MW01618298

Dedicated to
John Amos Nosler

April 4, 1913- October 10, 2010

Edited by
John R. Nosler

This book wouldn't have been possible without the diligent work performed by the following employees of Nosler, Inc.

Travis Furry
Eric Hansen
Mike Harris
Mike Lake
Dan Neumann
Mason Payer
Zach Waterman
Andy Nelson

MESSAGE FROM THE PRESIDENT

Looking Back...
Over the years, our Ballistics Department has acquired a collection that represents most of the reloading manuals that have been published by our competitors. While the value as a resource is significant, when we add-in the Nosler manuals, they also become a rich chronicle of cartridge, bullet and propellant development.

When I consider our own history, I remember compiling and editing Nosler Reloading Manual #1 on our dining room table. While that book has become somewhat of a collector's item, I chuckle at how basic the coverage was for the most popular cartridges of the day. That was 1976, and our entire Company fit nicely into a 6,000 square foot building. Since we continue to receive interesting questions about the previous six editions of our reloading guides, I wanted to share a little of the history:

Nosler Reloading Manual Number One
What most of our customers may not know about Number One is that there were actually "TWO" different books with the same title. Introduced in 1976, the first printing of Number One (234 pages) was dedicated to data for our flagship line of Partition bullets—with a few loads for our new line of Solid Base bullets. Shortly after introduction, we ran out of the first printing. This gave us an opportunity to add a few more Solid Base loads. This version of Number One (277 pages) has the same green cover with the same classic rendering of the moose standing at the edge of the water. The difference is the large red square on the front cover that reads "New Additional Solid Base Loading Data".

There was one more version of Number One. It was the first printing (234 pages) bound with a combination of buckram millinery and leather, and it came with a matching slip cover. With gold leaf embossing on the cover (and some personally signed by John A. Nosler), this was one for the collectors. There were only 1,000 of these special books produced, and only a few hundred that were signed.

Nosler Reloading Manual Number Two
Like Number One, Number Two was exclusively devoted to rifle data. Introduced in 1981, this book was an expanded version (309 pages) of Number One, with Solid Base bullets showing the most growth.

Nosler Reloading Manual Number Three
With the Company returned to Nosler family ownership, Number Three was introduced in 1989. At 516 pages, this was the first Nosler manual to have load data for handgun cartridges. In addition, Number Three was the only Nosler manual to show both Solid Base and Ballistic Tip load data.

At the urging of then Nosler Vice President, Chub Eastman, we began the tradition of asking prominent shooting sports personalities to write brief introductions to the cartridges. As it turned-out, the cartridge intros became one of the most popular parts of the Nosler books.

Nosler Reloading Guide Number Four
This book was first released in 1996. In recognition of the fact that hand-loaders are often great experimenters, we changed the title from Reloading Manual to Reloading Guide. At 722 pages, Number

MESSAGE FROM THE PRESIDENT

Four was the first edition to have our popular "Comments from the Lab" section.

Nosler Reloading Guide Number Five
Released in 2002, Number Five had 728 pages, and contained load data for Partition®, Ballistic Tip®, Combined Technology® (CT) Ballistic Silvertip®, CT Partition Gold®, CT Partition Gold Handgun®, and CT Failsafe bullets®.

Nosler Reloading Guide Number Six
Released in 2007, Number Six had new, up-to-date load data for most Nosler products. At 827 pages, Number Six also contained data for several new cartridge and bullet combinations.

As you can see, we have come a long way from the days when I laid out our first reloading manual. I am proud to say that the Nosler team has done another great job with book seven—and I know you will enjoy the read!

Good hunting and good shooting,

Bob Nosler

Bob Nosler
Tanzania 2011

Nosler®
UP FRONT

P.O. BOX 671
BEND, OREGON 97709
(800) 285-3701
NOSLER.COM

PRINTED FEBRUARY, 2013
PRINTED IN U.S.A.

THE LOAD DATA REPRESENTED IN THIS GUIDE SUPERSEDES ALL PREVIOUS DATA.

Table of Contents

Message from the President....................2

In Memoriam.......................................6

Index to Cartridge Reloading..............12

Foreword by Aaron Carter...................15

We Check Our Work..........................21

Bullet Cutaways..................................25

Brass Cutaways...................................38

Getting Ready to Reload......................40
- Safety....................................41
- Chamber Pressure.................44
- Cartridge Components..........45
- Reloading Equipment 51
- Case Preparation....................55
- Priming..................................58
- Powder Handling....................59
- Tools and Accessories.............61
- Step-by- Step Reloading.........63
- Tips on Bullet Seating............67

Basics of Bullet Stability.......................69

Rifle Cartridge Data............................74

Handgun Cartridge Data..................619

Appendix of Classic Cartridge Data....674

Rifle Drop Tables..............................737

Energy Tables...................................840

Glossary..844

By John R. Nosler

I think it is a well-known story, the legend of my grandfather, John Nosler, or "Big John" as we affectionately called him throughout his life. However, for anyone who is unfamiliar with the story, the brief would read as follows: John was a lifelong hunter, a lifelong innovator, and a man who questioned everything. I was named after him, and let me tell you; trying to live up to Big John as a namesake is not the easiest task. My grandfather had many important accomplishments, but he is best-known for inventing and developing the Partition® bullet—a bullet that has become the benchmark for an entire industry. While this feat may sound simple to some, inventing and developing the bullet (and the manufacturing systems required to produce it) was a major engineering accomplishment, all of which he did without a formal education. John was self-taught in the science of ballistics, self-taught in the art of manufacturing, and self-taught in business. What he did in his life was unselfish, courageous, and entrepreneurial. I don't have enough pages to give the entire account in the

detail it deserves, but thankfully the story has been recorded by Gary Lewis in his biography of John A. Nosler, *Going Ballistic*.

My grandfather and Gary spent hours on that project. I could hear him recounting all of the stories to Gary from my office, which was about 50 feet from his. His hearing was diminished from years of shooting without hearing protection, so he would always talk in a very loud voice. Even as he aged, his memories were sharp—being able to tell in great detail the events that shaped the company of today. I have told many people this, but I always marveled at the fact that my grandfather could visualize every concept in 3D. His favorite tool wasn't a computer, but a paper napkin and a black felt pen. With pen in hand, his ideas would flow onto the napkin like water. His life and the company he created make an interesting story. There were many paths to choose, several forks in the road, and a long list of people that influenced and inspired him along the way.

I often think about what would have been had my grandfather chosen a different path, or taken another road? What would our lives be like? He tried so hard to have someone else make the Partition®, what if had he sold the Partition patent to Winchester Ammunition, or taken up the offer to be a minority partner with Sierra Bullets? What would it be like had we not sold a majority interest to Leupold & Stevens back in 1969? I wonder, would I have been brought up in such a strong hunting heritage, raised to be a staunch supporter of our Second Amendment rights?

My fondest memories growing up involve days target shooting with my grandfather and father in a gravel pit or out in the BLM shooting cans. My first rifle was a Rogue Rifle Company Chipmunk. I still remember the day Big John came to the house to give it to me. We practiced in the back yard, how to hold it, how to use the peep sight, proper safety practices. I remember taking photos with him and the rifle, he was so proud and I was in complete awe. Those memories would be quite different had he chosen a different path in life, had he stayed a truck driver in California or never went on that now famous moose hunt in British Columbia. What if he had decided to hunt with his old 30-40 Krag or 30-30

Winchester? After all, without the 300 H&H Magnum, the thin jacketed bullets of the 1940's and that stubborn Canadian moose, the Nosler Partition bullet would have never been invented!

Moving through the timeline of our company history, it is easy to see the cause-and-effect results of my grandfather's decisions. A couple of difficult years brought scope-maker Leupold in as a majority shareholder in the old Nosler Partition Bullet Company. Because he sold to Leupold, we were able to bring on additional equipment and begin to market our product the way it needed to be. Proceeds from the sale to Leupold helped finance all of the impact extrusion manufacturing capabilities that are now a cornerstone of the Nosler factory. Through our partnership with Leupold, Nosler bullets came to be loaded into Federal Premium Ammunition, and the rest is history.

With the success of the "new" company we were catching the eye of talented people. People like my dad, Robert (Bob) Nosler. He was out of the Navy and my grandfather wanted him to join the company—but didn't want to force him to join the family business. He decided to let my father make that decision on his own. Thankfully, he did. Because of my father's sweat and blood, the family was able to purchase the company back from Leupold in 1988. Like dominos, this caused another crucial move, and his name was Chub Eastman. Chub was an important part of the Leupold/Nosler success, and when the family bought back the company, Chub came with us. Many other talented folks joined the new company, and soon we appointed our own Board of Directors to help guide our growth. Finally, we had control of our own destiny.

My dad's leadership led us to partner with Winchester®, then into new bullet designs such as AccuBond®, E-Tip®, BT Lead Free, and Custom Competition® just to name a few. Our biggest move was into Nosler Custom products. Rifles, Ammunition and Brass were a monumental step forward for this company. It couldn't have been done without us going through everything in our history. It made us who we are.

People often ask me if I am grateful for what my grandfather did. I can proudly say, not a day goes by that I don't think of him and say thanks for everything he and my grandmother sacrificed, and for taking the plunge. That is the American spirit, and that is what made this country great! So, I answer "of course I am grateful, who wouldn't be?" Fortunately, I was able to tell him how grateful I am many times before he passed in 2010. After reviewing my thoughts on paper for some time, I realized two absolute truths. First, opportunity presents itself, but only truly revolutionary individuals take that opportunity and run with it. Second, that all of the events leading to today were perfectly aligned. From my grandfather turning down an offer to join Sierra, to partnering with Leupold, to my dad deciding to join the company, to buying back our shares from Leupold, to creating a Board of Directors that helped guide us through uncertain times, to the advent of our ammunition and rifles, each event couldn't have happened without the one preceding it. It is all so eerily similar to a plan, a master plan, that all started with my grandfather recognizing the opportunity in front of him.

What Big John built is a legacy. We named one of our rifles in honor of him, the M48 Legacy. It was developed by our great team, but inspired by him. He had a lasting imprint on that firearm, and I was lucky enough to show it to him just two days before he died. He just smiled, and I could tell he was pleased.

That type of legacy is one that someday I can pass down to my children—if they choose to be in the company. My dad allowed me to choose, as my grandfather allowed him to choose, and so my children will be given the same choice. This is a family business, one that was built on the core values of faith, family, honesty, integrity, and quality. We will always strive to make a product that Big John would be proud of, and one that we are honored to put our name on.

I look forward to the day when my dad, Robert, can present his grandson, also named Robert, his first rifle and carry-on the legacy of Big John.

JOHN AMOS NOSLER
APRIL 4, 1913 - OCTOBER 10, 2010

By Chub Eastman

On October 10, 2010 at the age of 97 John Amos Nosler the founder of Nosler Inc. passed away peacefully at his home in Bend, Oregon. This date will be remembered by all as the passing of the last great icon of the shooting and reloading industry as we know it today.

John was born April 4, 1913 in Brawley, California. He was the type of person that only comes along once in a generation or so. John was the epitome of the American success story. With less than a formal high school education he was a self-taught metallurgical and mechanical genius that had the innate ability to consistently think outside the box to not only solve metal or mechanical problems but at the same time was always thinking of new products or ways to better the efficiency of production.

What started as a bullet failure on the side of a mud caked moose in Northern Canada became an obsession that he could design a bullet that would get the job done.

At the time, John had a trucking business delivering produce up and down the West Coast. Over the course of the next year the back of his mind was working on a bullet that would penetrate as it should to get the job done. While his trucks were making deliveries his spare time was spent at the milling machine and rifle range trying to come up with the right bullet.

The end product was tested the next hunting season with his hunting partner back in Canada looking for the one that got away. The result was two shots and two dead bull moose using the bullet John had designed and was christened the "Partition". Word got around as to what he had accomplished with his new bullet and the demand started to grow. He was far sighted enough to see there was a definite demand for a premium bullet that was reliable and the hunter could expect the same performance no matter what the hunting situation was. Consequently in 1948 John founded Nosler Partition Bullet Company and brought the new Partition bullet to market. The trucking business went by the wayside and his dream of being in the shooting sports industry became a reality.

John maintained his office at the Nosler facility and could be found most of the time cruising through the Engineering Dept. or in the manufacturing area. Always with a smile and a pocket full of Snickers candy bars he would pass out to anyone who looked hungry. His mind was always sharp as a tack and still looking for ways to improve production or new products.

In recognition of his contribution to the shooting sports, John was the unanimous recipient of the NRA's first prestigious 2007 Golden Bullseye Pioneer Award. There is also the NRA John A. Nosler Endowment Fund which is dedicated to the advancement and future of youth shooting programs.

A humble man, family man, and a man whose innate ability and talent to see the future has left a legacy that will be remembered by all who were privileged to have known him. No one is irreplaceable, but John A. Nosler comes very close.

Chub Eastman pictured with John A. Nosler

Grits Gresham 1922-2008
Author, Television Host, Former Shooting Editor of Sports Afield

Ron Mason 1937-2011
President, Federal Cartridge 1990-1998

Ian McMurchy 1944-2008
Author, Photographer, Big Game Specialist with Saskatchewan Department of Wildlife

RIFLE CARTRIDGES		Author	Page
17 Caliber	17 Remington Fireball	Mic McPherson	75
.172" Diameter	17 Remington	Layne Simpson	78
204 Caliber	204 Ruger	Bob Forker	81
.204" Diameter			
22 CALIBER	22 Hornet	John R. Nosler	85
.224" Diameter	221 Remington Fireball	Charles E. Petty	90
	222 Remington	Holt Bodinson	96
	223 Remington	Tom Gresham	102
	5.56x45 NATO	Patrick Sweeney	110
	222 Remington Magnum	Kevin Steele	114
	22 PPC-USA	Sam Fadala	120
	22 BR Remington	Chub Eastman	126
	225 Winchester	Appendix	676
	22-250 Remington	Brian Pearce	132
	22-250 Remington (FAST TWIST)	Ron Spomer	141
	22-250 Remington Ackley Imp	John Anderson	146
	220 Swift	Holt Bodinson	153
	223 Winchester Super Short Magnum (WSSM)	Gary Lewis	160
6mm CALIBER	6mm PPC	Jim Carmichel	167
.243" Diameter	6mm BR Remington	Appendix	679
	243 Winchester	John Barsness	173
	6mm Remington	Ed Timerson	181
	243 Winchester Super Short Magnum (WSSM)	Gary Lewis	188
	240 Weatherby Magnum	Greg Rodriguez	195
25 CALIBER	250-3000 Savage	Bob Nosler	203
.257" Diameter	257 Roberts	Dave Anderson	210
	257 Roberts Ackley Improved	Appendix	683
	25-06 Remington	Brandon Ray	217
	25 Winchester Super Short Magnum (WSSM)	Appendix	689
	257 Weatherby Magnum	Aaron Carter	224
6.5mm CALIBER	6.5mm x 55 Swedish	John Barsness	232
.264" Diameter	6.5 Grendel	Patrick Sweeney	239
	6.5 Creedmoor	Scott J. Rupp	244
	260 Remington	Don Polacek	251
	6.5-284 Norma	Layne Simpson	258
	6.5-06 A-Square	Bob Forker	265
	6.5 Remington Magnum	Lee J. Hoots	272
	264 Winchester Magnum	Craig Boddington	279
6.8 CALIBER	6.8 Remington SPC	Bill Wilson	286
.277" Diameter			
270 CALIBER	270 Winchester	John Zent	292
.277" Diameter	270 Winchester Short Magnum (WSM)	Mark Hampton	300
	270 Weatherby Magnum	Terry Wieland	308

RIFLE CARTRIDGES		Author	Page
7mm CALIBER	7mm-08 Remington	John Haviland	316
.284" Diameter	7x57mm Mauser	Dave Scovill	323
	280 Remington	John B. Snow	330
	284 Winchester	Ron Spomer	337
	280 Remington Ackley Improved	Bob Nosler	344
	7mm Remington Short Action Ultra Mag (SAUM)	Jon R. Sundra	351
	7mm Winchester Short Magnum (WSM)	Tom C. Tabor	358
	7mm Remington Magnum	Diana B. Rupp	365
	7mm Weatherby Magnum	Jeff H. Johnston	372
	7mm Shooting Times Westerner (STW)	John R. Nosler	379
	7mm Remington Ultra Magnum (RUM)	Art Wheaton	386
30 CALIBER	300 AAC Blackout	Iain Harrison	393
.308" Diameter	30-30 Winchester	Karen Mehall	398
	300 Savage	Wayne van Zwoll	402
	30/40 Krag	Appendix	695
	308 Marlin Express	Lane Pearce	408
	30 TC	Scott Haugen	412
	308 Winchester	Bob Robb	418
	30-06 Springfield	Wayne van Zwoll	427
	30-06 Springfield Ackley Improved	Appendix	700
	300 Holland & Holland (H&H) Magnum	Layne Simpson	437
	308 Norma Magnum	Appendix	707
	300 Winchester Magnum	James Brion	446
	300 Winchester Short Magnum (WSM)	Paul Fortino	455
	300 Remington Short Action Ultra Magnum	Chub Eastman	464
	300 Ruger Compact Magnum (RCM)	Jim Bequette	473
	300 Remington Ultra Magnum (RUM)	Ludo Wurfbain	482
	300 Weatherby Magnum	Joseph von Benedikt	491
	30-378 Weatherby Magnum	Aaron Carter	500
8mm CALIBER	8x57mm JS Mauser	Mike Venturino	509
.323" Diameter	8mm-06	Appendix	713
	325 Winchester Short Magnum (WSM)	Steve Gash	513
	8mm Remington Magnum	Craig Boddington	517
338 CALIBER	338 Federal	Jack Mitchell	521
.338" Diameter	338-06 A-Square	Steve Comus	526
	338 Winchester Magnum	Dave Petzal	532
	330 Dakota	Appendix	716
	340 Weatherby Magnum	Mike Harris	539
	338 Remington Ultra Magnum (RUM)	Mark A. Keefe IV	546
	338 Lapua Magnum	Adam Heggenstaller	553
	338-378 Weatherby Magnum	Joe Busalacchi	560
35 CALIBER	358 Winchester	Appendix	721
.358" Diameter	35 Whelen	Bryce Towsley	567
	350 Remington Magnum	Buck Pope	572
	358 Norma Magnum	Appendix	723
	358 Shooting Times Alaskan (STA)	Appendix	726

Caliber	Cartridge	Author	Page
9.3mm CALIBER	9.3x62 Mauser	Gary Lewis	576
.366" Diameter	9.3x74 R	Roland Zeitler	580
	9.3x64mm Brenneke	Appendix	729
375 CALIBER	375 Holland & Holland (H&H) Magnum	J. Scott Olmsted	584
.375" Diameter	375 Ruger	Phil Shoemaker	588
	375 Remington Ultra Magnum (RUM)	Art Wheaton	592
	378 Weatherby Magnum	Appendix	732
416 CALIBER	416 Remington Magnum	John R. Nosler	596
.416" Diameter	416 Rigby	Mark A. Keefe IV	599
	416 Weatherby Magnum	Appendix	735
	416 Ruger	Buck Pope	602
458 CALIBER	458 Winchester	Roger G. Roberts	605
.458" Diameter	458 Lott	Bob Nosler	608

LEVER RIFLE SECTION		Author	Page
44 CALIBER			
.429" Diameter	44 Remington Magnum	Roy Welch	611
45 CALIBER			
.458" Diameter	45-70 Gov't (Strong Actions Only)	Richard Mann	616

HANDGUN CARTRIDGES		Author	Page
9mm CALIBER	380 Auto (ACP)	Dave Workman	620
.355" Diameter	9mm Luger (Parabellum)	Iain Harrison	623
	357 Sig	Dave Workman	627
38 CALIBER	38 Special	Roy Huntington	631
.357" Diameter	357 Magnum	Brian Pearce	634
	357 Remington Maximum	Gary Lewis	637
10mm CALIBER	40 S&W	Massad Ayoob	640
.400" Diameter	10mm Auto	Mike McNett	646
41 CALIBER	41 Remington Magnum	Tom Gresham	652
.410" Diameter			
44 CALIBER	44 Special	John Taffin	655
.429" Diameter	44 Remington Magnum	Mark Hampton	659
45 CALIBER	45 Auto (ACP)	Clint Smith	664
.451" Diameter	45 Colt (Single Action Army & Replicas)	Charles E. Petty	668
	45 Colt (Ruger & T/C Contender & Encore)	Dave Scovill	671

FOREWORD
BY AARON CARTER

By Aaron M. Carter

It seems only yesterday when my maternal grandfather, Norman L. Yowell, and I—a just-turned teenager—visited a small sporting goods store in Winchester, Va. Truly selfless, he always supported my endeavors, from fishing and trapping to hunting and handloading. That particular day I was fulfilling the latter. The purpose: Nosler bullets.

Months prior, during the fall season a family friend and fellow handloader and I hunted together for whitetail deer, he with a 264 Win. Mag. and I a 30-30 Win. Although success eluded me, late that rainy evening he shot a mature doe. The Ballistic Tip's remarkable terminal performance, coupled with its stylish appearance, left an indelible impression, convincing me it was the bullet to use. It was my formal introduction to Nosler. The store's handloading equipment and components were to the immediate right upon entering. Behind the counter, prominently displayed, were the black boxes with red and green labels of Nosler's bullets. But, the one that caught my attention stated, "Nosler, 100 Solid Base Boat Tail, Cal. 270/130 Grain Ballistic Tip." As I stared intently, the clerk asked, "Is there something I can help you with?" "Can I see that box of Nosler Ballistic Tip bullets," I promptly replied.

Pointing to the box, he handed it to me; however, the label prevented me from seeing the contents. Without saying a word, the clerk cut both sides of the label, revealing the svelte, highly polished projectiles with sharp, yellow polymer tips. I was mesmerized; they were so different than any bullets I had seen to date. "I thought, why not make a real sharp, good-looking tip?," wrote John A. Nosler in Nosler Reloading Manual 6. "I thought that a plastic tip would be attractive, appealing, and give spectacular performance." They certainly caught my attention.

Although I tried to convince my grandfather to purchase the box, he wouldn't budge. His reasoning was valid, however. "You don't own a 270," he said. "Once you get the rifle then you can get the bullets." There was no harm in looking, right?

In subsequent weeks our visits to the sporting goods store became routine, as was revealing the contents of the Nosler boxes. Always 0.277" diameter, always Ballistic Tips. Looking back, I appreciate the patience of the clerk; he knew we weren't purchasing the bullets, yet gladly showed them.

Within a few years, though, I was gifted a 270 Win.—a Weatherby Vanguard topped with a 3-9X 40 mm Leupold riflescope. The yellow-polymer-tipped, 130-gr. projectiles I so desired to use complemented the setup nicely, and soon after the varmints, predators, turkeys and deer of Northern Virginia felt the impact, literally. The bullets delivered just as advertised, and have done so ever since, too. With my history with the Ballistic Tip, it's little wonder that, to me, it is the projectile that represents Nosler, not the Partition. Undoubtedly, the Partition, which was created in 1946, first field-tested in 1947 and introduced to consumers in 1948, was a game-changer with regard to bullet design, and more specifically, the concept of consistent, controlled expansion and deep penetration, but it's the Ballistic Tip that's been more widely copied—a simple measure of the design's effectiveness and popularity—and used. A cursory glance at the websites of large-scale, American bullet manufacturers reveals that most offer polymer-tipped bullets, be them for big game, varmints, tactical applications, competition or informal shooting. Remington's Bronze Point both predated and inspired the Ballistic Tip, but didn't attain the popularity or exhibit the influence that Nosler's bullet has. Interestingly, oftentimes sportsmen refer to polymer tipped bullets, regardless of manufacturer, as "Ballistic Tips." That speaks volumes.

The reason for its popularity is simple: performance. Unless hunting what is often referred to as "thick-skinned, heavily muscled" animals, the now-called "Ballistic Tip Hunting" is my go-to Nosler bullet. Opinions differ vastly on the subject of weight retention, expansion and penetration; however, for animals caribou size and smaller, I prefer to use a bullet that sheds considerable material during expansion and penetration, yet drives sufficient depth to disrupt the vital organs. Material "washed off" serves only to enhance the wounding capability. John A. Nosler and Roy Weatherby likewise subscribed to this philosophy.
Featuring a tapered, gilding-metal jacket and lead-alloy core with no mechanical bond, Ballistic Tip Hunting bullets fit the bill nicely, typically retaining 50 percent or less of their original weight. Despite this, the Solid Base and remaining core oftentimes exit the animal, or are found against the hide on the offside.
Proper penetration has never proved problematic.

The color-coded tip isn't only for appearance, though. Not only does it resist deformation in the magazine during setback, but, when combined with the bullet's boattail base, serves to enhance exterior ballistics. With their aerodynamic designs—thus higher ballistic coefficients—Ballistic Tips are conducive to long-range hunting and shooting, where less wind deflection, drop and more energy are invaluable. But, just as important is that, due to the Ballistic Tip's design, they expand at considerably lower velocities—down to 1,800 f.p.s.—than many bullets, an important consideration as shot distance increases. Terminal performance means nothing if the bullet cannot be accurately placed, though.

Due to the extremely tight tolerances held by Nosler, Ballistic Tip bullets are incredibly accurate, oftentimes equal to that of the company's Custom Competition match bullets—hence the reason I make heavy use of them to evaluate the accuracy potential of rifles and single-shot handguns chambered for rifle cartridges. Considering the complexity of the manufacturing process, that's a remarkable feat. When the size of a game animal increases to that of an elk or larger, or where deep penetration is paramount, the Partition is the best option from Nosler. Featuring a tapered jacket and dual lead-alloy cores separated by an integral cross-member or "partition," Nosler Partition bullets expand uniformly, resulting in the coveted "mushroom" profile while retaining 65 to 70 percent of their original weight. Although considered antiquated by some—due mostly to the current trend of near-100-percent weight retention projectiles—John A. Nosler believed that shedding material, and thus weight, was desirable.

Typically, Partitions leave small entry and exit holes on smaller or thin-skinned species, or are found against the offside hide on larger ones, though I've witnessed "extra-large" holes and wound channels. On the former, one might initially suspect bullet failure, that is until they realize the rapid expiring of the animal, and certainly once the organs are removed and the bullet's true terminal performance is observed. Recovered Partitions are remarkably consistent, exhibiting the classic mushroom shape with most, if not all, of the front core absent.

As good as Ballistic Tips are, impact velocities in excess of 3,000 f.p.s. are not its forte; with many modern cartridges, exceeding that number is achieved relatively easily. That's where the Partition excels. The lowest velocity at which the Partition expands is 1800 f.p.s., the same as the Ballistic Tip, but, due to its construction, and more specifically, its integral partition, Nosler lists its upper limit as "unlimited." No commercially available cartridge should make it "fail."

Contrary to some naysayers, I've found Partitions exhibit good accuracy; in fact, the smallest three-shot group at 100 yds. that I've shot to date was produced by the 60-gr. Partition loaded in 22-250 Rem. NoslerCustom ammunition, fired from a Nosler Model 48 Custom Varmint rifle. The group measured a scant 0.17" center to center.

One would think it would be difficult to improve upon either the Ballistic Tip or the Partition, but Nosler did just that by creating the AccuBond, which was introduced in 2003. Like the Ballistic Tip Hunting, the AccuBond has a tapered, copper-alloy jacket and lead-alloy core; however, it differs in that a proprietary process bonds the two, preventing core slippage for higher retained weights—upward of 70 percent—resulting in penetration depths similar to those achieved by the Partition. But, being profiled similar to the Ballistic Tip, with a polymer tip and boattail base, the AccuBond is preferable when long shots are possible. The AccuBond is also similar to the Partition in that it can withstand impacts in excess of the Ballistic Tip's upper limit, but it will also reliably expand at speeds as low as 1,800 f.p.s.—essentially an all-around bullet.

My initial field-test of the AccuBond came on the Dark Continent, during my first plains-game safari. During the seven-day hunt, 180-gr. AccuBonds fired from a 300 WSM-chambered Nosler Model 48 TGR accounted for seven animals, but three were especially representative of the bullet's toughness. The first, a quartering-toward blue wildebeest at 125 yards, proved the PH's longstanding belief inaccurate, as a single, well-placed AccuBond indeed dropped the bull in place and kept it there. Found behind the offside shoulder against the hide, the bullet weighed 110.6 grs.—61.4-percent weight retention—and expanded to 0.845".

Next came a once-elusive gemsbok, arguably among the toughest of the plains game. Although only 60 yards distant, because the bull spotted us before we did it, I had to "thread the needle" at the near-straight-away animal. Offering only a glimpse of its right shoulder, I aimed for the last rib in an effort to angle into the vital organs before striking the offside shoulder. The bullet impacted about one inch left of the point of aim, entering the front of the hindquarter and stopping only after penetrating the heart—amazing! Death was swift and the recovery distance short. The bullet retained 47 percent of its original weight and expanded to 0.568"—I'll take that any day.
Lastly came a 54" kudu. From a vantage point known as "Pride Rock" we spotted an old bull certainly short on time, a perfect candidate to remove from the herd. At 265 yards, the tracker and PH were a bit hesitant, but all worries were alleviated when the bull collapsed within seconds and 20 yards of the shot. Despite the kudu's body size, the bullet penetrated both shoulders and exited. What more can be expected?

Nosler's lead-core bullets are proven performers, but what about areas requiring a lead-free projectile, or those desiring near-100-percent weight retention? Nosler has you covered with the E-Tip. Because the E-Tip is made from solid gilding metal, the bullet offers 95-pecent-plus weight retention (typically only the tip is lost), resulting in deep penetration. The boatail base, polymer-tipped bullet's "E2" (Energy Expansion) cavity results in four distinct petals, and it reliably expands at velocities as low as 1,800 f.p.s., yet is rated by Nosler as having an "unlimited" threshold for upper velocities.

To test the E-Tip's durability, in 2008 I pitted it against nilgai, a free-ranging exotic in South Texas that is easily among, if not, the toughest species pound-for-pound found in the United States. Descriptors such as "like shooting a tractor tire" certainly apply to the nilgai; I've never encountered hide so thick, or lethally hit animals put down so faint a blood trail. It would be a good test. In the waning light on the first evening of the hunt, I crept within 49 yards of a large, feeding bull before it noticed me. Silhouetted by the setting sun, I had but seconds to deliver a bullet before the nilgai vanished. The purposefully neck-shot bull dropped instantly, but two "insurance" shots were still delivered to the chest when approaching. The two recovered .30-cal., 180-gr. bullets exhibited picture-perfect expansion, measuring 0.640" and 0.663" in diameter, and weighing 178.3 and 178.6 grs., respectively.

Noticing an absence of a non-expanding option to complement its larger-caliber Partitions, in 2008 introduced the Solid Dangerous Game. Machined from solid brass, the flat-nose, non-expanding bullets are designed to penetrate straight-line, as well as deliver a similar point of impact as Partitions of the same weight and caliber. Testing the Solids in 0.366" and 0.416" diameters has proven their designs nicely meet Nosler's design objectives.

Best known for its rifle bullets, and more specifically, hunting options, Nosler's Custom Competition match rifle and handgun bullets, as well as Sporting Handgun, are first-rate projectiles, too. I consider them among the most underrated for their given purposes, and they represent a tremendous value. For accuracy testing, competition or even high-volume plinking, Custom Competition bullets are hard to beat. Like Ballistic Tips, I make heavy use of them.

The 240-gr., 0.429"-diameter Sporting Handgun Revolver bullet in an MMP sabot, propelled by two Pyrodex pellets and a Winchester Triple-Seven primer, is my go-to load for my CVA Optima muzzleloader. Each year it accounts for multiple Virginia whitetails, and the recovery distance, as can be expected, is usually quite short. The sole bullet I've recovered from a medium-size doe shot head-on at 35 yds. expanded to 0.626" and retained 64 percent of its pre-expansion weight, despite a velocity well in excess for which it was designed.

Through my many years of hunting and evaluating firearms I can attest that I've yet to have, or see, an inaccurate Nosler bullet, or have had one "fail." The reason is simple: the hunting bullets are field-tested extensively by the Nosler family themselves, as well as employees who are diehard hunters, too. Besides evaluating afield, all lots are evaluated for accuracy (hunting and competition), and the hunting projectiles are further subjected to media that replicates animal matter they will encounter. If it doesn't meet the company's strict standards, it doesn't leave the factory, period. But it all comes down to the designs … and all Nosler bullets are thoughtfully developed. All of this inspires confidence.

On Oct. 10, 2010, John A. Nosler, founder of what is now called Nosler, Inc., died at age 97. Six days later my maternal grandfather passed away, too, at the age of 82. During this mournful time, John R. Nosler, Jr., and I communicated at-length frequently about our grandfathers and their generation, and that was when I observed first-hand the value of the "family" aspect of the family-owned business. That's one thing about Nosler, everyone is "down to earth," approachable and sincere. I appreciate and respect that. In summer 2011, upon finding a successful draw status for an elite Wyoming antelope unit, I thought what better way to honor my maternal grandfather, my paternal grandfather, George C. "Cliff" Carter, who passed away in 2003, and John A. Nosler, than to use my 30-'06 Sprg.-chambered Kimber Sonora—Cliff exclusively used a Remington Model 760 Gamemaster in 30-'06 Sprg.—with my preferred bullet for it, the 168-gr. Ballistic Tip, which consistently groups near one-half M.O.A.

After glassing more than four-dozen bucks during the first morning of my do-it-yourself hunt, I found the one I wanted among a large group of does with several inferior bucks on the herd's fringe. The shot was taken at 331 yds., where the bullet entered the base of the quartering-toward buck's neck and exited behind the shoulder on the opposite side. It traveled less than 50 yards before collapsing—perfect!

Walking toward the antelope, I couldn't help but think of my grandfathers, both of whom contributed to who, and what, I am. Thinking of the sporting goods store and the Ballistic Tips, I cracked a smile, but it was immediately followed by tears. It was at that moment that I realized that Nosler was more than just bullets.
Thank you.

Aaron Carter

BIOGRAPHY

Aaron Carter penned his first magazine article, "Gunning for Geese," at age 16 for NRA's Insights. Eight years later, after graduating college, he became assistant editor for NRA's newsstand monthly, Shooting Illustrated. Carter was soon promoted to associate editor, and later became senior associate editor for American Rifleman, where he is currently the managing editor.

In addition to his managerial duties, Carter writes firearm reviews and handloading columns, many of which have earned national and international recognition, including platinum and gold, from Hermes Creative Awards and MarCom Awards. He also contributes to "American Rifleman Television" and americanrifleman.org, as well as conducts handloading seminars during NRA's Annual Meetings & Exhibits.
Carter resides near his birthplace in Northern Virginia with his wife, Bonnie, and three sons, Gabriel, Colby and Symon. When not spending time with his family, which he tries to include in all his outdoor pursuits, he can be found "working up" handloads or using them afield. Thus far, he has hunted in 18 states, two Canadian provinces and South Africa, twice. Still, it's near home, the American West and Alaska he prefers hunting most. In addition to hunting, Carter enjoys running a long trapline, fishing and working on his small family farm.

WE CHECK OUR WORK

Here at Nosler, we have always adhered to the motto "Quality First." We take the quality of our product very seriously, and will never sacrifice quality for speed, quantity, or profit. If we make a batch of bullets, only to discover that there is something wrong, we scrap that batch and start over. We don't just assume that the bullets we make are accurate and consistent, we check and double check several times throughout each run of bullets that we make. Every batch of bullets receives a variety of spot-checks during the manufacturing process as well as being test-fired for accuracy throughout the run. Prior to packaging, every bullet that we make is visually inspected to ensure that it is free from flaws, blemishes, or defects. When you open a box of Nosler bullets, you can rest assured that the bullets you have purchased are of the finest quality and consistency that we can possibly produce.

Talent

At Nosler, we pride ourselves on our product knowledge. While not all employees are necessarily hunters or shooters, every employee does have a basic knowledge of the products that we produce.

Every new employee at Nosler, regardless of position, must complete bullet identification training. Classes are provided, and after 90 days on the job, every employee must pass our bullet identification test. There is a written portion of the test covering company history and basic bullet terminology. The heart of the test is the bullet identification section. Each candidate is given 30 bullets in unmarked paper cups, a scale, and a micrometer. The candidate must then identify each bullet by Product Line, Diameter, Caliber, Weight, Base Style, and Point Style. A minimum 90% score is required to pass the test.

Spot Checks

A variety of measurements and tests are performed on our bullets during the manufacturing process. These spot checks begin when we receive our raw materials, and continue until the bullets are packaged and ready to ship

A Gilding Metal slug, the beginning of a bullet jacket, is hardness tested after annealing to verify that the batch is soft enough to begin extruding into jackets.

On the production floor, completed bullets are weighed on highly sensitive scales to verify that they are within our weight tolerances.

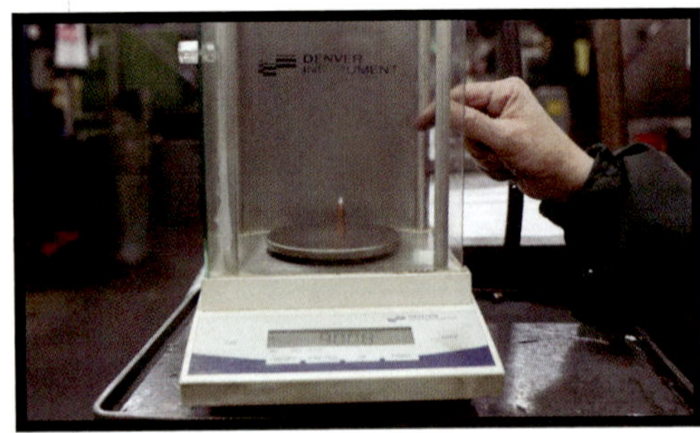

Periodically, sample bullets from each run are measured with a micrometer to verify that they are within our tolerances for diameter. A sample bullet is checked for runout near the base of the shank to insure that it is absolutely concentric. Concentricity, minimal to no runout, is a key feature of an accurate bullet. Here, the white polymer tip of this sample AccuBond® bullet is being measured for runout.

Visual Inspection

Prior to packaging, every bullet is visually inspected to ensure that it is free from flaws or visual imperfections. Bullets are fed by hand onto a moving belt, which spins the bullets along their length so that the entire surface of the bullet can be inspected. Any flaw that would affect the performance of the bullet, either in terms of accuracy or terminal performance, results in that bullet being rejected. Rejected bullets are de-milled, and then recycled as scrap. Any bullet showing cosmetic blemishes such as oxidation, discoloration, or minor scratches, is separated out to be sold as a "blem".

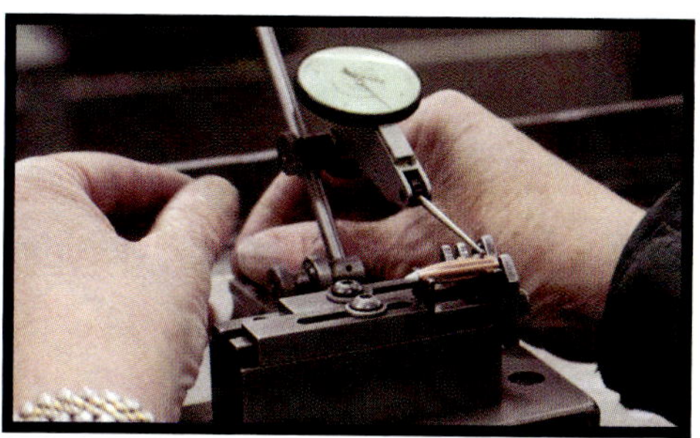

By visually inspecting every single bullet we make, we can be sure that the bullets we sell to our valued customers will always be of the upmost quality and consistency.

After washing and polishing, finished bullets are visually inspected one at a time by experienced members of our Packaging and Inspection department staff. The metal bar lying at an angle across the belt causes the bullets to spin along their length as they pass by, allowing the inspector to view the complete exterior of the bullet. Cotton gloves are worn by staff to prevent skin oils from tarnishing the bullets. Visual inspection is the final step of our thorough quality assurance process. After being inspected, bullets are packaged and shipped to customers.

Test Firing

While frequent measurements and weight checks help to ensure consistency, the true test of a bullet's quality is its accuracy. Throughout every run of bullets, production samples are pulled literally "hot of the press" and test-fired for accuracy in our underground ballistics-testing tunnel. Sample bullets are fired from test barrels that are clamped into a machine rest, mitigating the potential for human error. During the test-firing process, bullets are checked for accuracy, velocity, and pressure. Test firing is performed every day, on every batch of bullets that we make. By test-firing every batch of bullets we can be certain that the bullets you buy are consistent, accurate, and ready to perform whether on the range or in the field.

ISO Certification

In 2012, Nosler's Quality Management system was awarded ISO 9001:2008 certification accredited by DNV Certification B.V. ISO 9001:2008 certification recognizes that our policies and practices ensure consistent quality in the products that we manufacture. Certification provides objective confirmation that our reputation for quality and consistency can be continually relied upon by our distributors, dealers, and customers.

BULLET CUTAWAYS

- **Signature White Polymer Tip**
- **Tapered Copper-Alloy Gilding Metal Jacket**
- **Lead-Alloy Core**
- **Proprietary Bonding Process**
- **Boat Tail Design**

Unique White Polymer Tip:
Assures accuracy, smooth chambering and eliminates tip damage during recoil.

Tapered Copper-Alloy Jacket:
Lethal penetration and uniform expansion with maximum weight-retention.

Lead-Alloy Core:
Engineered for balanced expansion.

Proprietary Bonding Process:
Eliminates all possibility of component separation.

Boat Tail Design:
Boat tail configuration combines with the streamlined polymer tip for extreme long-range performance and easier loading.

Scan this QR code to go to www.Nosler.com and watch a video about the AccuBond®

Optimum Performance Velocity:
Minimum: 1800 fps (549 mps)
Maximum: Unlimited

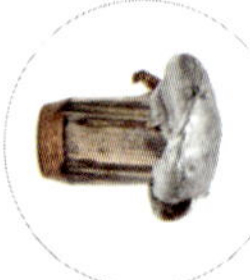
1800 FPS

2300 FPS

2600 FPS

2900 FPS

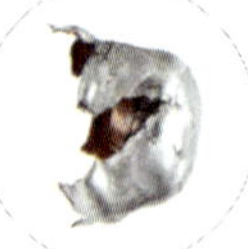
3200 FPS

PARTITION®

The Benchmark for Hunting Bullets Since 1948™

- **Broad Range of Animals**
- **Devastating Terminal Performance**
- **Deep Penetration**

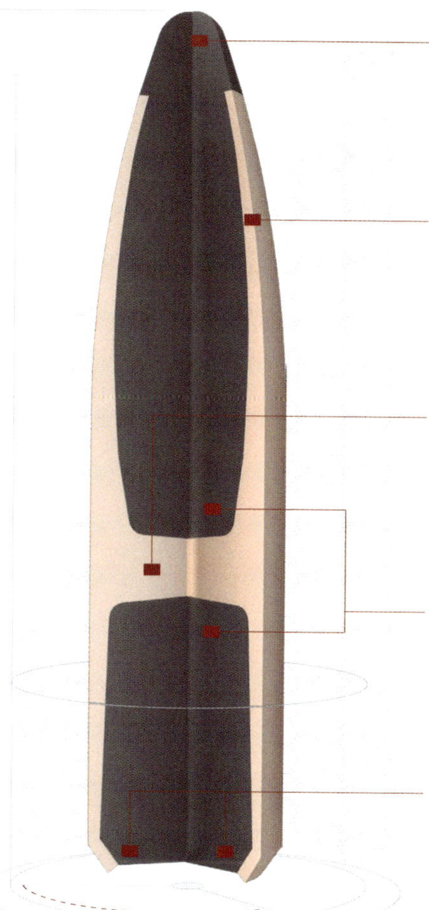

Nosler Engineering:
Nosler's special lead-alloy, dual-core provides superior mushrooming characteristics at virtually all impact velocities.

Fully Tapered Copper-Alloy Jacket:
Ruptures instantly at the thin jacket mouth, yet the gradual thickening along the bullet's axis controls expansion and curls the jacket uniformly outward at high and low velocities.

Nosler's Integral Partition:
Supports the expanded mushroom and retains the rear lead-alloy core. The enclosed rear core retains more than two-thirds of the original bullet weight for deep penetration.

Dual-Core Construction:
Every Partition® bullet in the Nosler® line delivers optimum length, weight and ogive design for maximum in-flight and terminal performance.

Special Crimp Lock:
Adds strength to resist deformation under the pressure of heavy magnums.

Scan this QR code to go to www.Nosler.com and watch a video about the Partition®

Optimum Performance Velocity:
Minimum: 1800 fps (549 mps)
Maximum: Unlimited

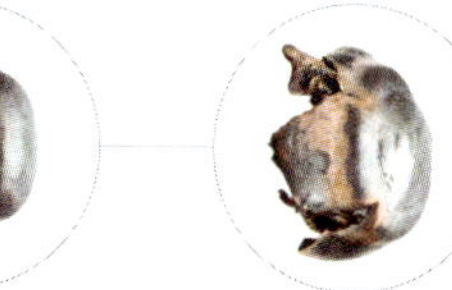

1800 FPS *2300 FPS* *2600 FPS* *2900 FPS* *3200 FPS*

THE ONLY BULLET THAT CAN BE CALLED™

BALLISTIC TIP®
HUNTING

6mm cal · 25 cal · 6.5mm cal · 270 cal · 7mm cal · 30 cal · 8mm

TIP COLOR BY CALIBER

- **Polymer Tip (Color-Coded by Caliber)**
- **Fully-Tapered Jacket and Special Lead-Alloy Core**
- **Ballistically Engineered Boat Tail Solid Base**
- **High Ballistic Coefficient**

Nosler Ballistic Tip®:
Streamlined polymer tip, color-coded by caliber, resists deformation in the magazine and initiates expansion upon impact.

Fully-Tapered Jacket and Special Lead-Alloy Core:
Allows controlled expansion and optimum weight retention at all practical velocity levels. The heavy jacketed base prevents bullet deformation during firing.

Ballistically Engineered Solid Base®:
Boat tail configuration combines with the streamlined polymer tip for extreme long-range performance and for easier loading.

Optimum Performance Velocity:
Minimum: 1800 fps (549 mps)
Maximum: 3200 fps (975 mps)

Scan this QR code to go to www.Nosler.com and watch a video about the Ballistic Tip®

1800 FPS

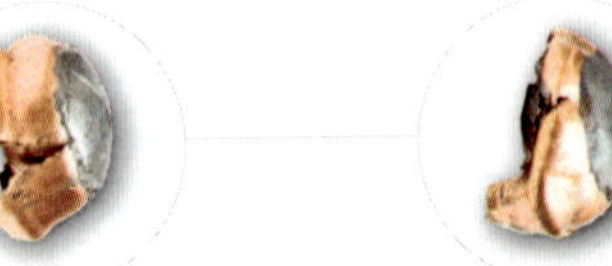

2600 FPS

2900 FPS

3200 FPS

- **Legendary Ballistic Tip® Performance**
- **Lubalox® Exterior Coating**
- **Defining Silver Polymer Tip**
- **Non-Moly Coating Which Allows More Shots Between Cleaning**

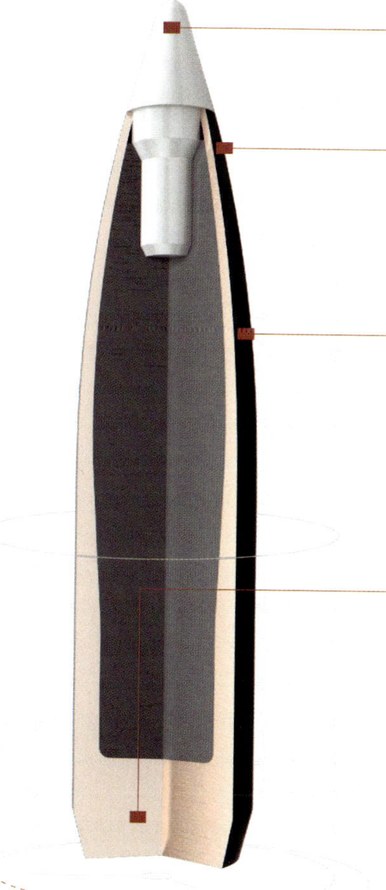

Polymer Silvertip:
Polymer Silvertip prevents deformation in the magazine.

Lubalox ®Exterior Coating:
Reduces fouling, pressure and friction between bullet and bore; provides longer barrel life, easier cleaning and enhanced accuracy.

Fully Tapered Jacket and Special Lead-Alloy Core:
Allows controlled expansion and optimum weight retention at all practical velocity levels. The heavy jacketed base prevents bullet deformation during firing.

Ballistically Engineered Solid Base®:
Boat tail configuration combines with the streamlined polymer tip for extreme long-range performance and easier loading.

Optimum Performance Velocity:
Minimum: 1800 fps (549 mps)
Maximum: 3200 fps (975 mps)

Scan this QR code to go to www.Nosler.com and watch a video about the Ballistic Silvertip®

1800 FPS

2600 FPS

2900 FPS

3200 FPS

- Polymer Tip
- Ultra-Thin Jacket Mouth
- Varmint Jacket Wall Design
- Boat Tail
- Long Range Capabilities

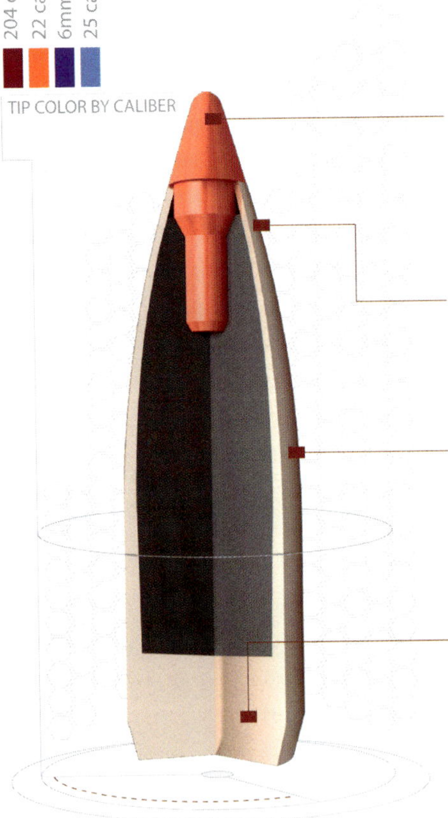

Nosler Ballistic Tip®:
Nestled in the jacket mouth is the streamlined polymer tip, color-coded by caliber.

Ultra-Thin Jacket Mouth:
Assures violent expansion at either end of the velocity scale.

Varmint Jacket Wall Design:
The uniform, gradual thickening at the bullet's mid-section is designed to keep the Ballistic Tip® Varmint together until impact-at any velocity.

Ballistically Engineered Solid Base®:
Boat tail configuration combines with the streamlined polymer tip for extreme long-range performance and easier loading.

Optimum Performance Velocity:
Minimum: 1600 fps (488 mps)
Maximum: Unlimited

Scan this QR code to go to www.Nosler.com and watch a video about the Ballistic Tip® Varmint

- Metallic-Colored Polymer Tip
- Explosive Expansion
- Larger Cavity
- Patent-Pending Fragmenting Copper Core Technology™
- Certified For Use In All Non-Lead Zones

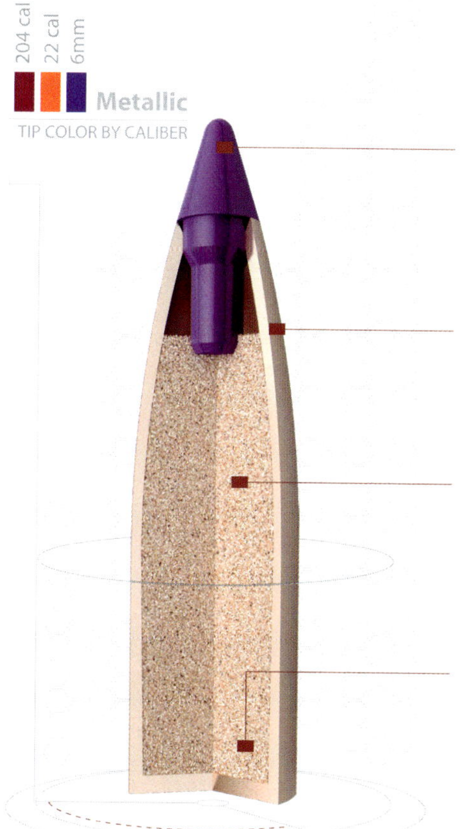

Unique Metallic- Colored Polymer Tip:
Activates immediate expansion upon contact at a wide range of velocities.

Ultra-Thin Copper Jacket:
Retains structural integrity at ultra-high velocities yet fragments completely upon impact.

Fragmenting Copper Core:
Allows rapid and absolute displacement of target. (Patent Pending)

Flat Base:
Ballistically engineered for ultra-high velocities and extreme accuracy.

Optimum Performance Velocity:
Minimum: 1600 fps (488 mps)
Maximum: Unlimited

Scan this QR code to go to www.Nosler.com and watch a video about the Ballistic Tip® Lead Free

- OD Green Polymer Tip
- E2 Cavity™
- Solid Gilding Metal Construction
- Lead-Free
- 95% Weight Retention

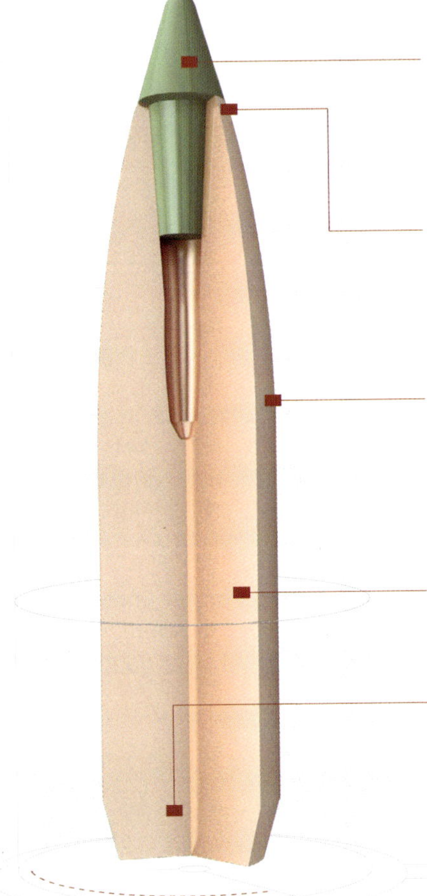

Signature OD Green Polymer Tip:
Nosler's advanced polymer tip initiates expansion while delivering unsurpassed penetration and weight retention.

ECR™:
The Expansion Control Ring insures controlled expansion at a wide range of velocities and conditions, making the E-Tip very versatile.

Patented E² Cavity™:
The exclusive Energy Expansion Cavity™ provides a dual-expansion chamber for amazing stopping power and uniform expansion with 95%+ weight retention for superior penetration.

Solid Copper-Alloy Construction:
Nosler's one-piece design meets all "lead-free" hunting regulations.

Boat Tail Design:
Precisely formed boat tail provides a more efficient flight profile for increased long-range performance and easier loading

Optimum Performance Velocity:
Minimum: 1800 fps (549 mps)
Maximum: Unlimited

Scan this QR code to go to www.Nosler.com and watch a video about the E-Tip™

1800 FPS

2600 FPS

2900 FPS

3200 FPS

- **Flat-Nose Design**
- **Solid™ Bullet Construction**
- **Concentric Design**
- **Designed for Same Point of Impact as Nosler Bullets**

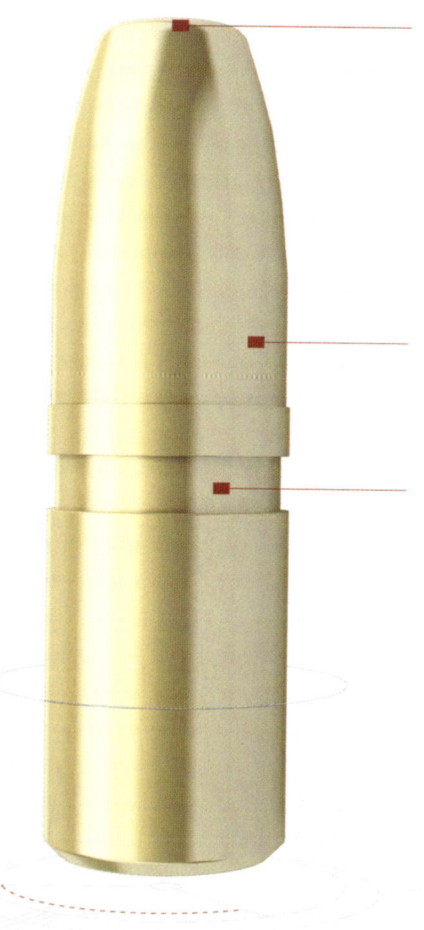

Flat-Nose Design:
Reduces deflection off of bone and muscle for deep penetration and an impressively straight wound channel.

Solid™ Bullet Construction:
Lead-free alloy delivers lethal penetration.

Concentric Design:
Seating groove produces optimal load versatility

Scan this QR code to go to www.Nosler.com and watch a video about the Solid™

Always Have a PH Back Up When Taking Dangerous Game

- Ultra-Thin Jacket
- Nosler Proven Accuracy
- Special Lead-Alloy Core
- Streamlined Flat-Base Design

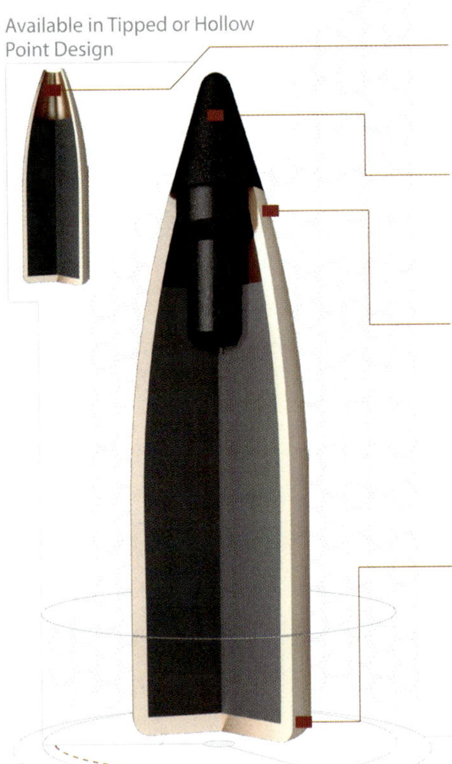

Hollow Point:
Designed for optimum flight and expansion characteristics

Metallic Black Polymer Tip:
Nestled in the jacket mouth is the streamlined polymer tip.

Ultra-Thin Jacket Mouth:
Assures violent expansion at either end of the velocity scale.

Flat Base:
Proven bench rest accuracy design.

Optimum Performance Velocity:
Minimum: 1600 fps (488 mps)
Maximum: Unlimited

Scan this QR code to go to www.Nosler.com and watch a video about the Varmageddon™

- Dedicated Machinery Means Uniform Consistency
- Production Lots Are Never Mixed
- Each Bullet is Visually Inspected
- Our Quality is Legendary!

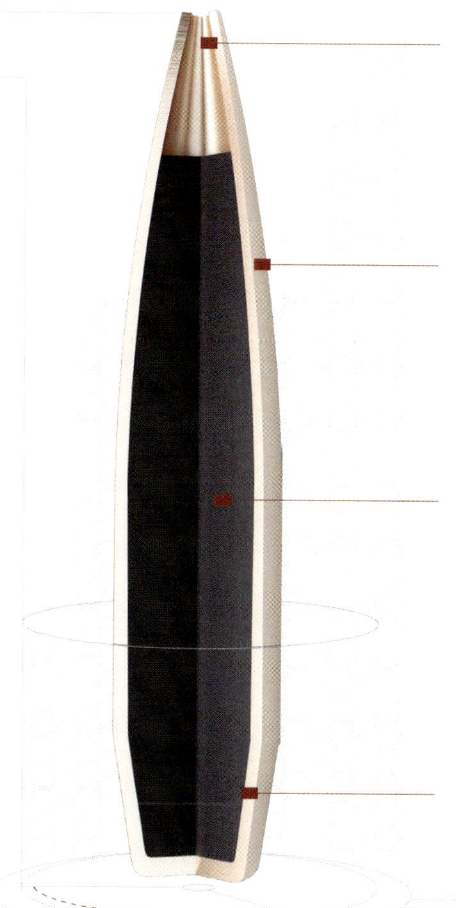

Nosler Hollow Point:
Provides small meplat for reduced drag and increased aerodynamic efficiency.

Custom Quality Jacket:
Engineered to exhibit unmatched accuracy.

Form-Fitted, Lead-Alloy Core:
For maximum bullet stability and balance.

Pronounced Boat Tail Design:
Provides excellent flight characteristics over a range of velocities.

Scan this QR code to go to www.Nosler.com and watch a video about the Custom Competition™

SPORTING™ REVOLVER HANDGUN

- Dedicated Machinery Means Uniform Consistency
- Production Lots Are Never Mixed
- Each Bullet is Visually Inspected
- Our Quality is Legendary!

Jacketed Soft Point
JSP

Jacketed Hollow Point
JHP

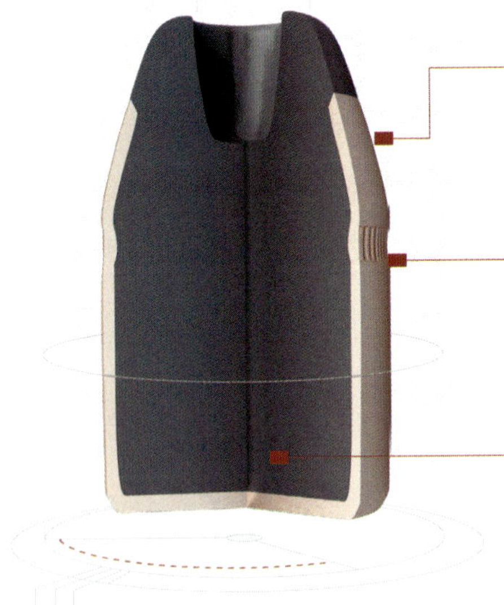

Tapered Copper-Alloy Jacket:
Extremely concentric, tapered jacket is built for maximum accuracy and enhanced terminal performance.

Well-Defined Cannelure:
Holds jacket and core together and permits a tight crimp that eliminates bullet movement within the case.

Form-Fitted Pure Lead Core:
Allows for reliable expansion and maximum accuracy across a broad velocity range.

Scan this QR code to go to www.Nosler.com and watch a video about the Sporting Handgun™

800 FPS

1100 FPS

1250 FPS

1350 FPS

SPORTING™ PISTOL HANDGUN

- **Dedicated Machinery Means Uniform Consistency**
- **Production Lots Are Never Mixed**
- **Each Bullet is Visually Inspected**
- **Our Quality is Legendary!**

Full Metal Jacket
FMJ

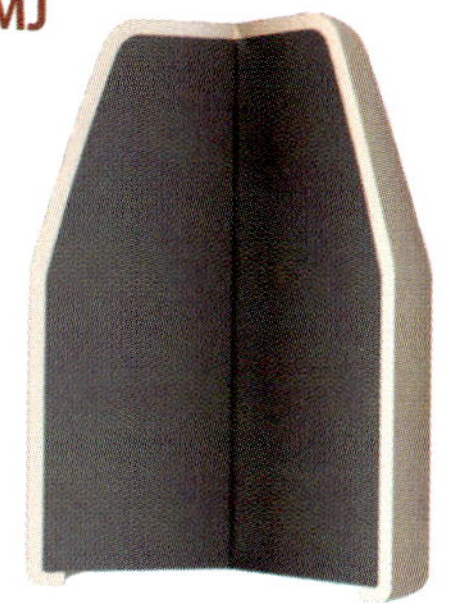

Jacketed Hollow Point
JHP

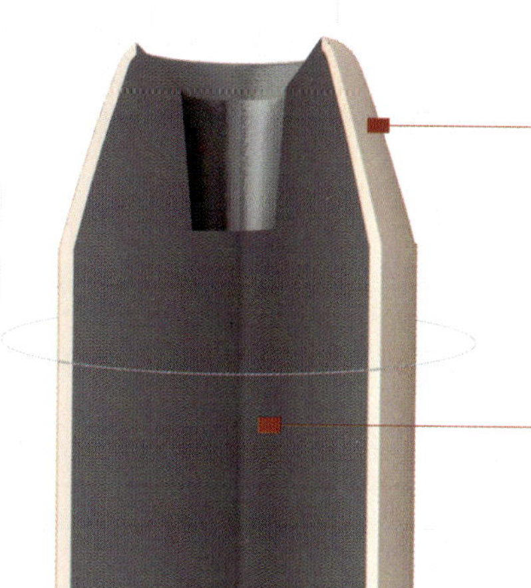

Tapered Copper- Alloy Jacket: Extremely concentric, tapered jacket is built for maximum accuracy and enhances terminal performance.

Form-Fitted Pure Lead Core: Allows for reliable expansion and maximum accuracy across a broad velocity range.

Scan this QR code to go to www.Nosler.com and watch a video about the Sporting Handgun™

800 FPS

1100 FPS

1250 FPS

1350 FPS

Nosler® BRASS

FULLY PREPPED, READY TO LOAD

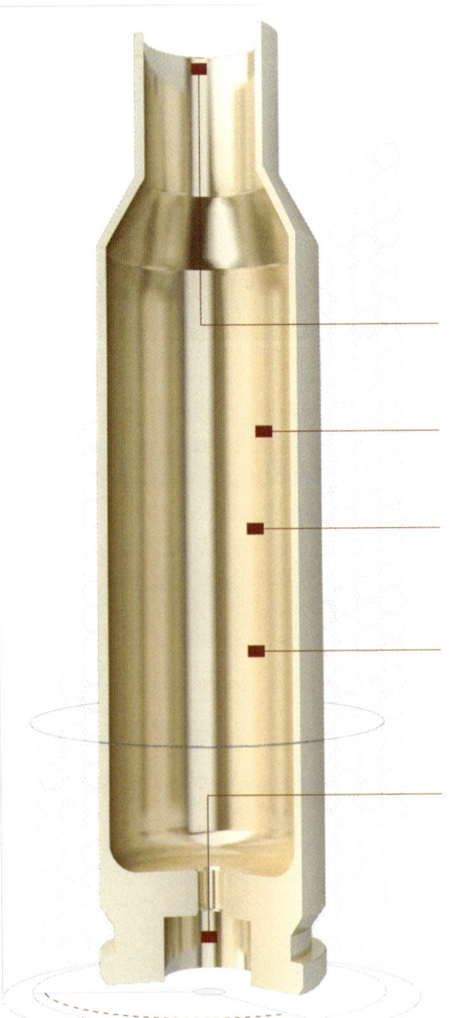

Case mouths are chamfered and deburred

NoslerCustom® Brass is hand-inspected and weight-sorted

Quality made, NoslerCustom® Brass is packaged in quantities of 25, 50 and 100 count boxes

Each piece of brass is full length sized and trimmed to proper length

Flash holes are deburred and checked for proper alignment

Each piece of NoslerCustom® cartridge brass bears the "Nosler" headstamp and is manufactured under the company's strict policy of "quality first." With uncompromising attention to detail, each round of NoslerCustom cartridge brass is made to precise dimensional standards and tolerances using top-grade materials to maximize accuracy and consistency while extending case life. Each piece of brass is completely prepped and ready to load. Case mouths are deburred and chamfered. Flash holes are deburred, and then checked for proper alignment. Every piece of brass is full-length sized, and then visually inspected to ensure that it is free from manufacturing flaws or blemishes. Each box of brass is weight-sorted so that every piece in a box is weight-matched within +/- 1/2 grain. NoslerCustom cartridge brass comes packaged in boxes of 25, 50 or 100 pieces.

Flash holes are deburred and checked for proper alignment

BEFORE | AFTER

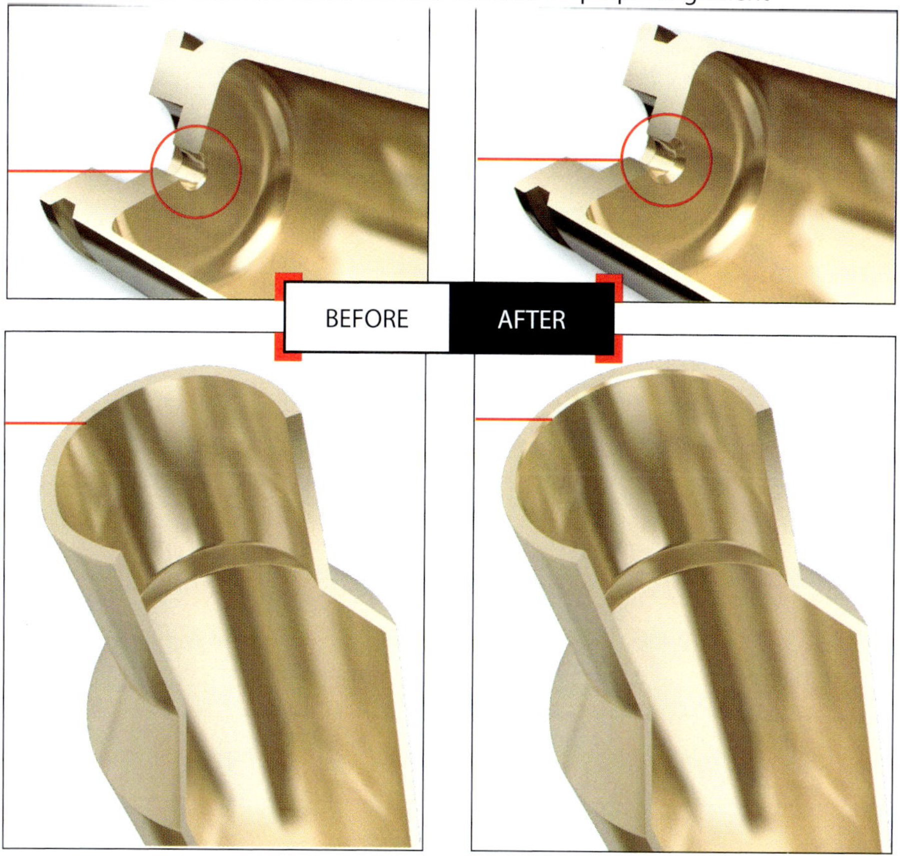

Case mouths are chamfered and deburred

GETTING READY TO RELOAD

BASIC RELOADING PRECAUTIONS

1. Modern ammunition uses smokeless powder as the energy source. Smokeless powder is much more powerful than black powder or Pyrodex®. Never substitute smokeless powder for black powder or Pyrodex® and never mix it with either.
2. Follow loading recommendations exactly. Don't substitute components for those listed. Start loading with the minimum powder charge listed in the data.
3. Never exceed manufacturers' reloading data. Excess pressures caused by excessive loads could severely damage a firearm and cause serious injury or death.
4. Understand what you are doing and why it must be done in a specific way.
5. Stay alert when reloading. Don't reload when distracted, disturbed or tired.
6. Set up a loading procedure and follow it. Don't vary your sequence of operations.
7. Set up your reloading bench where powder and primers will not be exposed to heat, sparks or flame.
8. DO NOT smoke while reloading.
9. ALWAYS wear safety glasses while reloading.
10. Keep everything out of the reach of small children.
11. Keep your reloading bench clean and uncluttered. Label components and reloads for easy identification.
12. Do not eat while handling lead.
13. NEVER try to dislodge a loaded cartridge that has become stuck in the chamber by impacting it with a cleaning rod. Have a competent gunsmith remove the round.

SMOKELESS POWDER

All smokeless powders obviously have to burn very fast, but handgun and shotgun powders must burn faster than rifle powders. You will readily note the differences in physical size and shape of various powders, but you cannot see differences in chemical composition that help to control the rate of burning. Burn rate is also affected by pressure, "hot primers," seating the bullet too deep, over crimping the case onto the bullet, tight gun chambers, oversize bullets, use of heavy shot loads and anything that increases friction or confinement of the powder will increase the pressure. Obviously this hobby requires attention to detail, patience and meticulousness to ensure the safety and quality of loads produced.

1. NEVER mix powders of different kinds.
2. Use the powder ONLY as recommended in manufacturer reloading manuals.
3. Store powder in a cool, dry place.
4. If you throw or measure powder charges by volume, check-weigh the charges every time you begin loading, occasionally during loading and when you finish.
5. Pour out only enough powder for the immediate work.
6. NEVER substitute smokeless powder for black powder or Pyrodex®.
7. Don't carry powder in your clothing. Wash your hands thoroughly after handling it.
8. Store powders only in original package. Don't repackage.

9. Keep powder containers tightly closed when not in use.
10. Specific powders are designed for specific uses. Don't use them for other purposes.
11. Smokeless powder is EXTREMELY FLAMMABLE. To dispose of deteriorated powders, follow recommendations in The Properties and Storage of Smokeless Powder—SAAMI Item #200 (http://www.saami.org).
12. Empty the powder measure back into the original powder container when through with a reloading session. DO NOT MIX POWDERS.
13. Clean up spilled powder with a brush and dustpan; do not use a vacuum cleaner because fire or explosion may result.

PRIMERS

Priming materials differ in brisance (initial explosive force) and in the amount of hot gas produced. Don't mix primers of different makes.
1. Don't decap live primers. Fire them in the appropriate gun—then decap.
2. Don't ream out or enlarge the flash hole in primer pockets. This can increase chamber pressure.
3. Over-ignition creates high chamber pressures. The best results are obtained by using the mildest primer consistent with good ignition.
4. Don't use primers you can't identify. Ask your local police or fire department to dispose of unidentifiable or non-serviceable primers.
5. Keep primers in the original packaging until used. Return unused primers to the same package. Don't dump together and store in bulk. There is risk of mass detonation if one is ignited.
6. If resistance to seating or feeding of primers is felt, STOP and investigate. DO NOT FORCE PRIMERS.
7. Store primers in a cool, dry place. High temperature, such as in a summer attic, causes them to deteriorate.
8. Don't handle primers with oily or greasy hands.
Oil contamination can affect ignitability.
9. There have been instances of "primer dusting" in the tubes of loading tools because of vibration. Clean the machines after each use.
10. Refer to SAAMI reprint "SPORTING AMMUNITION PRIMERS: Properties, Handling & Storage for Handloading." SAAMI Item #201 (http://www.saami.org)

HANDLING LEAD

Lead, a substance known to cause birth defects, reproductive harm and other serious physical injury, must be handled with extreme care. Handle lead bullets or lead shot only in a well-ventilated area and ALWAYS wash hands after handling lead and before eating. Discharging firearms in poorly ventilated areas, cleaning firearms, or handling ammunition also may result in exposure to lead. Have adequate ventilation at all times.

HANDLOADING RIFLE AND PISTOL CARTRIDGES

1. Examine cases before loading. Discard any that are not in good condition.
2. Put labels on boxes of loaded cartridges. Identify caliber, primer, powder and charge, bullet and weight, and date ofreloading.
3. In handgun cartridges, the seating depth of the bullet is extremely important. Handgun powders must burn very quickly because of the short barrel. They are sensitive to small changes in crimp, bullet hardness, bullet diameter, primer brisance and especially to bullet seating depth.
4. Check the overall length of the cartridge to be sure the bullet is seated properly.*
5. If you cast your own bullets, remember their hardness, diameter and lubrication will affect chamber pressure.
6. Plastic cases designed for practice loads (where the bullet is propelled by primer gas only) can't be used for full powder loads.
7. Consult the manufacturer regarding disposal of unserviceable ammunition. Ask your local police or fire department to dispose of small quantities.

*Accumulation of lead or grease in the bullet-seating tool may force the bullet in too far, which will increase chamber pressure. If the bullet isn't deep enough, it may engage the lands of the barrel when loaded. This will also increase the chamber pressure.

PREVENT MISSING AND DOUBLE CHARGES

1. It is easy to double charge if you are momentarily distracted. Use a depth gauge to check powder height in shell. A piece of doweling rod can be used as a depth gauge.
2. Observe the powder level of cases placed in the loading block. This is a way to discover any cases with missing or double powder charges.
3. Take care to operate progressive loaders as the manufacturer recommends. Don't back up the turret or jiggle the handle. Don't use a shell to catch the residue when cleaning out the powder train.

Chamber Pressure

The loading data published in this guide was developed and tested in rifles and handguns that are in excellent mechanical condition. A wide variety of factors can affect chamber pressure which is why we strongly recommend that you always begin loading with the minimum powder charge listed, and approach maximum loads with caution. As maximum loads are approached, even small increases in powder charge can cause large spikes in chamber pressure. Though this data has been developed for use in a wide variety of firearms, keep in mind that your particular firearm may develop high chamber pressures at powder charges lower than those listed as maximum. If at any time, a fired cartridge exhibits signs of high chamber pressure such as those listed, immediately cease firing all cartridges loaded with that powder charge, and reduce the powder charge accordingly. Cartridges that you suspect have been over-charged should be disassembled.

An example of a blown primer and ejector marks, due to high pressure.

Ejector Marks- Shiny marks on the base of a case head indicates that enough pressure was present to flow the metal of the head into the ejector recess of the bolt face. If these marks appear only on new cases, and if older cases using the same load show no ejector marks, it is possible that the new cases are from a production lot that was annealed to a softer condition. In either case, ejector marks are a definite sign of pressure and the load should be reduced accordingly.

Loose Primer Pockets- This condition is detected when new primers seat too easily. Cases with loose primer pockets have usually been exposed to high pressure loads and need to be discarded. Continued use may result in gas leakage around the primer.

Gas Leakage Around the Primer-This is indicated by a black sooty appearance around the primer and is a sign that the primer seats loosely in the primer pocket. Discard these cases and reduce the load to a lighter charge.

Flattened primer

Flattened Primer-Where other signs accompany this condition, it is due to excessive chamber pressure. However, flattened primers can also be an indication that the resizing die was improperly adjusted and has changed the shoulder position of the case. This causes excessive headspace and allows case stretching. The flattening is a result of the case moving forward slightly upon impact from the firing pin. Ignition shoves the primer firmly against the bolt face, and when the pressure from the powder ignition reaches its peak, the case head stretches until it contacts the bolt face, reseating the primer and giving it a flattened surface.

A cratered primer, note the pronounced ring around the firing pin indent.

Heavily Cratered or Extruded Primers-This condition should not be confused with the light cratering that can occur from a poorly fitted or badly worn firing pin. If other pressure signs are present, significant cratering should be regarded as a sign of excessive pressure.

Action Opens with Difficulty or Will Not Open-This is a definite sign of excessive pressure. If the bolt or lever offers resistance to opening or will not extract the case without forcing, it is an indication that the case was jammed tightly against the chamber wall from too much pressure.

Short Case Life-While this symptom can be caused by other factors such as chamber dimensions, excessive headspace or faulty resizing techniques, repeated heavy pressures can shorten case life.

Cartridge Components

Every metallic cartridge is made up of four main components: case, primer, powder, and bullet. There are a wide variety of cartridge components available, manufactured by many different companies. One of the enjoyable aspects of reloading is the ability to experiment with various component combinations searching for the perfect load. It is important to realize that components are designed for specific calibers and specific uses. Bullets designed for target shooting typically won't perform well on game. Understanding the intended use of a particular component will help you choose the right components for a cartridge's intended purpose.

Brass

Cartridge cases or "brass" as they are commonly called are the foundation of a cartridge. The case not only holds all of the components together, but also seals the chamber so that the pressure and gas from the burning powder drives the bullet forward down the barrel. When you think about it, the brass cartridge case is what reloading is all about. Bullets, primers, and powder can only be used once, but a cartridge case can be reloaded over and over again. Reloading wouldn't exist without the unique ability to save and reuse brass cases.

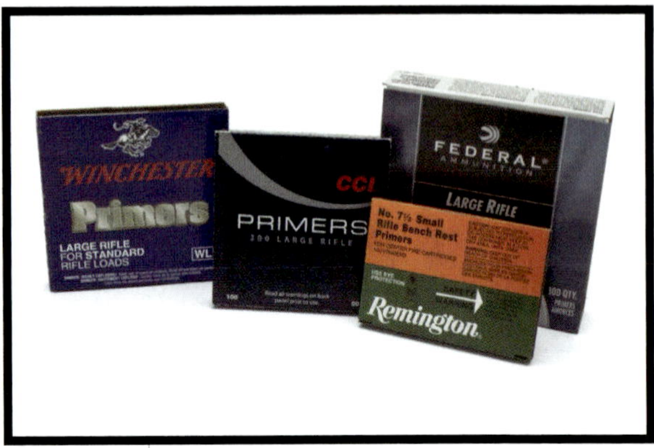

Cases are manufactured and designed for one specific cartridge. Some calibers can be loaded in cases formed from brass intended for other calibers, but the vast majority of reloading is done using brass made for the cartridge being loaded. Cases can be obtained either by buying packages of new unfired brass, or by reusing once-fired cases from commercially loaded ammunition. The advantage to buying new brass is that you can obtain larger quantities of brass from the same manufacturing lot. Using cases from the same lot helps to improve consistency which will in turn improve accuracy. At a minimum, only cases with the same head stamp should be used with a particular load. Cases from different manufacturers can have variations in internal capacity or volume due to differences in case wall thickness. Changes in internal case volume will result in variations in chamber pressure, so powder charges should be reduced and reevaluated when changing the brass used with a specific load. If a load is already at or near maximum pressure, loading it into a case with a smaller internal volume could result in an over pressure load. Keeping brass separated by head stamp not only promotes safety, but also helps to improve consistency and accuracy.

Primers

The primer is what ignites the powder charge, thereby firing the cartridge. The primer also seals the base of the case, containing the pressure needed to drive the bullet down the barrel. Using the proper primer for the cartridge is absolutely essential to loading safe and effective ammunition. Primers come in several sizes to match the different sizes of cartridge cases. Primers also have varying hardness's based upon their size and intended use. In general, rifle primers are made with harder cups and larger amounts of priming compound. The harder cup is necessary to contain the higher pressure of rifle loads, while the larger priming pellet is necessary to ignite the larger volume of powder. Pistol primers are typically softer and contain less priming compound due to the lower pressures and smaller powder charges associated with pistol cartridges. In all but a very few cases, pistol and rifle primers should never be used interchangeably. Though they may appear to be the same, they are not. Some primers come with a magnum designation for use

in magnum pistol and rifle cartridges. Magnum primers produce a longer, hotter flame which is necessary to sufficiently ignite the larger powder charges used in magnum cartridges. It is recommended to always use the type and size of primer listed in the load data that you are using.

Match grade primers are also available in both large and small rifle sizes. Match grade primers are simply manufactured to tighter tolerances. These tighter tolerances are meant to increase uniformity and consistency between primers, which should enhance accuracy. Whether or not a match primer will produce better results with your chosen load is something that can only be discovered through experimentation. Typically, standard, non-match primers will produce very accurate loads, so you may never have a need to try match primers.

Powder

Finding the powder and bullet combination that works best in your rifle is the heart of load development. Typically, there are a wide variety of powders that can be used in a particular cartridge. Picking which powder to try first can be difficult as there are so many good powders available to try. Nosler's Reloading Guides are extremely helpful as they list which powder produced the most accurate results in a particular cartridge during testing. Starting with the powder recommended as most accurate or "favorite" in the manual can greatly accelerate the load development process. Often, you may have a particular bullet that you want to use, so powder type, charge weight and cartridge overall length will be the variables that you adjust to find a load that will achieve the desired results with your chosen bullet. Always begin loading with the minimum load published.

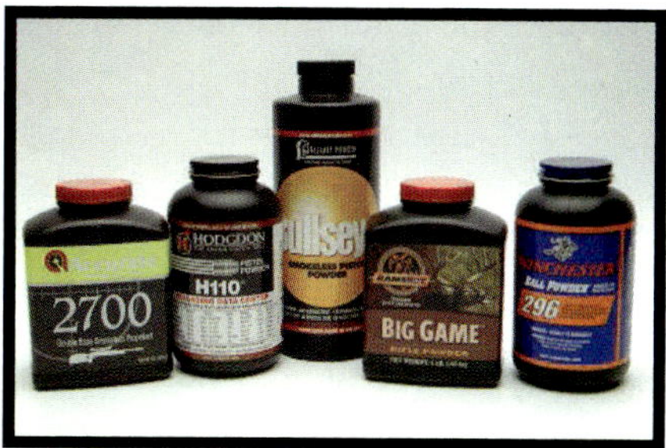

Modern smokeless powder can be either single or double-based. These terms refer to the chemical composition of the powder. Single-based powders are those made solely of nitrocellulose. Double –based powders are made up of both nitrocellulose and nitroglycerine. The individual granules of both types of powder are further coated with a variety of chemicals to control burn rate, increase storage life, decrease temperature sensitivity, decrease muzzle flash, and aid in powder handling. Some powders even contain copper inhibitors that are designed to limit barrel fouling. Whether a powder is single or double-based is usually of little

Example of the three types of powder: (L to R) Flake, Stick (Extruded), and Ball

concern to the handloader, and has no real bearing on how the powder may perform in a particular load.

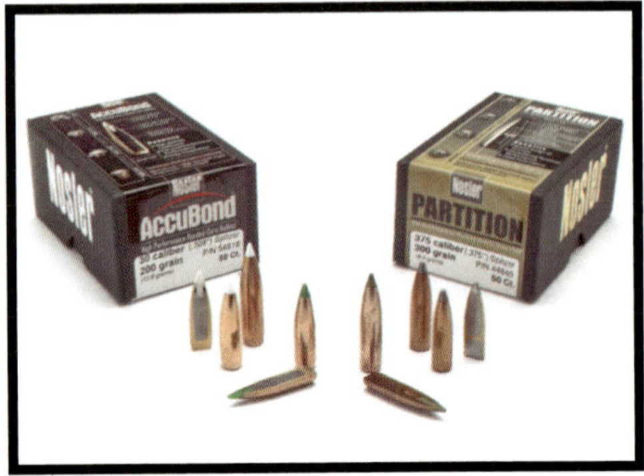

The three main types of powders are ball, stick, and flake powder. These terms refer to the shape of the individual granules of powder. The shape of the powder has little effect on how well it will perform in a particular load. The biggest difference between the shapes is how well they flow through powder measures. Ball powders tend to meter the best through powder measures simply due to the shape of the granules which allows them to flow through a measure very smoothly and consistently.

Powders also vary by their burn rate. Powders are typically rated by how fast or slow they burn. There is no calibrated rating scale for powder burn rate, but rather a broad range running from fast to slow, with powders placed along the scale according to their individual burn rates. A powder's individual burn rate will have some correlation to how it performs in different cartridges and different barrel lengths. In general, magnum cartridges and heavy bullets perform better when slower burning powders are used. Always use published load data from trustworthy sources. There is plenty of data available for a wide variety of powders no matter what cartridge you are loading. If there is no data available for a particular powder in your chosen cartridge, contact the powder manufacturer and ask for their recommendation. Realize however, if data isn't available for a certain powder in a specific cartridge, it is probably because that powder doesn't perform well in that cartridge.

Bullets

Of all the components of a cartridge, selecting the proper bullet for the intended purpose of the cartridge is the most critical. The purpose for which a cartridge is going to be used has little bearing on one's choice of powder, primer, or case, but bullets are designed and constructed for specific purposes. Hunting bullets are designed for maximum terminal performance on game. Match or target bullets are manufactured to provide the highest levels of accuracy with no regard to terminal performance.

Depending on the caliber being loaded, there may be dozens of different bullets to choose from. While one must only use bullets of the proper caliber, bullets of varying weights are usually available. Within a specific caliber, a reloader can choose from a wide variety of bullet brands, types, and weights. It is usually easiest to work backwards having first decided what the bullet will be used for, then choosing weight and brand. Or, you may simply decide that you want to use a particular brand and are willing to find a load that will work with the bullets made by a certain bullet company in your caliber.

Note the differences in construction and jacket thickness. L to R Varmint Bullet, Target Bullet, Big Game Hunting Bullet

Bullet Construction
To perform reliably on game, a hunting bullet needs to expand enough to transfer energy while retaining enough weight to ensure adequate penetration to reach the vitals after passing through muscle and bone. This necessity requires hunting bullets to be constructed with thicker jackets and either bonded or divided lead cores. Monolithic hunting bullets also perform very well as they combine controlled expansion with near one hundred percent weight retention.

With target bullets, the only concern is how the bullet performs in flight to the target. Once the bullet has struck the target, it has done its job and it doesn't matter how the bullet performs in regard to expansion and penetration. In general, target bullets have much thinner jackets, one-piece soft lead cores, and no bonding between the jacket and core. If a target bullet were to be used on game, it would likely expand and fragment far too quickly to achieve adequate penetration. Or, in the case of a full metal jacket or FMJ bullet, the bullet may simply pass through the animal without expanding, transferring little energy and causing minimal damage. While heavily constructed hunting bullets can and often do provide match grade accuracy, rarely do target bullets provide desirable terminal effects on big game.

For varmint hunting, rapid bullet expansion for complete energy transfer is desired. Varmint bullets are lightly constructed and designed to fragment easily upon impact providing complete energy transfer and explosive terminal performance. A bullet that will fragment readily is desirable for varmint shooting as misses are common and a bullet that will fragment quickly when it hits the ground or rocks is far less likely to skip or ricochet. Varmints are small targets, so high levels of accuracy are also desired.

Weight
Bullets of a given caliber are available in a variety of weights measured in grains. More common calibers will typically have a broader range of bullet weights available where less common calibers may only be available in one or two weights. Bullet weight has a large effect on velocity and recoil. A cartridge can only produce so much energy, and this energy can either push a light bullet fast or a heavier bullet more slowly. Heavy bullets will generally retain their energy better at long range while drifting less due to wind. Heavier bullets also penetrate better due to their greater mass and higher sectional densities. Lighter bullets typically expand more rapidly due to their higher velocities.

When choosing a hunting bullet, it is important to match the bullet weight to the game you will be pursuing. Generally, the bigger the game, the heavier the bullet you will want to use. Heavy bullets work fine on smaller game, but lighter bullets typically don't work as well on larger game. The caliber of the rifle you will be using also has an effect on what bullet weight will work for your intended game. If you are using a caliber that is considered very powerful for the game you are hunting, then your bullet weight will be less critical than if you are using a caliber that is on the lower end of the power scale. If you plan to use a smaller caliber for medium to large game, you should use the heaviest bullets available in that caliber.

Which weight bullet you choose will also be dictated by your individual rifle's performance. All rifles will prefer a particular bullet weight. The twist rate of your rifle's barrel will affect what weight bullets it will shoot the best. Most commercial rifle barrels are cut to a twist rate that will work well for the most common weights available in that caliber. If you start to experiment with bullets that are either very light or very heavy for caliber, you may experience performance issues due to twist rate. If you start loading much heavier bullets and accuracy drops off dramatically, it is likely that your twist rate is too slow to adequately stabilize the heavier bullets. In general, faster twist rates are necessary for stabilizing longer bullets. If you have a choice of twist rate, it is best to pick the twist rate that will stabilize the heaviest bullet you intend to use as lighter bullets will be stabilized by it as well. One problem to consider is that a twist rate which is too fast can cause light weight, lightly constructed varmint bullets to literally spin themselves apart in midair. As long as you are using bullets that are properly constructed for their intended use, it is more important to use the bullet that shoots the most accurately than to pick a bullet based solely on its weight.

(L to R) Varmint, Target, and three Hunting bullets. Notice the differences in construction.

Ballistic Coefficient

Ballistic coefficient is a numerical value which provides an indication of a bullet's ability to overcome air resistance as it travels through the air. A bullet's ballistic coefficient tells us how aerodynamic a bullet is. Ballistic coefficient is affected by a bullets point shape, weight, ogive, length, and base style. Lately, as interest in long range hunting and marksmanship has grown, shooters and reloaders have become more aware of and interested in the importance of ballistic coefficient. Ballistic coefficient is important as a bullet with a higher ballistic coefficient will retain its velocity and energy over a longer distance. This results in a "flatter" shooting bullet, meaning the bullet will drop less at a given range than a bullet of the same weight and muzzle velocity with a lower ballistic coefficient. Velocity retention results in greater energy on target. A bullet with a higher ballistic coefficient will also be less susceptible to the effects of wind drift.

Within normal hunting range, 300 yards or less, ballistic coefficient really isn't a large factor in bullet performance. The advantages of a higher ballistic coefficient simply cannot be realized until the target gets out to the 600 yard mark and beyond. Of all the qualities that make a good hunting bullet, a high ballistic coefficient is the least important.

Sectional Density

Sectional density is a mathematical value of a bullet's weight in relation to its diameter. The reason that sectional density is important to hunters is that it provides an indication of how well a bullet will penetrate. The higher a bullet's sectional density, the better its ability to overcome resistance when penetrating a solid medium. If velocity and construction are equal, the bullet with the higher sectional density will penetrate more deeply. Sectional density is part of the reason heavier bullets are recommended for heavier game. In a given caliber, the heaviest bullets will have the highest sectional densities, offering the greatest penetration potential. Using sectional density to compare bullets only works well if the bullets are of the same or similar caliber. It is inaccurate to use sectional density to compare two bullets of greatly different calibers. Bullet construction must also be the same when comparing sectional densities. A varmint bullet with a high sectional density still won't perform on large game as well as a heavily constructed hunting bullet with a lower sectional density.

Reloading Equipment

Example of a Single Stage press

Single Stage Press

The reloading press is one of the simplest yet most essential tools in the reloading process. Basically, all the press does is hold a die or dies which the casing to be loaded is forced into by the ram of the press. The press provides the leverage and force required to push a case into a die for resizing as well as the force necessary to push a bullet into the case neck. Presses range from very simple and basic single stage presses to large, complex progressive presses. A single stage press can only accommodate one die and one case at a time, so it can only perform one step of the reloading process at a time. Single stage presses are known for being extremely strong and rugged. Due to the fact that they are only performing one operation at a time, single stage presses are also very easy to set up and use. Single stage presses are ideal for someone learning to reload as there is only one operation happening at once and fewer variables to pay attention to. A single stage press can handle all the operations necessary to load a complete cartridge, and the only thing to be gained by using larger progressive presses is speed. The strength, simplicity, precision, and level of control make the single stage press the press type of choice for those shooters seeking to produce very consistent, highly accurate ammunition.

Turret Press

Turret presses are the next step up in size and production speed. A turret press is similar to a single stage press in that the ram still only holds one case, but the turret press will hold multiple dies in a rotating turret at the top of the press. The purpose of a turret press is to allow the handloader to perform multiple reloading operations in sequence without removing the case from the press. Since the turret holds multiple dies, the various steps of the process can be completed in sequence

Example of a Turret press

Example of a Progressive press

simply by rotating the turret and running the case back up into the next die. Turret presses also provide a time savings by eliminating the need to remove and replace individual dies between loading operations. It is possible to buy several turrets so that dies can be left set up in the turret, and caliber changes on the press can be achieved by simply changing the turret and the shellholder if necessary.

Progressive Press

The largest and most complex presses are progressive presses. Progressive presses are designed to hold multiple dies as well as multiple cases so that all steps of the reloading process can be performed simultaneously. With a progressive press, once the shellplate is full, every stroke of the ram produces a finished round of ammunition. Progressive presses greatly increase production speed by automating steps such as powder measuring that would otherwise have to be performed by hand. Also, once a case is in the shell plate, there is no need to remove and replace it between steps. Progressive presses are ideal for someone who needs to load a large number of the same cartridge quickly. Due to the complexity of progressive presses and the number of adjustments involved, they are not well suited to loading multiple calibers on the same press. Progressive presses are typically used by shooters who load for high-volume competition such as Cowboy Action or IPSC.

Dies

Dies are essential for resizing fired cases to the proper dimensions. This allows the case to fit into the chamber of a firearm as well as allow for seating a new bullet into the mouth of the case. There are a variety of dies available for performing various steps of the reloading process. Some dies perform multiple operations at once while others only perform one specific operation at a time.

Full Length Sizing Die

The main purpose of a full length sizing die is to return a fired case back to the specified dimensions for the cartridge.

Full Length die with Decapping/Expander Button assembly removed.

The Sporting Arms and Ammunition Manufacturers' Institute or SAAMI establishes industry standards for cartridge and firearm chamber dimensions. For a case to chamber easily in a firearm, its outside dimensions must be slightly smaller than the internal dimensions of the chamber. When a cartridge is fired, the brass case expands outward against the walls of the chamber. Brass has a unique ability to spring back to the shape it had before being expanded, but it doesn't quite spring back all the way to its original dimensions. To return the case to the proper dimensions for trouble free insertion and extraction, a full length sizing die is used to reshape the entire body and neck of the case.

A full length die will also contain the decapping rod and the sizing button. The decapping rod presses the spent primer out of the primer pocket by means of a small pin which extends down through the flash hole. The neck expander or sizing button expands the neck of the case back to the proper dimensions for gripping the bullet.

Small Base Die

A small base die is basically an extra tight, full length sizing die. A small base die will size brass down to the minimum SAAMI specification for that cartridge. Small base dies are typically used to size brass that will be fired in semi-automatic, pump, or lever-action rifles. Small base dies insure that brass is sized properly for reliable chambering and extraction in semi-automatic rifles. Reloaders who are experiencing poor functioning in their semi-automatic rifles with reloads may experience better performance through the use of a small base die. Small base dies are typically only available in calibers commonly chambered in semi-auto rifles.

Bullet Seating Die

The bullet seating die seats the bullet into the neck of the case and if desired, will also crimp the case mouth around the bullet. The bullet seating die is adjustable so that the seating depth of the bullet within the case can be varied. This is how changes to the overall length of the cartridge can be made.

The crimp is typically unnecessary for most bottleneck rifle cartridge reloading. Crimps are used on handgun cartridges and cartridges intended for use in semi-automatic or tubular magazine fed rifles. A crimp may also be applied to cartridges which produce heavy recoil to prevent bullet shift.

Neck Sizing Die

A neck sizing die is designed to only size the neck of the case. If a particular piece of brass is only going to be fired in one rifle, it need not be full length resized after every firing. Unnecessarily sizing brass will overwork it and contribute to shortened case life. Many shooters also believe that neck sizing only, produces more accurate ammunition. The drawback to using neck sizing only, is that after cases have been fired several times, they will be difficult or impossible to chamber as the body will have expanded too much to fit back into the chamber of your rifle. Like the full length die, the neck die also contains a decapping rod and neck expander.

Bushing Dies

A very specialized type of sizing die is a bushing die. Bushing dies use interchangeable bushing inserts to very precisely control the amount of sizing performed on the neck of a case. With traditional dies, there is no way to adjust the amount of sizing performed on the neck to account for varying brass thickness. With bushing dies, the neck sizing operation is performed by interchangeable bushings that come in a range of sizes. With these bushings, a reloader can closely adjust the amount of sizing performed on the neck of the case. Minimal neck sizing helps to preserve concentricity and case life while also providing greater control over neck tension. This promotes consistency which is a necessary part of accuracy. For most reloaders, this level of precision is unnecessary, but for those involved in long range competition, it is a necessity. Bushing dies are available for either neck or full length sizing.

Carbide Sizing Die

Special dies which feature a titanium carbide sizing ring are available for resizing straight wall rifle and pistol cartridges. In these dies, the resizing of the case is performed by the small carbide ring held within the body of the die. Carbide is much harder, smoother, and generates less friction than steel, so carbide sizing dies can be used without first lubricating the cases. Since most handgun reloading is a high-volume affair performed on progressive presses, carbide dies eliminate the need to lubricate all of the cases being loaded. Carbide dies also last longer due to the increased hardness of the carbide. Because of the higher material cost for carbide and the additional machining required, carbide dies typically cost twice the price of a traditional steel die.

Crimp Die

Some cartridges require the case mouth to be crimped into the cannelure or band of the bullet. The crimp prevents bullets from shifting within the case during recoil. Most bullet seating dies include an integral crimp ring, but crimping the case and seating the bullet in one step can result in bullet deformation when using soft lead pistol bullets. Better results and enhanced accuracy can be achieved by performing bullet seating and crimping in two separate steps. For this reason, separate crimp dies are available which only perform the crimp, and no other process. They are available in both taper and profile crimp styles. A taper crimp is used for cartridges that headspace on the case mouth while a roll crimp is used for rimmed cartridges which headspace on the rim of the case.

Expander Die

When reloading straight wall cases, a separate expander die must be used to expand the mouth of the case so that a bullet may be seated. An expander die does not size the outside of the case at all, but simply expands the case mouth slightly so that a new bullet can be inserted.

Decapping Die

A universal decapping die is designed to remove the spent primers from cases without sizing the case. As the die doesn't size the case at all, decapping dies are made to accommodate a wide range of case sizes and calibers. Universal decapping dies are used to remove spent primers before the case cleaning process to allow for cleaning of the primer pocket.

Die sets

Dies are available either individually or sold in sets. Dies are most commonly sold in sets of two or three. The most basic set of dies is a standard full length, two die set for reloading rifle cartridges. A two die set contains a full length sizing die and a bullet seating die. Deluxe three die rifle sets are also available that consist of a full length die, a neck sizing die, and a bullet seating die. Die sets for straight wall cartridges come with three dies, a full length sizing die, expander die, and bullet seating die. The more specialized dies such as body and crimp-only dies are sold individually.

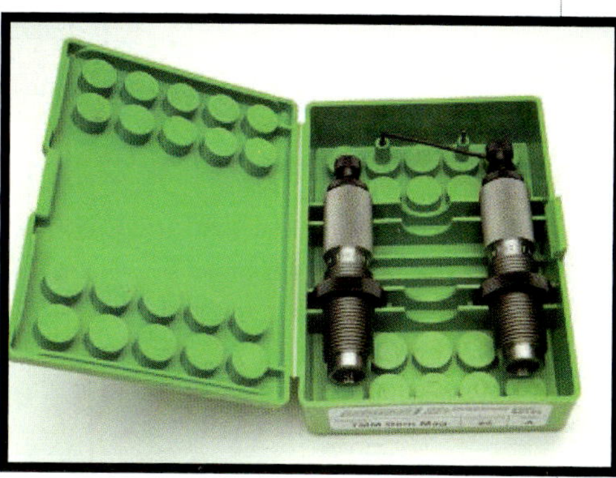

Case Preparation

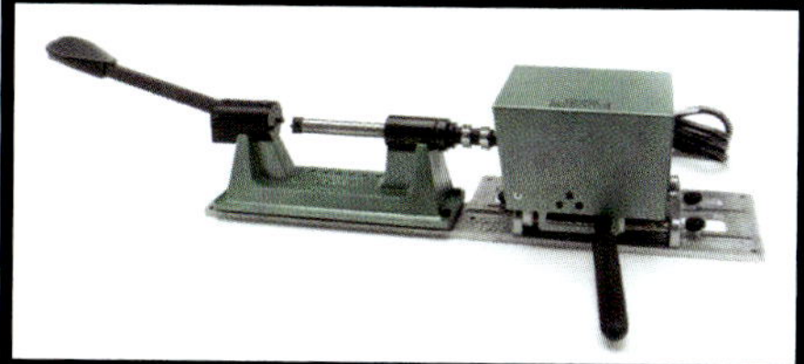

Case Trimmer

As cases are fired and resized, they will stretch and grow longer. The excess length of the case has only one direction to go and that is out the neck, so cases need to have their necks trimmed occasionally to keep the total case length within safe limits. A case that is too long will extend too far into the chamber of a rifle and effectively be crimped between the bullet and the chamber wall which will in turn create a pressure spike higher than the same load would with a case of the proper length. How much a case stretches is largely dependent on the case design and the pressure of the load. In general, high velocity, tapered, bottleneck cartridges will stretch their cases faster than low pressure, straight walled cartridges. It is important to measure your cases after resizing and trim the whole batch if you find any that are over SAAMI specified length.

There are a variety of different case trimmers available ranging from simple trim dies to electrically powered miniature lathes. The most common type of case trimmer resembles a miniature lathe. One end of the trimmer holds the case by its base while the other end holds a sharp cutting head. On some trimmers, the case rotates while the cutter remains stationary while on others the opposite is true. Trimmers are adjustable to accommodate a wide range of cases and trim lengths. Once the trimmer is set up, it will trim all of the cases to the same length.

Most case trimmers are powered by hand, but some feature electric motors to make the process quicker and easier. Many manual trimmers come with an adapter that allows them to be powered by a cordless drill or screwdriver. Whether you choose a manual or powered trimmer will be dictated by your preference and how many cases you need to trim at a time.

Example of a deburring tool

Deburring Tool

After cases have been trimmed, the mouth will be cut square leaving sharp burrs on the inside and outside of the case mouth. These burrs need to be removed so that they don't shave the bullet as it is seated, or hang up on the inside of the seating die or the rifle's chamber. A deburring tool is a simple hand-held cutter that removes the burr and slightly chamfers the inside and outside of the case mouth. The simplest and least expensive deburring tools are used by hand, but manually cranked and motor-driven deburring tools are available. As with a case trimmer, the type of deburring tool you choose will be determined by the number of cases you need to deburr at one time.

Primer Pocket Cleaner

A primer pocket cleaning tool uses a brush or sharp head to manually scrape residue from primer pockets after they have been deprimed. As with trimmers

Example of a primer pocket cleaner

and deburring tools, there are both manual and powered primer pocket cleaning tools available. Some reloaders believe that it is important to clean primer pockets between loadings, while some do it once every few loadings and others don't do it at all. The common thinking is that cleaning the residue from primer pockets helps the primers to seat better and more consistently. It is also thought to improve accuracy. Many reloaders clean rifle cartridge primer pockets, but don't clean primer pockets in handgun cases. Whether you choose to clean your primer pockets or not is up to you.

Case Neck Brush

Before resizing cases, it is a good idea to clean out the inside of the case neck. This is easily accomplished by the use of a case neck brush. Running a brush in and out of the case neck a few times will remove any excess powder residue or cleaning

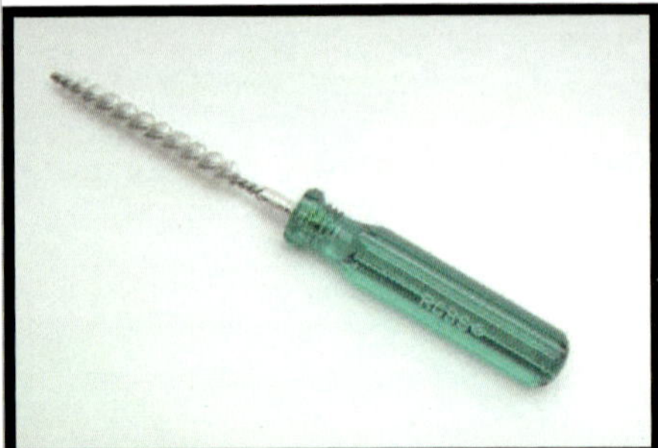

media that is left in the case neck. There are nylon brushes made specifically for this purpose, but the same result can be achieved, perhaps even better, by using a bore brush of the same caliber that you are loading. Case neck brushes can be operated by hand, used with a powered tool, or chucked in a drill and powered by the drill.

Case Tumbler

Cleaning cases between loadings is an important step in the reloading process. Clean cases resize easier and function in firearms more smoothly. Running clean cases into your dies will greatly enhance the longevity of your dies and help to prevent cases from getting stuck. There are a variety of ways to clean fired cases, but by far the most common is to tumble them with cleaning media in a vibratory case cleaner. Vibratory case cleaners, commonly called "tumblers," consist of a plastic bowl with lid mounted on top of

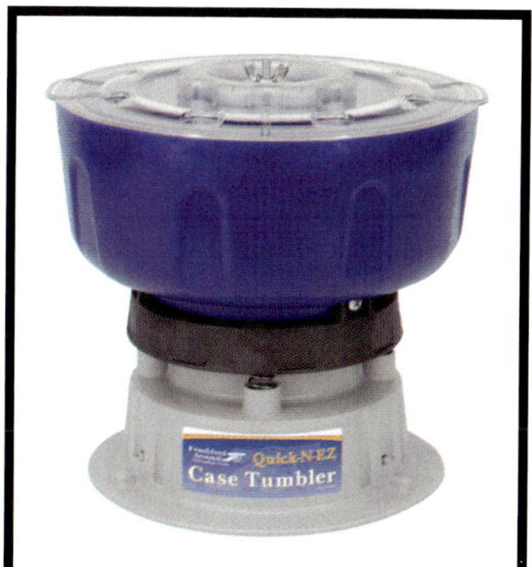

an electric motor. Cleaning media, such as ground corncobs or ground walnut shells, is placed in the bowl of the tumbler along with the brass to be cleaned. The motor then shakes the bowl in such a way that the cases and media are turned over and over which allows the media to scrub the cases clean. Depending on how dirty the cases are, cleaning in a tumbler can take anywhere from a half hour to several hours. One of the nicest things about a tumbler is that once you fill it up and turn it on, you can simply walk away and let it do its job. Various types of dry or liquid polish can also be added to the tumbling media which will speed up the process as well as make the cases extra clean and shiny. Cases need only be clean for proper function, but many loaders like to polish them simply for aesthetic reasons. After tumbling, cases can be removed from the media either by hand or through the use of a media separator which is similar to a colander and allows the media to fall out while keeping the cases inside. Once cases are separated from the media, they are immediately ready for reloading.

There are other ways to clean fired cases such as washing or cleaning in an ultrasonic cleaner. These methods are effective as well, but the main drawback to "wet" case cleaning methods is that the cases must then be allowed to dry before they are ready for reloading. Decapping cases before using wet cleaning methods both allows the primer pockets to be cleaned as well as allowing them to dry out faster after cleaning.

Case Lube

Case lube really isn't a piece of equipment, but it is a definite necessity for reloading. Sizing cases, particularly full-length sizing, creates a lot of friction as a case is forced into the die by the ram of the press. Without sufficient lubrication, a case can become stuck in the die rendering it unusable until the stuck case can be removed. Case lube reduces the friction between the case and die so that cases can easily be run in and out of the die. There are a variety of case lubes available on the market which can be sprayed on, applied with fingers, or applied with the use of a case lube pad. Using liquid case

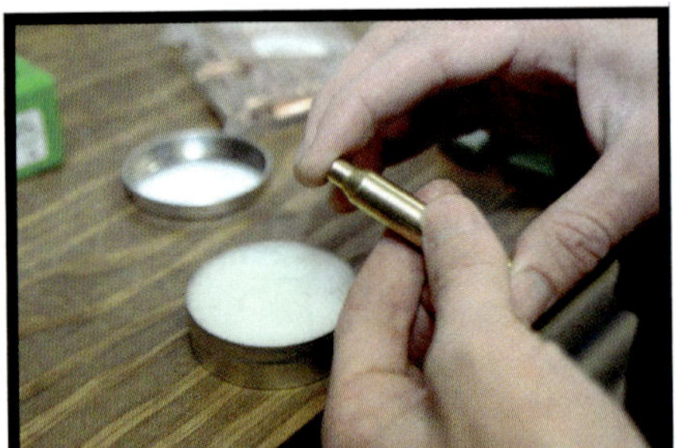

lubes that aren't of the aerosol spray type, requires the use of a lube pad. A lube pad is a dense pad to which lube is applied and then the cases to be lubricated are rolled across by hand. Which lube to use is a matter of personal preference and preferred application method. For lubing a large number of cases at once, spray lubes seem to be the quickest and most popular. **It is important not to lubricate the shoulder of the case as this can cause denting.**

Priming

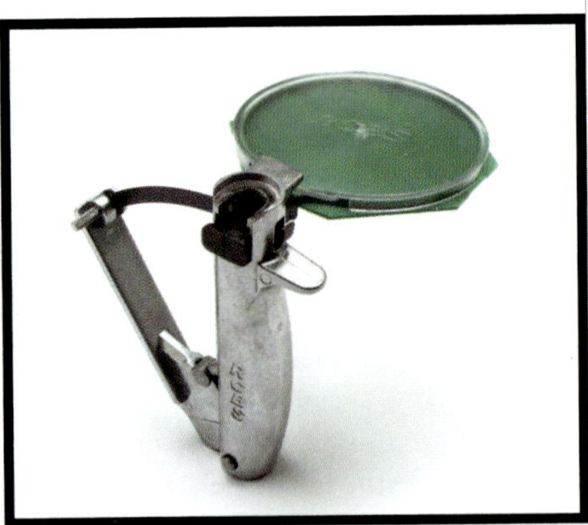

Priming Tool

A key operation in reloading is priming your brass by inserting a new primer into the primer pocket. Most presses have an attached priming arm, but many reloaders prefer to use a separate hand priming tool. Hand priming tools usually incorporate a priming tray that holds primers and feeds them into the priming tool one at a time. A benefit of the separate priming tool is that most eliminate the need to handle primers with your fingers, which could potentially contaminate them with oil from your skin. Larger turret and progressive presses usually incorporate some type of manual or automatic priming system.

Primer flipping tray

While not an essential piece of equipment, a primer flipping tray is a handy

tool to have on your loading bench. A primer flipping tray is simply a small, flat, plastic tray with grooves milled into its surface. When primers are dumped into this tray and the tray is shaken from side to side, the grooved floor of the tray catches the edge of the primers and flips them anvil side up. This is important as it allows you to easily pick up and place primers into a priming tool by hand with a minimum amount of handling. If you are going to be using a hand priming tool with a feed tray, a separate flipping tray is unnecessary. To minimize handling primers, always try to dump out only as many primers as you need for the batch of cases you are currently priming.

Powder Handling

Scale

An accurate scale is an absolute necessity for reloading. A scale is primarily used to weigh a powder charge before it is poured into a case. Even when a powder measure is used to meter out powder, a

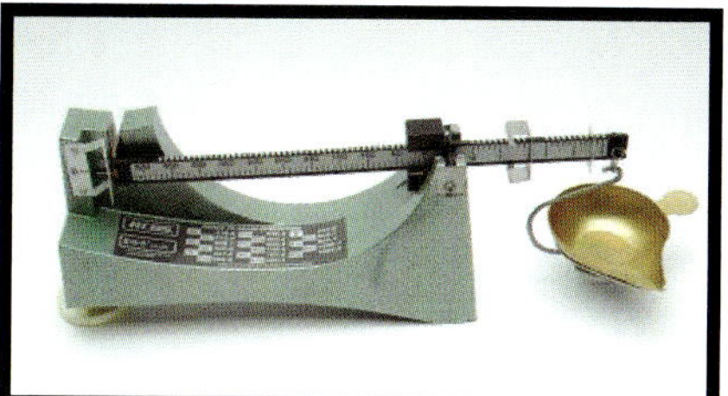

scale is necessary for initially setting up the powder measure to meter out the desired charge as well as for periodically check-weighing a charge thrown from

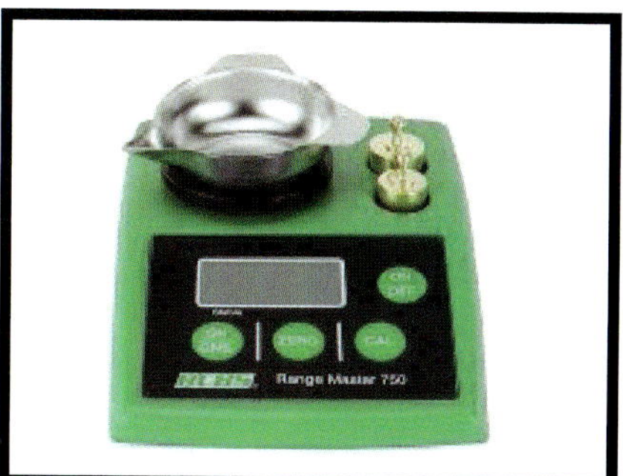

the powder measure to confirm that it is still metering properly. A scale can also be used to weigh bullets or brass when sorting components for uniformity. For a very high degree of accuracy, many reloaders will individually weigh each powder charge when loading rifle cartridges. This is the only way to be absolutely sure of loading the exact same powder charge in each cartridge. Individually weighing charges is also advised when loading at or near maximum load levels where even a small degree of overcharging could result in a high pressure situation.

The two main types of scales are mechanical and digital. Mechanical scales use a balancing beam and movable counterpoises to weigh items placed in a hanging pan. Digital scales are typically flat and have a digital read out which displays the weight of anything placed on the scale. Digital scales are somewhat faster to use, but their accuracy can be affected by a variety of factors such as low batteries or close proximity to other electronic devices. Mechanical scales are slightly slower to use, but their simplicity makes them less susceptible to interference from outside sources. Both types have their own advantages and disadvantages, so the choice of which to use comes down to personal preference. All scales must be zeroed before using, and should be checked for accuracy periodically by the use of a check weight. Whichever type of scale is used, it must be used on a flat, level surface to ensure consistent, accurate measurements. Reloading scales are so sensitive they can be affected by drafts and air movement, so they should be used in an area away from air vents or other sources of air flow.

Powder Measure

A powder measure is not a necessity for reloading, but it can certainly speed up

the process. A powder measure dispenses powder charges that have been measured by volume by means of an adjustable metering chamber. Measuring powder by volume isn't quite as accurate as measuring by weight on a scale, but it can be very close. In fact, once they are properly set up, high quality powder measures can repeatedly throw consistent powder charges again and again. The accuracy of powder measures is affected by the consistency of their use, so a consistent motion when operating the measure is critical for consistent results. Powder type can also affect the operation and consistency of a powder measure. Ball powders typically meter much more easily than stick powders simply due to the shape of the granules.

A powder measure and scale work together as a team and while a scale can be used without a measure, a measure cannot be set up without a scale. A scale is necessary to verify the weight of powder being metered out by the powder measure. Once the measure is set up, the scale is still used to periodically verify that a consistent weight of powder is being thrown by the measure.

The powder measure is also very useful when every charge is being weighed. For this, the measure is set up to throw a charge just slightly below the desired weight, then a trickler is used to bring the charge up to the desired weight on the scale. The powder measure speeds up this process by getting the charge close to the correct weight before being placed on the scale. A powder measure isn't necessary for weighing charges, but it can really speed things along.

Powder Trickler

A Trickler adds powder to the scale a few granules at a time.

A powder trickler is a simple device that allows you to add powder to your scale pan one or two granules at a time by simply turning a knob. Tricklers consist of a small hopper with an internally threaded tube running through the bottom of the hopper. The tube extends out either side of the hopper and is open on one end. The tube also has a hole in it so that powder can flow from the hopper into the tube. As the tube is turned, the threads inside move powder down its length and out the end very slowly. This tool allows you to slowly add very small amounts of powder to the pan on the scale when bringing a charge up to the desired weight. A trickler is a very handy tool as it is all but impossible to add such tiny amounts of powder any other way. The way in which the trickler adds powder so slowly makes it very controlled and easy to add just enough powder to the pan without adding too much.

Powder Funnel

Even large caliber case mouths are relatively small, so a powder funnel is absolutely essential for pouring powder into cases. Most powder funnels are made out of plastic and are designed to accommodate a range of calibers. The funnel

actually sits over the outside of the case mouth and has small steps cut into it to fit various common calibers. When purchasing a powder funnel, be sure that the caliber you intend to load is included in the range of the funnel you are going to buy. Caliber specific funnels are also available for those who load a large number of the same caliber, or simply desire the enhanced stability of a caliber specific funnel. A funnel isn't necessary for charging cases directly from a powder measure, but it is essential for filling cases from the pan of your scale.

Tools and Accessories

A Reloading Manual

Without a doubt, the most important piece of reloading equipment is a good reloading manual. A reloading manual is absolutely critical to loading safe and effective ammunition. Reloading manuals published by either bullet or powder manufacturers contain a wealth of data that is the result of many thousands of hours of research and development. To reload safely, it is necessary to use powder charges which will operate within acceptable pressure limits for your firearm. Powder charges that are either too heavy or too light can both result in unsafe situations. A reloading manual will provide you with the safe minimum and maximum load for a certain bullet/powder

combination in the cartridge that you are loading. Manuals also provide suggestions for which powders to use based on which ones produced the best results in the testing laboratory. Always use published data, and always begin loading with the minimum powder charge.

Loading Block

Loading blocks are flat plastic or wooden trays with holes in them designed to hold cases upright by the base. Loading blocks are very useful as they help to keep the reloading process organized. Cases can be kept organized by moving them from one end of the block to the other after a par-

ticular step has been performed. They are particularly useful when charging a large number of cases from a powder measure.

Calipers
Calipers are another tool that are an absolute necessity for reloading. Calipers allow you to measure items very accu-

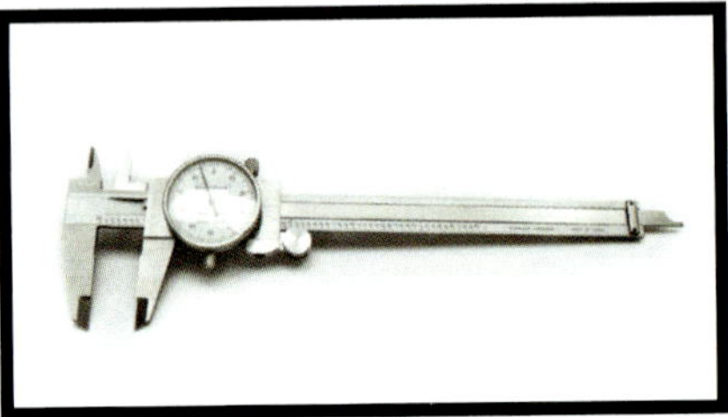

rately down to thousandths of an inch. In reloading, calipers are used to measure case lengths as well as overall cartridge length. Measuring case lengths between firings is important as cases that are too long can result in dangerously high pressure. It would be next to impossible to set up a bullet seating die properly without a set of calipers. There are two common types of calipers, digital and dial. Digital models require batteries and feature a digital read out for measurements. Dial calipers don't require batteries and feature a traditional dial read out for measurements. Both types work equally well, so picking one is just a matter of preference.

Stuck Case Removal Kit
Sooner or later, everyone gets a case stuck in a die. When a case becomes stuck in a die, the only safe way to remove it without damaging the die is to use a stuck case removal kit. Most kits are the same and include a drill bit, tap, threaded bolt, spacer block and Allen wrench. To remove a stuck case with such a kit, first the primer pocket is drilled out with the included drill bit. Next, use the tap to cut threads in the hole that was just drilled through the web of the case. Finally, place

the spacer block over the base of the die and head of the case, then thread the bolt into the case. As the bolt is turned in the case, its head will come to rest on the spacer block and the threads will pull the case out of the die. Be sure to clean the die afterwards as there will be brass shavings in it from the case removal process. A stuck case removal kit isn't a tool that you will need to use very often, but when you do need it, there is no other alternative.

Bullet Puller
Eventually, most reloaders end up with loaded rounds that for one reason or another they do not want to fire, but they want to salvage the components from. The only way to safely disassemble loaded rounds without firing them is to use a bullet puller. A bullet puller makes it possible to safely remove the bullet from a cartridge without damaging the bullet or the case. Bullets and brass can then be

reloaded if desired. Bullet pullers come in two styles, kinetic and collet. Kinetic bullet pullers look like a plastic hammer and use the force of inertia to pull bullets from cartridges. A cartridge is placed into the head of the puller, and an internal collet supports the cartridge case. When the puller is struck against a hard surface, the inertia of the bullet pulls it forward and out of the case. A collet or clamp style bullet puller looks like a die, and is used in conjunction with a reloading press. This type of puller clamps onto the bullet with a caliber-specific collet which holds the bullet while the ram of the press pulls the cartridge case down and away. The advantage of the inertia pullers is that they are inexpensive and can accommodate a wide range of cartridges. The drawback is that when a bullet is pulled, the powder spills out of the case all over the inside of the puller. The collet style pullers exert much more force, and don't spill the powder, but they require caliber specific collets that must be purchase separately.

Step By Step Reloading:

1. Clean and inspect fired cases. Discard any cases with split necks, heavy ejector marks on the case head, stretch marks ahead of the web area, or any other signs of damage.

2. Sort cases by headstamp, reloading cases from only one manufacturer at a time.

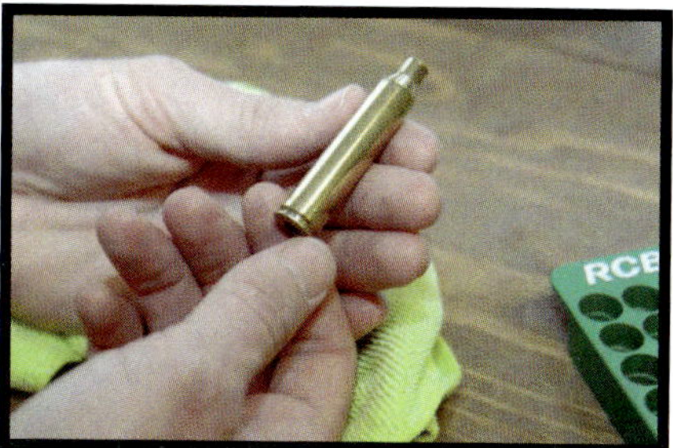

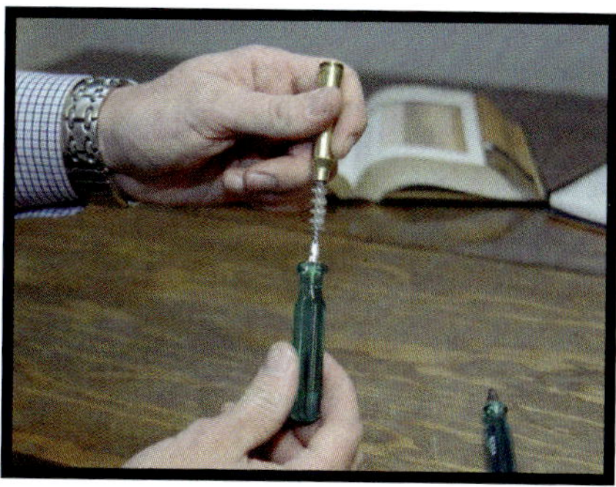

3. Using a case neck brush, clean the inside of the case neck to remove any powder residue.

4. Lightly lubricate cases with a lube pad, your fingers, or a spray lube. Be careful to avoid excess lubrication as it will result in small dents in the case. Do not lubricate the shoulder of the case.

Resizing

5. Insert a lubed case into the shellholder in the press holding the resizing die. Making sure that the case is fully

inserted into the shellholder, begin running the case into the die with a slow, deliberate pull of the operating handle.

6. You should feel the sizing, decapping, and interior neck sizing (expander ball) operations. Only moderate force is required. If you feel resistance as the case is being run into the die, back it out and apply some more lube to the body of the case with your fingers.

7. After sizing the cases, wipe the lube off with a rag, and place the sized cases in your loading block separate from the unsized cases.

8. Using a caliper, measure the cases' overall length to be sure they do not exceed SAAMI maximum length.

9. If the cases are too long, trim them to the proper length with a case trimmer.

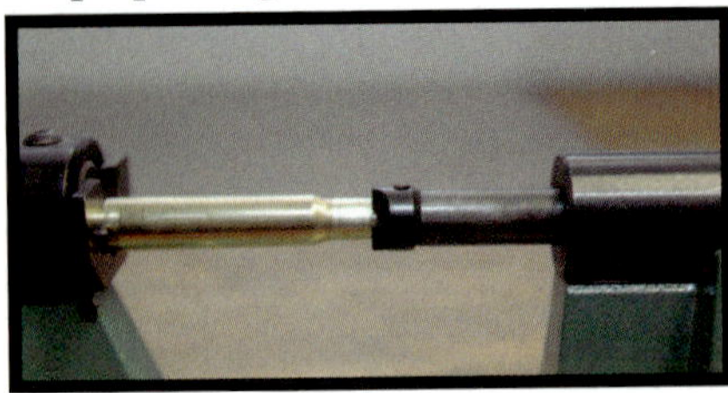

10. With a deburring tool, remove rough edges from the inside and outside of the case neck.

11. Remove any carbon from the primer pocket with a primer pocket cleaner.

12. If you are loading handgun or straight-walled rifle cartridges, the next operation is belling the case mouth. Most handgun die sets include a third die, called an expander die. Set up this die according to the manufacturer's instructions. Handgun case mouths must be expanded slightly to ease bullet seating and to prevent collapsing the case.

13. To prime your cases, remove the resizing die from the press. If your press has a priming arm, place a primer in the cup, then push it into the recess in the ram. Seat the primer with a firm, consistent pressure until you feel

the primer touch the bottom of the pocket. You should be able to feel slight resistance as the primer is seated in the primer pocket. If the primer seats without resistance, it is an indication that the primer pocket is worn and the case should be discarded. Otherwise, the primer may dislodge during firing.

14. If you're reloading in large batches, a priming tool will speed up the priming process. Priming tools, either bench-mounted or hand-held, have tubes or trays which feed the primers onto the priming arm, without having to handle them individually.

15. Check each primed case to be sure that the primer is fully seated. Primers should be slightly recessed below the surface of the case head.

Charging the cases

16. Charging cases with powder is one of the most important and critical operations in reloading. Great care must be taken to follow load data and drop the precise amount of powder specified in the data.

17. Once you have selected the proper load from the data, set up your reloading scale to weigh precisely the right number of grains. Remember, your powder scale will weigh powder down to the closest 1/10th of a grain. Use a powder trickler for adding minute amounts of powder to the scale pan.

18. Powder charges can either be weighed for every case, or dropped via the use of a powder measure. When using a powder measure, be sure to periodically check-weigh the amount of powder being dropped to verify that your loads are safe and consistent.

19. After you have charged a loading block full of cases, "eyeball" them to spot any missed or over-charged cases. Note: When reloading handgun cartridges, extra care must be taken against double charges since the volume of powder is quite small.

Bullet Seating-Rifle Cartridges

20. Install and adjust the bullet seating die according to the manufacturer's instructions. Most bullet seating dies provide for crimping the case into the bullet's cannelure groove, if one is present. Since most handloaders prefer not to crimp except where it is required, you should adjust the seating die to provide about 1/16" clearance between the

shellholder and the face of the die. This will assure no crimping will take place when the bullet is seated. Make sure the case is fully pushed into the shellholder to assure the case is properly aligned under the die.

21. Place the bullet in the case mouth and guide it up into the die while you pull the handle all the way down to seat the bullet to its predetermined depth. The cartridge is now complete.

Bullet Seating-Handgun Cartridges

22. Handgun bullet seating dies provide for crimping the case into the bullet's cannelure if one is present. "Roll" crimping is used for revolvers while "taper" crimping is used for pistols and bullets lacking a cannelure or groove. If you do not require crimping, adjust the seating die to provide 1/16" clearance between the shellholder and the face of the die. Bullets requiring a crimp should be seated down to the cann-

elure. Be sure that the case is seated all the way into the shellholder to assure proper alignment.

23. Place the bullet into the belled mouth of the case, then lower the handle all the way to complete bullet seating. Your cartridge is now complete.

24. When you have finished reloading your cartridges, identify them by date,

bullet, powder type, charge, primer and overall cartridge length. Keep different loads separate, and clearly labeled.

Tips on Bullet Seating

Before you adjust your seating plug, it is necessary to decide what O.A.C.L. you plan to use. In the reloading manual you will see a cartridge diagram which displays the S.A.A.M.I. established dimensions for the cartridge you are loading. This diagram will also list a maximum cartridge overall length. This length has more to do with cartridges being able to fit into a variety of different magazines, and function properly in all action types than anything else. After powder and bullet selection, the O.A.C.L. adjustment is a handloader's best tool for improving the accuracy of a cartridge. O.A.C.L. can be adjusted to fit your particular rifle. When you adjust O.A.C.L., you are changing the distance that the bullet has to "jump" before initially coming into contact with the lands of the barrel. Every rifle has a "sweet spot" where it will shoot the tightest groups. This best O.A.C.L. will be different for different bullets, so a little experimentation is necessary whenever you are changing loads. There are a few factors that must be considered that will affect how much you can adjust your O.A.C.L. for your rifle. First of all, you do not want to seat a bullet so far out that it is actually jammed into the lands when you chamber a cartridge. This can result in an increase in pressure as well as making rounds difficult to chamber. A cartridge that is too long may not fit into the magazine of your rifle, or won't function reliably in the action. Bullets must be seated deeply enough that the cartridge neck can grip the bullet sufficiently. The general rule of thumb is that a bullet should be seated at least as deep as its diameter for proper neck tension and alignment. Bullets that are seated too deeply can also cause an increase in pressure. In general, don't seat the bullets any deeper than necessary, and don't seat them any longer than what will fit into the magazine of your rifle.

To find the ideal O.A.C.L. for your rifle will require testing several loads at various lengths to see which one your rifle likes best. If you are just getting started, you have two options. You can either measure your rifle's "seating depth" or just use the recommended O.A.C.L. listed in your load data.

Determining the Proper Bullet Seating Depth for YOUR chamber

You will need:

1. A cartridge case that has been fired in your gun, and not resized.

2. A bullet of the type to be used, with a full, undamaged nose.

3. Calipers

4. A dark felt-tipped marker.

Step 1: Insert the bullet into the neck of the fired case. It should fall freely into the case, with no resistance.

Step 2: Remove the bullet from the fired case and press the case neck lightly against a flat surface to create a small indentation or flat surface in the case neck so that it will grip the bullet.

Step 3: Insert the bullet, base first, into the case so that the case just grips the bullet by itself. Just get the bullet started into the case, don't seat it too deeply.

Step 4: Completely color the bullet with the marker.

Step 5: Gently insert the case and bullet into the chamber of the firearm, and close the action. Do not pull the trigger.

Step 6: Carefully open the action and gently remove the case.

Step 7: Retrieve the bullet. It will either be stuck up in the lands of the barrel or still in the case. If the bullet is stuck in the lands, it can be removed by tapping the butt of the rifle on the ground. Or, it can be dislodged by gently pushing it out with a cleaning rod. If the bullet is still in the case, then gently remove it with your fingers, taking care not to mar the ink, and proceed to step 8.

Step 8: During Step 5, the lands will have contacted the bullet and pushed it back into the case, causing the case neck to scrape the ink off of the bearing surface of the bullet. Simply push the bullet into the case until the edge of the case neck is just to where the ink has been scraped off.

Step 9: Carefully measure the overall length of the dummy cartridge. This overall length is called your "rifle seating" depth. It is where the bullet contacts the lands of the barrel. This length is different for every different type of bullet, as it depends upon the shape of the ogive (the taper) and the meplat (the tip of the nose) of the bullet. This process should be repeated three or four times to obtain a consistent average.

Step 10:

a: Set your seating die to seat at a depth between .015" and .03" inches less than your rifle seating depth.

b: When loading our E-Tip® bullets, seat the bullets from .050" -.100" away from the lands for optimum accuracy.

c: Lightweight bullets may need to be seated further from the rifling. A depth of one bullet diameter inside the case neck gives good alignment and neck tension for ignition.

d: The overall length must be short enough to function through the magazine.

Technical Assistance

If you feel you need further assistance or instructions, please do not hesitate to contact our Customer Service department. We will be happy to assist you in any way possible, as well as answer any reloading questions you may have. We can be reached at 1-800-285-3701 or by email via our website Nosler.com.

Based on the number of calls we receive on the subject, there seems to be a lot of confusion out there regarding rates of twist and bullet stability. The primary factors in determining bullet stability are twist rate, bullet length, velocity and air density.

Twist Rate

Twist rate is generally expressed as two numbers separated by a hyphen. A 1-10" twist rate (one-in-ten), for example, simply means that the rifling makes one complete revolution for every ten inches of barrel. A twist rate of 1-14" would be called a "slower" twist rate (even though the numbers are higher) because it does not spin the bullet as fast as a 1-10" twist does. The faster the rate of twist, the better or more easily it will stabilize a bullet.

Bullet Length & Density

Bullet length and density is a key factor in bullet stability. The longer and less dense the bullet, the more difficult it is to stabilize. If you had a 30 caliber, 180-grain spitzer style (pointed) bullet made completely of polycarbonate, it would be an extremely long projectile (a couple of inches or more), and very difficult to stabilize. By contrast, a round nosed bullet of the same caliber and weight made completely of lead would be relatively short, denser and would be much easier to stabilize in the same barrel. This explains why certain lead free bullets are longer and sometimes more difficult to stabilize in standard twist barrels than lead core bullets.

Velocity

The faster a bullet goes, the easier it is to stabilize. It does not spin any more than what the twist rate is in the barrel, it simply generates more rotations per second. For example, a bullet coming out of a 1-10" twist barrel will make one complete revolution for every ten inches it travels, even after it leaves the barrel, regardless of velocity. A bullet that is traveling at 3400 FPS will cover ten inches much more quickly than one traveling 2400 FPS, and it can be said that the faster bullet is "spinning faster", but it is simply cranking out more RPM and still not turning more than once every ten inches.

Air Density

Air Density varies with factors such as temperatures and elevation. The lower the elevation (closer to sea level) and the lower the temperature, the denser the air becomes. The denser the air, the more difficult it is to stabilize a bullet. For example, it is much easier to stabilize a long bullet on a 90 degree day in Denver than it is to stabilize the same bullet at 20 degrees on the Alaskan coast.

Now, these are not the things that most shooters need to concern themselves with. Standard bullets coming from the factory are designed to stabilize in barrels with twist rates standard to their caliber.

A very common misconception is "over stability". Bullets do not become "destabilized" or "over stabilized" with faster twist rates, or greater velocities. They may not be quite as accurate as they were with slower twist rates or velocities, but they are stable. An unstable bullet may actually tip or tumble on its way to the target.

Powder Burn Rates: Fastest to Slowest

1) Norma R-1
2) Winchester WAALite
3) Vihtavuori N310
4) Accurate Nitro 100
5) Alliant e3
6) Hodgdon Titewad
7) Ramshot Competition
8) Scot Royal D
9) Alliant Red Dot
10) Alliant Promo
11) Winchester AA Plus
12) Hodgdon Clays
13) Alliant Clay Dot
14) IMR Hi-Skor 700-X
15) Alliant Bullseye
16) Hodgdon Titegroup
17) Alliant American Select
18) Accurate Solo 1000
19) Scot Red Diamond
20) Vectan AS
21) Alliant Green Dot
22) Winchester WST
23) IMR Trail Boss
24) Winchester Super Handicap
25) Hodgdon International
26) Accurate Solo 1250
27) IMR PB
28) Vihtavuori N320
29) Accurate No. 2
30) Ramshot Zip
31) IMR SR-7625
32) Vectan AO
33) Winchester 452AA
34) Scot 453
35) Hodgdon HP-38
36) Winchester 231
37) Alliant 20/28
38) Winchester 473AA
39) Hodgdon HS-5
40) Winchester WSL
41) Alliant Unique
42) Hodgdon Universal
43) Alliant Power Pistol
44) Vihtavuori N330
45) Alliant Herco
46) Winchester WSF
47) Winchester WAP
48) Vihtavuori N340
49) IMR Hi-Skor 800-X
50) IMR SR-4756
51) Scot Solo 1500
52) Ramshot True Blue
53) Accurate No. 5
54) Hodgdon HS-6
55) Winchester 540
56) Winchester AutoComp
57) Ramshot Silhouette
58) Vihtavuori 3N37
59) Vihtavuori N350
60) Hodgdon HS-7
61) Vihtavuori 3N38
62) Alliant Blue Dot
63) Accurate No. 7
64) Alliant Pro Reach
65) Hodgdon Longshot
66) Alliant 410
67) Alliant 2400
68) Ramshot Enforcer
69) Winchester 571
70) Vihtavuori N105
71) Accurate No. 9
72) Accurate 4100
73) Scot 4100
74) Alliant Steel
75) Norma R-123
76) Vihtavuori N110
77) Hodgdon Lil'Gun
78) Hodgdon H-110
79) Winchester 296
80) IMR-4227
81) Hodgdon H-4227
82) IMR SR-4759
83) Accurate 5744
84) Accurate 1680
85) Vectan SP-3
86) Winchester 680
87) Norma N200
88) Scot Brig 3032
89) Alliant Reloder 7
90) IMR-4198
91) Hodgdon H-4198
92) Vihtavuori N120
93) Hodgdon H-322
94) Scot Brig 4197
95) Accurate 2015BR
96) Alliant Reloder 10X
97) Vihtavuori N130
98) IMR-3031
99) Vihtavuori N133
100) Norma N-201
101) Scot Brig 322
102) Hodgdon Benchmark
103) Hodgdon H-335
104) Ramshot X-Terminator
105) Accurate 2230
106) Accurate 2460s
107) IMR-8208 XBR
108) Ramshot TAC
109) Hodgdon H-4895
110) Vihtavuori N530
111) IMR-4895
112) Scot Brig 4065
113) Vihtavuori N135
114) Alliant Reloder 12
115) Accurate 2495BR
116) Vectan 5000
117) IMR-4064
118) Norma N-202
119) Accurate 4064
120) Accurate 2520
121) Alliant Reloder 15
122) Norma N-203
123) Vihtavuori N140
124) Hodgdon Varget
125) IMR-4320
126) Winchester 748
127) Hodgdon BL-C(2)
128) Hodgdon CFE 223
129) Hodgdon LEVERevolution
130) Hodgdon H-380
131) IMR-4007 SSC
132) Ramshot Big Game
133) Vihtavuori N540
134) Winchester 760
135) Hodgdon H-414
136) Scot Brig 4351
137) Vihtavuori N150
138) Accurate 2700
139) Vectan 7000
140) IMR-4350
141) Hodgdon H-4350
142) Alliant Reloder 17
143) Accurate 4350
144) Norma N-204
145) Hodgdon Hybrid 100V
146) Vihtavuori N550
147) Alliant Reloder 19
148) IMR-4831
149) Ramshot Hunter
150) Scot Brig 4831
151) Norma N-205
152) Accurate 3100
153) Vihtavuori N160
154) Winchester WMR
155) Hodgdon H-4831 & H-4831SC
156) Hodgdon SUPERFORMANCE
157) Winchester Supreme 780
158) Norma MRP
159) Alliant Reloder 22
160) Vihtavuori N560
161) Vihtavuori N165
162) Winchester WXR
163) IMR-7828
164) Alliant Reloder 25
165) Vihtavuori N170
166) Winchester 785
167) Hodgdon H-450
168) Accurate Magpro
169) Hodgdon H-1000
170) Ramshot Magnum
171) Hodgdon Retumbo
172) Vihtavuori N570
173) Accurate 8700
174) Hodgdon H-870
175) Vihtavuori 24N41
176) Hodgdon H-50BMG
177) Hodgdon US-869
178) Vihtavuori 20N29

Powder Abbreviations for RG#7

Accurate Powder	Alliant Powder	Hodgdon	IMR	Norma	Ramshot	VihtaVuori	Winchester
A-1680	Varmint	Clays	700-X	Norma R1	Enforcer	Viht N110	WST
A-2015	2000-MR	Titegroup	TRAIL BOSS	Norma 200	Silhouette	Viht N120	WSF
A-2200	4000-MR	HP-38	PB	Norma 201	True Blue	Viht N130	Auto-Comp
A-2230	AR-Comp	Universal	SR7625	Norma 202	Zip	Viht N133	W231
A-2460	RL7	HS-6	800-X	Norma 203	X-Terminator	Viht N135	W296
A-2495	RL10x	H110	SR4756	Norma 203B	TAC	Viht N140	W748
A-2520	RL15	Lil'Gun	SR4759	Norma URP	Hunter	Viht N150	W760
A-2700	RL17	H4227	IMR 4227	Norma 204	Big Game	Viht N160	Supreme 780
A-4064	RL19	H4198	IMR 4198	Norma MRP	Magnum	Viht N165	
A-4350	RL22	Benchmark	IMR 3031	Norma MRP2		Viht N170	
A-5744	RL25	H322	IMR 8208			Viht N530	
MAGPRO	RL33	H335	XBR			Viht N540	
A-No.2	RL50	H4895	IMR 4895			Viht N550	
A-No.5	300-MP	Varget	IMR 4064			Viht N560	
A-No.7	2400	BL-C2	IMR 4320			Viht N570	
A-No.9	Bullseye	H380	IMR 4007			Viht N310	
A-4100	Power	Hybrid 100V	SSC			Viht N320	
A-3100	Pistol	H414	IMR 4350			Viht N32C	
	Unique	H4350	IMR 4831			Viht N330	
	American	H4831	IMR 7828			Viht N340	
	Select	H4831SC	IMR 7828			Viht N350	
	Blue Dot	H1000	SSC			Viht 3N37	
	Herco	Retumbo				Viht 3N38	
	Green Dot	H870				Viht N105	
	Red Dot	US 869				Viht 24N41	
		H50BMG				Viht 20N29	
		CFE 223					
		LVR					
		Suprform					

Discontinued Powders

Nosler Bullets Requiring Faster Twist Rates	
Bullet	**Minimum Required Twist Rate**
22 cal. 40gr Ballistic Tip Lead Free	1-12" Twist
22 cal. 60gr Partition	1-12" Twist
22 cal. 60gr Ballistic Tip	1-10" Twist
22 cal. 69gr Custom Competition	1-9" Twist
22 cal 77gr Custom Competition	1-8" Twist
22 cal 80gr Custom Competition	1-8" Twist
6mm 90gr E-Tip	1-9" Twist
6mm 105gr Custom Competition	1-8" Twist
6mm 107gr Custom Competition	1-8" Twist

Due to the long length and/or heavy for caliber bullets listed above, a faster twist-rate than SAAMI standard is required to reliably stabilize these bullets.

WARNING

RELOADING METALLIC CENTERFIRE CARTRIDGES CAN BE DANGEROUS!

Careless or incorrect handloading procedures can result in serious personal injury. Handloading should only be done after thorough and complete instruction. Since individual handloading procedures, operations and safety practices are beyond our control, we disclaim all liability for any damages which may result.

Reloading any cartridge requires a familiarity with safe, conservative reloading practices. When combined with personal discipline and attention to detail, such practices can help create a margin of safety for both the reloader and the shooter.

1. NEVER BEGIN LOADING WITH THE MAXIMUM POWDER CHARGE SHOWN IN THESE RELOADING DATA TABLES, REGARDLESS OF YOUR EXPERIENCE.

2. NEVER EXCEED MAXIMUM LOADS BECAUSE EXCESSIVE CHAMBER PRESSURE WILL RESULT.

3. DISCONTINUE FIRING AT THE FIRST SIGN OF EXCESSIVE PRESSURE OR UPON NOTICING A CHANGE IN THE REPORT OF THE FIREARM.

Lot-to-lot variations, design changes, or specification changes that are beyond our control may occur in certain components mentioned in this book. To safely use this or any other guide or manual, you must keep yourself well informed on the technical progress of reloading. It is our belief that the information offered will result in safe operation if the proper equipment and components are employed in a reasonable manner. Because Nosler, Inc. has no control over the actual components selected, the manner in which the components are assembled, or the condition or quality of firearms used, no responsibility, either expressed or implied is assumed for the use of this guide or for any other Nosler product. In no event shall Nosler, Inc. be liable for any damages resulting from the use of this book, whether in contract, tort or otherwise, including consequential damages.

Nosler® bullets contain lead, a chemical known to the State of California to cause birth defects, reproductive, or other physical harm. Wear gloves or wash hands thoroughly after handling.

(See Basic Rules for Reloading Safely, Pages 41-43.)

Load Data in this book supersedes all previous Nosler data.

RIFLE CARTRIDGE RELOADING DATA

17 REMINGTON FIREBALL

Standardized about 2007, the 17 Fireball is practically identical to the 17 Mach IV, a decades-old wildcat.

Predecessor of the 17 Fireball, the 17 Remington (standardized in 1971), was revolutionary — the first 17-caliber factory cartridge. Unfortunately, barrel, bullet, and propellant technology of that era limited 17 Remington performance. Many who purchased such a rifle discovered that thorough bore cleaning was necessary after firing only a few shots. Some rifles did far better, but the reputation swiftly spread that before one had fired 20 rounds the bore would foul so seriously that pressure would soar and accuracy would plummet. The stubby, hollow-point bullets presented another significant limitation; velocity dropped off so rapidly that it simply was not suitable for shots at distances where the then long-established king of varmint chamberings, the 222 Remington, still performed well. The finely granulated extruded propellant, designed to freely flow into 17 Remington cases, necessarily had an unusually heavy deterrent application and therefore generated relatively heavy fouling.

For most shooters, these limitations more than offset the advantages: miniscule recoil — see more hits and misses; high velocity — flat trajectory and explosive performance on close-range targets; and, limited ricochet danger — safe for use in more areas.

Modern barrels, bullets, and propellants change all that. Today the 17 Remington offers impressive usefulness but, in my opinion, the 17 Fireball does it one better. Despite a lower pressure limit and the fact that it has only three-fourths the case capacity, with maximum loads using the plastic-tipped Nosler Varmageddon, the 17 Fireball far outperforms any original 17 Remington factory load or handload — bullet design matters.

When hunting smaller species of vermin, the 17 Fireball is fully suitable for shots to 300 yards. For most of us, most of the time, that is about as far as we can expect to routinely make good hits. Therefore, the 17 Fireball

is the premier minimalist factory varmint chambering: it has modest recoil and muzzle blast; barrel heating is not a concern; barrel life is significantly greater than bigger varmint chamberings; component cost is relatively low; fouling is modest; and, with the right tools, these cartridges are a pleasure to make.

Most varminters will find that, most of the time, the 17 Fireball does most of what any larger varmint chambering will do; and, that it has significant advantages over most of those.

Mic McPherson

Mic McPherson is the author of Metallic Cartridge Handloading.

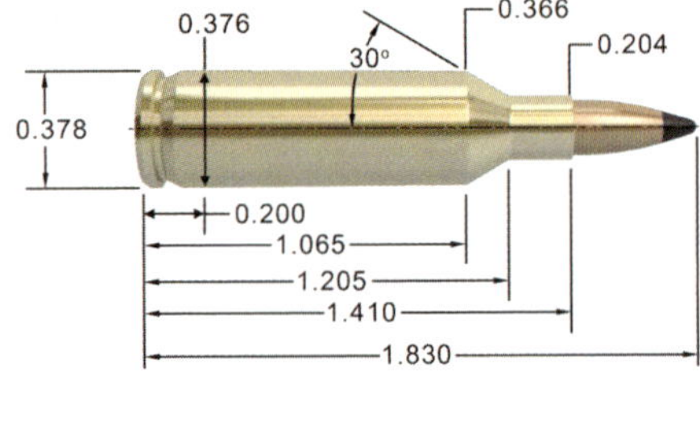

Maximum S.A.A.M.I. Overall Cartridge Length: 1.830"

BULLET CHOICES FOR THE 17 REMINGTON FIREBALL

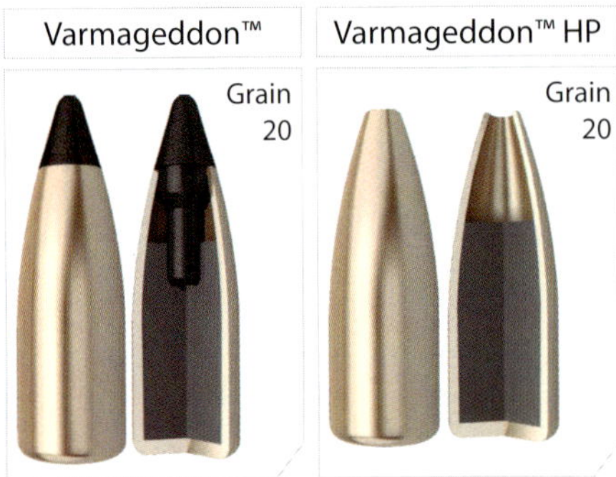

TECHNICAL INFORMATION

The 17 Remington Fireball is light on recoil, easy on powder and almost as fast as its big brother, the 17 Remington. Using a slightly modified 221 FB case it is similar to the popular 17 Mach IV wildcat cartridge. Due to the small bore diameter, clean burning powders are preferred to reduce fouling. Benchmark would be our second choice after IMR 8208 XBR for the Fireball.

17 Remington Fireball - 20 grain		MAXIMUM S.A.A.M.I. O.A.C.L.		1.830"
		TESTED O.A.C.L.	B.C.	S.D.
Varmageddon™	20gr. FBHP	1.780"	0.119	0.097
Varmageddon™	20gr. FB Tipped	1.800"	0.183	0.097

CASE TYPE:	Remington		PRIMER TYPE	Rem 7 1/2
CASE HOLDS:	21.7	Gr. WATER	BARREL Length/Make	26" Pac-Nor
			BARREL Twist	1-10"

POWDER TYPE	POWDER CHG. GRS.		MUZZLE VEL. F.P.S.	LOAD DENSITY (VOLUME)
RL7	16.5	MAX.	3970	83%
	15.5		3799	78%
	14.5 *		3630	73%
A-2015	18.5	MAX.	3991	97%
	17.5 *		3788	92%
	16.5		3611	86%
IMR 4198	16.5 *	MAX.	4053	86%
	15.5		3902	81%
	14.5		3651	76%
TAC	20.5	MAX.	4129	96%
	19.5		3942	92%
	18.5 *		3778	87%
H322	19.0	MAX.	4132	95%
	18.0 *		3943	90%
	17.0		3788	85%
IMR 8208 XBR	20.0 *	MAX.	4148	** 101%
Most Accurate Powder Tested	19.0		3977	96%
	18.0		3783	91%
Benchmark	20.0	MAX.	4246	100%
	19.0 *		4020	95%
	18.0		3820	90%

BC=Ballistic Coefficient SD=Sectional Density
*Most Accurate Load Tested **Compressed Load

Use Maximum Loads with Caution
Refer to page 73 for additional safety information

Layne Simpson

17 REMINGTON

Introduced in 1971 in the Remington Model 700 rifle, the 17 Remington shoots as flat as the 22-250 and burns about the same amount of powder as the 223 Remington. Muzzle blast is not at all bad. Recoil is so close to nonexistent, the shooter can observe his bullet striking yon varmint right there inside the scope, thereby eliminating the need for a spotter. An accurate rifle chambered for this cartridge will shoot five good bullets inside half an inch at 100 yards all day long.

Based on reloading die sales at RCBS, the 17 Remington is the most popular cartridge of its caliber, with the 17 Mach IV/17 Fireball behind it and the 17 Ackley Hornet in third place. While not as popular as it deserves to be, enough varmint shooters buy it each year to keep companies like Remington, CZ USA, Cooper, Sako and a few others producing rifles chambered for it.

Our biggest 17-caliber factory cartridge has its critics with the lazy ones repeating what was written decades ago rather than going to the trouble of actually loading the cartridge with a modern bullet and shooting it in a top-quality barrel with a smooth bore. Other handy items we now have that were not available back in the old days are really good cleaning rods, bore brushes and powder funnels of the proper size and solvents that dissolve bullet jacket fouling.

The 17 Remington is not a logical choice for a one-rifle varmint battery simply because its tiny bullet does not buck breezes as well as heavier bullets of larger calibers. But the wind does not always blow hard in varmint country and the shooter who decides to add a second rifle to his collection won't go wrong if one of them is chambered for this fun little cartridge.

Layne is Executive Field Editor for Shooting Times, RifleShooter and Guns & Ammo, and Senior Writer for BuckMasters GunHunter. He has also authored books on rifles, shotguns, handguns and hunting.

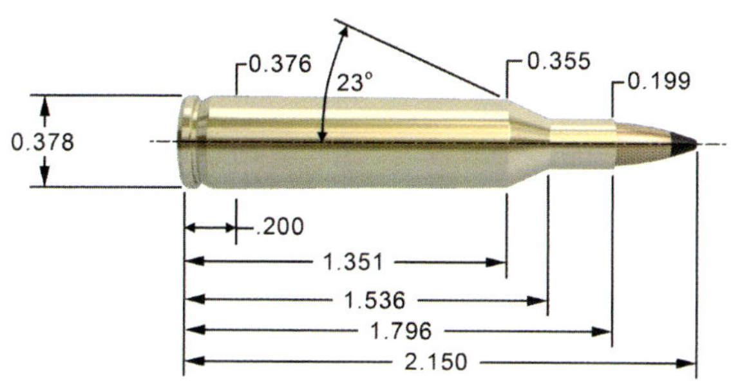

Maximum S.A.A.M.I. Overall Cartridge Length: 2.150″

BULLET CHOICES FOR THE 17 REMINGTON

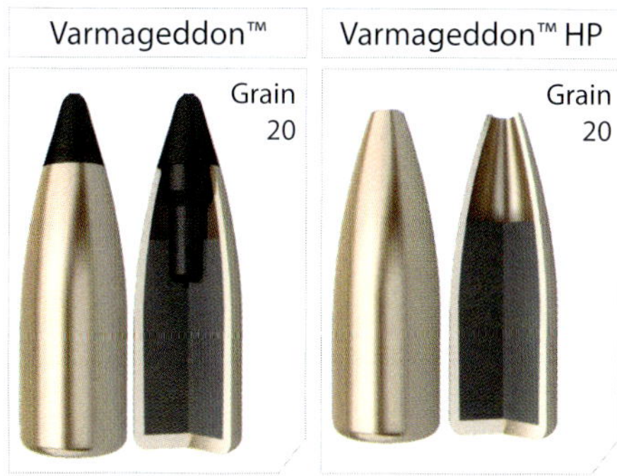

TECHNICAL INFORMATION

Based on a necked down 223 Remington case, it is generally chambered in bolt rifles, but could be used in AR platforms. An excellent varmint cartridge if the wind is not blowing too hard it shoots fast and flat with little noise or recoil. The explosiveness of the 17 cal. bullets make it a good choice for saving pelts on furbearers and reducing ricochets in populated areas. H322 would be a good second choice powder in this cartridge.

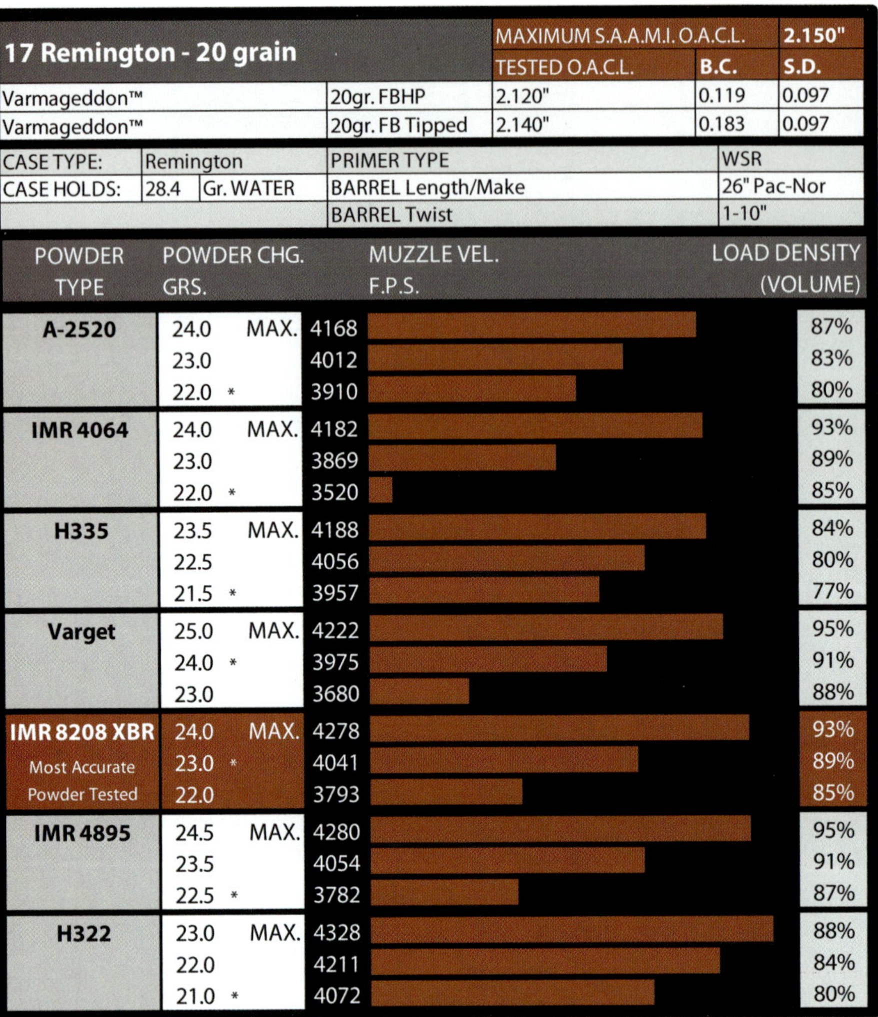

17 Remington - 20 grain		MAXIMUM S.A.A.M.I. O.A.C.L.		2.150"
		TESTED O.A.C.L.	B.C.	S.D.
Varmageddon™	20gr. FBHP	2.120"	0.119	0.097
Varmageddon™	20gr. FB Tipped	2.140"	0.183	0.097

CASE TYPE:	Remington		PRIMER TYPE	WSR
CASE HOLDS:	28.4	Gr. WATER	BARREL Length/Make	26" Pac-Nor
			BARREL Twist	1-10"

POWDER TYPE	POWDER CHG. GRS.		MUZZLE VEL. F.P.S.	LOAD DENSITY (VOLUME)
A-2520	24.0	MAX.	4168	87%
	23.0		4012	83%
	22.0 *		3910	80%
IMR 4064	24.0	MAX.	4182	93%
	23.0		3869	89%
	22.0 *		3520	85%
H335	23.5	MAX.	4188	84%
	22.5		4056	80%
	21.5 *		3957	77%
Varget	25.0	MAX.	4222	95%
	24.0 *		3975	91%
	23.0		3680	88%
IMR 8208 XBR Most Accurate Powder Tested	24.0	MAX.	4278	93%
	23.0 *		4041	89%
	22.0		3793	85%
IMR 4895	24.5	MAX.	4280	95%
	23.5		4054	91%
	22.5 *		3782	87%
H322	23.0	MAX.	4328	88%
	22.0		4211	84%
	21.0 *		4072	80%

BC=Ballistic Coefficient SD=Sectional Density
*Most Accurate Load Tested **Compressed Load

Use Maximum Loads with Caution
Refer to page 73 for additional safety information

204 RUGER

In the mid-1960's, I decided that the world needed a good 5mm centerfire cartridge. Starting with a 223 case and after a great deal of work, I produced a cartridge that I called the 5mm-223. P. O. Ackley made me a barrel and after making my own bullets and obtaining a set of custom dies, I had a functioning gun. It shot well, but the folks in the shooting world didn't even bother to cover their mouths when they yawned.

Now, jump about 50 years later. Hornady and Ruger, in 2004, introduced their 204 Ruger. Unlike my 5mm-223, the 204 Ruger took off. By 2011, it had climbed to 21st place on the reloading die sales list, a good indication of the popularity.

The base case for the 204 is said to be the 222 Remington Magnum. That may be true but the exact dimensions of a new factory cartridge are more often driven by manufacturing concerns than by any other factor.

My experience with the 204 is that it shoots extremely well and is a good in-between caliber between the .17's and the centerfire .22's. Half-inch, 5-shot groups at 100 yards are not uncommon, making the 204 an excellent choice for long-range (150 yards or more) shooting of ground squirrels and prairie dogs. Of course, it is useful at shorter ranges too, but if there are enough targets, the short range ones tend to get passed up.

Reloading this cartridge is no more difficult than any of the centerfire 22's. There are plenty of components and dies available. Small caliber guns seem to appeal to everyone. They are especially useful for graduating young shooters from their rimfire 22 guns to centerfire rifles.

While the 5mm-223 went exactly nowhere, the 204 Ruger is well on its way to being a small gun standard.

Bob Forker

Bob Forker is Technical Editor of Guns & Ammo Magazine

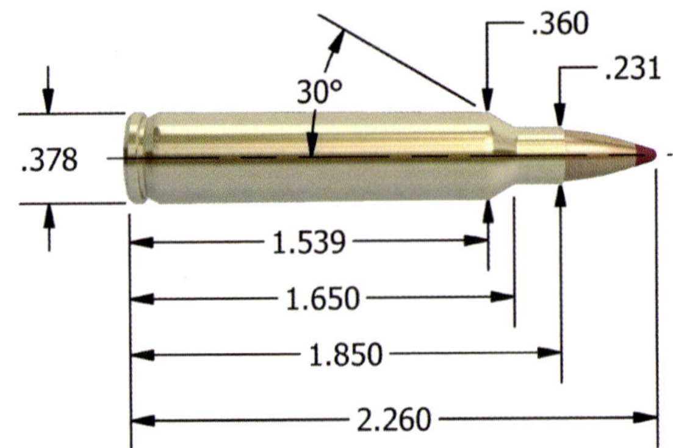

Maximum S.A.A.M.I. Overall Cartridge Length: 2.260"

BULLET CHOICES FOR THE 204 RUGER

TECHNICAL INFORMATION

The 204 Ruger was developed from the 222 Remington Magnum, which has the largest case capacity in the 222 family; providing for about 5% more case capacity than the most popular member of the family, the 223 Remington. The 204 Ruger has proven to be a very accurate and efficient cartridge. This is not surprising, considering that the first cartridge in the family, the 222 Remington, was a top benchrest competition cartridge for many years.

204 Ruger - 32/34 grain

Bullet		MAXIMUM S.A.A.M.I. O.A.C.L.		2.260"
		TESTED O.A.C.L.	B.C.	S.D.
Ballistic Tip® Lead Free	32gr. Spitzer	2.250"	0.228	0.110
Ballistic Tip® Varmint	32gr. Spitzer	2.250"	0.206	0.110
CT® Ballistic Silvertip®	32gr. Spitzer	2.250"	0.206	0.110
Varmageddon™	32gr. FBHP	2.250"	0.131	0.110
Varmageddon™	32gr. FB Tipped	2.250"	0.204	0.110
Nosler®	34gr. FBHP	2.220"	0.135	0.117

CASE TYPE:	Nosler		PRIMER TYPE	Fed. 205M
CASE HOLDS:	30.6	Gr. WATER	BARREL Length/Make	26" Pac-Nor
			BARREL Twist	1-12"

POWDER TYPE	POWDER CHG. GRS.		MUZZLE VEL. F.P.S.	LOAD DENSITY (VOLUME)
A-2015	24.5	MAX.	3812	94%
	23.5 *		3688	90%
	22.5		3554	87%
TAC	26.0	MAX.	3863	87%
	25.0 *		3800	83%
	24.0		3737	80%
Benchmark	26.5 *	MAX.	3930	94%
	25.5		3809	90%
	24.5		3702	87%
H322	25.5 *	MAX.	3996	90%
	24.5		3825	87%
	23.5		3682	83%
W748	28.5	MAX.	3998	97%
	27.5 *		3852	94%
	26.5		3710	90%
Varget	29.0	MAX.	4043	** 102%
	28.0 *		3921	99%
	27.0		3811	95%
H335	27.0 *	MAX.	4056	90%
	26.0		3915	86%
	25.0		3823	83%
IMR 4198 Most Accurate Powder Tested	23.0	MAX.	4060	85%
	22.0		3909	82%
	21.0 *		3760	78%
Varmint	29.5 *	MAX.	4087	97%
	28.5		3887	94%
	27.5		3688	91%
CFE 223	31.0	MAX.	4158	** 102%
	30.0		3994	99%
	29.0 *		3830	96%

BC=Ballistic Coefficient SD=Sectional Density
*Most Accurate Load Tested **Compressed Load

Use Maximum Loads with Caution
Refer to page 73 for additional safety information

204 Ruger - 40 grain		MAXIMUM S.A.A.M.I. O.A.C.L.		2.260"
		TESTED O.A.C.L.	B.C.	S.D.
Ballistic Tip® Varmint	40gr. Spitzer	2.250"	0.239	0.137

CASE TYPE:	Nosler		PRIMER TYPE	Fed. 205M
CASE HOLDS:	29.9	Gr. WATER	BARREL Length/Make	26" Pac-Nor
			BARREL Twist	1-12"

POWDER TYPE	POWDER CHG. GRS.		MUZZLE VEL. F.P.S.	LOAD DENSITY (VOLUME)
A-2015	24.0	MAX.	3480	91%
	23.0		3361	87%
	22.0 *		3221	84%
RL10x	23.5	MAX.	3545	88%
	22.5 *		3440	85%
	21.5		3320	81%
IMR 4895	27.0	MAX.	3545	99%
	26.0 *		3445	96%
	25.0		3352	92%
H335	25.0	MAX.	3565	85%
	24.0		3445	81%
	23.0 *		3311	78%
H322 Most Accurate Powder Tested	24.5	MAX.	3608	89%
	23.5		3473	85%
	22.5 *		3340	81%
Varget	27.5	MAX.	3648	99%
	26.5 *		3527	96%
	25.5		3407	92%
Benchmark	26.0	MAX.	3664	94%
	25.0		3527	90%
	24.0 *		3374	87%
W748	27.5	MAX.	3676	96%
	26.5 *		3534	92%
	25.5		3402	89%
Varmint	28.0	MAX.	3728	95%
	27.0		3584	91%
	26.0 *		3441	88%
CFE 223	29.5	MAX.	3815	100%
	28.5		3642	96%
	27.5 *		3471	93%

BC=Ballistic Coefficient SD=Sectional Density
*Most Accurate Load Tested **Compressed Load

Use Maximum Loads with Caution
Refer to page 73 for additional safety information

22 HORNET

Developed in the 1920's-the 22 Hornet was the brain-child of Colonel Townsend Whelen and Captain G.L. Wotkyns. The actual introduction of ammunition wasn't until 1930 by Winchester. The 22 Hornet is based on the 22 Winchester Centerfire (WCF) black powder case.

Hornets, like anything that have survived this long, have had a few generations of change. One is that the original rifle was based on the Springfield M1903 and the groove diameter was .222". The twist was a 1-16", far too slow by today's standards. Most rifles chambered today are .224" diameter and 1-14" providing a great small bore-higher velocity cartridge. The Hornet is very capable of delivering a knock-out punch on prairie dogs at 200-250 yards.

My first personal account of the 22 Hornet was in Wyoming in the 1980's. I was around 12 years old at the time. My grandfather contacted the folks at "Kimber of Oregon" and had their gunsmith build a matching set of rifles, one in 22 LR and the other in 22 Hornet as a gift to his grandson. Working the bolt and looking through the fixed Leupold 6 power at prairie dogs had a very natural feel to it. It forced you to concentrate on your breathing, but the benefit of the patience was very worthwhile. After shooting a full day with a 223 Remington and 22-250 Remington, the 22 Hornet was a gift from above. For a 12-year old boy, the ability to fire a shot and actually watch the end result through the scope was extremely rewarding.

The problem most Hornet's face is the magazine length versus bullet availability. I fortunately could load the 40-grain Ballistic Tip® in my Kimber. If you are one of the millions who own a Ruger 77, the use of tipped bullets presents a problem. This is why we have recently introduced the 35-grain Varmageddon Tipped bullet. We specifically addressed the magazine vs. bullet length issue.
Over the years, my Hornet has seen some miles, thousands of rounds, and unfortunately due to a mis-placed magazine- it is now a single shot. All of this just makes it more enjoyable to me as it has become one of my most cherished gifts and I think of my grandfather when I pop a prairie dog. It has in-fact turned into my go-to prairie dog gun when the shot is between 100-150 yards.

I can still hear "Big John" giggle at the sight of me sitting at a bench pulling the trigger.

John R. Nosler is the VP/General Manager of Nosler, Inc.

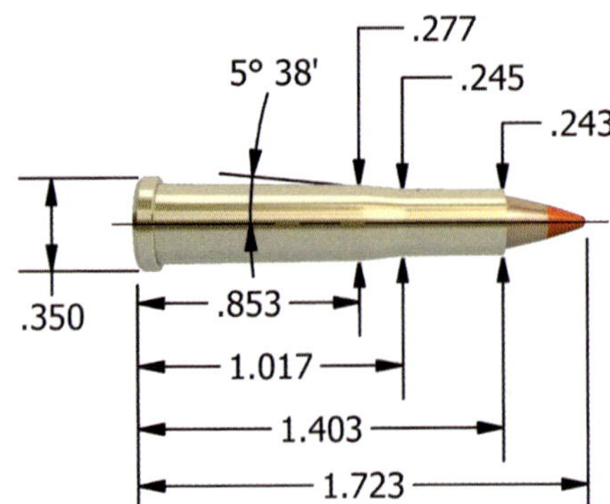

Maximum S.A.A.M.I. Overall Cartridge Length: 1.723"

BULLET CHOICES FOR THE 22 HORNET

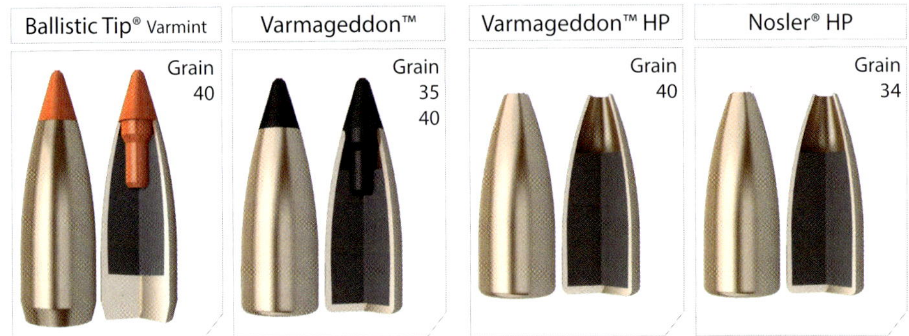

TECHNICAL INFORMATION

The 22 Hornet is an older cartridge that was originally adapted from the 22 WCF. Most of the early firearms (and even some modern guns) chambered for the Hornet have barrels with 1-16" twist rates and/or .223" groove diameters. All of our .22 caliber bullets are .224" diameter, so be sure the groove diameter of your gun is proper for the bullet you intend to use. Additionally, our 40-grain Ballistic Tip® bullets require 14" twist barrels to properly stabilize when fired at Hornet velocities. 50-grain and heavier bullets are not recommended for this cartridge.

22 Hornet - 34/35 grain		MAXIMUM S.A.A.M.I. O.A.C.L.		1.723"
		TESTED O.A.C.L.	B.C.	S.D.
Nosler®	34gr. FBHP	1.705"	0.113	0.097
Varmageddon™	35gr. FB Tipped	1.705"	0.120	0.100

CASE TYPE:	Remington		PRIMER TYPE	WSR
CASE HOLDS:	14.2	Gr. WATER	BARREL Length/Make	22" Lilja
			BARREL Twist	1-14"

POWDER TYPE	POWDER CHG. GRS.	MUZZLE VEL. F.P.S.	LOAD DENSITY (VOLUME)
Enforcer	10.0 * MAX.	2835	75%
	9.0	2671	67%
	8.0	2502	60%
2400	10.5 MAX.	2863	86%
	9.5	2745	78%
	8.5 *	2631	70%
Lil'Gun	10.0 * MAX.	3048	74%
	9.0	2802	67%
	8.0	2534	59%
W296	12.0 * MAX.	3075	86%
Most Accurate	11.0	2921	79%
Powder Tested	10.0	2755	72%
H110	12.5 MAX.	3154	89%
	11.5	2989	82%
	10.5 *	2817	75%

BC=Ballistic Coefficient SD=Sectional Density
*Most Accurate Load Tested **Compressed Load

Use Maximum Loads with Caution
Refer to page 73 for additional safety information

22 Hornet - 40 grain		MAXIMUM S.A.A.M.I. O.A.C.L.		1.723"
		TESTED O.A.C.L.	B.C.	S.D.
Ballistic Tip®	40gr. Spitzer	1.845"	0.221	0.114
Varmageddon™	40gr. FBHP	1.723"	0.158	0.114
Varmageddon™	40gr. FB Tipped	1.723"	0.211	0.114

CASE TYPE:	Remington		PRIMER TYPE	Rem 6 1/2
CASE HOLDS:	11.6	Gr. WATER	BARREL Length/Make	22" Lilja
			BARREL Twist	1-14"

POWDER TYPE	POWDER CHG. GRS.		MUZZLE VEL. F.P.S.		LOAD DENSITY (VOLUME)
IMR 4198	11.5	MAX.	2355	**	113%
	10.5		2245	**	103%
	9.5 *		2011		93%
Viht N120	12.0	MAX.	2528	**	126%
	11.0 *		2321	**	116%
	10.0		2154	**	105%
Enforcer	9.5	MAX.	2566		87%
	8.5 *		2347		78%
	7.5		2189		69%
A-1680 Most Accurate Powder Tested	13.0 *	MAX.	2578	**	118%
	12.0		2384	**	109%
	11.0		2192		100%
IMR 4227	12.0	MAX.	2627	**	121%
	11.0		2413	**	111%
	10.0 *		2233	**	101%
2400	10.0	MAX.	2639		100%
	9.0 *		2471		90%
	8.0		2312		80%
Lil'Gun	11.0 *	MAX.	2644		100%
	10.0		2466		91%
	9.0		2309		82%
Viht N110	10.0	MAX.	2660	**	113%
	9.0 *		2473	**	102%
	8.0		2292		91%
W296	11.0 *	MAX.	2689		97%
	10.0		2524		88%
	9.0		2424		79%
H110	11.5	MAX.	2794	**	101%
	10.5 *		2593		92%
	9.5		2399		83%

BC=Ballistic Coefficient SD=Sectional Density
*Most Accurate Load Tested **Compressed Load

Use Maximum Loads with Caution
Refer to page 73 for additional safety information

John A. Nosler and John R. Nosler on an Oregon mule deer hunt.

John A. Nosler circa 1969, shortly after the Nosler Partition Bullet Company became incorporated.

Charles E. Petty

221 REMINGTON FIREBALL

The 221 Remington Fireball started life in 1962 as a purpose built cartridge based on the 222 Magnum, for Remington's XP-100 handgun. Those were the early days of handgun hunting and even though the Fireball was a dandy little varmint round, shooters started doing all sorts of stuff to the XP-100 and it almost seemed as if the 221 was on the road to nowhere. When Remington dropped the XP, things looked really bad for the cartridge.

Even though you'll find it in the handgun data section of many manuals, it is alive and well in rifles. Almost all of my experience has been with one of Dan Cooper's dandy rifles. In fact, my affection for the cartridge is a direct result of an assignment to review a Cooper rifle and since I had never shot the cartridge in a rifle, it was an easy choice.

I learned a lot from that rifle. Foremost is that it is fun to shoot, doesn't beat up your ears or shoulder and doesn't need to be pushed to the speed of light to shoot well under half minute groups. One of the best loads used the 40gr Ballistic Tip; a combination that was later proven to be harmful to prairie dogs throughout the West.

Since the Fireball was new, there have been great improvements in components, especially powders. One of those is Hodgdon Lil'Gun. My friend Ron Reiber at Hodgdon suggested I try it. Of course, there was no data, so I ventured into never-never land and after looking at powders with similar burning rates, I put together several charges and went to the range. The chronograph gave nice numbers and bolt lift was easy. When the maximum charge loaded was fired, the velocity was almost 200 fps faster than anything published. Bolt lift was still easy, headstamps were not erased and the primers were still where they were supposed to be. I rushed home and called Reiber. An hour or so later, he called back with even better news. Not only were my velocities accurate, the pressure of my maximum load was well within industry standards.

So now I have to go back to the bench. Nosler has some new bullets to try and there might even be a better powder.

Charlie Petty

Charles is a contributing editor for Handloader and The American Rifleman.

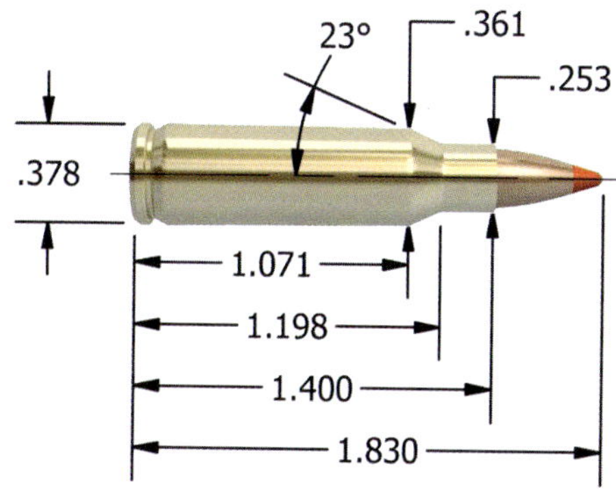

Maximum S.A.A.M.I. Overall Cartridge Length: 1.830"

BULLET CHOICES FOR THE 221 REMINGTON FIREBALL

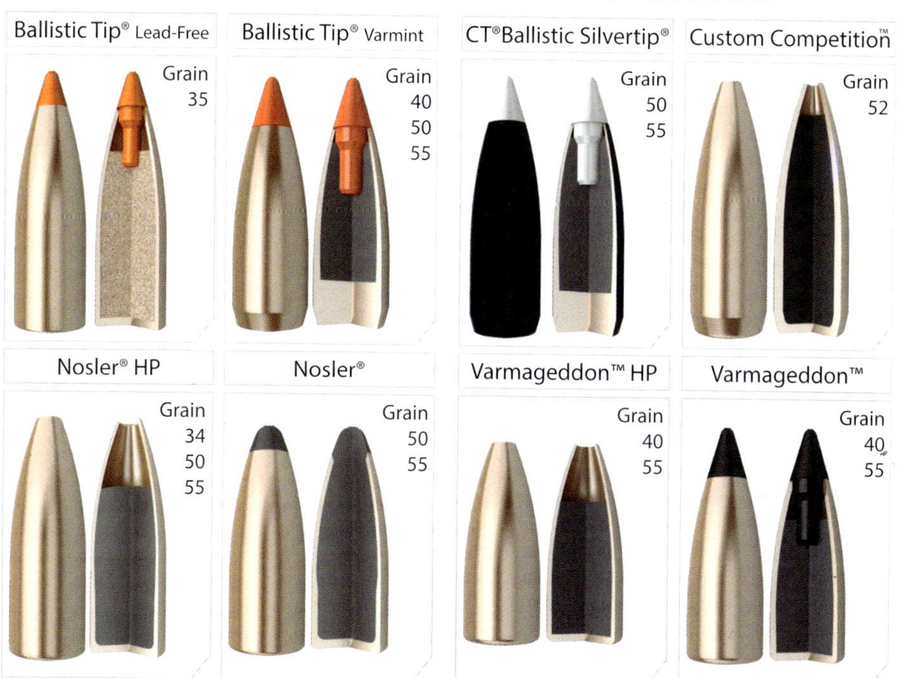

TECHNICAL INFORMATION

Remington originally developed the 221 Fireball for use in their XP-100 handguns. While it was developed for handgun use, the Fireball operates at rifle pressure levels. As such, small rifle primers should be used. The 221 Fireball is capable of very good accuracy and lethal performance on varmints at well over 200 yards, and is underrated as a rifle cartridge. It has become a favorite small varmint round among the Nosler staff.

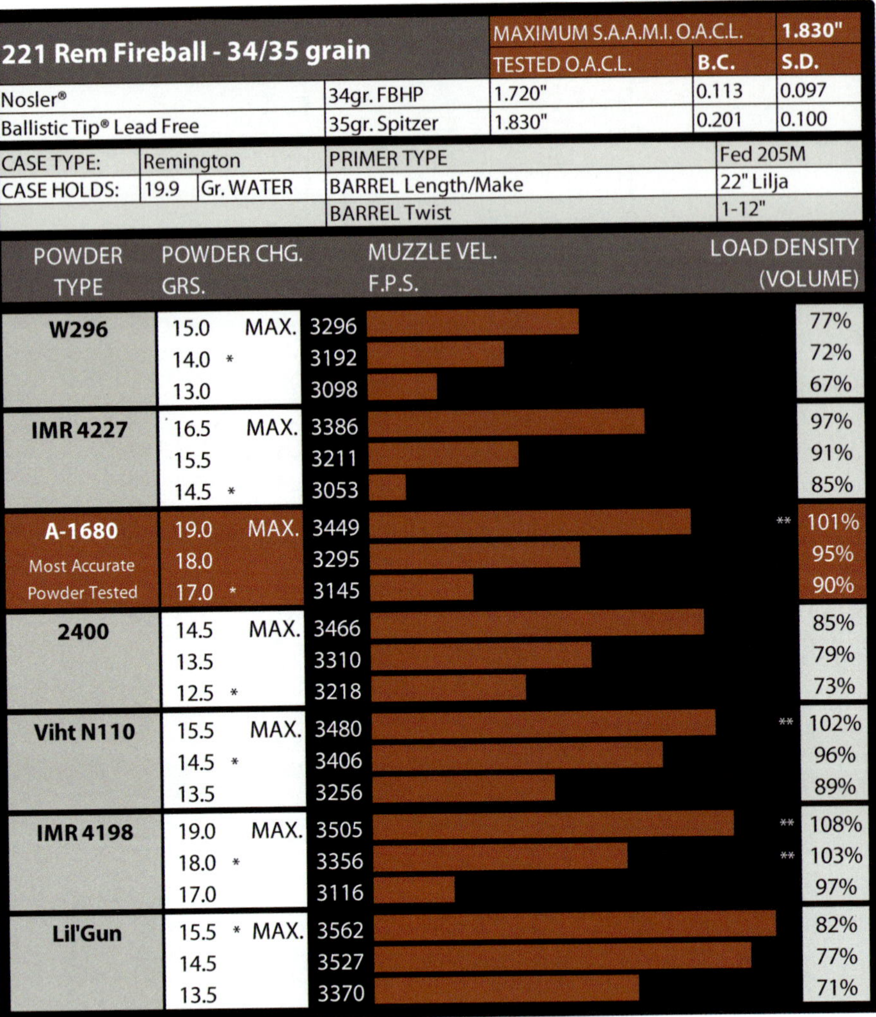

221 Rem Fireball - 34/35 grain		MAXIMUM S.A.A.M.I. O.A.C.L.		1.830"
		TESTED O.A.C.L.	B.C.	S.D.
Nosler®	34gr. FBHP	1.720"	0.113	0.097
Ballistic Tip® Lead Free	35gr. Spitzer	1.830"	0.201	0.100

CASE TYPE:	Remington	PRIMER TYPE	Fed 205M
CASE HOLDS:	19.9 Gr. WATER	BARREL Length/Make	22" Lilja
		BARREL Twist	1-12"

POWDER TYPE	POWDER CHG. GRS.		MUZZLE VEL. F.P.S.		LOAD DENSITY (VOLUME)
W296	15.0	MAX.	3296		77%
	14.0 *		3192		72%
	13.0		3098		67%
IMR 4227	16.5	MAX.	3386		97%
	15.5		3211		91%
	14.5 *		3053		85%
A-1680 Most Accurate Powder Tested	19.0	MAX.	3449	**	101%
	18.0		3295		95%
	17.0 *		3145		90%
2400	14.5	MAX.	3466		85%
	13.5		3310		79%
	12.5 *		3218		73%
Viht N110	15.5	MAX.	3480	**	102%
	14.5 *		3406		96%
	13.5		3256		89%
IMR 4198	19.0	MAX.	3505	**	108%
	18.0 *		3356	**	103%
	17.0		3116		97%
Lil'Gun	15.5 *	MAX.	3562		82%
	14.5		3527		77%
	13.5		3370		71%

BC=Ballistic Coefficient SD=Sectional Density
*Most Accurate Load Tested **Compressed Load

Use Maximum Loads with Caution
Refer to page 73 for additional safety information

221 Rem Fireball - 40 grain		MAXIMUM S.A.A.M.I. O.A.C.L.		1.830"
		TESTED O.A.C.L.	B.C.	S.D.
Ballistic Tip®	40gr. Spitzer	1.830"	0.221	0.114
Varmageddon™	40gr. FBHP	1.765"	0.158	0.114
Varmageddon™	40gr. FB Tipped	1.830"	0.211	0.114

CASE TYPE:	Remington		PRIMER TYPE	Fed 205M
CASE HOLDS:	20.0	Gr. WATER	BARREL Length/Make	22" Lilja
			BARREL Twist	1-12"

POWDER TYPE	POWDER CHG. GRS.		MUZZLE VEL. F.P.S.		LOAD DENSITY (VOLUME)
H110	14.0	MAX.	2977		71%
	13.0		2796		66%
	12.0 *		2615		61%
W296	14.0 *	MAX.	3031		71%
	13.0		2882		66%
	12.0		2732		61%
Enforcer	14.0	MAX.	3047		74%
	13.0 *		2890		69%
	12.0		2731		64%
IMR 4227	15.5 *	MAX.	3069		91%
	14.5		2865		85%
	13.5		2661		79%
H4198	18.0 *	MAX.	3176	**	101%
	17.0		2974		96%
	16.0		2772		90%
Viht N130	19.0	MAX.	3194	**	112%
	18.0		2996	**	106%
	17.0 *		2799		100%
A-2015	20.5	MAX.	3207	**	116%
	19.5		3009	**	111%
	18.5 *		2811	**	105%
A-1680	18.0 *	MAX.	3229		95%
	17.0		3023		89%
	16.0		2817		84%
RL 7 Most Accurate Powder Tested	19.0	MAX.	3242	**	103%
	18.0		3056		98%
	17.0 *		2869		92%
IMR 4198	18.0 *	MAX.	3259	**	102%
	17.0		3025		97%
	16.0		2792		91%

BC=Ballistic Coefficient SD=Sectional Density
*Most Accurate Load Tested **Compressed Load

Use Maximum Loads with Caution
Refer to page 73 for additional safety information

221 Rem Fireball - 50/52 grain		MAXIMUM S.A.A.M.I. O.A.C.L.		1.830"
		TESTED O.A.C.L.	B.C.	S.D.
Ballistic Tip®	50gr. Spitzer	1.830"	0.238	0.142
CT® Ballistic Silvertip®	50gr. Spitzer	1.830"	0.238	0.142
Nosler®	50gr. FBHP	1.830"	0.192	0.142
Nosler®	50gr. FBSP	1.830"	0.195	0.142
Custom Competition™	52gr. HPBT	1.830"	0.220	0.148

CASE TYPE:	Remington		PRIMER TYPE	Fed 205M
CASE HOLDS:	19.6	Gr. WATER	BARREL Length/Make	22" Lilja
			BARREL Twist	1-12"

POWDER TYPE	POWDER CHG. GRS.		MUZZLE VEL. F.P.S.	LOAD DENSITY (VOLUME)
Enforcer	13.5 *	MAX.	2733	73%
	12.5		2593	68%
	11.5		2436	62%
W296	14.0	MAX.	2768	72%
	13.0		2580	67%
	12.0 *		2392	62%
H110	14.0	MAX.	2815	73%
	13.0		2639	67%
	12.0 *		2463	62%
H4198 Most Accurate Powder Tested	17.0	MAX.	2852	97%
	16.0		2672	92%
	15.0 *		2491	86%
Viht N130	18.0	MAX.	2904	** 108%
	17.0		2724	** 102%
	16.0 *		2543	96%
IMR 4227	15.5	MAX.	2937	92%
	14.5		2750	87%
	13.5 *		2562	81%
RL 7	18.0 *	MAX.	2965	100%
	17.0		2780	94%
	16.0		2594	89%
A-1680	17.5	MAX.	3002	94%
	16.5		2806	89%
	15.5 *		2610	83%
A-2015	20.0 *	MAX.	3033	** 116%
	19.0		2863	** 110%
	18.0		2693	** 104%
IMR 4198	17.0	MAX.	3034	99%
	16.0		2853	93%
	15.0 *		2672	87%

BC=Ballistic Coefficient SD=Sectional Density
*Most Accurate Load Tested **Compressed Load

Use Maximum Loads with Caution
Refer to page 73 for additional safety information

221 Rem Fireball - 55 grain		MAXIMUM S.A.A.M.I. O.A.C.L.		1.830"
		TESTED O.A.C.L.	B.C.	S.D.
Ballistic Tip®	55gr. Spitzer	1.830"	0.267	0.157
CT® Ballistic Silvertip®	55gr. Spitzer	1.830"	0.267	0.157
Nosler®	55gr. FBSP	1.830"	0.210	0.157
Varmageddon™	55gr. FBHP	1.830"	0.218	0.157
Varmageddon™	55gr. FB Tipped	1.830"	0.255	0.157

CASE TYPE:	Remington	PRIMER TYPE	Fed 205M
CASE HOLDS:	19.4 Gr. WATER	BARREL Length/Make	22" Lilja
		BARREL Twist	1-12"

POWDER TYPE	POWDER CHG. GRS.	MUZZLE VEL. F.P.S.	LOAD DENSITY (VOLUME)
W296	13.5 * MAX.	2622	70%
	12.5	2454	65%
	11.5	2286	60%
H110	13.5 MAX.	2652	71%
	12.5	2465	65%
	11.5 *	2279	60%
IMR 4227	14.5 MAX.	2695	87%
	13.5	2524	81%
	12.5 *	2352	75%
H4198	16.5 MAX.	2700	96%
	15.5	2524	90%
	14.5 *	2349	84%
X-Terminator	20.5 MAX.	2754	** 107%
	19.5	2694	** 102%
	18.5 *	2561	97%
Viht N130	17.5 MAX.	2777	** 106%
	16.5	2623	100%
	15.5 *	2468	94%
A-1680	16.5 MAX.	2786	90%
	15.5	2620	84%
	14.5 *	2455	79%
RL 7	17.5 MAX.	2808	98%
	16.5 *	2646	92%
	15.5	2483	87%
IMR 4198	17.0 MAX.	2900	100%
	16.0	2723	94%
	15.0 *	2547	88%
A-2015	19.5 MAX.	2929	** 114%
Most Accurate	18.5	2767	** 108%
Powder Tested	17.5 *	2604	** 103%

BC=Ballistic Coefficient SD=Sectional Density
*Most Accurate Load Tested **Compressed Load

Use Maximum Loads with Caution
Refer to page 73 for additional safety information

Holt Bodinson

222 REMINGTON

Looking much like a scaled-down 30-06, the 222 Remington is a historically significant cartridge. The "Triple Deuce" was the first, new, commercial cartridge introduced after WWII and the precursor to the 222 Remington Magnum and the 223 Remington.

Designed by avid shooter, Mike Walker of Remington Arms, the 222 was introduced in 1950 to a hunting and competitive shooting public who were starved for something new after the war. Chambered initially in the Remington Model 722 with its 26", medium heavyweight barrel and superb adjustable trigger and then, only months later, in the inexpensive Model 340 Savage, the Triple Deuce was an overnight success.

Inherently accurate, the little cartridge was delivering sub-minute-of-angle groups in factory sporting rifles in 1950, a feat almost unheard of at the time. Varmint hunters loved its mild report and effectiveness out to 300 yards. The benchrest clan adopted the 222, and the cartridge dominated benchrest events for years. In 1952, Mike Walker's wife established the 100 yard, ten shot, world record group measuring 0.3268" with her 222.

My first true varmint rifle was a Model 722 in 222 bought at the full retail price of $87.50. Mounted with a Weaver 6X scope, it proved to be pure poison in the field. Foxes, coyotes, crows and 'chucks were in fatal trouble out to 320 yards (my longest shot). I never bought a factory box of ammunition for that rifle, rather I loaded all my ammunition using an old Herter's press and their Rube Goldberg derived Universal loading die.

Two shots with my 222 were memorable—a sunning fox in the snow at 200 yards that turned out to be two foxes lying back-to-back and a trophy roe deer in Germany.

Years later, I experimented with the 6mm-222 and the 6mm-223. The long-necked 222 case always proved to be the more accurate of the two. In fact, for pure accuracy, I'll take the 222 over the 223 any day.

Long live the Triple Deuce! A golden oldie, but not forgotten.

Holt is a Contributing Editor of GUNS magazine and the Ammunition, Ballistics and Components editor of GUN DIGEST

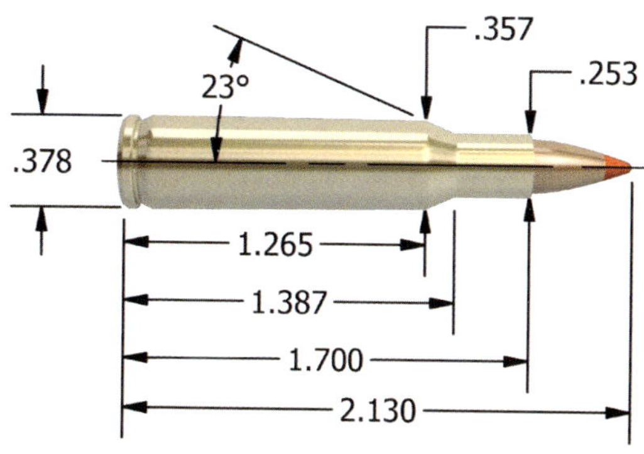

Maximum S.A.A.M.I. Overall Cartridge Length: 2.130″

BULLET CHOICES FOR THE 222 REMINGTON

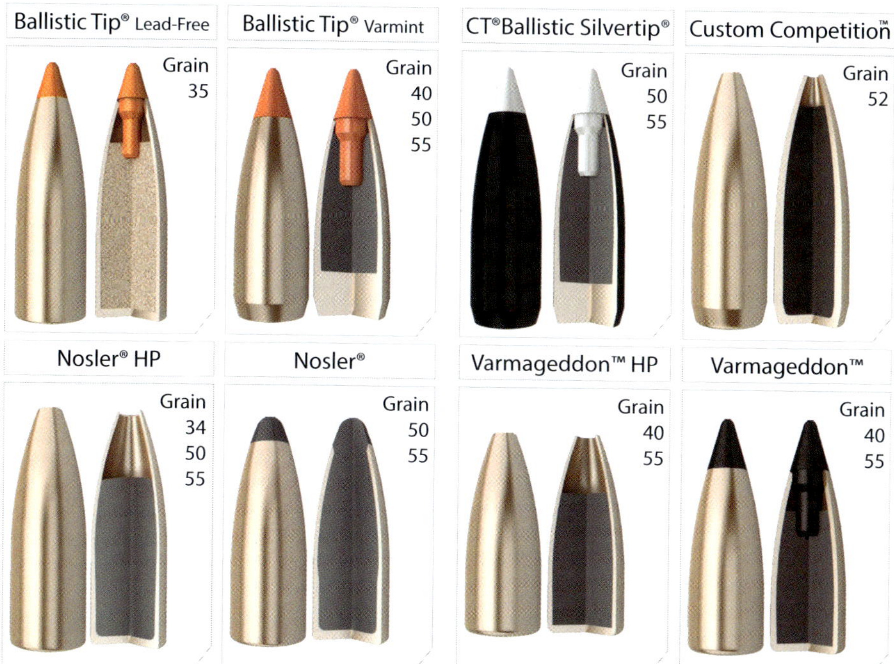

TECHNICAL INFORMATION

The "triple-deuce," as often called, is a superbly accurate cartridge. For optimum consistency, we recommend RL7, IMR 4198, and A-2015. Since this cartridge is usually teamed up with a 14" twist barrel, we do not recommend loading bullets that are heavier than 55 grains. Long, heavy bullets will require more "spin" to stabilize, and may actually "keyhole" at the target.

222 Remington - 34/35 grain

		MAXIMUM S.A.A.M.I. O.A.C.L.		2.130"
		TESTED O.A.C.L.	B.C.	S.D.
Nosler®	34gr. FBHP	2.020"	0.113	0.097
Ballistic Tip® Lead Free	35gr. Spitzer	2.130"	0.201	0.100

CASE TYPE:	Remington		PRIMER TYPE	Rem 7 1/2
CASE HOLDS:	23.6	Gr. WATER	BARREL Length/Make	20" Lilja
			BARREL Twist	1-14"

POWDER TYPE	POWDER CHG. GRS.		MUZZLE VEL. F.P.S.		LOAD DENSITY (VOLUME)
A-2015 Most Accurate Powder Tested	24.0	MAX.	3535	**	116%
	23.0	*	3435	**	111%
	22.0		3308	**	106%
Benchmark	25.0	* MAX.	3595	**	115%
	24.0		3454	**	110%
	23.0		3320	**	105%
Viht N133	23.0	* MAX.	3614	**	113%
	22.0		3368	**	108%
	21.0		3205	**	103%
N200	22.5	MAX.	3702		100%
	21.5	*	3508		96%
	20.5		3305		91%
H322	25.0	MAX.	3703	**	115%
	24.0		3570	**	110%
	23.0	*	3439	**	105%
A-1680	21.5	MAX.	3732		96%
	20.5		3593		91%
	19.5	*	3403		87%
RL 7	22.5	MAX.	3765	**	104%
	21.5		3563		99%
	20.5	*	3329		94%
IMR 4198	21.0	MAX.	3810	**	101%
	20.0		3605		96%
	19.0	*	3371		91%
H4198	21.5	MAX.	3813	**	102%
	20.5	*	3611		98%
	19.5		3360		93%

BC=Ballistic Coefficient SD=Sectional Density
*Most Accurate Load Tested **Compressed Load

Use Maximum Loads with Caution
Refer to page 73 for additional safety information

222 Remington - 40 grain

		MAXIMUM S.A.A.M.I. O.A.C.L.		2.130"
		TESTED O.A.C.L.	B.C.	S.D.
Ballistic Tip®	40gr. Spitzer	2.130"	0.221	0.114
Varmageddon™	40gr. FBHP	2.060"	0.158	0.114
Varmageddon™	40gr. FB Tipped	2.130"	0.211	0.114

CASE TYPE:	Remington		PRIMER TYPE	Rem 7 1/2
CASE HOLDS:	25.8	Gr. WATER	BARREL Length/Make	20" Lilja
			BARREL Twist	1-14"

POWDER TYPE	POWDER CHG. GRS.		MUZZLE VEL. F.P.S.		LOAD DENSITY (VOLUME)
X-Terminator	25.0	* MAX.	3321		98%
	24.0		3234		94%
	23.0		3146		91%
A-2230	25.0	* MAX.	3336		97%
	24.0		3217		93%
	23.0		3099		89%
Varmint	26.0	MAX.	3402	**	102%
	25.0		3277		98%
	24.0	*	3155		94%
RL7	22.0	* MAX.	3428		93%
	21.0		3306		88%
	20.0		3183		84%
RL10x	23.0	MAX.	3444		100%
	22.0	*	3238		96%
	21.0		3073		91%
H322	24.0	* MAX.	3490	**	101%
	23.0		3322		96%
	22.0		3154		92%
IMR 4198	21.0	* MAX.	3522		92%
	20.0		3314		88%
	19.0		3107		84%
H335	27.0	MAX.	3536	**	106%
	26.0	*	3416	**	102%
	25.0		3295		98%
A-2015 Most Accurate Powder Tested	24.5	MAX.	3540	**	108%
	23.5		3396	**	104%
	22.5	*	3252		99%
Viht N120	21.5	MAX.	3564	**	102%
	20.5		3422		97%
	19.5	*	3281		92%

BC=Ballistic Coefficient SD=Sectional Density
*Most Accurate Load Tested **Compressed Load

Use Maximum Loads with Caution
Refer to page 73 for additional safety information

222 Remington - 50/52 grain		MAXIMUM S.A.A.M.I. O.A.C.L.		2.130"
		TESTED O.A.C.L.	B.C.	S.D.
Ballistic Tip®	50gr. Spitzer	2.130"	0.238	0.142
CT® Ballistic Silvertip®	50gr. Spitzer	2.130"	0.238	0.142
Nosler®	50gr. FBHP	2.130"	0.192	0.142
Nosler®	50gr. FBSP	2.130"	0.195	0.142
Custom Competition™	52gr. HPBT	2.130"	0.220	0.148

CASE TYPE:	Remington		PRIMER TYPE	Rem 7 1/2
CASE HOLDS:	25.1	Gr. WATER	BARREL Length/Make	20" Lilja
			BARREL Twist	1-14"

POWDER TYPE	POWDER CHG. GRS.		MUZZLE VEL. F.P.S.	LOAD DENSITY (VOLUME)
X-Terminator	23.0	MAX.	3022	93%
	22.0 *		2916	89%
	21.0		2798	85%
RL10x	21.0 *	MAX.	3045	94%
	20.0		2874	90%
	19.0		2745	85%
W748	24.5	MAX.	3100	99%
	23.5		3020	95%
	22.5 *		2940	91%
Viht N120	19.5	MAX.	3140	95%
	18.5		3014	90%
	17.5 *		2889	85%
IMR 4198	20.0	MAX.	3172	91%
	19.0		3017	86%
	18.0 *		2862	81%
A-2460	24.5	MAX.	3198	98%
	23.5		3103	94%
	22.5 *		3008	90%
H322 Most Accurate Powder Tested	22.5	MAX.	3200	97%
	21.5		3090	93%
	20.5 *		2980	88%
CFE 223	26.5	MAX.	3202	** 107%
	25.5		3112	** 103%
	24.5 *		3020	99%
Varmint	25.5	MAX.	3246	** 103%
	24.5		3141	99%
	23.5 *		3035	95%
Viht N130	22.0	MAX.	3247	** 103%
	21.0		3102	98%
	20.0 *		2958	94%

BC=Ballistic Coefficient SD=Sectional Density
*Most Accurate Load Tested **Compressed Load

Use Maximum Loads with Caution
Refer to page 73 for additional safety information

222 Remington - 55 grain		MAXIMUM S.A.A.M.I. O.A.C.L.		2.130"
		TESTED O.A.C.L.	B.C.	S.D.
Ballistic Tip®	55gr. Spitzer	2.130"	0.267	0.157
CT® Ballistic Silvertip®	55gr. Spitzer	2.130"	0.267	0.157
Nosler®	55gr. FBHP	2.130"	0.210	0.157
Nosler®	55gr. FBSP	2.130"	0.218	0.157
Varmageddon™	55gr. FBHP	2.130"	0.210	0.157
Varmageddon™	55gr. FB Tipped	2.130"	0.255	0.157

CASE TYPE:	Remington		PRIMER TYPE	Rem 7 1/2
CASE HOLDS:	24.4	Gr. WATER	BARREL Length/Make	20" Lilja
			BARREL Twist	1-14"

POWDER TYPE	POWDER CHG. GRS.	MUZZLE VEL. F.P.S.	LOAD DENSITY (VOLUME)
X-Terminator	22.5 * MAX.	2913	94%
	21.5	2796	89%
	20.5	2661	85%
RL10x	20.5 MAX.	2923	94%
	19.5	2813	90%
	18.5 *	2658	85%
Viht N120	18.5 MAX.	2931	92%
	17.5	2818	87%
	16.5 *	2705	82%
W748	24.0 * MAX.	3012	99%
	23.0	2927	95%
	22.0	2842	91%
IMR 4198	19.5 * MAX.	3012	91%
	18.5	2897	86%
	17.5	2782	82%
IMR 4895	24.0 * MAX.	3028	** 108%
	23.0	2883	** 104%
	22.0	2738	99%
H335	24.5 * MAX.	3108	** 102%
	23.5	2983	98%
	22.5	2858	94%
H322 Most Accurate Powder Tested	22.0 * MAX.	3120	97%
	21.0	2990	93%
	20.0	2860	89%
Varmint	24.5 MAX.	3157	** 101%
	23.5	3043	97%
	22.5 *	2921	93%
CFE 223	26.0 MAX.	3205	** 108%
	25.0	3089	** 103%
	24.0 *	2968	99%

BC=Ballistic Coefficient SD=Sectional Density
*Most Accurate Load Tested **Compressed Load

Use Maximum Loads with Caution
Refer to page 73 for additional safety information

Nosler®

Tom Gresham

223 REMINGTON

From the military, to law enforcement, to competition, to self defense and hunting, the 223 Remington cartridge reigns. The 223, and its near-twin 5.56 NATO round, arrived in early 1964. Spawned by the military, the 223 grabbed the attention of varmint hunters, who soon dropped interest in the 222 Remington and 222 Remington Magnum -- rounds that flanked the 223 in power.

The U.S. military M-16s and M-4s are chambered for the 5.56x45mm NATO cartridge, which is almost the same as the 223. Due to differences in chamber dimensions and pressures, you should not fire 5.56x45mm ammunition in a rifle with a 223 Remington chamber. You can, however, fire 223 ammo in a 5.56 NATO rifle.

In civilian hands, the 223 Remington was usually chambered in bolt-action varmint rifles, and it was seen in fields where groundhogs or prairie dogs were the targets. In M-16s, it was used in the national matches at Camp Perry.

Gradually, citizens began buying AR-15s, and realized just how versatile that platform is. It was, ironically, an effort to ban AR-15s which caused the tsunami of interest in it, and the 223 Remington cartridge. The so-called "assault weapon" ban in 1994 awakened gun owners. They bought rifles which complied with the cosmetic requirements of this new law, and after this ban ended in 2004, the dam burst.

Today, hundreds of models of AR's are made by dozens of gun makers. From three-gun competition to serious hunting, the AR has become the most popular model firearm sold in the U.S.

Long range shooters think nothing of hitting targets at 600 yards with an accurate AR in 223. With the right bullets, this cartridge (where legal) can easily take game the size of big mule deer or even black bear.

Loading the 223 falls into two categories - economy and accuracy. Competitors want economy for loading thousands of rounds for practice, but they still need accuracy. Long range shooters want the longer bullets. Note that competition bullets such as the Nosler Custom Competition are made for accuracy - not terminal performance. In other words, they are not for hunting.

Hunters need look no further than the amazing Partition design. A 60gr Partition in the 223 gives quick expansion plus deep penetration.

The 223 Remington is the latest military cartridge to prove itself for hunting, competition, self-defense, law enforcement, and just pure fun shooting.

Tom Gresham

Tom Gresham is host of "Tom Gresham's Gun Talk," a nationally-syndicated talk radio program, as well as co-host of Gun Talk and Guns and Gear television series. He's also the co-author of "Weatherby: The Man, The Gun, The Legend."

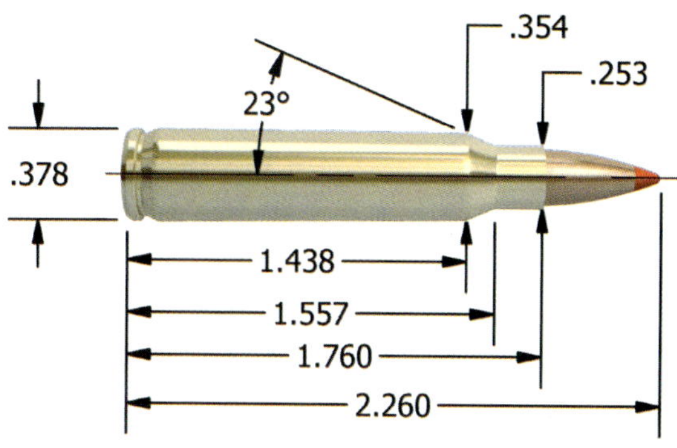

Maximum S.A.A.M.I. Overall Cartridge Length: 2.260"

BULLET CHOICES FOR THE 223 REMINGTON

Ballistic Tip® Lead-Free

Grain
35
40

Ballistic Tip® Varmint

Grain
40
50
55
60

CT® Ballistic Silvertip®

Grain
50
55

Custom Competition™

Grain
52

Nosler® HP

Grain
34
50
55

Nosler®

Grain
50
55

Varmageddon™ HP

Grain
40
55
62

Varmageddon™

Grain
40
55

Bonded Solid Base®

Grain
64

Partition®

Grain
60

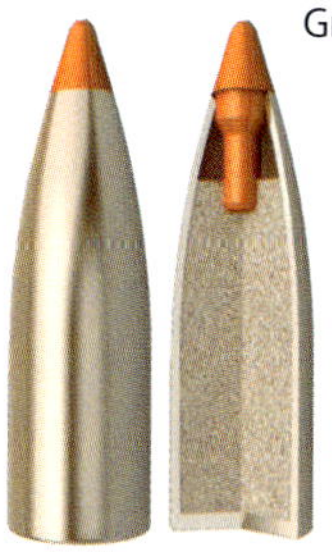

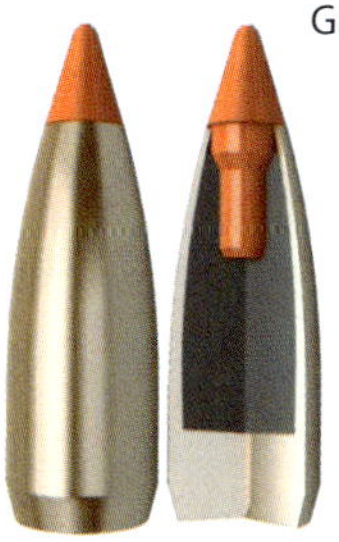

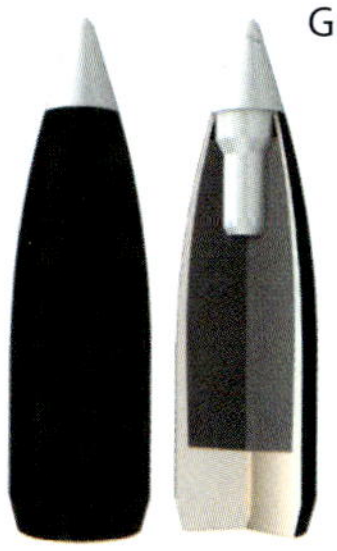

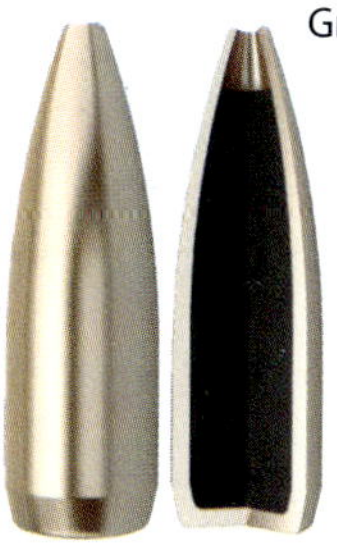

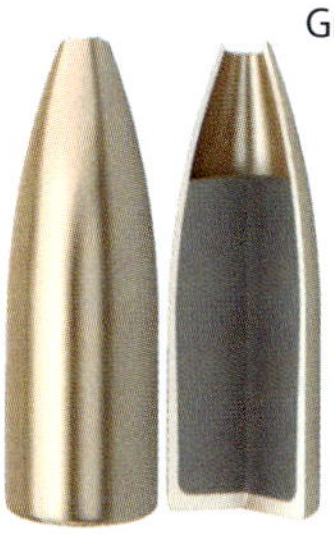

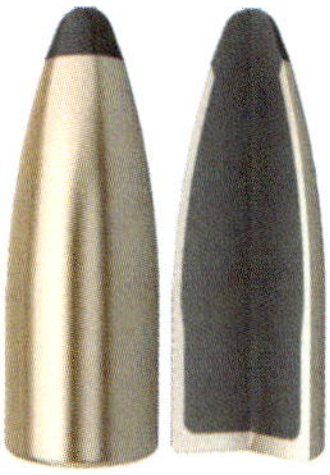

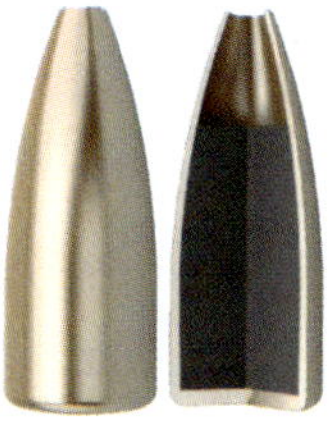

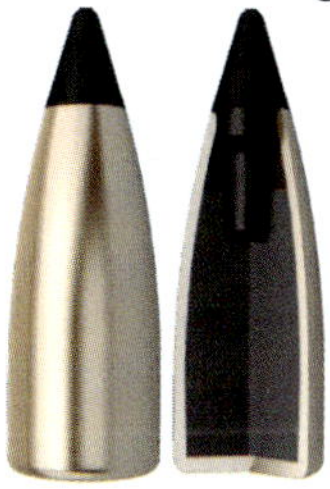

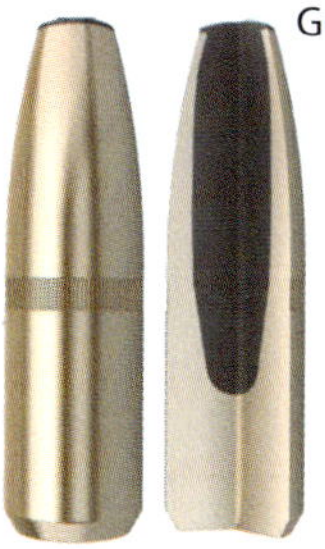

TECHNICAL INFORMATION

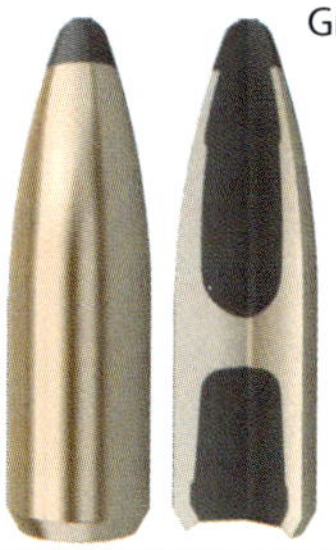

For Technical Information on the 223 Remington turn to page 109

223 Remington - 34/35 grain		MAXIMUM S.A.A.M.I. O.A.C.L.		2.260"
		TESTED O.A.C.L.	B.C.	S.D.
Nosler®	34gr. FBHP	2.080"	0.113	0.097
Ballistic Tip® Lead Free	35gr. Spitzer	2.260"	0.201	0.100

CASE TYPE:	Nosler		PRIMER TYPE	Rem 7 1/2
CASE HOLDS:	28.2	Gr. WATER	BARREL Length/Make	24" Lilja
			BARREL Twist	1-12"

POWDER TYPE	POWDER CHG. GRS.		MUZZLE VEL. F.P.S.		LOAD DENSITY (VOLUME)
A-5744	21.5	MAX.	3577	**	104%
	20.5		3450		100%
	19.5 *		3329		96%
RL 7	23.5	MAX.	3750		91%
	22.5		3598		87%
	21.5 *		3517		83%
H322	26.0	MAX.	3806		100%
	25.0		3730		96%
	24.0 *		3675		92%
TAC Most Accurate Powder Tested	28.5	MAX.	3832		90%
	27.5 *		3767		87%
	26.5		3720		83%
Viht N133	25.5	MAX.	3841		93%
	24.5 *		3672		89%
	23.5		3455		85%
IMR 4198	22.5 *	MAX.	3851	**	115%
	21.5		3682	**	111%
	20.5		3533	**	107%
H335	27.5	MAX.	3854	**	101%
	26.5 *		3743		97%
	25.5		3655		94%
X-Terminator	28.0	MAX.	3890		94%
	27.0 *		3796		90%
	26.0		3723		86%
RL10x	25.0	MAX.	3897	**	114%
	24.0 *		3701	**	110%
	23.0		3533	**	106%
A-2230	28.5	MAX.	3907		100%
	27.5		3872		96%
	26.5 *		3762		92%

BC=Ballistic Coefficient SD=Sectional Density
*Most Accurate Load Tested **Compressed Load

Use Maximum Loads with Caution
Refer to page 73 for additional safety information

223 Remington - 40 grain

		MAXIMUM S.A.A.M.I. O.A.C.L.		2.260"
		TESTED O.A.C.L.	B.C.	S.D.
Ballistic Tip®	40gr. Spitzer	2.260"	0.221	0.114
Ballistic Tip® Lead Free	40gr. Spitzer	2.260"	0.220	0.114
Varmageddon™	40gr. FBHP	2.125"	0.158	0.114
Varmageddon™	40gr. FB Tipped	2.225"	0.211	0.114

CASE TYPE:	Nosler		PRIMER TYPE	Rem 7 1/2
CASE HOLDS:	29.2	Gr. WATER	BARREL Length/Make	24" Lilja
			BARREL Twist	1-12"

POWDER TYPE	POWDER CHG. GRS.		MUZZLE VEL. F.P.S.	LOAD DENSITY (VOLUME)
Varget	27.0	MAX.	3383	100%
	26.0		3247	96%
	25.0 *		3111	93%
W748 Most Accurate Powder Tested	28.0 *	MAX.	3547	100%
	27.0		3426	96%
	26.0		3304	93%
H322	25.0	MAX.	3567	93%
	24.0		3442	89%
	23.0 *		3302	85%
RL7	23.5 *	MAX.	3614	87%
	22.5		3492	84%
	21.5		3370	80%
Varmint	29.0	MAX.	3642	100%
	28.0		3476	97%
	27.0 *		3311	93%
H335	27.5 *	MAX.	3681	96%
	26.5		3544	92%
	25.5		3406	89%
IMR 4198	23.0	MAX.	3682	90%
	22.0		3492	86%
	21.0 *		3302	82%
A-2015	25.5	MAX.	3796	99%
	24.5		3627	95%
	23.5 *		3458	91%
Viht N120	23.5	MAX.	3810	98%
	22.5		3677	94%
	21.5 *		3543	90%
Benchmark	28.0 *	MAX.	3860 **	104%
	27.0		3751	100%
	26.0		3602	96%

BC=Ballistic Coefficient SD=Sectional Density
*Most Accurate Load Tested **Compressed Load

Use Maximum Loads with Caution
Refer to page 73 for additional safety information

223 Remington - 50/52 grain		MAXIMUM S.A.A.M.I. O.A.C.L.		2.260"
		TESTED O.A.C.L.	B.C.	S.D.
Ballistic Tip®	50gr. Spitzer	2.260"	0.238	0.142
CT® Ballistic Silvertip®	50gr. Spitzer	2.260"	0.238	0.142
Nosler®	50gr. FBHP	2.230"	0.192	0.142
Nosler®	50gr. FBSP	2.200"	0.195	0.142
Custom Competition™	52gr. HPBT	2.260"	0.220	0.148

CASE TYPE:	Nosler		PRIMER TYPE	Rem 7 1/2
CASE HOLDS:	27.5	Gr. WATER	BARREL Length/Make	24" Lilja
			BARREL Twist	1-12"

POWDER TYPE	POWDER CHG. GRS.	MUZZLE VEL. F.P.S.	LOAD DENSITY (VOLUME)
Varget	26.0 * MAX.	3220	** 102%
	25.0	3092	98%
	24.0	2963	94%
IMR 4198	22.0 * MAX.	3230	91%
	21.0	3110	87%
	20.0	2990	83%
H335	26.0 MAX.	3260	96%
	25.0	3170	92%
	24.0 *	3080	89%
W748	26.5 * MAX.	3260	100%
	25.5	3150	97%
	24.5	3040	93%
Varmint	27.0 MAX.	3262	99%
	26.0	3119	96%
	25.0 *	2977	92%
CFE 223	29.0 MAX.	3379	** 107%
	28.0	3233	** 103%
	27.0 *	3088	99%
TAC	27.0 MAX.	3393	100%
	26.0	3238	96%
	25.0 *	3170	93%
Viht N133	25.0 MAX.	3412	** 106%
	24.0	3301	** 101%
	23.0 *	3190	97%
A-2460	28.0 MAX.	3497	** 103%
	27.0	3391	99%
	26.0 *	3285	96%
Benchmark	26.5 MAX.	3540	** 104%
Most Accurate	25.5 *	3411	100%
Powder Tested	24.5	3285	96%

BC=Ballistic Coefficient SD=Sectional Density
*Most Accurate Load Tested **Compressed Load

Use Maximum Loads with Caution
Refer to page 73 for additional safety information

223 Remington - 55 grain		MAXIMUM S.A.A.M.I. O.A.C.L.		2.260"
		TESTED O.A.C.L.	B.C.	S.D.
Ballistic Tip®	55gr. Spitzer	2.260"	0.267	0.157
CT® Ballistic Silvertip®	55gr. Spitzer	2.260"	0.267	0.157
Nosler®	55gr. FBHP	2.260"	0.210	0.157
Nosler®	55gr. FBSP	2.260"	0.218	0.157
Varmageddon™	55gr. FBHP	2.260"	0.210	0.157
Varmageddon™	55gr. FB Tipped	2.260"	0.255	0.157

CASE TYPE:	Nosler		PRIMER TYPE	Rem 7 1/2
CASE HOLDS:	27.4	Gr. WATER	BARREL Length/Make	24" Lilja
			BARREL Twist	1-12"

POWDER TYPE	POWDER CHG. GRS.		MUZZLE VEL. F.P.S.	LOAD DENSITY (VOLUME)
Varget	25.0	MAX.	3037	99%
	24.0		2921	95%
	23.0 *		2805	91%
H335 Most Accurate Powder Tested	25.0 *	MAX.	3140	93%
	24.0		3030	89%
	23.0		2920	85%
Varmint	26.0 *	MAX.	3140	96%
	25.0		3018	92%
	24.0		2897	88%
W748	26.0	MAX.	3140	99%
	25.0		3060	95%
	24.0 *		2980	91%
Viht N120	21.0 *	MAX.	3149	93%
	20.0		3020	89%
	19.0		2892	85%
IMR 4895	25.5 *	MAX.	3178	** 102%
	24.5		3083	98%
	23.5		2988	94%
Viht N135	25.0	MAX.	3195	** 106%
	24.0		3091	** 102%
	23.0 *		2987	98%
TAC	26.0	MAX.	3236	97%
	25.0		3096	93%
	24.0 *		2913	89%
CFE 223	28.0 *	MAX.	3292	** 103%
	27.0		3155	100%
	26.0		3018	96%
Benchmark	25.0 *	MAX.	3302	99%
	24.0		3194	95%
	23.0		3050	91%

BC=Ballistic Coefficient SD=Sectional Density
*Most Accurate Load Tested **Compressed Load

Use Maximum Loads with Caution
Refer to page 73 for additional safety information

223 Rem - 60/62 grain (fast twist)		MAXIMUM S.A.A.M.I. O.A.C.L.		2.260"
		TESTED O.A.C.L.	B.C.	S.D.
Ballistic Tip®	60gr. Spitzer	2.260"	0.270	0.171
Partition®	60gr. Spitzer	2.260"	0.228	0.171
Varmageddon™	62gr. FBHP	2.180"	0.251	0.176

CASE TYPE:	Nosler		PRIMER TYPE	Rem 7 1/2
CASE HOLDS:	27.6	Gr. WATER	BARREL Length/Make	20" Pac-Nor
			BARREL Twist	1-7"

POWDER TYPE	POWDER CHG. GRS.		MUZZLE VEL. F.P.S.		LOAD DENSITY (VOLUME)
Varget	24.5	MAX.	2995		96%
	23.5 *		2867		92%
	22.5		2769		88%
A-2460	23.0 *	MAX.	3022		84%
	22.0		2896		81%
	21.0		2799		77%
TAC	24.5	MAX.	3061		91%
	23.5 *		2983		87%
	22.5		2852		83%
H322	23.5 *	MAX.	3070		92%
	22.5		2965		88%
	21.5		2860		84%
RL15	25.5 *	MAX.	3078		96%
	24.5		2899		92%
	23.5		2809		89%
H335	25.0 *	MAX.	3102		92%
	24.0		3014		88%
	23.0		2909		85%
H4895	25.5 *	MAX.	3119	**	102%
	24.5		3080		98%
	23.5		2950		94%
IMR 4895	25.5 *	MAX.	3131	**	102%
	24.5		2980		98%
	23.5		2860		94%
Benchmark	23.5	MAX.	3136		92%
Most Accurate	22.5 *		3048		88%
Powder Tested	21.5		2932		84%
CFE 223	27.0 *	MAX.	3152		99%
	26.0		3037		95%
	25.0		2922		91%

BC=Ballistic Coefficient SD=Sectional Density
*Most Accurate Load Tested **Compressed Load

Use Maximum Loads with Caution
Refer to page 73 for additional safety information

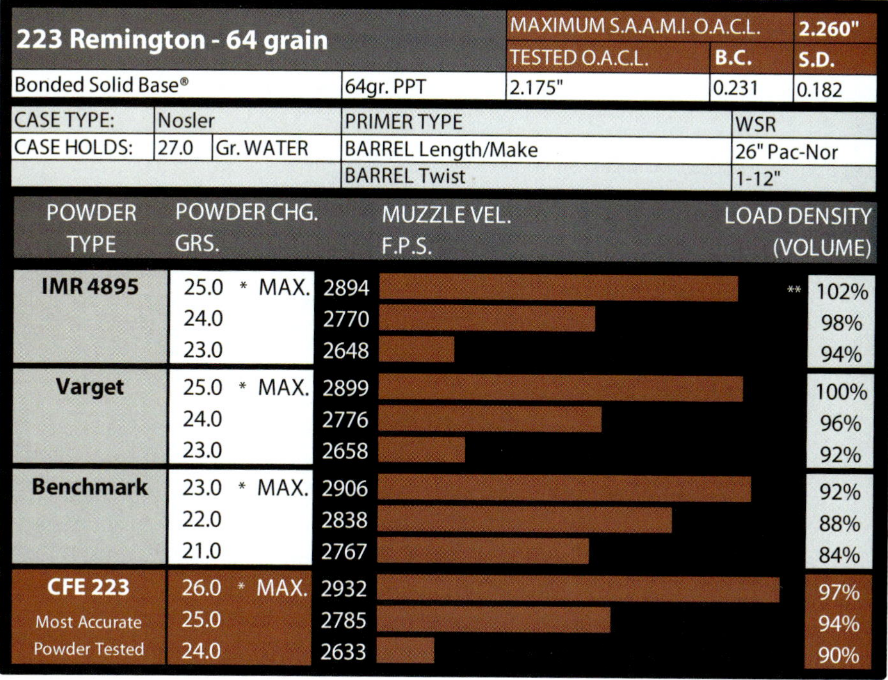

223 Remington - 64 grain				MAXIMUM S.A.A.M.I. O.A.C.L.		2.260"
				TESTED O.A.C.L.	B.C.	S.D.
Bonded Solid Base®			64gr. PPT	2.175"	0.231	0.182
CASE TYPE:	Nosler		PRIMER TYPE		WSR	
CASE HOLDS:	27.0	Gr. WATER	BARREL Length/Make		26" Pac-Nor	
			BARREL Twist		1-12"	

POWDER TYPE	POWDER CHG. GRS.	MUZZLE VEL. F.P.S.	LOAD DENSITY (VOLUME)
IMR 4895	25.0 * MAX.	2894	** 102%
	24.0	2770	98%
	23.0	2648	94%
Varget	25.0 * MAX.	2899	100%
	24.0	2776	96%
	23.0	2658	92%
Benchmark	23.0 * MAX.	2906	92%
	22.0	2838	88%
	21.0	2767	84%
CFE 223 Most Accurate Powder Tested	26.0 * MAX.	2932	97%
	25.0	2785	94%
	24.0	2633	90%

223 REMINGTON TECHNICAL INFORMATION

Simply a commercialized version of the 5.56x45mm NATO cartridge, the 223 Remington is without a doubt the single most popular centerfire varmint round in the civilized world. The loads listed here were developed using standard commercial brass. Military brass can have less case capacity because of heavier construction, and often yields higher pressures. We recommend caution when using military brass for the 223, and suggest starting at or below the minimum loads listed. Additionally, if the brass you are using had crimped primers, the crimp MUST be removed prior to re-priming. If you are loading for a semi-auto and find that crimping the bullet is necessary, we recommend using a taper crimp since there is no crimping groove on any of our .22 caliber products. Crimping with a standard seating die (roll crimping) can adversely affect accuracy. The standard 223 Remington barrel twist rate is 1-12". This twist rate will not properly stabilize bullets longer than our 60-grain Partition®.

BC=Ballistic Coefficient SD=Sectional Density
*Most Accurate Load Tested **Compressed Load

Use Maximum Loads with Caution
Refer to page 73 for additional safety information

Patrick Sweeney

5.56X45 NATO

Unlike the 223 Remington, the 5.56 is a no-hold's-barred effort at making a varmint cartridge into a defensive caliber. In order to do this, it is necessary to fire a 5.56 round (or a loaded-to-5.56 specs round) in a 5.56 chamber. The two are not interchangeable. Firing a 5.56 in a 223 Remington chamber leads to excessive pressure. This is not enough extra pressure to disassemble a rifle in ballistic time frames, but it does lead to loosened primer pockets and "popped" primers, which in a self-loading rifle can lead to problems.

Loaded to its potential, the 5.56 can launch heavier-than-customary bullets at impressive speeds, which explains its attractiveness. It also brings with it the need to attend to every detail; consistent and closely-monitored powder charge weights, neck tension and crimp, and cases trimmed to the minimum. It is also advisable to use cannelured bullets, to avoid bullet set-back on feeding, as the full-case powder charges common here can be unforgiving of error.

Developing and firing a regular diet of 5.56-level loads is not the time nor place to be using your miscellaneous range-found brass, nor the high-mileage cases that have already been trimmed three or four times.

Some do not wish to give up the perceived accuracy advantage of the .223's short leade, while gaining the velocity of the 5.56. Accuracy here is more a matter of proper rifle assembly, and not simply of the length of leade. NRA High Power shooters have been using rifles with 5.56-length leades (and then some) for many years now, and if a High Master is happy with the accuracy he gets, the rest of us really don't have much cause for complaint.

Even so, running a cartridge, any cartridge, right to the "red line" on every shot is not a formula for longevity. If you decide you have to get the performance of the 5.56, do so with the knowledge that you will experience a rifle with a shorter accurate service life, and in the AR, shorten bolt life to some extent as well. It

is entirely accepted to load your ammunition "down" to 223 Remington performance for practice and plinking, and save the more-strenuous loads for hunting, match use and other tasks.

That said, I have found I don't really wear out brass loaded to 5.56 specs at anything like the rate I expect, as it is usually lost to the terrain, left on ranges in "lost brass" matches, or otherwise irretrievable.

Patrick Sweeney

Currently the Handguns Editor of Guns and Ammo magazine, Patrick Sweeney is an author and gunsmith best known for his many books on gunsmithing.

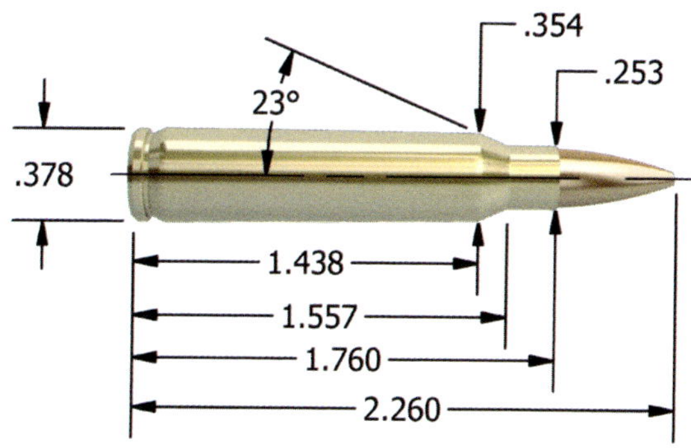

Maximum S.A.A.M.I. Overall Cartridge Length: 2.260"

BULLET CHOICES FOR THE 5.56X45 NATO

Custom Competition™

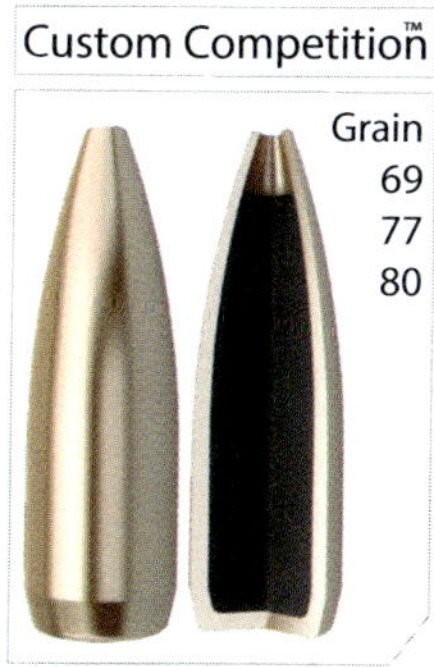

TECHNICAL INFORMATION

The 5.56x45mm NATO cartridge was developed for military use over 50 years ago, and became known as the 223 Remington only when introduced as a commercial round by Remington. Military use notwithstanding, the "5.56" enjoys widespread civilian use in AR-15 style rifles, and is the mainstay of service rifle match competition. Because of the fast-twist rifling used in the 5.56, it will stabilize bullets of up to 80-grains when fired from a 1-8" or faster barrel. Our 80-grain Custom Competition bullet is an excellent choice for extremely long-range shooting, but must be fired as a single shot cartridge (this bullet cannot be loaded to magazine length). For magazine length loading in the 5.56, the 69-grain and 77-grain Custom Competition products are excellent choices. If your brass originally had crimped primers, the crimp MUST be removed prior to re-priming. Also, as it is usually necessary to crimp the bullet in place when loading for a semi-auto use, we recommend using a taper crimp. Crimping with a standard seating die (roll crimping) on bullets without a crimping groove can adversely affect accuracy. The data listed in this section is safe for use in both 223 Remington and 5.56x45mm cartridges and chambers.

5.56x45mm NATO - 69 grain

		MAXIMUM S.A.A.M.I. O.A.C.L.		2.260"
		TESTED O.A.C.L.	B.C.	S.D.
Custom Competition™	69gr. HPBT	2.260"	0.305	0.196

CASE TYPE:	Nosler		PRIMER TYPE	WSR
CASE HOLDS:	27.5	Gr. WATER	BARREL Length/Make	20" Pac-Nor
			BARREL Twist	1-7"

POWDER TYPE	POWDER CHG. GRS.		MUZZLE VEL. F.P.S.	LOAD DENSITY (VOLUME)
RL15	24.5	MAX.	2786	93%
	23.5 *		2641	89%
	22.5		2495	85%
Viht N540	25.5	MAX.	2794	** 103%
	24.5		2665	99%
	23.5 *		2457	95%
TAC	24.5	MAX.	2815	91%
	23.5		2709	87%
	22.5 *		2535	83%
IMR 4895	23.5	MAX.	2818	94%
	22.5 *		2648	90%
	21.5		2531	86%
H4895	24.0	MAX.	2820	96%
	23.0		2649	92%
	22.0 *		2538	88%
Varget	25.0	MAX.	2835	98%
	24.0		2750	94%
	23.0 *		2603	90%
Viht N135	23.5	MAX.	2853	99%
	22.5		2733	95%
	21.5 *		2606	91%
H335	24.0	MAX.	2858	89%
	23.0		2747	85%
	22.0 *		2610	81%
A-2230 Most Accurate Powder Tested	23.5 *	MAX.	2870	87%
	22.5		2784	83%
	21.5		2720	80%
W748	26.0	MAX.	2957	98%
	25.0		2841	95%
	24.0 *		2723	91%

BC=Ballistic Coefficient SD=Sectional Density
*Most Accurate Load Tested **Compressed Load

Use Maximum Loads with Caution
Refer to page 73 for additional safety information

5.56x45mm NATO - 77/80 grain		MAXIMUM S.A.A.M.I. O.A.C.L.		2.260"
		TESTED O.A.C.L.	B.C.	S.D.
Custom Competition™	77gr. HPBT	2.260"	0.340	0.219
Custom Competition™	80gr. HPBT	2.360"	0.415	0.228

CASE TYPE:	Nosler		PRIMER TYPE	WSR
CASE HOLDS:	26.5	Gr. WATER	BARREL Length/Make	20" Pac-Nor
			BARREL Twist	1-7"

POWDER TYPE	POWDER CHG. GRS.		MUZZLE VEL. F.P.S.		LOAD DENSITY (VOLUME)
A-2015	21.0	MAX.	2554		90%
	20.0 *		2449		86%
	19.0		2344		81%
H335	23.0	MAX.	2601		88%
	22.0 *		2461		84%
	21.0		2376		80%
Viht N540	24.5	MAX.	2604	**	103%
	23.5 *		2466		99%
	22.5		2352		94%
W748	24.0	MAX.	2611		94%
	23.0 *		2480		90%
	22.0		2392		86%
Viht N140	23.5	MAX.	2636	**	103%
	22.5 *		2526		99%
	21.5		2424		94%
TAC	23.5	MAX.	2658		90%
	22.5 *		2530		87%
	21.5		2413		83%
Varget	24.0	MAX.	2658		98%
	23.0		2547		94%
	22.0 *		2454		90%
IMR 4064	23.5	MAX.	2659		97%
	22.5 *		2583		93%
	21.5		2458		89%
RL15	24.0	MAX.	2685		94%
Most Accurate	23.0 *		2566		90%
Powder Tested	22.0		2439		86%
H4895	23.5 *	MAX.	2692		97%
	22.5		2594		93%
	21.5		2468		89%

BC=Ballistic Coefficient SD=Sectional Density
*Most Accurate Load Tested **Compressed Load

Use Maximum Loads with Caution
Refer to page 73 for additional safety information

Kevin Steele

222 REMINGTON MAGNUM

I'm convinced that the 222 Remington Magnum was a victim of either bad timing or wishful thinking on the part of Remington. Introduced in 1958, the 222 Rem Mag was intended to increase the velocity and range of Remington's already successful 222 Remington launched eight years previously. Originally developed as a potential military cartridge in tandem with a rifle designed by Springfield Armory, the 222 Rem Mag was fated to enjoy an all too brief period of limited popularity.

The cartridge case of the 222 Rem Mag is slightly longer than the 222 Rem with subsequently more propellant capacity and thus capable of attaining higher velocities. This proved popular with varmint hunters of the era, who readily snapped up various rifles from Remington. But, the ultimate 222 Rem Mag in my opinion was the Sako L-461 Vixen. In fact, it was the first center-fire rifle I ever shot back in the early 1960's as my father, an avid woodchuck hunter, was in love with both the 222 Rem Mag and his little Finnish Vixen.

Logic tells me that Remington introduced the 222 Remington Magnum commercially, believing that the Springfield experimental assault rifle which also chambered the round would eventually be adopted by the U.S. Armed Forces. At the time that would have seemed a "no brainer" to the Remington brass, as Springfield Armory had been producing rifles for the U.S. military since 1816.

What Remington didn't plan on was a young engineer named Eugene Stoner who submitted his own assault rifle design chambered for another high-velocity, 22 center-fire developed by Bob Hutton, the technical editor at the time of Guns & Ammo magazine.

For five or six years, the 222 Rem Mag enjoyed a brief period of popularity, but in 1964 the writing was on the wall when the military adopted what would become known as the M16 rifle and its Hutton-designed cartridge…the soon to be 223 Remington.

Ballistically, the 222 Rem Mag is a near duplicate of the 223 Rem. Nosler Custom offers loaded ammunition today, but to my knowledge no major rifle manufacturer is currently chambering rifles for the 222 Magnum. It's true that I retain a bit of sentiment for the 222 Remington Magnum, but then, I also have a soft spot for the 225 Winchester as well!

Kevin is the publisher of Petersen's HUNTING and the producer and co-host of Petersen's HUNTING Adventures on The Sportsman Channel. He is also publisher and editor-in-chief emeritus of Guns & Ammo.

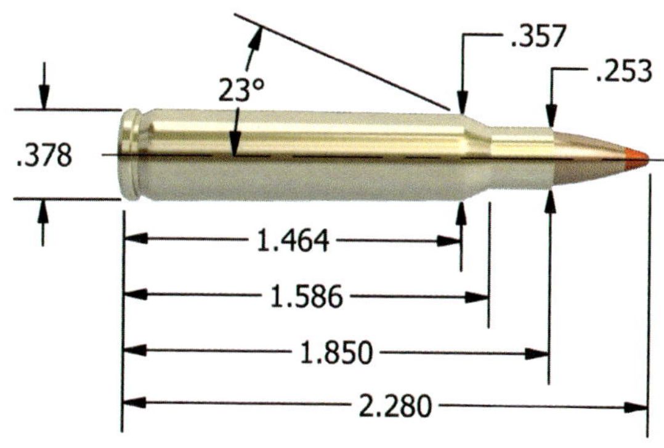

Maximum S.A.A.M.I. Overall Cartridge Length: 2.280"

BULLET CHOICES FOR THE 222 REMINGTON MAGNUM

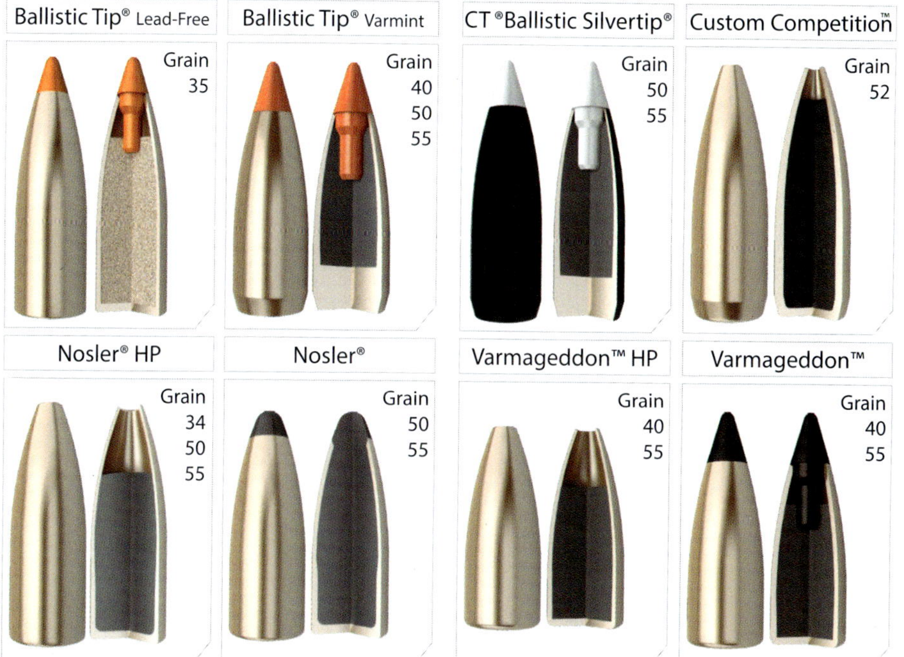

TECHNICAL INFORMATION

Developed to answer the call for more velocity than the standard 222 Remington, the 222 Remington Magnum is capable of slightly outperforming the 223 Remington when loaded with heavier bullets. This is due to its almost 5% greater case capacity.

222 Rem Mag - 34/35 grain		MAXIMUM S.A.A.M.I. O.A.C.L.		2.280"
		TESTED O.A.C.L.	B.C.	S.D.
Nosler®	34gr. FBHP	2.170"	0.113	0.097
Ballistic Tip® Lead Free	35gr. Spitzer	2.280"	0.201	0.100

CASE TYPE:	Remington		PRIMER TYPE	Rem 7 1/2
CASE HOLDS:	30.5	Gr. WATER	BARREL Length/Make	24" Lilja
			BARREL Twist	1-14"

POWDER TYPE	POWDER CHG. GRS.		MUZZLE VEL. F.P.S.	LOAD DENSITY (VOLUME)
A-2015	26.5	MAX.	3661	99%
	25.5 *		3514	95%
	24.5		3378	91%
Benchmark	27.5	MAX.	3678	97%
	26.5		3520	94%
	25.5 *		3331	90%
W748	29.5	MAX.	3696	** 101%
	28.5 *		3617	97%
	27.5		3422	94%
H322	26.5	MAX.	3704	94%
	25.5		3510	90%
	24.5 *		3339	87%
X-Terminator	27.5	MAX.	3723	92%
	26.5 *		3631	88%
	25.5		3510	85%
N201	29.0	MAX.	3729	** 101%
	28.0		3628	98%
	27.0 *		3497	94%
RL 7	24.5	MAX.	3768	87%
	23.5 *		3575	84%
	22.5		3424	80%
H335	29.0 *	MAX.	3797	97%
	28.0		3668	93%
	27.0		3552	90%
IMR 4198	23.0	MAX.	3810	86%
Most Accurate	22.0 *		3600	82%
Powder Tested	21.0		3444	78%

BC=Ballistic Coefficient SD=Sectional Density
*Most Accurate Load Tested **Compressed Load

Use Maximum Loads with Caution
Refer to page 73 for additional safety information

222 Rem Mag - 40 grain

		MAXIMUM S.A.A.M.I. O.A.C.L.		2.280"
		TESTED O.A.C.L.	B.C.	S.D.
Ballistic Tip®	40gr. Spitzer	2.280"	0.221	0.114
Varmageddon™	40gr. FBHP	2.210"	0.158	0.114
Varmageddon™	40gr. FB Tipped	2.280"	0.211	0.114

CASE TYPE:	Remington		PRIMER TYPE	Rem 7 1/2
CASE HOLDS:	30.5	Gr. WATER	BARREL Length/Make	24" Lilja
			BARREL Twist	1-14"

POWDER TYPE	POWDER CHG. GRS.		MUZZLE VEL. F.P.S.		LOAD DENSITY (VOLUME)
Benchmark	26.5	MAX.	3468		94%
	25.5	*	3258		90%
	24.5		3078		87%
N201	28.0	MAX.	3494		98%
	27.0	*	3370		94%
	26.0		3221		91%
X-Terminator	26.5	MAX.	3508		88%
	25.5		3425		85%
	24.5	*	3310		82%
IMR 4198	23.5	MAX.	3640		88%
	22.5		3455		84%
	21.5	*	3270		80%
W748 Most Accurate Powder Tested	29.5	* MAX.	3657	**	101%
	28.5		3528		97%
	27.5		3399		94%
Varget	30.0	* MAX.	3691	**	106%
	29.0		3551	**	103%
	28.0		3412		99%
RL7	25.0	* MAX.	3765		89%
	24.0		3600		86%
	23.0		3435		82%
A-2015	27.0	* MAX.	3771	**	101%
	26.0		3620		97%
	25.0		3469		93%
Viht N133	27.5	* MAX.	3834	**	105%
	26.5		3708	**	101%
	25.5		3581		97%
H335	29.5	MAX.	3852		98%
	28.5		3742		95%
	27.5	*	3632		92%

BC=Ballistic Coefficient SD=Sectional Density
*Most Accurate Load Tested **Compressed Load

Use Maximum Loads with Caution
Refer to page 73 for additional safety information

222 Rem Mag - 50/52 grain		MAXIMUM S.A.A.M.I. O.A.C.L.		2.280"
		TESTED O.A.C.L.	B.C.	S.D.
Ballistic Tip®	50gr. Spitzer	2.280"	0.238	0.142
CT® Ballistic Silvertip®	50gr. Spitzer	2.280"	0.238	0.142
Nosler®	50gr. FBHP	2.280"	0.192	0.142
Nosler®	50gr. FBSP	2.280"	0.195	0.142
Custom Competition™	52gr. HPBT	2.280"	0.220	0.148

CASE TYPE:	Remington		PRIMER TYPE	Rem 7 1/2
CASE HOLDS:	28.7	Gr. WATER	BARREL Length/Make	24" Lilja
			BARREL Twist	1-14"

POWDER TYPE	POWDER CHG. GRS.		MUZZLE VEL. F.P.S.	LOAD DENSITY (VOLUME)
W748	27.0	MAX.	3164	98%
	26.0		3038	94%
	25.0 *		2989	91%
IMR 4198	21.0	MAX.	3190	83%
	20.0		3080	79%
	19.0 *		2970	75%
X-Terminator	25.5	MAX.	3262	90%
	24.5 *		3138	87%
	23.5		3006	83%
Benchmark	25.5	MAX.	3272	96%
	24.5		3076	92%
	23.5 *		2935	89%
N201	27.0	MAX.	3304	100%
	26.0		3195	96%
	25.0 *		3015	93%
IMR 3031	25.0	MAX.	3328	98%
	24.0		3223	94%
	23.0 *		3118	91%
BL-C2	26.5 *	MAX.	3331	94%
	25.5		3228	90%
	24.5		3126	87%
H335 Most Accurate Powder Tested	27.5 *	MAX.	3332	97%
	26.5		3157	94%
	25.5		2982	90%
IMR 4320	27.5	MAX.	3368 **	103%
	26.5		3263	99%
	25.5 *		3158	96%
IMR 4895	26.0 *	MAX.	3370	100%
	25.0		3200	96%
	24.0		3030	92%

BC=Ballistic Coefficient SD=Sectional Density
*Most Accurate Load Tested **Compressed Load

Use Maximum Loads with Caution
Refer to page 73 for additional safety information

222 Rem Mag - 55 grain

Bullet	Type	TESTED O.A.C.L. (MAXIMUM S.A.A.M.I. O.A.C.L. 2.280")	B.C.	S.D.
Ballistic Tip®	55gr. Spitzer	2.280"	0.267	0.157
CT® Ballistic Silvertip®	55gr. Spitzer	2.280"	0.267	0.157
Nosler®	55gr. FBSP	2.280"	0.218	0.157
Varmageddon™	55gr. FBHP	2.280"	0.210	0.157
Varmageddon™	55gr. FB Tipped	2.280"	0.255	0.157

CASE TYPE:	Remington		PRIMER TYPE	Rem 7 1/2
CASE HOLDS:	28.6	Gr. WATER	BARREL Length/Make	24" Lilja
			BARREL Twist	1-14"

POWDER TYPE	POWDER CHG. GRS.		MUZZLE VEL. F.P.S.	LOAD DENSITY (VOLUME)
RL10x	21.5	MAX.	2896	84%
	20.5 *		2792	81%
	19.5		2721	77%
X-Terminator Most Accurate Powder Tested	24.0	MAX.	3009	85%
	23.0		2881	82%
	22.0 *		2747	78%
Norma 201	26.0 *	MAX.	3135	97%
	25.0		3006	93%
	24.0		2888	89%
Benchmark	25.0	MAX.	3141	95%
	24.0		2990	91%
	23.0 *		2860	87%
IMR 4198	21.0	MAX.	3180	83%
	20.0		3070	79%
	19.0 *		2960	75%
IMR 4320	26.0 *	MAX.	3200	98%
	25.0		3070	94%
	24.0		2940	90%
H335	27.0 *	MAX.	3230	96%
	26.0		3090	92%
	25.0		2950	89%
BL-C2	26.0	MAX.	3310	92%
	25.0		3210	89%
	24.0 *		3110	85%
IMR 3031	25.0 *	MAX.	3330	99%
	24.0		3180	95%
	23.0		3030	91%
IMR 4895	26.0 *	MAX.	3380	100%
	25.0		3260	96%
	24.0		3140	92%

BC=Ballistic Coefficient SD=Sectional Density
*Most Accurate Load Tested **Compressed Load

Use Maximum Loads with Caution
Refer to page 73 for additional safety information

Sam Fadala

22 PPC-USA

Lou Palmisano was a vascular surgeon of note in the NY/NJ area, a graduate of the esteemed University of Bologna Medical College (Italy) and inventor of medical tools. A colleague informed me that on the morning train to school, Lou read his medical books, tearing out each page as he finished. He had understood and memorized all salient points. Along with medicine, Palmisano had a burning interest in precision shooting. One of his early ideas was spinning 22 Long Rifle cases during manufacture to evenly distribute priming mixture around the rim, since any part of the rim could be struck by the firing pin. But, his chief goal was a supreme benchrest cartridge. When he saw the 220 Russian, his dream was on the road.

The dream went from paper to reality at the hands of talented machinist Ferris Pindell just as Galileo turned Hans Lippershey's 1608 idea into a workable telescope in 1609. Lou called upon ballistician Dan Pawlak to run sophisticated tests to examine the efficacy of his short/fat case design—now with 10% body taper and 30-degree shoulder angle. Pawlak found that the bullet exited the barrel of the test rifle 640 microseconds after peak pressure, leaving only 8,000 psi muzzle pressure from about 55,000, indicating efficient consumption of the powder charge. Dan wrote, "The significance of this cartridge is surely underestimated." Pawlak also concluded that "The harmonics—vibrations—when firing the cartridge were conducive to uniformity and consequent high accuracy."

The 22 PPC (Palmisano-Pindell-Cartridge) debuted in the August 1974 issue of Precision Shooting Magazine, and in the 1975 Super Shoot, Lou Palmisano won the 100 yard Heavy Varmint Class Aggregate with a 0.265" center-to-center group. In 1976, a 14-year old boy emerged victorious over 80 men, winning the 200 yard 25-shot aggregate Benchrest Shooters National Championship with a group of 0.3280". His name was David Palmisano, Lou's son. Between 1981 and 1985, 116 International Benchrest Shooters (IBS) records were broken—by the 22 PPC.

Lou joined me at my Wyoming home for a long weekend of prairie dog control that lasted three weeks. Later, I met him in NYC, where we drove to his Pennsylvania farm for more shooting. If we did not achieve five-shot groups of 0.25" center to center at 100 yards, we blamed range conditions, faulty ambient temperatures, anti-accuracy gremlins hovering over the bench, and occasionally ourselves. The 22 PPC, while challenged by the 6mm PPC and others, remains a cartridge of terrific accuracy potential at the bench and in the varmint field.

Sam Fadala

Sam Fadala is a freelance writer, and the author of many books on firearms and hunting.

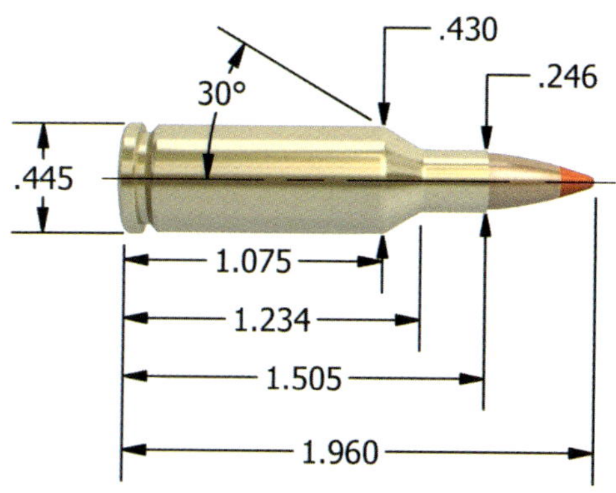

Maximum S.A.A.M.I. Overall Cartridge Length: 1.960"

BULLET CHOICES FOR THE 22 PPC-USA

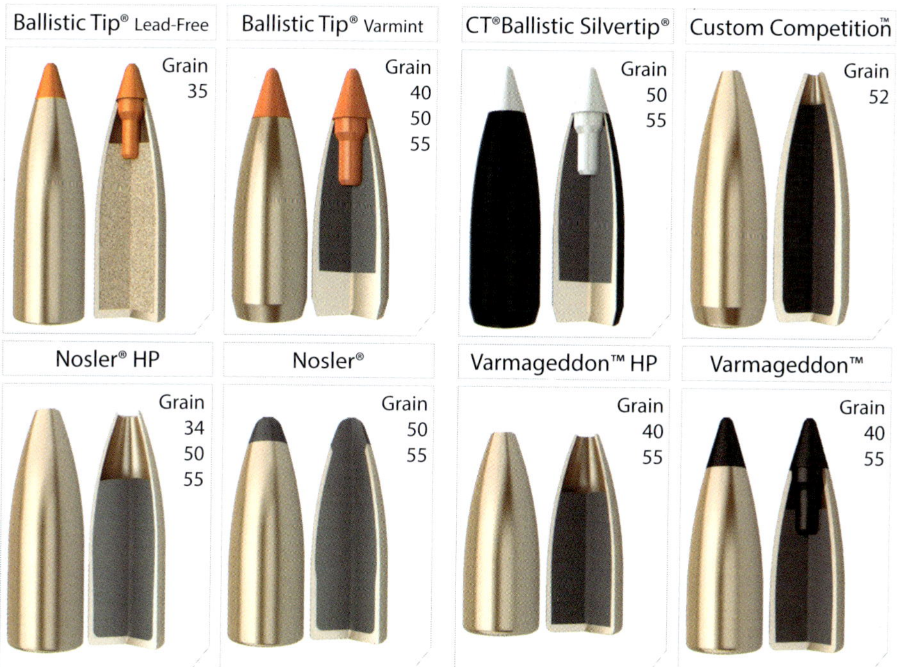

TECHNICAL INFORMATION

In terms of accuracy, this cartridge is tough to beat! For our data, we used production 22 PPC-USA brass from Sako with the necks "outside turned" to fit our chamber. The loads listed here will produce similar pressures in a chamber with standard neck dimensions. Hodgdon's Varget gave excellent results with all bullet weights.

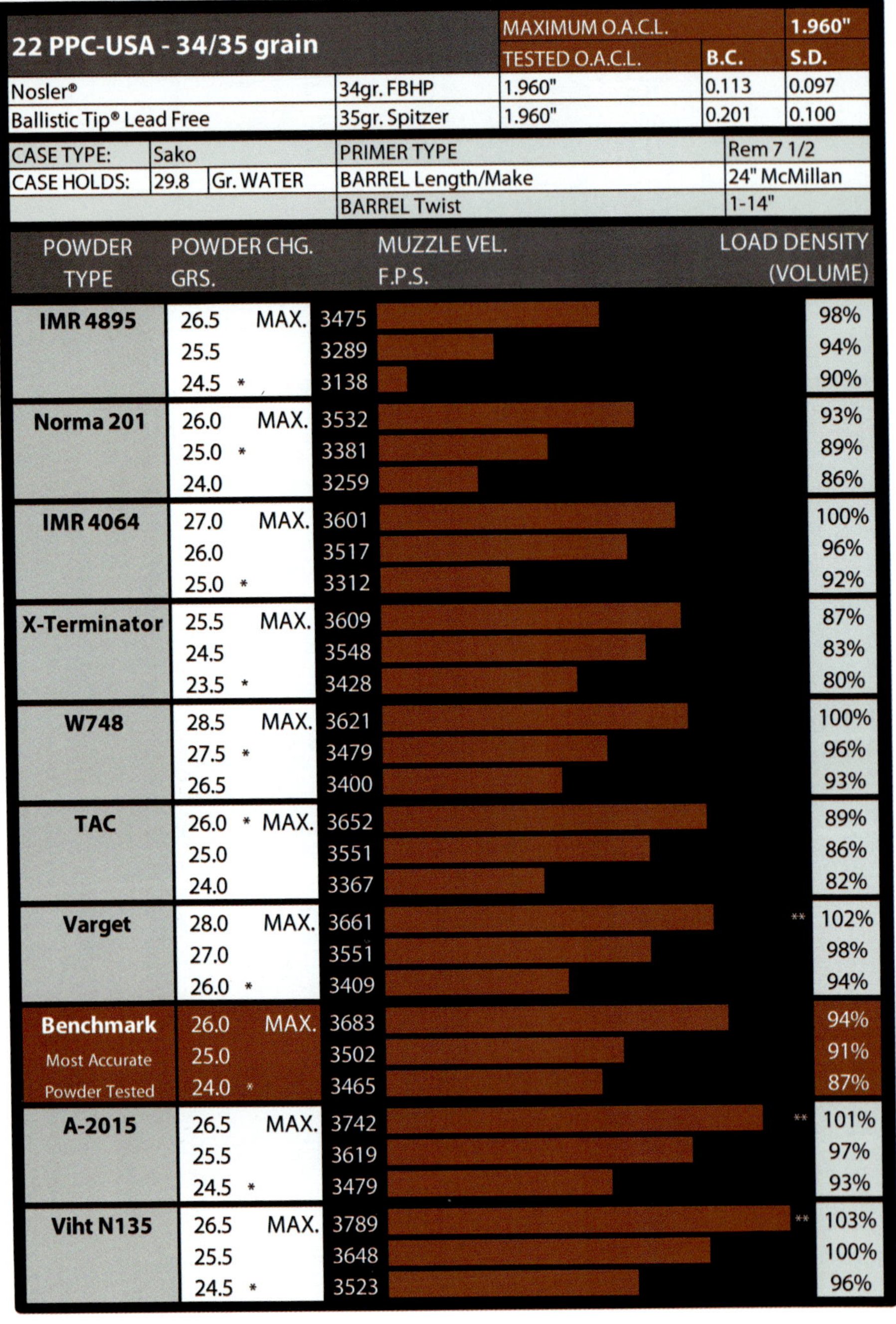

22 PPC-USA - 34/35 grain		MAXIMUM O.A.C.L.		1.960"
		TESTED O.A.C.L.	B.C.	S.D.
Nosler®	34gr. FBHP	1.960"	0.113	0.097
Ballistic Tip® Lead Free	35gr. Spitzer	1.960"	0.201	0.100

CASE TYPE:	Sako		PRIMER TYPE	Rem 7 1/2
CASE HOLDS:	29.8	Gr. WATER	BARREL Length/Make	24" McMillan
			BARREL Twist	1-14"

POWDER TYPE	POWDER CHG. GRS.		MUZZLE VEL. F.P.S.	LOAD DENSITY (VOLUME)
IMR 4895	26.5	MAX.	3475	98%
	25.5		3289	94%
	24.5 *		3138	90%
Norma 201	26.0	MAX.	3532	93%
	25.0 *		3381	89%
	24.0		3259	86%
IMR 4064	27.0	MAX.	3601	100%
	26.0		3517	96%
	25.0 *		3312	92%
X-Terminator	25.5	MAX.	3609	87%
	24.5		3548	83%
	23.5 *		3428	80%
W748	28.5	MAX.	3621	100%
	27.5 *		3479	96%
	26.5		3400	93%
TAC	26.0 *	MAX.	3652	89%
	25.0		3551	86%
	24.0		3367	82%
Varget	28.0	MAX.	3661	** 102%
	27.0		3551	98%
	26.0 *		3409	94%
Benchmark Most Accurate Powder Tested	26.0	MAX.	3683	94%
	25.0		3502	91%
	24.0 *		3465	87%
A-2015	26.5	MAX.	3742	** 101%
	25.5		3619	97%
	24.5 *		3479	93%
Viht N135	26.5	MAX.	3789	** 103%
	25.5		3648	100%
	24.5 *		3523	96%

BC=Ballistic Coefficient SD=Sectional Density
*Most Accurate Load Tested **Compressed Load

Use Maximum Loads with Caution
Refer to page 73 for additional safety information

22 PPC-USA - 40 grain

		MAXIMUM O.A.C.L.		1.960"
		TESTED O.A.C.L.	B.C.	S.D.
Ballistic Tip®	40gr. Spitzer	1.960"	0.221	0.114
Varmageddon™	40gr. FBHP	1.960"	0.158	0.114
Varmageddon™	40gr. FB Tipped	1.960"	0.211	0.114

CASE TYPE:	Sako		PRIMER TYPE	Fed 205M
CASE HOLDS:	30.1	Gr. WATER	BARREL Length/Make	24" Lilja
			BARREL Twist	1-14"

POWDER TYPE	POWDER CHG. GRS.	MUZZLE VEL. F.P.S.	LOAD DENSITY (VOLUME)
Viht N135	27.5 * MAX.	3577	** 106%
	26.5	3452	** 102%
	25.5	3327	99%
Varget Most Accurate Powder Tested	29.0 * MAX.	3611	** 104%
	28.0	3476	** 101%
	27.0	3340	97%
W748	30.5 MAX.	3611	** 106%
	29.5	3481	** 102%
	28.5 *	3350	99%
Benchmark	28.0 MAX.	3661	** 101%
	27.0 *	3472	97%
	26.0	3371	93%
A-2230	28.5 MAX.	3690	97%
	27.5	3559	93%
	26.5 *	3429	90%
H322	27.0 * MAX.	3726	97%
	26.0	3594	93%
	25.0	3462	90%
H4198	23.5 MAX.	3726	88%
	22.5 *	3611	84%
	21.5	3476	80%
A-2015	27.0 MAX.	3745	** 102%
	26.0	3606	98%
	25.0 *	3468	94%
IMR 4198	24.5 MAX.	3785	92%
	23.5	3641	89%
	22.5 *	3497	85%
RL7	25.5 MAX.	3790	92%
	24.5	3645	88%
	23.5 *	3500	85%

BC=Ballistic Coefficient SD=Sectional Density
*Most Accurate Load Tested **Compressed Load

Use Maximum Loads with Caution
Refer to page 73 for additional safety information

22 PPC-USA - 50/52 grain		MAXIMUM O.A.C.L.		1.960"
		TESTED O.A.C.L.	B.C.	S.D.
Ballistic Tip®	50gr. Spitzer	1.960"	0.238	0.142
CT® Ballistic Silvertip®	50gr. Spitzer	1.960"	0.238	0.142
Nosler®	50gr. FBHP	1.960"	0.192	0.142
Nosler®	50gr. FBSP	1.960"	0.195	0.142
Custom Competition™	52gr. HPBT	1.960"	0.220	0.148

CASE TYPE:	Sako		PRIMER TYPE	Fed 205M
CASE HOLDS:	29.6	Gr. WATER	BARREL Length/Make	24" Lilja
			BARREL Twist	1-14"

POWDER TYPE	POWDER CHG. GRS.	MUZZLE VEL. F.P.S.	LOAD DENSITY (VOLUME)
Benchmark Most Accurate Powder Tested	26.0 MAX.	3335	95%
	25.0	3170	91%
	24.0 *	3011	88%
H322	25.0 * MAX.	3360	91%
	24.0	3260	88%
	23.0	3160	84%
Varget	28.0 * MAX.	3369	** 102%
	27.0	3255	99%
	26.0	3142	95%
W748	29.0 * MAX.	3388	** 102%
	28.0	3263	99%
	27.0	3138	95%
IMR 4895	29.0 MAX.	3420	** 108%
	28.0	3286	** 104%
	27.0 *	3151	100%
A-2460	28.0 * MAX.	3480	96%
	27.0	3360	92%
	26.0	3240	89%
A-2015	26.0 MAX.	3503	100%
	25.0	3355	96%
	24.0 *	3206	92%
Viht N135	27.5 MAX.	3505	** 108%
	26.5	3390	** 104%
	25.5 *	3274	100%
RL7	25.0 * MAX.	3512	92%
	24.0	3397	88%
	23.0	3282	84%
IMR 4198	24.0 MAX.	3527	92%
	23.0	3380	88%
	22.0 *	3233	84%

BC=Ballistic Coefficient SD=Sectional Density
*Most Accurate Load Tested **Compressed Load

Use Maximum Loads with Caution
Refer to page 73 for additional safety information

22 PPC-USA - 55 grain

Bullet	Type	MAXIMUM O.A.C.L. / TESTED O.A.C.L.	B.C.	S.D.
		MAXIMUM O.A.C.L.		1.960"
		TESTED O.A.C.L.	B.C.	S.D.
Ballistic Tip®	55gr. Spitzer	1.960"	0.267	0.157
CT® Ballistic Silvertip®	55gr. Spitzer	1.960"	0.267	0.157
Nosler®	55gr. FBSP	1.960"	0.210	0.157
Varmageddon™	55gr. FBHP	1.960"	0.218	0.157
Varmageddon™	55gr. FB Tipped	1.960"	0.255	0.157

CASE TYPE:	Sako		PRIMER TYPE	Fed 205M
CASE HOLDS:	29.2	Gr. WATER	BARREL Length/Make	24" Lilja
			BARREL Twist	1-14"

POWDER TYPE	POWDER CHG. GRS.		MUZZLE VEL. F.P.S.		LOAD DENSITY (VOLUME)
IMR 4895	26.0 *	MAX.	3140		98%
	25.0		3010		94%
	24.0		2880		90%
W748	28.5 *	MAX.	3252	**	102%
	27.5		3137		98%
	26.5		3022		95%
A-2520	28.0	MAX.	3312		99%
	27.0		3187		95%
	26.0 *		3062		92%
RL7	24.0	MAX.	3320		89%
	23.0		3210		86%
	22.0 *		3100		82%
Viht N135	27.5	MAX.	3352	**	110%
	26.5		3240	**	106%
	25.5 *		3127	**	102%
Varget Most Accurate Powder Tested	28.0 *	MAX.	3361	**	104%
	27.0		3236		100%
	26.0		3110		96%
IMR 4198	23.5 *	MAX.	3362		91%
	22.5		3231		88%
	21.5		3099		84%
A-2495	27.0	MAX.	3391	**	105%
	26.0		3268	**	101%
	25.0 *		3145		97%
H335	28.0	MAX.	3440		97%
	27.0		3340		94%
	26.0 *		3240		90%
RL15	29.0	MAX.	3448	**	103%
	28.0		3321		100%
	27.0 *		3195		96%

BC=Ballistic Coefficient SD=Sectional Density
*Most Accurate Load Tested **Compressed Load

Use Maximum Loads with Caution
Refer to page 73 for additional safety information

Chub Eastman

22 BR REMINGTON

A few prairie dog seasons ago, one of my favorite long-range varmint rifles finally gave up the ghost, after many years of reliable service. Accuracy went from bragging size, bug-hole groups to an inch or a little more which is more than respectable for a hunting rifle. However, for a varmint rifle where accuracy on Coke can size targets at a couple of hundred yards is paramount, that's not good enough.

In a conversation with the head ballistician at Nosler, I casually asked what cartridge they used to test for accuracy with the .224" diameter bullets. His answer was the 22 BR. When asked why the 22 BR, he said "the case was very efficient and gave great accuracy with most powders used." Velocities were close to the 22-250 and cases lasted considerably longer than most other high velocity .224 cartridges.

This all made sense to me and would be an easy conversion. My rifle was a single shot Remington and all that was needed to change from 22-250 to the 22 BR was the barrel and chamber.

If there is a down side to the 22 BR, it is the fact you have to make the brass for it. 6mm BR cases are sized down and neck turned which can be a pain in the butt. However, when finished, they will last longer than most any cartridge I have ever loaded. I have some cases that have been shot at least a dozen times and the primer pockets are still good.

The other advantage learned in the field for the first time was the barrel does not heat up as quickly as with most other varmint cartridges that push bullets to or past the 3600 fps mark.

The 22 BR is an off-shoot of technology developed by the benchrest fraternity. To them, short and fat relates to accuracy. For varmint hunters, accuracy is also one of the high priorities. All I know is, it will vaporize a prairie dog with a 50gr Ballistic tip at 250 yards to 300 yards, not heat the barrel as fast, is easy to load and cases last forever.

I almost wish I would have shot the 22-250 barrel out sooner than I did.

Chub Eastman

Chub Eastman is the Rifle Field Editor for Sports Afield Magazine

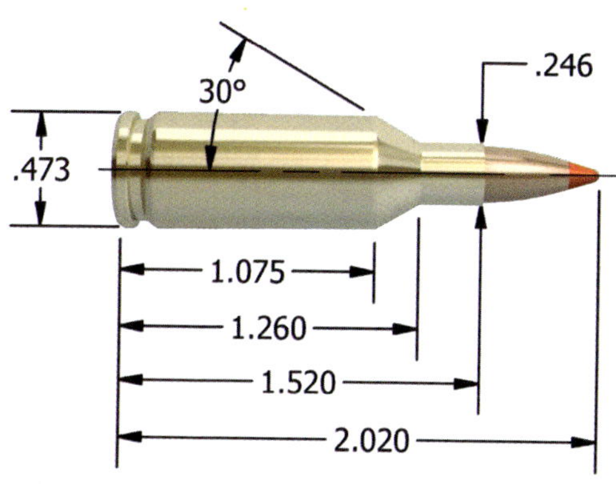

Maximum S.A.A.M.I. Overall Cartridge Length: 2.020"

BULLET CHOICES FOR THE 22 BR REMINGTON

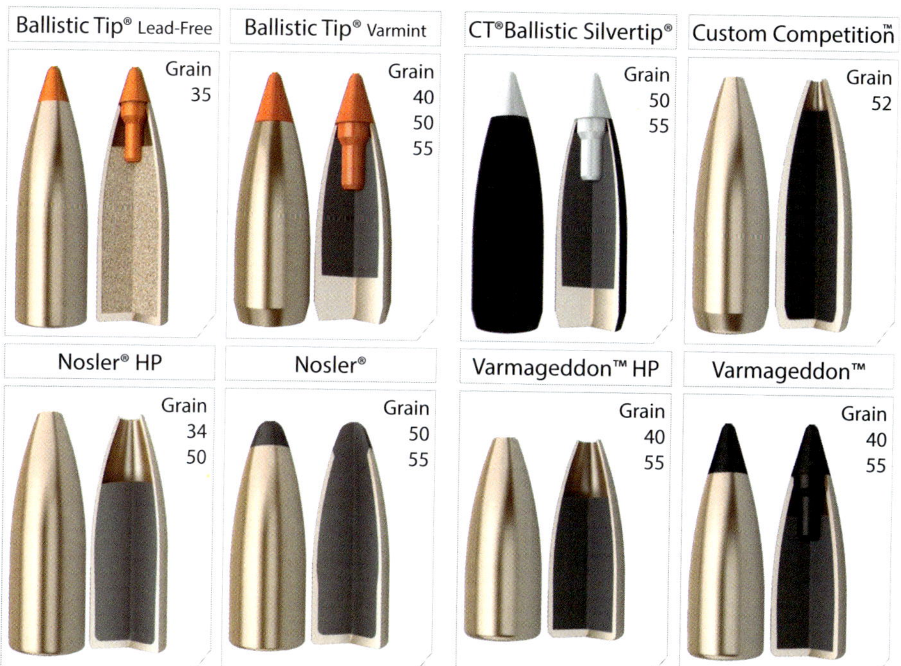

TECHNICAL INFORMATION

If there is a 22 caliber cartridge that is as accurate as the 22 PPC, this is it! At present, there is no commercial source for 22 BR brass, so start with brass from another member of the BR family. We used Remington 6mm BR brass necked down, trimmed, and outside neck turned for the development of these loads. For production testing, we have had our best results with RL7 and W748.

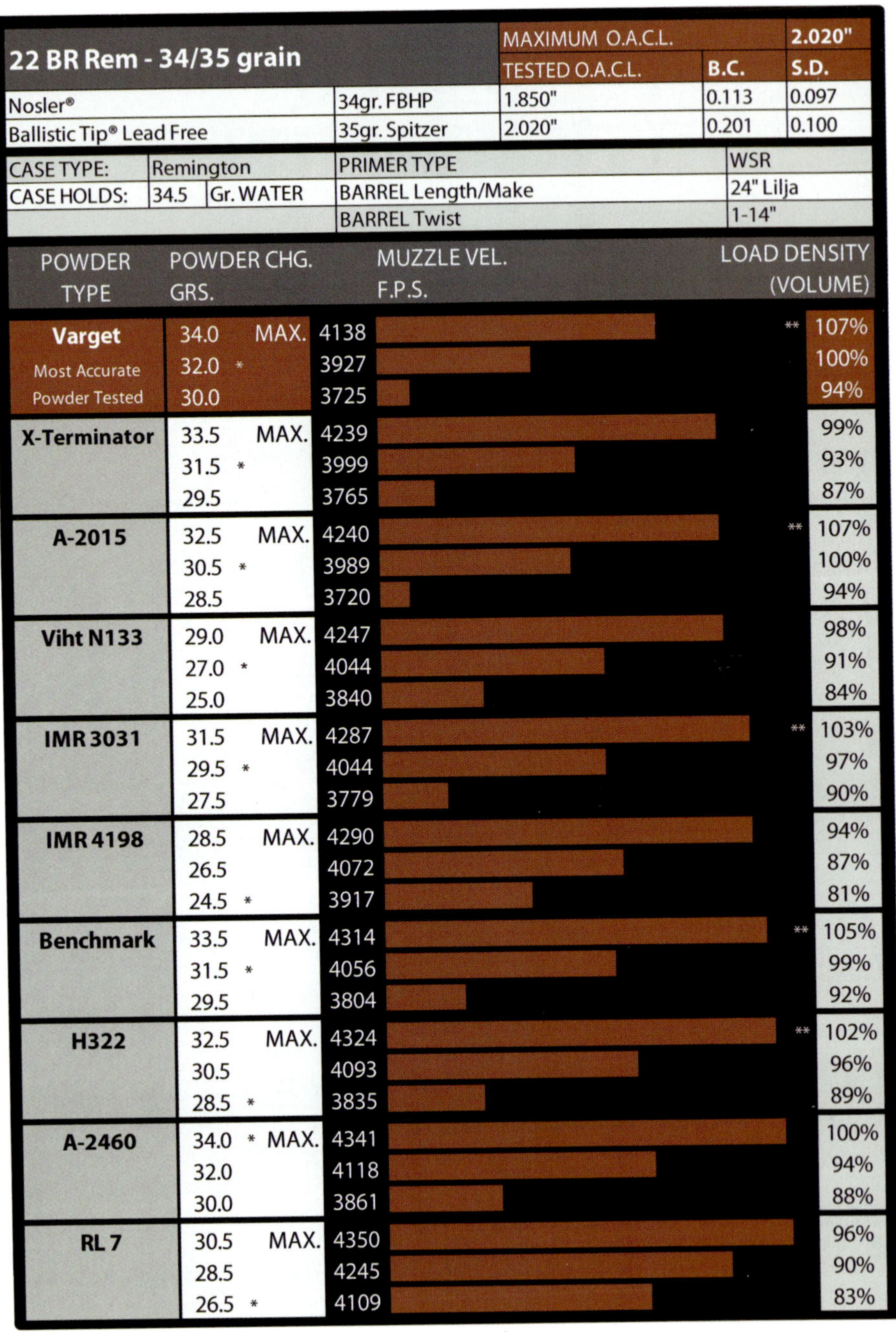

22 BR Rem - 34/35 grain		MAXIMUM O.A.C.L.		2.020"
		TESTED O.A.C.L.	B.C.	S.D.
Nosler®	34gr. FBHP	1.850"	0.113	0.097
Ballistic Tip® Lead Free	35gr. Spitzer	2.020"	0.201	0.100

CASE TYPE:	Remington	PRIMER TYPE	WSR
CASE HOLDS:	34.5 Gr. WATER	BARREL Length/Make	24" Lilja
		BARREL Twist	1-14"

POWDER TYPE	POWDER CHG. GRS.		MUZZLE VEL. F.P.S.		LOAD DENSITY (VOLUME)
Varget Most Accurate Powder Tested	34.0	MAX.	4138	**	107%
	32.0	*	3927		100%
	30.0		3725		94%
X-Terminator	33.5	MAX.	4239		99%
	31.5	*	3999		93%
	29.5		3765		87%
A-2015	32.5	MAX.	4240	**	107%
	30.5	*	3989		100%
	28.5		3720		94%
Viht N133	29.0	MAX.	4247		98%
	27.0	*	4044		91%
	25.0		3840		84%
IMR 3031	31.5	MAX.	4287	**	103%
	29.5	*	4044		97%
	27.5		3779		90%
IMR 4198	28.5	MAX.	4290		94%
	26.5		4072		87%
	24.5	*	3917		81%
Benchmark	33.5	MAX.	4314	**	105%
	31.5	*	4056		99%
	29.5		3804		92%
H322	32.5	MAX.	4324	**	102%
	30.5		4093		96%
	28.5	*	3835		89%
A-2460	34.0	* MAX.	4341		100%
	32.0		4118		94%
	30.0		3861		88%
RL 7	30.5	MAX.	4350		96%
	28.5		4245		90%
	26.5	*	4109		83%

BC=Ballistic Coefficient SD=Sectional Density
*Most Accurate Load Tested **Compressed Load

Use Maximum Loads with Caution
Refer to page 73 for additional safety information

22 BR Rem - 40 grain

		MAXIMUM O.A.C.L.		2.020"
		TESTED O.A.C.L.	B.C.	S.D.
Ballistic Tip®	40gr. Spitzer	2.020"	0.221	0.114
Varmageddon™	40gr. FBHP	1.890"	0.158	0.114
Varmageddon™	40gr. FB Tipped	1.990"	0.211	0.114

CASE TYPE:	Remington		PRIMER TYPE	WSR
CASE HOLDS:	35.5	Gr. WATER	BARREL Length/Make	24" Lilja
			BARREL Twist	1-14"

POWDER TYPE	POWDER CHG. GRS.		MUZZLE VEL. F.P.S.		LOAD DENSITY (VOLUME)
Viht N133	28.5	MAX.	3655		93%
	26.5		3412		87%
	24.5 *		3169		80%
BL-C2	36.0 *	MAX.	3795	**	103%
	34.0		3516		97%
	32.0		3236		92%
IMR 4198	28.0	MAX.	3841		90%
	26.0		3508		83%
	24.0 *		3174		77%
IMR 3031	32.0	MAX.	3877	**	102%
	30.0		3496		95%
	28.0 *		3116		89%
Varget	34.0	MAX.	3881	**	104%
	32.0		3603		97%
	30.0 *		3326		91%
W748 Most Accurate Powder Tested	36.0 *	MAX.	3965	**	106%
	34.0		3726		100%
	32.0		3486		94%
A-2015	31.5	MAX.	3986	**	101%
	29.5		3718		94%
	27.5 *		3450		88%
H322	32.0	MAX.	3987		97%
	30.0		3728		91%
	28.0 *		3468		85%
RL7	30.5	MAX.	4028		93%
	28.5		3665		87%
	26.5 *		3303		81%
A-2230	34.5	MAX.	4053		99%
	32.5		3687		93%
	30.5 *		3320		88%

BC=Ballistic Coefficient SD=Sectional Density
*Most Accurate Load Tested **Compressed Load

Use Maximum Loads with Caution
Refer to page 73 for additional safety information

22 BR Rem - 50/52 grain

		MAXIMUM O.A.C.L.		2.020"
		TESTED O.A.C.L.	B.C.	S.D.
Ballistic Tip®	50gr. Spitzer	2.020"	0.238	0.142
CT® Ballistic Silvertip®	50gr. Spitzer	2.020"	0.238	0.142
Nosler®	50gr. FBHP	2.000"	0.192	0.142
Nosler®	50gr. FBSP	1.970"	0.195	0.142
Custom Competition™	52gr. HPBT	2.020"	0.220	0.148

CASE TYPE:	Remington		PRIMER TYPE	WSR
CASE HOLDS:	34.5	Gr. WATER	BARREL Length/Make	24" Lilja
			BARREL Twist	1-14"

POWDER TYPE	POWDER CHG. GRS.		MUZZLE VEL. F.P.S.		LOAD DENSITY (VOLUME)
RL7 Most Accurate Powder Tested	27.5 *	MAX.	3500		87%
	25.5		3227		80%
	23.5		2953		74%
A-2230	31.0 *	MAX.	3536		92%
	29.0		3280		86%
	27.0		3024		80%
A-2015	30.0	MAX.	3541		99%
	28.0		3243		92%
	26.0 *		2944		86%
H322	30.5	MAX.	3556		96%
	28.5		3362		89%
	26.5 *		3169		83%
IMR 3031	30.5 *	MAX.	3593		100%
	28.5		3302		93%
	26.5		3012		87%
BL-C2	35.0	MAX.	3642	**	103%
	33.0		3431		97%
	31.0 *		3221		91%
Varget	32.5	MAX.	3664	**	102%
	30.5		3417		96%
	28.5 *		3170		89%
Viht N135	32.0	MAX.	3669	**	108%
	30.0		3446	**	101%
	28.0 *		3222		94%
IMR 4895	32.5	MAX.	3681	**	104%
	30.5		3373		97%
	28.5 *		3065		91%
W748	34.5	MAX.	3700	**	104%
	32.5		3450		98%
	30.5 *		3201		92%

BC=Ballistic Coefficient SD=Sectional Density
*Most Accurate Load Tested **Compressed Load

Use Maximum Loads with Caution
Refer to page 73 for additional safety information

22 BR Rem - 55 grain

Bullet		MAXIMUM O.A.C.L.		2.020"
		TESTED O.A.C.L.	B.C.	S.D.
Ballistic Tip®	55gr. Spitzer	2.020"	0.267	0.157
CT® Ballistic Silvertip®	55gr. Spitzer	2.020"	0.267	0.157
Nosler®	55gr. FBSP	2.020"	0.218	0.157
Varmageddon™	55gr. FBHP	2.020"	0.210	0.157
Varmageddon™	55gr. FB Tipped	2.020"	0.255	0.157

CASE TYPE:	Remington		PRIMER TYPE	WSR
CASE HOLDS:	34.5	Gr. WATER	BARREL Length/Make	24" Lilja
			BARREL Twist	1-14"

POWDER TYPE	POWDER CHG. GRS.		MUZZLE VEL. F.P.S.		LOAD DENSITY (VOLUME)
RL7	27.5	MAX.	3371		87%
	25.5		3088		80%
	23.5 *		2806		74%
BL-C2	33.5 *	MAX.	3396		99%
	31.5		3191		93%
	29.5		2987		87%
H322	29.5	MAX.	3435		92%
	27.5		3183		86%
	25.5 *		2931		80%
A-2015	29.5	MAX.	3474		97%
	27.5		3202		91%
	25.5 *		2930		84%
IMR 3031	30.0 *	MAX.	3496		98%
	28.0		3203		92%
	26.0		2908		85%
Viht N135	31.5	MAX.	3526	**	106%
	29.5		3306		99%
	27.5 *		3087		93%
IMR 4895	32.0	MAX.	3542	**	102%
	30.0		3304		96%
	28.0 *		2992		89%
Varget	32.0	MAX.	3544		100%
	30.0		3323		94%
	28.0 *		3102		88%
A-2460	31.5	MAX.	3553		92%
	29.5		3290		86%
	27.5 *		3027		81%
W748 Most Accurate Powder Tested	35.0	MAX.	3680	**	106%
	33.0		3448		100%
	31.0 *		3214		94%

BC=Ballistic Coefficient SD=Sectional Density
*Most Accurate Load Tested **Compressed Load

Use Maximum Loads with Caution
Refer to page 73 for additional safety information

Brian Pearce

22-250 REMINGTON

The "22-250" or "22 Varminter" started life as a wildcat during the 1930's with various experimenters necking the 250 Savage case to 22 caliber, each having only slight dimensional differences. Those that were willing to have a rifle built and assemble quality handloads were generally thrilled with the results. The savvy folks at Remington recognized the cartridge's potential and in 1965 began offering the 22-250 Remington as an industry standard cartridge. It was an instant success and quickly climbed high on the list of most reloaded cartridges. Now, nearly half a century later, it remains in high demand and is one of the most popular sporting rifle cartridges. It has been forty years since I fired my first 22-250. At that time, I marveled at the speed that the 55-grain bullet, with a muzzle velocity of around 3800 fps, struck the target several hundred yards down range, with practically no holdover and very little recoil. I still do.

Over the past four decades, I have relied on the 22-250 for many varmint shoots and it is an honest 500-yard cartridge and even beyond on lucky days, combined with correct bullets and a good rifle. It has served well in taking many ranch and barnyard pests, requiring very little holdover on long shots.

Most important, it hits coyotes hard. In late winter, when my expensive registered cows are calving, I move them to a pasture close to the house, where I keep a close eye on them in the event that one has trouble or if a predator moves in. We have lost many young calves to coyotes. Resting by the front door is an accurate bolt action rifle, naturally chambered in 22-250. When the big western Idaho coyotes get brave and move in for a kill, the crosshairs are held steady and the 55-grain Nosler Ballistic Tip is launched at 3800 fps. Hair flies and the problem resolved!

As to accuracy, I always marvel at how easy it is to develop handloads that simply stack into a ragged hole at 100 yards. There are many excellent powders with correct burn rates and when combined with the legendary Nosler Ballistic Tip bullet, match grade accuracy along with pest destroying field performance can be expected.

Brian Pearce

Brian Pearce is a contributing editor for Rifle and Handloader magazines.

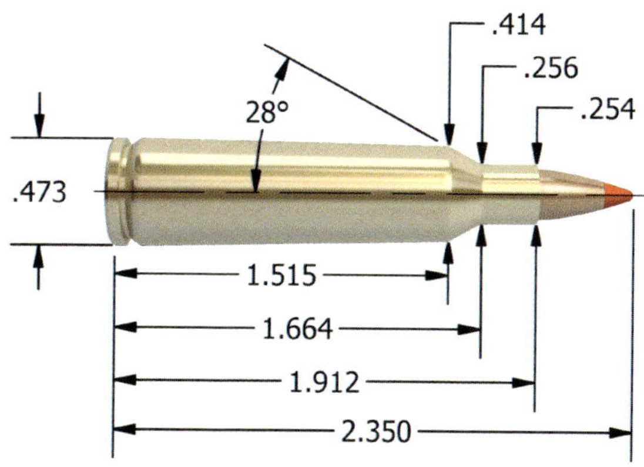

Maximum S.A.A.M.I. Overall Cartridge Length: 2.350"

BULLET CHOICES FOR THE 22-250 REMINGTON

Ballistic Tip® Lead-Free

Grain
35

Ballistic Tip® Varmint

Grain
40
50
55
60

CT® Ballistic Silvertip®

Grain
50
55

Custom Competition™

Grain
52

Nosler® HP

Grain
34
50
55

Nosler®

Grain
50
55

Varmageddon™ HP

Grain
40
55
62

Varmageddon™

Grain
40
55

Bonded Solid Base®

Grain
64

Partition®

Grain
60

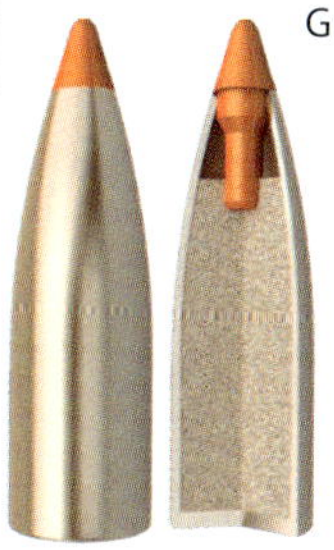

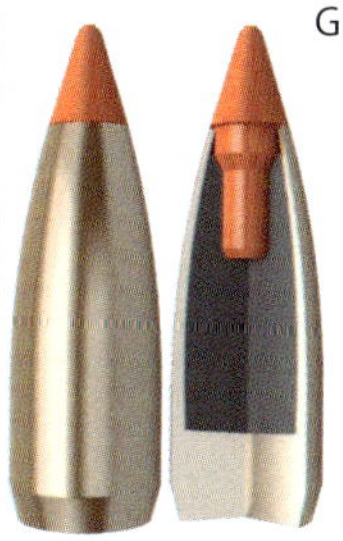

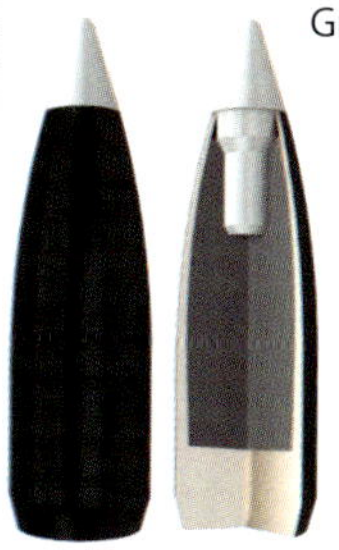

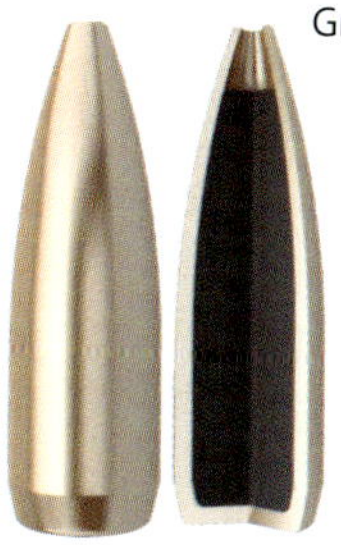

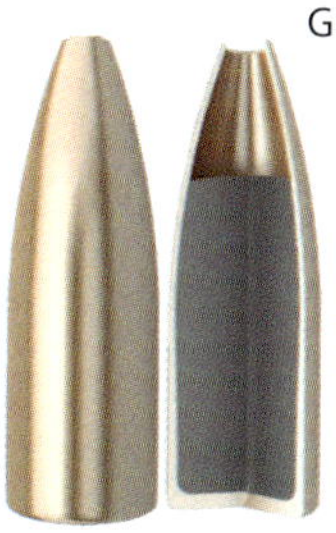

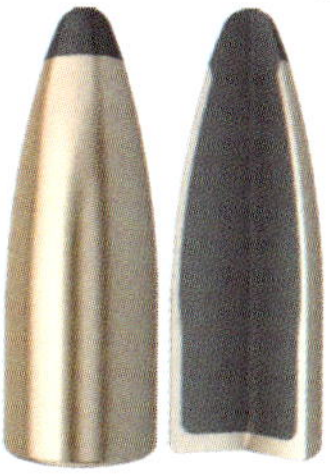

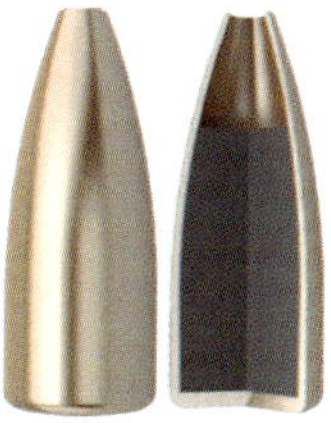

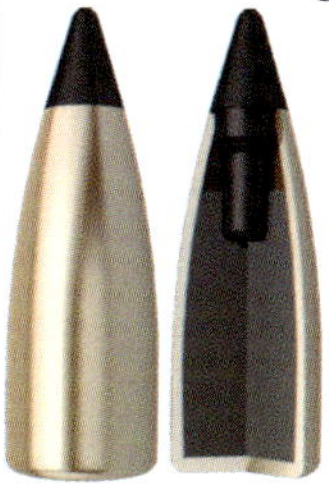

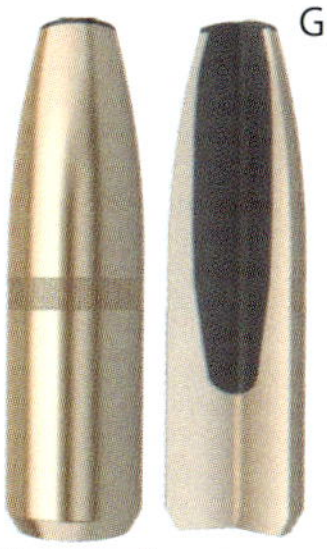

TECHNICAL INFORMATION

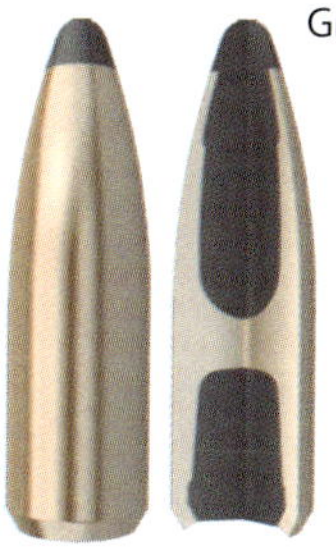

For Technical information on the 22-250 Remington turn to page 139.

22-250 Rem - 34/35 grain		MAXIMUM S.A.A.M.I. O.A.C.L.		2.350"
		TESTED O.A.C.L.	B.C.	S.D.
Nosler®	34gr. FBHP	2.350"	0.113	0.097
Ballistic Tip® Lead Free	35gr. Spitzer	2.350"	0.201	0.100

CASE TYPE:	Nosler		PRIMER TYPE	WLR
CASE HOLDS:	40.9	Gr. WATER	BARREL Length/Make	24" Lilja
			BARREL Twist	1-14"

POWDER TYPE	POWDER CHG. GRS.		MUZZLE VEL. F.P.S.		LOAD DENSITY (VOLUME)
RL15	39.0	MAX.	4344		99%
	37.0		4112		94%
	35.0 *		3820		89%
H335	38.0	MAX.	4397		94%
	36.0		4212		89%
	34.0 *		3992		84%
A-2520	39.0	MAX.	4405		98%
	37.0		4224		93%
	35.0 *		4021		88%
IMR 4320	40.0	MAX.	4430	**	105%
	38.0		4238		100%
	36.0 *		3950		95%
Varget	40.0	MAX.	4458	**	106%
	38.0		4294		100%
	36.0 *		4105		95%
W748	40.0	MAX.	4466	**	102%
Most Accurate	38.0		4277		97%
Powder Tested	36.0 *		4116		92%
IMR 4064	38.0 *	MAX.	4480	**	102%
	36.0		4321		97%
	34.0		3957		91%
IMR 4895	39.0	MAX.	4492	**	105%
	37.0 *		4252		99%
	35.0		4019		94%
TAC	38.5	MAX.	4545		96%
	36.5		4406		91%
	34.5 *		4277		86%

BC=Ballistic Coefficient SD=Sectional Density
*Most Accurate Load Tested **Compressed Load

Use Maximum Loads with Caution
Refer to page 73 for additional safety information

22-250 Rem - 40 grain		MAXIMUM S.A.A.M.I. O.A.C.L.		2.350"
		TESTED O.A.C.L.	B.C.	S.D.
Ballistic Tip®	40gr. Spitzer	2.350"	0.221	0.114
Varmageddon™	40gr. FBHP	2.280"	0.158	0.114
Varmageddon™	40gr. FB Tipped	2.350"	0.211	0.114

CASE TYPE:	Nosler		PRIMER TYPE	WLR
CASE HOLDS:	43.0	Gr. WATER	BARREL Length/Make	24" Lilja
			BARREL Twist	1-14"

POWDER TYPE	POWDER CHG. GRS.		MUZZLE VEL. F.P.S.		LOAD DENSITY (VOLUME)
IMR 4895	36.0	* MAX.	3961		92%
	34.0		3678		87%
	32.0		3395		82%
RL15	37.0	* MAX.	3990		90%
	35.0		3711		85%
	33.0		3431		80%
W760	41.0	* MAX.	3997		100%
	39.0		3750		95%
	37.0		3503		90%
Viht N140	37.5	MAX.	4006	**	101%
	35.5		3763		96%
	33.5	*	3519		91%
A-2460	35.5	MAX.	4050		83%
	33.5		3833		79%
	31.5	*	3616		74%
H380	41.0	* MAX.	4070		100%
	39.0		3776		95%
	37.0		3482		91%
Varmint	39.0	* MAX.	4089		92%
	37.0		3880		87%
	35.0		3694		82%
Varget	38.0	MAX.	4100		96%
	36.0		3863		91%
	34.0	*	3625		85%
TAC	37.5	MAX.	4181		89%
Most Accurate	35.5		3942		84%
Powder Tested	33.5	*	3723		79%
CFE 223	42.5	MAX.	4325		100%
	40.5		4101		95%
	38.5	*	3896		90%

BC=Ballistic Coefficient SD=Sectional Density
*Most Accurate Load Tested **Compressed Load

Use Maximum Loads with Caution
Refer to page 73 for additional safety information

22-250 Rem - 50/52 grain		MAXIMUM S.A.A.M.I. O.A.C.L.		2.350"
		TESTED O.A.C.L.	B.C.	S.D.
Ballistic Tip®	50gr. Spitzer	2.350"	0.238	0.142
CT®Ballistic Silvertip®	50gr. Spitzer	2.350"	0.238	0.142
Nosler®	50gr. FBHP	2.350"	0.192	0.142
Nosler®	50gr. FBSP	2.350"	0.195	0.142
Custom Competition™	52gr. HPBT	2.350"	0.220	0.148

CASE TYPE:	Nosler		PRIMER TYPE	WLR
CASE HOLDS:	42.6	Gr. WATER	BARREL Length/Make	24" Lilja
			BARREL Twist	1-14"

POWDER TYPE	POWDER CHG. GRS.		MUZZLE VEL. F.P.S.	LOAD DENSITY (VOLUME)
IMR 4064	34.5	MAX.	3690	89%
	32.5		3470	84%
	30.5 *		3250	79%
Viht N140	36.0 *	MAX.	3736	98%
	34.0		3500	93%
	32.0		3265	87%
A-2230	35.5	MAX.	3740	85%
	33.5		3500	80%
	31.5 *		3260	75%
Varget	36.0 *	MAX.	3794	91%
	34.0		3604	86%
	32.0		3415	81%
IMR 4350	40.0	MAX.	3800	99%
	38.0		3590	94%
	36.0 *		3380	89%
IMR 4895	34.0 *	MAX.	3802	88%
	32.0		3577	83%
	30.0		3352	77%
Varmint	37.5	MAX.	3805	89%
	35.5		3603	84%
	33.5 *		3421	79%
Big Game	40.5	MAX.	3849	100%
	38.5		3729	95%
	36.5 *		3528	90%
H380	41.0	MAX.	3897	** 101%
Most Accurate	39.0		3742	96%
Powder Tested	37.0 *		3547	91%
CFE 223	40.0	MAX.	3975	95%
	38.0		3762	90%
	36.0 *		3573	85%

BC=Ballistic Coefficient SD=Sectional Density
*Most Accurate Load Tested **Compressed Load

Use Maximum Loads with Caution
Refer to page 73 for additional safety information

22-250 Rem - 55 grain		MAXIMUM S.A.A.M.I. O.A.C.L.		2.350"
		TESTED O.A.C.L.	B.C.	S.D.
Ballistic Tip®	55gr. Spitzer	2.350"	0.267	0.157
CT® Ballistic Silvertip®	55gr. Spitzer	2.350"	0.267	0.157
Nosler®	55gr. FBSP	2.350"	0.218	0.157
Varmageddon™	55gr. FBHP	2.350"	0.210	0.157
Varmageddon™	55gr. FB Tipped	2.350"	0.255	0.157

CASE TYPE:	Nosler		PRIMER TYPE	WLR
CASE HOLDS:	42.4	Gr. WATER	BARREL Length/Make	24" Lilja
			BARREL Twist	1-14"

POWDER TYPE	POWDER CHG. GRS.		MUZZLE VEL. F.P.S.	LOAD DENSITY (VOLUME)
Varmint	36.0 *	MAX.	3584	86%
	34.0		3420	81%
	32.0		3386	76%
A-2460	34.5	MAX.	3598	82%
	32.5		3383	77%
	30.5 *		3168	73%
IMR 4895	34.0 *	MAX.	3602	88%
	32.0		3357	83%
	30.0		3112	78%
IMR 4064	34.5	MAX.	3603	89%
	32.5		3407	84%
	30.5 *		3211	79%
IMR 4350	39.0 *	MAX.	3608	97%
	37.0		3423	92%
	35.0		3238	87%
Varget	35.0	MAX.	3635	89%
	33.0		3454	84%
	31.0 *		3273	79%
IMR 4320	35.5 *	MAX.	3690	90%
	33.5		3480	85%
	31.5		3270	80%
Big Game	39.5	MAX.	3724	98%
Most Accurate	37.5 *		3568	93%
Powder Tested	35.5		3411	88%
H380	39.5 *	MAX.	3736	98%
	37.5		3591	93%
	35.5		3381	88%
CFE 223	39.0	MAX.	3803	93%
	37.0		3604	88%
	35.0 *		3408	83%

BC=Ballistic Coefficient SD=Sectional Density
*Most Accurate Load Tested **Compressed Load

Use Maximum Loads with Caution
Refer to page 73 for additional safety information

22-250 Rem - 60 grain		MAXIMUM S.A.A.M.I. O.A.C.L.		2.350"
		TESTED O.A.C.L.	B.C.	S.D.
Ballistic Tip®	60gr. Spitzer	2.350"	0.270	0.171
Partition®	60gr. Spitzer	2.350"	0.228	0.171
Varmageddon™	62gr. FBHP	2.322"	0.251	0.176

CASE TYPE:	Nosler		PRIMER TYPE	WLR
CASE HOLDS:	41.0	Gr. WATER	BARREL Length/Make	24" Wiseman
			BARREL Twist	1-8"

POWDER TYPE	POWDER CHG. GRS.		MUZZLE VEL. F.P.S.	LOAD DENSITY (VOLUME)
IMR 4064	33.0 *	MAX.	3500	88%
	31.0		3310	83%
	29.0		3145	78%
Viht N150	33.5	MAX.	3500	96%
	31.5 *		3375	90%
	29.5		3250	85%
Varget	34.0 *	MAX.	3525	90%
	32.0		3380	84%
	30.0		3270	79%
IMR 4895	33.0	MAX.	3540	88%
	31.0 *		3354	83%
	29.0		3164	78%
RL15	34.0	MAX.	3540	86%
	32.0		3370	81%
	30.0 *		3220	76%
IMR 4350	38.5	MAX.	3560	99%
	36.5		3380	94%
	34.5 *		3200	89%
A-2520	34.0	MAX.	3570	85%
	32.0 *		3430	80%
	30.0		3280	75%
H414	38.5	MAX.	3600	98%
	36.5 *		3405	93%
	34.5		3230	88%
H380 Most Accurate Powder Tested	38.5	MAX.	3603	99%
	36.5 *		3441	94%
	34.5		3257	89%
Hunter	41.0	MAX.	3612 **	105%
	39.0		3478	100%
	37.0 *		3275	95%

BC=Ballistic Coefficient SD=Sectional Density
*Most Accurate Load Tested **Compressed Load

Use Maximum Loads with Caution
Refer to page 73 for additional safety information

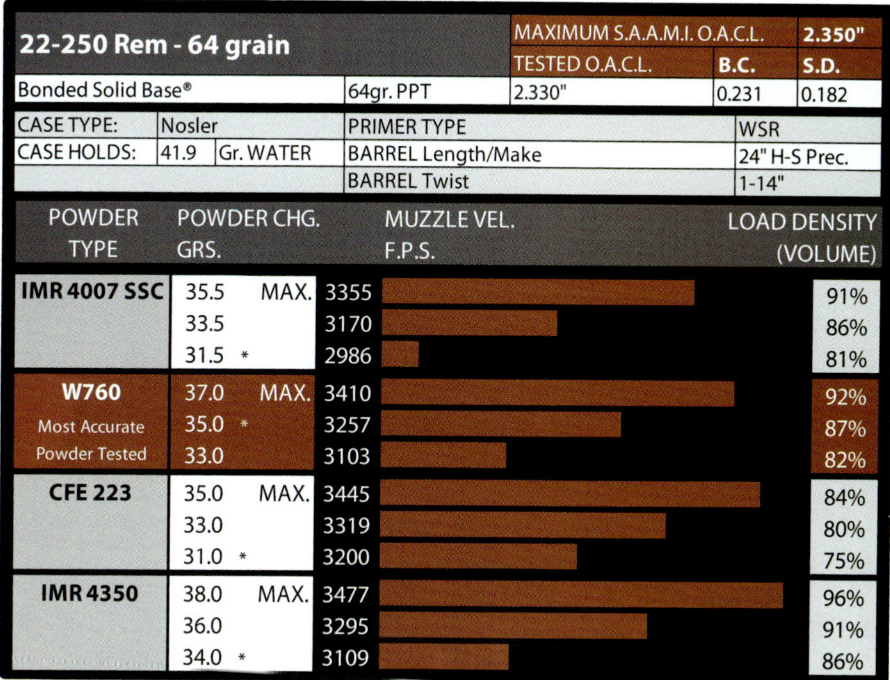

22-250 Rem - 64 grain			MAXIMUM S.A.A.M.I. O.A.C.L.		2.350"
			TESTED O.A.C.L.	B.C.	S.D.
Bonded Solid Base®		64gr. PPT	2.330"	0.231	0.182

CASE TYPE:	Nosler		PRIMER TYPE	WSR
CASE HOLDS:	41.9	Gr. WATER	BARREL Length/Make	24" H-S Prec.
			BARREL Twist	1-14"

POWDER TYPE	POWDER CHG. GRS.		MUZZLE VEL. F.P.S.	LOAD DENSITY (VOLUME)
IMR 4007 SSC	35.5	MAX.	3355	91%
	33.5		3170	86%
	31.5	*	2986	81%
W760 Most Accurate Powder Tested	37.0	MAX.	3410	92%
	35.0	*	3257	87%
	33.0		3103	82%
CFE 223	35.0	MAX.	3445	84%
	33.0		3319	80%
	31.0	*	3200	75%
IMR 4350	38.0	MAX.	3477	96%
	36.0		3295	91%
	34.0	*	3109	86%

TECHNICAL INFORMATION

Due to the tapered body of the 22-250, cases are prone to stretch when full-length sized between firings. Case life can be extended dramatically by using a neck sizing die, or a full-length die adjusted to size only the neck portion of the case. IMR4895 and H380 are excellent powders for the 22-250 Remington. Capable of stabilizing bullets as long as our 55-grain Ballistic Tip® and 60-grain Partition®, the 22-250 uses a standard rifling twist rate of 1-14". In recent years, there has been significant interest in load data for fast-twist barrels and heavy weight bullets. Note: in some elevations/temperatures, the 60gr Partition may need a 1-12" twist to reliably stabilize.

BC=Ballistic Coefficient SD=Sectional Density
*Most Accurate Load Tested **Compressed Load

Use Maximum Loads with Caution
Refer to page 73 for additional safety information

John A. Nosler and Bob Nosler performing a routine quality check circa 2005.

22-250 REMINGTON (FAST-TWIST)

The 22-250 Remington was one of the best 20th century cartridges, but it's time for a twist. A fast twist, which will make this venerable cartridge even better.

A century has nearly flowed under the bridge since Charles Newton created the first commercial cartridge that broke the 3,000 fps barrier, but his 250-3000 Savage is better known in its subsequent configuration – the 22-250 Remington released in 1967.

Squeezing the neck of that little case to .224" was a brilliant idea. It put 3,000 fps in the rear view mirror real quick. Today's powders help the 22-250 spit 40-grain Ballistic Tips over 4,180 fps and 55-grain BT's over 3,700 fps. Fun. But -- not all the fun possible.

The 22-250 case has always had the horsepower to drive stouter, heavier, more wind resistant bullets fast enough to make them deadly, effective on not only long-range varmints, but mid-range pronghorns and deer. Hunters in Texas, South Dakota, Idaho and many other states may legally use the 22-250 for big game. I once put a 60-grain .224 Partition through a big whitetail buck, breaking shoulder, spine and hip before exiting. But I used a 1-8 twist custom barrel to do so. Traditional 1-14 or 1-12 rifling twist rates won't stabilize most bullets heavier than 60 grains.

Fast twist barrels are the wave of the future, especially as longer, sleek, wind resistant, high B.C. bullets come on line. With Nosler's 69-, 77- and 80-grain Custom Competition bullets, a fast twist 22-250 Rem. becomes a mild-recoiling, long-range bell ringer matching the trajectory curves of many 6.5mm and 7mm rounds. As more and more fast twist 22-250s are built, I'm hoping Nosler will give us 70-grain and heavier AccuBonds and Ballistic Tips, too. With a fast twist barrel, the BT Lead-Free .224 bullets can be stretched to 50-grains, increasing their downrange punch on varmints and coyotes.

What's not to like about a fast twist 22-250 barrel? The 22-250 Remington of old was a great round, but the fast twist version is even better.

Ron Spomer

Ron Spomer is the Travel Field Editor for Sports Afield magazine as well as a contributing editor for Wolfe Publications.

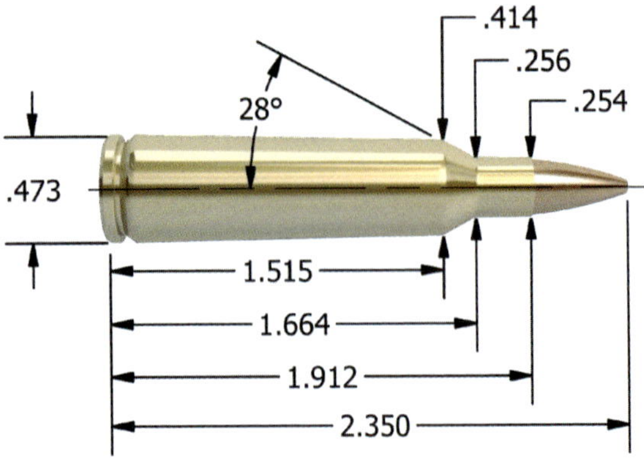

BULLET CHOICES FOR THE 22-250 REMINGTON (FAST TWIST)

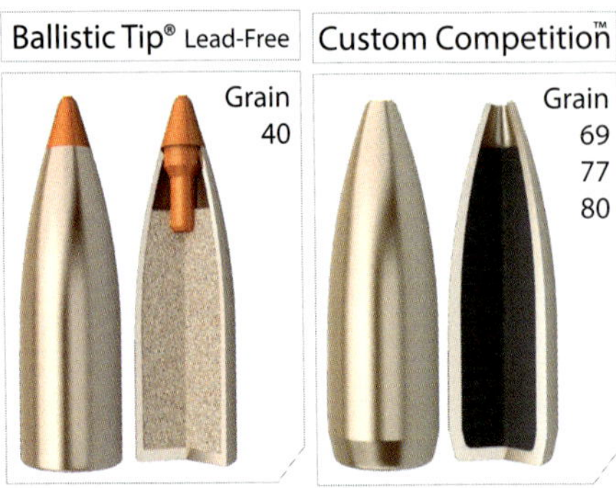

TECHNICAL INFORMATION

In recent years, there has been significant interest in load data for fast-twist barrels and heavy weight bullets. The load data developed for the 40 grain BTLF as well as the bullets weighing 60 grains or more, was fired from a 24" Wiseman barrel with 1-8" twist. The longer length of our Custom Competition bullets weighing 69 grains and higher require that they be loaded longer than the S.A.A.M.I. maximum C.O.A.L. of 2.350". We achieved best performance loading our Custom Competition bullets to a C.O.A.L. of 2.600".

22-250 Rem 40 gr BTLF (Fast Twist)		MAXIMUM S.A.A.M.I. O.A.C.L.		2.350"
		TESTED O.A.C.L.	B.C.	S.D.
Ballistic Tip® Lead Free	40gr. Spitzer	2.350"	0.220	0.114

CASE TYPE:	Nosler		PRIMER TYPE	WLR
CASE HOLDS:	43.0	Gr. WATER	BARREL Length/Make	24" Wiseman
			BARREL Twist	1-8"

POWDER TYPE	POWDER CHG. GRS.		MUZZLE VEL. F.P.S.		LOAD DENSITY (VOLUME)
RL15	38.0	MAX.	4112		92%
	36.0		3897		87%
	34.0 *		3677		82%
IMR 4320	36.0	MAX.	4121		90%
	34.0 *		3893		85%
	32.0		3668		80%
W748	38.0	MAX.	4196		92%
	36.0		3999		87%
	34.0 *		3718		82%
IMR 4895 Most Accurate Powder Tested	37.5	MAX.	4211		96%
	35.5 *		3960		91%
	33.5		3724		86%
Viht N140	37.5 *	MAX.	4224	**	101%
	35.5		4006		96%
	33.5		3796		91%
Viht N135	35.5	MAX.	4248		96%
	33.5		4014		91%
	31.5 *		3764		85%
Varget	38.5	MAX.	4252		97%
	36.5		4079		92%
	34.5 *		3856		87%
A-2460	36.5	MAX.	4287		86%
	34.5		4073		81%
	32.5 *		3911		76%
TAC	37.5	MAX.	4300		89%
	35.5 *		4179		84%
	33.5		4045		79%
RL10x	34.5	MAX.	4334		90%
	32.5 *		4128		85%
	30.5		3898		80%

BC=Ballistic Coefficient SD=Sectional Density
*Most Accurate Load Tested **Compressed Load

Use Maximum Loads with Caution
Refer to page 73 for additional safety information

22-250 Rem - 69 grain (Fast Twist)		MAXIMUM S.A.A.M.I. O.A.C.L.		2.350"
		TESTED O.A.C.L.	B.C.	S.D.
Custom Competition™	69gr. HPBT	2.600"	0.305	0.196

CASE TYPE:	Nosler		PRIMER TYPE	WLR
CASE HOLDS:	41.3	Gr. WATER	BARREL Length/Make	24" Wiseman
			BARREL Twist	1-8"

POWDER TYPE	POWDER CHG. GRS.	MUZZLE VEL. F.P.S.	LOAD DENSITY (VOLUME)
IMR 4350	37.0 * MAX.	3176	95%
	35.0	3000	90%
	33.0	2847	85%
A-2015	30.0 MAX.	3220	83%
	28.0 *	3051	77%
	26.0	2876	72%
Varget	32.5 MAX.	3282	85%
	30.5 *	3113	80%
	28.5	2941	75%
Viht N540 Most Accurate Powder Tested	33.5 * MAX.	3283	90%
	31.5	3104	85%
	29.5	2980	79%
RL15	33.0 MAX.	3305	83%
	31.0	3144	78%
	29.0 *	2982	73%
Big Game	36.0 MAX.	3316	92%
	34.0	3145	87%
	32.0 *	2995	82%
IMR 4064	32.5 MAX.	3337	86%
	30.5	3144	81%
	28.5 *	2938	76%
W760	36.5 MAX.	3341	93%
	34.5 *	3149	87%
	32.5	3002	82%
H414	37.0 MAX.	3411	93%
	35.0	3190	88%
	33.0 *	3015	83%
Viht N550	36.0 MAX.	3414	97%
	34.0 *	3153	91%
	32.0	3000	86%

BC=Ballistic Coefficient SD=Sectional Density
*Most Accurate Load Tested **Compressed Load

Use Maximum Loads with Caution
Refer to page 73 for additional safety information

22-250 Rem - 77/80 gr (Fast Twist)

		MAXIMUM S.A.A.M.I. O.A.C.L.		2.350"
		TESTED O.A.C.L.	B.C.	S.D.
Custom Competition™	77gr. HPBT	2.600"	0.340	0.219
Custom Competition™	80gr. HPBT	2.600"	0.415	0.228

CASE TYPE:	Nosler		PRIMER TYPE	WLR
CASE HOLDS:	42.1	Gr. WATER	BARREL Length/Make	24" Wiseman
			BARREL Twist	1-8"

POWDER TYPE	POWDER CHG GRS.		MUZZLE VEL. F.P.S.	LOAD DENSITY (VOLUME)
Viht N160 Most Accurate Powder Tested	35.0	MAX.	2979	92%
	33.0		2845	87%
	31.0 *		2740	82%
H4831sc	38.0	MAX.	3083	94%
	36.0		2902	89%
	34.0 *		2750	84%
IMR 4350	36.0	MAX.	3106	90%
	34.0		2949	85%
	32.0 *		2782	80%
RL22	38.0	MAX.	3128	98%
	36.0		2946	93%
	34.0 *		2779	88%
H4350	36.0	MAX.	3131	90%
	34.0		3001	85%
	32.0 *		2836	80%
IMR 4831	37.0 *	MAX.	3137	93%
	35.0		2925	88%
	33.0		2790	83%
H414	35.5	MAX.	3143	88%
	33.5 *		2963	83%
	31.5		2819	78%
W760	36.0	MAX.	3167	90%
	34.0 *		3003	85%
	32.0		2858	80%
RL19	38.0	MAX.	3187	98%
	36.0 *		2991	93%
	34.0		2774	88%

BC=Ballistic Coefficient SD=Sectional Density
*Most Accurate Load Tested **Compressed Load

Use Maximum Loads with Caution
Refer to page 73 for additional safety information

John Anderson

22-250 REMINGTON ACKLEY IMPROVED

This improved "wildcat" cartridge is like most others in that the case is blown out to a nearly straight body and a sharper shoulder. The result is higher velocity, and more energy, than the factory round produces, generally about 100 fps more velocity, depending on powder. Such increase in performance can be enough to justify the cost of "improvement" in the eyes of many riflemen.

I live in the great plains of South Dakota and commissioned gunsmith Darrell Holland in Oregon to build me a 22-250 AI as a coyote calling rifle where often the shots are long and you want all the punch you can get at extended ranges. Darrell used a Remington Model 700 action, a Pac-Nor barrel, and a McMillan stock to produce a very accurate rifle and one that's reasonably light for carrying purposes. Not only is this rifle just what I wanted for predator calling, it has accounted for more than its share of prairie dogs at extended ranges.

When Darrell finished with the rifle, it weighed 7.8 pounds, without scope, and it has proven to be very accurate. For the purposes for which this rifle was built, and the wide open spaces where I live and hunt, a scope of at least 4.5-14x or even more magnification would be appropriate.

This cartridge also offers the advantage of four different styles of Nosler bullets of five different weights, and will launch each of them at velocities higher than provided by the standard 22-250. With such bullet choices, a hunter can put together a wide range of handloads that are appropriate for critters from ground squirrels to coyotes, and for ranges out to 300 yards or more even when shooting at the small squirrels. For larger targets like a coyote, I would have no hesitation firing at a standing target at longer ranges, when using a solid rest and the appropriate Nosler bullet.

John Anderson

John Anderson
Editor, The Varmint Hunter Magazine

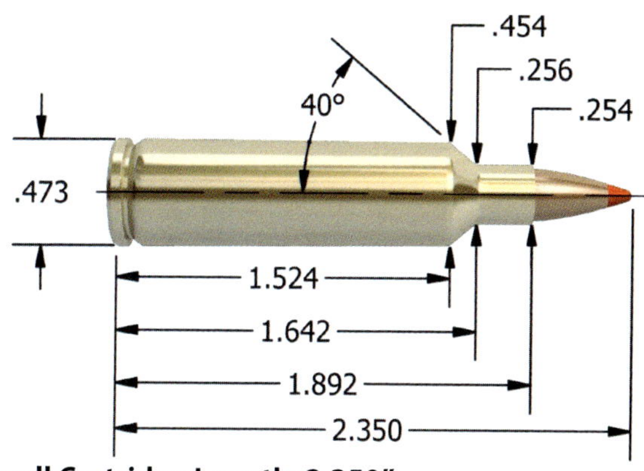

Maximum Overall Cartridge Length: 2.350"

BULLET CHOICES FOR THE 22-250 REMINGTON ACKLEY IMPROVED

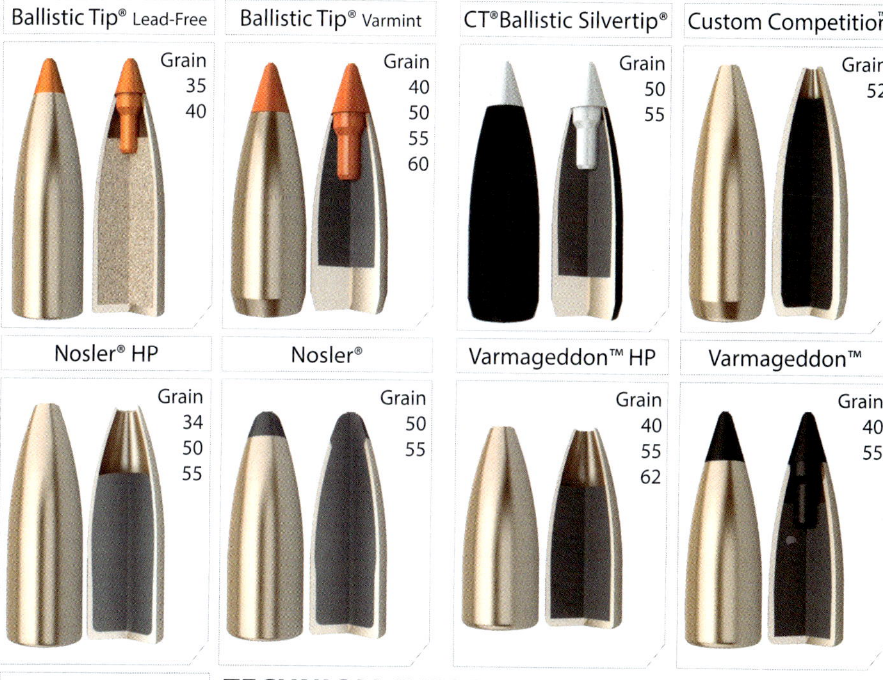

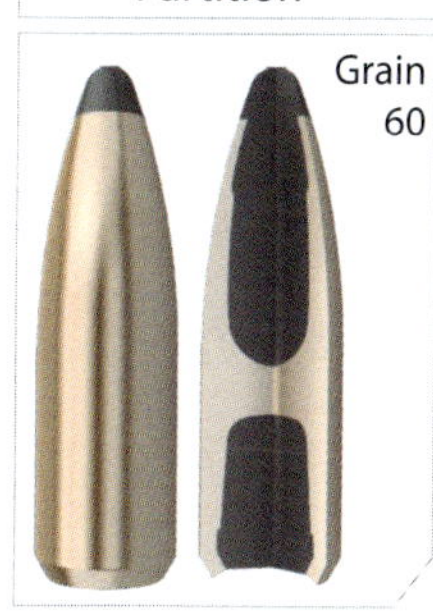

TECHNICAL INFORMATION

To fire-form for the .22-250 Ackley Improved, we suggest the following procedure:

- Start with standard .22-250 Remington cases.
- Select a light load of relatively fast powder (IMR4895 is great) from the standard. 22-250 Remington data.
- Seat the bullet well out into the rifling so it makes good contact with the lands when the cartridge is chambered. This holds the case head against the bolt face and eliminates case stretching in the web area, a cause of case head separation. We recommend using 50-55 grain bullets. Lighter bullets are generally too short to reach the lands.
- After fire-forming, follow load data for the 22-250 Remington Ackley improved.

22-250 Rem A.I. - 34/35 grain		MAXIMUM O.A.C.L.		2.350"
		TESTED O.A.C.L.	B.C.	S.D.
Nosler®	34gr. FBHP	2.230"	0.113	0.097
Ballistic Tip® Lead Free	35gr. Spitzer	2.350"	0.201	0.100

CASE TYPE:	Nosler		PRIMER TYPE	Rem 9 1/2
CASE HOLDS:	46.3	Gr. WATER	BARREL Length/Make	26" Shilen
			BARREL Twist	1-14"

POWDER TYPE	POWDER CHG. GRS.		MUZZLE VEL. F.P.S.		LOAD DENSITY (VOLUME)
Big Game	45.0	MAX.	4285	**	102%
	43.0 *		4102		98%
	41.0		3919		93%
H380	46.5	MAX.	4328	**	106%
	44.5		4154	**	101%
	42.5 *		3980		97%
RL15	42.5	MAX.	4456		96%
	40.5 *		4288		91%
	38.5		4120		87%
Varget Most Accurate Powder Tested	42.0	MAX.	4459		98%
	40.0		4257		93%
	38.0 *		4055		89%
Viht N140	41.0	MAX.	4468	**	103%
	39.0 *		4265		98%
	37.0		4062		93%
IMR 4895	42.0 *	MAX.	4556		100%
	40.0		4315		95%
	38.0		4074		90%
IMR 4064	41.0	MAX.	4570		97%
	39.0 *		4339		93%
	37.0		4108		88%
W748	43.0 *	MAX.	4583		97%
	41.0		4419		92%
	39.0		4255		88%
Viht N135	40.5	MAX.	4654	**	102%
	38.5		4458		97%
	36.5 *		4262		92%
IMR 3031	39.5	MAX.	4760		96%
	37.5		4515		92%
	35.5 *		4270		87%

BC=Ballistic Coefficient SD=Sectional Density
*Most Accurate Load Tested **Compressed Load

Use Maximum Loads with Caution
Refer to page 73 for additional safety information

22-250 Rem A.I. - 40 grain

		MAXIMUM O.A.C.L.		2.350"
		TESTED O.A.C.L.	B.C.	S.D.
Ballistic Tip® Lead-Free	40gr. Spitzer	2.350"	0.220	0.114
Ballistic Tip®	40gr. Spitzer	2.350"	0.221	0.114
Varmageddon™	40gr. FBHP	2.280"	0.158	0.114
Varmageddon™	40gr. FB Tipped	2.350"	0.211	0.114

CASE TYPE:	Nosler		PRIMER TYPE	Rem 9 1/2
CASE HOLDS:	48.0	Gr. WATER	BARREL Length/Make	26" Shilen
			BARREL Twist	1-14"

POWDER TYPE	POWDER CHG. GRS.		MUZZLE VEL. F.P.S.		LOAD DENSITY (VOLUME)
IMR 4895	41.0	* MAX.	4124		94%
	39.0		3883		89%
	37.0		3642		85%
H380	46.5	MAX.	4129	**	102%
	44.5		3955		98%
	42.5	*	3781		93%
RL15	41.5	MAX.	4181		90%
	39.5	*	4013		86%
	37.5		3763		81%
Viht N135	39.5	MAX.	4199		96%
	37.5		4003		91%
	35.5	*	3806		86%
Big Game	44.0	MAX.	4206		96%
	42.0	*	4023		92%
	40.0		3826		88%
Varget	41.0	MAX.	4209		92%
	39.0		4007		88%
	37.0	*	3800		83%
Viht N140	40.0	MAX.	4209		97%
	38.0	*	4009		92%
	36.0		3840		87%
W748	42.0	* MAX.	4217		91%
	40.0		4053		87%
	38.0		3888		82%
IMR 4064	40.0	* MAX.	4298		92%
	38.0		4067		87%
	36.0		3833		82%
IMR 3031	38.5	MAX.	4338		91%
Most Accurate	36.5	*	4093		86%
Powder Tested	34.5		3869		81%

BC=Ballistic Coefficient SD=Sectional Density
*Most Accurate Load Tested **Compressed Load

Use Maximum Loads with Caution
Refer to page 73 for additional safety information

Nosler

22-250 Rem A.I. - 50/52 grain		MAXIMUM O.A.C.L.		2.350"
		TESTED O.A.C.L.	B.C.	S.D.
Ballistic Tip®	50gr. Spitzer	2.350"	0.238	0.142
CT® Ballistic Silvertip®	50gr. Spitzer	2.350"	0.238	0.142
Nosler®	50gr. FBHP	2.350"	0.192	0.142
Nosler®	50gr. FBSP	2.350"	0.195	0.142
Custom Competition™	52gr. HPBT	2.350"	0.220	0.148

CASE TYPE:	Nosler		PRIMER TYPE	Rem 9 1/2
CASE HOLDS:	47.0	Gr. WATER	BARREL Length/Make	26" Shilen
			BARREL Twist	1-14"

POWDER TYPE	POWDER CHG. GRS.		MUZZLE VEL. F.P.S.	LOAD DENSITY (VOLUME)
Viht N140	38.0	MAX.	3805	94%
	36.0		3603	89%
	34.0 *		3402	84%
Viht N540	39.0	MAX.	3829	92%
	37.0 *		3651	87%
	35.0		3479	83%
Big Game	42.0	MAX.	3846	94%
	40.0		3671	90%
	38.0 *		3533	85%
IMR 4064	38.0	MAX.	3858	89%
	36.0 *		3614	84%
	34.0		3392	79%
RL15	40.0	MAX.	3877	89%
	38.0		3702	84%
	36.0 *		3527	80%
IMR 3031	36.5	MAX.	3884	88%
	34.5 *		3648	83%
	32.5		3466	78%
W760	43.0	MAX.	3909	96%
	41.0		3696	91%
	39.0 *		3520	87%
Varget Most Accurate Powder Tested	40.0 *	MAX.	3922	92%
	38.0		3730	87%
	36.0		3539	83%
IMR 4895	40.0 *	MAX.	3951	94%
	38.0		3745	89%
	36.0		3539	84%

BC=Ballistic Coefficient SD=Sectional Density
*Most Accurate Load Tested **Compressed Load

Use Maximum Loads with Caution
Refer to page 73 for additional safety information

22-250 Rem A.I. - 55 grain		MAXIMUM O.A.C.L.		2.350"
		TESTED O.A.C.L.	B.C.	S.D.
Ballistic Tip®	55gr. Spitzer	2.350"	0.267	0.157
CT® Ballistic Silvertip®	55gr. Spitzer	2.350"	0.267	0.157
Nosler®	55gr. FBHP	2.350"	0.210	0.157
Nosler®	55gr. FBSP	2.350"	0.218	0.157
Varmageddon™	55gr. FBHP	2.350"	0.210	0.157
Varmageddon™	55gr. FB Tipped	2.350"	0.255	0.157

CASE TYPE:	Nosler		PRIMER TYPE	Rem 9 1/2
CASE HOLDS:	47.0	Gr. WATER	BARREL Length/Make	26" Shilen
			BARREL Twist	1-14"

POWDER TYPE	POWDER CHG. GRS.		MUZZLE VEL. F.P.S.	LOAD DENSITY (VOLUME)
IMR 3031	35.5	MAX.	3647	85%
Most Accurate Powder Tested	33.5	*	3462	81%
	31.5		3272	76%
IMR 4320	38.5	MAX.	3676	88%
	36.5	*	3522	84%
	34.5		3202	79%
Varget	38.0	MAX.	3722	87%
	36.0	*	3565	83%
	34.0		3388	78%
Big Game	41.5	MAX.	3727	93%
	39.5		3590	88%
	37.5	*	3418	84%
W760	42.0	MAX.	3735	94%
	40.0		3527	89%
	38.0	*	3386	85%
IMR 4895	37.0	MAX.	3745	87%
	35.0	*	3562	82%
	33.0		3394	77%
H414	44.5	MAX.	3753	99%
	42.5		3552	94%
	40.5	*	3351	90%
Viht N140	39.0	MAX.	3753	96%
	37.0		3597	92%
	35.0	*	3441	87%
IMR 4064	39.0	* MAX.	3787	91%
	37.0		3603	87%
	35.0		3419	82%
RL15	38.5	MAX.	3788	85%
	36.5		3642	81%
	34.5	*	3497	76%

BC=Ballistic Coefficient SD=Sectional Density
*Most Accurate Load Tested **Compressed Load

Use Maximum Loads with Caution
Refer to page 73 for additional safety information

22-250 Rem A.I. (Fast Twist) - 60/62 gr.

Bullet		Tested O.A.C.L. (Maximum O.A.C.L. 2.350")	B.C.	S.D.
Ballistic Tip®	60gr. Spitzer	2.350"	0.270	0.171
Partition®	60gr. Spitzer	2.350"	0.228	0.171
Varmageddon™	62gr. FBHP	2.303"	0.251	0.176

CASE TYPE:	Nosler		PRIMER TYPE	Rem 9 1/2
CASE HOLDS:	48.0	Gr. WATER	BARREL Length/Make	26" Shilen
			BARREL Twist	1-8"

POWDER TYPE	POWDER CHG. GRS.		MUZZLE VEL. F.P.S.	LOAD DENSITY (VOLUME)
H4831SC	44.0	MAX.	3528	95%
	42.0		3356	91%
	40.0 *		3219	87%
IMR 4895 Most Accurate Powder Tested	35.5 *	MAX.	3552	81%
	33.5		3381	77%
	31.5		3214	72%
IMR 4350	41.0	MAX.	3564	90%
	39.0 *		3394	86%
	37.0		3257	82%
IMR 4064	35.5	MAX.	3572	81%
	33.5 *		3394	77%
	31.5		3264	72%
Big Game	39.0	MAX.	3578	86%
	37.0		3364	81%
	35.0 *		3185	77%
IMR 4831	42.0	MAX.	3596	93%
	40.0 *		3428	89%
	38.0		3256	84%
Viht N540	36.5 *	MAX.	3622	84%
	34.5		3496	80%
	32.5		3375	75%
Varget	36.5 *	MAX.	3626	82%
	34.5		3492	78%
	32.5		3365	73%
RL15	36.5	MAX.	3638	79%
	34.5 *		3465	75%
	32.5		3299	71%

BC=Ballistic Coefficient SD=Sectional Density
*Most Accurate Load Tested **Compressed Load

Use Maximum Loads with Caution
Refer to page 73 for additional safety information

Holt Bodinson

220 SWIFT

Blasting onto the varmint hunting stage in 1935 in Winchester's Model 54, the 220 Swift blew 48-grain pills down range at an unheard of muzzle velocity of 4,140 fps out of a factory 26" barrel. It stunned the shooting world. Not only was the cartridge highly accurate, but it was capable of drilling 3/8" diameter holes through 1/2" armor plate at 30 feet that 30-06 armor piercing rounds would only dimple.

The Swift is a sensational varmint cartridge with a trajectory so flat, it has to be experienced to be believed. With proper bullets, like Nosler's 60gr Partition and judicious shot placement, the Swift is suitable for deer size game as well.

Being the Champ of anything also brings with the title the inevitable batch of sideline critics who enjoy stirring the pot of public controversy. It was charged that the Swift ruined barrels within 1,000 rounds. It was charged that it was an inflexible cartridge to load, cursed with wild pressure excursions. It was charged that the case stretched and thickened so quickly that cases had to be trashed after two reloads. Of course, at the mere mention of its use on deer sized game, one could be stoned. Poppycock!

My first Swift was, and still is, a stock Model 70, bought new in 1960 for $150. The Winchester delivered sub-minute-of-angle groups from the get-go, and the only modification I made to the rifle was to replace the factory trigger with a Canjar set trigger. The Swift powder of the day was IMR 4064 until ball powders like H380, H414 and AA2520 began making serious inroads. Just by chance, I had bought 1,500 Sisk 49-grain Express bullets at a going-out-of-business sale. The Winchester Swift and the old, tiny tipped Express bullet thrived on each other and together rewarded me with many pleasant days afield hunting varmints and deer.

Years later, I added a Ruger Number One in 220 Swift to my battery. Like the Model 70, it shot like a house of fire and had a distinct appetite for Norma's hot loading.

Blessed as we are with today's 50gr Ballistic Tips, 52gr Custom Competition match bullets and the long-ranging 55gr Ballistic Tips, the Swift just keeps getting better every year.

Try one!

Holt is a Contributing Editor of GUNS magazine and the Ammunition, Ballistics and Components editor of GUN DIGEST

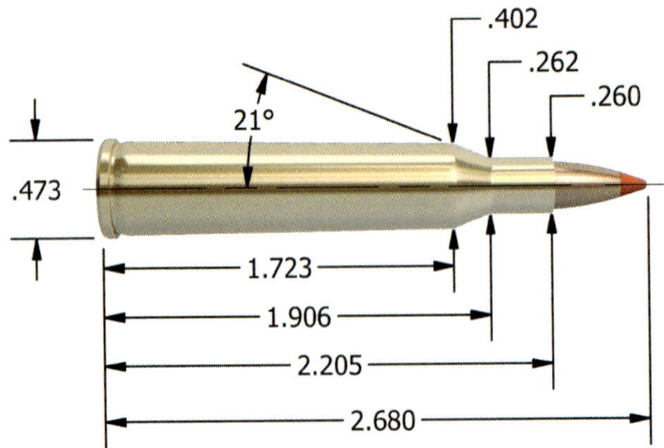

Maximum S.A.A.M.I. Overall Cartridge Length: 2.680"

BULLET CHOICES FOR THE 220 SWIFT

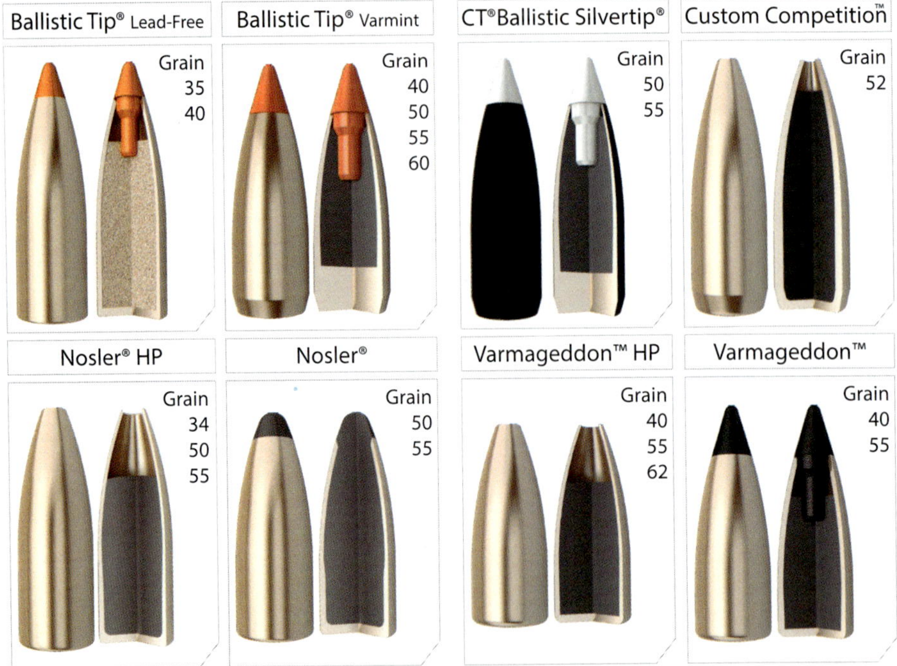

TECHNICAL INFORMATION

The 220 Swift, like the .22-250, is prone to case stretching due to its shallow shoulder angle and long, tapering shape. This stretching can result in neck thickening and premature case-head separation. Case life can be greatly increased by using a neck die, or a full-length die adjusted to partially size the case. If barrel heat is kept at a minimum, and proper cleaning procedures are followed, we have found barrels chambered for the "Swift" to have a surprisingly long life.

220 Swift - 34/35 grain

		MAXIMUM S.A.A.M.I. O.A.C.L.		2.680"
		TESTED O.A.C.L.	B.C.	S.D.
Nosler®	34gr. FBHP	2.530"	0.113	0.097
Ballistic Tip® Lead Free	35gr. Spitzer	2.670"	0.201	0.100

CASE TYPE:	Winchester		PRIMER TYPE	CCI 200
CASE HOLDS:	46.3	Gr. WATER	BARREL Length/Make	26" H-S Prec.
			BARREL Twist	1-14"

POWDER TYPE	POWDER CHG. GRS.		MUZZLE VEL. F.P.S.		LOAD DENSITY (VOLUME)
IMR 4007 SSC	44.0	MAX.	4286	**	102%
Most Accurate Powder Tested	42.0 *		4090		97%
	40.0		3843		93%
A-2520	40.0	MAX.	4326		89%
	38.0 *		4165		85%
	36.0		3971		80%
Varget	40.5 *	MAX.	4337		95%
	38.5		4092		90%
	36.5		3817		85%
H414	45.5	MAX.	4348	**	102%
	43.5		4199		98%
	41.5 *		4028		93%
Viht N150	40.5	MAX.	4350	**	103%
	38.5 *		4165		98%
	36.5		4057		93%
RL15	41.5	MAX.	4358		93%
	39.5 *		4109		89%
	37.5		3935		84%
H380	44.5	MAX.	4379	**	101%
	42.5		4275		97%
	40.5 *		4106		92%
W760	45.0	MAX.	4386	**	102%
	43.0		4219		97%
	41.0 *		3935		93%
H4895	39.5	MAX.	4417		94%
	37.5 *		4131		89%
	35.5		3945		84%
IMR 3031	37.5 *	MAX.	4423		92%
	35.5		4164		87%
	33.5		3953		82%

BC=Ballistic Coefficient SD=Sectional Density
*Most Accurate Load Tested **Compressed Load

Use Maximum Loads with Caution
Refer to page 73 for additional safety information

220 Swift - 40 grain		MAXIMUM S.A.A.M.I. O.A.C.L.		2.680"
		TESTED O.A.C.L.	B.C.	S.D.
Ballistic Tip® Lead-Free	40gr. Spitzer	2.680"	0.220	0.114
Ballistic Tip®	40gr. Spitzer	2.670"	0.221	0.114
Varmageddon™	40gr. FBHP	2.570"	0.158	0.114
Varmageddon™	40gr. FB Tipped	2.670"	0.211	0.114

CASE TYPE:	Winchester		PRIMER TYPE	CCI 200
CASE HOLDS:	48.0	Gr. WATER	BARREL Length/Make	26" H-S Prec.
			BARREL Twist	1-14"

POWDER TYPE	POWDER CHG GRS.		MUZZLE VEL. F.P.S.	LOAD DENSITY (VOLUME)
Viht N140	38.5	MAX.	3942	93%
	36.5		3692	88%
	34.5 *		3442	84%
IMR 4064	38.0	MAX.	3949	87%
	36.0		3695	82%
	34.0 *		3441	78%
IMR 3031	36.0	MAX.	4005	85%
	34.0		3684	80%
	32.0 *		3364	75%
Viht N150	39.5	MAX.	4050	97%
	37.5		3825	92%
	35.5 *		3600	87%
Varget	39.0	MAX.	4067	88%
	37.0		3823	83%
	35.0 *		3579	79%
W760	44.0 *	MAX.	4160	96%
	42.0		3898	92%
	40.0		3636	87%
RL15	39.0 *	MAX.	4165	85%
	37.0		3892	80%
	35.0		3618	76%
H380 Most Accurate Powder Tested	44.0 *	MAX.	4168	96%
	42.0		3991	92%
	40.0		3814	88%
H414	44.0 *	MAX.	4203	95%
	42.0		3976	91%
	40.0		3750	87%
H4895	39.0	MAX.	4203	89%
	37.0		3990	85%
	35.0 *		3792	80%

BC=Ballistic Coefficient SD=Sectional Density
*Most Accurate Load Tested **Compressed Load

Use Maximum Loads with Caution
Refer to page 73 for additional safety information

220 Swift - 50/52 grain		MAXIMUM S.A.A.M.I. O.A.C.L.		2.680"
		TESTED O.A.C.L.	B.C.	S.D.
Ballistic Tip®	50gr. Spitzer	2.670"	0.238	0.142
CT® Ballistic Silvertip®	50gr. Spitzer	2.670"	0.238	0.142
Nosler®	50gr. FBHP	2.670"	0.192	0.142
Nosler®	50gr. FBSP	2.670"	0.195	0.142
Custom Competition™	52gr. HPBT	2.670"	0.220	0.148

CASE TYPE:	Winchester		PRIMER TYPE	CCI 200
CASE HOLDS:	46.8	Gr. WATER	BARREL Length/Make	26" H-S Prec.
			BARREL Twist	1-14"

POWDER TYPE	POWDER CHG. GRS.		MUZZLE VEL. F.P.S.	LOAD DENSITY (VOLUME)
Viht N140	37.0	MAX.	3676	92%
	35.0		3430	87%
	33.0 *		3185	82%
RL15	37.0 *	MAX.	3754	82%
	35.0		3548	78%
	33.0		3343	73%
Varget	36.0 *	MAX.	3787	83%
	34.0		3598	79%
	32.0		3409	74%
H414	42.0	MAX.	3845	93%
	40.0		3645	89%
	38.0 *		3445	85%
IMR 4831	44.0	MAX.	3870	100%
	42.0		3640	95%
	40.0 *		3410	91%
IMR 4350	42.0 *	MAX.	3888	95%
	40.0		3683	90%
	38.0		3478	86%
Viht N150	39.0	MAX.	3890	98%
	37.0		3686	93%
	35.0 *		3482	88%
H380	43.0	MAX.	3900	97%
	41.0		3760	92%
	39.0 *		3620	88%
A-2520 Most Accurate Powder Tested	38.5	MAX.	3930	85%
	36.5		3750	80%
	34.5 *		3570	76%
W760	43.0	MAX.	3988	96%
	41.0		3792	92%
	39.0 *		3595	87%

BC=Ballistic Coefficient SD=Sectional Density
*Most Accurate Load Tested **Compressed Load

Use Maximum Loads with Caution
Refer to page 73 for additional safety information

220 Swift - 55 grain

220 Swift - 55 grain		MAXIMUM S.A.A.M.I. O.A.C.L.		2.680"
		TESTED O.A.C.L.	B.C.	S.D.
Ballistic Tip®	55gr. Spitzer	2.670"	0.267	0.157
CT® Ballistic Silvertip®	55gr. Spitzer	2.670"	0.267	0.157
Nosler®	55gr. FBHP	2.670"	0.210	0.157
Nosler®	55gr. FBSP	2.670"	0.218	0.157
Varmageddon™	55gr. FBHP	2.670"	0.210	0.157
Varmageddon™	55gr. FB Tipped	2.670"	0.255	0.157

CASE TYPE:	Winchester		PRIMER TYPE	CCI 200
CASE HOLDS:	46.4	Gr. WATER	BARREL Length/Make	26" H-S Prec.
			BARREL Twist	1-14"

POWDER TYPE	POWDER CHG. GRS.		MUZZLE VEL. F.P.S.	LOAD DENSITY (VOLUME)
Varget	35.0 *	MAX.	3587	82%
	33.0		3415	77%
	31.0		3243	72%
H414	40.5	MAX.	3661	91%
	38.5		3437	86%
	36.5 *		3213	82%
RL15	37.0 *	MAX.	3662	83%
	35.0		3575	79%
	33.0		3468	74%
Viht N150	38.0 *	MAX.	3736	96%
	36.0		3550	91%
	34.0		3364	86%
IMR 4064	38.0 *	MAX.	3767	90%
	36.0		3533	85%
	34.0		3299	81%
H380	40.0	MAX.	3770	91%
	38.0		3660	86%
	36.0 *		3550	82%
A-2520	37.5 *	MAX.	3780	83%
	35.5		3630	79%
	33.5		3480	74%
IMR 4320	37.0	MAX.	3792	86%
	35.0		3637	81%
	33.0 *		3482	76%
W760	42.0	MAX.	3830	95%
	40.0		3628	90%
	38.0 *		3427	86%
IMR 4831	44.0 *	MAX.	3852	** 101%
Most Accurate	42.0		3677	96%
Powder Tested	40.0		3482	92%

BC=Ballistic Coefficient SD=Sectional Density
*Most Accurate Load Tested **Compressed Load

Use Maximum Loads with Caution
Refer to page 73 for additional safety information

220 Swift - 60 grain

220 Swift - 60 grain		MAXIMUM S.A.A.M.I. O.A.C.L.		2.680"
		TESTED O.A.C.L.	B.C.	S.D.
Ballistic Tip®	60gr. Spitzer	2.680"	0.270	0.171
Partition®	60gr. Spitzer	2.680"	0.228	0.171
Varmageddon™	62gr. FBHP	2.615"	0.251	0.176

CASE TYPE:	Winchester		PRIMER TYPE	CCI 200
CASE HOLDS:	47.0	Gr. WATER	BARREL Length/Make	26" H-S Prec.
			BARREL Twist	1-14"

POWDER TYPE	POWDER CHG. GRS.		MUZZLE VEL. F.P.S.	LOAD DENSITY (VOLUME)
RL15	35.5	MAX.	3598	79%
	33.5		3459	74%
	31.5 *		3320	70%
IMR 4320	36.0	MAX.	3605	82%
	34.0 *		3458	78%
	32.0		3248	73%
Big Game	38.0 *	MAX.	3615	85%
	36.0		3488	81%
	34.0		3376	76%
IMR 4350 Most Accurate Powder Tested	39.5 *	MAX.	3630	89%
	37.5		3508	84%
	35.5		3330	80%
W760	39.5	MAX.	3636	88%
	37.5 *		3526	84%
	35.5		3406	79%
RL22	43.0	MAX.	3646	99%
	41.0		3490	95%
	39.0 *		3351	90%
H414	40.5	MAX.	3676	90%
	38.5 *		3536	85%
	36.5		3396	81%
RL19	43.0 *	MAX.	3685	99%
	41.0		3576	95%
	39.0		3446	90%
IMR 4831	43.0 *	MAX.	3715	97%
	41.0		3576	93%
	39.0		3456	88%

BC=Ballistic Coefficient SD=Sectional Density
*Most Accurate Load Tested **Compressed Load

Use Maximum Loads with Caution
Refer to page 73 for additional safety information

Gary Lewis

223 WINCHESTER SUPER SHORT MAGNUM

The promise of the Super Short Magnum concept was high velocity, and nowhere was it more successful than in the 223 Winchester Super Short Magnum envelope.

The case was developed from the 300 Winchester Short Magnum, shortened and necked down to marry the diminutive .224" projectile. With velocities as high as 4,400 feet per second, given the bullet and powder combination, this 223 is a screamer.

It is a good choice for prairie dogs, rockchucks, woodchucks and coyotes at long range. Thanks to its lightweight offerings, the 223 WSSM is easy on recoil and is an option when outfitting a young person for their first deer hunt.

When loaded with a 60-grain Nosler Partition, it becomes a rifle to be depended upon to put meat in the freezer if the bullet is placed properly.

But, the cartridge really shines when varmints are in the crosshair. One year, my friend Brian Smith, from Alabama, invited me to hunt his home turf, a hunt club carved out of beaver marsh and pines near Tuscaloosa. One of the rifles he put at my disposal was a bull-barreled Winchester, chambered in 223 WSSM.

While I hunted deer in the mornings and evenings with a 7mm Remington Magnum, I opted to carry the 223 WSSM on forays for varmints in the early afternoons. We were on the lookout for raccoons, armadillos and various other critters that prowl the edge habitat.

Watching the tree line one lazy afternoon, I spied an armadillo in the short grass. The distance I guessed at close to 300, but there was no time for the rangefinder. Instead, I took a rest and squeezed and the 50-grain Ballistic Tip screamed across the field and anchored the armored burrower in his tracks.

Later on that trip, the same gun accounted for a bobcat and other varmints. If a flat-shooting small game rifle is in your future, it would be hard to find a better choice than the 223 WSSM.

Gary, with John Nosler, is the author of John Nosler – Going Ballistic. Lewis makes his home in Bend, Oregon

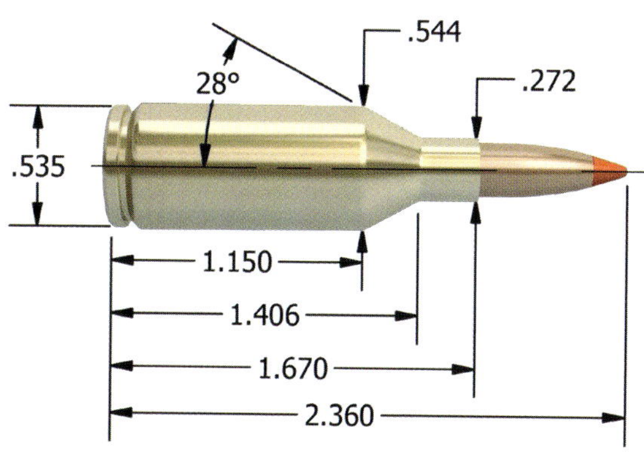

Maximum S.A.A.M.I. Overall Cartridge Length: 2.360"

BULLET CHOICES FOR THE 223 WINCHESTER SUPER SHORT MAGNUM

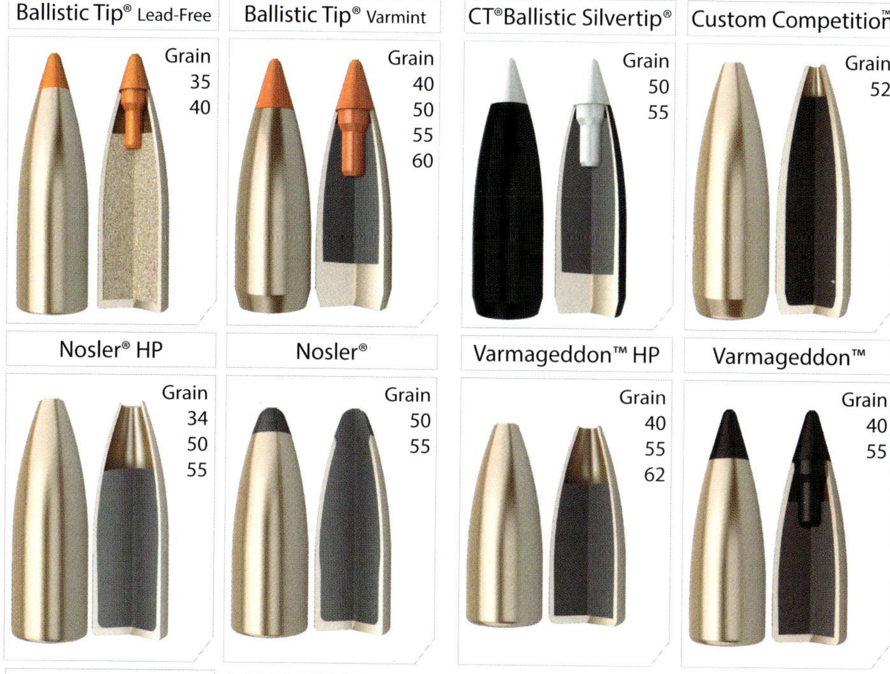

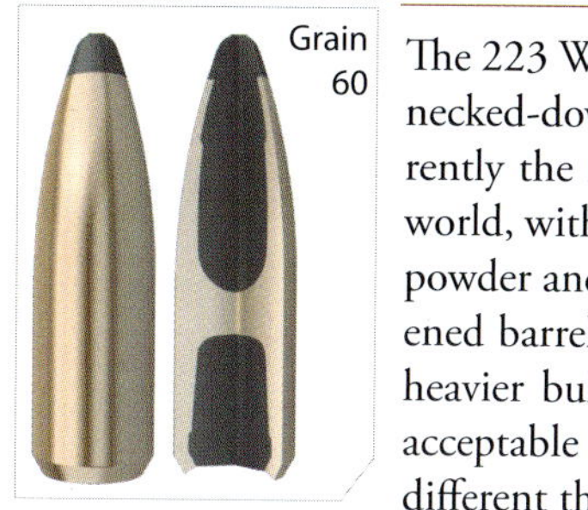

TECHNICAL INFORMATION

The 223 WSSM was introduced by Winchester as a shortened and necked-down version of the very sucessful 300 WSM. It is currently the fastest commercially produced 22 caliber round in the world, with velocities as high as 4,400 fps achievable with the right powder and bullet choices. Due to these extreme velocities, shortened barrel life can result depending on your load. Loaded with heavier bullets and/or at lower velocities, barrel life can be very acceptable if your barrel is well cared for. These concerns are no different than for other high-velocity cartridges.

223 WSSM - 34/35 grain		MAXIMUM S.A.A.M.I. O.A.C.L.		2.360"
		TESTED O.A.C.L.	B.C.	S.D.
Nosler®	34gr. FBHP	1.990"	0.113	0.097
Ballistic Tip® Lead Free	35gr. Spitzer	2.200"	0.201	0.100

CASE TYPE:	Winchester	PRIMER TYPE	WLR
CASE HOLDS:	52.6 Gr. WATER	BARREL Length/Make	24" Wiseman
		BARREL Twist	1-12"

POWDER TYPE	POWDER CHG. GRS.		MUZZLE VEL. F.P.S.		LOAD DENSITY (VOLUME)
W760	52.0	MAX.	4443	**	104%
	50.0		4326		100%
	48.0 *		4174		96%
A-2495	44.5	MAX.	4448		96%
	42.5		4311		92%
	40.5 *		4178		87%
H414 Most Accurate Powder Tested	52.0	MAX.	4450	**	103%
	50.0		4307		99%
	48.0 *		4213		95%
Varget	45.5	MAX.	4479		94%
	43.5 *		4309		89%
	41.5		4176		85%
A-2520	46.5	MAX.	4488		91%
	44.5		4371		87%
	42.5 *		4184		83%
H380	50.5	MAX.	4508	**	101%
	48.5		4366		97%
	46.5 *		4230		93%
TAC	44.5 *	MAX.	4515		86%
	42.5		4349		82%
	40.5		4210		79%
IMR 4064	44.5	MAX.	4546		93%
	42.5 *		4423		89%
	40.5		4264		85%
RL15	47.0 *	MAX.	4547		93%
	45.0		4420		89%
	43.0		4361		85%
H4895	44.5	MAX.	4568		93%
	42.5		4373		89%
	40.5 *		4186		85%

BC=Ballistic Coefficient SD=Sectional Density
*Most Accurate Load Tested **Compressed Load

Use Maximum Loads with Caution
Refer to page 73 for additional safety information

223 WSSM - 40 grain		MAXIMUM S.A.A.M.I. O.A.C.L.		2.360"
		TESTED O.A.C.L.	B.C.	S.D.
Ballistic Tip®	40gr. Spitzer	2.170"	0.221	0.114
Ballistic Tip® Lead Free	40gr. Spitzer	2.170"	0.220	0.114
Varmageddon™	40gr. FBHP	2.030"	0.158	0.114
Varmageddon™	40gr. FB Tipped	2.130"	0.211	0.114

CASE TYPE:	Winchester		PRIMER TYPE	WLR
CASE HOLDS:	52.5	Gr. WATER	BARREL Length/Make	24" Wiseman
			BARREL Twist	1-12"

POWDER TYPE	POWDER CHG. GRS.		MUZZLE VEL. F.P.S.		LOAD DENSITY (VOLUME)
H380 Most Accurate Powder Tested	50.0	MAX.	4251		100%
	48.0		4115		96%
	46.0 *		3940		92%
Varget	45.0 *	MAX.	4260		93%
	43.0		4092		89%
	41.0		3910		84%
H4895	43.5	MAX.	4280		91%
	41.5 *		4105		87%
	39.5		3920		83%
BL-C2	46.5	MAX.	4294		90%
	44.5		4148		86%
	42.5 *		4023		82%
A-2495	43.5	MAX.	4296		94%
	41.5 *		4140		90%
	39.5		4002		85%
W760	51.0	MAX.	4315	**	102%
	49.0		4171		98%
	47.0 *		4020		94%
Viht N150	46.0	MAX.	4316	**	103%
	44.0 *		4179		99%
	42.0		4081		94%
H414	51.0	MAX.	4323	**	101%
	49.0		4146		97%
	47.0 *		3986		93%
RL15	45.5	MAX.	4350		90%
	43.5		4133		86%
	41.5 *		3963		82%
IMR 4064	46.0	MAX.	4411		96%
	44.0 *		4245		92%
	42.0		4064		88%

BC=Ballistic Coefficient SD=Sectional Density
*Most Accurate Load Tested **Compressed Load

Use Maximum Loads with Caution
Refer to page 73 for additional safety information

223 WSSM - 50/52 grain

223 WSSM - 50/52 grain		MAXIMUM S.A.A.M.I. O.A.C.L.		2.360"
		TESTED O.A.C.L.	B.C.	S.D.
Ballistic Tip®	50gr. Spitzer	2.270"	0.238	0.142
CT® Ballistic Silvertip®	50gr. Spitzer	2.270"	0.238	0.142
Nosler®	50gr. FBHP	2.140"	0.192	0.142
Nosler®	50gr. FBSP	2.100"	0.195	0.142
Custom Competition™	52gr. HPBT	2.270"	0.220	0.148

CASE TYPE:	Winchester		PRIMER TYPE	WLR
CASE HOLDS:	51.8	Gr. WATER	BARREL Length/Make	24" Wiseman
			BARREL Twist	1-12"

POWDER TYPE	POWDER CHG. GRS.		MUZZLE VEL. F.P.S.	LOAD DENSITY (VOLUME)
A-2495	39.5	MAX.	3848	87%
	37.5 *		3686	82%
	35.5		3518	78%
Viht N150	40.5 *	MAX.	3882	92%
	38.5		3743	87%
	36.5		3589	83%
IMR 4350	48.0	MAX.	3917	98%
	46.0 *		3775	94%
	44.0		3600	90%
RL15 Most Accurate Powder Tested	42.5	MAX.	3932	85%
	40.5 *		3767	81%
	38.5		3621	77%
H4350	47.0	MAX.	3935	96%
	45.0		3817	92%
	43.0 *		3706	88%
H414	47.5	MAX.	3950	96%
	45.5 *		3791	91%
	43.5		3671	87%
Varget	42.5 *	MAX.	3954	89%
	40.5		3816	85%
	38.5		3660	80%
Big Game	46.0	MAX.	3976	93%
	44.0 *		3838	89%
	42.0		3704	85%
H4895	41.5	MAX.	3994	88%
	39.5		3874	84%
	37.5 *		3703	80%
W760	47.5	MAX.	4000	96%
	45.5 *		3817	92%
	43.5		3703	88%

BC=Ballistic Coefficient SD=Sectional Density
*Most Accurate Load Tested **Compressed Load

Use Maximum Loads with Caution
Refer to page 73 for additional safety information

223 WSSM - 55 grain		MAXIMUM S.A.A.M.I. O.A.C.L.		2.360"
		TESTED O.A.C.L.	B.C.	S.D.
Ballistic Tip®	55gr. Spitzer	2.280"	0.267	0.157
CT® Ballistic Silvertip®	55gr. Spitzer	2.280"	0.267	0.157
Nosler®	55gr. FBHP	2.180"	0.210	0.157
Nosler®	55gr. FBSP	2.160"	0.218	0.157
Varmageddon™	55gr. FBHP	2.135"	0.210	0.157
Varmageddon™	55gr. FB Tipped	2.215"	0.255	0.157

CASE TYPE:	Winchester		PRIMER TYPE	WLR
CASE HOLDS:	51.4	Gr. WATER	BARREL Length/Make	24" Wiseman
			BARREL Twist	

POWDER TYPE	POWDER CHG. GRS.		MUZZLE VEL. F.P.S.	LOAD DENSITY (VOLUME)
A-2495	38.5	MAX.	3668	85%
	36.5 *		3513	81%
	34.5		3346	76%
Viht N150	39.5	MAX.	3730	90%
	37.5 *		3592	86%
	35.5		3451	81%
Big Game	44.5	MAX.	3790	91%
	42.5		3643	87%
	40.5 *		3531	83%
Varget Most Accurate Powder Tested	41.0	MAX.	3828	86%
	39.0		3685	82%
	37.0 *		3546	78%
RL15	41.5 *	MAX.	3848	84%
	39.5		3726	80%
	37.5		3569	76%
IMR 4350	46.5	MAX.	3856	96%
	44.5 *		3703	92%
	42.5		3567	87%
H4895	40.0	MAX.	3862	86%
	38.0 *		3721	81%
	36.0		3580	77%
H4350	46.0	MAX.	3874	95%
	44.0		3757	91%
	42.0 *		3618	86%
H414	46.5	MAX.	3882	94%
	44.5		3782	90%
	42.5 *		3668	86%
W760	46.5	MAX.	3896	95%
	44.5 *		3779	91%
	42.5		3655	87%

BC=Ballistic Coefficient SD=Sectional Density
*Most Accurate Load Tested **Compressed Load

Use Maximum Loads with Caution
Refer to page 73 for additional safety information

223 WSSM - 60 grain		MAXIMUM S.A.A.M.I. O.A.C.L.		2.360"
		TESTED O.A.C.L.	B.C.	S.D.
Ballistic Tip®	60gr. Spitzer	2.135"	0.270	0.171
Partition®	60gr. Spitzer	2.135"	0.228	0.171
Varmageddon™	62gr. FBHP	2.080"	0.251	0.176

CASE TYPE:	Winchester		PRIMER TYPE	WLR
CASE HOLDS:	51.0	Gr. WATER	BARREL Length/Make	24" Wiseman
			BARREL Twist	1-12"

POWDER TYPE	POWDER CHG. GRS.		MUZZLE VEL. F.P.S.	LOAD DENSITY (VOLUME)
Varget	38.0	MAX.	3635	81%
	36.0		3475	76%
	34.0 *		3345	72%
H4350	41.5	MAX.	3636	86%
	39.5 *		3524	82%
	37.5		3410	78%
IMR 4350	42.0	MAX.	3640	87%
	40.0		3455	83%
	38.0 *		3307	79%
RL15	38.5	MAX.	3647	79%
	36.5		3507	75%
	34.5 *		3340	70%
H4895	37.5	MAX.	3649	81%
	35.5 *		3504	76%
	33.5		3354	72%
H414	43.0	MAX.	3669	88%
	41.0		3559	84%
	39.0 *		3409	80%
H380 Most Accurate Powder Tested	43.0	MAX.	3698	89%
	41.0 *		3559	85%
	39.0		3425	80%
W760	43.5	MAX.	3703	89%
	41.5		3591	85%
	39.5 *		3479	81%
IMR 4064	40.0	MAX.	3721	86%
	38.0 *		3570	82%
	36.0		3436	78%
Big Game	44.5	MAX.	3755	92%
	42.5		3630	88%
	40.5 *		3488	84%

BC=Ballistic Coefficient SD=Sectional Density
*Most Accurate Load Tested **Compressed Load

Use Maximum Loads with Caution
Refer to page 73 for additional safety information

6MM PPC-USA

"THE MOST ACCURATE CARTRIDGE IN THE WORLD" was boldly emblazoned on a truck that for over two decades was often seen at rifle tournaments, gun shows or wherever accuracy minded shooters were likely to gather. The truck's owner was Dr. Lou Palmisano, a fanatical firearms and cartridge experimenter who, with equally consummate shooter Ferris Pindell, combined their know-how and names to a couple of cartridges they called the PPCs. Of these, the 6mm version of the PPC duo was destined to become an overwhelming favorite of accuracy buffs and overwhelmingly live up to Palmisano's promotion of it being the world's "Most Accurate."

It's no secret that the PPC line of cartridges were born of a previously obscure round known as the 220 Russian (AKA 5.6X39 Vostok, a stepchild of the 7.62X39 of AK-47 fame.) Ferris and "Doc", both of whom I knew well and shot with for many years, apparently took one look at the cute little Russian case and knew they had discovered what they and other bench shooters had been searching for. They reckoned that by blowing out the body of the case slightly and sharpening the shoulder angle, the resulting powder capacity would be like Goldilocks' Baby Bear, not too much and not too little, but just right! That was back in the '70's and since then the 6PPC has proven to be so just right that it has accounted for virtually all current benchrest records, plus winning untold numbers of national and international tournaments. No other centerfire cartridge has so totally dominated any modern shooting sport.

Palmisano and Pindell are gone now; Ferris in 2011 and Lou a few months later, but their names and legacy will long endure in The World's Most Accurate Cartridge.

The rifle I'm holding with Ferris Pindell (about 2008) is the same 6mm PPC caliber rifle I used to set a new World's Record in March of 2012. The new record is for the Light Varmint Class Benchrest Rifle (10.5 lbs max with scope) with a 100 yard aggregate group average size of .1441" for five 5-shot groups. The previous record of .1500" had stood for 18 years.

Retired Shooting Editor of Outdoor Life, Jim Carmichel is the current World's Record holder in Light Varmint Class Benchrest Rifle competition.

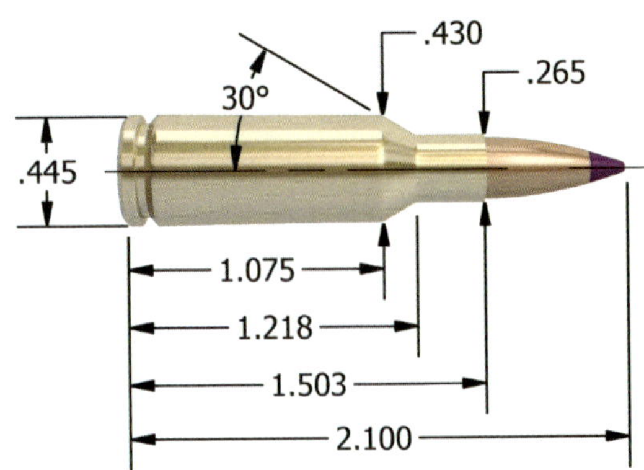

Maximum Overall Cartridge Length: 2.100"

BULLET CHOICES FOR THE 6MM PPC-USA

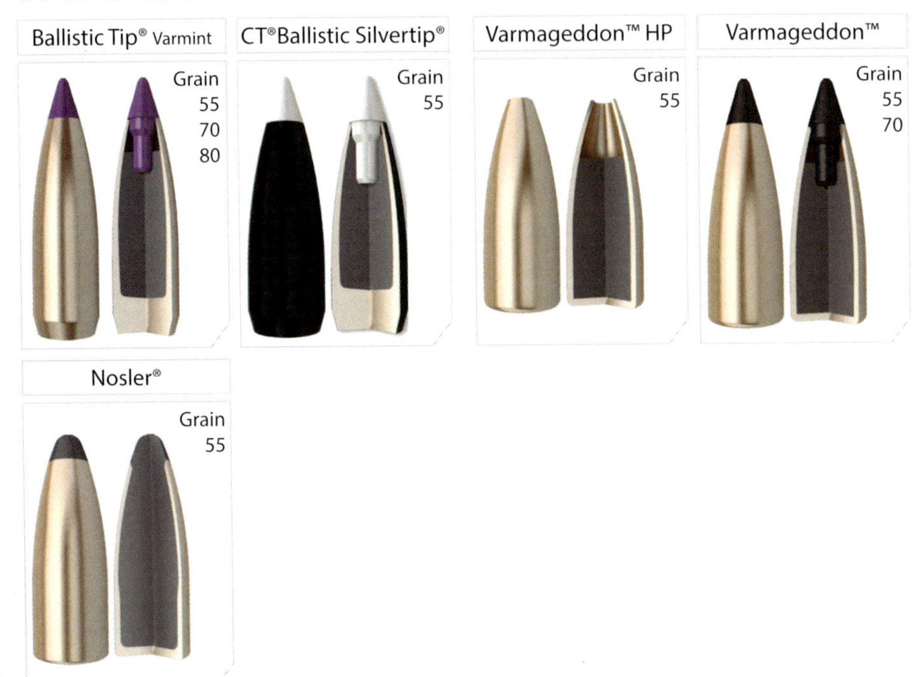

TECHNICAL INFORMATION

Our 6mm PPC data was worked-up using standard production Sako brass with the necks "outside turned" to fit our tighter, bench rest chamber. The loads listed here will create similar pressures in chambers with standard neck dimensions. Our 80-grain Ballistic Tip® is the heaviest Nosler bullet that will properly stabilize in the 1-14" twist of the 6mm PPC.

6mm PPC USA - 55 grain		MAXIMUM O.A.C.L.		2.100"
		TESTED O.A.C.L.	B.C.	S.D.
Ballistic Tip®	55gr. Spitzer	2.060"	0.276	0.133
CT® Ballistic Silvertip®	55gr. Spitzer	2.060"	0.276	0.133
Nosler®	55gr. FBSP	1.990"	0.203	0.133
Varmageddon™	55gr. FBHP	1.910"	0.192	0.133
Varmageddon™	55gr. FB Tipped	1.990"	0.252	0.133

CASE TYPE:	Sako		PRIMER TYPE	Rem 7 1/2
CASE HOLDS:	30.5	Gr. WATER	BARREL Length/Make	22" Wiseman
			BARREL Twist	1-14"

POWDER TYPE	POWDER CHG. GRS.	MUZZLE VEL. F.P.S.	LOAD DENSITY (VOLUME)
A-2015	27.0 * MAX.	3227	** 101%
	26.0	3073	97%
	25.0	2918	93%
Viht N120	24.0 * MAX.	3266	96%
	23.0	3138	92%
	22.0	3010	88%
W748	31.0 * MAX.	3281	** 106%
	30.0	3161	** 102%
	29.0	3042	99%
X-terminator	29.0 MAX.	3294	97%
	28.0 *	3180	93%
	27.0	3050	90%
Viht N130	26.5 * MAX.	3320	** 102%
	25.5	3188	98%
	24.5	3056	95%
H322 Most Accurate Powder Tested	28.5 * MAX.	3348	** 101%
	27.5	3208	97%
	26.5	3069	94%
A-2230	30.0 * MAX.	3357	100%
	29.0	3216	97%
	28.0	3075	94%
RL10x	27.5 MAX.	3362	** 101%
	26.5	3197	98%
	25.5 *	3096	94%
RL7	26.5 * MAX.	3395	94%
	25.5	3258	91%
	24.5	3120	87%
IMR 4198	25.5 MAX.	3428	95%
	24.5	3280	91%
	23.5 *	3134	88%

BC=Ballistic Coefficient SD=Sectional Density
*Most Accurate Load Tested **Compressed Load

Use Maximum Loads with Caution
Refer to page 73 for additional safety information

6mm PPC USA - 70 grain

		MAXIMUM O.A.C.L.		2.100"
		TESTED O.A.C.L.	B.C.	S.D.
Ballistic Tip®	70gr. Spitzer	2.060"	0.310	0.169
Varmageddon™	70gr. FB Tipped	2.060"	0.305	0.169

CASE TYPE:	Sako		PRIMER TYPE	Rem 7 1/2
CASE HOLDS:	29.4	Gr. WATER	BARREL Length/Make	22" Wiseman
			BARREL Twist	1-14"

POWDER TYPE	POWDER CHG. GRS.		MUZZLE VEL. F.P.S.		LOAD DENSITY (VOLUME)
Viht N120	22.5 *	MAX.	2925		93%
	21.5		2812		89%
	20.5		2698		85%
W748	28.5 *	MAX.	2960	**	101%
	27.5		2850		97%
	26.5		2740		94%
Viht N130	25.0 *	MAX.	3019		100%
	24.0		2896		96%
	23.0		2774		92%
A-2015	27.0	MAX.	3055	**	104%
	26.0		2937		100%
	25.0 *		2836		97%
RL7	24.5	MAX.	3082		91%
	23.5		2957		87%
	22.5 *		2832		83%
H322 Most Accurate Powder Tested	25.5	MAX.	3088		94%
	24.5		2983		90%
	23.5 *		2878		86%
RL10x	25.5	MAX.	3112		97%
	24.5		2961		94%
	23.5 *		2829		90%
H335	28.5 *	MAX.	3130		98%
	27.5		3030		95%
	26.5		2930		92%
A-2230	29.5 *	MAX.	3192	**	102%
	28.5		3097		99%
	27.5		3002		95%
A-2460	29.5	MAX.	3250	**	101%
	28.5		3160		98%
	27.5 *		3070		94%

BC=Ballistic Coefficient SD=Sectional Density
*Most Accurate Load Tested **Compressed Load

Use Maximum Loads with Caution
Refer to page 73 for additional safety information

6mm PPC USA - 80 grain

			MAXIMUM O.A.C.L.		2.100"
			TESTED O.A.C.L.	B.C.	S.D.
Ballistic Tip®		80gr. Spitzer	2.060"	0.329	0.194

CASE TYPE:	Sako		PRIMER TYPE	Rem 7 1/2
CASE HOLDS:	28.3	Gr. WATER	BARREL Length/Make	22" Wiseman
			BARREL Twist	1-14"

POWDER TYPE	POWDER CHG. GRS.		MUZZLE VEL. F.P.S.		LOAD DENSITY (VOLUME)
A-2015	25.0	MAX.	2655		100%
	24.0 *		2599		96%
	23.0		2498		92%
Varget	28.0	MAX.	2694	**	107%
	27.0		2606	**	103%
	26.0 *		2498		99%
H322 Most Accurate Powder Tested	25.0 *	MAX.	2722		96%
	24.0		2633		92%
	23.0		2559		88%
RL7	24.0	MAX.	2760		92%
	23.0 *		2657		88%
	22.0		2556		84%
Viht N130	24.0	MAX.	2774		100%
	23.0 *		2694		96%
	22.0		2606		91%
IMR 4198	22.5	MAX.	2776		90%
	21.5		2662		86%
	20.5 *		2560		82%
H335	27.0	MAX.	2846		97%
	26.0 *		2711		93%
	25.0		2607		90%
TAC	28.0 *	MAX.	2861	**	101%
	27.0		2798		97%
	26.0		2723		94%
Benchmark	26.5	MAX.	2862	**	101%
	25.5		2773		97%
	24.5 *		2670		94%
A-2460	28.0	MAX.	2890		100%
	27.0		2780		96%
	26.0 *		2704		93%

BC=Ballistic Coefficient SD=Sectional Density
*Most Accurate Load Tested **Compressed Load

Use Maximum Loads with Caution
Refer to page 73 for additional safety information

Always at the drawing board. John A. Nosler circa 1970.

John A. Nosler testing bullets in wet newspaper circa 1972.

243 WINCHESTER

A smart aleck once remarked that the only hunters who have trouble killing deer with the 243 Winchester are gun writers. There might be some truth to this. Quite a few magazine articles suggest the .243 is "marginal" for big bucks, while out in the real world, .243-toting hunters slay millions of deer each year.

For years, my wife Eileen avoided the .243 because of its reputation as a women's and kid's cartridge. Then one day, she picked up one of my recently acquired rifles, a lightweight Husqvarna .243, finding it fit her perfectly. Two months later a big whitetail buck charged out of some river bottom brush after a hot doe. Eileen put a 100-grain Nosler Partition in the right place, and now yet another deer hunter is thoroughly pleased with her marginal deer rifle—and her husband's looking for another .243.

John and his wife Eileen Clarke live in Montana, where they hunt and run their website www.riflesandrecipes.com

.454
.276
20°
.473
1.560
1.804
2.045
2.710

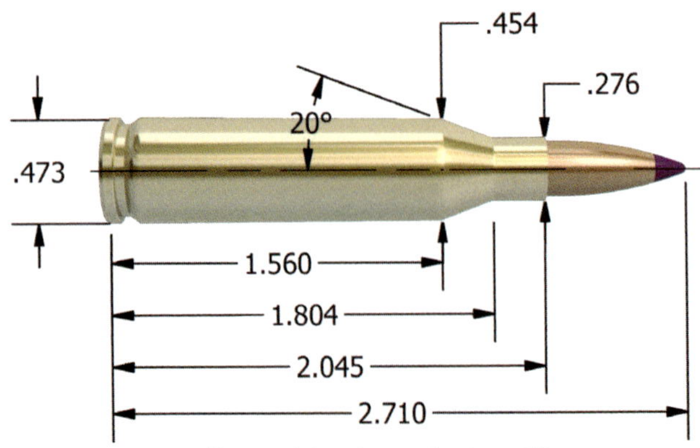

Maximum S.A.A.M.I. Overall Cartridge Length: 2.710"

BULLET CHOICES FOR THE 243 WINCHESTER

AccuBond®

Grain
90

Ballistic Tip®

Grain
55
70
80
90
95

CT®Ballistic Silvertip®

Grain
55
95

Ballistic Tip® Lead-Free

Grain
55

Custom Competition™

Grain
105
107

E-Tip®

Grain
90

Nosler®

Grain
55

Partition®

Grain
85
95
100

Varmageddon™ HP

Grain
55

Varmageddon™

Grain
55
70

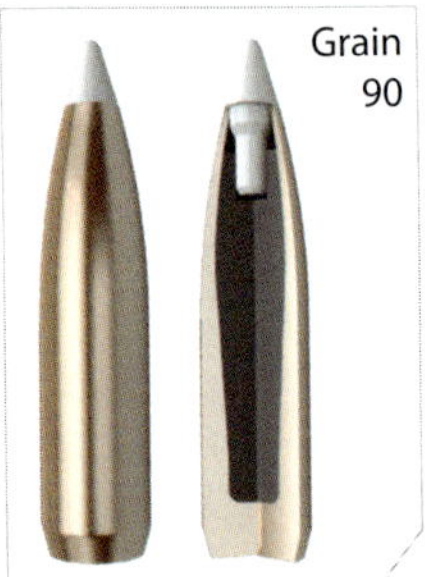

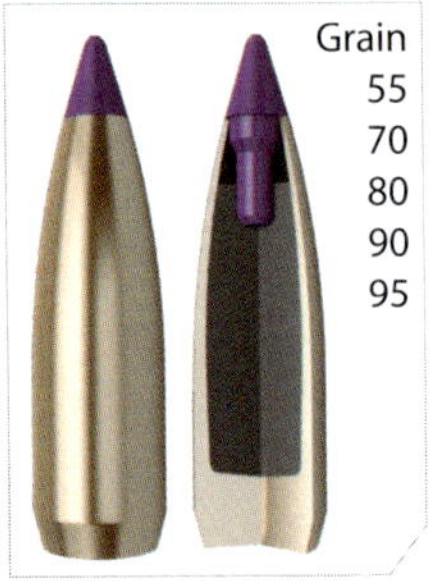

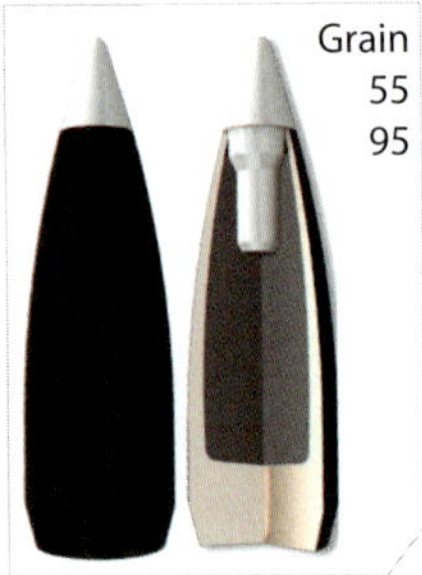

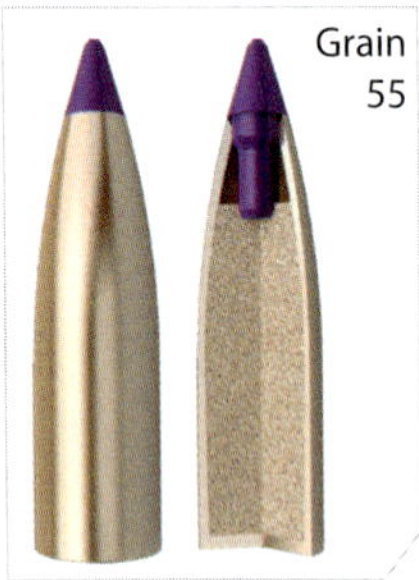

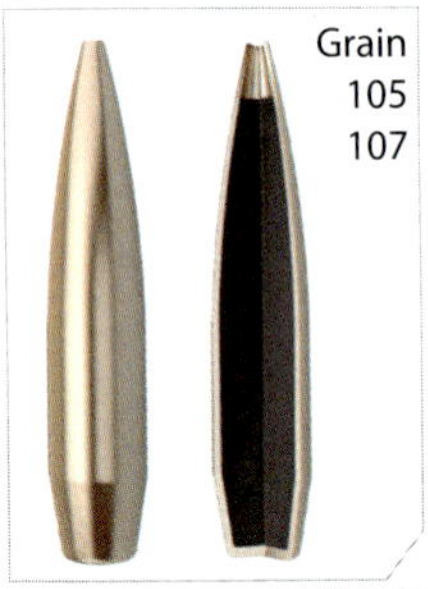

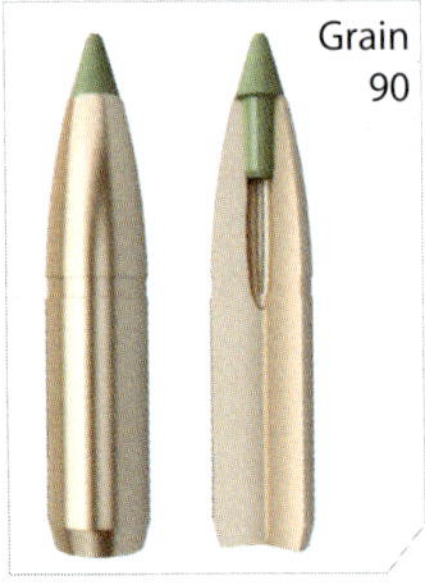

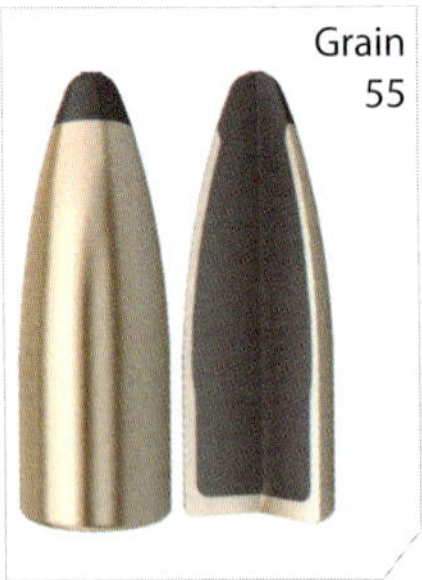

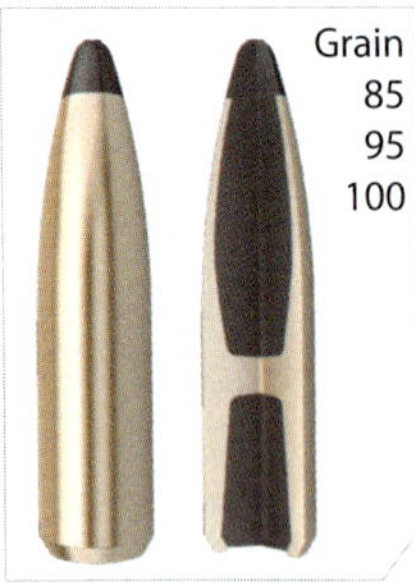

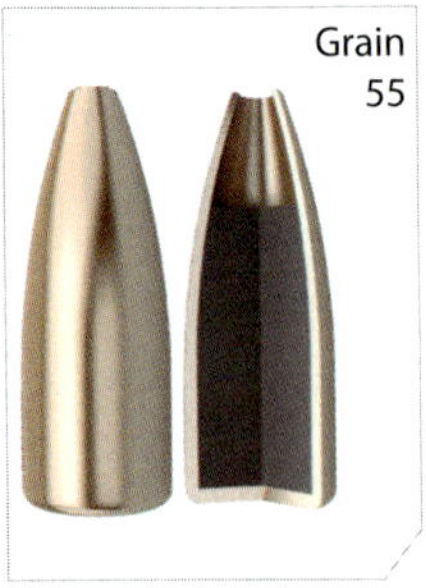

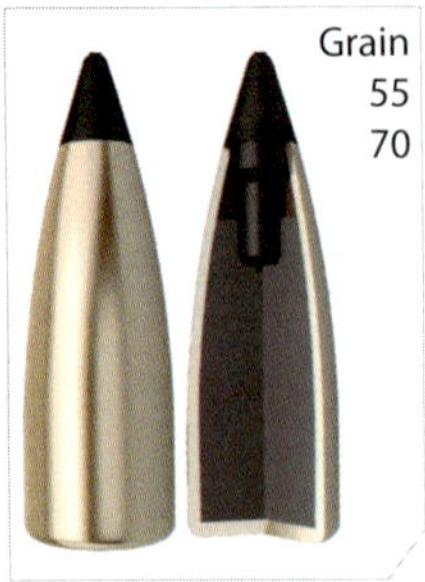

TECHNICAL INFORMATION

The 243 Winchester tolerates a wide variety of bullet/propellant combinations, making it an easy cartridge to successfully reload. With Nosler's 55, 70 or 80gr bullets, the 243 Winchester is a proven varminter. Step up to 85, 90, 95 or 100gr AccuBonds, E-Tips or Partitions to make the 243 Winchester a potent deer or antelope setup. Note: in some elevations/temperatures, the 90gr E-Tip may need a 1-9" twist to reliably stabilize.

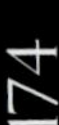

243 Winchester - 55 grain		MAXIMUM S.A.A.M.I. O.A.C.L.		2.710"
		TESTED O.A.C.L.	B.C.	S.D.
Ballistic Tip®	55gr. Spitzer	2.610"	0.276	0.133
Ballistic Tip® Lead Free	55gr. Spitzer	2.610"	0.288	0.133
CT® Ballistic Silvertip®	55gr. Spitzer	2.610"	0.276	0.133
Nosler®	55gr. FBSP	2.450"	0.203	0.133
Varmageddon™	55gr. FBHP	2.510"	0.192	0.133
Varmageddon™	55gr. FB Tipped	2.590"	0.252	0.133

CASE TYPE:	Nosler		PRIMER TYPE	Rem 9 1/2
CASE HOLDS:	53.9	Gr. WATER	BARREL Length/Make	24" Lilja
			BARREL Twist	1-10"

POWDER TYPE	POWDER CHG. GRS.	MUZZLE VEL. F.P.S.	LOAD DENSITY (VOLUME)
Viht N150	45.0 * MAX.	3872	98%
	43.0	3716	94%
	41.0	3560	89%
W760 Most Accurate Powder Tested	50.0 * MAX.	3901	97%
	48.0	3737	93%
	46.0	3573	89%
IMR 3031	42.0 MAX.	3927	88%
	40.0 *	3797	84%
	38.0	3607	80%
IMR 4895	44.5 * MAX.	3935	91%
	42.5	3735	87%
	40.5	3535	83%
Varget	45.5 MAX.	3941	91%
	43.5	3758	87%
	41.5 *	3575	83%
IMR 4320	45.5 * MAX.	3942	91%
	43.5	3743	87%
	41.5	3544	83%
IMR 4064	45.5 * MAX.	3970	93%
	43.5	3759	89%
	41.5	3549	85%
RL15	45.0 MAX.	3993	87%
	43.0	3784	83%
	41.0 *	3576	79%
Big Game	48.5 MAX.	4032	95%
	46.5	3860	91%
	44.5 *	3575	87%

BC=Ballistic Coefficient SD=Sectional Density
*Most Accurate Load Tested **Compressed Load

Use Maximum Loads with Caution
Refer to page 73 for additional safety information

243 Winchester - 70 grain			MAXIMUM S.A.A.M.I. O.A.C.L.		2.710"
			TESTED O.A.C.L.	B.C.	S.D.
Ballistic Tip®		70gr. Spitzer	2.680"	0.310	0.169
Varmageddon™		70gr. FB Tipped	2.680"	0.334	0.169

CASE TYPE:	Nosler		PRIMER TYPE	Rem 9 1/2
CASE HOLDS:	50.8	Gr. WATER	BARREL Length/Make	24" Lilja
			BARREL Twist	1-10"

POWDER TYPE	POWDER CHG. GRS.		MUZZLE VEL. F.P.S.	LOAD DENSITY (VOLUME)
IMR 3031	38.0	MAX.	3410	85%
	36.0		3250	80%
	34.0 *		3093	76%
H380	44.0	MAX.	3440	91%
	42.0		3300	87%
	40.0 *		3160	83%
H335	38.0	MAX.	3460	76%
	36.0		3330	72%
	34.0 *		3202	68%
Viht N160	47.0	MAX.	3463	** 103%
	45.0		3315	98%
	43.0 *		3167	94%
IMR 4064	41.5 *	MAX.	3478	90%
	39.5		3323	85%
	37.5		3168	81%
IMR 4895	40.0 *	MAX.	3483	87%
	38.0		3318	82%
	36.0		3160	78%
Big Game	45.5	MAX.	3600	94%
	43.5 *		3481	90%
	41.5		3312	86%
IMR 4350	47.0 *	MAX.	3610	98%
	45.0		3430	94%
	43.0		3250	90%
Varget Most Accurate Powder Tested	42.0 *	MAX.	3616	89%
	40.0		3477	85%
	38.0		3338	81%
H414	47.5 *	MAX.	3630	97%
	45.5		3470	93%
	43.5		3310	89%

BC=Ballistic Coefficient SD=Sectional Density
*Most Accurate Load Tested **Compressed Load

Use Maximum Loads with Caution
Refer to page 73 for additional safety information

243 Winchester - 80 grain			MAXIMUM S.A.A.M.I. O.A.C.L.		2.710"
			TESTED O.A.C.L.	B.C.	S.D.
Ballistic Tip®		80gr. Spitzer	2.680"	0.329	0.194
CASE TYPE:	Nosler		PRIMER TYPE		Rem 9 1/2
CASE HOLDS:	51.5	Gr. WATER	BARREL Length/Make		24" Lilja
			BARREL Twist		1-10"

POWDER TYPE	POWDER CHG. GRS.		MUZZLE VEL. F.P.S.	LOAD DENSITY (VOLUME)
H380	42.5	MAX.	3291	87%
Most Accurate	40.5		3169	83%
Powder Tested	38.5 *		3041	79%
Big Game	42.5	MAX.	3318	87%
	40.5		3183	83%
	38.5 *		3032	79%
IMR 4895	39.0 *	MAX.	3321	83%
	37.0		3185	79%
	35.0		3045	75%
Viht N160	46.0	MAX.	3368	99%
	44.0		3246	95%
	42.0 *		3107	91%
Varget	40.0	MAX.	3371	84%
	38.0		3253	80%
	36.0 *		3112	76%
IMR 4064	40.0 *	MAX.	3393	85%
	38.0		3252	81%
	36.0		3124	77%
RL19	47.0	MAX.	3400	99%
	45.0		3272	95%
	43.0 *		3137	91%
W760	45.0 *	MAX.	3431	91%
	43.0		3284	87%
	41.0		3171	83%
IMR 4350	45.0 *	MAX.	3438	92%
	43.0		3289	88%
	41.0		3168	84%

BC=Ballistic Coefficient SD=Sectional Density
*Most Accurate Load Tested **Compressed Load

Use Maximum Loads with Caution
Refer to page 73 for additional safety information

243 Winchester - 85/90 grain		MAXIMUM S.A.A.M.I. O.A.C.L.		2.710"
		TESTED O.A.C.L.	B.C.	S.D.
Partition®	85gr. Spitzer	2.680"	0.315	0.206
AccuBond®	90gr. Spitzer	2.680"	0.376	0.218
Ballistic Tip®	90gr. Spitzer	2.680"	0.365	0.218
E-Tip®	90gr. Spitzer	2.680"	0.403	0.218

Due to internal construction differences, always begin with starting loads when using E-Tip® products.

CASE TYPE:	Nosler		PRIMER TYPE	Rem 9 1/2
CASE HOLDS:	52.8	Gr. WATER	BARREL Length/Make	24" Lilja
			BARREL Twist	1-10"

POWDER TYPE	POWDER CHG. GRS.		MUZZLE VEL. F.P.S.	LOAD DENSITY (VOLUME)
RL19	43.5	* MAX.	3107	90%
	41.5		2981	85%
	39.5		2855	81%
IMR 4064	37.0	MAX.	3139	77%
	35.0		2974	73%
	33.0	*	2809	69%
W760	42.0	MAX.	3139	83%
	40.0		3054	79%
	38.0	*	2969	75%
IMR 4895	37.0	MAX.	3145	77%
	35.0		2995	73%
	33.0	*	2845	69%
Viht N160	44.0	* MAX.	3172	93%
	42.0		3064	88%
	40.0		2952	84%
Hunter	44.5	MAX.	3109	89%
	42.5	*	3011	85%
	40.5		2915	81%
H380	42.0	* MAX.	3232	84%
	40.0		3077	80%
	38.0		2922	76%
IMR 4350	43.5	MAX.	3240	87%
	41.5		3085	83%
	39.5	*	2930	79%
IMR 4831	44.5	* MAX.	3240	90%
Most Accurate	42.5		3120	86%
Powder Tested	40.5		3000	82%

BC=Ballistic Coefficient SD=Sectional Density
*Most Accurate Load Tested **Compressed Load

Use Maximum Loads with Caution
Refer to page 73 for additional safety information

243 Winchester - 95/100 grain		MAXIMUM S.A.A.M.I. O.A.C.L.		2.710"
		TESTED O.A.C.L.	B.C.	S.D.
Ballistic Tip®	95gr. Spitzer	2.680"	0.379	0.230
CT® Ballistic Silvertip®	95gr. Spitzer	2.680"	0.379	0.230
Partition®	95gr. Spitzer	2.680"	0.365	0.230
Partition®	100gr. Spitzer	2.680"	0.384	0.242

CASE TYPE:	Nosler		PRIMER TYPE	Rem 9 1/2
CASE HOLDS:	50.0	Gr. WATER	BARREL Length/Make	24" Lilja
			BARREL Twist	1-10"

POWDER TYPE	POWDER CHG. GRS.		MUZZLE VEL. F.P.S.	LOAD DENSITY (VOLUME)
IMR 4895	35.5	MAX.	2857	78%
	33.5		2722	74%
	31.5 *		2587	69%
IMR 3031	35.0 *	MAX.	2910	79%
	33.0		2803	75%
	31.0		2699	70%
IMR 4831	42.0 *	MAX.	2931	89%
	40.0		2830	85%
	38.0		2729	81%
H380	40.0 *	MAX.	2968	84%
	38.0		2843	80%
	36.0		2718	76%
RL22	42.5 *	MAX.	3019	92%
	40.5		2922	88%
	38.5		2820	84%
IMR 4350	41.5 *	MAX.	3060	88%
	39.5		2985	84%
	37.5		2909	79%
Hunter	44.0	MAX.	3063	93%
	42.0		2968	88%
	40.0 *		2824	84%
IMR 7828	45.5 *	MAX.	3123	97%
	43.5		2998	93%
	41.5		2873	88%
Viht N560 Most Accurate Powder Tested	44.5 *	MAX.	3144	99%
	42.5		3024	94%
	40.5		2906	90%

BC=Ballistic Coefficient SD=Sectional Density
*Most Accurate Load Tested **Compressed Load

Use Maximum Loads with Caution
Refer to page 73 for additional safety information

243 Win - 105/107 grain (Fast Twist)		MAXIMUM S.A.A.M.I. O.A.C.L.		2.710"
		TESTED O.A.C.L.	B.C.	S.D.
Custom Competition™	105gr. HPBT	2.710"	0.517	0.254
Custom Competition™	107gr. HPBT	2.710"	0.525	0.259

CASE TYPE:	Nosler		PRIMER TYPE	Fed 210
CASE HOLDS:	48.0	Gr. WATER	BARREL Length/Make	24" H-S Prec.
			BARREL Twist	1-8"

POWDER TYPE	POWDER CHG. GRS.	MUZZLE VEL. F.P.S.	LOAD DENSITY (VOLUME)
RL22	42.0 * MAX.	2791	95%
	40.0	2680	91%
	38.0	2577	86%
IMR 7828 SSC	43.0 MAX.	2798	92%
	41.0	2641	88%
	39.0 *	2497	84%
H1000	46.0 MAX.	2872	100%
	44.0 *	2780	96%
	42.0	2655	92%
IMR 4350	40.5 * MAX.	2893	89%
	38.5	2745	85%
	36.5	2648	80%
H4831 Most Accurate Powder Tested	43.5 MAX.	2899	94%
	41.5 *	2780	90%
	39.5	2652	86%
Retumbo	48.0 MAX.	2912	** 105%
	46.0 *	2791	** 101%
	44.0	2656	96%
Hunter	42.5 MAX.	2915	93%
	40.5 *	2775	89%
	38.5	2687	84%
Supreme 780	45.5 * MAX.	2978	99%
	43.5	2841	95%
	41.5	2733	91%

BC=Ballistic Coefficient SD=Sectional Density
*Most Accurate Load Tested **Compressed Load

Use Maximum Loads with Caution
Refer to page 73 for additional safety information

6MM REMINGTON

An old favorite, I've known the 6mm Remington since way back when it was the 244 Remington. In the early days, I preferred it as a varmint round. Those early Remington rifles with long, twenty-six inch barrels and one-in-twelve twist rate, were oh-so fast and accurate with light varmint bullets.

In the mid-nineteen seventies, a company transfer moved me to Green River, Wyoming. Early on, most of my hunting was with a 264 Winchester Magnum. A few new hunting friends were hunting deer and antelope with heavy barreled Ruger M77V Rifles in 6mm Remington. On the open sagebrush plains of Wyoming, shooting was a short walk from trucks, so a heavy barreled rifle was not a handicap.

I decided to also purchase one of these M77V Rugers, but could find none locally, and had to order one. It took a long time coming in, but proved its worth with a variety of bullets, as almost all of my shooting is with handloads. About this time, Nosler introduced the Solid Base bullet line. Proving to be very accurate, they soon became my favorite hunting bullets for everything from prairie dogs to deer.

Nine years later I moved back to the east coast, and to this day I still have that Ruger. The only way I would part with it is to give it to my son. It now carries a beautiful Myrtle wood stock supplied by old pen pal Dick Simmons of Washington State. It soon proved very accurate with the newer Nosler Ballistic Tip bullets. When Nosler introduced the ultra light fifty-five grain Ballistic Tip, I first thought it too light for the 6mm. But Chub Eastman assured me at very high velocity they were flat shooting and accurate.

He sent some to try out, and I quickly worked up handloads for these little bullets. At close to four thousand feet per second, I was getting one hundred yard, three shot groups in the one-half inch range. I was very impressed, and had found a load to rival my 220 Swift. On the heavy side, the 90 and 95-grain Ballistic Tips proved their worth on deer. This Remington round has always struggled to survive, and seems to be fading fast in today's market. But, if one handloads and can obtain a good 6mm rifle, the rewards are worth the effort.

Ed Timerson

Ed is an inveterate shooting experimenter and writer.

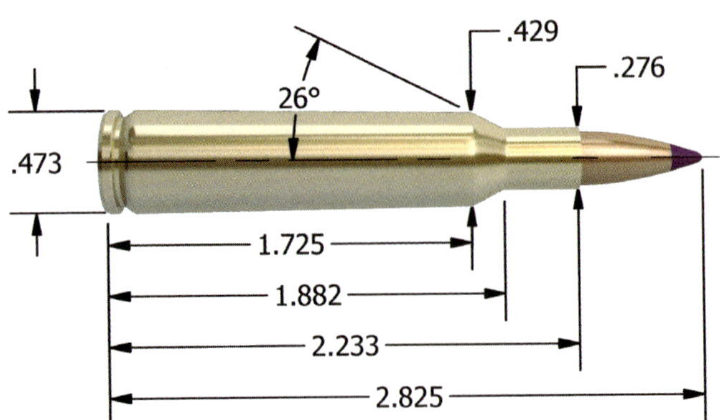

Maximum S.A.A.M.I. Overall Cartridge Length: 2.825"

BULLET CHOICES FOR THE 6MM REMINGTON

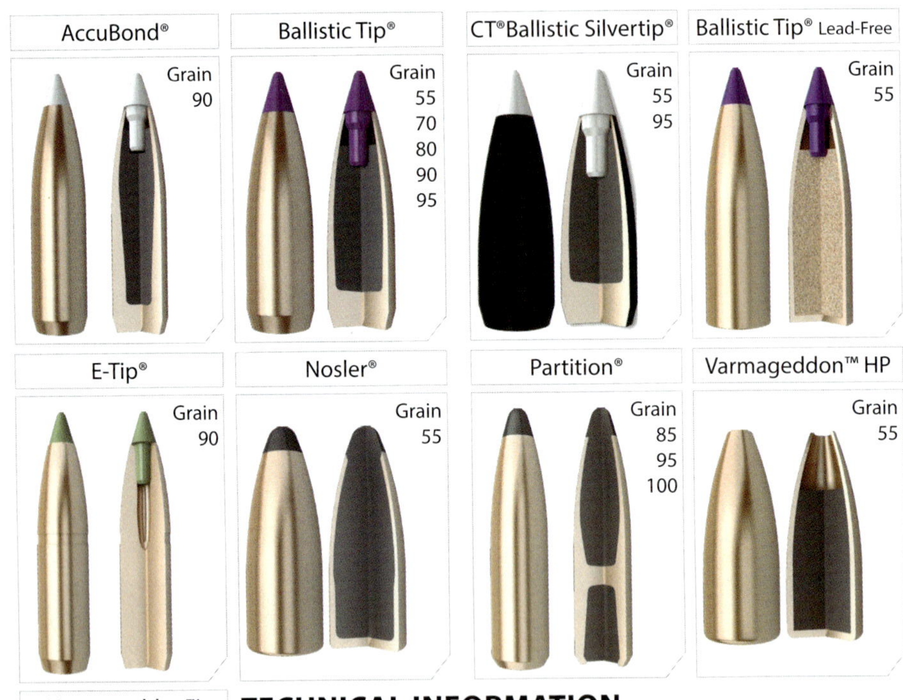

Varmageddon™

Grain

55

70

TECHNICAL INFORMATION

Originally dubbed the 244 Remington, the first production rifles chambered for this cartridge were equipped with 1-12" twist barrels. The combination proved to be less popular than Remington expected, and was soon reintroduced as the 6mm Remington. This time, the cartridge was matched with faster 1-9" twist barrels to properly stabilize the more popular hunting weight bullets. Like the 243 Winchester, the 6mm Remington is a versatile cartridge that, with proper bullet selection, can be used for everything from varmints, to deer and antelope.

6mm Remington - 55 grain

		MAXIMUM S.A.A.M.I. O.A.C.L.		2.825"
		TESTED O.A.C.L.	B.C.	S.D.
Ballistic Tip®	55gr. Spitzer	2.770"	0.276	0.133
Ballistic Tip® Lead Free	55gr. Spitzer	2.770"	0.288	0.133
CT® Ballistic Silvertip®	55gr. Spitzer	2.770"	0.276	0.133
Nosler®	55gr. FBSP	2.620"	0.203	0.133
Varmageddon™	55gr. FBHP	2.640"	0.192	0.133
Varmageddon™	55gr. FB Tipped	2.720"	0.252	0.133

CASE TYPE:	Remington		PRIMER TYPE	Fed 210M
CASE HOLDS:	55.5	Gr. WATER	BARREL Length/Make	24" Shilen
			BARREL Twist	1-10"

POWDER TYPE	POWDER CHG. GRS.		MUZZLE VEL. F.P.S.		LOAD DENSITY (VOLUME)
IMR 4831	52.0	MAX.	3710		100%
	50.0		3540		96%
	48.0 *		3370		92%
IMR 4350	52.0	MAX.	3808		99%
	50.0		3636		95%
	48.0 *		3464		92%
RL19	53.0	MAX.	3812	**	104%
	51.0		3629		100%
	49.0 *		3445		96%
H414	53.0 *	MAX.	3961		99%
	51.0		3760		96%
	49.0		3560		92%
Viht N150	47.5 *	MAX.	3975	**	101%
	45.5		3782		96%
	43.5		3589		92%
RL15	46.5	MAX.	3987		87%
	44.5		3754		84%
	42.5 *		3520		80%
Varget	47.0 *	MAX.	4002		92%
	45.0		3831		88%
	43.0		3661		84%
H380	53.0 *	MAX.	4019	**	101%
	51.0		3885		97%
	49.0		3751		93%
W760 Most Accurate Powder Tested	53.0 *	MAX.	4077		100%
	51.0		3875		96%
	49.0		3674		92%
Big Game	52.0	MAX.	4086		99%
	50.0		3950		95%
	48.0 *		3809		91%

BC=Ballistic Coefficient SD=Sectional Density
*Most Accurate Load Tested **Compressed Load

Use Maximum Loads with Caution
Refer to page 73 for additional safety information

6mm Remington - 70 grain

		MAXIMUM S.A.A.M.I. O.A.C.L.		2.825"
		TESTED O.A.C.L.	B.C.	S.D.
Ballistic Tip®	70gr. Spitzer	2.770"	0.310	0.169
Varmageddon™	70gr. FB Tipped	2.770"	0.305	0.169

CASE TYPE:	Remington		PRIMER TYPE	Fed 210M
CASE HOLDS:	53.2	Gr. WATER	BARREL Length/Make	24" Shilen
			BARREL Twist	1-10"

POWDER TYPE	POWDER CHG. GRS.		MUZZLE VEL. F.P.S.		LOAD DENSITY (VOLUME)
H414	47.0	MAX.	3469		92%
	45.0		3294		88%
	43.0 *		3119		84%
IMR 3031	40.0 *	MAX.	3503		85%
	38.0		3368		81%
	36.0		3243		76%
IMR 4895	40.0 *	MAX.	3511		83%
	38.0		3362		78%
	36.0		3218		74%
H380	45.0	MAX.	3515		89%
	43.0		3385		85%
	41.0 *		3238		81%
RL19	50.5 *	MAX.	3554	**	103%
	48.5		3387		99%
	46.5		3221		95%
IMR 4320	42.0	MAX.	3569		85%
	40.0		3386		81%
	38.0 *		3203		77%
Viht N160	50.0	MAX.	3582	**	104%
Most Accurate	48.0		3436		100%
Powder Tested	46.0 *		3291		96%
IMR 4350	48.0	MAX.	3596		95%
	46.0		3455		91%
	44.0 *		3312		88%
IMR 4831	50.5 *	MAX.	3620	**	101%
	48.5		3473		97%
	46.5		3306		93%
Big Game	48.0	MAX.	3648		95%
	46.0		3531		91%
	44.0 *		3422		87%

BC=Ballistic Coefficient SD=Sectional Density
*Most Accurate Load Tested **Compressed Load

Use Maximum Loads with Caution
Refer to page 73 for additional safety information

6mm Remington - 80 grain

		MAXIMUM S.A.A.M.I. O.A.C.L.	2.825"
		TESTED O.A.C.L.	B.C.	S.D.
Ballistic Tip®	80gr. Spitzer	2.770"	0.329	0.194

CASE TYPE:	Remington		PRIMER TYPE	Fed 210M
CASE HOLDS:	52.4	Gr. WATER	BARREL Length/Make	24" Shilen
			BARREL Twist	1-10"

POWDER TYPE	POWDER CHG. GRS.		MUZZLE VEL. F.P.S.	LOAD DENSITY (VOLUME)
IMR 4895	39.5	MAX.	3322	83%
	37.5		3211	79%
	35.5 *		3107	74%
RL19	49.0	MAX.	3370	** 102%
	47.0		3181	97%
	45.0 *		3012	93%
IMR 3031	39.0	MAX.	3382	84%
	37.0		3204	80%
	35.0 *		3086	75%
Varget	42.0 *	MAX.	3384	87%
	40.0		3261	83%
	38.0		3121	78%
RL15	42.0 *	MAX.	3430	83%
	40.0		3288	80%
	38.0		3162	76%
IMR 4831	48.0 *	MAX.	3447	97%
	46.0		3280	93%
	44.0		3077	89%
IMR 4064 Most Accurate Powder Tested	42.0	MAX.	3451	88%
	40.0		3311	84%
	38.0 *		3179	80%
Big Game	47.0	MAX.	3458	94%
	45.0		3370	90%
	43.0 *		3234	86%
IMR 4350	47.0	MAX.	3472	95%
	45.0		3281	91%
	43.0 *		3111	87%
H414	46.0	MAX.	3472	91%
	44.0		3323	87%
	42.0 *		3150	83%

BC=Ballistic Coefficient SD=Sectional Density
*Most Accurate Load Tested **Compressed Load

Use Maximum Loads with Caution
Refer to page 73 for additional safety information

6mm Remington - 85/90 grain		MAXIMUM S.A.A.M.I. O.A.C.L.		2.825"
		TESTED O.A.C.L.	B.C.	S.D.
Partition®	85gr. Spitzer	2.770"	0.315	0.206
AccuBond®	90gr. Spitzer	2.770"	0.376	0.218
Ballistic Tip®	90gr. Spitzer	2.770"	0.365	0.218
E-Tip®	90gr. Spitzer	2.770"	0.403	0.218

Due to internal construction differences, always begin with starting loads when using E-Tip® products.

CASE TYPE:	Remington		PRIMER TYPE	Fed 210M
CASE HOLDS:	52.0	Gr. WATER	BARREL Length/Make	24" Shilen
			BARREL Twist	1-10"

POWDER TYPE	POWDER CHG. GRS.		MUZZLE VEL. F.P.S.	LOAD DENSITY (VOLUME)
IMR 3031	37.5	MAX.	3178	81%
	35.5		3043	77%
	33.5 *		2908	73%
Varget	39.0 *	MAX.	3200	81%
	37.0		3056	77%
	35.0		2909	73%
H380	41.5	MAX.	3210	84%
	39.5		3060	80%
	37.5 *		2910	76%
Big Game	44.5	MAX.	3214	90%
	42.5 *		3174	86%
	40.5		3048	82%
IMR 4064	39.5	MAX.	3223	83%
	37.5		3073	79%
	35.5 *		2923	75%
IMR 4350	45.0 *	MAX.	3288	92%
	43.0		3123	88%
	41.0		2958	83%
IMR 4831	47.5 *	MAX.	3300	97%
Most Accurate	45.5		3140	93%
Powder Tested	43.5		2980	89%
Viht N160	47.0	MAX.	3319	100%
	45.0		3184	96%
	43.0 *		3050	92%
RL19	49.0 *	MAX.	3370 **	102%
	47.0		3200	98%
	45.0		3030	94%

BC=Ballistic Coefficient SD=Sectional Density
*Most Accurate Load Tested **Compressed Load

Use Maximum Loads with Caution
Refer to page 73 for additional safety information

6mm Remington - 95/100 grain

Bullet		TESTED O.A.C.L.	B.C.	S.D.
MAXIMUM S.A.A.M.I. O.A.C.L.				2.825"
Ballistic Tip®	95gr. Spitzer	2.770"	0.379	0.230
CT® Ballistic Silvertip®	95gr. Spitzer	2.770"	0.379	0.230
Partition®	95gr. Spitzer	2.770"	0.365	0.230
Partition®	100gr. Spitzer	2.770"	0.384	0.242

CASE TYPE:	Remington		PRIMER TYPE	Fed 210M
CASE HOLDS:	51.8	Gr. WATER	BARREL Length/Make	24" Shilen
			BARREL Twist	1-10"

POWDER TYPE	POWDER CHG. GRS.		MUZZLE VEL. F.P.S.	LOAD DENSITY (VOLUME)
H380	37.0	MAX.	2858	75%
	35.0		2714	71%
	33.0 *		2560	67%
IMR 4895	36.5 *	MAX.	2933	77%
	34.5		2803	73%
	32.5		2667	69%
IMR 3031	36.0	MAX.	2981	79%
	34.0		2864	74%
	32.0 *		2726	70%
Viht N160	44.0	MAX.	3048	94%
	42.0		2913	90%
	40.0 *		2779	86%
IMR 4350	43.0 *	MAX.	3071	88%
	41.0		2947	84%
	39.0		2833	80%
Big Game	43.0	MAX.	3096	87%
	41.0		2993	83%
	39.0 *		2888	79%
IMR 7828	49.0	MAX.	3140	** 101%
	47.0		2984	97%
	45.0 *		2821	92%
RL19 Most Accurate Powder Tested	46.0	MAX.	3159	97%
	44.0		3004	92%
	42.0 *		2856	88%
RL22	48.0 *	MAX.	3261	** 101%
	46.0		3100	97%
	44.0		2942	92%

BC=Ballistic Coefficient SD=Sectional Density
*Most Accurate Load Tested **Compressed Load

Use Maximum Loads with Caution
Refer to page 73 for additional safety information

Gary Lewis

243 WINCHESTER SUPER SHORT MAGNUM

There has long been a place in my heart for the .243 and the 6mm calibers. It was with a 243 that I tagged my first mule deer buck and what is still my biggest blacktail to date. And it was with a 243 that I have had the pleasure to watch several young people bag their first deer. Yes, we handloaded those bullets and yes, they were Nosler Partitions, the 100-grain Spitzers and, more often, the 95-grainSpitzers.

We think of this class of cartridges as good guns with which to start new hunters, but there is a place in the rifle rack and, indeed, in the pickup, for a gun that can reach out hundreds of yards to tag a loping, over-confident coyote.

Once, a rancher of my acquaintance said he wanted to show me his favorite critter-gitter, I wasn't surprised when he pulled out a beat-up Ruger bolt-action chambered in 243. I imagine he has added to the score since then, but he was well up in the triple digits with confirmed properly managed coyotes when last I talked to him.

Did I think the 243 Winchester could be improved upon? It never entered my mind. Fortunately, the wheels were in motion to put more fps behind my favorite bullets.

With a shortened 300 WSM, necked down to .243, the 243 WSSM was born and hunters begin to realize 4,000 feet per second and beyond with 55-grain Ballistic Tips. Where a 243 Winchester is a good choice for pronghorn, blacktails, whitetails and other thin-skinned, light-boned big game, the 243 WSSM is all that and more.

To my way of thinking, the 243s and the 6mms, with their relatively light weights, high accuracy potential and minimal recoil come into their own on hunts for coyote, bobcat and predators up to and including mountain lion. If the gun is doing double duty for big game and predators, my choice is either the 90-grain white-tipped AccuBond or the 95-grain Partition.

If this is strictly a coyote gun, I would opt for purple-tipped Ballistic Tip weighing in at 55, 70 or 90 grains. For government-mandated lead-free environments, the 55-grain Nosler BT Lead-Free or 90-grain E-Tip would get the nod.

Gary A. Lewis

Gary, with John Nosler, is the author of John Nosler – Going Ballistic. Lewis makes his home in Bend, Oregon.

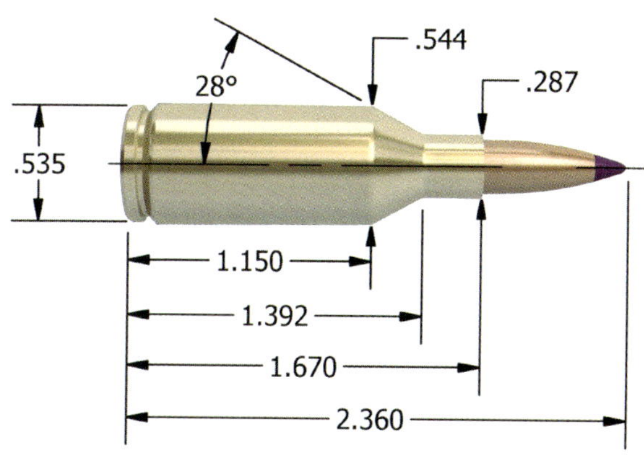

Maximum S.A.A.M.I. Overall Cartridge Length: 2.360"

BULLET CHOICES FOR THE 243 WINCHESTER SUPER SHORT MAGNUM

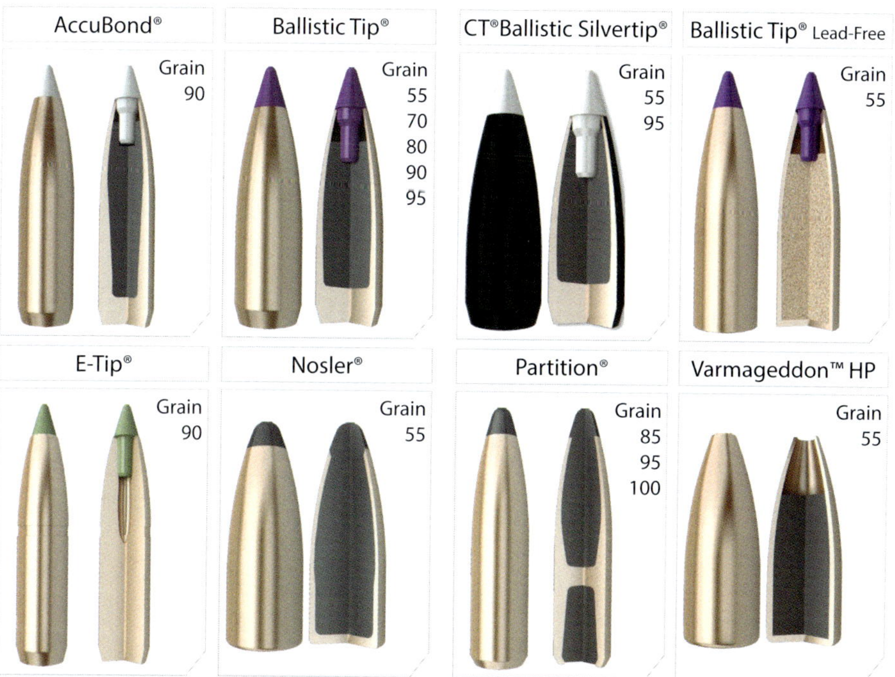

Varmageddon™

TECHNICAL INFORMATION

The 243 WSSM is a fairly new cartridge, using a shortened 300 Winchester Short Magnum case necked down to accept a .243"/6mm diameter bullet. Due to the extremely high velocities achievable with this cartridge, shortened barrel life can result depending on your load. Loaded with heavier bullets and/or at lower velocities, barrel life can be very acceptable if your barrel is well cared for. These concerns are no different than for other high-velocity cartridges.

243 WSSM - 55 grain

		MAXIMUM S.A.A.M.I. O.A.C.L.		2.360"
		TESTED O.A.C.L.	B.C.	S.D.
Ballistic Tip®	55gr. Spitzer	2.260"	0.276	0.133
Ballistic Tip® Lead Free	55gr. Spitzer	2.260"	0.288	0.133
CT® Ballistic Silvertip®	55gr. Spitzer	2.260"	0.276	0.133
Nosler®	55gr. FBSP	2.055"	0.203	0.133
Varmageddon™	55gr. FBHP	2.075"	0.192	0.133
Varmageddon™	55gr. FB Tipped	2.160"	0.252	0.133

CASE TYPE:	Winchester		PRIMER TYPE	WLR
CASE HOLDS:	52.3	Gr. WATER	BARREL Length/Make	24" Wiseman
			BARREL Twist	1-10"

POWDER TYPE	POWDER CHG. GRS.		MUZZLE VEL. F.P.S.		LOAD DENSITY (VOLUME)
Viht N150	48.0	MAX.	4082	**	108%
	46.0 *		3939	**	103%
	44.0		3830		99%
H380	51.5	MAX.	4109	**	104%
	49.5		4012		100%
	47.5 *		3889		96%
Varget	47.0 *	MAX.	4125		97%
	45.0		3980		93%
	43.0		3851		89%
A-2495	46.0	MAX.	4147		100%
	44.0		3982		96%
	42.0 *		3833		91%
H414 Most Accurate Powder Tested	53.0 *	MAX.	4150	**	106%
	51.0		4025	**	102%
	49.0		3826		98%
RL15	47.0	MAX.	4161		94%
	45.0 *		4002		90%
	43.0		3828		86%
W760	53.0	MAX.	4162	**	106%
	51.0		4042	**	102%
	49.0 *		3915		98%
H4895	46.0	MAX.	4165		97%
	44.0 *		4020		92%
	42.0		3878		88%
IMR 4895	46.0	MAX.	4196		97%
	44.0 *		4027		92%
	42.0		3880		88%
BL-C2	49.5 *	MAX.	4198		96%
	47.5		4028		92%
	45.5		3900		88%

BC=Ballistic Coefficient SD=Sectional Density
*Most Accurate Load Tested **Compressed Load

Use Maximum Loads with Caution
Refer to page 73 for additional safety information

243 WSSM - 70 grain		MAXIMUM S.A.A.M.I. O.A.C.L.		2.360"
		TESTED O.A.C.L.	B.C.	S.D.
Ballistic Tip®	70gr. Spitzer	2.260"	0.310	0.169
Varmageddon™	70gr. FB Tipped	2.060"	0.305	0.169

CASE TYPE:	Winchester		PRIMER TYPE	WLR
CASE HOLDS:	51.0	Gr. WATER	BARREL Length/Make	24" Wiseman
			BARREL Twist	1-10"

POWDER TYPE	POWDER CHG. GRS.	MUZZLE VEL. F.P.S.	LOAD DENSITY (VOLUME)
A-2495	42.0 * MAX.	3678	94%
	40.0	3524	89%
	38.0	3396	85%
RL15	43.5 * MAX.	3724	89%
	41.5	3596	85%
	39.5	3473	81%
IMR 4350	49.0 MAX.	3739	** 102%
	47.0 *	3611	98%
	45.0	3448	93%
Varget	43.5 MAX.	3744	92%
	41.5 *	3614	88%
	39.5	3480	84%
H4895	42.5 MAX.	3756	92%
	40.5	3613	87%
	38.5 *	3488	83%
H380	48.5 MAX.	3771	100%
	46.5 *	3637	96%
	44.5	3518	92%
H414	49.5 MAX.	3772	** 101%
	47.5 *	3647	97%
	45.5	3519	93%
H4350	49.0 * MAX.	3785	** 102%
Most Accurate	47.0	3672	98%
Powder Tested	45.0	3538	93%
Viht N160	52.0 * MAX.	3803	** 113%
	50.0	3672	** 109%
	48.0	3547	** 105%
W760	49.5 MAX.	3805	** 102%
	47.5 *	3666	98%
	45.5	3530	93%

243 WSSM - 80 grain		MAXIMUM S.A.A.M.I. O.A.C.L.		2.360"
		TESTED O.A.C.L.	B.C.	S.D.
Ballistic Tip®	80gr. Spitzer	2.260"	0.329	0.194

CASE TYPE:	Winchester		PRIMER TYPE	WLR
CASE HOLDS:	50.5	Gr. WATER	BARREL Length/Make	24" Wiseman
			BARREL Twist	1-10"

POWDER TYPE	POWDER CHG. GRS.	MUZZLE VEL. F.P.S.	LOAD DENSITY (VOLUME)
RL15	41.0 MAX.	3460	85%
	39.0	3295	80%
	37.0 *	3186	76%
Varget	41.0 MAX.	3465	88%
	39.0	3319	83%
	37.0 *	3195	79%
H4895 Most Accurate Powder Tested	41.0 MAX.	3509	89%
	39.0 *	3389	85%
	37.0	3248	81%
IMR 4350	47.5 MAX.	3527	100%
	45.5	3398	95%
	43.5 *	3258	91%
H380	46.5 MAX.	3549	97%
	44.5 *	3412	93%
	42.5	3305	89%
H414	47.5 MAX.	3550	98%
	45.5	3429	94%
	43.5 *	3325	90%
Viht N160	49.0 MAX.	3560	** 108%
	47.0 *	3432	** 103%
	45.0	3314	99%
W760	47.5 * MAX.	3578	98%
	45.5	3439	94%
	43.5	3339	90%
H4350	47.5 MAX.	3579	100%
	45.5	3458	95%
	43.5 *	3351	91%

BC=Ballistic Coefficient SD=Sectional Density
*Most Accurate Load Tested **Compressed Load

Use Maximum Loads with Caution
Refer to page 73 for additional safety information

243 WSSM - 85/90 grain		MAXIMUM S.A.A.M.I. O.A.C.L.		2.360"
		TESTED O.A.C.L.	B.C.	S.D.
Partition®	85gr. Spitzer	2.260"	0.315	0.206
AccuBond®	90gr. Spitzer	2.260"	0.376	0.218
Ballistic Tip®	90gr. Spitzer	2.260"	0.365	0.218
E-Tip®	90gr. Spitzer	2.260"	0.403	0.218

Due to internal construction differences, always begin with starting loads when using E-Tip® products.

CASE TYPE:	Winchester		PRIMER TYPE	WLR
CASE HOLDS:	49.6	Gr. WATER	BARREL Length/Make	24" Wiseman
			BARREL Twist	1-10"

POWDER TYPE	POWDER CHG. GRS.		MUZZLE VEL. F.P.S.		LOAD DENSITY (VOLUME)
H4895	38.5	MAX.	3242		85%
	36.5		3129		81%
	34.5 *		3000		76%
RL15	39.5	MAX.	3255		83%
	37.5		3146		79%
	35.5 *		3027		75%
Varget	39.5	MAX.	3258		86%
	37.5 *		3142		82%
	35.5		3013		77%
H4831SC	47.5	MAX.	3260		100%
	45.5		3139		96%
	43.5 *		3031		91%
IMR 4350	45.0	MAX.	3320		97%
	43.0		3205		92%
	41.0 *		3071		88%
H4350	44.0 *	MAX.	3332		94%
	42.0		3215		90%
	40.0		3093		85%
Viht N160	47.0	MAX.	3341	**	105%
Most Accurate	45.0 *		3227	**	101%
Powder Tested	43.0		3124		96%
H414	45.5 *	MAX.	3354		96%
	43.5		3235		91%
	41.5		3109		87%
W760	45.0	MAX.	3358		95%
	43.0		3240		91%
	41.0 *		3124		87%

BC=Ballistic Coefficient SD=Sectional Density
*Most Accurate Load Tested **Compressed Load

Use Maximum Loads with Caution
Refer to page 73 for additional safety information

243 WSSM - 95/100 gr		MAXIMUM S.A.A.M.I. O.A.C.L.		2.360"
		TESTED O.A.C.L.	B.C.	S.D.
Ballistic Tip®	95gr. Spitzer	2.260"	0.379	0.230
CT® Ballistic Silvertip®	95gr. Spitzer	2.260"	0.379	0.230
Partition®	95gr. Spitzer	2.260"	0.365	0.230
Partition®	100gr. Spitzer	2.260"	0.384	0.242

CASE TYPE:	Winchester		PRIMER TYPE	WLR
CASE HOLDS:	49.5	Gr. WATER	BARREL Length/Make	24" Wiseman
			BARREL Twist	1-10"

POWDER TYPE	POWDER CHG. GRS.		MUZZLE VEL. F.P.S.	LOAD DENSITY (VOLUME)
Varget	36.5	MAX.	3007	80%
	34.5		2889	75%
	32.5 *		2775	71%
RL15 Most Accurate Powder Tested	37.0	MAX.	3010	78%
	35.0		2919	74%
	33.0 *		2796	69%
H4831SC	44.5	MAX.	3065	94%
	42.5 *		2966	89%
	40.5		2877	85%
IMR 4831	44.0	MAX.	3131	95%
	42.0		3003	90%
	40.0 *		2896	86%
H414	42.5 *	MAX.	3132	89%
	40.5		3019	85%
	38.5		2908	81%
W760	43.0	MAX.	3143	91%
	41.0 *		3031	87%
	39.0		2907	83%
H4350	42.0 *	MAX.	3148	90%
	40.0		3048	86%
	38.0		2940	81%
IMR 4350	44.0	MAX.	3177	94%
	42.0		3032	90%
	40.0 *		2914	86%
Viht N165	50.0	MAX.	3222	** 112%
	48.0 *		3126	** 108%
	46.0		2984	** 103%

BC=Ballistic Coefficient SD=Sectional Density
*Most Accurate Load Tested **Compressed Load

Use Maximum Loads with Caution
Refer to page 73 for additional safety information

Greg Rodriguez

240 WEATHERBY MAGNUM

Our fastest commercial 6mm never interested me until a hunting client showed up at the ranch with one. I didn't guide him, but I had the pleasure of watching him from a hilltop as he stalked a big buck in a pear flat late in the day. I enjoyed watching him slither into range, and I had my binocular glued on the big ten pointer when a massive shockwave pulsed out of its shoulder. A millisecond later, two things happened: the buck dropped in its tracks, and I was sold on the 240 Weatherby.

I've owned several 240 Weatherby rifles over the years, but the Weatherby Mark V Ultra Lightweight I bought a few years ago is so accurate I'll never part with it. That incredible accuracy and its flat trajectory have combined to help me pile up an impressive amount of predators and big game at some pretty long ranges, including a coyote at 515 yards. I'd studied its ballistics enough that its long range capabilities didn't surprise me, but I was surprised by its ability to drop big-bodied deer and hogs in their tracks.

I've shot a dozen whitetails and a bunch of feral hogs with the 240 Weatherby. Our Texas deer aren't very big, but only one of those deer managed more than a few steps after the shot. It drops big pigs like the 250-pound bruiser I took with mine last year just as easily. My bullet took that porker right on the shoulder and dropped it where it stood. It didn't exit, but the bullet liquefied its heart and destroyed its lungs.

Its performance on pigs prompted me to test the 240 Weatherby on an elk that had moved onto the ranch from a nearby exotic ranch. That bull got on my bad side when it started destroying pricey deer feeders and fences. I found it standing in some thick brush, 80 yards away. It was almost chest-on, so I held right where the neck and shoulder meet and touched the trigger. The trouble-maker dropped in its tracks, deader than a hammer.

I would never call it an elk rifle, but it performed so well on that bull and on several big pigs that I wouldn't hesitate to ventilate any deer, hog or pronghorn on the planet with my flat-shooting 240 Weatherby.

Greg is the Shooting Editor at Shooting Times Magazine and a contributing editor at Petersen's Hunting and Guns & Ammo Magazines. He is also the host of A Rifleman's Journal, which airs on The Sportsman Channel.

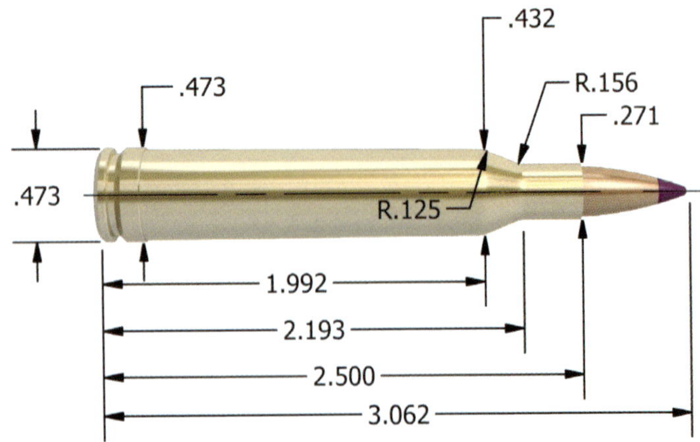

Maximum S.A.A.M.I. Overall Cartridge Length: 3.062"

BULLET CHOICES FOR THE 240 WEATHERBY MAGNUM

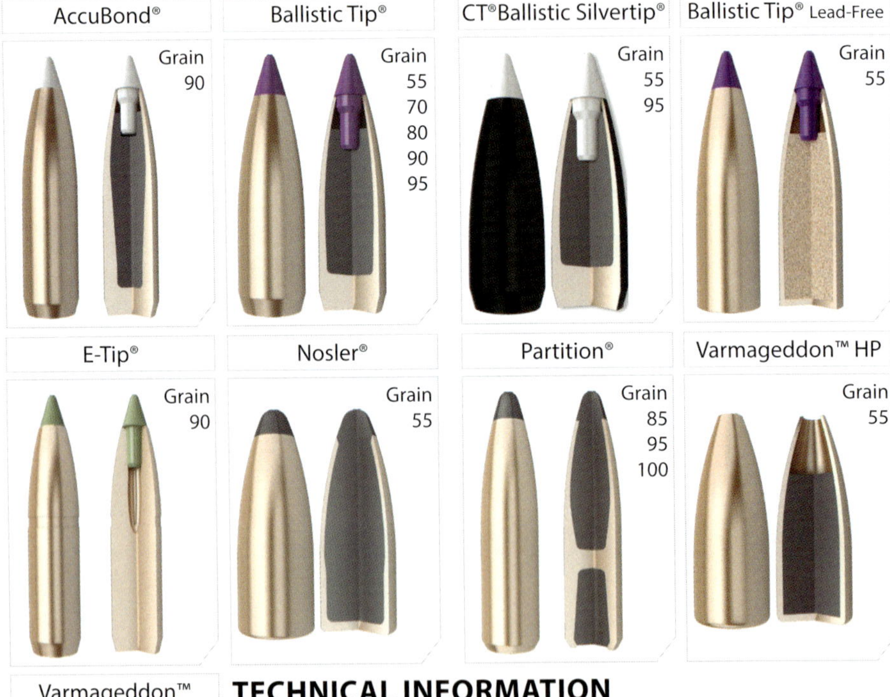

Varmageddon™

Grain
55
70

TECHNICAL INFORMATION

For technical information on the 240 Weatherby Magnum turn to page 202

240 Weatherby - 55 grain		MAXIMUM S.A.A.M.I. O.A.C.L.		3.062"
		TESTED O.A.C.L.	B.C.	S.D.
Ballistic Tip®	55gr. Spitzer	3.060"	0.276	0.133
Ballistic Tip® Lead Free	55gr. Spitzer	3.060"	0.288	0.133
CT® Ballistic Silvertip®	55gr. Spitzer	3.060"	0.276	0.133
Nosler®	55gr. FBSP	2.890"	0.203	0.133
Varmageddon™	55gr. FBHP	2.905"	0.192	0.133
Varmageddon™	55gr. FB Tipped	2.990"	0.252	0.133

CASE TYPE:	Weatherby		PRIMER TYPE	Fed 210M
CASE HOLDS:	63.5	Gr. WATER	BARREL Length/Make	24" Lilja
			BARREL Twist	1-10"

POWDER TYPE	POWDER CHG. GRS.		MUZZLE VEL. F.P.S.	LOAD DENSITY (VOLUME)
H380	52.5	MAX.	3786	87%
	50.5		3638	84%
	48.5 *		3542	80%
H414	55.0	MAX.	3866	90%
	53.0		3679	87%
	51.0 *		3512	84%
IMR 4350	57.0	MAX.	3965	95%
	55.0		3781	92%
	53.0 *		3617	88%
Viht N150	52.0	MAX.	4038	96%
	50.0		3891	93%
	48.0 *		3744	89%
A-2495	49.5	MAX.	4057	89%
	47.5		3978	85%
	45.5 *		3771	81%
IMR 4064	50.0 *	MAX.	4065	87%
Most Accurate	48.0		3839	83%
Powder Tested	46.0		3612	80%
Viht N540	52.5	MAX.	4092	92%
	50.5		3911	88%
	48.5 *		3727	85%
Big Game	54.0	MAX.	4092	90%
	52.0 *		3981	86%
	50.0		3835	83%
Varget	51.5	MAX.	4095	88%
	49.5 *		3954	84%
	47.5		3753	81%
RL15	51.5	MAX.	4099	84%
	49.5		3928	81%
	47.5 *		3765	78%

BC=Ballistic Coefficient SD=Sectional Density
*Most Accurate Load Tested **Compressed Load

Use Maximum Loads with Caution
Refer to page 73 for additional safety information

240 Weatherby - 70 grain

		MAXIMUM S.A.A.M.I. O.A.C.L.		3.062"
		TESTED O.A.C.L.	B.C.	S.D.
Ballistic Tip®	70gr. Spitzer	3.060"	0.310	0.169
Varmageddon™	70gr. FB Tipped	3.060"	0.305	0.169

CASE TYPE:	Weatherby		PRIMER TYPE	Fed 210M
CASE HOLDS:	61.4	Gr. WATER	BARREL Length/Make	24" Lilja
			BARREL Twist	1-10"

POWDER TYPE	POWDER CHG. GRS.		MUZZLE VEL. F.P.S.	LOAD DENSITY (VOLUME)
Viht N150	46.0 *	MAX.	3542	88%
	44.0		3412	84%
	42.0		3284	80%
H4831SC	56.0 *	MAX.	3595	95%
	54.0		3435	92%
	52.0		3310	88%
RL19	55.0	MAX.	3642	97%
	53.0		3479	94%
	51.0 *		3317	90%
H380 Most Accurate Powder Tested	51.5	MAX.	3663	88%
	49.5		3595	85%
	47.5 *		3480	81%
IMR 4831	55.0 *	MAX.	3692	95%
	53.0		3517	92%
	51.0		3342	88%
IMR 4350	53.0 *	MAX.	3718	91%
	51.0		3543	88%
	49.0		3368	84%
H414	53.0 *	MAX.	3750	90%
	51.0		3600	87%
	49.0		3450	83%
W760	53.0	MAX.	3750	90%
	51.0 *		3655	87%
	49.0		3555	84%
Hunter	57.0	MAX.	3772	98%
	55.0		3689	94%
	53.0 *		3642	91%

BC=Ballistic Coefficient SD=Sectional Density
*Most Accurate Load Tested **Compressed Load

Use Maximum Loads with Caution
Refer to page 73 for additional safety information

240 Weatherby - 80 grain			MAXIMUM S.A.A.M.I. O.A.C.L.		3.062"
			TESTED O.A.C.L.	B.C.	S.D.
Ballistic Tip®		80gr. Spitzer	3.060"	0.329	0.194

CASE TYPE:	Weatherby		PRIMER TYPE	Fed 210M
CASE HOLDS:	60.1	Gr. WATER	BARREL Length/Make	24" Lilja
			BARREL Twist	1-10"

POWDER TYPE	POWDER CHG. GRS.		MUZZLE VEL. F.P.S.	LOAD DENSITY (VOLUME)
RL19	54.0	MAX.	3415	98%
	52.0		3266	94%
	50.0	*	3074	90%
Viht N150	47.0	MAX.	3419	92%
	45.0		3315	88%
	43.0	*	3189	84%
Hunter	54.0	MAX.	3519	95%
	52.0		3485	91%
	50.0	*	3400	88%
H4831SC Most Accurate Powder Tested	56.0	MAX.	3532	97%
	54.0	*	3403	94%
	52.0		3275	90%
Viht N160	53.5	MAX.	3534	99%
	51.5		3431	95%
	49.5	*	3273	92%
W760	51.0	MAX.	3534	89%
	49.0	*	3452	85%
	47.0		3338	82%
H414	51.5	MAX.	3560	89%
	49.5		3438	86%
	47.5	*	3341	82%
IMR 4350	52.5	MAX.	3562	92%
	50.5	*	3403	89%
	48.5		3230	85%
IMR 4831	54.0	MAX.	3579	96%
	52.0		3440	92%
	50.0	*	3272	89%

BC=Ballistic Coefficient SD=Sectional Density
*Most Accurate Load Tested **Compressed Load

Use Maximum Loads with Caution
Refer to page 73 for additional safety information

240 Weatherby - 85/90 grain

		MAXIMUM S.A.A.M.I. O.A.C.L.		3.062"
		TESTED O.A.C.L.	B.C.	S.D.
Partition®	85gr. Spitzer	3.060"	0.315	0.206
AccuBond®	90gr. Spitzer	3.060"	0.376	0.218
Ballistic Tip®	90gr. Spitzer	3.060"	0.365	0.218
E-Tip®	90gr. Spitzer	3.060"	0.403	0.218

Due to internal construction differences, always begin with starting loads when using E-Tip® products.

CASE TYPE:	Weatherby		PRIMER TYPE	Fed 210M
CASE HOLDS:	60.4	Gr. WATER	BARREL Length/Make	24" Lilja
			BARREL Twist	1-10"

POWDER TYPE	POWDER CHG. GRS.		MUZZLE VEL. F.P.S.	LOAD DENSITY (VOLUME)
H414	48.5	MAX.	3275	84%
	46.5		3204	80%
	44.5 *		3066	77%
W760	49.0	MAX.	3306	85%
	47.0		3249	81%
	45.0 *		3112	78%
Magnum	59.5	MAX.	3309	100%
	57.5 *		3205	97%
	55.5		3110	93%
H1000	58.5	MAX.	3333	** 101%
	56.5 *		3250	98%
	54.5		3152	94%
RL19	52.5	MAX.	3443	94%
	50.5		3308	91%
	48.5 *		3174	87%
Viht N160 Most Accurate Powder Tested	52.0	MAX.	3445	96%
	50.0		3313	92%
	48.0 *		3181	88%
IMR 4350	51.0	MAX.	3448	89%
	49.0		3303	86%
	47.0 *		3158	82%
H4831	54.0 *	MAX.	3453	93%
	52.0		3363	90%
	50.0		3273	86%
IMR 4831	53.0 *	MAX.	3488	93%
	51.0		3343	90%
	49.0		3198	86%

BC=Ballistic Coefficient SD=Sectional Density
*Most Accurate Load Tested **Compressed Load

Use Maximum Loads with Caution
Refer to page 73 for additional safety information

240 Weatherby - 95/100 grain		MAXIMUM S.A.A.M.I. O.A.C.L.		3.062"
		TESTED O.A.C.L.	B.C.	S.D.
Ballistic Tip®	95gr. Spitzer	3.060"	0.379	0.230
CT® Ballistic Silvertip®	95gr. Spitzer	3.060"	0.379	0.230
Partition®	95gr. Spitzer	3.060"	0.365	0.230
Partition®	100gr. Spitzer	3.060"	0.384	0.242

CASE TYPE:	Weatherby		PRIMER TYPE	Fed 210M
CASE HOLDS:	58.8	Gr. WATER	BARREL Length/Make	24" Lilja
			BARREL Twist	1-10"

POWDER TYPE	POWDER CHG. GRS.	MUZZLE VEL. F.P.S.	LOAD DENSITY (VOLUME)
IMR 4350	47.0 * MAX.	3070	85%
	45.0	2930	81%
	43.0	2790	77%
Viht N160	48.0 * MAX.	3160	91%
	46.0	3065	87%
	44.0	2970	83%
W760	47.0 MAX.	3180	84%
	45.0 *	3114	80%
	43.0	2940	77%
H1000	56.5 MAX.	3220	** 101%
	54.5	3136	97%
	52.5 *	3019	93%
RL19	51.0 MAX.	3242	94%
	49.0	3140	91%
	47.0 *	3038	87%
H4831SC	52.0 * MAX.	3247	92%
	50.0	3152	89%
	48.0	3077	85%
Magnum	61.0 MAX.	3247	** 105%
Most Accurate	59.0 *	3195	** 102%
Powder Tested	57.0	3108	98%
IMR 7828	53.5 * MAX.	3260	97%
	51.5	3130	93%
	49.5	3000	90%
RL22	53.0 * MAX.	3352	98%
	51.0	3217	94%
	49.0	3082	91%

BC=Ballistic Coefficient SD=Sectional Density
*Most Accurate Load Tested **Compressed Load

Use Maximum Loads with Caution
Refer to page 73 for additional safety information

TECHNICAL INFORMATION

As with most Weatherby Magnums, there are a few points to keep in mind when reloading this cartridge:

• The chambers have "freebore," meaning they have a longer throat. For this reason, it is generally not possible to seat a bullet close to, or in contact with the lands (rifling).

• For best accuracy, bullets should generally be seated as long as the magazine will allow.

• This is a belted case, but the case head is the same size as that of a standard .30-06 Springfield.

• As with all belted magnums, we strongly recommend three-shot groups for accuracy testing

• We have found little difference in performance between "standard" or "magnum" primers in this particular cartridge.

250-3000 SAVAGE

I was 9, well really 9½, when I was invited to go on my first real hunting trip with my dad. Up to this point in my life, hunting with Dad meant a series of short, day hunts around our home in Ashland, Oregon—but, this time was going to be different. This time we would be camping near the top of the legendary Steens Mountain, one of Oregon's most productive mule deer hunting areas of the 1950's. While I had been tagging along from the time I was old enough to walk, this hunt was different—I had my very first real rifle, a 250-3000 Savage 99E. Dad had cut the stock to fit me, and mounted a little 2 ¾ power, Bear Cub on the receiver. No longer did I have to carry a Daisy BB gun on my hunts with Dad. I was excited and ready! For the next few years, that little rifle and I piled-up a truck load of mule deer.

The 250 Savage (also known as the 250-3000 Savage) was designed by Charles Newton and introduced in 1915. At the time, it was the undisputed velocity king and, when loaded with an 87-grain bullet, the little cartridge was the first to break the 3,000 fps barrier.

The original Savage rifles were fitted with a 1-14" twist barrel, which was just right for the stubby, round nose bullets of the day. Fast-forward to the 1970's and 1980's, and the slow twist rate would become a real problem for modern, high ballistic coefficient bullets.
Thompson Center is one of many new-generation companies to recognize the advantages of a faster twist, and offer their 250 Savage Contender barrel with a 1-10" twist rate. This gives new life to the cartridge, and allows for reliable stability for everything up to and including the 115-grain Ballistic Tip and the 120-grain Partition.

As a hunting cartridge, the 250 Savage is a good choice for deer-sized game. However, deer should be considered the upper limit for everyone except the more accomplished hunters. While a well placed shot will put an elk in the freezer, the Sunday Hunter may need a little more knock-down power than the old Savage can deliver.

"9 year old Bob Nosler shot his first buck using Nosler Partition 100 grain bullet in his 250 Savage."

Bob Nosler

Bob Nosler is the CEO/President of Nosler, Inc.

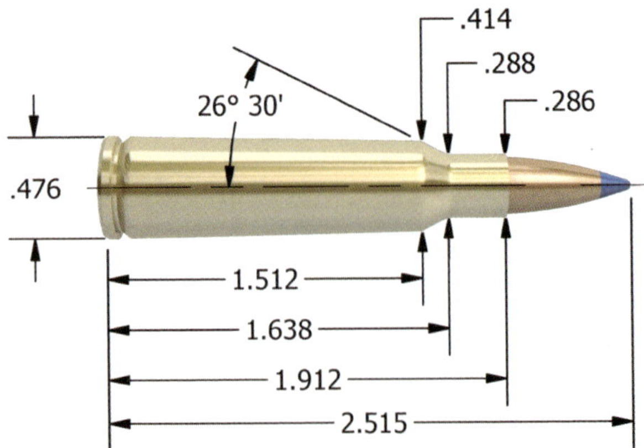

Maximum S.A.A.M.I. Overall Cartridge Length: 2.515"

BULLET CHOICES FOR THE 250-3000 SAVAGE

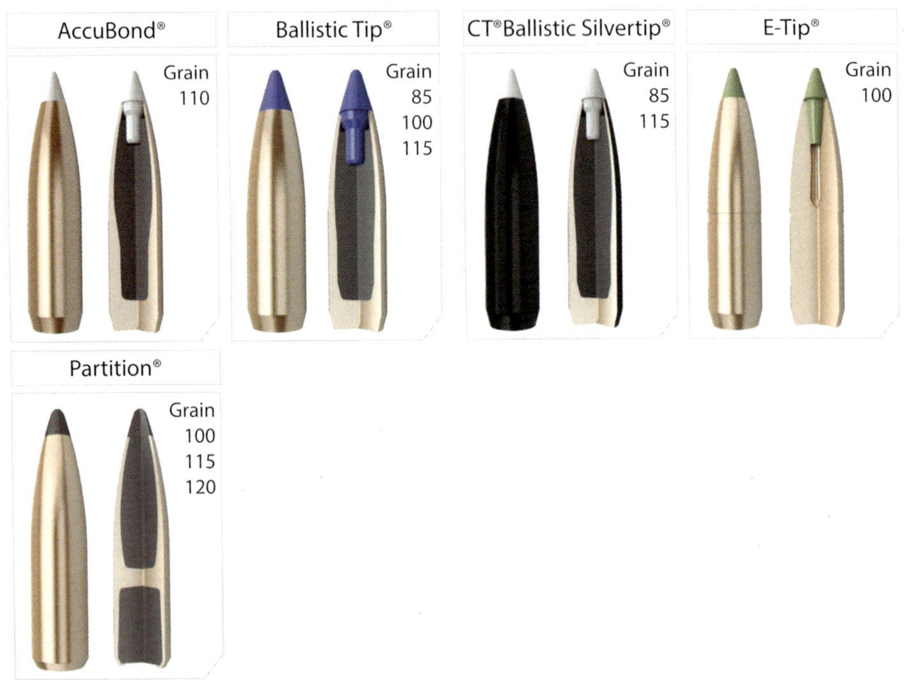

TECHNICAL INFORMATION

The Savage Model 99 was the original rifle chambered for the 250 Savage. Many older Model 99' s have a 1-14" twist barrel, and will not stabilize any of our .25 caliber bullets. If your rifle has the new, standard 1-10" twist barrel, it will stabilize all Nosler .25 caliber bullets. The 85-grain Ballistic Tip is a good choice for varmints, but for deer we recommend one of our heavier, 100-grain bullets. As always, when dealing with a lever action rifle, full-length case resizing will yield the best results.

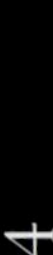

250 Savage - 85 Grain

		MAXIMUM S.A.A.M.I. O.A.C.L.		2.515"
		TESTED O.A.C.L.	B.C.	S.D.
Ballistic Tip®	85gr. Spitzer	2.515"	0.329	0.183
CT® Ballistic Silvertip®	85gr. Spitzer	2.515"	0.329	0.183

CASE TYPE:	Winchester		PRIMER TYPE	CCI BR-2
CASE HOLDS:	42.7	Gr. WATER	BARREL Length/Make	22" Hart
			BARREL Twist	1-10"

POWDER TYPE	POWDER CHG. GRS.		MUZZLE VEL. F.P.S.		LOAD DENSITY (VOLUME)
W760	39.0	* MAX.	2900		96%
	37.0		2790		91%
	35.0		2680		86%
W748	36.0	* MAX.	2932		88%
	34.0		2797		83%
	32.0		2662		78%
A-2520	35.0	MAX.	2980		85%
	33.0		2840		80%
	31.0	*	2700		75%
H414	41.0	* MAX.	3000		100%
	39.0		2860		95%
	37.0		2720		90%
H380	41.0	* MAX.	3002	**	101%
	39.0		2867		96%
	37.0		2732		91%
IMR 4895	35.0	MAX.	3018		90%
	33.0		2853		85%
	31.0	*	2688		80%
RL15	36.0	MAX.	3031		88%
	34.0	*	2847		83%
	32.0		2691		78%
Varget	36.0	MAX.	3036		91%
Most Accurate	34.0		2900		86%
Powder Tested	32.0	*	2700		81%
Viht N150	36.5	* MAX.	3057	**	101%
	34.5		2909		95%
	32.5		2742		90%
Big Game	40.0	MAX.	3089		99%
	38.0		2950		94%
	36.0	*	2802		89%

BC=Ballistic Coefficient SD=Sectional Density
*Most Accurate Load Tested **Compressed Load

Use Maximum Loads with Caution
Refer to page 73 for additional safety information

250 Savage - 100 Grain		MAXIMUM S.A.A.M.I. O.A.C.L.		2.515"
		TESTED O.A.C.L.	B.C.	S.D.
Ballistic Tip®	100gr. Spitzer	2.515"	0.393	0.216
E-Tip®	100gr. Spitzer	2.515"	0.409	0.216
Due to internal construction differences, always begin with starting loads when using E-Tip® products.				
Partition®	100gr. Spitzer	2.515"	0.377	0.216

CASE TYPE:	Winchester		PRIMER TYPE	CCI BR-2
CASE HOLDS:	41.2	Gr. WATER	BARREL Length/Make	22" Hart
			BARREL Twist	1-10"

POWDER TYPE	POWDER CHG. GRS.		MUZZLE VEL. F.P.S.	LOAD DENSITY (VOLUME)
IMR 4350	38.0 *	MAX.	2700	98%
	36.0		2540	92%
	34.0		2380	87%
A-2520	34.0	MAX.	2778	85%
	32.0		2643	80%
	30.0 *		2508	75%
Viht N150 Most Accurate Powder Tested	34.5	MAX.	2783	99%
	32.5 *		2660	93%
	30.5		2516	87%
IMR 4320	35.0	MAX.	2788	91%
	33.0		2663	86%
	31.0 *		2538	81%
RL19	41.0	MAX.	2798	** 108%
	39.0		2655	** 103%
	37.0 *		2497	98%
H380	38.5 *	MAX.	2798	98%
	36.5		2673	93%
	34.5		2548	88%
Varget	34.0	MAX.	2808	89%
	32.0 *		2673	84%
	30.0		2525	79%
W760	38.0	MAX.	2830	97%
	36.0		2690	91%
	34.0 *		2550	86%
Hunter	43.5	MAX.	2911	** 111%
	41.5		2834	** 106%
	39.5 *		2643	** 101%
H4895	35.0 *	MAX.	2952	93%
	33.0		2837	88%
	31.0		2722	83%

BC=Ballistic Coefficient SD=Sectional Density
*Most Accurate Load Tested **Compressed Load

Use Maximum Loads with Caution
Refer to page 73 for additional safety information

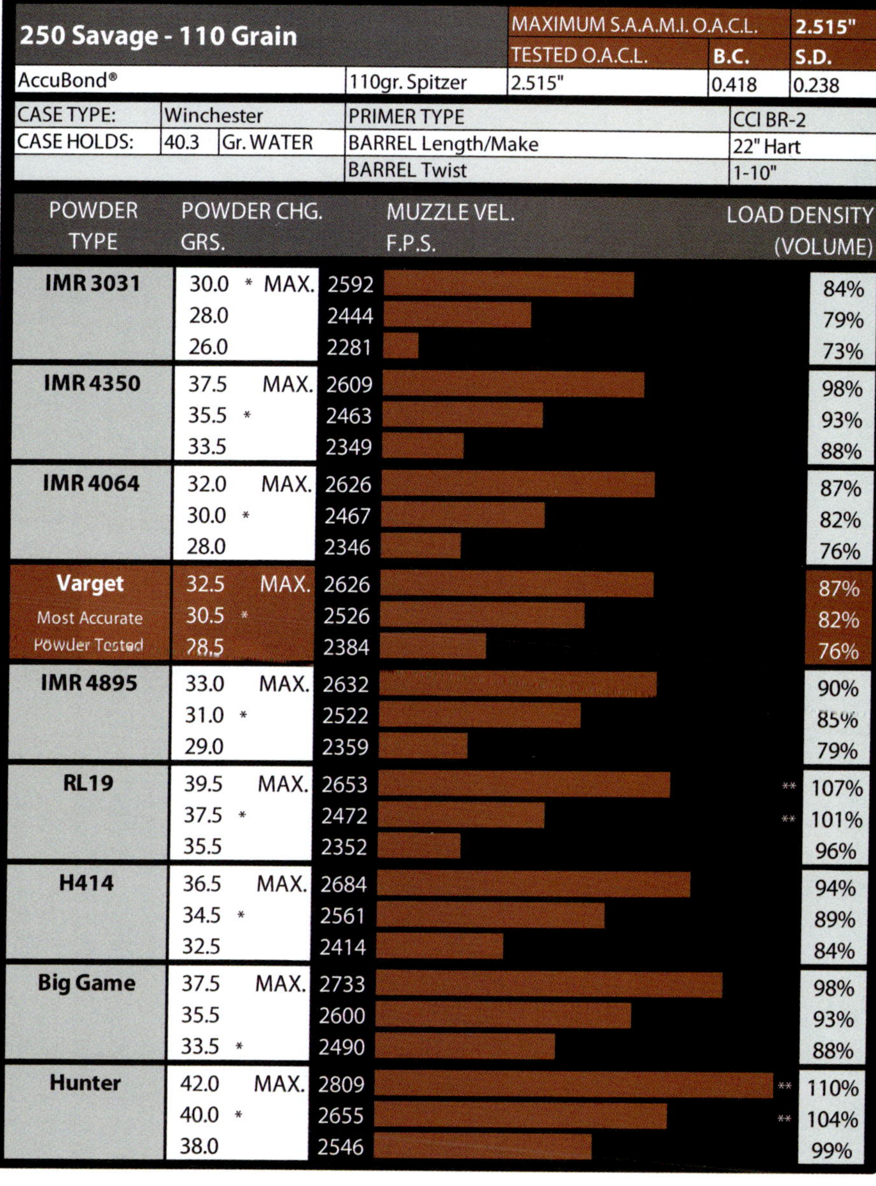

250 Savage - 110 Grain

		MAXIMUM S.A.A.M.I. O.A.C.L.		2.515"
		TESTED O.A.C.L.	B.C.	S.D.
AccuBond®	110gr. Spitzer	2.515"	0.418	0.238

CASE TYPE:	Winchester		PRIMER TYPE	CCI BR-2
CASE HOLDS:	40.3	Gr. WATER	BARREL Length/Make	22" Hart
			BARREL Twist	1-10"

POWDER TYPE	POWDER CHG. GRS.		MUZZLE VEL. F.P.S.		LOAD DENSITY (VOLUME)
IMR 3031	30.0 *	MAX.	2592		84%
	28.0		2444		79%
	26.0		2281		73%
IMR 4350	37.5	MAX.	2609		98%
	35.5 *		2463		93%
	33.5		2349		88%
IMR 4064	32.0	MAX.	2626		87%
	30.0 *		2467		82%
	28.0		2346		76%
Varget Most Accurate Powder Tested	32.5	MAX.	2626		87%
	30.5 *		2526		82%
	28.5		2384		76%
IMR 4895	33.0	MAX.	2632		90%
	31.0 *		2522		85%
	29.0		2359		79%
RL19	39.5	MAX.	2653	**	107%
	37.5 *		2472	**	101%
	35.5		2352		96%
H414	36.5	MAX.	2684		94%
	34.5 *		2561		89%
	32.5		2414		84%
Big Game	37.5	MAX.	2733		98%
	35.5		2600		93%
	33.5 *		2490		88%
Hunter	42.0	MAX.	2809	**	110%
	40.0 *		2655	**	104%
	38.0		2546		99%

BC=Ballistic Coefficient SD=Sectional Density
*Most Accurate Load Tested **Compressed Load

Use Maximum Loads with Caution
Refer to page 73 for additional safety information

250 Savage - 115 Grain		MAXIMUM S.A.A.M.I. O.A.C.L.		2.515"
		TESTED O.A.C.L.	B.C.	S.D.
Ballistic Tip®	115gr. Spitzer	2.515"	0.453	0.249
CT® Ballistic Silvertip®	115gr. Spitzer	2.515"	0.453	0.249
Partition®	115gr. Spitzer	2.515"	0.389	0.249

CASE TYPE:	Winchester		PRIMER TYPE	CCI BR-2
CASE HOLDS:	40.2	Gr. WATER	BARREL Length/Make	22" Hart
			BARREL Twist	1-10"

POWDER TYPE	POWDER CHG. GRS.		MUZZLE VEL. F.P.S.		LOAD DENSITY (VOLUME)
IMR 3031	30.0	MAX.	2484		84%
	28.0		2321		79%
	26.0 *		2157		73%
IMR 4064	32.0	MAX.	2509		87%
	30.0		2360		82%
	28.0 *		2211		77%
IMR 4895	31.0 *	MAX.	2510		85%
	29.0		2365		79%
	27.0		2220		74%
RL19	38.5	MAX.	2547	**	104%
	36.5		2428		99%
	34.5 *		2294		93%
IMR 4350	36.0 *	MAX.	2548		95%
	34.0		2376		89%
	32.0		2204		84%
IMR 4320 Most Accurate Powder Tested	33.0	MAX.	2584		88%
	31.0		2444		83%
	29.0 *		2304		78%
H414	36.0	MAX.	2615		93%
	34.0 *		2465		88%
	32.0		2345		83%
W760	37.0 *	MAX.	2627		96%
	35.0		2496		91%
	33.0		2365		86%
Hunter	39.5	MAX.	2654	**	103%
	37.5		2520		98%
	35.5 *		2422		93%

BC=Ballistic Coefficient SD=Sectional Density
*Most Accurate Load Tested **Compressed Load

Use Maximum Loads with Caution
Refer to page 73 for additional safety information

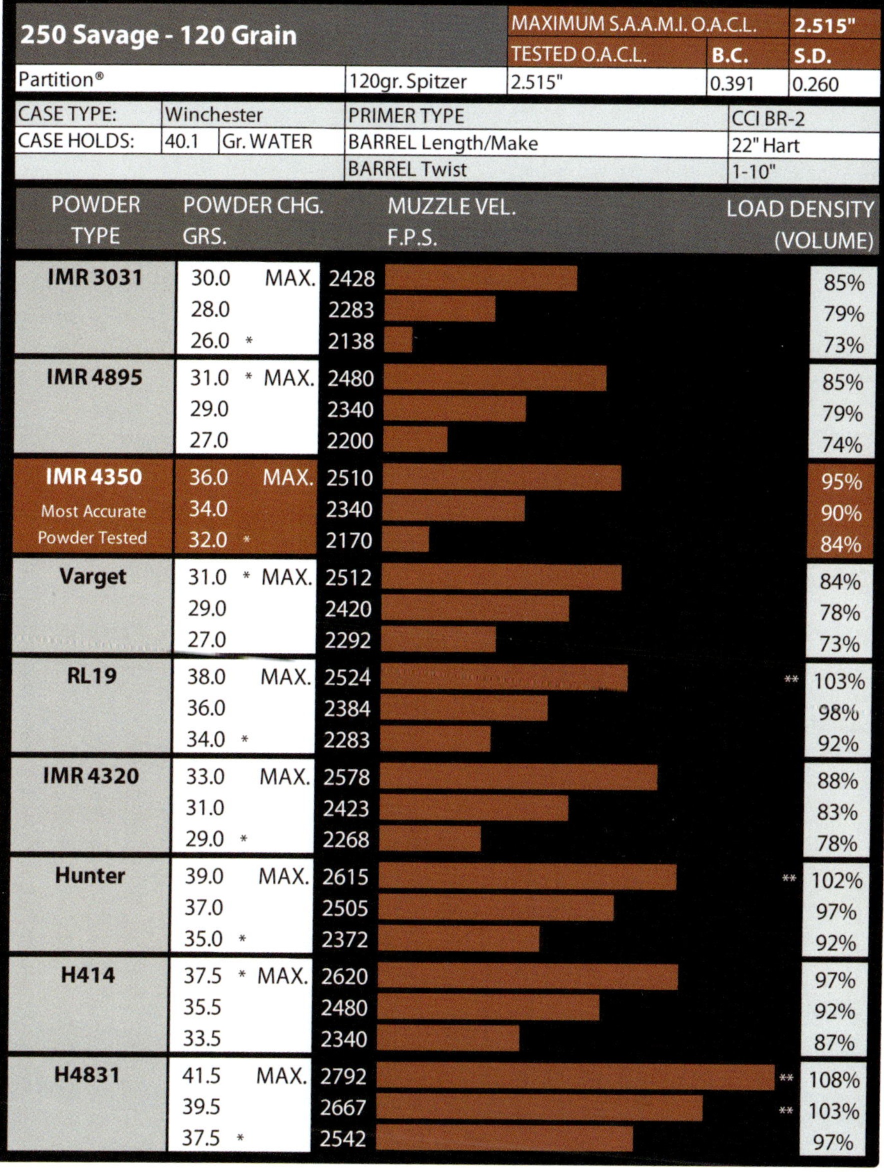

250 Savage - 120 Grain

		MAXIMUM S.A.A.M.I. O.A.C.L.		2.515"
		TESTED O.A.C.L.	B.C.	S.D.
Partition®	120gr. Spitzer	2.515"	0.391	0.260

CASE TYPE:	Winchester		PRIMER TYPE	CCI BR-2
CASE HOLDS:	40.1	Gr. WATER	BARREL Length/Make	22" Hart
			BARREL Twist	1-10"

POWDER TYPE	POWDER CHG. GRS.		MUZZLE VEL. F.P.S.		LOAD DENSITY (VOLUME)
IMR 3031	30.0	MAX.	2428		85%
	28.0		2283		79%
	26.0 *		2138		73%
IMR 4895	31.0 *	MAX.	2480		85%
	29.0		2340		79%
	27.0		2200		74%
IMR 4350 Most Accurate Powder Tested	36.0	MAX.	2510		95%
	34.0		2340		90%
	32.0 *		2170		84%
Varget	31.0 *	MAX.	2512		84%
	29.0		2420		78%
	27.0		2292		73%
RL19	38.0	MAX.	2524	**	103%
	36.0		2384		98%
	34.0 *		2283		92%
IMR 4320	33.0	MAX.	2578		88%
	31.0		2423		83%
	29.0 *		2268		78%
Hunter	39.0	MAX.	2615	**	102%
	37.0		2505		97%
	35.0 *		2372		92%
H414	37.5 *	MAX.	2620		97%
	35.5		2480		92%
	33.5		2340		87%
H4831	41.5	MAX.	2792	**	108%
	39.5		2667	**	103%
	37.5 *		2542		97%

BC=Ballistic Coefficient SD=Sectional Density
*Most Accurate Load Tested **Compressed Load

Use Maximum Loads with Caution
Refer to page 73 for additional safety information

Dave Anderson

257 ROBERTS

The 257 Roberts should be among the most popular cartridges in America. Its power and trajectory are ideal for deer, antelope, black bear, and an occasional varmint hunt for 'chucks or coyotes.

It has a mild report, long barrel life, and moderate recoil. Light recoil makes a rifle more pleasant to shoot, meaning more practice, more confidence, and better shot placement. This is not exactly breaking news. Author James Fenimore Cooper made the point in a novel written nearly two centuries ago.

His character "Hawkeye" says to his Mohican companion, "You are wasteful of your powder, and the kick of the rifle disconcerts your aim! Little powder, light lead, and a long arm…you waste the kernels by over-charging, and a kicking rifle never carries a true bullet."

Try and convince a new hunter. "What about elk and moose? What about the big bears? More power! I need more power!" So they get a 300 magnum, or a light 30-06, or worse yet a light 300 magnum. They don't shoot nearly enough or well enough because the darn rifle is loud, expensive, and unpleasant to shoot.

My own Roberts is one I put together around 1980, using a 24" Douglas Premium barrel on a pre-'64 Winchester 70 action. I've used it mainly on whitetails, plus a few mule deer, antelope, and a couple of unlucky coyotes. Most were taken with either the Nosler 100gr Solid Base or with a favorite .257-cal game bullet, the 115gr Partition. To this day, it has never fired a factory cartridge.

I don't recall ever firing a second shot. Most were taken at moderate ranges, 100 – 200 yards, but I remember one pronghorn shot in his bed at a bit over 300 yards. Another memorable shot was a three-point whitetail buck trotting briskly along broadside. I sat down, swung the reticle ahead and fired with the rifle swinging. The buck ran fifty yards and fell, shot through both lungs at about 325 yards.

Long ago, N.H. Roberts wrote "This cartridge [.257 Roberts] will prove a very fine, super-accurate killer of deer, antelope, and similar

big game…" Well, Ned, your cartridge may not have achieved "270 /30-06" levels of popularity but 80 years later, rifles are still being made for it. "Little powder, light lead, and a long arm…" Good advice 200 years ago and good advice today.

Dave Anderson

Dave Anderson is a lifelong hunter, competition shooter, gun collector and hand loader. Currently he is a columnist on rifle and handgun shooting for GUNS Magazine and American Handgunner."

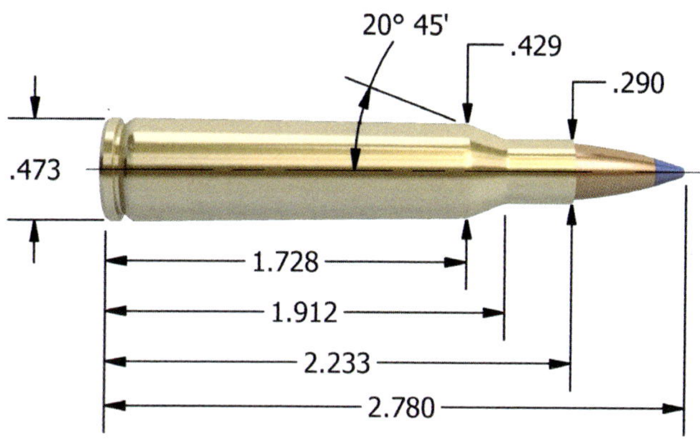

Maximum S.A.A.M.I. Overall Cartridge Length: 2.780"

BULLET CHOICES FOR THE 257 ROBERTS

TECHNICAL INFORMATION

The 257 Roberts is one of two .25 caliber cartridges used for daily production testing in the Nosler Ballistics Lab. Either the 100gr. Ballistic Tip, 100gr. Partition, or the 110gr. AccuBond would be our pick for deer-sized game. IMR 4350 is an excellent powder for our hunting weight bullets. For varmints, our 85-grain Ballistic Tip works very well with faster powders such as RL15 and H380.

257 Roberts - 85 grain

257 Roberts - 85 grain		MAXIMUM S.A.A.M.I. O.A.C.L.		2.780"
		TESTED O.A.C.L.	B.C.	S.D.
Ballistic Tip®	85gr. Spitzer	2.750"	0.329	0.183
CT® Ballistic Silvertip®	85gr. Spitzer	2.750"	0.329	0.183

CASE TYPE:	Winchester		PRIMER TYPE	CCI 200
CASE HOLDS:	53.6	Gr. WATER	BARREL Length/Make	24" Lilja
			BARREL Twist	1-10"

POWDER TYPE	POWDER CHG. GRS.		MUZZLE VEL. F.P.S.	LOAD DENSITY (VOLUME)
A-2520	38.5 *	MAX.	3090	74%
	36.5		2960	70%
	34.5		2830	66%
Varget	41.0	MAX.	3226	83%
	39.0		3133	79%
	37.0 *		3015	75%
H380	45.5	MAX.	3258	89%
	43.5		3143	85%
	41.5 *		3028	82%
BL-C2	43.0 *	MAX.	3270	81%
	41.0		3150	78%
	39.0		3030	74%
IMR 4350	47.5	MAX.	3272	94%
	45.5		3087	90%
	43.5 *		2902	86%
Big Game	45.5	MAX.	3280	89%
	43.5 *		3165	85%
	41.5		3046	82%
IMR 4895	42.0 *	MAX.	3352	86%
	40.0		3197	82%
	38.0		3042	78%
IMR 4064	43.5 *	MAX.	3360	89%
	41.5		3210	85%
	39.5		3060	81%
Viht N160	50.5 *	MAX.	3373	** 105%
	48.5		3232	** 101%
	46.5		3092	96%
RL15 Most Accurate Powder Tested	43.5		3381	88%
	41.5		3231	84%
	39.5 *		3082	80%

BC=Ballistic Coefficient SD=Sectional Density
*Most Accurate Load Tested **Compressed Load

Use Maximum Loads with Caution
Refer to page 73 for additional safety information

257 Roberts - 100 grain

		MAXIMUM S.A.A.M.I. O.A.C.L.		2.780"
		TESTED O.A.C.L.	B.C.	S.D.
Ballistic Tip®	100gr. Spitzer	2.750"	0.393	0.216
E-Tip®	100gr. Spitzer	2.750"	0.409	0.216
Due to internal construction differences, always begin with starting loads when using E-Tip® products.				
Partition®	100gr. Spitzer	2.750"	0.377	0.216

CASE TYPE:	Winchester		PRIMER TYPE	CCI 200
CASE HOLDS:	52.2	Gr. WATER	BARREL Length/Make	24" Lilja
			BARREL Twist	1-10"

POWDER TYPE	POWDER CHG. GRS.		MUZZLE VEL. F.P.S.	LOAD DENSITY (VOLUME)
A-2520	37.5	* MAX.	2862	74%
	35.5		2737	70%
	33.5		2612	66%
IMR 4320	37.0	* MAX.	2950	76%
	35.0		2810	72%
	33.0		2670	68%
IMR 4350 Most Accurate Powder Tested	43.5	MAX.	3007	88%
	41.5		2913	84%
	39.5	*	2820	80%
H4831SC	48.0	MAX.	3009	96%
	46.0		2922	92%
	44.0	*	2824	88%
Varget	39.0	MAX.	3016	81%
	37.0	*	2904	77%
	35.0		2758	72%
RL19	48.0	MAX.	3052	100%
	46.0		2918	96%
	44.0	*	2827	92%
IMR 4064	38.0	MAX.	3070	80%
	36.0		2940	76%
	34.0	*	2810	72%
H380	40.5	* MAX.	3078	82%
	38.5		2963	78%
	36.5		2848	74%
IMR 4831	45.5	* MAX.	3092	93%
	43.5		2947	89%
	41.5		2802	85%
IMR 4895	37.0	* MAX.	3120	78%
	35.0		2970	74%
	33.0		2820	69%

BC=Ballistic Coefficient SD=Sectional Density
*Most Accurate Load Tested **Compressed Load

Use Maximum Loads with Caution
Refer to page 73 for additional safety information

257 Roberts - 110 grain

257 Roberts - 110 grain		MAXIMUM S.A.A.M.I. O.A.C.L.		2.780"
		TESTED O.A.C.L.	B.C.	S.D.
AccuBond®	110gr. Spitzer	2.750"	0.418	0.238

CASE TYPE:	Winchester		PRIMER TYPE	CCI 200
CASE HOLDS:	51.6	Gr. WATER	BARREL Length/Make	24" Lilja
			BARREL Twist	1-10"

POWDER TYPE	POWDER CHG. GRS.		MUZZLE VEL. F.P.S.	LOAD DENSITY (VOLUME)
H4895	35.5	MAX.	2790	76%
	33.5	*	2670	71%
	31.5		2562	67%
Varget	37.0	MAX.	2821	78%
	35.0	*	2673	73%
	33.0		2546	69%
IMR 4350	43.0	MAX.	2866	88%
	41.0		2762	84%
	39.0	*	2645	80%
H4831SC	46.0	MAX.	2868	93%
	44.0	*	2799	89%
	42.0		2678	85%
H414	41.5	MAX.	2888	84%
	39.5	*	2786	80%
	37.5		2653	76%
RL22	46.0	MAX.	2900	97%
	44.0	*	2750	93%
	42.0		2670	88%
Viht N160	44.0	MAX.	2913	95%
	42.0	*	2842	90%
	40.0		2698	86%
Big Game	42.0	MAX.	2920	86%
	40.0		2802	82%
	38.0	*	2676	78%
RL19	47.0	MAX.	2940	99%
Most Accurate	45.0	*	2819	95%
Powder Tested	43.0		2709	91%

BC=Ballistic Coefficient SD=Sectional Density
*Most Accurate Load Tested **Compressed Load

Use Maximum Loads with Caution
Refer to page 73 for additional safety information

257 Roberts - 115 grain

		MAXIMUM S.A.A.M.I. O.A.C.L.		2.780"
		TESTED O.A.C.L.	B.C.	S.D.
Ballistic Tip®	115gr. Spitzer	2.750"	0.453	0.249
CT® Ballistic Silvertip®	115gr. Spitzer	2.750"	0.453	0.249
Partition®	115gr. Spitzer	2.750"	0.389	0.249

CASE TYPE:	Winchester		PRIMER TYPE	CCI 200
CASE HOLDS:	51.6	Gr. WATER	BARREL Length/Make	24" Lilja
			BARREL Twist	1-10"

POWDER TYPE	POWDER CHG. GRS.		MUZZLE VEL. F.P.S.	LOAD DENSITY (VOLUME)
IMR 3031	33.5	* MAX.	2626	73%
	31.5		2511	69%
	29.5		2397	65%
IMR 4320	34.5	* MAX.	2710	72%
	32.5		2532	68%
	30.5		2354	64%
H4895	35.0	MAX.	2725	75%
	33.0		2593	70%
	31.0	*	2463	66%
Big Game	40.5	MAX.	2740	83%
	38.5		2638	79%
	36.5	*	2542	74%
IMR 4064	35.0	* MAX.	2785	75%
	33.0		2654	70%
	31.0		2523	66%
IMR 4350 Most Accurate Powder Tested	40.0	* MAX.	2806	82%
	38.0		2667	78%
	36.0		2527	74%
IMR 4895	35.0	* MAX.	2807	75%
	33.0		2669	70%
	31.0		2530	66%
W760	40.0	MAX.	2809	81%
	38.0		2686	77%
	36.0	*	2563	73%
H380	38.0	MAX.	2816	78%
	36.0		2714	73%
	34.0	*	2612	69%
IMR 4831	42.5	MAX.	2827	88%
	40.5		2703	83%
	38.5	*	2579	79%

BC=Ballistic Coefficient SD=Sectional Density
*Most Accurate Load Tested **Compressed Load

Use Maximum Loads with Caution
Refer to page 73 for additional safety information

257 Roberts - 120 grain

		MAXIMUM S.A.A.M.I. O.A.C.L.		2.780"
		TESTED O.A.C.L.	B.C.	S.D.
Partition®	120gr. Spitzer	2.750"	0.391	0.260

CASE TYPE:	Winchester		PRIMER TYPE	CCI 200
CASE HOLDS:	51.4	Gr. WATER	BARREL Length/Make	24" Lilja
			BARREL Twist	1-10"

POWDER TYPE	POWDER CHG. GRS.		MUZZLE VEL. F.P.S.	LOAD DENSITY (VOLUME)
H4895	34.0 *	MAX.	2660	73%
	32.0		2538	68%
	30.0		2413	64%
IMR 4320	34.5 *	MAX.	2690	72%
	32.5		2510	68%
	30.5		2330	64%
Big Game	39.0	MAX.	2732	80%
	37.0		2674	76%
	35.0 *		2526	72%
IMR 4064	35.0 *	MAX.	2750	75%
	33.0		2630	71%
	31.0		2510	66%
IMR 4895	35.0	MAX.	2760	75%
	33.0		2630	71%
	31.0 *		2500	66%
IMR 4350 Most Accurate Powder Tested	40.0 *	MAX.	2780	82%
	38.0		2640	78%
	36.0		2500	74%
H4350	42.0 *	MAX.	2782	86%
	40.0		2677	82%
	38.0		2572	78%
H380	38.0	MAX.	2790	78%
	36.0		2690	74%
	34.0 *		2590	70%
IMR 4831	42.5	MAX.	2800	88%
	40.5		2670	84%
	38.5 *		2540	80%
IMR 7828	47.5 *	MAX.	2810	98%
	45.5		2720	94%
	43.5		2590	90%

BC=Ballistic Coefficient SD=Sectional Density
*Most Accurate Load Tested **Compressed Load

Use Maximum Loads with Caution
Refer to page 73 for additional safety information

25-06 REMINGTON

For deer and antelope-sized game, my first choice is the 25-06 Remington. In the average 8 pound rifle, the recoil is very manageable. Comparatively speaking, my shoulder tells me it kicks only a bit harder than a similar-weight gun in the mild 243 Win. Ballistically, it shoots similar to the 270 Winchester, but the 25-06 is easier on the shoulder. That means, at least for me, I shoot it more accurately.

I recall a 15-inch antelope buck in the Texas Panhandle. One of his does picked me off and it was time to shoot. Shooting off the sticks at 185 yards, I steadied the crosswire of the Leupold variable. When I tugged the trigger on the Browning X-Bolt, he dropped hard.

In Oklahoma, I stalked a frenzied mob of whitetail bucks chasing a hot doe in the CRP grass. When a white-horned 8-point hesitated at 100 yards, I squeezed. He dropped with no further advancement. The rest of the bucks continued the chase like nothing happened. I used the same gun, scope and ammo from the antelope hunt.

The largest animal I've ever taken with the 25-06 was a monstrous 300-pound aoudad ram in the rugged breaks of the Palo Duro Canyon. Shooting prone off the canyon rim with a single-shot Thompson Center Encore rifle, the big ram sucked up the lead from the first shot at 150 yards. He ran 20 yards and stopped. I broke the gun open, dropped a fresh shell in, then hit him mid-body with missile number two. He piled up 50 yards later.

The first bullet hit in the crease behind his right shoulder at a quartering away angle, exiting out the front of his chest. The second round was unnecessary, but a follow-up pill is always advisable, particularly on tough-as-hell aoudad sheep. His horns weren't bad either, 31-inches plus on each side.

In addition to being just about perfect for deer-sized game, the 25-06 makes a solid choice for someone picking one cartridge to pull double duty on both deer and predators. One afternoon, using my deer rifle, I shot a double on winter coyotes. I don't mean a double as in two shots; I mean two dogs with one bullet! Neither dog went more than three steps.

Out of my 25-06, I like bullets in the 115 to 120-grain weight.

If I had to choose only one, the 25-06 would be my caliber of choice for deer and antelope.

Brandon Ray

Brandon Ray lives with his wife, daughter and six dogs on a ranch in the Texas Panhandle. More than 700 of his articles have been published in outdoor publications.

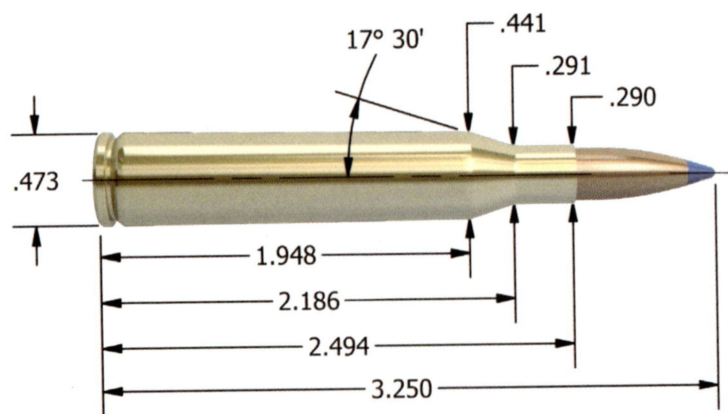

Maximum S.A.A.M.I. Overall Cartridge Length: 3.250"

BULLET CHOICES FOR THE 25-06 REMINGTON

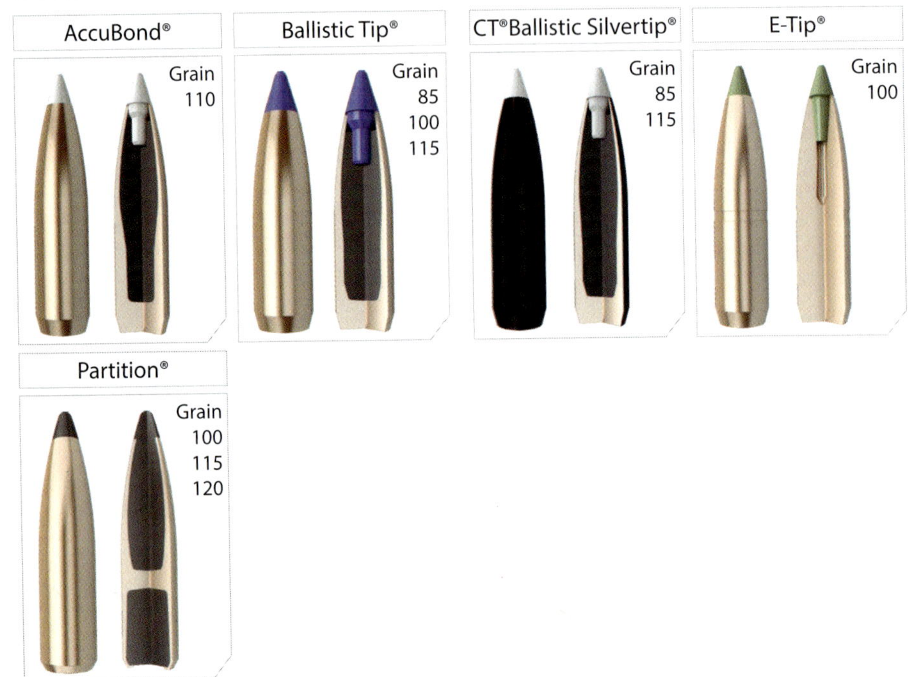

TECHNICAL INFORMATION

The 25-06 Remington is our standard cartridge for daily .25 caliber production testing. It is a terrific long-range varmint cartridge when loaded with our 85gr. Ballistic Tip, and it becomes an excellent choice for deer and antelope with any of our heavier .25 caliber bullets. Both 4350 and RL19 are great choices for the 25-06 Remington.

25-06 Remington - 85 grain

		MAXIMUM S.A.A.M.I. O.A.C.L.		3.250"
		TESTED O.A.C.L.	B.C.	S.D.
Ballistic Tip®	85gr. Spitzer	3.200"	0.329	0.183
CT® Ballistic Silvertip®	85gr. Spitzer	3.200"	0.329	0.183

CASE TYPE:	Winchester		PRIMER TYPE	WLR
CASE HOLDS:	65.8	Gr. WATER	BARREL Length/Make	24" Wiseman
			BARREL Twist	1-10"

POWDER TYPE	POWDER CHG. GRS.		MUZZLE VEL. F.P.S.	LOAD DENSITY (VOLUME)
H4831SC	55.5	MAX.	3408	88%
	53.5		3331	85%
	51.5	*	3257	82%
IMR 4064	47.0	MAX.	3410	78%
	45.0		3315	75%
	43.0	*	3220	72%
IMR 4895	46.0	MAX.	3450	77%
	44.0		3305	73%
	42.0	*	3160	70%
IMR 4350	53.5	MAX.	3450	86%
	51.5		3313	83%
	49.5	*	3175	80%
Varget	47.0	MAX.	3541	77%
	45.0		3401	74%
	43.0	*	3288	71%
RL15	48.0	* MAX.	3548	76%
	46.0		3420	73%
	44.0		3309	70%
Big Game	51.5	MAX.	3564	82%
	49.5		3457	79%
	47.5	*	3365	76%
Viht N160	57.0	MAX.	3581	96%
	55.0		3475	93%
	53.0	*	3369	89%
RL19	57.0	* MAX.	3600	94%
Most Accurate	55.0		3480	91%
Powder Tested	53.0		3360	88%

BC=Ballistic Coefficient SD=Sectional Density
*Most Accurate Load Tested **Compressed Load

Use Maximum Loads with Caution
Refer to page 73 for additional safety information

25-06 Remington - 100 grain

25-06 Remington - 100 grain		MAXIMUM S.A.A.M.I. O.A.C.L.		3.250"
		TESTED O.A.C.L.	B.C.	S.D.
Ballistic Tip®	100gr. Spitzer	3.200"	0.393	0.216
CT® Ballistic Silvertip®	100gr. Spitzer	3.200"	0.393	0.216
E-Tip®	100gr. Spitzer	3.170"	0.409	0.216
Due to internal construction differences, always begin with starting loads when using E-Tip® products.				
Partition®	100gr. Spitzer	3.200"	0.377	0.216

CASE TYPE:	Winchester		PRIMER TYPE	WLR
CASE HOLDS:	64.2	Gr. WATER	BARREL Length/Make	24" Wiseman
			BARREL Twist	1-10"

POWDER TYPE	POWDER CHG. GRS.		MUZZLE VEL. F.P.S.	LOAD DENSITY (VOLUME)
IMR 4320	42.0 *	MAX.	3118	70%
	40.0		2983	67%
	38.0		2848	64%
IMR 4064	41.5	MAX.	3142	71%
	39.5		2987	68%
	37.5 *		2832	64%
IMR 7828	56.5	MAX.	3252	94%
	54.5		3127	90%
	52.5 *		3002	87%
Magnum	62.0	MAX.	3265	98%
	60.0 *		3172	95%
	58.0		3079	92%
IMR 4831	53.5	MAX.	3290	89%
	51.5		3194	85%
	49.5 *		3099	82%
H4350	53.5	MAX.	3318	88%
	51.5		3143	85%
	49.5 *		2968	82%
Viht N160	54.0	MAX.	3343	93%
	52.0		3234	90%
	50.0 *		3125	87%
IMR 4350	51.0 *	MAX.	3352	84%
	49.0		3247	81%
	47.0		3142	77%
RL19 Most Accurate Powder Tested	54.5	MAX.	3361	92%
	52.5 *		3254	89%
	50.5		3147	86%

BC=Ballistic Coefficient SD=Sectional Density
*Most Accurate Load Tested **Compressed Load

Use Maximum Loads with Caution
Refer to page 73 for additional safety information

25-06 Remington - 110 grain		MAXIMUM S.A.A.M.I. O.A.C.L.		3.250"
		TESTED O.A.C.L.	B.C.	S.D.
AccuBond®	110gr. Spitzer	3.200"	0.418	0.238

CASE TYPE:	Winchester		PRIMER TYPE	WLR
CASE HOLDS:	63.3	Gr. WATER	BARREL Length/Make	24" Wiseman
			BARREL Twist	1-10"

POWDER TYPE	POWDER CHG. GRS.		MUZZLE VEL. F.P.S.	LOAD DENSITY (VOLUME)
W760	46.5	MAX.	3078	81%
	44.5 *		2997	77%
	42.5		2884	74%
H4831SC	51.0 *	MAX.	3131	84%
	49.0		2993	81%
	47.0		2950	77%
IMR 4831 Most Accurate Powder Tested	51.0	MAX.	3177	86%
	49.0		3081	82%
	47.0 *		2996	79%
IMR 4350	49.5	MAX.	3183	83%
	47.5 *		3092	79%
	45.5		2955	76%
Magnum	61.0	MAX.	3198	98%
	59.0		3092	95%
	57.0 *		2981	91%
MAGPRO	56.0	MAX.	3212	91%
	54.0		3129	88%
	52.0 *		3035	85%
RL19	52.0	MAX.	3213	89%
	50.0		3109	86%
	48.0 *		3044	82%
Viht N560	52.5	MAX.	3217	92%
	50.5 *		3121	89%
	48.5		3027	85%
Retumbo	59.0 *	MAX.	3267	98%
	57.0		3206	95%
	55.0		3121	91%

BC=Ballistic Coefficient SD=Sectional Density
*Most Accurate Load Tested **Compressed Load

Use Maximum Loads with Caution
Refer to page 73 for additional safety information

25-06 Remington - 115 grain		MAXIMUM S.A.A.M.I. O.A.C.L.		3.250"
		TESTED O.A.C.L.	B.C.	S.D.
Ballistic Tip®	115gr. Spitzer	3.200"	0.453	0.249
CT® Ballistic Silvertip®	115gr. Spitzer	3.200"	0.453	0.249
Partition®	115gr. Spitzer	3.200"	0.389	0.249

CASE TYPE:	Winchester		PRIMER TYPE	WLR
CASE HOLDS:	63.3	Gr. WATER	BARREL Length/Make	24" Wiseman
			BARREL Twist	1-10"

POWDER TYPE	POWDER CHG. GRS.		MUZZLE VEL. F.P.S.	LOAD DENSITY (VOLUME)
IMR 4064	40.0	MAX.	2871	69%
	38.0		2744	66%
	36.0 *		2617	62%
RL19	50.0	MAX.	3007	86%
	48.0		2909	82%
	46.0 *		2810	79%
IMR 7828	54.0	MAX.	3030	91%
	52.0		2880	87%
	50.0 *		2730	84%
Viht N560	49.5	MAX.	3050	87%
	47.5 *		2967	83%
	45.5		2915	80%
Viht N165	50.5	MAX.	3070	89%
	48.5		2970	85%
	46.5 *		2869	82%
IMR 4350 Most Accurate Powder Tested	49.0 *	MAX.	3095	82%
	47.0		2986	79%
	45.0		2877	75%
IMR 4831	52.0 *	MAX.	3116	87%
	50.0		3024	84%
	48.0		2933	81%
Magnum	60.5	MAX.	3150	97%
	58.5		3033	94%
	56.5 *		2927	91%
Retumbo	57.5	MAX.	3170	96%
	55.5		3083	92%
	53.5 *		3012	89%

BC=Ballistic Coefficient SD=Sectional Density
*Most Accurate Load Tested **Compressed Load

Use Maximum Loads with Caution
Refer to page 73 for additional safety information

25-06 Remington - 120 grain

25-06 Remington - 120 grain		MAXIMUM S.A.A.M.I. O.A.C.L.		3.250"
		TESTED O.A.C.L.	B.C.	S.D.
Partition®	120gr. Spitzer	3.200"	0.391	0.260

CASE TYPE:	Winchester		PRIMER TYPE	WLR
CASE HOLDS:	63.5	Gr. WATER	BARREL Length/Make	24" Wiseman
			BARREL Twist	1-10"

POWDER TYPE	POWDER CHG. GRS.		MUZZLE VEL. F.P.S.	LOAD DENSITY (VOLUME)
IMR 4064	40.0	MAX.	2870	69%
	38.0		2740	66%
	36.0 *		2610	62%
IMR 4320	41.0 *	MAX.	2888	69%
	39.0		2773	66%
	37.0		2658	63%
RL19	50.0	MAX.	2990	86%
	48.0		2890	82%
	46.0 *		2790	79%
Viht N165	50.5	MAX.	3007	88%
	48.5		2907	85%
	46.5 *		2807	81%
IMR 7828	51.0	MAX.	3030	90%
	52.0		2880	87%
	50.0 *		2730	84%
IMR 4831	50.0 *	MAX.	3040	84%
	48.0		2910	80%
	46.0		2742	77%
IMR 4350	49.0 *	MAX.	3080	82%
	47.0		2970	78%
	45.0		2860	75%
Magnum	60.0	MAX.	3096	96%
	58.0 *		3012	93%
	56.0		2910	90%
Retumbo	57.5	MAX.	3175	95%
Most Accurate	55.5		3112	92%
Powder Tested	53.5 *		3038	89%

BC=Ballistic Coefficient SD=Sectional Density
*Most Accurate Load Tested **Compressed Load

Use Maximum Loads with Caution
Refer to page 73 for additional safety information

Aaron Carter

257 WEATHERBY MAGNUM

Unveiled in 1948, the .257 Weatherby Magnum was Roy Weatherby's personal favorite and, to date, remains the second most popular cartridge in Weatherby's line-up; the reasons for such are simple—high performance without objectionable recoil. At its debut, the cartridge's closest competitor, at least commercially, was the 257 Roberts, which it bested by upward of 600fps. Even today, it still reigns supreme in the realm of "quarter bores". Despite utilizing technologically advanced propellants, the 25-'06 Rem.—its chief competition—is approximately 200fps slower when bullets of identical weights are used. The 257's additional velocity translates to less drop and wind deflection (and thus compensation) and more energy downrange; however, felt recoil is comparable to that of popular chamberings such as the 280 Rem., and the 30-'06 Sprg.

Because of the 257 Wby Mag's considerable velocities, bullet selection is critical. Unless nuisance species or small predators are the intended quarry, where minimal weight retention is preferred, controlled-expansion bullets are recommended. This is especially true when game animals at the cartridge's upper limit—such as elk and moose—are pursued, or simply for "insurance" in case less-than-perfect shot angles are encountered—both of which demand deep penetration.

For me, the "go-to" bullet for the 257 Wby Mag is the 115gr Partition. I prefer material to "wash off" during expansion/penetration, thereby enhancing the bullet's wounding capability; however, for larger species, such as elk, deep penetration is necessary, too. Partitions fit this philosophy perfectly. Typical weight retention is 65 to 70 percent, which, combined with a not-too-large frontal diameter, enables it to attain depths sufficient to quickly down big game. Animals I've shot with the aforementioned Partition required little, if any, tracking. No bullets were recovered, either.

Concerning accuracy, from my 24"-barreled, Weatherby Vanguard Series 2 Synthetic rifle, the 115gr Partition—seated to give a 3.250" routinely prints 100-yd., three-shot groups measuring 1/2" to 3/4" center to center. The smallest group to-date was 0.493".

For hunters craving even higher velocities, yet still want controlled-expansion projectiles, there's the 100gr Partition, 100gr E-Tip, and 110gr AccuBond. According to Nosler's reloading data, the former can be propelled in excess of 3,600 f.p.s, and the latter near 3,500 f.p.s.—talk about flat-shooting!

Given its level of performance, there's little wonder as to why the 257 Wby Mag remains so popular. It is capable of handling all North American species, except the largest bruins, has excellent external ballistics, delivers impressive terminal performance, yet the recoil is quite manageable for most shooters. What's not to like?

Aaron Carter

Aaron Carter, Managing Editor, American Rifleman

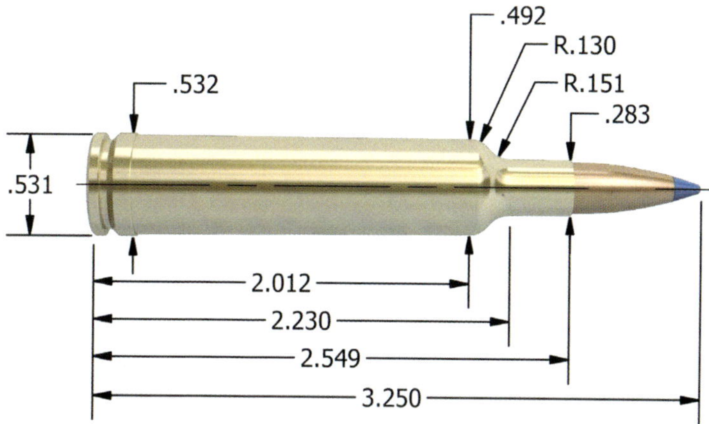

Maximum S.A.A.M.I. Overall Cartridge Length: 3.250"

BULLET CHOICES FOR THE 257 WEATHERBY MAGNUM

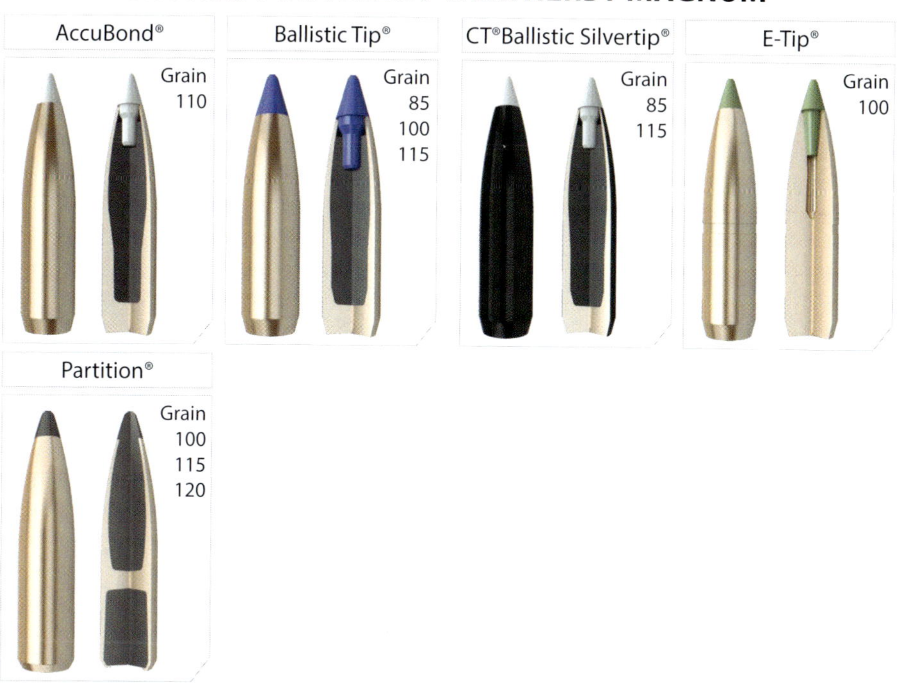

TECHNICAL INFORMATION

As with most Weatherby Magnums, there are a few points to keep in mind when loading this cartridge.

- The chambers have "freebore", meaning they have a longer throat. For this reason, it is generally not possible to seat a bullet close to, or in contact with the lands (rifling).
- For best accuracy, bullets should be seated as long as the magazine will allow.
- Weatherby rifles of German manufacture (pre-1972) may have barrels with 1-12" twist rates. These rifles will not reliably stabilize bullets larger than 100 grains.

257 Weatherby Magnum - 85 grain		MAXIMUM S.A.A.M.I. O.A.C.L.		3.250"
		TESTED O.A.C.L.	B.C.	S.D.
Ballistic Tip®	85gr. Spitzer	3.240"	0.329	0.183
CT® Ballistic Silvertip®	85gr. Spitzer	3.240"	0.329	0.183

CASE TYPE:	Nosler		PRIMER TYPE	Fed 215
CASE HOLDS:	84.8	Gr. WATER	BARREL Length/Make	26" Lilja
			BARREL Twist	1-10"

POWDER TYPE	POWDER CHG. GRS.		MUZZLE VEL. F.P.S.	LOAD DENSITY (VOLUME)
H4831	72.0	MAX.	3675	88%
	70.0		3552	86%
	68.0 *		3428	84%
IMR 7828	73.5	MAX.	3698	92%
	71.5		3546	90%
	69.5 *		3393	87%
Hunter	68.0	MAX.	3715	84%
	66.0 *		3614	82%
	64.0		3584	79%
Viht N160	70.0	MAX.	3764	92%
	68.0		3642	89%
	66.0 *		3519	86%
W760	68.0	MAX.	3769	84%
	66.0		3739	81%
	64.0 *		3642	79%
RL22	73.0 *	MAX.	3780	94%
	71.0		3617	91%
	69.0		3455	88%
H4350	68.0	MAX.	3788	85%
	66.0		3711	82%
	64.0 *		3643	80%
IMR 4350	69.5	MAX.	3793	87%
	67.5		3668	84%
	65.5 *		3543	82%
RL19	73.0	MAX.	3832	94%
Most Accurate	71.0 *		3745	91%
Powder Tested	69.0		3617	88%

BC=Ballistic Coefficient SD=Sectional Density
*Most Accurate Load Tested **Compressed Load

Use Maximum Loads with Caution
Refer to page 73 for additional safety information

257 Weatherby Magnum - 100 grain

		MAXIMUM S.A.A.M.I. O.A.C.L.		3.250"
		TESTED O.A.C.L.	B.C.	S.D.
Ballistic Tip®	100gr. Spitzer	3.240"	0.393	0.216
E-Tip®	100gr. Spitzer	3.240"	0.409	0.216
Due to internal construction differences, always begin with starting loads when using E-Tip® products.				
Partition®	100gr. Spitzer	3.240"	0.377	0.216

CASE TYPE:	Nosler		PRIMER TYPE	Fed 215
CASE HOLDS:	81.8	Gr. WATER	BARREL Length/Make	26" Lilja
			BARREL Twist	1-10"

POWDER TYPE	POWDER CHG. GRS.		MUZZLE VEL. F.P.S.	LOAD DENSITY (VOLUME)
H4831	68.0	MAX.	3446	87%
	66.0		3357	84%
	64.0	*	3268	81%
Hunter	65.5	MAX.	3482	84%
	63.5	*	3400	82%
	61.5		3333	79%
Viht N160	65.0	MAX.	3500	88%
	63.0		3404	86%
	61.0	*	3297	83%
RL25 Most Accurate Powder Tested	75.0	MAX.	3512	100%
	73.0		3411	97%
	71.0	*	3330	94%
IMR 4350	64.0	* MAX.	3532	83%
	62.0		3408	80%
	60.0		3294	78%
H1000	75.5	MAX.	3543	97%
	73.5	*	3470	94%
	71.5		3403	92%
RL19	69.0	MAX.	3569	92%
	67.0		3470	89%
	65.0	*	3359	86%
RL22	71.0	MAX.	3574	94%
	69.0		3498	92%
	67.0	*	3379	89%
IMR 4831	68.0	* MAX.	3576	88%
	66.0		3455	86%
	64.0		3333	83%

BC=Ballistic Coefficient SD=Sectional Density
*Most Accurate Load Tested **Compressed Load

Use Maximum Loads with Caution
Refer to page 73 for additional safety information

257 Weatherby Magnum - 110 grain		MAXIMUM S.A.A.M.I. O.A.C.L.		3.250"
		TESTED O.A.C.L.	B.C.	S.D.
AccuBond®	110gr. Spitzer	3.240"	0.418	0.238

CASE TYPE:	Nosler		PRIMER TYPE	Fed 215
CASE HOLDS:	81.0	Gr. WATER	BARREL Length/Make	26" Lilja
			BARREL Twist	1-10"

POWDER TYPE	POWDER CHG. GRS.		MUZZLE VEL. F.P.S.	LOAD DENSITY (VOLUME)
IMR 4831	65.0	MAX.	3282	85%
	63.0 *		3198	83%
	61.0		3109	80%
Hunter	63.0 *	MAX.	3304	82%
	61.0		3221	79%
	59.0		3135	77%
Viht N165	69.5	MAX.	3342	95%
	67.5 *		3251	93%
	65.5		3174	90%
IMR 4350	64.0	MAX.	3375	84%
Most Accurate	62.0		3291	81%
Powder Tested	60.0 *		3200	78%
H4831SC	68.5	MAX.	3386	88%
	66.5 *		3301	86%
	64.5		3228	83%
RL19	67.0	MAX.	3387	90%
	65.0		3312	87%
	63.0 *		3253	85%
H1000	74.0	MAX.	3446	96%
	72.0		3378	93%
	70.0 *		3302	90%
RL22	69.5 *	MAX.	3471	93%
	67.5		3381	91%
	65.5		3298	88%
IMR 7828	71.0	MAX.	3480	93%
	69.0 *		3380	91%
	67.0		3262	88%

BC=Ballistic Coefficient SD=Sectional Density
*Most Accurate Load Tested **Compressed Load

Use Maximum Loads with Caution
Refer to page 73 for additional safety information

257 Weatherby Magnum - 115 grain

		MAXIMUM S.A.A.M.I. O.A.C.L.	3.250"
		TESTED O.A.C.L.	B.C.	S.D.

Bullet	Type	TESTED O.A.C.L.	B.C.	S.D.
Ballistic Tip®	115gr. Spitzer	3.240"	0.453	0.249
CT® Ballistic Silvertip®	115gr. Spitzer	3.240"	0.453	0.249
Partition®	115gr. Spitzer	3.240"	0.389	0.249

MAXIMUM S.A.A.M.I. O.A.C.L.: 3.250"

CASE TYPE:	Nosler		PRIMER TYPE	Fed 215
CASE HOLDS:	81.0	Gr. WATER	BARREL Length/Make	26" Lilja
			BARREL Twist	1-10"

POWDER TYPE	POWDER CHG. GRS.		MUZZLE VEL. F.P.S.	LOAD DENSITY (VOLUME)
IMR 4350	59.0 *	MAX.	3115	77%
	57.0		3034	74%
	55.0		2952	72%
Hunter	61.5	MAX.	3198	80%
	59.5 *		3052	77%
	57.5		2905	75%
Viht N165	66.0	MAX.	3214	91%
	64.0		3125	88%
	62.0 *		3034	85%
RL19	65.0	MAX.	3256	87%
	63.0 *		3182	85%
	61.0		3115	82%
IMR 4831	64.0 *	MAX.	3281	84%
	62.0		3184	81%
	60.0		3086	79%
RL22	67.0	MAX.	3285	90%
	65.0 *		3203	87%
	63.0		3144	85%
RL25	72.0	MAX.	3299	97%
Most Accurate	70.0 *		3190	94%
Powder Tested	68.0		3068	91%
H1000	71.5	MAX.	3302	92%
	69.5 *		3221	90%
	67.5		3142	87%
IMR 7828	69.0	MAX.	3433	91%
	67.0 *		3342	88%
	65.0		3237	85%

BC=Ballistic Coefficient SD=Sectional Density
*Most Accurate Load Tested **Compressed Load

Use Maximum Loads with Caution
Refer to page 73 for additional safety information

257 Weatherby Magnum - 120 grain		MAXIMUM S.A.A.M.I. O.A.C.L.		3.250"
		TESTED O.A.C.L.	B.C.	S.D.
Partition®	120gr. Spitzer	3.240"	0.391	0.260

CASE TYPE:	Nosler		PRIMER TYPE	Fed 215
CASE HOLDS:	81.2	Gr. WATER	BARREL Length/Make	26" Lilja
			BARREL Twist	1-10"

POWDER TYPE	POWDER CHG. GRS.		MUZZLE VEL. F.P.S.	LOAD DENSITY (VOLUME)
IMR 4350	59.0 *	MAX.	3090	77%
	57.0		3010	74%
	55.0		2930	72%
Viht N165	66.0	MAX.	3183	90%
	64.0		3094	88%
	62.0 *		3005	85%
Hunter	62.0	MAX.	3186	80%
	60.0 *		3143	78%
	58.0		3030	75%
RL25 Most Accurate Powder Tested	71.0 *	MAX.	3225	95%
	69.0		3130	92%
	67.0		3037	90%
H1000	70.0	MAX.	3235	90%
	68.0 *		3139	88%
	66.0		3071	85%
IMR 4831	64.0	MAX.	3258	84%
	62.0		3153	81%
	60.0 *		3048	79%
RL19	65.0 *	MAX.	3260	87%
	63.0		3175	84%
	61.0		3089	82%
RL22	67.0	MAX.	3285	90%
	65.0 *		3192	87%
	63.0		3109	84%
IMR 7828	69.0	MAX.	3402	90%
	67.0 *		3312	88%
	65.0		3206	85%

BC=Ballistic Coefficient SD=Sectional Density
*Most Accurate Load Tested **Compressed Load

Use Maximum Loads with Caution
Refer to page 73 for additional safety information

90 year-old John A. Nosler on an Oregon mule deer hunt.

John Barsness

6.5MM X 55 SWEDISH

The 6.5x55 Mauser appeared in 1891, a joint Norwegian-Swedish development for use in each country's military rifles. If imitation is the sincerest form of flattery, the 6.5x55 is one of the most flattered cartridges in the world. The long list of imitations stretches from the 256 Newton to the 6.5 Creedmoor, but the 6.5x55 just keeps trucking along, appearing in new rifles every year.

The reasons are simple. The "Swede" combines light recoil and fine accuracy with a wide array of ballistically-efficient bullets suitable for everything from coyotes to moose. Plus, the cartridge functions perfectly in any bolt action that works with the 30-06. Mine is a custom rifle built on a commercial FN Mauser action, and it groups 140-grain Nosler Partitions into much less than an inch at 2700 fps, a combination suitable for just about any non-dangerous big game, at any sane range. No wonder the 6.5x55 is still kicking—though, as always, lightly.

John and his wife Eileen Clarke live in Montana, where they hunt and run their website www.riflesandrecipes.com

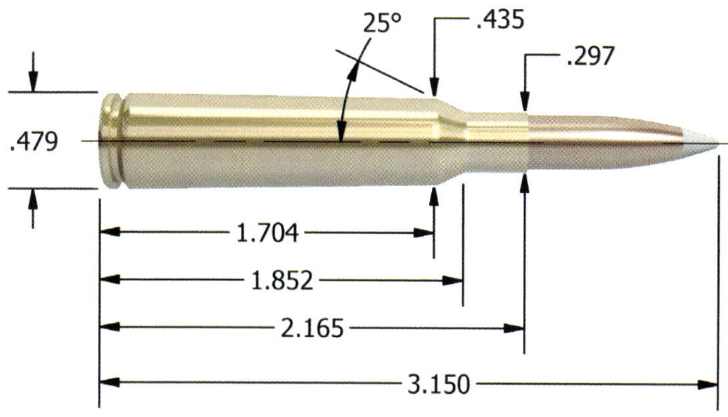

Maximum S.A.A.M.I. Overall Cartridge Length: 3.150"

BULLET CHOICES FOR THE 6.5MM X 55 SWEDISH

TECHNICAL INFORMATION

In a high-quality rifle, accuracy of the "Swede" is excellent, and its ability to handle long, 140-grain bullets makes it a great choice for all but the largest of North American big game species. These loads are intended for use only with new firearms in good condition. If you are using an older military rifle, or are unsure as to the condition of your firearm, it must be thoroughly inspected by a competent gunsmith before use.

6.5mm x 55 Swedish - 100 grain		MAXIMUM S.A.A.M.I. O.A.C.L.		3.150"
		TESTED O.A.C.L.	B.C.	S.D.
Ballistic Tip®	100gr. Spitzer	3.025"	0.350	0.205
Partition®	100gr. Spitzer	3.025"	0.326	0.205

CASE TYPE:	Norma		PRIMER TYPE	Rem. 9 1/2
CASE HOLDS:	55.3	Gr. WATER	BARREL Length/Make	23" Lilja
			BARREL Twist	1-9"

POWDER TYPE	POWDER CHG. GRS.		MUZZLE VEL. F.P.S.		LOAD DENSITY (VOLUME)
Varget Most Accurate Powder Tested	43.5 *	MAX.	3094		85%
	41.5		2985		81%
	39.5		2897		77%
H4831	51.0	MAX.	3117		96%
	49.0		2999		92%
	47.0 *		2882		89%
W760	47.5	MAX.	3118		90%
	45.5		3002		86%
	43.5 *		2885		82%
Viht N150	42.0 *	MAX.	3120		89%
	40.0		3010		85%
	38.0		2899		81%
IMR 4350	48.0	MAX.	3127		92%
	46.0		2973		88%
	44.0 *		2819		84%
Big Game	47.0	MAX.	3150		89%
	45.0 *		3035		86%
	43.0		2932		82%
RL19	51.5 *	MAX.	3163	**	101%
	49.5		3018		97%
	47.5		2874		93%
Viht N160	50.5	MAX.	3169	**	101%
	48.5		3042		97%
	46.5 *		2915		93%
RL15	43.5	MAX.	3188		82%
	41.5		3066		78%
	39.5 *		2944		74%

BC=Ballistic Coefficient SD=Sectional Density
*Most Accurate Load Tested **Compressed Load

Use Maximum Loads with Caution
Refer to page 73 for additional safety information

6.5mm x 55 Swedish - 120 grain

		MAXIMUM S.A.A.M.I. O.A.C.L.		3.150"
		TESTED O.A.C.L.	B.C.	S.D.
Ballistic Tip®	120gr. Spitzer	3.025"	0.458	0.246
E-Tip®	120gr. Spitzer	2.975"	0.497	0.246

Due to internal construction differences, always begin with starting loads when using E-Tip® products.

CASE TYPE:	Norma		PRIMER TYPE	Rem. 9 1/2
CASE HOLDS:	54.0	Gr. WATER	BARREL Length/Make	23" Lilja
			BARREL Twist	1-9"

POWDER TYPE	POWDER CHG. GRS.		MUZZLE VEL. F.P.S.	LOAD DENSITY (VOLUME)
A-2520	38.0	MAX.	2770	73%
	36.0		2670	69%
	34.0	*	2570	65%
Viht N150	38.0	MAX.	2791	83%
	36.0		2689	78%
	34.0	*	2587	74%
Varget	41.0	MAX.	2828	82%
	39.0	*	2737	78%
	37.0		2612	74%
Viht N160	46.5	MAX.	2837	96%
	44.5		2708	92%
	42.5	*	2579	87%
Hunter	47.0	MAX.	2852	92%
	45.0		2771	88%
	43.0	*	2662	84%
W760	45.5	* MAX.	2881	88%
Most Accurate	43.5		2754	84%
Powder Tested	41.5		2630	80%
IMR 4350	45.5	* MAX.	2902	89%
	43.5		2737	85%
	41.5		2572	81%
H4350	47.0	MAX.	3000	92%
	45.0		2870	88%
	43.0	*	2740	84%
RL19	48.5	MAX.	3002	98%
	46.5		2867	94%
	44.5	*	2732	90%

BC=Ballistic Coefficient SD=Sectional Density
*Most Accurate Load Tested **Compressed Load

Use Maximum Loads with Caution
Refer to page 73 for additional safety information

6.5mm x 55 Swedish - 123/125 grain

		MAXIMUM S.A.A.M.I. O.A.C.L.		3.150"
		TESTED O.A.C.L.	B.C.	S.D.
Custom Competition™	123gr. HPBT	3.025"	0.510	0.252
Partition®	125gr. Spitzer	3.025"	0.449	0.256

CASE TYPE:	Nosler		PRIMER TYPE	Rem. 9 1/2
CASE HOLDS:	53.4	Gr. WATER	BARREL Length/Make	23" Lilja
			BARREL Twist	1-9"

POWDER TYPE	POWDER CHG. GRS.		MUZZLE VEL. F.P.S.	LOAD DENSITY (VOLUME)
H4895	37.0	MAX.	2730	76%
	35.0		2620	72%
	33.0 *		2510	68%
Hunter	46.0	MAX.	2802	91%
	44.0		2652	87%
	42.0 *		2535	83%
Viht N560	48.5	MAX.	2819	** 101%
	46.5		2690	97%
	44.5 *		2598	93%
W760	45.0	MAX.	2819	88%
	43.0 *		2723	84%
	41.0		2636	80%
Varget	40.0 *	MAX.	2877	81%
	38.0		2775	77%
	36.0		2724	73%
IMR 4350 Most Accurate Powder Tested	45.5	MAX.	2910	90%
	43.5		2750	86%
	41.5 *		2590	82%
IMR 4831	47.0 *	MAX.	2940	94%
	45.0		2810	90%
	43.0		2680	86%
RL19	47.5	MAX.	2950	97%
	45.5		2850	93%
	43.5 *		2750	89%
H4350	45.5	MAX.	2954	90%
	43.5		2873	86%
	41.5 *		2741	82%

BC=Ballistic Coefficient SD=Sectional Density
*Most Accurate Load Tested **Compressed Load

Use Maximum Loads with Caution
Refer to page 73 for additional safety information

6.5mm x 55 Swedish - 130 grain

		MAXIMUM S.A.A.M.I. O.A.C.L.		3.150"
		TESTED O.A.C.L.	B.C.	S.D.
AccuBond®	130gr. Spitzer	3.025"	0.488	0.266

CASE TYPE:	Norma		PRIMER TYPE	Rem. 9 1/2
CASE HOLDS:	52.0	Gr. WATER	BARREL Length/Make	23" Lilja
			BARREL Twist	1-9"

POWDER TYPE	POWDER CHG. GRS.		MUZZLE VEL. F.P.S.		LOAD DENSITY (VOLUME)
IMR 4350	43.5	MAX.	2631		89%
	41.5		2508		84%
	39.5 *		2410		80%
Magnum	52.0 *	MAX.	2640	**	102%
	50.0		2518		98%
	48.0		2378		94%
H4831SC	48.0 *	MAX.	2691		96%
	46.0		2595		92%
	44.0		2496		88%
H4350	43.5	MAX.	2719		89%
	41.5 *		2610		84%
	39.5		2522		80%
Viht N550	43.0	MAX.	2737		92%
	41.0 *		2632		88%
	39.0		2503		83%
W760	44.0 *	MAX.	2753		89%
	42.0		2655		85%
	40.0		2546		81%
RL19	47.5	MAX.	2775		99%
Most Accurate	45.5 *		2642		95%
Powder Tested	43.5		2506		91%
Viht N560	48.0	MAX.	2800	**	103%
	46.0		2649		98%
	44.0 *		2536		94%
Hunter	46.5	MAX.	2801		94%
	44.5 *		2730		90%
	42.5		2608		86%

BC=Ballistic Coefficient SD=Sectional Density
*Most Accurate Load Tested **Compressed Load

Use Maximum Loads with Caution
Refer to page 73 for additional safety information

6.5mm x 55 Swedish - 140 grain		MAXIMUM S.A.A.M.I. O.A.C.L.		3.150"
		TESTED O.A.C.L.	B.C.	S.D.
AccuBond®	140gr. Spitzer	3.025"	0.509	0.287
Ballistic Tip®	140gr. Spitzer	3.025"	0.509	0.287
Custom Competition™	140gr. HPBT	3.025"	0.529	0.287
Partition®	140gr. Spitzer	3.025"	0.490	0.287

CASE TYPE:	Norma		PRIMER TYPE	Rem. 9 1/2
CASE HOLDS:	50.2	Gr. WATER	BARREL Length/Make	23" Lilja
			BARREL Twist	1-9"

POWDER TYPE	POWDER CHG. GRS.		MUZZLE VEL. F.P.S.		LOAD DENSITY (VOLUME)
Viht N150	36.5	MAX.	2494		86%
	34.5 *		2382		81%
	32.5		2273		76%
Viht N160	41.0	MAX.	2509		91%
	39.0		2413		86%
	37.0 *		2316		82%
Magnum	51.5	MAX.	2597	**	104%
	49.5 *		2471		100%
	47.5		2379		96%
Viht N560	46.0	MAX.	2690	**	102%
	44.0		2582		97%
	42.0 *		2479		93%
IMR 4350	43.0	MAX.	2730		91%
	41.0		2620		86%
	39.0 *		2510		82%
RL22	46.5 *	MAX.	2740	**	101%
Most Accurate	44.5		2650		96%
Powder Tested	42.5		2560		92%
IMR 4831	44.5	MAX.	2762		94%
	42.5		2657		90%
	40.5 *		2552		86%
H4831	47.0	MAX.	2778		98%
	45.0		2663		93%
	43.0 *		2548		89%
H1000	52.0	MAX.	2790	**	108%
	50.0		2690	**	104%
	48.0 *		2590		100%

BC=Ballistic Coefficient SD=Sectional Density
*Most Accurate Load Tested **Compressed Load

Use Maximum Loads with Caution
Refer to page 73 for additional safety information

Patrick Sweeney

6.5 GRENDEL

In an earlier era, we would probably be describing the 6.5 Grendel as the "6.5-7.62X39 Ackley Improved." Which, were we to do so, would overlook the technical hurdles that Bill Alexander had to overcome in order to see his brainchild to fruition.

It is true that the Grendel has, as its parent case, the 7.62X39. But, it uses a small rifle primer, and it took a lot of work to determine just where the shoulder had to be, the neck length, and body taper, so the Grendel would feed reliably out of an AR-15 magazine. While many other rifles can be chambered in 6.5 Grendel, it was designed for the AR platform.

Part of the shoulder location and neck length calculus that went into its design was the existence and utility of low-drag 6.5 millimeter bullets. In order to use long bullets with high BCs in the strictly limited space of an AR magazine, called for some juggling. Bill juggled his variables until he came up with an amazing long-range performer.

Yes, you can hurl some 6.5 bullets faster, but the high BC of long-range match bullets means they slow down at a lesser rate, and it is not uncommon for these bullets to have retained more velocity out to the target than lighter bullets that started out faster, some much faster.

The free-floated barrel of the AR, combined with a top-quality barrel, means the Grendel can deliver accuracy that back in Ackley's time would have been unheard of out of a self-loading rifle.

The vagaries of the military procurement system almost certainly means we will be using M4s, chambered in 5.56, until armies can transition to "phased plasma rifles" phasers, lasers, or something-yet-to-be-invented. That does not, however, mean that a savvy shooter, and a detail-oriented reloader, can't take advantage of the advances that inventors like Bill have produced.

At the max, the Grendel is an amazing mix of accuracy and controllability. At one demonstration, I recently won an impromptu contest among other gun writers; I scored five hits on a steel plate with an Alexander Arms in 6.5 Grendel in just over seven seconds. The plate: 640 yards away.

If you start loading the 6.5 Grendel, you won't regret it.

Currently the Handguns Editor of Guns and Ammo magazine, Patrick Sweeney is an author and gunsmith best known for his many books on gunsmithing.

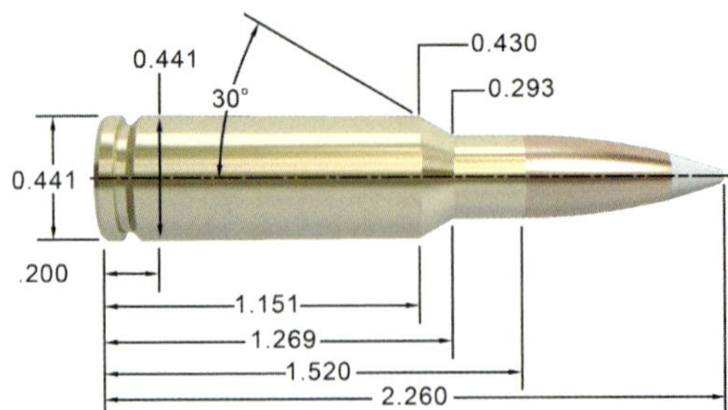

Maximum S.A.A.M.I. Overall Cartridge Length: 2.260"

BULLET CHOICES FOR THE 6.5 GRENDEL

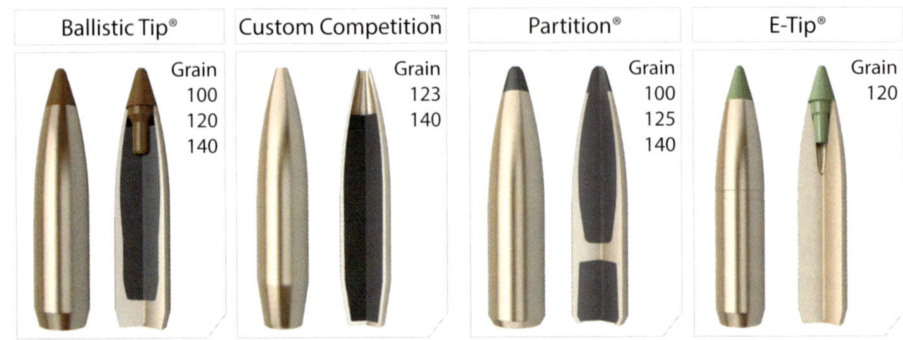

TECHNICAL INFORMATION

Designed for the AR platform and having a greater case capacity than the 223 Remington family of cartridges, the Grendel is an excellent small to medium game cartridge. A Ballistic Tip for deer and a Partition for tougher species such as hogs will do nicely. RL15 and Benchmark are powders which also do a fine job with the Grendel in the accuracy department.

6.5 Grendel - 100 grain		MAXIMUM S.A.A.M.I. O.A.C.L.		2.260"
		TESTED O.A.C.L.	B.C.	S.D.
Ballistic Tip®	100gr. Spitzer	2.245"	0.350	0.205
Partition®	100gr. Spitzer	2.200"	0.326	0.205

CASE TYPE:	Hornady		PRIMER TYPE	Rem 7 1/2
CASE HOLDS:	33.0	Gr. WATER	BARREL Length/Make	24" Pac-Nor
			BARREL Twist	1-8"

POWDER TYPE	POWDER CHG. GRS.		MUZZLE VEL. F.P.S.	LOAD DENSITY (VOLUME)
RL10x	25.0	MAX.	2632	85%
	24.0 *		2559	82%
	23.0		2486	78%
H335	28.5	MAX.	2695	88%
	27.5 *		2570	85%
	26.5		2445	82%
A-2460	28.5	MAX.	2717	87%
	27.5		2639	84%
	26.5 *		2561	81%
TAC	29.5 *	MAX.	2742	91%
	28.5		2653	88%
	27.5		2564	85%
Benchmark	28.5 *	MAX.	2742	93%
	27.5		2673	90%
	26.5		2604	87%
IMR 8208 XBR	29.0	MAX.	2745	97%
Most Accurate Powder Tested	28.0 *		2628	93%
	27.0		2511	90%

BC=Ballistic Coefficient SD=Sectional Density
*Most Accurate Load Tested **Compressed Load

Use Maximum Loads with Caution
Refer to page 73 for additional safety information

6.5 Grendel - 120 grain		MAXIMUM S.A.A.M.I. O.A.C.L.		2.260"
		TESTED O.A.C.L.	B.C.	S.D.
Ballistic Tip®	120gr. Spitzer	2.245"	0.458	0.246
E-Tip®	120gr. Spitzer	2.215"	0.497	0.246

Due to internal construction differences, always begin with starting loads when using E-Tip® products.

CASE TYPE:	Hornady		PRIMER TYPE	Rem 7 1/2
CASE HOLDS:	29.0	Gr. WATER	BARREL Length/Make	24" Pac-Nor
			BARREL Twist	1-8"

POWDER TYPE	POWDER CHG. GRS.		MUZZLE VEL. F.P.S.		LOAD DENSITY (VOLUME)
Benchmark	26.5 *	MAX.	2431		99%
	25.5		2338		95%
	24.5		2245		91%
IMR 8208 XBR	27.5	MAX.	2478	**	104%
	26.5 *		2360		100%
	25.5		2242		97%
A-2460	27.0	MAX.	2490		94%
	26.0 *		2378		91%
	25.0		2266		87%
TAC Most Accurate Powder Tested	28.0	MAX.	2493		99%
	27.0 *		2379		95%
	26.0		2265		91%
RL15	30.0	MAX.	2520	**	108%
	29.0 *		2430	**	104%
	28.0		2344	**	101%

BC=Ballistic Coefficient SD=Sectional Density
*Most Accurate Load Tested **Compressed Load

Use Maximum Loads with Caution
Refer to page 73 for additional safety information

6.5 Grendel - 123/125 grain		MAXIMUM S.A.A.M.I. O.A.C.L.		2.260"
		TESTED O.A.C.L.	B.C.	S.D.
Custom Competition™	123gr. HPBT	2.250"	0.510	0.252
Partition®	125gr. Spitzer	2.240"	0.449	0.256

CASE TYPE:	Hornady		PRIMER TYPE	Rem 7 1/2
CASE HOLDS:	30.0	Gr. WATER	BARREL Length/Make	24" Pac-Nor
			BARREL Twist	1-8"

POWDER TYPE	POWDER CHG. GRS.		MUZZLE VEL. F.P.S.	LOAD DENSITY (VOLUME)
RL15	28.0	* MAX.	2398	97%
	27.0		2305	94%
	26.0		2212	90%
IMR 4895	28.0	* MAX.	2414	** 103%
	27.0		2346	99%
	26.0		2278	95%
IMR 8208 XBR Most Accurate Powder Tested	27.0	MAX.	2420	99%
	26.0	*	2324	95%
	25.0		2228	92%
TAC	27.5	MAX.	2438	94%
	26.5	*	2334	90%
	25.5		2230	87%
A-2520	28.5	* MAX.	2484	98%
	27.5		2392	95%
	26.5		2302	91%

BC=Ballistic Coefficient SD=Sectional Density
*Most Accurate Load Tested **Compressed Load

Use Maximum Loads with Caution
Refer to page 73 for additional safety information

J. Scott Rupp

6.5 CREEDMOOR

A few years ago I had the privilege of being one of the first to see and shoot the 6.5 Creedmoor, which was developed by Joe Thielen and Dave Emary at Hornady. It was originally designed as a round for across-the-course NRA Highpower competition and NRA Long Range competition. I was impressed by how accurate it was and how flat it shot—and how little recoil it generated. My first thought was, "This would make a great hunting round."

A year or so later, Hornady added two hunting loads to the 6.5 Creedmoor lineup, and Ruger began chambering rifles (as have Thompson/Center and Savage since then). About that same time, I got an invite to go on a mule deer hunt, and it took me all of two seconds to order a Ruger No. 1 in the new chambering. On that hunt I killed a buck at a ranged 400 yards, the bullet passing completely through the deer's chest. He dropped in his tracks. When I got home, I sent Ruger a check for the rifle.

Why am I so smitten by it? Whereas the 260 Remington is a necked-down 308 Winchester, the 6.5 Creedmoor is based on the 30 TC—a shorter, more efficient case. And, while the .260 can typically get 100 fps more velocity than the Creedmoor due to its larger case capacity, at comparable velocities the 6.5 gets the job done with less powder, so it's more economical to load and the lighter powder charge translates to less recoil.

Most hunters are already aware of the 6.5mm (.264 inch) bullet's advantages, chief among them are excellent ballistic coefficients and, especially, superior sectional densities. A 130-grain 6.5 has an SD of .266, which bests such popular hunting bullets as the 130-grain .270 (.261), 140-grain 7mm (.248) and 150-grain .308 (.226). Couple that with a high BC (a 130-grain AccuBond in 6.5 has a BC of .488) and you've got a bullet that will drift less in the wind and lose less velocity downrange—and will get the job done when it strikes its target.

All this performance can be housed in a short-action rifle, and the two Rugers and one Savage I've shot it in have proven to be accurate. In short, I think the 6.5 Creedmoor is the best medium-power, all-around big game cartridge to come down the pike in quite a while. It will do anything that justifiably more popular cartridges can do, just with less recoil than most of them. That's a winner in my book.

Scott Rupp

J. Scott Rupp is editor in chief of Rifle-Shooter and Handguns magazines.

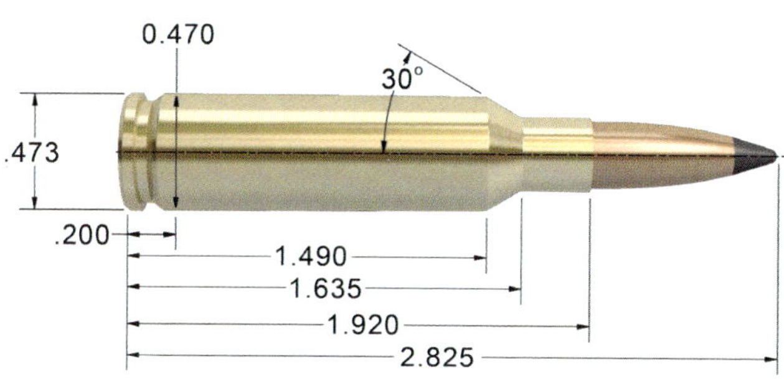

Maximum S.A.A.M.I. Overall Cartridge Length: 2.825"

BULLET CHOICES FOR THE 6.5 CREEDMOOR

Partition®

Grain
100
125
140

6.5 Creedmoor - 100 grain

		MAXIMUM S.A.A.M.I. O.A.C.L.		2.825"
		TESTED O.A.C.L.	B.C.	S.D.
Ballistic Tip®	100gr. Spitzer	2.775"	0.350	0.205
Partition®	100gr. Spitzer	2.740"	0.326	0.205

CASE TYPE:	Hornady		PRIMER TYPE	Fed 210
CASE HOLDS:	50.9	Gr. WATER	BARREL Length/Make	24" Pac-Nor
			BARREL Twist	1-8"

POWDER TYPE	POWDER CHG. GRS.		MUZZLE VEL. F.P.S.	LOAD DENSITY (VOLUME)
IMR 8208 XBR	39.5	MAX.	3129	85%
	37.5		3090	81%
	35.5 *		2935	77%
Varget	42.0	MAX.	3160	89%
	40.0 *		3057	85%
	38.0		2924	81%
RL15 Most Accurate Powder Tested	42.0	MAX.	3195	86%
	40.0		3078	82%
	38.0 *		2984	78%
A-2520	42.0	MAX.	3206	85%
	40.0 *		3087	81%
	38.0		2984	77%
TAC	40.0	MAX.	3212	80%
	38.0		3042	76%
	36.0 *		2934	72%
W748	43.0	MAX.	3231	88%
	41.0 *		3108	84%
	39.0		2992	80%
Big Game	47.0	MAX.	3256	97%
	45.0 *		3161	93%
	43.0		2978	89%
IMR 4007 SSC	46.5	MAX.	3272	98%
	44.5		3149	94%
	42.5 *		3042	89%

BC=Ballistic Coefficient SD=Sectional Density
*Most Accurate Load Tested **Compressed Load

Use Maximum Loads with Caution
Refer to page 73 for additional safety information

6.5 Creedmoor - 120 grain		MAXIMUM S.A.A.M.I. O.A.C.L.		2.825"
		TESTED O.A.C.L.	B.C.	S.D.
Ballistic Tip®	120gr. Spitzer	2.775"	0.458	0.246
E-Tip®	120gr. Spitzer	2.775"	0.497	0.246

Due to internal construction differences, always begin with starting loads when using E-Tip® products.

CASE TYPE:	Hornady		PRIMER TYPE	Fed 210
CASE HOLDS:	48.4	Gr. WATER	BARREL Length/Make	24" Pac-Nor
			BARREL Twist	1-8"

POWDER TYPE	POWDER CHG. GRS.		MUZZLE VEL. F.P.S.		LOAD DENSITY (VOLUME)
IMR 4895 Most Accurate Powder Tested	39.5	* MAX.	2945		90%
	37.5		2821		85%
	35.5		2702		81%
Varget	40.0	MAX.	2951		89%
	38.0	*	2823		85%
	36.0		2711		80%
W760	44.5	MAX.	2973		96%
	42.5		2866		92%
	40.5	*	2728		88%
IMR 4007 SSC	43.5	MAX.	2975		96%
	41.5		2841		92%
	39.5	*	2732		87%
Big Game	44.5	MAX.	2980		97%
	42.5	*	2878		92%
	40.5		2761		88%
H4350	45.0	MAX.	3022		98%
	43.0	*	2894		94%
	41.0		2788		90%
RL17	44.5	MAX.	3060		95%
	42.5	*	2896		91%
	40.5		2780		87%
Hunter	48.5	MAX.	3068	**	105%
	46.5		2959	**	101%
	44.5	*	2843		97%

BC=Ballistic Coefficient SD=Sectional Density
*Most Accurate Load Tested **Compressed Load

Use Maximum Loads with Caution
Refer to page 73 for additional safety information

6.5 Creedmoor - 123/125 grain		MAXIMUM S.A.A.M.I. O.A.C.L.		2.825"
		TESTED O.A.C.L.	B.C.	S.D.
Custom Competition™	123gr. HPBT	2.775"	0.510	0.252
Partition®	125gr. Spitzer	2.790"	0.449	0.256

CASE TYPE:	Hornady		PRIMER TYPE	Fed 210
CASE HOLDS:	49.5	Gr. WATER	BARREL Length/Make	24" Pac-Nor
			BARREL Twist	1-8"

POWDER TYPE	POWDER CHG. GRS.		MUZZLE VEL. F.P.S.	LOAD DENSITY (VOLUME)
Varget	39.0	MAX.	2855	85%
	37.0		2763	81%
	35.0 *		2644	76%
IMR 4895	39.0	MAX.	2893	87%
	37.0 *		2781	82%
	35.0		2668	78%
IMR 4007 SSC Most Accurate Powder Tested	43.0	MAX.	2901	93%
	41.0 *		2803	89%
	39.0		2705	84%
Big Game	43.5 *	MAX.	2907	93%
	41.5		2790	88%
	39.5		2685	84%
W760	44.0	MAX.	2927	93%
	42.0		2821	89%
	40.0 *		2703	85%
RL15	40.5	MAX.	2937	85%
	38.5		2828	81%
	36.5 *		2711	77%
H4350	44.5	MAX.	2985	95%
	42.5 *		2877	91%
	40.5		2772	87%
RL17	44.5	MAX.	3001	93%
	42.5 *		2881	89%
	40.5		2759	85%

BC=Ballistic Coefficient SD=Sectional Density
*Most Accurate Load Tested **Compressed Load

Use Maximum Loads with Caution
Refer to page 73 for additional safety information

6.5 Creedmoor - 130 grain			MAXIMUM S.A.A.M.I. O.A.C.L.		2.825"
			TESTED O.A.C.L.	B.C.	S.D.
AccuBond®		130gr. Spitzer	2.775"	0.488	0.266

CASE TYPE:	Hornady		PRIMER TYPE	Fed 210
CASE HOLDS:	47.1	Gr. WATER	BARREL Length/Make	24" Pac-Nor
			BARREL Twist	1-8"

POWDER TYPE	POWDER CHG. GRS.		MUZZLE VEL. F.P.S.	LOAD DENSITY (VOLUME)
RL15 Most Accurate Powder Tested	39.0 *	MAX.	2810	86%
	37.0		2687	82%
	35.0		2581	77%
Varget	38.5	MAX.	2812	88%
	36.5 *		2699	84%
	34.5		2586	79%
W748	39.5	MAX.	2822	87%
	37.5		2693	83%
	35.5 *		2572	79%
Big Game	43.0 *	MAX.	2839	96%
	41.0		2741	92%
	39.0		2631	87%
H4350	43.0	MAX.	2865	97%
	41.0 *		2751	92%
	39.0		2639	88%
W760	44.0	MAX.	2906	98%
	42.0		2775	93%
	40.0 *		2670	89%
Hybrid 100V	43.5	MAX.	2935	95%
	41.5		2800	91%
	39.5 *		2665	86%
RL17	43.5	MAX.	2953	98%
	41.5		2783	94%
	39.5 *		2660	89%

BC=Ballistic Coefficient SD=Sectional Density
*Most Accurate Load Tested **Compressed Load

Use Maximum Loads with Caution
Refer to page 73 for additional safety information

6.5 Creedmoor - 140 grain		MAXIMUM S.A.A.M.I. O.A.C.L.		2.825"
		TESTED O.A.C.L.	B.C.	S.D.
AccuBond®	140gr. Spitzer	2.805"	0.509	0.287
Ballistic Tip®	140gr. Spitzer	2.805"	0.509	0.287
Custom Competition™	140gr. HPBT	2.775"	0.529	0.287
Partition®	140gr. Spitzer	2.790"	0.490	0.287

CASE TYPE:	Hornady		PRIMER TYPE	Fed 210
CASE HOLDS:	46.8	Gr. WATER	BARREL Length/Make	24" Pac-Nor
			BARREL Twist	1-8"

POWDER TYPE	POWDER CHG. GRS.		MUZZLE VEL. F.P.S.	LOAD DENSITY (VOLUME)
IMR 8208 XBR	33.5	MAX.	2510	79%
Most Accurate Powder Tested	31.5	*	2392	74%
	29.5		2268	69%
IMR 4895	36.0	* MAX.	2620	85%
	34.0		2492	80%
	32.0		2384	75%
RL15	36.5	* MAX.	2635	81%
	34.5		2502	77%
	32.5		2369	72%
Hybrid 100V	40.0	* MAX.	2658	88%
	38.0		2547	84%
	36.0		2435	79%
IMR 4007 SSC	40.0	MAX.	2668	92%
	38.0		2556	87%
	36.0	*	2426	82%
W760	40.5	MAX.	2672	91%
	38.5	*	2543	86%
	36.5		2436	82%
H4350	41.0	MAX.	2699	93%
	39.0		2571	88%
	37.0	*	2485	84%
Hunter	44.0	MAX.	2730	99%
	42.0		2603	94%
	40.0	*	2479	90%

BC=Ballistic Coefficient SD=Sectional Density
*Most Accurate Load Tested **Compressed Load

Use Maximum Loads with Caution
Refer to page 73 for additional safety information

Donald Polacek

260 REMINGTON

The 260 Remington was originally conceived by Handloader's own noted gunwriter Ken Waters in 1956. It was a wildcat called the 263 Express. P.O. Ackley stated that it was a 308 Winchester necked down to accept .264 inch bullets. While this will work fine, it was easier for Ken to neck up the 243 Winchester. The cartridge was largely forgotten until Jim Carmichael revisited this wildcat. While 6.5mm cartridges are widely popular in Europe, they never really caught on here in the U.S.A. This changed in 1997 when Remington officially adopted the 260 into its product line.

The 260 Remington is known for its deep penetration and moderate recoil. When using 100-grain Ballistic Tips, the 260 Remington is perfect for varmint hunting. Move up to a 140-grain Nosler Partition, and you are ready to take down large game. With this bullet you can drop an elk in its tracks.

This cartridge is perfect for new shooters. Many kids hunting at the age of 10 have trouble with recoil. The 260 Remington has light recoil making it a pleasure for young shooters.

My son Jeremiah took his first elk at age 11. He was small for his age, but the recoil did not bother him at all. He became quite proficient at the range. The elk was taken at 85 yards, and it dropped in its tracks. I was amazed at the deep penetration and the wound channel. Two years later, Jeremiah took his first deer. Again, one shot dropped it where it stood.

The 260 Remington is ideal for a child's first rifle. They will not be afraid of the recoil, and with practice, it will deliver great accuracy and deep penetration. When coyotes enter the yard, my son grabs for his 260 Remington. Most of the time, they don't get away.

Donald Polacek
President
Wolfe Publishing Co.

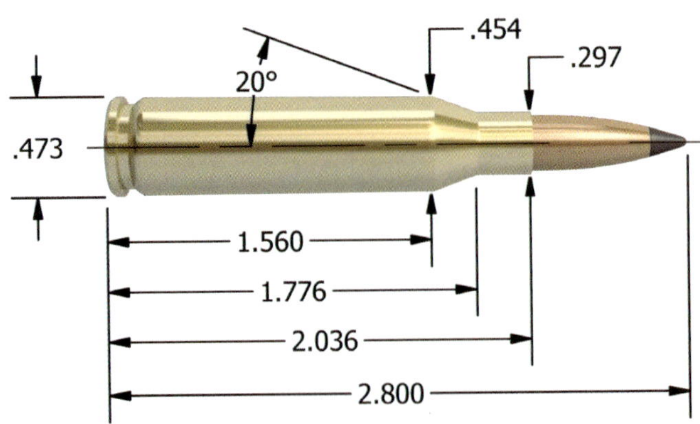

Maximum S.A.A.M.I. Overall Cartridge Length: 2.800"

BULLET CHOICES FOR THE 260 REMINGTON

TECHNICAL INFORMATION

Shortly after this cartridge was introduced, Nosler designers created a new Partition weight that generated optimum velocity for hunting applications. The 100gr. Partition Spitzer soon became a favorite of whitetail hunters throughout the U.S. Like its .308 cousins, the .260 is easy to load and will handle a wide variety of bullet/powder combinations.

260 Remington - 100 grain

260 Remington - 100 grain		MAXIMUM S.A.A.M.I. O.A.C.L.		2.800"
		TESTED O.A.C.L.	B.C.	S.D.
Ballistic Tip®	100gr. Spitzer	2.800"	0.350	0.205
Partition®	100gr. Spitzer	2.800"	0.326	0.205

CASE TYPE:	Nosler		PRIMER TYPE	Rem 9 1/2
CASE HOLDS:	50.2	Gr. WATER	BARREL Length/Make	24" Wiseman
			BARREL Twist	1-9"

POWDER TYPE	POWDER CHG. GRS.		MUZZLE VEL. F.P.S.		LOAD DENSITY (VOLUME)
A-2460	39.0	MAX.	3156		78%
	37.0		2990		74%
	35.0 *		2815		70%
Varget	43.0	MAX.	3230		93%
	41.0 *		3105		88%
	39.0		3000		84%
RL15	43.0	MAX.	3250		89%
	41.0 *		3110		85%
	39.0		3005		81%
Viht N540	43.0 *	MAX.	3260		95%
	41.0		3110		91%
	39.0		3000		86%
IMR 4350	47.5	MAX.	3265		100%
	45.5		3075		96%
	43.5 *		2985		92%
H4350	48.5	MAX.	3277	**	102%
	46.5		3162		98%
	44.5 *		3041		94%
IMR 4064 Most Accurate Powder Tested	43.0	MAX.	3290		94%
	41.0 *		3135		90%
	39.0		3005		85%
Big Game	47.5	MAX.	3314		100%
	45.5		3186		95%
	43.5 *		3068		91%
W760	48.0 *	MAX.	3330		100%
	46.0		3200		96%
	44.0		3095		92%
H414	49.0	MAX.	3365	**	102%
	47.0		3240		98%
	45.0 *		3110		93%

BC–Ballistic Coefficient SD=Sectional Density
*Most Accurate Load Tested **Compressed Load

Use Maximum Loads with Caution
Refer to page 73 for additional safety information

260 Remington - 120 grain		MAXIMUM S.A.A.M.I. O.A.C.L.		2.800"
		TESTED O.A.C.L.	B.C.	S.D.
Ballistic Tip®	120gr. Spitzer	2.800"	0.458	0.246
E-Tip®	120gr. Spitzer	2.770"	0.497	0.246

Due to internal construction differences, always begin with starting loads when using E-Tip® products.

CASE TYPE:	Nosler		PRIMER TYPE	Rem 9 1/2
CASE HOLDS:	47.8	Gr. WATER	BARREL Length/Make	24" Wiseman
			BARREL Twist	1-9"

POWDER TYPE	POWDER CHG. GRS.		MUZZLE VEL. F.P.S.		LOAD DENSITY (VOLUME)
Varget	40.0	MAX.	2870		90%
	38.0		2760		86%
	36.0	*	2640		81%
RL15	40.0	* MAX.	2895		87%
	38.0		2770		83%
	36.0		2690		78%
IMR 4064	39.0	* MAX.	2898		90%
	37.0		2775		85%
	35.0		2655		80%
H380	44.5	* MAX.	2944		98%
	42.5		2829		94%
	40.5		2714		89%
IMR 4350	45.5	MAX.	2979	**	101%
	43.5	*	2867		96%
	41.5		2796		92%
RL19	47.0	* MAX.	2980	**	107%
	45.0		2860	**	102%
	43.0		2780		98%
H414	45.0	* MAX.	3005		98%
	43.0		2864		94%
	41.0		2724		89%
W760	45.0	* MAX.	3016		99%
	43.0		2846		94%
	41.0		2747		90%
Hunter	48.0	MAX.	3043	**	106%
Most Accurate	46.0		2959	**	101%
Powder Tested	44.0	*	2877		97%
Viht N560	48.0	MAX.	3049	**	112%
	46.0		2914	**	107%
	44.0	*	2772	**	102%

BC=Ballistic Coefficient SD=Sectional Density
*Most Accurate Load Tested **Compressed Load

Use Maximum Loads with Caution
Refer to page 73 for additional safety information

260 Remington - 123/125 grain		MAXIMUM S.A.A.M.I. O.A.C.L.		2.800"
		TESTED O.A.C.L.	B.C.	S.D.
Custom Competition™	123gr. HPBT	2.800"	0.510	0.252
Partition®	125gr. Spitzer	2.800"	0.449	0.256

CASE TYPE:	Nosler		PRIMER TYPE	Rem 9 1/2
CASE HOLDS:	47.8	Gr. WATER	BARREL Length/Make	24" Wiseman
			BARREL Twist	1-9"

POWDER TYPE	POWDER CHG. GRS.		MUZZLE VEL. F.P.S.	LOAD DENSITY (VOLUME)
Varget	37.5	MAX.	2780	85%
	35.5 *		2656	80%
	33.5		2558	76%
IMR 4350	43.0	MAX.	2829	95%
	41.0 *		2718	91%
	39.0		2621	86%
H380	42.0 *	MAX.	2831	92%
	40.0		2710	88%
	38.0		2643	84%
W760	43.0 *	MAX.	2849	92%
	41.0		2719	88%
	39.0		2616	84%
Hunter	45.0	MAX.	2882	94%
	43.0 *		2807	90%
	41.0		2759	85%
RL15	40.0 *	MAX.	2902	99%
	38.0		2800	95%
	36.0		2691	90%
IMR 4064	39.0 *	MAX.	2903	87%
	37.0		2778	83%
	35.0		2653	78%
Viht N560	47.0	MAX.	2970	90%
	45.0		2846	85%
	43.0 *		2732	80%
RL19	47.0 *	MAX.	2980 **	109%
Most Accurate	45.0		2860 **	105%
Powder Tested	43.0		2780	100%

BC=Ballistic Coefficient SD=Sectional Density
*Most Accurate Load Tested **Compressed Load

Use Maximum Loads with Caution
Refer to page 73 for additional safety information

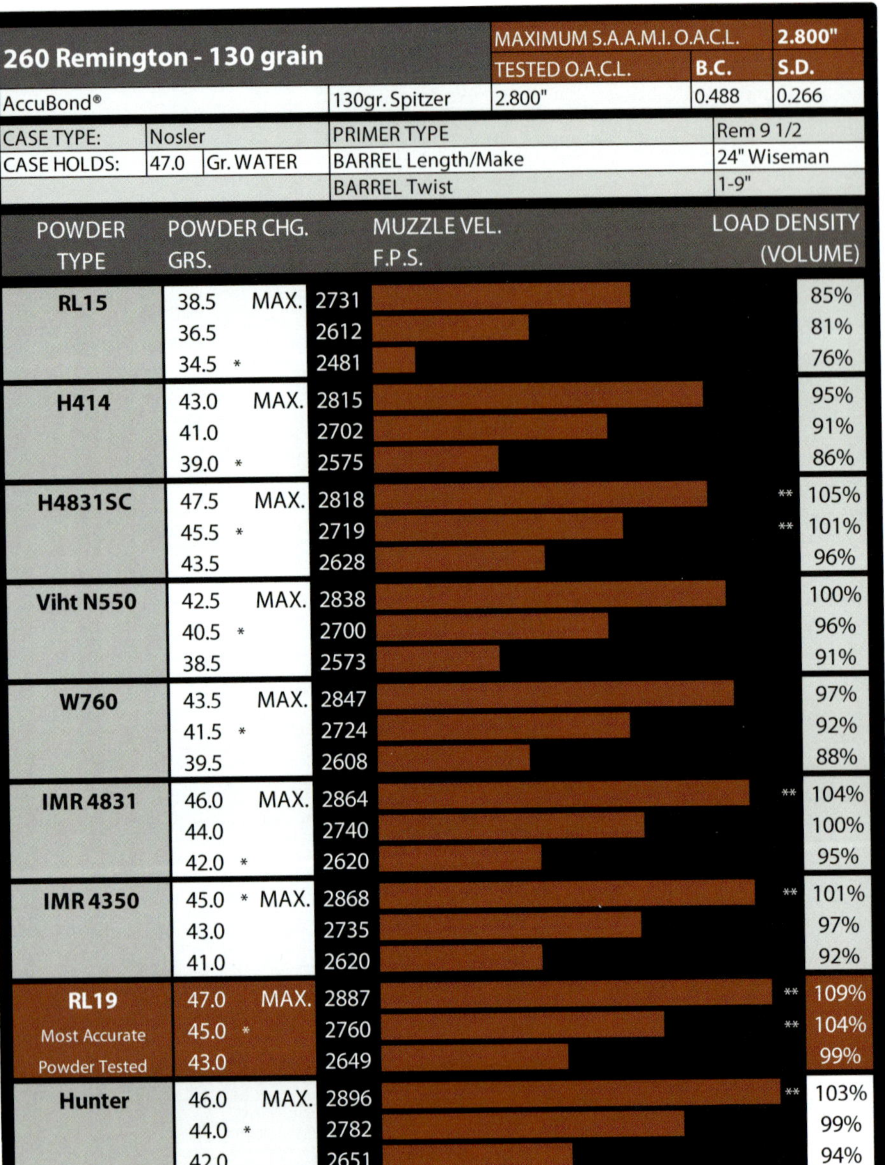

260 Remington - 130 grain

			MAXIMUM S.A.A.M.I. O.A.C.L.		2.800"
			TESTED O.A.C.L.	B.C.	S.D.
AccuBond®		130gr. Spitzer	2.800"	0.488	0.266

CASE TYPE:	Nosler		PRIMER TYPE	Rem 9 1/2
CASE HOLDS:	47.0	Gr. WATER	BARREL Length/Make	24" Wiseman
			BARREL Twist	1-9"

POWDER TYPE	POWDER CHG. GRS.		MUZZLE VEL. F.P.S.		LOAD DENSITY (VOLUME)
RL15	38.5	MAX.	2731		85%
	36.5		2612		81%
	34.5 *		2481		76%
H414	43.0	MAX.	2815		95%
	41.0		2702		91%
	39.0 *		2575		86%
H4831SC	47.5	MAX.	2818	**	105%
	45.5 *		2719	**	101%
	43.5		2628		96%
Viht N550	42.5	MAX.	2838		100%
	40.5 *		2700		96%
	38.5		2573		91%
W760	43.5	MAX.	2847		97%
	41.5 *		2724		92%
	39.5		2608		88%
IMR 4831	46.0	MAX.	2864	**	104%
	44.0		2740		100%
	42.0 *		2620		95%
IMR 4350	45.0 *	MAX.	2868	**	101%
	43.0		2735		97%
	41.0		2620		92%
RL19	47.0	MAX.	2887	**	109%
Most Accurate	45.0 *		2760	**	104%
Powder Tested	43.0		2649		99%
Hunter	46.0	MAX.	2896	**	103%
	44.0 *		2782		99%
	42.0		2651		94%

BC=Ballistic Coefficient SD=Sectional Density
*Most Accurate Load Tested **Compressed Load

Use Maximum Loads with Caution
Refer to page 73 for additional safety information

260 Remington - 140 grain

		MAXIMUM S.A.A.M.I. O.A.C.L.		2.800"
		TESTED O.A.C.L.	B.C.	S.D.
AccuBond®	140gr. Spitzer	2.800"	0.509	0.287
Ballistic Tip®	140gr. Spitzer	2.800"	0.509	0.287
Custom Competition™	140gr. HPBT	2.800"	0.529	0.287
Partition®	140gr. Spitzer	2.800"	0.490	0.287

CASE TYPE:	Nosler		PRIMER TYPE	Rem 9 1/2
CASE HOLDS:	47.4	Gr. WATER	BARREL Length/Make	24" Wiseman
			BARREL Twist	1-9"

POWDER TYPE	POWDER CHG. GRS.		MUZZLE VEL. F.P.S.		LOAD DENSITY (VOLUME)
IMR 4064	37.0	MAX.	2683		86%
	35.0	*	2558		81%
	33.0		2452		77%
H4831SC Most Accurate Powder Tested	46.0	MAX.	2774	**	101%
	44.0	*	2655		97%
	42.0		2565		92%
H1000	50.0	MAX.	2785	**	110%
	48.0		2686	**	106%
	46.0	*	2593	**	102%
IMR 4350	43.0	MAX.	2787		96%
	41.0		2650		92%
	39.0	*	2528		87%
RL19	45.0	MAX.	2797	**	103%
	43.0		2676		99%
	41.0	*	2552		94%
Hunter	45.0	MAX.	2811		100%
	43.0		2713		95%
	41.0	*	2618		91%
IMR 4831	44.0	MAX.	2820		99%
	42.0		2696		94%
	40.0	*	2566		90%
IMR 7828	46.5	MAX.	2822	**	104%
	44.5		2677		100%
	42.5	*	2569		95%
Viht N560	45.5	MAX.	2830	**	107%
	43.5	*	2725	**	102%
	41.5		2570		97%

BC=Ballistic Coefficient SD=Sectional Density
*Most Accurate Load Tested **Compressed Load

Use Maximum Loads with Caution
Refer to page 73 for additional safety information

Layne Simpson

6.5-284 NORMA

Soon after the introduction of the 284 Winchester in 1963, wildcatters across this great land shouted with glee as they pounced on its case and necked it both up and down. Once all the dust had settled, it was clear that the 6mm-284 was the one that every varmint shooter and deer hunter lusted for while the .25-284 stood in a distant second place. Fast-forward close to half a century and while we don't hear a lot about those two today, a littermate of larger caliber that few paid attention to in the old days is now getting its share of attention both here and abroad.

Back then the 6.5-284 was known as a short-action cartridge and it remained so until long-distance target shooters began to seat bullets further out of the case and use it in rifles with longer actions. That prompted the Swedish firm of Norma to adopt the cartridge in 2001 and call it the 6.5-284 Norma. Whereas maximum overall cartridge length for the old wildcat was usually held at around 2.800 inches, the factory cartridge is at 3.228 inches. This means a number of things, two of which should be kept in mind; the 6.5-284 Norma is too long for a short-action rifle and maximum loads published for it are generally too hot for a rifle chambered for the old wildcat.

I have not hunted with the 6.5-284 Norma all that much but I have used it in an incredibly accurate Cooper Model 22 to take my best-ever whitetail deer and pronghorn antelope. I have not tried this cartridge in a great many different rifles but those I have shot had one thing in common---they were exceptionally accurate. One of my favorite loads is the 125-grain Partition pushed along at 3100 fps by Reloder 22 powder. I also like the 130-grain AccuBond.

Layne is Senior Field Editor for Shooting Times, RifleShooter and Guns & Ammo, and Senior Writer for BuckMasters GunHunter. He has also authored books on rifles, shotguns, handguns and hunting.

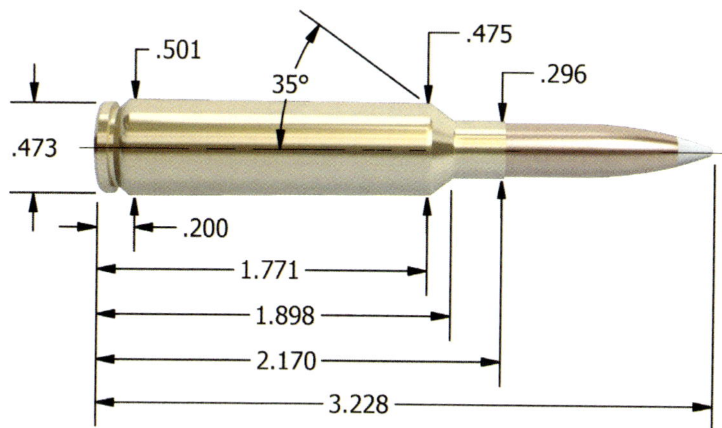

Maximum Overall Cartridge Length: 3.228"

BULLET CHOICES FOR THE 6.5-284 NORMA

TECHNICAL INFORMATION

The data presented in this book was developed in a true 6.5-284 Norma chamber. Since the 6.5-284 has a long history as a "wildcat" cartridge, variations in chamber dimensions are likely. The other most common wildcat variation of this cartridge is the 6.5mm-284 Winchester. The Winchester chambers are frequently short-throated, and **WILL** yield significantly higher pressures when compared with the 6.5mm-284 Norma. We suggest starting with the mildest load listed, and working up in half-grain increments. Always watch for visible signs of pressure as you increase your charge weight. NEVER exceed the maximum loads listed in this data.

6.5-284 Norma - 100 grain

		MAXIMUM O.A.C.L.		3.228"
		TESTED O.A.C.L.	B.C.	S.D.
Ballistic Tip®	100gr. Spitzer	2.980"	0.350	0.205
Partition®	100gr. Spitzer	2.980"	0.326	0.205

CASE TYPE:	Nosler		PRIMER TYPE	Fed 210M
CASE HOLDS:	60.8	Gr. WATER	BARREL Length/Make	26" Wiseman
			BARREL Twist	1-9"

POWDER TYPE	POWDER CHG. GRS.		MUZZLE VEL. F.P.S.	LOAD DENSITY (VOLUME)
RL15	48.0 *	MAX.	3342	82%
	46.0		3281	79%
	44.0		3187	75%
Varget	48.0 *	MAX.	3358	85%
	46.0		3272	82%
	44.0		3139	78%
H4831SC	57.5	MAX.	3365	99%
	55.5 *		3264	95%
	53.5		3183	92%
IMR 4350	53.0 *	MAX.	3380	92%
	51.0		3258	89%
	49.0		3160	85%
W760	52.5 *	MAX.	3399	90%
	50.5		3325	87%
	48.5		3232	84%
RL19	58.0	MAX.	3430	** 104%
	56.0 *		3349	100%
	54.0		3224	97%
Viht N160 Most Accurate Powder Tested	57.0	MAX.	3469	** 104%
	55.0 *		3369	** 101%
	53.0		3352	97%
H4350	55.0	MAX.	3470	96%
	53.0 *		3350	92%
	51.0		3246	89%
Hunter	57.5	MAX.	3480	100%
	55.5 *		3357	96%
	53.5		3236	93%

BC=Ballistic Coefficient SD=Sectional Density
*Most Accurate Load Tested **Compressed Load

Use Maximum Loads with Caution
Refer to page 73 for additional safety information

6.5-284 Norma - 120 grain		MAXIMUM O.A.C.L.		3.228"
		TESTED O.A.C.L.	B.C.	S.D.
Ballistic Tip®	120gr. Spitzer	2.980"	0.458	0.246
E-Tip®	120gr. Spitzer	2.980"	0.497	0.246

Due to internal construction differences, always begin with starting loads when using E-Tip® products.

CASE TYPE:	Nosler		PRIMER TYPE	Fed 210M
CASE HOLDS:	59.0	Gr. WATER	BARREL Length/Make	26" Wiseman
			BARREL Twist	1-9"

POWDER TYPE	POWDER CHG. GRS.		MUZZLE VEL. F.P.S.	LOAD DENSITY (VOLUME)
H4831SC	54.0	MAX.	3008	95%
	52.0		2898	92%
	50.0 *		2795	88%
RL15	45.0 *	MAX.	3011	79%
	43.0		2926	76%
	41.0		2829	72%
Varget	46.0	MAX.	3046	84%
	44.0 *		2966	81%
	42.0		2856	77%
H414	49.5 *	MAX.	3063	87%
	47.5		2973	84%
	45.5		2879	80%
W760	49.5	MAX.	3070	88%
	47.5 *		2971	84%
	45.5		2870	81%
IMR 4350	50.5	MAX.	3078	91%
	48.5		2962	87%
	46.5 *		2867	83%
Viht N160	53.0	MAX.	3133	100%
	51.0		3019	96%
	49.0 *		2909	92%
Hunter	53.5	MAX.	3150	95%
	51.5		3030	92%
	49.5 *		2938	88%
RL19 Most Accurate Powder Tested	54.5	MAX.	3175	100%
	52.5		3052	97%
	50.5 *		2946	93%

BC=Ballistic Coefficient SD=Sectional Density
*Most Accurate Load Tested **Compressed Load

Use Maximum Loads with Caution
Refer to page 73 for additional safety information

6.5-284 Norma - 123/125 grain		MAXIMUM O.A.C.L.		3.228"
		TESTED O.A.C.L.	B.C.	S.D.
Custom Competition™	123gr. HPBT	2.980"	0.326	0.205
Partition®	125gr. Spitzer	2.980"	0.350	0.205

CASE TYPE:	Nosler		PRIMER TYPE	Fed 210M
CASE HOLDS:	59.7	Gr. WATER	BARREL Length/Make	26" Wiseman
			BARREL Twist	1-9"

POWDER TYPE	POWDER CHG. GRS.		MUZZLE VEL. F.P.S.		LOAD DENSITY (VOLUME)
H4831SC	53.5 *	MAX.	3029		93%
	51.5		2940		90%
	49.5		2851		86%
Hunter	52.0	MAX.	3041		92%
	50.0 *		2972		88%
	48.0		2878		85%
W760	49.0	MAX.	3050		86%
	47.0		2963		82%
	45.0 *		2881		79%
IMR 4350	50.0	MAX.	3052		89%
	48.0		2910		85%
	46.0 *		2828		82%
Viht N160	52.0	MAX.	3066		97%
	50.0		2964		93%
	48.0 *		2854		89%
Viht N165 Most Accurate Powder Tested	56.5 *	MAX.	3076	**	105%
	54.5		2976	**	101%
	52.5		2840		98%
IMR 4831	51.5	MAX.	3106		92%
	49.5		2935		88%
	47.5 *		2816		85%
RL19	54.0 *	MAX.	3112		98%
	52.0		3009		95%
	50.0		2915		91%
RL22	54.5	MAX.	3122		99%
	52.5		2982		96%
	50.5 *		2893		92%

BC=Ballistic Coefficient SD=Sectional Density
*Most Accurate Load Tested **Compressed Load

Use Maximum Loads with Caution
Refer to page 73 for additional safety information

6.5-284 Norma - 130 grain

		MAXIMUM O.A.C.L.		3.228"
		TESTED O.A.C.L.	B.C.	S.D.
AccuBond®	130gr. Spitzer	2.980"	0.488	0.266

CASE TYPE:	Nosler		PRIMER TYPE	Fed 210M
CASE HOLDS:	59.4	Gr. WATER	BARREL Length/Make	26" Wiseman
			BARREL Twist	1-9"

POWDER TYPE	POWDER CHG. GRS.		MUZZLE VEL. F.P.S.	LOAD DENSITY (VOLUME)
H4831SC	51.5	MAX.	2927	90%
	49.5 *		2829	90%
	47.5		2776	87%
H414	48.0 *	MAX.	2979	84%
	46.0		2893	81%
	44.0		2806	77%
W760	48.5	MAX.	2987	85%
Most Accurate	46.5		2882	82%
Powder Tested	44.5 *		2797	78%
Viht N160	50.5	MAX.	3004	94%
	48.5		2892	91%
	46.5 *		2808	87%
IMR 4350	50.0	MAX.	3023	89%
	48.0		2897	86%
	46.0 *		2812	82%
RL22	54.0	MAX.	3053	99%
	52.0 *		2940	95%
	50.0		2832	91%
IMR 4831	52.0	MAX.	3070	93%
	50.0 *		2956	90%
	48.0		2849	86%
Hunter	51.5	MAX.	3092	91%
	49.5 *		2939	88%
	47.5		2842	84%
RL19	53.5	MAX.	3096	98%
	51.5 *		2985	94%
	49.5		2880	91%

BC=Ballistic Coefficient SD=Sectional Density
*Most Accurate Load Tested **Compressed Load

Use Maximum Loads with Caution
Refer to page 73 for additional safety information

6.5-284 Norma - 140 grain		MAXIMUM O.A.C.L.		3.228"
		TESTED O.A.C.L.	B.C.	S.D.
AccuBond®	140gr. Spitzer	2.980"	0.509	0.287
Ballistic Tip®	140gr. Spitzer	2.980"	0.509	0.287
Custom Competition™	140gr. HPBT	2.980"	0.529	0.287
Partition®	140gr. Spitzer	2.980"	0.490	0.287

CASE TYPE:	Nosler		PRIMER TYPE	Fed 210M
CASE HOLDS:	58.4	Gr. WATER	BARREL Length/Make	26" Wiseman
			BARREL Twist	1-9"

POWDER TYPE	POWDER CHG. GRS.		MUZZLE VEL. F.P.S.	LOAD DENSITY (VOLUME)
H4831SC	49.0 *	MAX.	2769	87%
	47.0		2682	84%
	45.0		2601	80%
W760 Most Accurate Powder Tested	46.5	MAX.	2809	83%
	44.5		2694	80%
	42.5 *		2593	76%
H414	46.5 *	MAX.	2814	83%
	44.5		2725	79%
	42.5		2620	76%
Viht N160	48.5	MAX.	2816	92%
	46.5 *		2718	88%
	44.5		2611	85%
IMR 4350	48.0	MAX.	2874	87%
	46.0		2733	83%
	44.0 *		2656	80%
Hunter	50.0	MAX.	2884	90%
	48.0		2785	87%
	46.0 *		2676	83%
IMR 4831	50.0	MAX.	2924	91%
	48.0		2777	87%
	46.0 *		2673	84%
RL22	52.5	MAX.	2925	98%
	50.5 *		2761	94%
	48.5		2676	90%
RL19	52.0	MAX.	2953	97%
	50.0 *		2847	93%
	48.0		2746	89%

BC=Ballistic Coefficient SD=Sectional Density
*Most Accurate Load Tested **Compressed Load

Use Maximum Loads with Caution
Refer to page 73 for additional safety information

6.5-06 A-SQUARE

This is an interesting cartridge. Right after the .30-06 was adopted as the standard U. S. Army infantry rifle cartridge, Charles Newton necked the .30-06 cartridge to a variety of different calibers. One of those calibers was called the .256 Newton. Despite the .256 designation, the .256 Newton used .264 (6.5mm) diameter bullets (go figure). Western Cartridge made .256 Newton ammunition up until about 1938.

While the .256 Newton didn't have enough buyers to stay on the factory ammo list, the wildcatters renamed the cartridge the 6.5-06 (undoubtedly with slight dimensional differences since there are seldom two wildcats exactly alike) and it maintained favor with plenty of shooters. One of the advantages of the 6.5 bullets that were available at the time was a ballistic coefficient superior to most any other bullet. The 6.5-06 found a good home with long-range shooters.

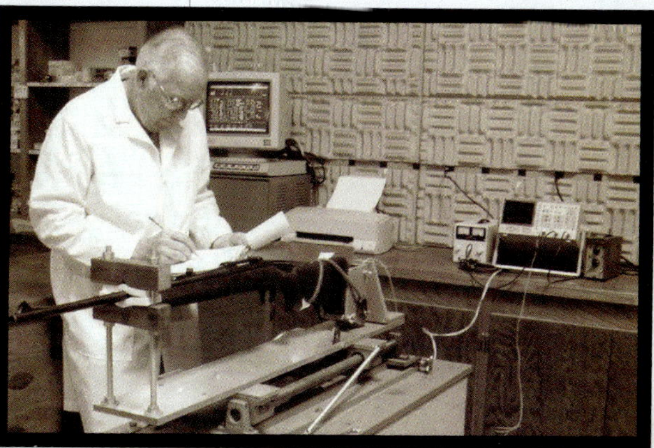

Before WWII the metric calibers found little favor in the U. S. While the 6.5 bullets were excellent, the 6.5 calibers never really took off. Since then, led mostly by Remington starting with the 7mm Remington Magnum and continuing with the 6mm Remington, the metric designations have become common. Winchester seems to have an aversion to metric names and that may partially explain that they promoted the .270 instead of a 6.5mm version of the .30-06.

Here are two other cartridges that should be mentioned along with the 6.5-06. They are the 6.5-.284 and the 6.5 Remington Magnum. Both have exactly the same powder capacity as the 6.5-06 and at the same chamber pressures will produce essentially the same performance. The 6.5 Remington Magnum got a bum rap right from the beginning and soon fell off the charts, but has been revived recently. The wildcat 6.5-.284 has established itself as a premier long-range target cartridge but it, in-fact, does nothing the 6.5-06 won't do.

With lighter bullets the 6.5-06 makes an excellent medium to heavy varmint cartridge.

Bob Forker

Bob Forker is Technical Editor for Guns & Ammo Magazine.

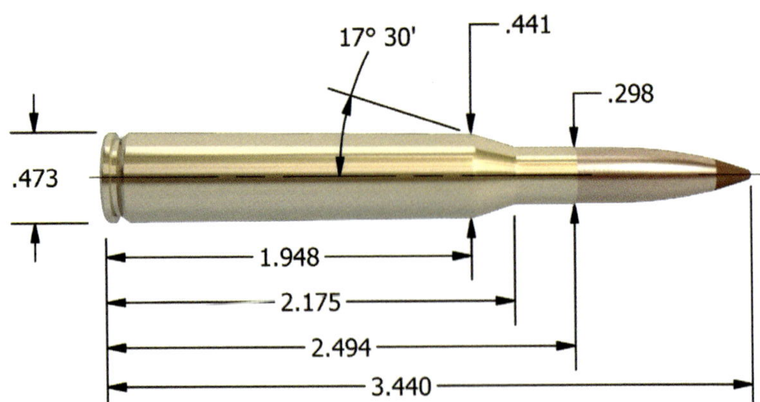

Maximum S.A.A.M.I. Overall Cartridge Length: 3.440"

BULLET CHOICES FOR THE 6.5-06 A-SQUARE

TECHNICAL INFORMATION

Since the 6.5mm-06 A-Square is a true "wildcat" cartridge, variations in chamber dimensions are likely. We suggest starting with the mildest load listed, and working up in half-grain increments. Always watch for visible signs of pressure as you increase your charge weight. NEVER exceed the maximum loads listed in this data. S.A.A.M.I. approved this cartridge in 1999.

6.5mm-06 A-Square - 100 grain

		MAXIMUM S.A.A.M.I. O.A.C.L.		3.440"
		TESTED O.A.C.L.	B.C.	S.D.
Ballistic Tip®	100gr. Spitzer	3.250"	0.350	0.205
Partition®	100gr. Spitzer	3.250"	0.326	0.205

CASE TYPE:	Winchester		PRIMER TYPE	WLR
CASE HOLDS:	65.7	Gr. WATER	BARREL Length/Make	24" Lilja
			BARREL Twist	1-9"

POWDER TYPE	POWDER CHG. GRS.		MUZZLE VEL. F.P.S.	LOAD DENSITY (VOLUME)
IMR 4350 Most Accurate Powder Tested	55.5	MAX.	3367	89%
	53.5		3229	86%
	51.5 *		3090	83%
IMR 7828	60.0 *	MAX.	3413	97%
	58.0		3268	94%
	56.0		3134	91%
RL19	58.0 *	MAX.	3420	96%
	56.0		3287	93%
	54.0		3154	89%
Viht N165	60.5	MAX.	3422	** 102%
	58.5		3301	99%
	56.5 *		3179	96%
H4831	60.0	MAX.	3425	95%
	58.0		3309	92%
	56.0 *		3193	89%
W760	53.5	MAX.	3520	85%
	51.5		3428	82%
	49.5 *		3319	79%
H4350	54.0	MAX.	3532	87%
	52.0 *		3447	84%
	50.0		3380	81%
Viht N160	56.0	MAX.	3533	95%
	54.0		3429	91%
	52.0 *		3328	88%
Hunter	57.0	MAX.	3547	91%
	55.0		3451	88%
	53.0 *		3324	85%

BC=Ballistic Coefficient SD=Sectional Density
*Most Accurate Load Tested **Compressed Load

Use Maximum Loads with Caution
Refer to page 73 for additional safety information

6.5mm-06 A-Square - 120 grain		MAXIMUM S.A.A.M.I. O.A.C.L.		3.440"
		TESTED O.A.C.L.	B.C.	S.D.
Ballistic Tip®	120gr. Spitzer	3.250"	0.458	0.246
E-Tip®	120gr. Spitzer	3.250"	0.497	0.246

Due to internal construction differences, always begin with starting loads when using E-Tip® products.

CASE TYPE:	Winchester		PRIMER TYPE	WLR
CASE HOLDS:	63.3	Gr. WATER	BARREL Length/Make	24" Lilja
			BARREL Twist	1-9"

POWDER TYPE	POWDER CHG. GRS.		MUZZLE VEL. F.P.S.	LOAD DENSITY (VOLUME)
H4831 Most Accurate Powder Tested	56.5	MAX.	2938	93%
	54.5		2813	90%
	52.5 *		2588	86%
IMR 7828	57.5 *	MAX.	3070	97%
	55.5		2920	93%
	53.5		2770	90%
Viht N160	52.0 *	MAX.	3118	91%
	50.0		3035	88%
	48.0		2940	84%
RL19	54.0 *	MAX.	3144	93%
	52.0		3031	89%
	50.0		2919	86%
H4350	50.0	MAX.	3179	84%
	48.0 *		3105	80%
	46.0		2996	77%
Hunter	53.0	MAX.	3215	88%
	51.0		3122	85%
	49.0 *		3008	81%
IMR 4350	52.5 *	MAX.	3248	88%
	50.5		3086	84%
	48.5		2930	81%
Viht N560	55.0	MAX.	3282	97%
	53.0		3170	93%
	51.0 *		3068	90%
RL22	55.5	MAX.	3294	95%
	53.5		3201	92%
	51.5 *		3061	88%

BC=Ballistic Coefficient SD=Sectional Density
*Most Accurate Load Tested **Compressed Load

Use Maximum Loads with Caution
Refer to page 73 for additional safety information

6.5mm-06 A-Square - 123/125 grain		MAXIMUM S.A.A.M.I. O.A.C.L.		3.440"
		TESTED O.A.C.L.	B.C.	S.D.
Custom Competition™	123gr. HPBT	3.250"	0.510	0.252
Partition®	125gr. Spitzer	3.250"	0.449	0.256

CASE TYPE:	Winchester		PRIMER TYPE	WLR
CASE HOLDS:	64.6	Gr. WATER	BARREL Length/Make	24" Lilja
			BARREL Twist	1-9"

POWDER TYPE	POWDER CHG. GRS.		MUZZLE VEL. F.P.S.	LOAD DENSITY (VOLUME)
IMR 4831	50.0	MAX.	3000	82%
	48.0		2942	79%
	46.0 *		2848	76%
Viht N160	47.0	MAX.	3012	81%
	45.0 *		2908	77%
	43.0		2801	74%
Hunter	49.0	MAX.	3048	80%
	47.0		2945	77%
	45.0 *		2849	73%
Viht N165	54.5	MAX.	3049	94%
	52.5		2954	90%
	50.5 *		2859	87%
IMR 7828	53.5	MAX.	3061	88%
	51.5		2970	85%
	49.5 *		2864	82%
RL19	53.5	MAX.	3070	90%
Most Accurate	51.5		2960	87%
Powder Tested	49.5 *		2850	83%
H4831SC	50.0	MAX.	3078	81%
	48.0		3022	77%
	46.0 *		2961	74%
IMR 4350	48.0	MAX.	3088	79%
	46.0		2982	75%
	44.0 *		2871	72%
RL22	54.5 *	MAX.	3110	92%
	52.5		3010	88%
	50.5		2910	85%

BC=Ballistic Coefficient SD=Sectional Density
*Most Accurate Load Tested **Compressed Load

Use Maximum Loads with Caution
Refer to page 73 for additional safety information

6.5mm-06 A-Square - 130 grain		MAXIMUM S.A.A.M.I. O.A.C.L.		3.440"
		TESTED O.A.C.L.	B.C.	S.D.
AccuBond®	130gr. Spitzer	3.250"	0.488	0.266

CASE TYPE:	Norma		PRIMER TYPE	Rem 9 1/2
CASE HOLDS:	60.8	Gr. WATER	BARREL Length/Make	24" Lilja
			BARREL Twist	1-9"

POWDER TYPE	POWDER CHG. GRS.		MUZZLE VEL. F.P.S.	LOAD DENSITY (VOLUME)
Hunter	49.0	MAX.	2960	85%
	47.0 *		2852	81%
	45.0		2779	78%
Viht N165	52.5	MAX.	3031	96%
	50.5 *		2924	92%
	48.5		2839	89%
IMR 4350	49.5	MAX.	3040	87%
	47.5		2947	83%
	45.5 *		2825	80%
H4831SC	52.0 *	MAX.	3058	89%
	50.0		2981	86%
	48.0		2902	82%
IMR 4831	50.0	MAX.	3074	87%
	48.0		2959	84%
	46.0 *		2862	80%
IMR 7828	52.5	MAX.	3078	92%
	50.5 *		2959	88%
	48.5		2874	85%
RL19	52.0	MAX.	3089	93%
	50.0		3013	89%
	48.0 *		2899	86%
RL22	53.0	MAX.	3121	95%
	51.0		3040	91%
	49.0 *		2937	88%
Viht N560	53.0	MAX.	3151	97%
Most Accurate	51.0 *		3024	93%
Powder Tested	49.0		2912	90%

BC=Ballistic Coefficient SD=Sectional Density
*Most Accurate Load Tested **Compressed Load

Use Maximum Loads with Caution
Refer to page 73 for additional safety information

6.5mm-06 A-Square - 140 grain

Bullet	Type	MAXIMUM S.A.A.M.I. O.A.C.L.		3.440"
		TESTED O.A.C.L.	B.C.	S.D.
AccuBond®	140gr. Spitzer	3.250"	0.509	0.287
Ballistic Tip®	140gr. Spitzer	3.250"	0.509	0.287
Custom Competition™	140gr. HPBT	3.250"	0.529	0.287
Partition®	140gr. Spitzer	3.250"	0.490	0.287

CASE TYPE:	Winchester		PRIMER TYPE	WLR
CASE HOLDS:	62.0	Gr. WATER	BARREL Length/Make	24" Lilja
			BARREL Twist	1-9"

POWDER TYPE	POWDER CHG. GRS.		MUZZLE VEL. F.P.S.	LOAD DENSITY (VOLUME)
Viht N160	47.0	MAX.	2773	84%
	45.0		2691	81%
	43.0	*	2608	77%
H4831	49.5	* MAX.	2800	83%
	47.5		2710	80%
	45.5		2620	76%
IMR 4831	48.0	MAX.	2800	82%
	46.0		2720	79%
	44.0	*	2640	75%
H4350 Most Accurate Powder Tested	48.5	MAX.	2808	83%
	46.5		2723	79%
	44.5	*	2638	76%
Hunter	47.0	MAX.	2832	80%
	45.0	*	2748	76%
	43.0		2652	73%
RL22	50.5	* MAX.	2860	89%
	48.5		2760	85%
	46.5		2660	82%
IMR 4350	46.5	MAX.	2869	79%
	44.5		2771	76%
	42.5	*	2682	73%
RL19	48.5	MAX.	2902	85%
	46.5	*	2781	82%
	44.5		2712	78%
Viht N165	50.5	* MAX.	2906	91%
	48.5		2811	87%
	46.5		2735	83%

BC=Ballistic Coefficient SD=Sectional Density
*Most Accurate Load Tested **Compressed Load

Use Maximum Loads with Caution
Refer to page 73 for additional safety information

Lee J. Hoots

6.5 REMINGTON MAGNUM

Four years after the introduction of the iconic 7mm Remington Magnum and Model 700 rifle in 1962, Remington followed up with a commercial flop, the short action Model 600 Magnum carbine and the belted 6.5 Remington Magnum cartridge.

First introduced in 1965 with the potent 350 Remington Magnum cartridge, the rifle had an 18 1/2-inch barrel, full-length rib, a tall, unbecoming front sight blade and a laminated stock. The 6.5 version showed up the following year, but the carbine and its mini-magnums were doomed from the outset. At the time, mule deer numbers were peaking and elk hunting was increasing in popularity. Hunters exploring beyond the family farm in search of adventure promised in sporting magazines of the day wanted proven traditional cartridges like the 270 Winchester or a new belted magnum like Remington's 7mm to shoot across the open West. They also wanted optical sights and trim sporter rifles with 22- or 24-inch barrels. Apparently they weren't at all interested in carbines or .264-caliber bullets.

The 600 Magnum was revamped in 1968 as the Model 660 Magnum with a 20-inch barrel sans rib and with a more tasteful front sight, and the 6.5 and 350 Magnums were added to the Model 700 line in 1969. The 660 disappeared within a couple of years and the cartridges didn't hang on much longer. That's too bad.

The 6.5 Remington Magnum's original loading included a 120-grain SP CoreLokt at an advertized 3,030 fps, but it is unclear if Remington used an 18 1/2-inch barrel for this figure. Still, it is plenty of velocity for the open country and enough bullet weight for deer and pronghorn.

Handloaded 120-grain Nosler Ballistic Tips in my Remington Custom Shop rifle with a 24-inch barrel seem about on par with Remington's first load. With a moderate powder charge producing good accuracy and a muzzle velocity of 3,287 fps, the combination was

quite adequate on an old California buck shot a couple of years back. The 6.5 Remington Magnum will shoot nearly as fast with the 125-grain Partition, and with a 130-grain AccuBond or 140-grain Partition, its ballistics fairly equal that of the 270 Winchester given identical barrel lengths. What's more, .264-caliber bullets have a higher ballistic coefficient. Brass availability is sketchy today for the 6.5 Remington Magnum but can be made from 350 Remington Magnum brass in a pinch—or various other belted magnums in desperation.

Lee J. Hoots is the editor of Successful Hunter magazine

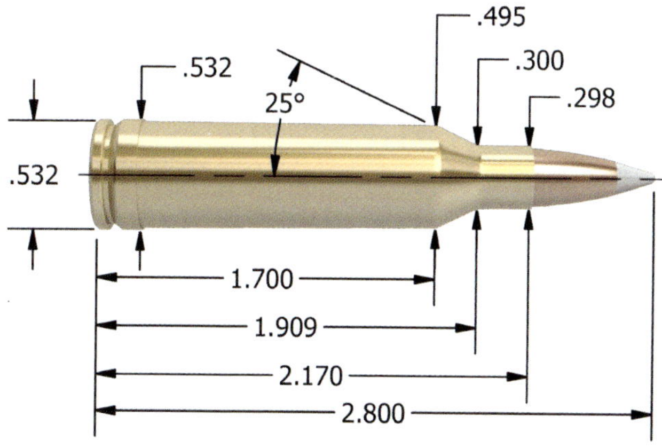

Maximum S.A.A.M.I. Overall Cartridge Length: 2.800"

BULLET CHOICES FOR THE 6.5 REMINGTON MAGNUM

TECHNICAL INFORMATION

One of the original "short magnums" (the other being the 350 Remington Magnum), the 6.5mm Rem Mag was hindered by the short barrel length on the Remington Model 600 it was introduced for. Chambered in a standard length rifle barrel, this cartridge becomes a viable contender in the growing 6.5mm arena.

6.5 Remington Magnum - 100 Grain		MAXIMUM S.A.A.M.I. O.A.C.L.		2.800"
		TESTED O.A.C.L.	B.C.	S.D.
Ballistic Tip®	100gr. Spitzer	2.800"	0.350	0.205
Partition®	100gr. Spitzer	2.800"	0.326	0.205

CASE TYPE:	Remington		PRIMER TYPE	Rem 9 1/2M
CASE HOLDS:	64.7	Gr. WATER	BARREL Length/Make	24" Pac-Nor
			BARREL Twist	1-9"

POWDER TYPE	POWDER CHG. GRS.		MUZZLE VEL. F.P.S.		LOAD DENSITY (VOLUME)
H4831SC Most Accurate Powder Tested	65.0	MAX.	3500	**	105%
	63.0		3389	**	101%
	61.0	*	3295		98%
Big Game	58.0	MAX.	3543		94%
	56.0		3431		91%
	54.0	*	3319		88%
W760	59.5	MAX.	3556		96%
	57.5	*	3466		93%
	55.5		3305		90%
H414	59.0	MAX.	3562		95%
	57.0	*	3443		92%
	55.0		3330		89%
Viht N550	58.0	MAX.	3566		100%
	56.0	*	3428		96%
	54.0		3293		93%
RL19	65.0	MAX.	3567	**	109%
	63.0	*	3426	**	106%
	61.0		3321	**	102%
Hunter	61.0	MAX.	3571		99%
	59.0	*	3418		96%
	57.0		3331		93%
IMR 4831	63.0	MAX.	3584	**	104%
	61.0	*	3443	**	100%
	59.0		3342		97%
H4350	60.5	MAX.	3585		99%
	58.5	*	3479		96%
	56.5		3369		92%
IMR 4350	61.5		3598	**	101%
	59.5	*	3487		97%
	57.5		3366		94%

BC=Ballistic Coefficient SD=Sectional Density
*Most Accurate Load Tested **Compressed Load

Use Maximum Loads with Caution
Refer to page 73 for additional safety information

6.5 Remington Magnum - 120 Grain

		MAXIMUM S.A.A.M.I. O.A.C.L.		2.800"
		TESTED O.A.C.L.	B.C.	S.D.
Ballistic Tip®	120gr. Spitzer	2.800"	0.458	0.246
E-Tip®	120gr. Spitzer	2.770"	0.497	0.246

Due to internal construction differences, always begin with starting loads when using E-Tip® products.

CASE TYPE:	Remington		PRIMER TYPE	Rem 9 1/2M
CASE HOLDS:	61.0	Gr. WATER	BARREL Length/Make	24" Pac-Nor
			BARREL Twist	1-9"

POWDER TYPE	POWDER CHG. GRS.		MUZZLE VEL. F.P.S.		LOAD DENSITY (VOLUME)
Hunter	58.0 *	MAX.	3260		100%
	56.0		3217		97%
	54.0		3104		93%
H4831SC	61.5	MAX.	3263	**	105%
	59.5 *		3177	**	102%
	57.5		3103		98%
W760	56.0 *	MAX.	3264		96%
	54.0		3172		93%
	52.0		3055		89%
IMR 4350	57.0	MAX.	3277		99%
Most Accurate Powder Tested	55.0		3194		95%
	53.0 *		3088		92%
H4350	57.0	MAX.	3286		99%
	55.0		3228		95%
	53.0 *		3110		92%
IMR 4831	59.0	MAX.	3299	**	103%
	57.0 *		3187		99%
	55.0		3094		96%
Viht N550	55.5 *	MAX.	3305	**	101%
	53.5		3173		97%
	51.5		3082		94%
RL19	60.5	MAX.	3323	**	108%
	58.5 *		3212	**	104%
	56.5		3131	**	101%
RL22	62.5	MAX.	3326	**	111%
	60.5 *		3212	**	108%
	58.5		3103	**	104%

BC=Ballistic Coefficient SD=Sectional Density
*Most Accurate Load Tested **Compressed Load

Use Maximum Loads with Caution
Refer to page 73 for additional safety information

6.5 Remington Mag - 123/125 Grain		MAXIMUM S.A.A.M.I. O.A.C.L.		2.800"
		TESTED O.A.C.L.	B.C.	S.D.
Custom Competition™	123gr. HPBT	2.800"	0.510	0.252
Partition®	125gr. Spitzer	2.800"	0.449	0.256

CASE TYPE:	Remington		PRIMER TYPE	Rem 9 1/2M
CASE HOLDS:	62.8	Gr. WATER	BARREL Length/Make	24" Pac-Nor
			BARREL Twist	1-9"

POWDER TYPE	POWDER CHG. GRS.		MUZZLE VEL. F.P.S.		LOAD DENSITY (VOLUME)
W760	54.5	MAX.	3161		91%
	52.5		3053		88%
	50.5	*	2955		84%
H4831SC Most Accurate Powder Tested	58.5	MAX.	3165		97%
	56.5		3082		94%
	54.5	*	2983		90%
Hunter	56.5	MAX.	3177		95%
	54.5		3063		91%
	52.5	*	2962		88%
IMR 4350	55.0	MAX.	3180		93%
	53.0		3074		89%
	51.0	*	2980		86%
IMR 4831	57.0	MAX.	3217		97%
	55.0		3120		93%
	53.0	*	3020		90%
RL19	59.0	MAX.	3260	**	102%
	57.0	*	3176		99%
	55.0		3075		95%
IMR 7828	61.5	MAX.	3266	**	104%
	59.5		3188	**	101%
	57.5	*	3080		97%
RL22	61.0		3273	**	106%
	59.0	*	3170	**	102%
	57.0		3082		99%

BC=Ballistic Coefficient SD=Sectional Density
*Most Accurate Load Tested **Compressed Load

Use Maximum Loads with Caution
Refer to page 73 for additional safety information

6.5 Remington Magnum - 130 Grain

AccuBond®	130gr. Spitzer	MAXIMUM S.A.A.M.I. O.A.C.L.		2.800"
		TESTED O.A.C.L.	B.C.	S.D.
		2.800"	0.488	0.266

CASE TYPE:	Remington		PRIMER TYPE	Rem 9 1/2M
CASE HOLDS:	60.1	Gr. WATER	BARREL Length/Make	24" Pac-Nor
			BARREL Twist	1-9"

POWDER TYPE	POWDER CHG. GRS.		MUZZLE VEL. F.P.S.		LOAD DENSITY (VOLUME)
W760	52.5	MAX.	3007		91%
	50.5	*	2925		88%
	48.5		2849		85%
Hunter	56.5	MAX.	3102		99%
	54.5		3026		95%
	52.5	*	2979		92%
H4831SC	57.5	MAX.	3103		100%
	55.5		3044		96%
	53.5	*	2950		93%
IMR 4350	54.0	MAX.	3104		95%
	52.0		2999		92%
	50.0	*	2882		88%
IMR 4831	56.0	MAX.	3136		99%
Most Accurate	54.0		3057		96%
Powder Tested	52.0	*	2953		92%
RL19	57.5	MAX.	3141	**	104%
	55.5		3040		100%
	53.5	*	2935		97%
RL22	59.5	MAX.	3170	**	108%
	57.5		3095	**	104%
	55.5	*	2999		100%
IMR 7828	60.5		3180	**	107%
	58.5	*	3096	**	104%
	56.5		3015		100%

BC=Ballistic Coefficient SD=Sectional Density
*Most Accurate Load Tested **Compressed Load

Use Maximum Loads with Caution
Refer to page 73 for additional safety information

6.5 Remington Magnum - 140 Grain		MAXIMUM S.A.A.M.I. O.A.C.L.		2.800"
		TESTED O.A.C.L.	B.C.	S.D.
AccuBond®	140gr. Spitzer	2.800"	0.509	0.287
Ballistic Tip®	140gr. Spitzer	2.800"	0.509	0.287
Custom Competition™	140gr. HPBT	2.800"	0.529	0.287
Partition®	140gr. Spitzer	2.800"	0.490	0.287

CASE TYPE:	Remington		PRIMER TYPE	Rem 9 1/2M
CASE HOLDS:	59.3	Gr. WATER	BARREL Length/Make	24" Pac-Nor
			BARREL Twist	1-9"

POWDER TYPE	POWDER CHG. GRS.		MUZZLE VEL. F.P.S.		LOAD DENSITY (VOLUME)
W760	50.5	MAX.	2900		89%
	48.5	*	2810		86%
	46.5		2725		82%
Hunter	55.0	MAX.	2958		98%
	53.0		2906		94%
	51.0	*	2801		91%
IMR 4350	53.0	MAX.	2961		95%
	51.0	*	2874		91%
	49.0		2768		87%
H4831SC	57.0	MAX.	2979		100%
	55.0		2908		97%
	53.0	*	2835		93%
RL19 Most Accurate Powder Tested	55.5	MAX.	2997	**	102%
	53.5		2908		98%
	51.5	*	2801		94%
IMR 4831	55.0	MAX.	2997		99%
	53.0	*	2909		95%
	51.0		2833		91%
RL22	57.5	MAX.	3042	**	105%
	55.5	*	2956	**	102%
	53.5		2877		98%
IMR 7828	58.5	MAX.	3059	**	105%
	56.5	*	2983	**	101%
	54.5		2877		98%

BC=Ballistic Coefficient SD=Sectional Density
*Most Accurate Load Tested **Compressed Load

Use Maximum Loads with Caution
Refer to page 73 for additional safety information

264 WINCHESTER MAGNUM

The 264 Winchester Magnum and its companion introduction, the 338 Winchester Magnum, hit the scene together in 1958, pretty much at the beginning of what we now think of as the first "magnum craze." Based on the 300 H&H case shortened, blown out, and appropriately necked, they were the second and third of Winchester's quartet of 30-06-length belted magnums, following the 458 Winchester Magnum (1956) and preceding the 300 Winchester Magnum (1963). The .264 was introduced in a 26-inch-barreled version of the Winchester Model 70 called the "Westerner," while the .338's Model 70 was dubbed the "Alaskan."

Interestingly, the .264 initially took off like a rocket, while the .338 plodded along and caught on very slowly. For many years now, the situation has been reversed; the .338 is an established favorite for large North American game while the .264 barely hangs on in production rifles and factory loads. The .264 probably suffers from the 6.5mm curse: No 6.5mm cartridge has achieved lasting popularity in North America. The .264 also had some issues. Especially with the powders available 50 years ago, it was somewhat overbore capacity, and achieved an early reputation for rapid throat erosion. Also, to be honest, there was a bit of blue sky in the initial published ballistics!

Even so, I've always had a soft spot for the .264. I got my first one in 1965, and in those innocent pre-chronograph days, I thought it was magic! More recently I've returned to the .264, using it for sheep and goats as well as open country deer hunting. Loaded properly, the .264 is a bit faster and flatter-shooting than the 270 Winchester, and it offers the high Sectional Density and Ballistic Coefficients inherent in the typically heavy-for-caliber 6.5mm bullets.

Nosler has long been a leader in developing 6.5mm hunting bullets, offering a good selection in the Ballistic Tip, Partition, and AccuBond lines. In recent years, the 130gr AccuBond has become one of my favorites. Loaded to about 3100 fps in my 26-inch-barrel .264, this bullet has accounted for one of my best whitetails as well as my best mule deer in many a year…along with several Asian sheep and ibex. Regrettably, I don't think the .264 Winchester Magnum will ever again become extremely popular, but it remains a marvelous choice for medium-sized game in big country.

Craig Boddington

Author and Television personality Craig Boddington is Executive Field Editor, Intermedia Outdoor Group as well as a Contributing Editor to Sports Afield.

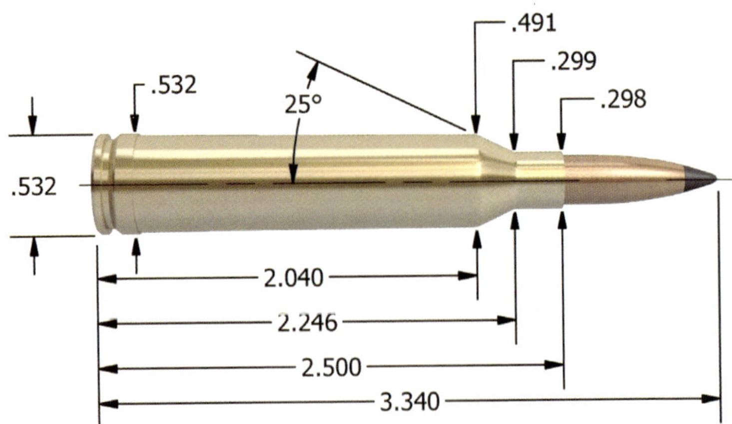

Maximum S.A.A.M.I. Overall Cartridge Length: 3.340"

BULLET CHOICES FOR THE 264 WINCHESTER MAGNUM

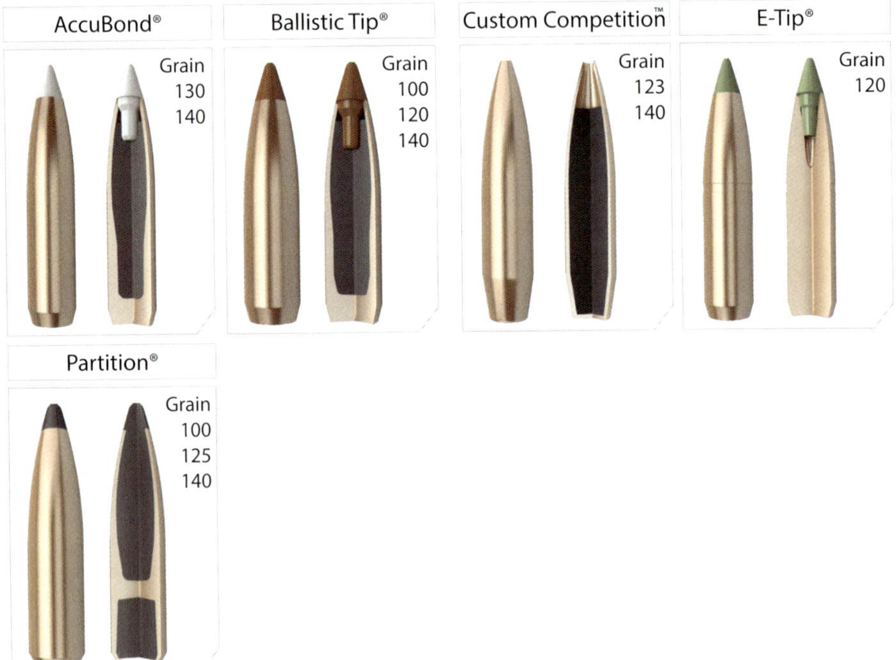

TECHNICAL INFORMATION

The 264 Winchester Magnum, like many cartridges with big cases and skinny bullets, has developed a reputation as a "barrel burner". We have found that surprisingly long barrel life can be expected, even with full power loads, provided the shooter takes care not to overheat the barrel and employs frequent bore scrubbings.

264 Winchester Magnum - 100 grain

		MAXIMUM S.A.A.M.I. O.A.C.L.		3.340"
		TESTED O.A.C.L.	B.C.	S.D.
Ballistic Tip®	100gr. Spitzer	3.190"	0.350	0.205
Partition®	100gr. Spitzer	3.190"	0.326	0.205

CASE TYPE:	Nosler		PRIMER TYPE	Rem 9 1/2M
CASE HOLDS:	82.0	Gr. WATER	BARREL Length/Make	24" Wiseman
			BARREL Twist	1-9"

POWDER TYPE	POWDER CHG. GRS.		MUZZLE VEL. F.P.S.	LOAD DENSITY (VOLUME)
Hunter	64.5 *	MAX.	3482	83%
	62.5		3456	80%
	60.5		3336	78%
IMR 4350 Most Accurate Powder Tested	62.5	MAX.	3487	81%
	60.5		3412	78%
	58.5 *		3297	75%
IMR 7828	69.0	MAX.	3494	90%
	67.0		3412	87%
	65.0 *		3330	84%
RL22	67.5	MAX.	3516	89%
	65.5		3457	87%
	63.5 *		3398	84%
H4350	63.5	MAX.	3528	82%
	61.5		3462	79%
	59.5 *		3361	77%
H4831SC	69.0	MAX.	3548	88%
	67.0 *		3440	85%
	65.0		3366	83%
W760	63.5	MAX.	3552	81%
	61.5		3472	79%
	59.5 *		3330	76%
Viht N160	66.0	MAX.	3561	89%
	64.0 *		3471	87%
	62.0		3399	84%
RL19	68.5 *	MAX.	3570	91%
	66.5		3481	88%
	64.5		3312	85%

BC=Ballistic Coefficient SD=Sectional Density
*Most Accurate Load Tested **Compressed Load

Use Maximum Loads with Caution
Refer to page 73 for additional safety information

264 Winchester Magnum - 120 grain		MAXIMUM S.A.A.M.I. O.A.C.L.		3.340"
		TESTED O.A.C.L.	B.C.	S.D.
Ballistic Tip®	120gr. Spitzer	3.190"	0.458	0.246
E-Tip®	120gr. Spitzer	3.190"	0.497	0.246

Due to internal construction differences, always begin with starting loads when using E-Tip® products.

CASE TYPE:	Nosler		PRIMER TYPE	Rem 9 1/2M
CASE HOLDS:	78.9	Gr. WATER	BARREL Length/Make	24" Wiseman
			BARREL Twist	1-9"

POWDER TYPE	POWDER CHG. GRS.		MUZZLE VEL. F.P.S.	LOAD DENSITY (VOLUME)
IMR 4320	53.0	MAX.	3100	72%
	51.0		2990	70%
	49.0 *		2880	67%
W760	58.0 *	MAX.	3114	77%
	56.0		3053	74%
	54.0		2949	72%
IMR 4064	54.0	MAX.	3140	75%
	52.0		3060	72%
	50.0 *		2980	70%
IMR 4831	62.0 *	MAX.	3162	84%
	60.0		3057	81%
	58.0		2952	78%
RL22	62.5	MAX.	3187	86%
	60.5		3099	83%
	58.5 *		3011	81%
Hunter	60.0	MAX.	3195	80%
	58.0		3111	77%
	56.0 *		3003	75%
IMR 4350 Most Accurate Powder Tested	60.0 *	MAX.	3202	80%
	58.0		3127	78%
	56.0		3052	75%
H4350	54.5	MAX.	3203	73%
	52.5 *		3128	70%
	50.5		3047	68%
RL19	64.0	MAX.	3309	88%
	62.0		3211	85%
	60.0 *		3123	83%

BC=Ballistic Coefficient SD=Sectional Density
*Most Accurate Load Tested **Compressed Load

Use Maximum Loads with Caution
Refer to page 73 for additional safety information

264 Winchester Mag - 123/125 grain

		MAXIMUM S.A.A.M.I. O.A.C.L.		3.340"
		TESTED O.A.C.L.	B.C.	S.D.
Custom Competition™	123gr. HPBT	3.290"	0.510	0.252
Partition®	125gr. Spitzer	3.190"	0.449	0.256

CASE TYPE:	Nosler		PRIMER TYPE	WLRM
CASE HOLDS:	81.2	Gr. WATER	BARREL Length/Make	24" H-S Prec.
			BARREL Twist	1-9"

POWDER TYPE	POWDER CHG. GRS.		MUZZLE VEL. F.P.S.	LOAD DENSITY (VOLUME)
IMR 4350	57.5	* MAX.	3097	75%
	55.5		3022	72%
	53.5		2947	70%
H4350	55.5	MAX.	3120	72%
	53.5	*	3030	70%
	51.5		2960	67%
IMR 4831	61.5	* MAX.	3130	81%
	59.5		2995	78%
	57.5		2885	75%
W760	56.5	MAX.	3139	73%
	54.5		3029	70%
	52.5	*	2945	68%
Hunter	59.0	MAX.	3158	76%
	57.0		3090	74%
	55.0	*	3031	71%
RL19	61.0	MAX.	3222	82%
	59.0	*	3118	79%
	57.0		3040	76%
MAGPRO	70.0	MAX.	3245	89%
	68.0	*	3168	86%
	66.0		3111	84%
Magnum	70.0	MAX.	3246	88%
	68.0		3183	85%
	66.0	*	3128	83%
Supreme 780	67.0	* MAX.	3298	86%
Most Accurate	65.0		3220	84%
Powder Tested	63.0		3120	81%

BC–Ballistic Coefficient SD=Sectional Density
*Most Accurate Load Tested **Compressed Load

Use Maximum Loads with Caution
Refer to page 73 for additional safety information

264 Winchester Magnum - 130 grain		MAXIMUM S.A.A.M.I. O.A.C.L.		3.340"
		TESTED O.A.C.L.	B.C.	S.D.
AccuBond®	130gr. Spitzer	3.190"	0.488	0.266

CASE TYPE:	Nosler		PRIMER TYPE	Rem 9 1/2M
CASE HOLDS:	78.0	Gr. WATER	BARREL Length/Make	24" Wiseman
			BARREL Twist	1-9"

POWDER TYPE	POWDER CHG. GRS.		MUZZLE VEL. F.P.S.	LOAD DENSITY (VOLUME)
W760	54.0	MAX.	2981	72%
	52.0		2892	70%
	50.0 *		2814	67%
Hunter	55.0	MAX.	2986	74%
	53.0 *		2869	72%
	51.0		2800	69%
IMR 4350 Most Accurate Powder Tested	55.0	MAX.	3042	75%
	53.0		2944	72%
	51.0 *		2866	69%
H4831SC	57.5	MAX.	3053	77%
	55.5 *		2969	74%
	53.5		2875	71%
Viht N560	59.5	MAX.	3105	85%
	57.5		3017	82%
	55.5 *		2929	79%
RL19	58.5 *	MAX.	3117	82%
	56.5		3023	79%
	54.5		2955	76%
IMR 4831	58.5	MAX.	3141	80%
	56.5 *		3032	77%
	54.5		2963	74%
IMR 7828	61.0	MAX.	3153	83%
	59.0		3050	80%
	57.0 *		2977	78%
RL22	61.0	MAX.	3166	85%
	59.0		3086	82%
	57.0 *		3008	79%

BC=Ballistic Coefficient SD=Sectional Density
*Most Accurate Load Tested **Compressed Load

Use Maximum Loads with Caution
Refer to page 73 for additional safety information

264 Winchester Magnum - 140 grain		MAXIMUM S.A.A.M.I. O.A.C.L.		3.340"
		TESTED O.A.C.L.	B.C.	S.D.
AccuBond®	140gr. Spitzer	3.190"	0.509	0.287
Ballistic Tip®	140gr. Spitzer	3.190"	0.509	0.287
Custom Competition™	140gr. HPBT	3.190"	0.529	0.287
Partition®	140gr. Spitzer	3.190"	0.490	0.287

CASE TYPE:	Winchester		PRIMER TYPE	Rem 9 1/2M
CASE HOLDS:	79.6	Gr. WATER	BARREL Length/Make	24" Wiseman
			BARREL Twist	1-9"

POWDER TYPE	POWDER CHG. GRS.		MUZZLE VEL. F.P.S.	LOAD DENSITY (VOLUME)
IMR 4320	50.0 *	MAX.	2848	68%
	48.0		2763	65%
	46.0		2678	62%
IMR 4064	51.0	MAX.	2860	70%
	49.0		2790	68%
	47.0 *		2720	65%
Hunter	54.0	MAX.	2908	71%
	52.0		2845	69%
	50.0 *		2751	66%
RL22 Most Accurate Powder Tested	59.0	MAX.	2929	81%
	57.0		2821	78%
	55.0 *		2712	75%
H4350	52.0	MAX.	2930	69%
	50.0 *		2852	66%
	48.0		2775	64%
W760	54.5 *	MAX.	2942	72%
	52.5		2861	69%
	50.5		2793	66%
IMR 4831	59.0	MAX.	2970	79%
	57.0		2870	76%
	55.0 *		2770	74%
IMR 4350	57.0 *	MAX.	2980	76%
	55.0		2920	73%
	53.0		2860	70%
RL19	57.5	MAX.	3021	79%
	55.5 *		2954	76%
	53.5		2869	73%

BC=Ballistic Coefficient SD=Sectional Density
*Most Accurate Load Tested **Compressed Load

Use Maximum Loads with Caution
Refer to page 73 for additional safety information

Bill Wilson

6.8 REMINGTON SPC

These days, the popularity of AR platform rifles and carbines is staggering. This is due in part to their tremendous versatility. No other long gun platform is even close to the AR when it comes to being able to customize to your specific needs and wants. Not only can you have the specifications you want, you now have a substantial choice of cartridges suitable for a wide variety of applications.

I'm often asked "what's the best all-around AR cartridge?" and my answer is always the 6.8 SPC. While no cartridge is perfect for everything or everyone, the 6.8 sure fits the bill when you need a semi-auto for tactical use, personal defense, range work and hunting. It's accurate, low on recoil and packs a substantial down-range punch. It's a proven performer on game up to mule deer size when loaded with premium bullets like the excellent Nosler AccuBond.

Its versatility is aided by the availability of factory ammunition from at least eight different manufacturers and a large selection of bullets designed specifically for the cartridge. Nosler alone has Custom Competition match grade target bullets, as well as the proven AccuBond for medium size game. The 110gr AccuBond with its excellent accuracy and optimum terminal performance is my bullet of choice for deer hunting. We have to do a lot of deer culling on the ranch for our management plan and the 6.8 loaded with 110gr AccuBonds has been an extremely reliable performer.

The 6.8 is also a terrific choice for women and young shooters due to its mild recoil. The ability to fit rifles with butt stocks that permit adjustable lengths-of-pull also make the AR perfect for those of small stature. Another plus is the fact that the 6.8 SPC is an easy caliber to handload for, which allows you to shoot a lot more for less money. For extreme versatility you simply can't beat a quality AR chambered in the 6.8 SPC.

Bill Wilson

Bill Wilson is the founder of Wilson Combat.

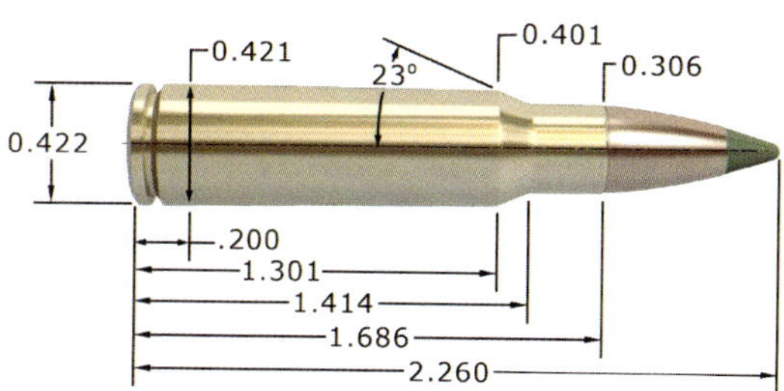

Maximum S.A.A.M.I. Overall Cartridge Length: 2.260"

BULLET CHOICES FOR THE 6.8 REMINGTON SPC

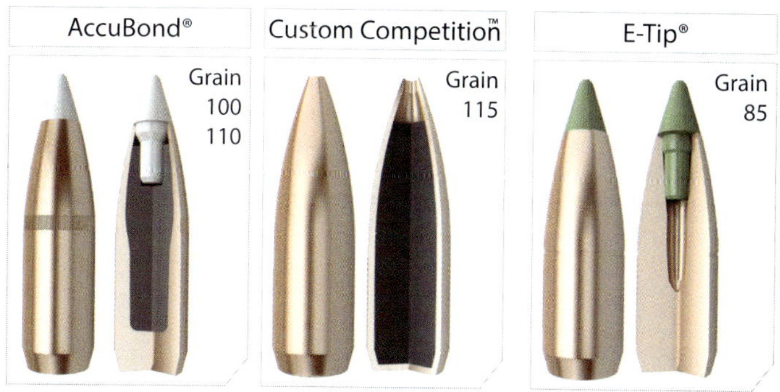

TECHNICAL INFORMATION

The 6.8 Remington Special Purpose Cartridge was designed by Remington Arms and the USAMU & USSOC as a replacement for the 5.56x45 NATO cartridge in AR/M-16 rifles. Developed from the 30 Remington rimless cartridge case shortened to function at 2.260" with a .277" dia. bullet, the performance is between the 5.56 and 7.62 NATO rounds. Nosler has developed a line of 6.8 bullets to take maximum advantage of the cartridge's potential from targets to hunting medium game. While there are other variants of the cartridge developed by custom rifle makers, the load data herein is for the SAAMI standard 6.8 SPC cartridge. IMR3031 and H322 powders provide excellent performance with all bullet weights.

6.8 Remington SPC - 85 grain		MAXIMUM S.A.A.M.I. O.A.C.L.		2.260"
		TESTED O.A.C.L.	B.C.	S.D.
E-Tip®	85gr. Spitzer	2.225"	0.273	0.158

Due to internal construction differences, always begin with starting loads when using E-Tip® products.

CASE TYPE:	Remington		PRIMER TYPE	Rem 9 1/2
CASE HOLDS:	30.5	Gr. WATER	BARREL Length/Make	20" Pac-Nor
			BARREL Twist	1-10"

POWDER TYPE	POWDER CHG. GRS.		MUZZLE VEL. F.P.S.		LOAD DENSITY (VOLUME)
Benchmark	32.0	MAX.	2881	**	113%
	31.0		2793	**	110%
	30.0 *		2723	**	106%
RL10x	29.0	MAX.	2896	**	107%
	28.0 *		2809	**	103%
	27.0		2732		99%
A-2230	32.5	MAX.	2902	**	109%
	31.5		2831	**	105%
	30.5 *		2760	**	102%
RL7	27.5 *	MAX.	2917		98%
	26.5		2822		94%
	25.5		2725		91%
IMR 4198	26.5 *	MAX.	2924		99%
	25.5		2829		95%
	24.5		2727		91%
X-Terminator	33.0	MAX.	2924	**	110%
	32.0		2843	**	107%
	31.0 *		2754	**	103%
H322 Most Accurate Powder Tested	31.0 *	MAX.	2940	**	110%
	30.0		2869	**	106%
	29.0		2777	**	103%
H4198	28.5	MAX.	2985	**	105%
	27.5		2912	**	101%
	26.5 *		2839		98%

BC=Ballistic Coefficient SD=Sectional Density
*Most Accurate Load Tested **Compressed Load

Use Maximum Loads with Caution
Refer to page 73 for additional safety information

6.8 Remington SPC - 100 grain		MAXIMUM S.A.A.M.I. O.A.C.L.		2.260"
		TESTED O.A.C.L.	B.C.	S.D.
AccuBond®	100gr. Spitzer	2.245"	0.323	0.186

CASE TYPE:	Remington	PRIMER TYPE	Rem 9 1/2
CASE HOLDS:	28.6 Gr. WATER	BARREL Length/Make	20" Pac-Nor
		BARREL Twist	1-10"

POWDER TYPE	POWDER CHG. GRS.		MUZZLE VEL. F.P.S.		LOAD DENSITY (VOLUME)
IMR 4198	24.0	MAX.	2660		95%
	23.0		2562		91%
	22.0 *		2446		87%
H335	30.5	MAX.	2684	**	108%
	29.5 *		2617	**	105%
	28.5		2534	**	101%
RL7	26.5	MAX.	2693	**	101%
	25.5		2598		97%
	24.5 *		2485		93%
X-Terminator	31.0	MAX.	2716	**	110%
	30.0		2638	**	106%
	29.0 *		2597	**	103%
RL10x	28.0	MAX.	2736	**	110%
	27.0		2670	**	106%
	26.0 *		2570	**	102%
IMR 3031	29.0	MAX.	2751	**	115%
	28.0		2673	**	111%
	27.0 *		2564	**	107%
Benchmark Most Accurate Powder Tested	31.0	MAX.	2760	**	117%
	30.0		2683	**	113%
	29.0 *		2569	**	110%
H322	29.5	MAX.	2765	**	112%
	28.5		2697	**	108%
	27.5 *		2591	**	104%

BC=Ballistic Coefficient SD=Sectional Density
*Most Accurate Load Tested **Compressed Load

Use Maximum Loads with Caution
Refer to page 73 for additional safety information

6.8 Remington SPC - 110 grain

		MAXIMUM S.A.A.M.I. O.A.C.L.		2.260"
		TESTED O.A.C.L.	B.C.	S.D.
AccuBond®	110gr. Spitzer	2.245"	0.370	0.205

CASE TYPE:	Remington		PRIMER TYPE	Rem 9 1/2
CASE HOLDS:	27.5	Gr. WATER	BARREL Length/Make	20" Pac-Nor
			BARREL Twist	1-10"

POWDER TYPE	POWDER CHG. GRS.		MUZZLE VEL. F.P.S.		LOAD DENSITY (VOLUME)
IMR 4198	23.0 *	MAX.	2474		95%
	22.0		2438		91%
	21.0		2342		87%
H322	28.0	MAX.	2531	**	110%
	27.0 *		2451	**	106%
	26.0		2381	**	102%
X-Terminator Most Accurate Powder Tested	29.5	MAX.	2532	**	109%
	28.5 *		2453	**	105%
	27.5		2428	**	102%
H335	29.0	MAX.	2534	**	107%
	28.0 *		2435	**	103%
	27.0		2351		100%
RL10x	26.5	MAX.	2540	**	108%
	25.5		2462	**	104%
	24.5 *		2361		100%
Benchmark	29.0	MAX.	2540	**	114%
	28.0 *		2450	**	110%
	27.0		2358	**	106%
IMR 3031	27.5	MAX.	2540	**	113%
	26.5		2473	**	109%
	25.5 *		2384	**	105%
A-2230	29.0	MAX.	2543	**	108%
	28.0		2480	**	104%
	27.0 *		2423		100%

BC=Ballistic Coefficient SD=Sectional Density
*Most Accurate Load Tested **Compressed Load

Use Maximum Loads with Caution
Refer to page 73 for additional safety information

6.8 Remington SPC - 115 grain		MAXIMUM S.A.A.M.I. O.A.C.L.		2.260"
		TESTED O.A.C.L.	B.C.	S.D.
Custom Competition™	115gr. HPBT	2.230"	0.375	0.214

CASE TYPE:	Remington		PRIMER TYPE	Rem 9 1/2
CASE HOLDS:	29.5	Gr. WATER	BARREL Length/Make	20" Pac-Nor
			BARREL Twist	1-10"

POWDER TYPE	POWDER CHG. GRS.		MUZZLE VEL. F.P.S.		LOAD DENSITY (VOLUME)
RL10x	25.0	MAX.	2387		95%
	24.0 *		2285		91%
	23.0		2201		88%
IMR 4198	22.5	MAX.	2397		87%
	21.5 *		2293		83%
	20.5		2224		79%
H335	28.5	MAX.	2446		98%
	27.5		2341		95%
	26.5 *		2238		91%
IMR 3031 Most Accurate Powder Tested	27.0	MAX.	2460	**	103%
	26.0 *		2346		100%
	25.0		2237		96%
X-Terminator	29.0 *	MAX.	2544		100%
	28.0		2506		96%
	27.0		2400		93%
Benchmark	29.0	MAX.	2520	**	106%
	28.0 *		2427	**	103%
	27.0		2324		99%
H322	27.5	MAX.	2556	**	101%
	26.5		2491		97%
	25.5 *		2401		93%

BC=Ballistic Coefficient SD=Sectional Density
*Most Accurate Load Tested **Compressed Load

Use Maximum Loads with Caution
Refer to page 73 for additional safety information

John Zent

270 WINCHESTER

Though it's not unusual for a gun magazine guy to hunt with a loaner rifle, I was having a hard time choosing from the tired lot offered by my Mexican outfitter. Then came the camp old-timer bearing another option. He introduced himself as a former ranch owner and long-time Tucson resident, and was way more fluent in English than anyone else I'd met. The old gent handed me a nice Winchester M70 carbine in 270 Win., along with a box of ammo. That made the choice easy.

Later he confided: "O'Connor told me to get one of these [rifles];" and yes, he was talking about famed gunwriter Jack O'Connor, Mr. .270 himself. The legendary columnist is universally credited with popularizing what ranks as America's No. 3 bestselling caliber, and no other cartridge has been so closely linked to one person.

While never shy about stating his preference for the .270, O'Connor was usually fair-minded about grouping it with ballistically similar rounds, and so perhaps he is given too much credit. With its stellar 90-year track record on nearly all of our antlered and horned game, the 270 Win. was bound to catch on and hold on.

Upon introduction in 1925, the .270 offered sizzle. Here was a medium bore that could shoot big-game loads faster than 3000 fps, and even with heavier bullets, velocities ranged as high as 2800 fps. Although bullet weights were fairly limited, commonly available loads neatly addressed the priorities—130-grain for deer, 150- or 160-grain for elk. For years, the .270's long-range trajectory was the best in class, and easing holdover guesswork helped to account for a lot of dead game. Just as important to many shooters was lighter recoil than the 30-calibers, especially when magnums came along. The ammo the old man handed me that day turned out to be an assortment of leftovers from past hunters. Most plentiful was an old-hat, round-nosed-bullet load, and while that would never have been my preference, it did shoot reasonably well from the carbine. And so off I trudged into the desert, hoping I wouldn't have to shoot long-range.

As it played out, the hunt required heck of a lot of walking and tracking before yielding a shot inside 100 yards. The short-barreled Model 70 and its .270 round-nosed ammo proved perfect. If ever it seemed that fate chose to hand me the proper shooting gear, that was it.

John Zent is the editorial director for NRA Publications.

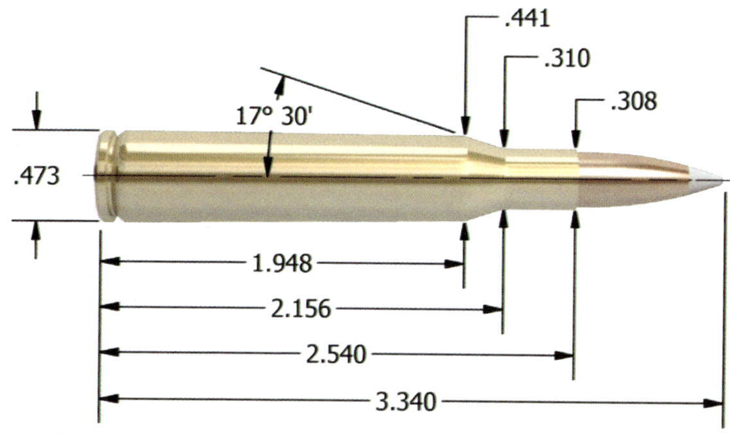

Maximum S.A.A.M.I. Overall Cartridge Length: 3.340"

BULLET CHOICES FOR THE 270 WINCHESTER

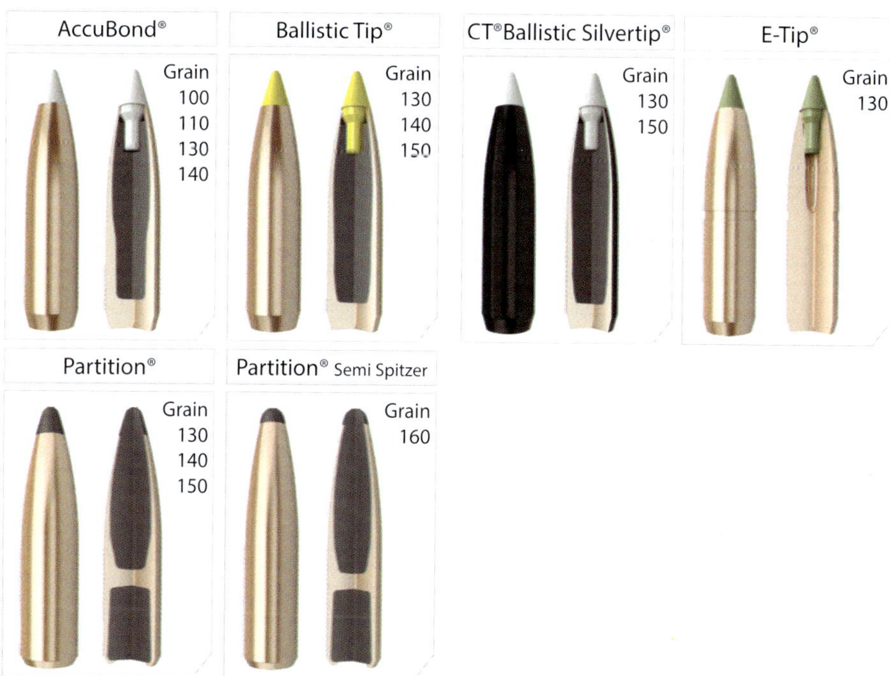

TECHNICAL INFORMATION

The 270 Winchester is clearly one of the shooting sport's premier big game cartridges. At the range and in the field, this cartridge works well with a variety of bullet /powder combinations. Our favorite powders for the .270 include 4350, 4831, and RL19.

270 Winchester - 100 grain			MAXIMUM S.A.A.M.I. O.A.C.L.		3.340"
			TESTED O.A.C.L.	B.C.	S.D.
AccuBond®		100gr. Spitzer	3.275"	0.323	0.186
CASE TYPE:	Nosler		PRIMER TYPE		WLR
CASE HOLDS:	64.7	Gr. WATER	BARREL Length/Make		26" Pac-Nor
			BARREL Twist		1-10"

POWDER TYPE	POWDER CHG. GRS.		MUZZLE VEL. F.P.S.	LOAD DENSITY (VOLUME)
RL15 Most Accurate Powder Tested	54.0	MAX.	3451	87%
	52.0 *		3322	84%
	50.0		3198	80%
Varget	54.0	MAX.	3492	90%
	52.0		3358	87%
	50.0 *		3267	84%
IMR 4350	60.5 *	MAX.	3528	99%
	58.5		3420	96%
	56.5		3239	92%
W760	58.0	MAX.	3550	94%
	56.0 *		3455	91%
	54.0		3339	87%

BC=Ballistic Coefficient SD=Sectional Density
*Most Accurate Load Tested **Compressed Load

Use Maximum Loads with Caution
Refer to page 73 for additional safety information

270 Winchester - 110 grain		MAXIMUM S.A.A.M.I. O.A.C.L.		3.340"
		TESTED O.A.C.L.	B.C.	S.D.
AccuBond®	110gr. Spitzer	3.275"	0.370	0.205

CASE TYPE:	Nosler		PRIMER TYPE	WLR
CASE HOLDS:	64.6	Gr. WATER	BARREL Length/Make	24" Shilen
			BARREL Twist	1-10"

POWDER TYPE	POWDER CHG. GRS.		MUZZLE VEL. F.P.S.	LOAD DENSITY (VOLUME)
H4895	48.0	MAX.	3280	82%
	46.0 *		3170	78%
	44.0		3082	75%
Viht N150	50.0	MAX.	3312	91%
	48.0 *		3209	87%
	46.0		3115	84%
H380	53.0	MAX.	3345	86%
	51.0 *		3254	83%
	49.0		3177	80%
Varget	50.5	MAX.	3371	85%
	48.5 *		3252	81%
	46.5		3155	78%
IMR 4320	50.5	MAX.	3376	84%
	48.5		3248	81%
	46.5 *		3157	77%
RL15	51.0	MAX.	3380	82%
	49.0 *		3262	79%
	47.0		3159	76%
IMR 4064 Most Accurate Powder Tested	51.0	MAX.	3411	87%
	49.0 *		3261	83%
	47.0		3157	80%
IMR 4350	57.0 *	MAX.	3437	93%
	55.0		3321	90%
	53.0		3175	87%
W760	56.0	MAX.	3470	91%
	54.0		3371	88%
	52.0 *		3216	84%
N 204	57.0	MAX.	3503	95%
	55.0 *		3350	92%
	53.0		3225	88%

BC=Ballistic Coefficient SD=Sectional Density
*Most Accurate Load Tested **Compressed Load

Use Maximum Loads with Caution
Refer to page 73 for additional safety information

270 Winchester - 130 grain		MAXIMUM S.A.A.M.I. O.A.C.L.		3.340"
		TESTED O.A.C.L.	B.C.	S.D.
AccuBond®	130gr. Spitzer	3.320"	0.435	0.242
Ballistic Tip®	130gr. Spitzer	3.320"	0.433	0.242
CT® Ballistic Silvertip®	130gr. Spitzer	3.320"	0.433	0.242
E-Tip®	130gr. Spitzer	3.300"	0.459	0.242
Due to internal construction differences, always begin with starting loads when using E-Tip® products.				
Partition®	130gr. Spitzer	3.320"	0.416	0.242

CASE TYPE:	Nosler		PRIMER TYPE	Fed 210
CASE HOLDS:	64.6	Gr. WATER	BARREL Length/Make	24" Shilen
			BARREL Twist	1-10"

POWDER TYPE	POWDER CHG. GRS.		MUZZLE VEL. F.P.S.	LOAD DENSITY (VOLUME)
IMR 4320	47.0	MAX.	2928	78%
	45.0		2833	75%
	43.0 *		2738	72%
H380	50.0	MAX.	2932	81%
	48.0		2847	78%
	46.0 *		2762	75%
IMR 4895	46.0 *	MAX.	2960	78%
	44.0		2870	75%
	42.0		2780	71%
H414	54.5 *	MAX.	3062	88%
	52.5		2937	85%
	50.5		2812	81%
IMR 4350	55.0 *	MAX.	3078	90%
	53.0		2953	87%
	51.0		2828	84%
Hunter	57.5	MAX.	3093	94%
	55.5		2981	90%
	53.5 *		2892	87%
RL22	58.0 *	MAX.	3100	98%
	56.0		2950	94%
	54.0		2800	91%
Viht N160	55.0	MAX.	3102	95%
	53.0		3004	91%
	51.0 *		2907	88%
H4831SC Most Accurate Powder Tested	59.0	MAX.	3124	95%
	57.0		2997	92%
	55.0 *		2909	89%
W760	54.0 *	MAX.	3158	88%
	52.0		3050	84%
	50.0		2946	81%

BC=Ballistic Coefficient SD=Sectional Density
*Most Accurate Load Tested **Compressed Load

Use Maximum Loads with Caution
Refer to page 73 for additional safety information

270 Winchester - 140 grain		MAXIMUM S.A.A.M.I. O.A.C.L.		3.340"
		TESTED O.A.C.L.	B.C.	S.D.
AccuBond®	140gr. Spitzer	3.320"	0.496	0.261
Ballistic Tip®	140gr. Spitzer	3.320"	0.456	0.261
Partition®	140gr. Spitzer	3.320"	0.432	0.261

CASE TYPE:	Nosler		PRIMER TYPE	Fed 210
CASE HOLDS:	63.7	Gr. WATER	BARREL Length/Make	24" Shilen
			BARREL Twist	1-10"

POWDER TYPE	POWDER CHG. GRS.		MUZZLE VEL. F.P.S.	LOAD DENSITY (VOLUME)
RL15	47.0 *	MAX.	2832	77%
	45.0		2727	74%
	43.0		2622	70%
H414	51.5	MAX.	2838	84%
	49.5		2693	81%
	47.5 *		2548	78%
IMR 4350	52.5	MAX.	2858	87%
	50.5		2723	84%
	48.5 *		2588	81%
IMR 4831	53.5	MAX.	2910	89%
Most Accurate	51.5		2790	86%
Powder Tested	49.5 *		2670	83%
Hunter	56.5	MAX.	2971	93%
	54.5		2888	90%
	52.5 *		2782	87%
IMR 7828	57.5	MAX.	2977	96%
	55.5		2870	93%
	53.5 *		2763	89%
H4831SC	58.0 *	MAX.	3000	95%
	56.0		2883	92%
	54.0		2774	88%
RL19	55.0	MAX.	3003	94%
	53.0		2891	90%
	51.0 *		2779	87%
W760	53.0 *	MAX.	3018	87%
	51.0		2932	84%
	49.0		2835	81%

BC=Ballistic Coefficient SD=Sectional Density
*Most Accurate Load Tested **Compressed Load

Use Maximum Loads with Caution
Refer to page 73 for additional safety information

270 Winchester - 150 grain		MAXIMUM S.A.A.M.I. O.A.C.L.		3.340"
		TESTED O.A.C.L.	B.C.	S.D.
Ballistic Tip®	150gr. Spitzer	3.320"	0.496	0.279
CT® Ballistic Silvertip®	150gr. Spitzer	3.320"	0.496	0.279
Partition®	150gr. Spitzer	3.320"	0.465	0.279

CASE TYPE:	Nosler		PRIMER TYPE	Fed 210
CASE HOLDS:	62.8	Gr. WATER	BARREL Length/Make	24" Shilen
			BARREL Twist	1-10"

POWDER TYPE	POWDER CHG. GRS.		MUZZLE VEL. F.P.S.	LOAD DENSITY (VOLUME)
H4350 Most Accurate Powder Tested	52.0 *	MAX.	2782	88%
	50.0		2657	84%
	48.0		2532	81%
W760	50.0	MAX.	2810	83%
	48.0		2740	80%
	46.0 *		2670	77%
IMR 4320	45.0 *	MAX.	2818	77%
	43.0		2723	74%
	41.0		2628	70%
Viht N165	54.0	MAX.	2842	96%
	52.0		2752	92%
	50.0 *		2662	88%
IMR 7828	57.5	MAX.	2862	97%
	55.5		2737	94%
	53.5 *		2612	91%
RL19	55.0	MAX.	2892	95%
	53.0		2777	92%
	51.0 *		2662	88%
RL22	56.5	MAX.	2902	98%
	54.5		2787	94%
	52.5 *		2672	91%
H4831	55.0	MAX.	2905	91%
	53.0		2817	88%
	51.0 *		2728	85%
MAGPRO	61.5	MAX.	2913	** 101%
	59.5 *		2838	98%
	57.5		2735	94%

BC=Ballistic Coefficient SD=Sectional Density
*Most Accurate Load Tested **Compressed Load

Use Maximum Loads with Caution
Refer to page 73 for additional safety information

270 Winchester - 160 grain			MAXIMUM S.A.A.M.I. O.A.C.L.		3.340"
			TESTED O.A.C.L.	B.C.	S.D.
Partition®		160gr. Semi Spitz	3.320"	0.434	0.298
CASE TYPE:	Nosler		PRIMER TYPE	Fed 210	
CASE HOLDS:	63.8	Gr. WATER	BARREL Length/Make	24" Shilen	
			BARREL Twist	1-10"	

POWDER TYPE	POWDER CHG. GRS.	MUZZLE VEL. F.P.S.	LOAD DENSITY (VOLUME)
H4350	51.5 * MAX.	2672	85%
	49.5	2557	82%
	47.5	2442	79%
IMR 7828	57.0 * MAX.	2732	95%
	55.0	2637	92%
	53.0	2542	88%
RL19 Most Accurate Powder Tested	53.5 * MAX.	2750	91%
	51.5	2620	88%
	49.5	2490	84%
IMR 4350	51.0 MAX.	2758	85%
	49.0	2643	81%
	47.0 *	2528	78%
Viht N165	52.0 MAX.	2760	91%
	50.0	2688	87%
	48.0 *	2615	84%
H4831SC	54.5 * MAX.	2777	89%
	52.5	2678	86%
	50.5	2593	82%
MAGPRO	60.0 MAX.	2794	97%
	58.0	2672	94%
	56.0 *	2578	90%
IMR 4831	52.0 * MAX.	2820	87%
	50.0	2722	83%
	48.0	2624	80%
RL22	56.0 * MAX.	2828	95%
	54.0	2703	92%
	52.0	2578	89%

BC=Ballistic Coefficient SD=Sectional Density
*Most Accurate Load Tested **Compressed Load

Use Maximum Loads with Caution
Refer to page 73 for additional safety information

Mark Hampton

270 WINCHESTER SHORT MAGNUM

For those of us plagued with an emotional attachment to .277 caliber firearms, it was a glorious day back in 2002 when Winchester introduced the 270 Winchester Short Magnum (WSM), a necked-down version of their 300 WSM. This new, fat, stubby cartridge was intended to achieve magnum performance in short action rifles, a ballistic improvement over the classic 270 Winchester. The cartridge quickly became an admirable long range candidate for hunting wide open spaces where extended shots are the norm.

After shooting several different rifles in 270 WSM, I have found one common denominator- superb accuracy. Perhaps I have just been lucky, but minute-of-angle accuracy is common and one-half MOA is not uncommon. Not only does the 270 WSM function well in short action rifles, it also achieves impressive performance in short barrels. Currently, I am working with an H-S Precision handgun shooting both Nosler's 130 grain Ballistic Tip and Accubond bullets. Even with the short barrel and handgun scope, groups less than one inch are obtained consistently from one hundred yards.

The 270 WSM with Nosler's 130-grain BT makes a superlative cartridge for deer and antelope. The flat trajectory characteristics provide long range effectiveness in open country. Two days before this piece was written, I took a mature free-range west Texas aoudad with the 270 WSM and 130-grain AccuBond from 240 yards. A cross-canyon, one shot affair on a challenging ram confirmed the efficiency of cartridge and premium bullet. Performance was undeniably flawless! I will be using Nosler's AccuBond on an upcoming hunt in Mozambique for larger plains game. If penetration is a concern on elk-sized game, then Nosler's 150-grain Partition is just the ticket. Most assuredly, the 270 WSM is a great all-around hunting cartridge that will handle a multitude of big game pursuits.

There are several short, fat cartridges running around the block. The jury is still out on a verdict rendering which ones will survive. Even though I won't be here fifty years from now, I'd be willing to bet a cooler full of antelope tenderloin the 270 WSM will be going strong.

Mark Hampton

Mark Hampton has written numerous articles and published two books related to hunting. He has hunted on six continents in thirty countries and taken over 160 different species of big game with a handgun alone. Mark is considered one of the world's top handgun hunters.

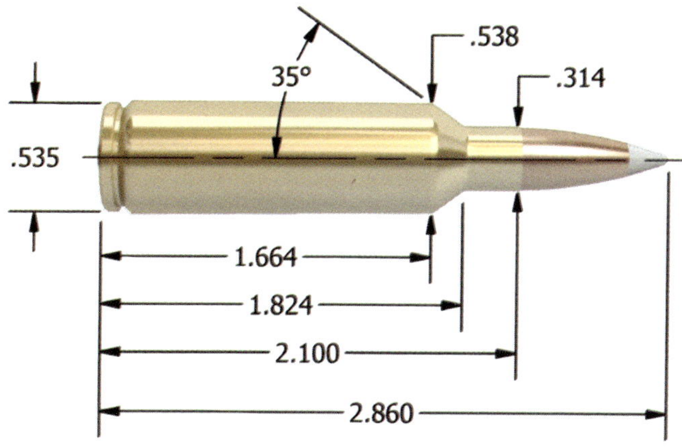

Maximum S.A.A.M.I. Overall Cartridge Length: 2.860"

BULLET CHOICES FOR THE 270 WINCHESTER SHORT MAGNUM

TECHNICAL INFORMATION

Capable of near-Weatherby performance, the 270 WSM is a true short action magnum. This cartridge will typically yield uniform pressures and velocities, while providing excellent accuracy with a variety of bullets. Our favorite powders are 4350, 4831, and RL19.

270 WSM - 100 grain

270 WSM - 100 grain		MAXIMUM S.A.A.M.I. O.A.C.L.		2.860"
		TESTED O.A.C.L.	B.C.	S.D.
AccuBond®	100gr. Spitzer	2.760"	0.323	0.186

CASE TYPE:	Winchester		PRIMER TYPE	WLRM
CASE HOLDS:	75.1	Gr. WATER	BARREL Length/Make	26" Pac-Nor
			BARREL Twist	1-10"

POWDER TYPE	POWDER CHG. GRS.		MUZZLE VEL. F.P.S.	LOAD DENSITY (VOLUME)
H4831SC	71.0	MAX.	3638	98%
	69.0		3546	96%
	67.0 *		3464	93%
W760	65.0	MAX.	3679	91%
	63.0		3620	88%
	61.0 *		3544	85%
RL15	61.0	MAX.	3724	85%
	59.0		3653	82%
	57.0 *		3602	79%
IMR 4350 Most Accurate Powder Tested	68.0	MAX.	3762	96%
	66.0		3624	93%
	64.0 *		3537	90%

BC=Ballistic Coefficient SD=Sectional Density
*Most Accurate Load Tested **Compressed Load

Use Maximum Loads with Caution
Refer to page 73 for additional safety information

270 WSM - 110 grain			MAXIMUM S.A.A.M.I. O.A.C.L.		2.860"
			TESTED O.A.C.L.	B.C.	S.D.
AccuBond®		110gr. Spitzer	2.760"	0.370	0.205

CASE TYPE:	Winchester		PRIMER TYPE	WLRM
CASE HOLDS:	74.5	Gr. WATER	BARREL Length/Make	24" Wiseman
			BARREL Twist	1-10"

POWDER TYPE	POWDER CHG. GRS.		MUZZLE VEL. F.P.S.	LOAD DENSITY (VOLUME)
H1000	73.0	MAX.	3482	** 103%
	71.0		3369	100%
	69.0	*	3278	97%
H4831SC	67.0	MAX.	3542	94%
	65.0		3448	91%
	63.0	*	3349	88%
H4350 Most Accurate Powder Tested	62.0	MAX.	3560	88%
	60.0		3456	85%
	58.0	*	3369	82%
RL19	68.0	MAX.	3635	99%
	66.0		3514	96%
	64.0	*	3421	93%

BC=Ballistic Coefficient SD=Sectional Density
*Most Accurate Load Tested **Compressed Load

Use Maximum Loads with Caution
Refer to page 73 for additional safety information

270 WSM - 130 grain		MAXIMUM S.A.A.M.I. O.A.C.L.		2.860"
		TESTED O.A.C.L.	B.C.	S.D.
AccuBond®	130gr. Spitzer	2.850"	0.435	0.242
Ballistic Tip®	130gr. Spitzer	2.850"	0.433	0.242
CT® Ballistic Silvertip®	130gr. Spitzer	2.850"	0.433	0.242
E-Tip®	130gr. Spitzer	2.820"	0.459	0.242
Due to internal construction differences, always begin with starting loads when using E-Tip® products.				
Partition®	130gr. Spitzer	2.850"	0.416	0.242

CASE TYPE:	Winchester		PRIMER TYPE	WLRM
CASE HOLDS:	73.0	Gr. WATER	BARREL Length/Make	24" Wiseman
			BARREL Twist	1-10" 4 groove

POWDER TYPE	POWDER CHG. GRS.		MUZZLE VEL. F.P.S.	LOAD DENSITY (VOLUME)
H4831SC	66.5	MAX.	3239	95%
	64.5		3149	92%
	62.5 *		3065	89%
W760	59.0 *	MAX.	3255	85%
	57.0		3203	82%
	55.0		3128	79%
IMR 4350 Most Accurate Powder Tested	63.0	MAX.	3294	91%
	61.0		3141	88%
	59.0 *		3051	86%
RL19	65.5	MAX.	3307	98%
	63.5 *		3167	95%
	61.5		3056	92%
Magnum	73.0	MAX.	3308	** 102%
	71.0		3242	99%
	69.0 *		3159	96%
IMR 4831	61.0	MAX.	3330	89%
	59.0		3234	86%
	57.0 *		3112	83%
IMR 7828	65.5	MAX.	3385	95%
	63.5		3293	93%
	61.5 *		3171	90%
MAGPRO	72.0	MAX.	3389	** 102%
	70.0 *		3317	99%
	68.0		3240	96%
RL22	65.0	MAX.	3396	97%
	63.0 *		3301	94%
	61.0		3222	91%

BC=Ballistic Coefficient SD=Sectional Density
*Most Accurate Load Tested **Compressed Load

Use Maximum Loads with Caution
Refer to page 73 for additional safety information

270 WSM - 140 grain

		MAXIMUM S.A.A.M.I. O.A.C.L.		2.860"
		TESTED O.A.C.L.	B.C.	S.D.
AccuBond®	140gr. Spitzer	2.850"	0.496	0.261
Ballistic Tip®	140gr. Spitzer	2.850"	0.456	0.261
Partition®	140gr. Spitzer	2.850"	0.432	0.261

CASE TYPE:	Winchester		PRIMER TYPE	WLRM
CASE HOLDS:	72.4	Gr. WATER	BARREL Length/Make	24" Wiseman
			BARREL Twist	1-10" 4 groove

POWDER TYPE	POWDER CHG. GRS.		MUZZLE VEL. F.P.S.	LOAD DENSITY (VOLUME)
H4831SC	62.5	MAX.	3108	90%
	60.5		3038	87%
	58.5 *		2949	84%
IMR 4350	60.5	MAX.	3154	88%
	58.5 *		3049	86%
	56.5		2938	83%
IMR 7828	61.0 *	MAX.	3161	90%
	59.0		3047	87%
	57.0		2950	84%
H1000	65.0 *	MAX.	3165	94%
	63.0		3086	91%
	61.0		3025	88%
RL22	61.0	MAX.	3182	92%
	59.0 *		3102	89%
	57.0		3012	86%
RL19	63.5	MAX.	3186	95%
	61.5		3114	92%
	59.5 *		3022	89%
Magnum	71.0	MAX.	3200	100%
	69.0		3116	97%
	67.0 *		3026	94%
Viht N170	67.0	MAX.	3209	** 103%
	65.0 *		3117	100%
	63.0		3074	97%
MAGPRO	69.5	MAX.	3237	99%
Most Accurate	67.5 *		3181	96%
Powder Tested	65.5		3092	93%

BC=Ballistic Coefficient SD=Sectional Density
*Most Accurate Load Tested **Compressed Load

Use Maximum Loads with Caution
Refer to page 73 for additional safety information

270 WSM - 150 grain

		MAXIMUM S.A.A.M.I. O.A.C.L.		2.860"
		TESTED O.A.C.L.	B.C.	S.D.
Ballistic Tip®	150gr. Spitzer	2.850"	0.496	0.279
CT® Ballistic Silvertip®	150gr. Spitzer	2.850"	0.496	0.279
Partition®	150gr. Spitzer	2.850"	0.465	0.279

CASE TYPE:	Winchester		PRIMER TYPE	WLRM
CASE HOLDS:	71.8	Gr. WATER	BARREL Length/Make	24" Wiseman
			BARREL Twist	1-10" 4 groove

POWDER TYPE	POWDER CHG. GRS.		MUZZLE VEL. F.P.S.		LOAD DENSITY (VOLUME)
H4831SC Most Accurate Powder Tested	60.0	MAX.	2966		87%
	58.0		2886		84%
	56.0	*	2803		81%
IMR 4350	58.0	* MAX.	2992		85%
	56.0		2894		83%
	54.0		2813		80%
RL19	60.5	MAX.	3007		92%
	58.5	*	2916		89%
	56.5		2834		86%
H1000	63.5	* MAX.	3068		93%
	61.5		3003		90%
	59.5		2903		87%
Magnum	69.0	MAX.	3089		98%
	67.0	*	2989		95%
	65.0		2888		92%
IMR 7828	61.5	MAX.	3117		91%
	59.5	*	3024		88%
	57.5		2949		85%
Viht N170	67.0	* MAX.	3147	**	104%
	65.0		3076	**	101%
	63.0		3006		97%
RL22	61.0	MAX.	3155		92%
	59.0		3069		89%
	57.0	*	2965		86%
MAGPRO	68.0	MAX.	3187		98%
	66.0	*	3122		95%
	64.0		3022		92%

BC=Ballistic Coefficient SD=Sectional Density
*Most Accurate Load Tested **Compressed Load

Use Maximum Loads with Caution
Refer to page 73 for additional safety information

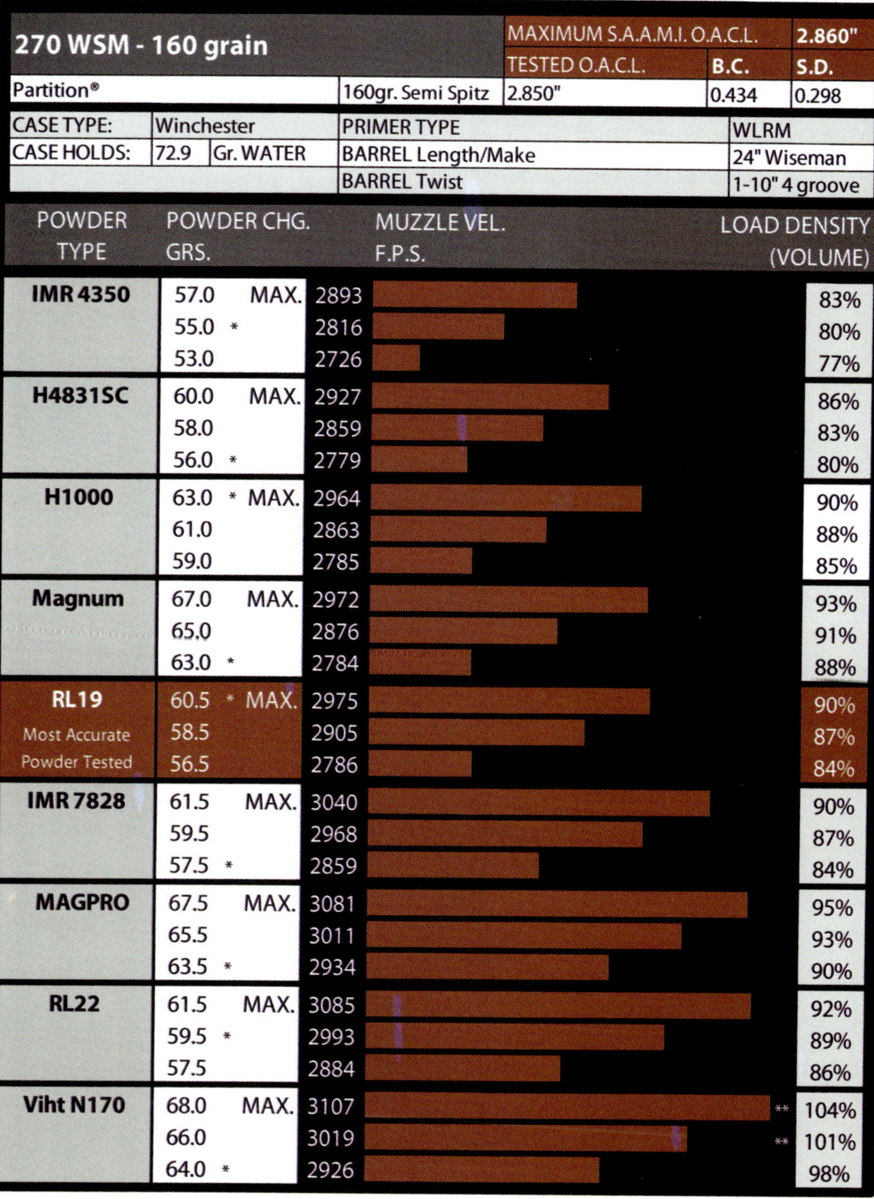

270 WSM - 160 grain		MAXIMUM S.A.A.M.I. O.A.C.L.		2.860"
		TESTED O.A.C.L.	B.C.	S.D.
Partition®	160gr. Semi Spitz	2.850"	0.434	0.298

CASE TYPE:	Winchester		PRIMER TYPE	WLRM
CASE HOLDS:	72.9	Gr. WATER	BARREL Length/Make	24" Wiseman
			BARREL Twist	1-10" 4 groove

POWDER TYPE	POWDER CHG. GRS.		MUZZLE VEL. F.P.S.		LOAD DENSITY (VOLUME)
IMR 4350	57.0	MAX.	2893		83%
	55.0 *		2816		80%
	53.0		2726		77%
H4831SC	60.0	MAX.	2927		86%
	58.0		2859		83%
	56.0 *		2779		80%
H1000	63.0 *	MAX.	2964		90%
	61.0		2863		88%
	59.0		2785		85%
Magnum	67.0	MAX.	2972		93%
	65.0		2876		91%
	63.0 *		2784		88%
RL19 Most Accurate Powder Tested	60.5 *	MAX.	2975		90%
	58.5		2905		87%
	56.5		2786		84%
IMR 7828	61.5	MAX.	3040		90%
	59.5		2968		87%
	57.5 *		2859		84%
MAGPRO	67.5	MAX.	3081		95%
	65.5		3011		93%
	63.5 *		2934		90%
RL22	61.5	MAX.	3085		92%
	59.5 *		2993		89%
	57.5		2884		86%
Viht N170	68.0	MAX.	3107	**	104%
	66.0		3019	**	101%
	64.0 *		2926		98%

BC=Ballistic Coefficient SD=Sectional Density
*Most Accurate Load Tested **Compressed Load

Use Maximum Loads with Caution
Refer to page 73 for additional safety information

Terry Wieland

270 WEATHERBY MAGNUM

The 270 Weatherby Magnum was the first of Roy Weatherby's high-velocity cartridges based on the 375 H&H case, shortened, blown out, and necked down. It was designed in 1943 when Weatherby was a wildcatter living in Kansas.

At the time, the 270 Winchester was the hottest hunting cartridge around, so it was a logical caliber for Weatherby to begin with. His new magnum added up to 300 fps with every bullet weight. Because the 270 Weatherby was the same overall length as the 270 Winchester, it was an easy matter to rechamber existing rifles, especially the Winchester Model 70, to the more powerful cartridge, and much of Weatherby's early business consisted of just that.

Ballistically, the 270 Weatherby is a serious step up from the 270 Winchester. Where that cartridge is on the light side for elk, the Weatherby round, loaded with a tough 150-grain bullet, is more than adequate for anything up to moose. As for trajectory, the .270 is the original flat-shooting magnum, with a point-blank range of up to 350 yards.

When it was introduced, the 270 Weatherby was handicapped by a lack of bullets that could withstand impact at high velocity. John Nosler remedied this with the development of the Partition bullet. The Partition's excellent accuracy and legendary terminal performance on game really allowed the 270 Weatherby to perform to its full potential. The Nosler Partition was the first premium game bullet to be loaded in Weatherby's factory ammunition, and it is still offered today.

Although it is pushing 70 years old, the 270 Weatherby is still the top performer in its caliber, improving further with every genuine development in game bullets and smokeless powders. Delivering almost 3400 fps with a 130-grain bullet, and 3200 with the 150-grain, the .270 is an excellent all-around choice for North American hunting, as well as for all African plains game.

If it has a handicap, it's the fact that rifles need 26-inch barrels to deliver this potential, and so they are long and heavy for backpacking or mountain work. As a pronghorn rifle, however, for mule deer at long range or elk across a canyon, the 270 Weatherby remains one of the most accurate and flattest shooting cartridges around, with more than ample power.

Terry Wieland

Terry Wieland is the author of several books on guns and hunting as well as a writer for Gray's Sporting Journal and Shooting Sportsman.

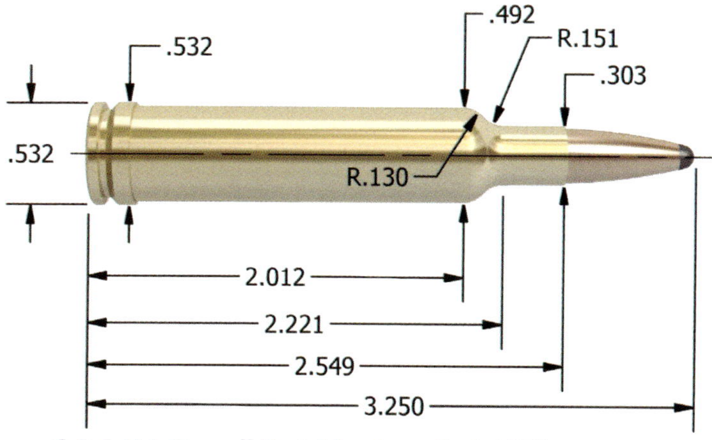

Maximum S.A.A.M.I. Overall Cartridge Length: 3.250"

BULLET CHOICES FOR THE 270 WEATHERBY MAGNUM

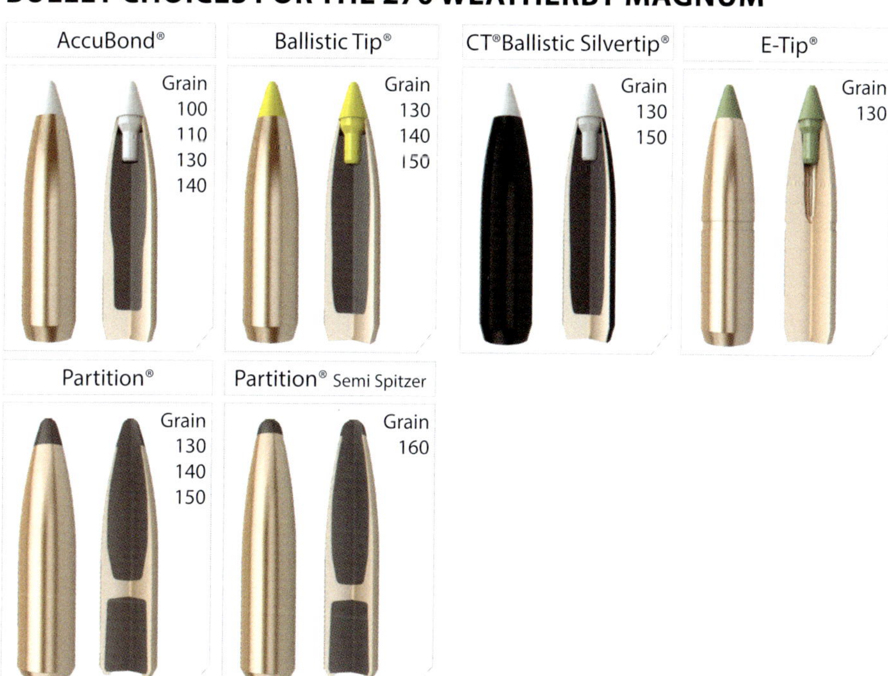

TECHNICAL INFORMATION

As with most Weatherby Magnums, there are a few points to keep in mind when reloading for this cartridge:

- Factory chambers have "freebore", meaning they have a long throat. For this reason it is generally not possible to seat a bullet close to, or in contact with the lands (rifling).

- Weatherby rifles of German manufacture (pre-1972) may have barrels with 1-12" twist rates. These rifles may not reliably stabilize bullets larger than 130 grains.

270 Weatherby Magnum - 100 grain		MAXIMUM S.A.A.M.I. O.A.C.L.		3.250"
		TESTED O.A.C.L.	B.C.	S.D.
AccuBond®	100gr. Spitzer	3.200"	0.323	0.186

CASE TYPE:	Nosler		PRIMER TYPE	WLRM
CASE HOLDS:	83.1	Gr. WATER	BARREL Length/Make	26" Lilja
			BARREL Twist	1-10"

POWDER TYPE	POWDER CHG. GRS.		MUZZLE VEL. F.P.S.	LOAD DENSITY (VOLUME)
IMR 7828 SSC	76.0	MAX.	3603	97%
	74.0		3474	95%
	72.0 *		3362	92%
MagPro	80.5	MAX.	3638	100%
	78.5		3576	97%
	76.5 *		3507	95%
H4831SC	76.5 *	MAX.	3644	96%
	74.5		3568	93%
	72.5		3461	91%
RL22 Most Accurate Powder Tested	76.5	MAX.	3723	100%
	74.5		3587	97%
	72.5 *		3419	95%

BC=Ballistic Coefficient SD=Sectional Density
*Most Accurate Load Tested **Compressed Load

Use Maximum Loads with Caution
Refer to page 73 for additional safety information

270 Weatherby Magnum - 110 grain		MAXIMUM S.A.A.M.I. O.A.C.L.		3.250"
		TESTED O.A.C.L.	B.C.	S.D.
AccuBond®	110gr. Spitzer	3.200"	0.370	0.205

CASE TYPE:	Weatherby		PRIMER TYPE	WLRM
CASE HOLDS:	82.0	Gr. WATER	BARREL Length/Make	26" Lilja
			BARREL Twist	1-10"

POWDER TYPE	POWDER CHG. GRS.		MUZZLE VEL. F.P.S.	LOAD DENSITY (VOLUME)
RL22	74.5	MAX.	3501	99%
	72.5 *		3358	96%
	70.5		3259	93%
H4831SC	74.0 *	MAX.	3520	94%
	72.0		3413	91%
	70.0		3294	89%
IMR 4350	70.0	MAX.	3531	90%
Most Accurate	68.0		3386	88%
Powder Tested	66.0 *		3255	85%

BC=Ballistic Coefficient SD=Sectional Density
*Most Accurate Load Tested **Compressed Load

Use Maximum Loads with Caution
Refer to page 73 for additional safety information

270 Weatherby Magnum - 130 grain		MAXIMUM S.A.A.M.I. O.A.C.L.		3.250"
		TESTED O.A.C.L.	B.C.	S.D.
AccuBond®	130gr. Spitzer	3.230"	0.435	0.242
Ballistic Tip®	130gr. Spitzer	3.230"	0.433	0.242
CT® Ballistic Silvertip®	130gr. Spitzer	3.230"	0.433	0.242
E-Tip®	130gr. Spitzer	3.230"	0.459	0.242
Due to internal construction differences, always begin with starting loads when using E-Tip® products.				
Partition®	130gr. Spitzer	3.230"	0.416	0.242

CASE TYPE:	Weatherby		PRIMER TYPE	Fed 215
CASE HOLDS:	81.7	Gr. WATER	BARREL Length/Make	26" Lilja
			BARREL Twist	1-10"

POWDER TYPE	POWDER CHG. GRS.		MUZZLE VEL. F.P.S.	LOAD DENSITY (VOLUME)
IMR 4831	69.0	* MAX.	3240	90%
	67.0		3130	87%
	65.0		3020	85%
H4831	72.0	* MAX.	3260	92%
	70.0		3110	89%
	68.0		2960	87%
MAGPRO	76.0	MAX.	3295	96%
	74.0	*	3231	93%
	72.0		3111	91%
Viht N560	71.0	MAX.	3309	97%
	69.0	*	3214	94%
	67.0		3139	91%
IMR 4350	68.0	* MAX.	3350	88%
	66.0		3250	85%
	64.0		3150	83%
RL25	75.0	* MAX.	3350	100%
	73.0		3256	97%
	71.0		3147	94%
RL19	71.0	* MAX.	3414	94%
Most Accurate	69.0		3318	92%
Powder Tested	67.0		3221	89%
RL22	73.0	* MAX.	3458	97%
	71.0		3386	94%
	69.0		3263	92%
IMR 7828	74.5	* MAX.	3463	97%
	72.5		3402	94%
	70.5		3264	92%

BC=Ballistic Coefficient SD=Sectional Density
*Most Accurate Load Tested **Compressed Load

Use Maximum Loads with Caution
Refer to page 73 for additional safety information

270 Weatherby Magnum - 140 grain		MAXIMUM S.A.A.M.I. O.A.C.L.		3.250"
		TESTED O.A.C.L.	B.C.	S.D.
AccuBond®	140gr. Spitzer	3.230"	0.496	0.261
Ballistic Tip®	140gr. Spitzer	3.230"	0.456	0.261
Partition®	140gr. Spitzer	3.230"	0.432	0.261

CASE TYPE:	Weatherby		PRIMER TYPE	Fed 215
CASE HOLDS:	79.0	Gr. WATER	BARREL Length/Make	26" Lilja
			BARREL Twist	1-10"

POWDER TYPE	POWDER CHG. GRS.		MUZZLE VEL. F.P.S.	LOAD DENSITY (VOLUME)
H4831	71.0	* MAX.	3080	94%
	69.0		2980	91%
	67.0		2880	88%
RL22	68.5	* MAX.	3116	94%
	66.5		3039	91%
	64.5		2961	89%
IMR 4350	66.5	* MAX.	3138	89%
	64.5		3043	86%
	62.5		2948	84%
IMR 4831	68.5	MAX.	3155	92%
	66.5		3045	90%
	64.5	*	2935	87%
Magnum	77.0	MAX.	3160	99%
	75.0		3089	96%
	73.0	*	2981	94%
MagPro	74.0	* MAX.	3162	97%
	72.0		3044	94%
	70.0		2996	91%
RL25	73.0	* MAX.	3193	100%
	71.0		3143	98%
	69.0		3021	95%
IMR 7828	71.0	MAX.	3260	96%
	69.0		3155	93%
	67.0	*	3053	90%
Viht N560	70.5	* MAX.	3293	99%
Most Accurate	68.5		3170	96%
Powder Tested	66.5		3093	94%

BC=Ballistic Coefficient SD=Sectional Density
*Most Accurate Load Tested **Compressed Load

Use Maximum Loads with Caution
Refer to page 73 for additional safety information

270 Weatherby Magnum - 150 grain

		MAXIMUM S.A.A.M.I. O.A.C.L.		3.250"
		TESTED O.A.C.L.	B.C.	S.D.
Ballistic Tip®	150gr. Spitzer	3.230"	0.496	0.279
CT® Ballistic Silvertip®	150gr. Spitzer	3.230"	0.496	0.279
Partition®	150gr. Spitzer	3.230"	0.465	0.279

CASE TYPE:	Weatherby		PRIMER TYPE	Fed 215
CASE HOLDS:	79.4	Gr. WATER	BARREL Length/Make	26" Lilja
			BARREL Twist	1-10"

POWDER TYPE	POWDER CHG. GRS.		MUZZLE VEL. F.P.S.	LOAD DENSITY (VOLUME)
RL22	66.5	MAX.	2988	91%
	64.5		2901	88%
	62.5	*	2815	86%
MAGPRO	70.0	MAX.	3031	91%
	68.0	*	2913	88%
	66.0		2812	86%
IMR 4831	67.0	* MAX.	3060	90%
	65.0		2950	87%
	63.0		2840	84%
IMR 4350	65.0	* MAX.	3060	87%
	63.0		2990	84%
	61.0		2920	81%
H4831	69.0	MAX.	3068	91%
	67.0		2943	88%
	65.0	*	2818	85%
Retumbo	75.0	MAX.	3091	99%
	73.0	*	3039	97%
	71.0		2965	94%
RL25 Most Accurate Powder Tested	72.0	* MAX.	3109	99%
	70.0		3040	96%
	68.0		2924	93%
Viht N560	69.5	* MAX.	3190	97%
	67.5		3103	94%
	65.5		2998	92%
IMR 7828	70.5	MAX.	3207	94%
	68.5		3118	92%
	66.5	*	3033	89%

BC=Ballistic Coefficient SD=Sectional Density
*Most Accurate Load Tested **Compressed Load

Use Maximum Loads with Caution
Refer to page 73 for additional safety information

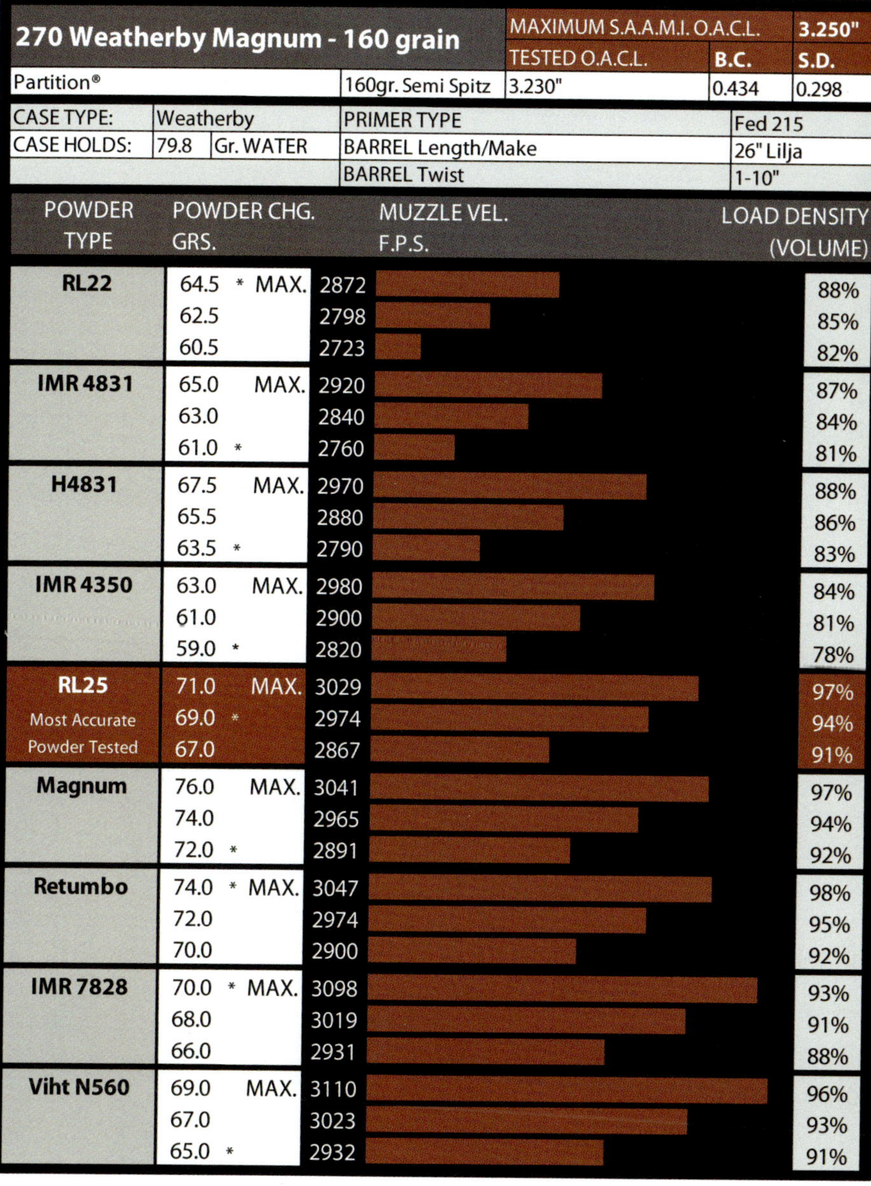

270 Weatherby Magnum - 160 grain		MAXIMUM S.A.A.M.I. O.A.C.L.		3.250"
		TESTED O.A.C.L.	B.C.	S.D.
Partition®	160gr. Semi Spitz	3.230"	0.434	0.298

CASE TYPE:	Weatherby		PRIMER TYPE	Fed 215
CASE HOLDS:	79.8	Gr. WATER	BARREL Length/Make	26" Lilja
			BARREL Twist	1-10"

POWDER TYPE	POWDER CHG. GRS.		MUZZLE VEL. F.P.S.	LOAD DENSITY (VOLUME)
RL22	64.5 *	MAX.	2872	88%
	62.5		2798	85%
	60.5		2723	82%
IMR 4831	65.0	MAX.	2920	87%
	63.0		2840	84%
	61.0 *		2760	81%
H4831	67.5	MAX.	2970	88%
	65.5		2880	86%
	63.5 *		2790	83%
IMR 4350	63.0	MAX.	2980	84%
	61.0		2900	81%
	59.0 *		2820	78%
RL25 Most Accurate Powder Tested	71.0	MAX.	3029	97%
	69.0 *		2974	94%
	67.0		2867	91%
Magnum	76.0	MAX.	3041	97%
	74.0		2965	94%
	72.0 *		2891	92%
Retumbo	74.0 *	MAX.	3047	98%
	72.0		2974	95%
	70.0		2900	92%
IMR 7828	70.0 *	MAX.	3098	93%
	68.0		3019	91%
	66.0		2931	88%
Viht N560	69.0	MAX.	3110	96%
	67.0		3023	93%
	65.0 *		2932	91%

BC=Ballistic Coefficient SD=Sectional Density
*Most Accurate Load Tested **Compressed Load

Use Maximum Loads with Caution
Refer to page 73 for additional safety information

John Haviland

7MM-08 REMINGTON

The 7mm-08 Remington reigns around our house. Over the last 23 years, my three sons have used various 7mm-08 rifles to shoot antelope, black bears, elk and whitetail and mule deer. My wife even used the cartridge in a Winchester Model 70 Featherweight to kill a bull moose. Once in awhile the Winchester is left unattended and I steal away with it to the mountains. It kills elk and deer just as quickly as my old 30-06.

I've handloaded 120 to 175-grain bullets in the 7mm-08 for family hunts over the years, but finally settled on 140-grain bullets for all big game hunting. With a muzzle velocity of nearly 2,900 fps from a 22-inch barrel, that bullet weight has the same trajectory as a 30-06 shooting 150-grain bullets at maximum velocity. However, the 7mm-08 generates nearly 20 percent less recoil, even though it is chambered in a lighter and more compact rifle than the 30-06.

That comfortable recoil allows us to shoot our 7mm-08s during the summer months so we are skillful at placing a single bullet in the correct spot on game come hunting season. That ability to fire a bullet in exactly the right spot is essential no matter what cartridge a hunter shoots, even one with magnum as its last name.

Last fall at sundown on opening day of elk season, I heard my son fire his 7mm-08 from the far side of a mountain. Most of two hours after dark, I met Thomas along the trail leading off the mountain. "We got one," he said, hefting his 7-08. After all, the 7mm-08 is a member of the family.

John Haviland

John Haviland is a Contributing Editor to Wolfe Publications

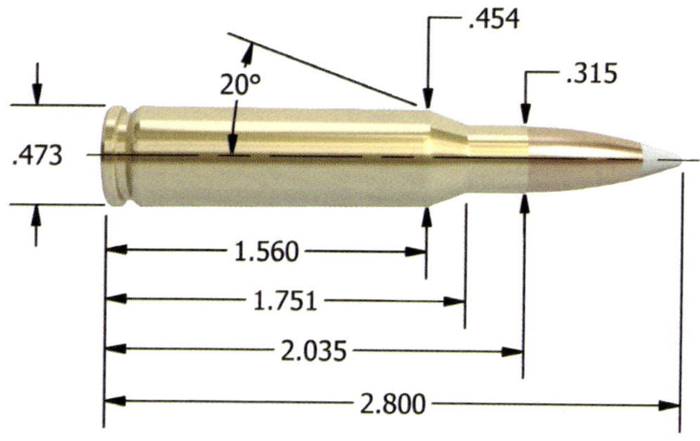

Maximum S.A.A.M.I. Overall Cartridge Length: 2.800"

BULLET CHOICES FOR THE 7MM-08 REMINGTON

TECHNICAL INFORMATION

Like its 308 Winchester parent, the 7mm-08 is an excellent cartridge, one that performs well with a wide variety of bullet/powder combinations. We have had good results with 4350, H-380, and Big Game powders. Some 7mm-08 chambers have short throats, so when seating bullets make sure they do not contact the rifling.

7mm-08 Remington - 120 grain		MAXIMUM S.A.A.M.I. O.A.C.L.		2.800"
		TESTED O.A.C.L.	B.C.	S.D.
Ballistic Tip®	120gr. Spitzer	2.800"	0.417	0.213

CASE TYPE:	Nosler		PRIMER TYPE	WLR
CASE HOLDS:	50.4	Gr. WATER	BARREL Length/Make	26" Wiseman
			BARREL Twist	1-9"

POWDER TYPE	POWDER CHG. GRS.		MUZZLE VEL. F.P.S.		LOAD DENSITY (VOLUME)
IMR 4895	43.0	MAX.	3040		94%
	41.0		2950		89%
	39.0 *		2860		85%
IMR 4350 Most Accurate Powder Tested	49.0	MAX.	3051	**	103%
	47.0		2942		99%
	45.0 *		2809		94%
IMR 4064	43.0	MAX.	3052		94%
	41.0		2917		89%
	39.0 *		2782		85%
H4895	42.5	MAX.	3090		93%
	40.5		2960		88%
	38.5 *		2830		84%
Viht N150	45.5 *	MAX.	3111	**	106%
	43.5		2994	**	102%
	41.5		2877		97%
RL15	44.0	MAX.	3118		91%
	42.0		2983		87%
	40.0 *		2848		83%
Varget	45.0	MAX.	3139		97%
	43.0 *		3018		92%
	41.0		2895		88%
H414	50.5 *	MAX.	3180	**	104%
	48.5		3060		100%
	46.5		2940		96%
W760	52.0	MAX.	3233	**	108%
	50.0		3122	**	104%
	48.0 *		3015		100%
Big Game	52.0	MAX.	3262	**	109%
	50.0 *		3190	**	104%
	48.0		3090		100%

BC=Ballistic Coefficient SD=Sectional Density
*Most Accurate Load Tested **Compressed Load

Use Maximum Loads with Caution
Refer to page 73 for additional safety information

7mm-08 Remington - 140 grain		MAXIMUM S.A.A.M.I. O.A.C.L.		2.800"
		TESTED O.A.C.L.	B.C.	S.D.
AccuBond®	140gr. Spitzer	2.800"	0.485	0.248
Ballistic Tip®	140gr. Spitzer	2.800"	0.485	0.248
CT® Ballistic Silvertip®	140gr. Spitzer	2.800"	0.485	0.248
E-Tip®	140gr. Spitzer	2.770"	0.489	0.248
Due to internal construction differences, always begin with starting loads when using E-Tip® products.				
Partition®	140gr. Spitzer	2.800"	0.434	0.248

CASE TYPE:	Nosler		PRIMER TYPE	WLR
CASE HOLDS:	48.6	Gr. WATER	BARREL Length/Make	26" Wiseman
			BARREL Twist	1-9"

POWDER TYPE	POWDER CHG. GRS.		MUZZLE VEL. F.P.S.	LOAD DENSITY (VOLUME)
Viht N150	41.0	MAX.	2739	99%
	39.0		2626	94%
	37.0 *		2513	90%
H335	40.0 *	MAX.	2742	84%
	38.0		2597	79%
	36.0		2452	75%
A-2520	41.5	MAX.	2806	88%
	39.5		2670	84%
	37.5 *		2550	80%
RL15	41.5	MAX.	2827	89%
	39.5		2724	85%
	37.5 *		2596	80%
H414	46.0	MAX.	2828	99%
	44.0		2683	94%
	42.0 *		2538	90%
IMR 4350	46.0 *	MAX.	2840	100%
	44.0		2720	96%
	42.0		2600	91%
Varget	42.0	MAX.	2849	93%
	40.0		2718	89%
	38.0 *		2610	85%
IMR 4895	41.0	MAX.	2862	93%
	39.0		2737	88%
	37.0 *		2612	84%
RL19	49.0	MAX.	2922	** 110%
	47.0		2807	** 105%
	45.0 *		2692	** 101%
Big Game	47.5	MAX.	2953	** 103%
Most Accurate	45.5		2832	99%
Powder Tested	43.5 *		2760	94%

BC=Ballistic Coefficient SD=Sectional Density
*Most Accurate Load Tested **Compressed Load

Use Maximum Loads with Caution
Refer to page 73 for additional safety information

7mm-08 Remington - 150 grain		MAXIMUM S.A.A.M.I. O.A.C.L.		2.800"
		TESTED O.A.C.L.	B.C.	S.D.
Ballistic Tip®	150gr. Spitzer	2.800"	0.493	0.266
CT® Ballistic Silvertip®	150gr. Spitzer	2.800"	0.493	0.266
E-Tip®	150gr. Spitzer	2.770"	0.498	0.266
Due to internal construction differences, always begin with starting loads when using E-Tip® products.				
Partition®	150gr. Spitzer	2.800"	0.456	0.266

CASE TYPE:	Nosler		PRIMER TYPE	WLR
CASE HOLDS:	47.8	Gr. WATER	BARREL Length/Make	26" Wiseman
			BARREL Twist	1-9"

POWDER TYPE	POWDER CHG. GRS.		MUZZLE VEL. F.P.S.		LOAD DENSITY (VOLUME)
IMR 4350	44.5	MAX.	2692		99%
	42.5		2567		94%
	40.5 *		2442		90%
IMR 4831	46.0 *	MAX.	2710	**	102%
	44.0		2580		98%
	42.0		2450		93%
Varget	41.0	MAX.	2747		93%
	39.0		2630		88%
	37.0 *		2505		84%
IMR 4895	40.0 *	MAX.	2750		92%
	38.0		2620		87%
	36.0		2490		83%
H380 Most Accurate Powder Tested	45.5 *	MAX.	2750		100%
	43.5		2640		96%
	41.5		2530		91%
Viht N160	48.0	MAX.	2788	**	112%
	46.0		2680	**	107%
	44.0 *		2571	**	102%
H414	46.0	MAX.	2800		100%
	44.0		2660		96%
	42.0 *		2520		92%
RL15	41.5	MAX.	2802		90%
	39.5		2657		86%
	37.5 *		2512		82%
RL19	48.5	MAX.	2862	**	110%
	46.5		2757	**	106%
	44.5 *		2652	**	101%
Hunter	49.5	MAX.	2869	**	109%
	47.5		2801	**	105%
	45.5 *		2748		100%

BC=Ballistic Coefficient SD=Sectional Density
*Most Accurate Load Tested **Compressed Load

Use Maximum Loads with Caution
Refer to page 73 for additional safety information

7mm-08 Remington - 160 grain

		MAXIMUM S.A.A.M.I. O.A.C.L.		2.800"
		TESTED O.A.C.L.	B.C.	S.D.
AccuBond®	160gr. Spitzer	2.800"	0.531	0.283
Partition®	160gr. Spitzer	2.800"	0.475	0.283

CASE TYPE:	Nosler		PRIMER TYPE	WLR
CASE HOLDS:	47.6	Gr. WATER	BARREL Length/Make	26" Wiseman
			BARREL Twist	1-9"

POWDER TYPE	POWDER CHG. GRS.		MUZZLE VEL. F.P.S.		LOAD DENSITY (VOLUME)
IMR 4895	38.0	MAX.	2570		88%
	36.0		2450		83%
	34.0 *		2330		78%
RL15 Most Accurate Powder Tested	39.0	MAX.	2593		85%
	37.0 *		2520		81%
	35.0		2402		77%
Varget	39.0	MAX.	2594		89%
	37.0 *		2492		84%
	35.0		2380		79%
H414	44.5 *	MAX.	2650		97%
	42.5		2520		93%
	40.5		2390		89%
IMR 4350	44.0	MAX.	2650		98%
	42.0		2550		93%
	40.0 *		2450		89%
IMR 4831	46.0	MAX.	2680	**	103%
	44.0		2560		98%
	42.0 *		2440		94%
H4350	44.5	MAX.	2682		99%
	42.5		2557		94%
	40.5 *		2432		90%
Viht N160	47.0	MAX.	2723	**	110%
	45.0		2617	**	105%
	43.0 *		2511		100%
Hunter	49.0	MAX.	2762	**	108%
	47.0 *		2683	**	104%
	45.0		2623		100%
RL19	47.5	MAX.	2780	**	108%
	45.5		2660	**	104%
	43.5 *		2540		99%

BC=Ballistic Coefficient SD=Sectional Density
*Most Accurate Load Tested **Compressed Load

Use Maximum Loads with Caution
Refer to page 73 for additional safety information

7mm-08 Remington - 175 grain		MAXIMUM S.A.A.M.I. O.A.C.L.		2.800"
		TESTED O.A.C.L.	B.C.	S.D.
Partition®	175gr. Spitzer	2.800"	0.519	0.310

CASE TYPE:	Nosler		PRIMER TYPE	WLR
CASE HOLDS:	47.2	Gr. WATER	BARREL Length/Make	26" Wiseman
			BARREL Twist	1-9"

POWDER TYPE	POWDER CHG. GRS.		MUZZLE VEL. F.P.S.		LOAD DENSITY (VOLUME)
IMR 4895	37.0	MAX.	2472		86%
	35.0		2357		81%
	33.0 *		2242		77%
Viht N160 Most Accurate Powder Tested	43.5	MAX.	2479	**	102%
	41.5		2383		98%
	39.5 *		2288		93%
H414	42.0	MAX.	2488		93%
	40.0		2363		88%
	38.0 *		2238		84%
H4831	45.0	MAX.	2510		99%
	43.0		2416		95%
	41.0 *		2311		90%
W760	41.5 *	MAX.	2534		92%
	39.5		2425		88%
	37.5		2330		83%
IMR 4350	43.0 *	MAX.	2540		96%
	41.0		2428		92%
	39.0		2312		87%
IMR 4831	45.0	MAX.	2612	**	101%
	43.0		2497		97%
	41.0 *		2382		92%
RL19	46.0	MAX.	2614	**	106%
	44.0 *		2488	**	101%
	42.0		2392		97%
Hunter	46.0	MAX.	2623	**	103%
	44.0		2545		98%
	42.0 *		2439		94%

BC=Ballistic Coefficient SD=Sectional Density
*Most Accurate Load Tested **Compressed Load

Use Maximum Loads with Caution
Refer to page 73 for additional safety information

7X57MM MAUSER

Introduced in the Model 1892 Mauser in that same year, the 7x57mm Mauser is in its 120th year as this is written, quite a milestone considering the upstart competition that has dominated the 7mm lineup since the advent of the 7mm Remington Magnum in 1962.

From the outset, the 7x57mm Mauser was designed around a 173- to 175-grain roundnose bullet, both solids and softpoints, that averaged a respectable 2,300 fps, depending on barrel length. While standard military chambers designed for the long, roundnose, 175-grain bullets had a .495-inch leade, sporting rifle manufacturers reduced the leade by about half, .25 to .265 inch, in the late 1970s and early 1980s to accommodate more streamlined 130- to 162-grain spitzers that could be seated out to 3.065 inches in standard 30-06 length bolt actions. And, up until 1962, owing to the proliferation of surplus military rifles, all 7mm bullets in this country were designed to expand and penetrate at prudent velocities when fired from the Mauser; aka, the 275 Rigby.

As a rule, optimum velocities with 120-, 139/140- and 160-grain bullets average around 3,000, 2,800 and 2,700 fps, respectively, in typical 22-inch barrels in modern rifles such as the Winchester Model 70, Remington 700 and Ruger M77. Folks could debate whether or not velocities might be increased a bit, but for all practical purposes, the 7x57mm Mauser performance is similar to the 280 and 7mm-08 Remingtons and 284 and 270 Winchesters. Assuming the proper bullet is used for the task, and proper placement, they are all peas in a pod.

Like any cartridge, however, success of the 7x57mm Mauser was, and still is, based on its bullets, and one of the great stories in 7mm bullet design is rooted in the 1946 failure of a 30-caliber bullet in John Nosler's 300 H&H on a Canadian moose that ultimately led to the creation of the Nosler Partition Bullet Co. in 1948. Introduced in 1955, the 7mm 140-, 150- and 160-grain Partitions, and the more recent 140- and 160-grain AccuBonds, will likely secure a successful future for the 7x57mm Mauser for generations to come. As our late friend, writer and African professional hunter, Finn Aagaard, once advised, "the Partition helped to establish the 7mm Mauser as one of the great sporting cartridges our time."

Dave Scovill is the Editor in Chief of Successful Hunter, Rifle, and Handloader magazines.

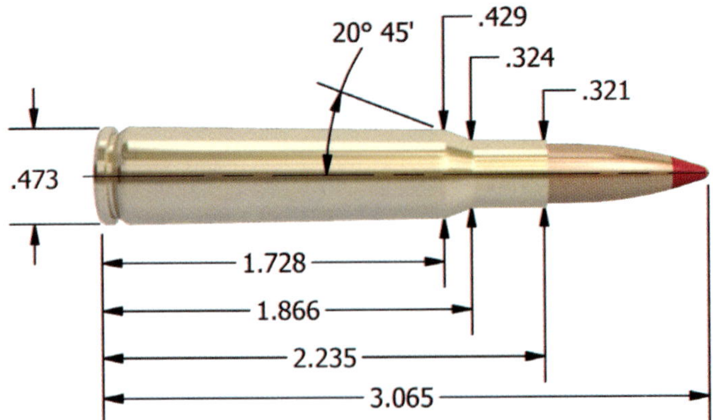

Maximum S.A.A.M.I. Overall Cartridge Length: 3.065"

BULLET CHOICES FOR THE 7X57MM MAUSER

TECHNICAL INFORMATION

Developed for military use in 1892, the 7mm Mauser is one of the oldest cartridges still in sporting use. This load data is intended for use only with new firearms in good condition. If you are using an older military rifle, or are unsure as to the condition of your firearm, it must be thoroughly inspected by a competent gunsmith before use.

7x57 Mauser - 120 grain

Ballistic Tip®	120gr. Spitzer	MAXIMUM S.A.A.M.I. O.A.C.L.		3.065"
		TESTED O.A.C.L.	B.C.	S.D.
		3.045"	0.417	0.213

CASE TYPE:	Winchester	PRIMER TYPE	WLR
CASE HOLDS:	54.3 Gr. WATER	BARREL Length/Make	22" Lilja
		BARREL Twist	1-9"

POWDER TYPE	POWDER CHG. GRS.		MUZZLE VEL. F.P.S.		LOAD DENSITY (VOLUME)
IMR 4831	52.0 *	MAX.	2830	**	102%
	50.0		2750		98%
	48.0		2670		94%
H380	47.5	MAX.	2922		92%
	45.5		2797		88%
	43.5 *		2672		84%
BL-C2	44.5 *	MAX.	2928		83%
	42.5		2793		79%
	40.5		2658		76%
Varget Most Accurate Powder Tested	45.0	MAX.	2928		90%
	43.0 *		2824		86%
	41.0		2717		82%
IMR 4064	47.0	MAX.	3000		95%
	45.0		2870		91%
	43.0 *		2740		87%
RL15	44.5 *	MAX.	3001		85%
	42.5		2931		82%
	40.5		2861		78%
IMR 4320	47.0 *	MAX.	3008		93%
	45.0		2883		89%
	43.0		2758		85%
IMR 4895	45.0	MAX.	3020		91%
	43.0		2880		87%
	41.0 *		2740		83%
Hunter	53.5	MAX.	3040	**	104%
	51.5 *		2921		100%
	49.5		2787		96%
Viht N150	46.5	MAX.	3047	**	101%
	44.5		2944		96%
	42.5 *		2841		92%

BC=Ballistic Coefficient SD=Sectional Density
*Most Accurate Load Tested **Compressed Load

Use Maximum Loads with Caution
Refer to page 73 for additional safety information

7x57 Mauser - 140 grain		MAXIMUM S.A.A.M.I. O.A.C.L.		3.065"
		TESTED O.A.C.L.	B.C.	S.D.
AccuBond®	140gr. Spitzer	3.065"	0.485	0.248
Ballistic Tip®	140gr. Spitzer	3.065"	0.485	0.248
CT® Ballistic Silvertip®	140gr. Spitzer	3.065"	0.485	0.248
E-Tip®	140gr. Spitzer	3.015"	0.489	0.248
Due to internal construction differences, always begin with starting loads when using E-Tip® products.				
Partition®	140gr. Spitzer	3.065"	0.434	0.248

CASE TYPE:	Winchester		PRIMER TYPE	WLR
CASE HOLDS:	52.5	Gr. WATER	BARREL Length/Make	22" Lilja
			BARREL Twist	1-9"

POWDER TYPE	POWDER CHG. GRS.		MUZZLE VEL. F.P.S.	LOAD DENSITY (VOLUME)
RL15	42.0	MAX.	2693	83%
	40.0 *		2597	79%
	38.0		2498	75%
IMR 4064	42.0 *	MAX.	2712	88%
	40.0		2607	84%
	38.0		2502	80%
IMR 4895	40.0 *	MAX.	2720	84%
	38.0		2590	80%
	36.0		2460	75%
IMR 4831	51.0 *	MAX.	2762	** 103%
	49.0		2637	99%
	47.0		2512	95%
IMR 4320	43.0	MAX.	2768	88%
	41.0		2643	84%
	39.0 *		2518	80%
Hunter	49.5	MAX.	2768	99%
	47.5 *		2682	95%
	45.5		2593	91%
Viht N150	43.5	MAX.	2779	97%
	41.5		2686	93%
	39.5 *		2594	89%
H414 Most Accurate Powder Tested	49.0	MAX.	2790	97%
	47.0		2700	93%
	45.0 *		2610	89%
IMR 4350	49.0 *	MAX.	2820	99%
	47.0		2680	95%
	45.0		2540	91%
RL19	51.5	MAX.	2892	** 107%
	49.5		2787	** 102%
	47.5 *		2683	98%

BC=Ballistic Coefficient SD=Sectional Density
*Most Accurate Load Tested **Compressed Load

Use Maximum Loads with Caution
Refer to page 73 for additional safety information

7x57 Mauser - 150 grain		MAXIMUM S.A.A.M.I. O.A.C.L.		3.065"
		TESTED O.A.C.L.	B.C.	S.D.
Ballistic Tip®	150gr. Spitzer	3.045"	0.493	0.266
CT® Ballistic Silvertip®	150gr. Spitzer	3.045"	0.493	0.266
E-Tip®	150gr. Spitzer	3.015"	0.498	0.266
Due to internal construction differences, always begin with starting loads when using E-Tip® products.				
Partition®	150gr. Spitzer	3.045"	0.456	0.266

CASE TYPE:	Winchester		PRIMER TYPE	WLR
CASE HOLDS:	51.4	Gr. WATER	BARREL Length/Make	22" Lilja
			BARREL Twist	1-9"

POWDER TYPE	POWDER CHG. GRS.		MUZZLE VEL. F.P.S.		LOAD DENSITY (VOLUME)
H4831SC Most Accurate Powder Tested	49.0	MAX.	2570		99%
	47.0		2482		95%
	45.0 *		2408		91%
IMR 4895	38.5	MAX.	2590		82%
	36.5		2410		78%
	34.5 *		2230		74%
IMR 4064	41.0 *	MAX.	2600		88%
	39.0		2490		83%
	37.0		2380		79%
H414	48.5	MAX.	2610		98%
	46.5		2570		94%
	44.5 *		2530		90%
IMR 4320	42.0	MAX.	2612		88%
	40.0		2517		84%
	38.0 *		2422		79%
W760	46.0 *	MAX.	2633		94%
	44.0		2548		90%
	42.0		2452		86%
Hunter	48.5	MAX.	2686		99%
	46.5		2592		95%
	44.5 *		2519		91%
IMR 4350	48.0	MAX.	2730		99%
	46.0		2580		95%
	44.0 *		2430		91%
Viht N160	50.0	MAX.	2770	**	108%
	48.0		2670	**	104%
	46.0 *		2569		99%
RL19	51.0	MAX.	2831	**	108%
	49.0		2718	**	104%
	47.0 *		2605		99%

BC=Ballistic Coefficient SD=Sectional Density
*Most Accurate Load Tested **Compressed Load

Use Maximum Loads with Caution
Refer to page 73 for additional safety information

7x57 Mauser - 160 grain		MAXIMUM S.A.A.M.I. O.A.C.L.		3.065"
		TESTED O.A.C.L.	B.C.	S.D.
AccuBond®	160gr. Spitzer	3.045"	0.531	0.283
Partition®	160gr. Spitzer	3.045"	0.475	0.283

CASE TYPE:	Winchester		PRIMER TYPE	WLR
CASE HOLDS:	50.6	Gr. WATER	BARREL Length/Make	22" Lilja
			BARREL Twist	1-9"

POWDER TYPE	POWDER CHG. GRS.		MUZZLE VEL. F.P.S.		LOAD DENSITY (VOLUME)
IMR 4895	38.0	MAX.	2448		83%
	36.0		2333		78%
	34.0 *		2218		74%
IMR 4064	39.0	MAX.	2450		85%
	37.0		2360		80%
	35.0 *		2270		76%
W760	43.0	MAX.	2480		89%
	41.0		2405		85%
	39.0 *		2272		81%
H414	45.0 *	MAX.	2490		93%
	43.0		2380		89%
	41.0		2270		84%
Hunter	46.0 *	MAX.	2527		96%
	44.0		2448		92%
	42.0		2318		87%
IMR 4320	41.0	MAX.	2580		87%
	39.0		2470		83%
	37.0 *		2360		79%
IMR 4350 Most Accurate Powder Tested	46.0 *	MAX.	2622		96%
	44.0		2517		92%
	42.0		2412		88%
RL19	47.5	MAX.	2643	**	102%
	45.5		2532		98%
	43.5 *		2421		93%
IMR 4831	49.0	MAX.	2648	**	103%
	47.0		2523		99%
	45.0 *		2398		95%
Viht N160	48.5	MAX.	2686	**	106%
	46.5		2603	**	102%
	44.5 *		2520		98%

BC=Ballistic Coefficient SD=Sectional Density
*Most Accurate Load Tested **Compressed Load

Use Maximum Loads with Caution
Refer to page 73 for additional safety information

7x57 Mauser - 175 grain

		MAXIMUM S.A.A.M.I. O.A.C.L.		3.065"
		TESTED O.A.C.L.	B.C.	S.D.
Partition®	175gr. Spitzer	3.045"	0.519	0.310

CASE TYPE:	Winchester		PRIMER TYPE	WLR
CASE HOLDS:	48.7	Gr. WATER	BARREL Length/Make	22" Lilja
			BARREL Twist	1-9"

POWDER TYPE	POWDER CHG. GRS.		MUZZLE VEL. F.P.S.	LOAD DENSITY (VOLUME)
W760	42.0	MAX.	2374	77%
	40.0	*	2296	73%
	38.0		2206	70%
Hunter	44.5	MAX.	2394	82%
	42.5		2321	78%
	40.5	*	2236	75%
IMR 4895	37.0	MAX.	2420	71%
	35.0		2270	67%
	33.0	*	2120	64%
IMR 4064	39.0	* MAX.	2438	75%
	37.0		2303	71%
	35.0		2168	67%
H414	44.0	MAX.	2460	80%
	42.0		2390	77%
	40.0	*	2320	73%
IMR 4320	40.0	* MAX.	2480	75%
	38.0		2400	72%
	36.0		2320	68%
IMR 4831	47.0	* MAX.	2498	88%
	45.0		2373	84%
	43.0		2248	80%
IMR 4350	45.0	MAX.	2520	84%
	43.0		2430	80%
	41.0	*	2340	76%
RL22	47.5	* MAX.	2539	91%
	45.5		2440	87%
	43.5		2340	83%
Viht N165	50.5	* MAX.	2574	98%
Most Accurate	48.5		2482	95%
Powder Tested	46.5		2391	91%

BC=Ballistic Coefficient SD=Sectional Density
*Most Accurate Load Tested **Compressed Load

Use Maximum Loads with Caution
Refer to page 73 for additional safety information

John B. Snow

280 REMINGTON

I've always had a weakness for underdogs, which is part of the reason I'm such a fan of the 280 Remington. When this cartridge was first announced back in 1957, it got a chilly reception from both gun writers and the shooting public. It was like a weed growing between two mighty oaks—the 30-06 and the 270 Winchester—and as such, it wasn't accorded much respect or given long-term prospects for survival.
After all, what did it have to offer that those undisputed lords among our big-game cartridges didn't?

To make matters worse, the fledgling .280 was burdened with rumors about it being purposely underpowered and then went through a series of name changes, first as the 7mm-06 and then as the 7mm Express, before going back to the 280 Rem.
Such a rough start on life would have doomed a lesser round.

Who saved the .280 from an early grave? Handloaders did, that's who. What they discovered was that the spectrum of 7mm bullets performed remarkably well when paired with the .280's case and could handle everything from varmints to moose with the right combination of projectile and powder.

Conventional wisdom held that 130- and 140-grain bullets were best on deer-sized game, while the stout 160-grainers were the way to go for elk or moose. This is still a valid rule of thumb but bullets like the AccuBond have opened our eyes to other possibilities.

In recent years, I find myself returning time and again to the 140-grain AccuBond in my .280. I've used it to take numerous head of game, large and small, in Alaska, Canada and the Lower 48. The accuracy and terminal ballistics it delivers, thanks to its aerodynamic profile and bonded construction, makes it difficult to justify using any other rifle or cartridge when I hunt.

Of course, it is worth tinkering with multiple bullet weights and styles with a cartridge as versatile as the .280. Do so and I doubt you'll be disappointed.

It might have started as an underdog, but the .280's grown beyond that now. Its star-crossed origins are part of its charm but the real appeal of the .280 lies in its performance. As an all-around hunting cartridge it is second to none.

John B. Snow

John B. Snow is the Shooting Editor of Outdoor Life

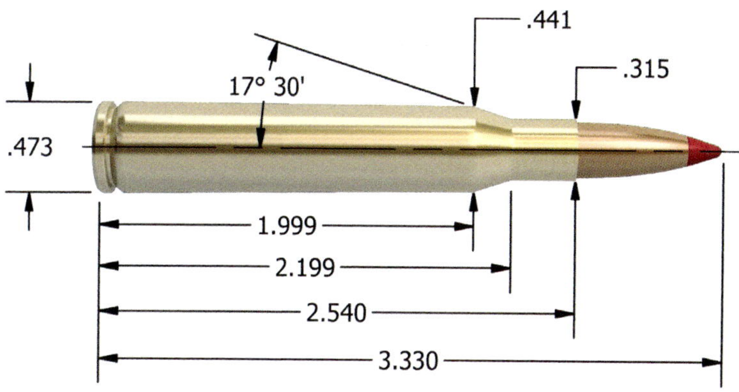

Maximum S.A.A.M.I. Overall Cartridge Length: 3.330"

BULLET CHOICES FOR THE 280 REMINGTON

TECHNICAL INFORMATION

Although the 280 Remington was derived from the .30-06-case, it has a slightly different shoulder location and will not chamber in either .30-06 or 270 Winchester rifles. Because the .280 can generate impressive velocities and still maintain good accuracy, it is one of our primary cartridges for 7mm production testing. For propellants, RL19 or 4350 will usually give very good results.

280 Remington - 120 grain

Ballistic Tip®	120gr. Spitzer	MAXIMUM S.A.A.M.I. O.A.C.L.		3.330"
		TESTED O.A.C.L.	B.C.	S.D.
		3.330"	0.417	0.213

CASE TYPE:	Nosler		PRIMER TYPE	Fed 210
CASE HOLDS:	65.4	Gr. WATER	BARREL Length/Make	26" Lilja
			BARREL Twist	1-9"

POWDER TYPE	POWDER CHG. GRS.		MUZZLE VEL. F.P.S.	LOAD DENSITY (VOLUME)
IMR 4831	57.0 *	MAX.	3080	100%
	55.0		2930	97%
	53.0		2780	93%
IMR 4064	48.0	MAX.	3138	81%
	46.0		2990	77%
	44.0 *		2773	74%
H4895	47.5	MAX.	3198	80%
	45.5		3103	76%
	43.5 *		3008	73%
RL15	50.0 *	MAX.	3240	80%
	48.0		3150	76%
	46.0		3060	73%
IMR 4350	56.0	MAX.	3243	91%
	54.0		3065	87%
	52.0 *		3008	84%
RL19	57.5	MAX.	3250	87%
	55.5		3130	84%
	53.5 *		3010	81%
Hunter	56.5	MAX.	3257	91%
	54.5		3177	88%
	52.5 *		3061	85%
Viht N160	59.0 *	MAX.	3265	96%
	57.0		3154	92%
	55.0		3043	89%
W760 Most Accurate Powder Tested	54.5 *	MAX.	3286	87%
	52.5		3193	84%
	50.5		3100	81%

BC=Ballistic Coefficient SD=Sectional Density
*Most Accurate Load Tested **Compressed Load

Use Maximum Loads with Caution
Refer to page 73 for additional safety information

280 Remington - 140 grain

		MAXIMUM S.A.A.M.I. O.A.C.L.		3.330"
		TESTED O.A.C.L.	B.C.	S.D.
AccuBond®	140gr. Spitzer	3.330"	0.485	0.248
Ballistic Tip®	140gr. Spitzer	3.330"	0.485	0.248
CT® Ballistic Silvertip®	140gr. Spitzer	3.330"	0.485	0.248
E-Tip®	140gr. Spitzer	3.300"	0.489	0.248
Due to internal construction differences, always begin with starting loads when using E-Tip® products.				
Partition®	140gr. Spitzer	3.300"	0.434	0.248

CASE TYPE:	Nosler		PRIMER TYPE	Fed 210
CASE HOLDS:	62.6	Gr. WATER	BARREL Length/Make	26" Lilja
			BARREL Twist	1-9"

POWDER TYPE	POWDER CHG. GRS.		MUZZLE VEL. F.P.S.	LOAD DENSITY (VOLUME)
IMR 4895	44.0 *	MAX.	2840	77%
	42.0		2710	74%
	40.0		2580	70%
Varget	45.0	MAX.	2853	78%
	43.0		2747	74%
	41.0 *		2658	71%
Viht N160	54.0	MAX.	2929	96%
	52.0		2820	92%
	50.0 *		2711	89%
IMR 4350	52.0	MAX.	2947	88%
	50.0 *		2834	85%
	48.0		2751	81%
Big Game	50.5	MAX.	2974	85%
	48.5		2872	82%
	46.5 *		2744	78%
H4831SC	57.0	MAX.	3002	95%
	55.0 *		2923	92%
	53.0		2826	88%
W760	52.0	MAX.	3056	87%
	50.0		2966	84%
	48.0 *		2860	80%
IMR 4831	56.0	MAX.	3070	95%
	54.0		2952	92%
	52.0 *		2850	88%
RL19	57.0	MAX.	3152	99%
Most Accurate	55.0		3027	95%
Powder Tested	53.0 *		2902	92%

BC=Ballistic Coefficient SD=Sectional Density
*Most Accurate Load Tested **Compressed Load

Use Maximum Loads with Caution
Refer to page 73 for additional safety information

280 Remington - 150 grain		MAXIMUM S.A.A.M.I. O.A.C.L.		3.330"
		TESTED O.A.C.L.	B.C.	S.D.
Ballistic Tip®	150gr. Spitzer	3.330"	0.493	0.266
CT® Ballistic Silvertip®	150gr. Spitzer	3.330"	0.493	0.266
E-Tip®	150gr. Spitzer	3.300"	0.498	0.266
Due to internal construction differences, always begin with starting loads when using E-Tip® products.				
Partition®	150gr. Spitzer	3.330"	0.456	0.266

CASE TYPE:	Nosler		PRIMER TYPE	Fed 210
CASE HOLDS:	62.6	Gr. WATER	BARREL Length/Make	26" Lilja
			BARREL Twist	1-9"

POWDER TYPE	POWDER CHG. GRS.	MUZZLE VEL. F.P.S.	LOAD DENSITY (VOLUME)
IMR 4895	43.0 * MAX.	2718	75%
	41.0	2593	72%
	39.0	2468	68%
IMR 4320	44.0 * MAX.	2778	76%
	42.0	2653	72%
	40.0	2528	69%
H414	48.5 MAX.	2850	81%
	46.5 *	2752	77%
	44.5	2670	74%
IMR 4350 Most Accurate Powder Tested	51.5 * MAX.	2870	87%
	49.5	2757	84%
	47.5	2653	80%
Viht N165	57.0 * MAX.	2923	** 101%
	55.0	2815	98%
	53.0	2712	94%
IMR 4831	55.0 * MAX.	2966	93%
	53.0	2850	90%
	51.0	2734	87%
RL22	56.0 * MAX.	2982	97%
	54.0	2877	94%
	52.0	2772	90%
IMR 7828	59.0 MAX.	2990	100%
	57.0	2881	97%
	55.0 *	2777	93%
RL19	55.0 MAX.	2995	95%
	53.0	2862	92%
	51.0 *	2716	89%

BC=Ballistic Coefficient SD=Sectional Density
*Most Accurate Load Tested **Compressed Load

Use Maximum Loads with Caution
Refer to page 73 for additional safety information

280 Remington - 160 grain

		MAXIMUM S.A.A.M.I. O.A.C.L.		3.330"
		TESTED O.A.C.L.	B.C.	S.D.
AccuBond®	160gr. Spitzer	3.330"	0.531	0.283
Partition®	160gr. Spitzer	3.330"	0.475	0.283

CASE TYPE:	Nosler		PRIMER TYPE	Fed 210
CASE HOLDS:	61.5	Gr. WATER	BARREL Length/Make	26" Lilja
			BARREL Twist	1-9"

POWDER TYPE	POWDER CHG. GRS.		MUZZLE VEL. F.P.S.	LOAD DENSITY (VOLUME)
IMR 7828	54.5	* MAX.	2752	94%
	52.5		2659	91%
	50.5		2571	87%
H414	47.5	* MAX.	2762	80%
	45.5		2663	77%
	43.5		2578	74%
Viht N165	55.0	MAX.	2788	99%
	53.0		2697	96%
	51.0	*	2607	92%
H4831SC	54.0	MAX.	2792	91%
	52.0		2724	88%
	50.0	*	2621	85%
IMR 4350	50.5	* MAX.	2810	87%
	48.5		2700	83%
	46.5		2590	80%
RL19 Most Accurate Powder Tested	53.0	MAX.	2824	94%
	51.0	*	2710	90%
	49.0		2632	87%
Hunter	51.5	MAX.	2830	88%
	49.5	*	2728	85%
	47.5		2653	81%
RL22	54.0	MAX.	2873	95%
	52.0	*	2789	92%
	50.0		2690	88%
IMR 4831	54.0	MAX.	2888	93%
	52.0		2783	90%
	50.0	*	2678	86%
MAGPRO	61.5	MAX.	2929 **	103%
	59.5	*	2865	100%
	57.5		2782	96%

BC=Ballistic Coefficient SD=Sectional Density
*Most Accurate Load Tested **Compressed Load

Use Maximum Loads with Caution
Refer to page 73 for additional safety information

280 Remington - 175 grain		MAXIMUM S.A.A.M.I. O.A.C.L.		3.330"
		TESTED O.A.C.L.	B.C.	S.D.
Partition®	175gr. Spitzer	3.330"	0.519	0.310

CASE TYPE:	Nosler		PRIMER TYPE	Fed 210
CASE HOLDS:	61.1	Gr. WATER	BARREL Length/Make	26" Lilja
			BARREL Twist	1-9"

POWDER TYPE	POWDER CHG. GRS.		MUZZLE VEL. F.P.S.	LOAD DENSITY (VOLUME)
IMR 7828	53.0	MAX.	2627	92%
	51.0		2531	89%
	49.0 *		2444	85%
IMR 4350	49.0	MAX.	2630	85%
	47.0		2527	81%
	45.0 *		2416	78%
Magnum	58.5	MAX.	2631	97%
	56.5		2593	94%
	54.5 *		2482	91%
Viht N165 Most Accurate Powder Tested	53.0	MAX.	2661	96%
	51.0		2564	93%
	49.0 *		2468	89%
RL19	51.5 *	MAX.	2707	92%
	49.5		2613	88%
	47.5		2510	85%
RL22	52.5	MAX.	2717	93%
	50.5 *		2606	90%
	48.5		2498	86%
H1000	56.5	MAX.	2724	97%
	54.5		2651	93%
	52.5 *		2580	90%
MAGPRO	58.5	MAX.	2755	99%
	56.5		2690	95%
	54.5 *		2630	92%
IMR 4831	52.5 *	MAX.	2760	91%
	50.5		2653	88%
	48.5		2546	84%

BC=Ballistic Coefficient SD=Sectional Density
*Most Accurate Load Tested **Compressed Load

Use Maximum Loads with Caution
Refer to page 73 for additional safety information

284 WINCHESTER

I missed the 284 Winchester its first time around, partly because I was only 10-years old and partly because nearly everyone else missed it, too.

This was no fault of the cartridge, which, as planned by its designers, does replicate 270 Win. performance (and 280 Rem. for that matter) in a short-action. The fault was chambering this efficient case in the then new Model 88 lever-action and Model 100 semi-auto rifles. Their actions were too short to accommodate long, efficient .284 bullets without shoving them deep into powder space. In those days, the M70 wasn't made in a true short-action length, so there was no benefit in chambering it for the .284. If folks wanted .270 performance in a bolt action, they bought their M70s in 270 Win.

End of story. Well, not quite. Wildcatters to the rescue.

That independent breed of cat that hangs around garage gunrooms took note of the efficient, fat, straight-walled, sharp-shouldered 284 Win. case and proceeded to subject it to the usual ups and downs. Up to .338, down to .243. At 6.5mm, they hit the magic number. Long-range target shooters kept the .284 case alive even as M88 and M100 rifles died on the vine. Then came a West Virginia gunmaker with wild ideas about efficiency. Melvin Forbes engineered his Ultra Light Arms Model 20 bolt-action from the ground up to maximize efficiency. He shrunk the rifle without shrinking performance, and he did it first by wrapping everything around the 284 Winchester.

Here was a 4.75-pound bolt-action rifle designed to milk the promise of the 284 Win. The action and magazine were long enough to permit seating pointy, efficient .284 bullets long in the case allowing for more powder capacity.

Critics claim the 284 Winchester's rebated rim and 35-degree shoulder compromise smooth cycling. They forgot to inform my rifle -- not that it often needs to cycle quickly. In its first hunting season it barked 11 times and collected 10 head of game, including a 6x7 bull elk, out to 411 yards. The extra round was to insure a dead mountain goat didn't kick himself off a ledge. Other than that, a single 140-grain Partition has usually been enough.

Ron Spomer

Ron Spomer is the Travel Field Editor for Sports Afield magazine as well as a contributing editor for Wolfe Publications.

.475
.320
35°
.473
1.775
1.885
2.170
2.800

Maximum S.A.A.M.I. Overall Cartridge Length: 2.800"

BULLET CHOICES FOR THE 284 WINCHESTER

284 Winchester - 120 grain		MAXIMUM S.A.A.M.I. O.A.C.L.		2.800"
		TESTED O.A.C.L.	B.C.	S.D.
Ballistic Tip®	120gr. Spitzer	2.800"	0.417	0.213

CASE TYPE:	Winchester		PRIMER TYPE	Fed 210
CASE HOLDS:	61.3	Gr. WATER	BARREL Length/Make	24" Lilja
			BARREL Twist	1-9"

POWDER TYPE	POWDER CHG. GRS.		MUZZLE VEL. F.P.S.		LOAD DENSITY (VOLUME)
W760	53.0	* MAX.	3036		91%
	51.0		2906		87%
	49.0		2776		84%
IMR 4350	55.0	MAX.	3046		95%
	53.0		2933		91%
	51.0	*	2820		88%
Viht N160 Most Accurate Powder Tested	56.0	MAX.	3061	**	102%
	54.0		2949		98%
	52.0	*	2837		94%
H4831	59.0	MAX.	3062		100%
	57.0		2960		97%
	55.0	*	2860		93%
RL15	48.5	MAX.	3064		82%
	46.5		2929		79%
	44.5	*	2794		76%
RL19	57.5	* MAX.	3101	**	102%
	55.5		2986		98%
	53.5		2870		95%
H4350	57.0	MAX.	3231		98%
	55.0		3142		95%
	53.0	*	3028		91%
Hunter	62.0	MAX.	3295	**	106%
	60.0	*	3255	**	103%
	58.0		3160		100%

BC=Ballistic Coefficient SD=Sectional Density
*Most Accurate Load Tested **Compressed Load

Use Maximum Loads with Caution
Refer to page 73 for additional safety information

284 Winchester - 140 grain		MAXIMUM S.A.A.M.I. O.A.C.L.		2.800"
		TESTED O.A.C.L.	B.C.	S.D.
AccuBond®	140gr. Spitzer	2.800"	0.485	0.248
Ballistic Tip®	140gr. Spitzer	2.800"	0.485	0.248
CT® Ballistic Silvertip®	140gr. Spitzer	2.800"	0.485	0.248
E-Tip®	140gr. Spitzer	2.780"	0.489	0.248
Due to internal construction differences, always begin with starting loads when using E-Tip® products.				
Partition®	140gr. Spitzer	2.800"	0.434	0.248

CASE TYPE:	Winchester		PRIMER TYPE	Fed 210
CASE HOLDS:	61.3	Gr. WATER	BARREL Length/Make	24" Lilja
			BARREL Twist	1-9"

POWDER TYPE	POWDER CHG. GRS.		MUZZLE VEL. F.P.S.	LOAD DENSITY (VOLUME)
RL19	55.0	MAX.	2836	98%
	53.0		2715	94%
	51.0	*	2595	90%
W760	50.5	MAX.	2837	86%
	48.5		2711	83%
	46.5	*	2586	79%
IMR 4064	46.5	MAX.	2849	83%
	44.5		2710	80%
	42.5	*	2572	76%
H4350	53.0	MAX.	2856	91%
	51.0		2716	88%
	49.0	*	2576	85%
RL15	48.0	MAX.	2873	82%
Most Accurate	46.0		2748	78%
Powder Tested	44.0	*	2623	75%
H4831SC	58.5	MAX.	2974	99%
	56.5	*	2882	96%
	54.5		2803	93%
Hunter	56.5	MAX.	3012	97%
	54.5	*	2902	94%
	52.5		2844	90%

BC=Ballistic Coefficient SD=Sectional Density
*Most Accurate Load Tested **Compressed Load

Use Maximum Loads with Caution
Refer to page 73 for additional safety information

284 Winchester - 150 grain

284 Winchester - 150 grain		MAXIMUM S.A.A.M.I. O.A.C.L.		2.800"
		TESTED O.A.C.L.	B.C.	S.D.
Ballistic Tip®	150gr. Spitzer	2.800"	0.493	0.266
CT® Ballistic Silvertip®	150gr. Spitzer	2.800"	0.493	0.266
E-Tip®	150gr. Spitzer	2.780"	0.498	0.266
Due to internal construction differences, always begin with starting loads when using E-Tip® products.				
Partition®	150gr. Spitzer	2.800"	0.456	0.266

CASE TYPE:	Winchester		PRIMER TYPE	Fed 210
CASE HOLDS:	57.6	Gr. WATER	BARREL Length/Make	24" Lilja
			BARREL Twist	1-9"

POWDER TYPE	POWDER CHG. GRS.		MUZZLE VEL. F.P.S.		LOAD DENSITY (VOLUME)
H4831 Most Accurate Powder Tested	56.0	* MAX.	2730	**	101%
	54.0		2623		98%
	52.0		2515		94%
RL15	46.0	* MAX.	2736		83%
	44.0		2627		80%
	42.0		2518		76%
IMR 4350	49.5	MAX.	2741		91%
	47.5	*	2647		87%
	45.5		2552		84%
W760	50.0	MAX.	2759		91%
	48.0		2636		87%
	46.0	*	2514		84%
RL19	54.5	MAX.	2780	**	103%
	52.5		2668		99%
	50.5	*	2556		95%
Big Game	51.0	* MAX.	2870		93%
	49.0		2758		90%
	47.0		2677		86%
Viht N550	51.0	MAX.	2879		98%
	49.0	*	2769		95%
	47.0		2671		91%

BC=Ballistic Coefficient SD=Sectional Density
*Most Accurate Load Tested **Compressed Load

Use Maximum Loads with Caution
Refer to page 73 for additional safety information

284 Winchester - 160 grain		MAXIMUM S.A.A.M.I. O.A.C.L.		2.800"
		TESTED O.A.C.L.	B.C.	S.D.
AccuBond®	160gr. Spitzer	2.800"	0.531	0.283
Partition®	160gr. Spitzer	2.800"	0.475	0.283

CASE TYPE:	Winchester		PRIMER TYPE	Fed 210
CASE HOLDS:	58.0	Gr. WATER	BARREL Length/Make	24" Lilja
			BARREL Twist	1-9"

POWDER TYPE	POWDER CHG. GRS.		MUZZLE VEL. F.P.S.	LOAD DENSITY (VOLUME)
RL15 Most Accurate Powder Tested	44.5 *	MAX.	2592	80%
	42.5		2490	76%
	40.5		2387	73%
IMR 4064	44.5	MAX.	2629	84%
	42.5		2506	81%
	40.5 *		2383	77%
IMR 4350	48.5	MAX.	2636	88%
	46.5		2505	85%
	44.5 *		2373	81%
H4350	51.5	MAX.	2689	94%
	49.5		2566	90%
	47.5 *		2443	87%
RL19	54.0	MAX.	2748	** 101%
	52.0		2642	97%
	50.0 *		2536	94%
H4831SC	55.5	MAX.	2757	100%
	53.5 *		2681	96%
	51.5		2615	92%
Hunter	55.0	MAX.	2816	100%
	53.0		2749	96%
	51.0 *		2646	93%

BC=Ballistic Coefficient SD=Sectional Density
*Most Accurate Load Tested **Compressed Load

Use Maximum Loads with Caution
Refer to page 73 for additional safety information

284 Winchester - 175 grain

		MAXIMUM S.A.A.M.I. O.A.C.L.		2.800"
		TESTED O.A.C.L.	B.C.	S.D.
Partition®	175gr. Spitzer	2.800"	0.519	0.310

CASE TYPE:	Winchester		PRIMER TYPE	Fed 210
CASE HOLDS:	57.0	Gr. WATER	BARREL Length/Make	24" Lilja
			BARREL Twist	1-9"

POWDER TYPE	POWDER CHG. GRS.		MUZZLE VEL. F.P.S.	LOAD DENSITY (VOLUME)
RL15	43.5	MAX.	2511	79%
Most Accurate Powder Tested	41.5		2416	76%
	39.5 *		2321	72%
IMR 4064	42.0 *	MAX.	2512	81%
	40.0		2369	77%
	38.0		2226	73%
Viht N560	48.5 *	MAX.	2551	95%
	46.5		2382	91%
	44.5		2327	87%
IMR 4831	49.0 *	MAX.	2560	91%
	47.0		2478	88%
	45.0		2396	84%
IMR 4350	47.5	MAX.	2588	88%
	45.5		2470	84%
	43.5 *		2352	81%
RL19	52.0	MAX.	2629	99%
	50.0		2540	95%
	48.0 *		2451	92%
Hunter	51.0	MAX.	2641	94%
	49.0 *		2539	90%
	47.0		2409	87%

BC=Ballistic Coefficient SD=Sectional Density
*Most Accurate Load Tested **Compressed Load

Use Maximum Loads with Caution
Refer to page 73 for additional safety information

Bob Nosler

280 REMINGTON ACKLEY IMPROVED

For more than four decades, shooters and hunters have argued about the merits of P.O. Ackley's "improvements" to various cartridges. One thing is certain, in the 1950's shooters were dazzled by the idea that a necked-down '06 case could push a 150-grain bullet at 3,000 fps—all without raising the pressure! But, how does it stack-up today? With a 150-grain bullet, the 280 Ackley delivers 100-150 fps more velocity than the standard 280 Remington, and gives up approximately 100 – 125 fps to its belted magnum cousins, the 7mm Remington Magnum and the 7mm Remington Short-Action Ultra Magnum.

While the 280 Ackley was an important advancement for the time, attention faded when Remington introduced the 7mm Remington Magnum in 1962. This versatile hunting cartridge overshadowed the original 280 Remington to a point that, in 1979, Remington actually re-introduced the 280 Remington as the 7mm Express Remington. The idea fizzled, and in 1981 Remington changed the name back to the 280 Remington. Personally, I believe the better choice would have been for Remington to simply adopt the Ackley version!

In spite of the diverse opinions about the merits of the cartridge, the 280 Ackley remains an impressive hunting round—especially when typical North American big game is the quarry. In fact, the elk in the picture was taken using a Nosler Model 48 rifle chambered in 280 Ackley, and loaded with 140-grain Nosler AccuBond® bullets.

The Sporting Arms and Ammunition Manufacturers Institute (SAAMI) is the official U.S. governing body that approves cartridge and chamber dimensions, as well as establishes critical breach pressure tolerances for the shooting industry. I am proud that Nosler was able to bring the 280 Ackley to the SAAMI Technical Committee for standardization and approval. This action, while long overdue, legitimized the cartridge and elevated it out of the "Wildcat" category. During the process, some had suggested that we should even change the name to the "280 Nosler", but we opted to keep the original name as a tribute to one of the great "experimenters" of the 1950's.

Since the 280 Ackley Improved was adopted in 2007, both Nosler and Kimber chamber models in their popular hunting rifles. In addition, Nosler Trophy Grade ammunition and Nosler Custom brass bearing the "280 Ack Imp" headstamp are available for hunters and reloaders.

Bob Nosler

Bob Nosler is the CEO/President of Nosler, Inc.

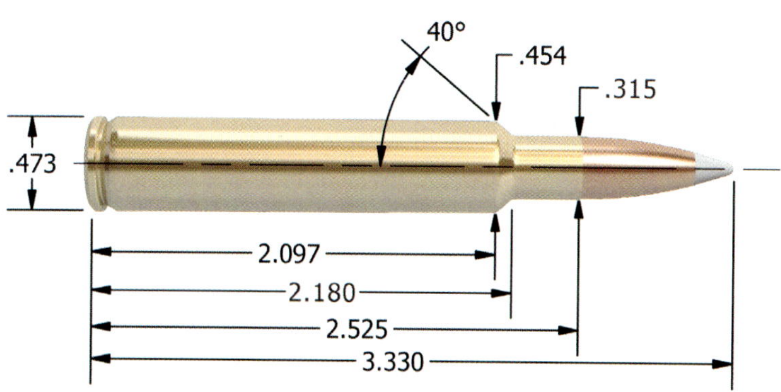

Maximum S.A.A.M.I. Overall Cartridge Length: 3.330"

BULLET CHOICES FOR THE 280 REMINGTON ACKLEY IMPROVED

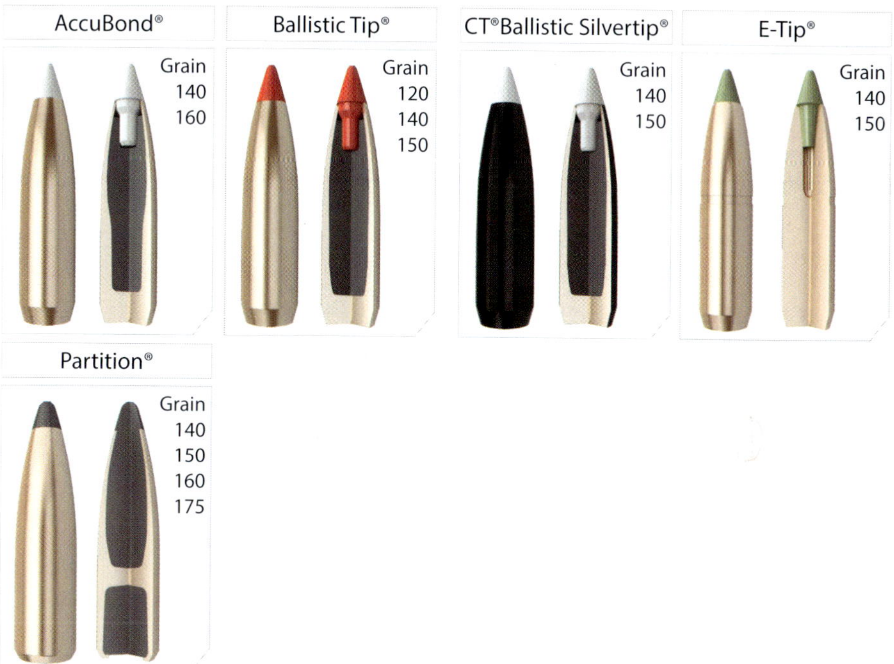

TECHNICAL INFORMATION

While Nosler offers pre-formed 280 Ackley brass, you may want to form your own. If so, we suggest the following procedure:

- Start with standard .280 Remington cases.
- Select a light load of a relatively fast powder (IMR4895 is great) from the standard .280 Remington data.
- Seat the bullet well into the rifling so that it makes good contact with the lands when the cartridge is chambered. This holds the case against the bolt face and eliminates case stretching in the web area, a cause of case head separation.
- After fire-forming, follow load data for the .280 Remington Ackley Improved.

280 Remington A.I. - 120 grain		MAXIMUM S.A.A.M.I. O.A.C.L.		3.330"
		TESTED O.A.C.L.	B.C.	S.D.
Ballistic Tip®	120gr. Spitzer	3.330"	0.417	0.213

CASE TYPE:	Nosler		PRIMER TYPE	Fed 210
CASE HOLDS:	68.0	Gr. WATER	BARREL Length/Make	26" Wiseman
			BARREL Twist	1-9"

POWDER TYPE	POWDER CHG. GRS.		MUZZLE VEL. F.P.S.		LOAD DENSITY (VOLUME)
H4831SC Most Accurate Powder Tested	65.0	MAX.	3331		100%
	63.0		3227		97%
	61.0	*	3124		93%
W760	59.5	MAX.	3347		92%
	57.5		3245		89%
	55.5	*	3143		85%
RL19	63.0	MAX.	3365	**	101%
	61.0		3253		98%
	59.0	*	3140		94%
IMR 4350	60.5	MAX.	3366		94%
	58.5		3250		91%
	56.5	*	3135		88%
Viht N160	63.0	MAX.	3376	**	103%
	61.0		3268		100%
	59.0	*	3160		96%
H414	60.0	MAX.	3396		92%
	58.0		3294		89%
	56.0	*	3192		86%

BC=Ballistic Coefficient SD=Sectional Density
*Most Accurate Load Tested **Compressed Load

Use Maximum Loads with Caution
Refer to page 73 for additional safety information

280 Remington A.I. - 140 grain		MAXIMUM S.A.A.M.I. O.A.C.L.		3.330"
		TESTED O.A.C.L.	B.C.	S.D.
AccuBond®	140gr. Spitzer	3.330"	0.485	0.248
Ballistic Tip®	140gr. Spitzer	3.330"	0.485	0.248
CT® Ballistic Silvertip®	140gr. Spitzer	3.330"	0.485	0.248
E-Tip®	140gr. Spitzer	3.300"	0.489	0.248
Due to internal construction differences, always begin with starting loads when using E-Tip® products.				
Partition®	140gr. Spitzer	3.300"	0.434	0.248

CASE TYPE:	Nosler		PRIMER TYPE	Fed 210
CASE HOLDS:	65.4	Gr. WATER	BARREL Length/Make	26" Wiseman
			BARREL Twist	1-9"

POWDER TYPE	POWDER CHG. GRS.	MUZZLE VEL. F.P.S.	LOAD DENSITY (VOLUME)
Viht N160	59.5 MAX.	3157	** 101%
	57.5	3100	98%
	55.5 *	3037	94%
H4831SC	62.5 MAX.	3179	100%
	60.5	3099	96%
	58.5 *	3031	93%
IMR 4350	58.0 MAX.	3197	94%
	56.0	3097	91%
	54.0 *	2995	88%
IMR 4831 Most Accurate Powder Tested	60.0 * MAX.	3222	98%
	58.0	3115	94%
	56.0	3006	91%
IMR 7828	64.5 * MAX.	3241	** 105%
	62.5	3138	** 102%
	60.5	3023	98%
Viht N560	63.5 MAX.	3253	** 108%
	61.5 *	3146	** 104%
	59.5	3047	** 101%
RL19	63.0 MAX.	3257	** 105%
	61.0 *	3127	** 101%
	59.0	3018	98%
MRP	64.0 MAX.	3259	** 109%
	62.0 *	3176	** 105%
	60.0	3076	** 102%
RL22	64.0 * MAX.	3265	** 106%
	62.0	3153	** 103%
	60.0	3037	100%

BC=Ballistic Coefficient SD=Sectional Density
*Most Accurate Load Tested **Compressed Load

Use Maximum Loads with Caution
Refer to page 73 for additional safety information

280 Remington A.I. - 150 grain		MAXIMUM S.A.A.M.I. O.A.C.L.		3.330"
		TESTED O.A.C.L.	B.C.	S.D.
Ballistic Tip®	150gr. Spitzer	3.330"	0.493	0.266
CT® Ballistic Silvertip®	150gr. Spitzer	3.330"	0.493	0.266
E-Tip®	150gr. Spitzer	3.300"	0.498	0.266
Due to internal construction differences, always begin with starting loads when using E-Tip® produ				
Partition®	150gr. Spitzer	3.330"	0.456	0.266

CASE TYPE:	Nosler		PRIMER TYPE	Fed 210
CASE HOLDS:	64.8	Gr. WATER	BARREL Length/Make	26" Wiseman
			BARREL Twist	1-9"

POWDER TYPE	POWDER CHG. GRS.		MUZZLE VEL. F.P.S.		LOAD DENSITY (VOLUME)
H4831SC Most Accurate Powder Tested	60.5 *	MAX.	2994		97%
	58.5		2889		94%
	56.5		2776		91%
IMR 4350	56.5	MAX.	3014		92%
	54.5 *		2912		89%
	52.5		2794		86%
Viht N165	63.0 *	MAX.	3055	**	108%
	61.0		2956	**	105%
	59.0		2857	**	101%
RL22	61.5	MAX.	3095	**	103%
	59.5		2988		100%
	57.5 *		2880		96%
IMR 7828	63.0	MAX.	3107	**	103%
	61.0		3015		100%
	59.0 *		2923		97%

BC=Ballistic Coefficient SD=Sectional Density
*Most Accurate Load Tested **Compressed Load

Use Maximum Loads with Caution
Refer to page 73 for additional safety information

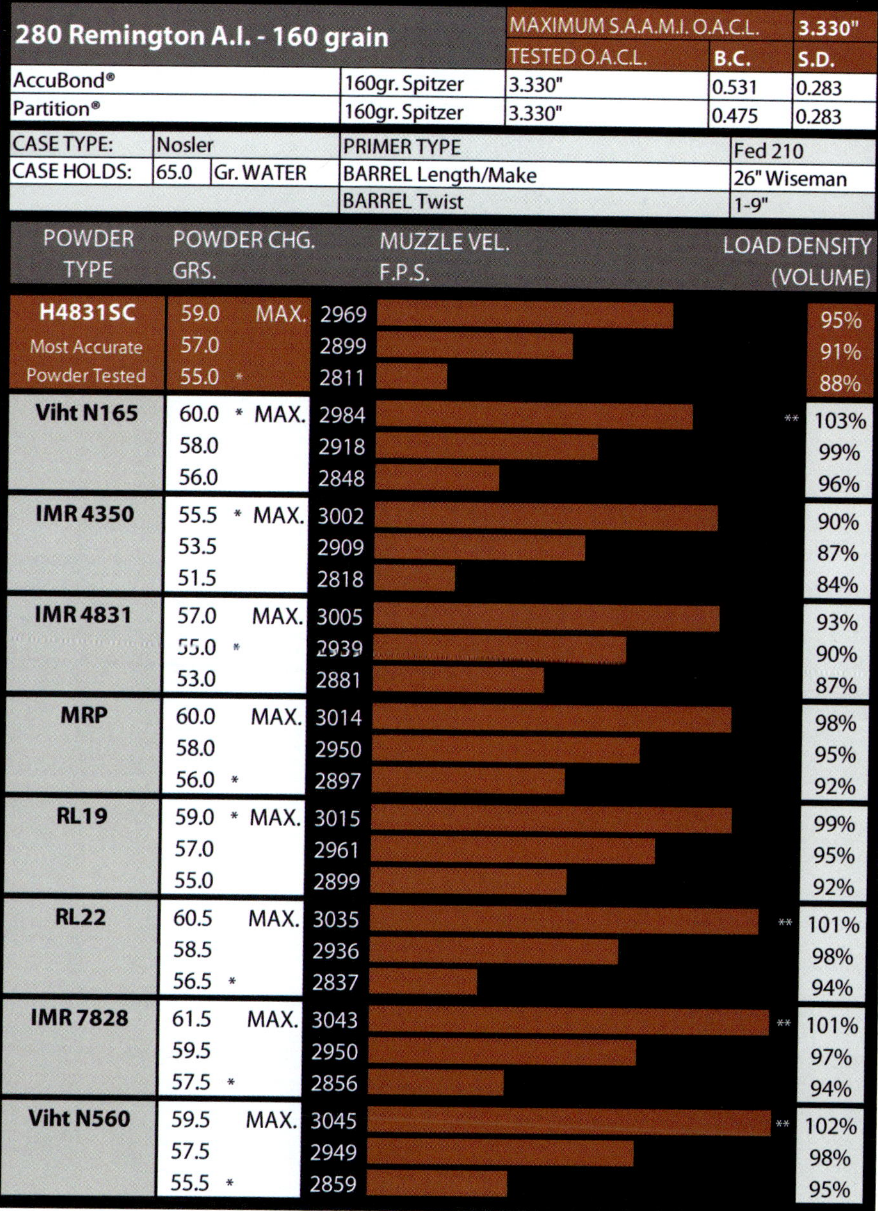

280 Remington A.I. - 160 grain		MAXIMUM S.A.A.M.I. O.A.C.L.		3.330"
		TESTED O.A.C.L.	B.C.	S.D.
AccuBond®	160gr. Spitzer	3.330"	0.531	0.283
Partition®	160gr. Spitzer	3.330"	0.475	0.283

CASE TYPE:	Nosler		PRIMER TYPE	Fed 210
CASE HOLDS:	65.0	Gr. WATER	BARREL Length/Make	26" Wiseman
			BARREL Twist	1-9"

POWDER TYPE	POWDER CHG. GRS.		MUZZLE VEL. F.P.S.		LOAD DENSITY (VOLUME)
H4831SC Most Accurate Powder Tested	59.0	MAX.	2969		95%
	57.0		2899		91%
	55.0 *		2811		88%
Viht N165	60.0 *	MAX.	2984	**	103%
	58.0		2918		99%
	56.0		2848		96%
IMR 4350	55.5 *	MAX.	3002		90%
	53.5		2909		87%
	51.5		2818		84%
IMR 4831	57.0	MAX.	3005		93%
	55.0 *		2939		90%
	53.0		2881		87%
MRP	60.0	MAX.	3014		98%
	58.0		2950		95%
	56.0 *		2897		92%
RL19	59.0 *	MAX.	3015		99%
	57.0		2961		95%
	55.0		2899		92%
RL22	60.5	MAX.	3035	**	101%
	58.5		2936		98%
	56.5 *		2837		94%
IMR 7828	61.5	MAX.	3043	**	101%
	59.5		2950		97%
	57.5 *		2856		94%
Viht N560	59.5	MAX.	3045	**	102%
	57.5		2949		98%
	55.5 *		2859		95%

BC=Ballistic Coefficient SD=Sectional Density
*Most Accurate Load Tested **Compressed Load

Use Maximum Loads with Caution
Refer to page 73 for additional safety information

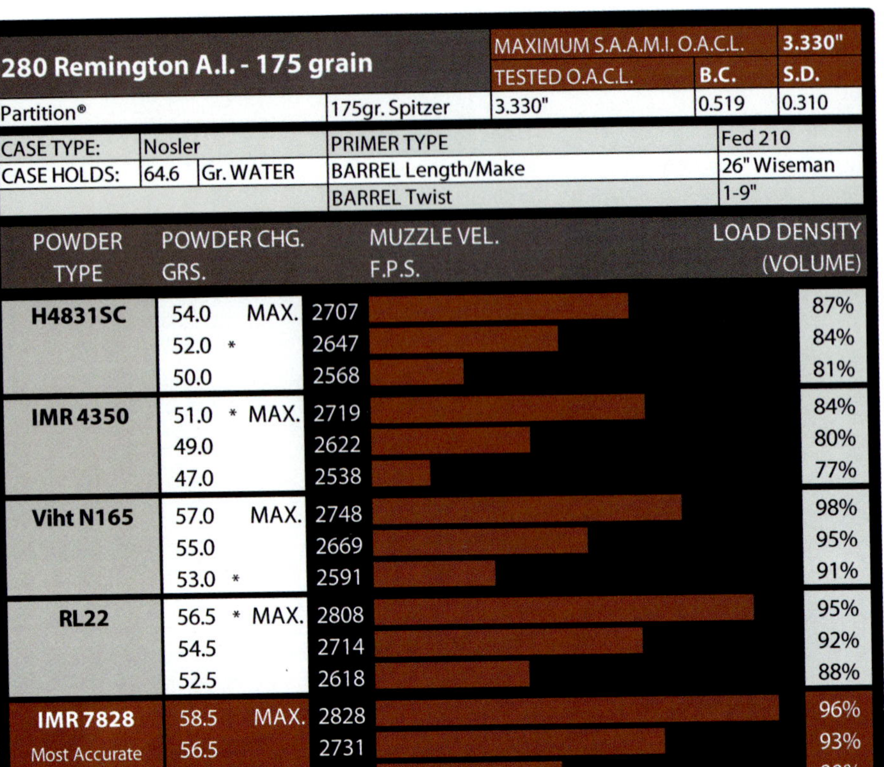

280 Remington A.I. - 175 grain			MAXIMUM S.A.A.M.I. O.A.C.L.		3.330"
			TESTED O.A.C.L.	B.C.	S.D.
Partition®		175gr. Spitzer	3.330"	0.519	0.310
CASE TYPE:	Nosler	PRIMER TYPE		Fed 210	
CASE HOLDS:	64.6 Gr. WATER	BARREL Length/Make		26" Wiseman	
		BARREL Twist		1-9"	

POWDER TYPE	POWDER CHG. GRS.		MUZZLE VEL. F.P.S.	LOAD DENSITY (VOLUME)
H4831SC	54.0	MAX.	2707	87%
	52.0 *		2647	84%
	50.0		2568	81%
IMR 4350	51.0 *	MAX.	2719	84%
	49.0		2622	80%
	47.0		2538	77%
Viht N165	57.0	MAX.	2748	98%
	55.0		2669	95%
	53.0 *		2591	91%
RL22	56.5 *	MAX.	2808	95%
	54.5		2714	92%
	52.5		2618	88%
IMR 7828 Most Accurate Powder Tested	58.5	MAX.	2828	96%
	56.5		2731	93%
	54.5 *		2644	90%

BC=Ballistic Coefficient SD=Sectional Density
*Most Accurate Load Tested **Compressed Load

Use Maximum Loads with Caution
Refer to page 73 for additional safety information

7MM REM SA ULTRA MAG

The 7mm Remington Short Action Ultra Mag (SAUM) is an excellent cartridge, but it, along with its sibling, the 300 SAUM, were among the worst commercial failures in the history of modern rifle cartridges. Within six years of their introduction, not even Remington was chambering for either round. It had nothing to do with ballistic merit, but rather timing and competition in the marketplace. They were Remington's answer to the 300 Winchester Short Magnum (WSM), which was introduced in 2001. The 300 WSM created such an intense interest among both the firearms media and consumers that Remington couldn't even wait until the following year to introduce their own version of a short-action magnum cartridge. "Big Green" not only introduced a .300, but a 7mm version as well.

Like the WSM's, the SAUM's are 404 Jeffery-based beltless cartridges having almost, but not quite, the same internal capacity as the standard-length 7mm Remington Magnum, but also claimed to match the performance of the 7mm WSM, which also has about 5 percent more powder capacity. Now, 5 percent is not a lot, but handloaders know, all other things being equal, that the larger case has the potential for more velocity. So when those who were smitten by the short magnum bug started shopping, they overwhelmingly chose the Winchester version over the Remington. Nevertheless, the 7mm SAUM is an excellent cartridge and the real world difference between it, the WSM and Remington's own 7 Mag is purely academic.

I've used this cartridge to take mule deer and pronghorn using a Remington Model Seven Magnum, but would not hesitate to use it on all but a few critters in the world. On deer-sized game, it really shines with a 140gr Ballistic Tip or CT Ballistic Silvertip. If seating depth is restricted to factory cartridge length, bullets of more than 150 grains need to be seated rather deeply and I don't like to infringe on usable powder space to that extent. So for larger, soft-skinned game, I would choose the 150gr Partition or E-Tip.

Jon R. Sundra

Jon R. Sundra is the Field Editor for Safari Magazine

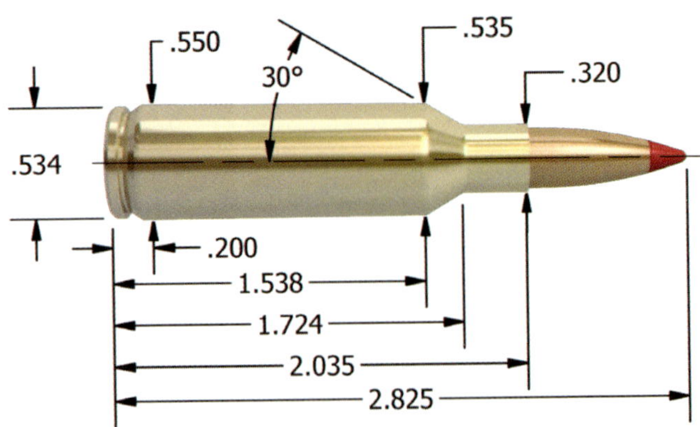

Maximum S.A.A.M.I. Overall Cartridge Length: 2.825"

BULLET CHOICES FOR THE 7MM REM SA ULTRA MAG

TECHNICAL INFORMATION

Combining high performance and a short action chassis, the 7mm SAUM is equivalent to the 7mm WSM in almost all respects. We observed excellent performance with 4350 and 4831 powders.

7mm Rem SA Ultra Mag - 120 grain		MAXIMUM S.A.A.M.I. O.A.C.L.		2.825"
		TESTED O.A.C.L.	B.C.	S.D.
Ballistic Tip®	120gr. Spitzer	2.800"	0.417	0.213

CASE TYPE:	Nosler		PRIMER TYPE	Rem 9 1/2M
CASE HOLDS:	69.2	Gr. WATER	BARREL Length/Make	24" Wiseman
			BARREL Twist	1-9 1/2"

POWDER TYPE	POWDER CHG. GRS.		MUZZLE VEL. F.P.S.	LOAD DENSITY (VOLUME)
RL19	65.5	MAX.	3358	** 103%
	63.5		3235	100%
	61.5 *		3128	97%
IMR 4350 Most Accurate Powder Tested	64.0	MAX.	3407	98%
	62.0 *		3318	95%
	60.0		3198	92%
H414	64.5	MAX.	3424	97%
	62.5		3348	94%
	60.5 *		3246	91%

BC=Ballistic Coefficient SD=Sectional Density
*Most Accurate Load Tested **Compressed Load

Use Maximum Loads with Caution
Refer to page 73 for additional safety information

7mm Rem SA Ultra Mag - 140 grain		MAXIMUM S.A.A.M.I. O.A.C.L.		2.825"
		TESTED O.A.C.L.	B.C.	S.D.
AccuBond®	140gr. Spitzer	2.800"	0.485	0.248
Ballistic Tip®	140gr. Spitzer	2.800"	0.485	0.248
CT® Ballistic Silvertip®	140gr. Spitzer	2.800"	0.485	0.248
E-Tip®	140gr. Spitzer	2.770"	0.489	0.248
Due to internal construction differences, always begin with starting loads when using E-Tip® products.				
Partition®	140gr. Spitzer	2.800"	0.434	0.248

CASE TYPE:	Nosler		PRIMER TYPE	Rem 9 1/2M
CASE HOLDS:	67.1	Gr. WATER	BARREL Length/Make	24" Wiseman
			BARREL Twist	1-9 1/2"

POWDER TYPE	POWDER CHG. GRS.		MUZZLE VEL. F.P.S.	LOAD DENSITY (VOLUME)
H414	60.5	MAX.	3133	94%
	58.5		3030	91%
	56.5 *		2942	88%
IMR 4350 Most Accurate Powder Tested	60.0 *	MAX.	3146	95%
	58.0		3035	91%
	56.0		2943	88%
RL19	62.0	MAX.	3147	100%
	60.0		3048	97%
	58.0 *		2949	94%

BC=Ballistic Coefficient SD=Sectional Density
*Most Accurate Load Tested **Compressed Load

Use Maximum Loads with Caution
Refer to page 73 for additional safety information

7mm Rem SA Ultra Mag - 150 grain		MAXIMUM S.A.A.M.I. O.A.C.L.		2.825"
		TESTED O.A.C.L.	B.C.	S.D.
Ballistic Tip®	150gr. Spitzer	2.800"	0.493	0.266
CT® Ballistic Silvertip®	150gr. Spitzer	2.800"	0.493	0.266
E-Tip®	150gr. Spitzer	2.770"	0.498	0.266
Due to internal construction differences, always begin with starting loads when using E-Tip® products.				
Partition®	150gr. Spitzer	2.800"	0.456	0.266

CASE TYPE:	Nosler		PRIMER TYPE	Rem 9 1/2M
CASE HOLDS:	65.8	Gr. WATER	BARREL Length/Make	24" Wiseman
			BARREL Twist	1-9 1/2"

POWDER TYPE	POWDER CHG. GRS.		MUZZLE VEL. F.P.S.	LOAD DENSITY (VOLUME)
H4831SC Most Accurate Powder Tested	62.5	MAX.	3037	99%
	60.5	*	2946	96%
	58.5		2848	93%
RL19	60.5	* MAX.	3044	100%
	58.5		2925	97%
	56.5		2812	93%
IMR 4350	58.5	* MAX.	3047	94%
	56.5		2883	91%
	54.5		2786	88%

BC=Ballistic Coefficient SD=Sectional Density
*Most Accurate Load Tested **Compressed Load

Use Maximum Loads with Caution
Refer to page 73 for additional safety information

7mm Rem SA Ultra Mag - 160 grain		MAXIMUM S.A.A.M.I. O.A.C.L.		2.825"
		TESTED O.A.C.L.	B.C.	S.D.
AccuBond®	160gr. Spitzer	2.800"	0.531	0.283
Partition®	160gr. Spitzer	2.800"	0.475	0.283

CASE TYPE:	Nosler		PRIMER TYPE	Fed 215
CASE HOLDS:	65.4	Gr. WATER	BARREL Length/Make	24" Wiseman
			BARREL Twist	1-9 1/2"

POWDER TYPE	POWDER CHG. GRS.		MUZZLE VEL. F.P.S.	LOAD DENSITY (VOLUME)
H4831SC Most Accurate Powder Tested	59.5 *	MAX.	2901	95%
	57.5		2835	92%
	55.5		2750	88%
IMR 4350	56.0	MAX.	2917	91%
	54.0		2829	87%
	52.0 *		2732	84%
RL22	59.0	MAX.	2942	98%
	57.0		2896	95%
	55.0 *		2845	91%

BC=Ballistic Coefficient SD=Sectional Density
*Most Accurate Load Tested **Compressed Load

Use Maximum Loads with Caution
Refer to page 73 for additional safety information

7mm Rem SA Ultra Mag - 175 grain		MAXIMUM S.A.A.M.I. O.A.C.L.		2.825"
		TESTED O.A.C.L.	B.C.	S.D.
Partition®	175gr. Spitzer	2.800"	0.519	0.310

CASE TYPE:	Nosler		PRIMER TYPE	Rem 9 1/2
CASE HOLDS:	65.0	Gr. WATER	BARREL Length/Make	24" Wiseman
			BARREL Twist	1-9 1/2"

POWDER TYPE	POWDER CHG. GRS.	MUZZLE VEL. F.P.S.	LOAD DENSITY (VOLUME)
H4831SC Most Accurate Powder Tested	58.5 * MAX.	2788	94%
	56.5	2721	91%
	54.5	2642	87%
IMR 4350	55.5 * MAX.	2810	90%
	53.5	2749	87%
	51.5	2654	84%
RL22	58.0 * MAX.	2811	99%
	56.0	2730	95%
	54.0	2658	92%

BC=Ballistic Coefficient SD=Sectional Density
*Most Accurate Load Tested **Compressed Load

Use Maximum Loads with Caution
Refer to page 73 for additional safety information

Thomas C. Tabor

7MM WINCHESTER SHORT MAGNUM (WSM)

Just when I thought virtually everything had been done in the area of new cartridge development, Winchester decided to take a walk on the short and fat side with their short magnums. The objective was to develop a series of cartridges that would be short enough to be used in short action rifles, but would be capable of achieving magnum level performance. Through a joint venture between Winchester Ammunition and Browning Arms, that goal was achieved in 2001. In my opinion, one of the most impressive of those is the 7mm WSM. Obviously, dramatic changes were necessary to form the 7mm WSM from its 404 Jeffery parent case, but the end result was a cartridge that in some instances actually outperforms its much larger counterpart - the 7mm Rem. Mag. And those results were achieved from a cartridge case that is 0.40-inch shorter and only falls 0.3% shy in its powder capacity.

Of course, I couldn't control the urge to give the 7mm WSM a try and eventually it wound up accompanying me on hunts spanning three continents, beginning with deer hunting in my home state of Montana. Over a period of several years, many mulies were taken, sometimes at distances out beyond 400 yards. I became so enamored with the results that I eventually took it on safari to South Africa with me, where a variety of plains game species fell victim to its charm. And even later on, I joined forces with the Jawoyn Territory game ranger in the Northern Territory of Australia to assist him on a feral animal eradication project. This gave me the opportunity to see how the cartridge would perform on a wide variety of different species, including wild hogs, donkeys and even on one occasion, an Asian water buffalo that was approaching a ton in weight. Of course I wouldn't recommend the 7mm WSM as a viable buffalo cartridge, but I found the little 140-grain bullet sent directly into the brain at an estimated 125 yards did a fine job dispatching the critter.

My experience with the 7mm WSM has involved bullets ranging in weight from 140-grain up to 160-grain and all performed their duties exceptionally well. Nosler offers a wide variety of 7mm bullets to choose from, but my personal favorite would likely be the 140-grain Partition, but the 140-grain AccuBond is also a superb choice.

Thomas C. Tabor is Contributing Editor, Varmint Hunter Magazine and Contributing Editor, Gun Digest

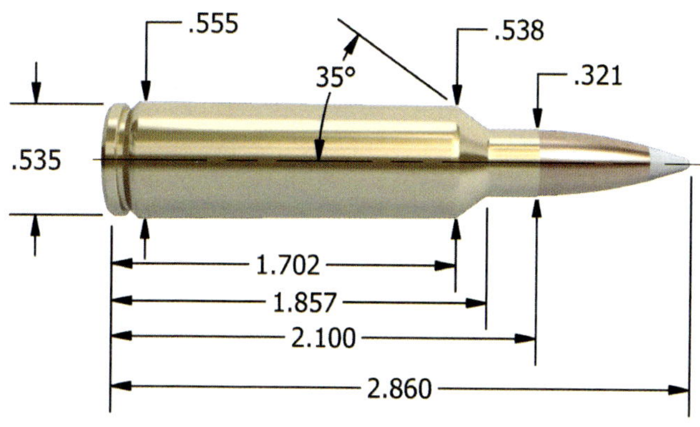

Maximum S.A.A.M.I. Overall Cartridge Length: 2.860"

BULLET CHOICES FOR THE 7MM WINCHESTER SHORT MAGNUM (WSM)

TECHNICAL INFORMATION

All of the WSM cartridges are based on the 404 Jeffery non-belted magnum cartridge, and have been shortened to fit a short action. The 7mm WSM has a case shoulder location that is slightly longer than its 30 caliber parent or 270 caliber child. This is to prevent the 7mm from being chambered in either of the other two WSMs. Our favorite powders are 4350 and RL19.

7mm WSM - 120 grain		MAXIMUM S.A.A.M.I. O.A.C.L.		2.860"
		TESTED O.A.C.L.	B.C.	S.D.
Ballistic Tip®	120gr. Spitzer	2.850"	0.417	0.213

CASE TYPE:	Winchester		PRIMER TYPE	WLRM
CASE HOLDS:	76.8	Gr. WATER	BARREL Length/Make	24" Wiseman
			BARREL Twist	1-9 1/2"

POWDER TYPE	POWDER CHG. GRS.		MUZZLE VEL. F.P.S.		LOAD DENSITY (VOLUME)
Varget	58.5 *	MAX.	3351		82%
	56.5		3252		80%
	54.5		3166		77%
H4831SC	71.0 *	MAX.	3360		96%
	69.0		3292		94%
	67.0		3201		91%
H414	65.5	MAX.	3387		89%
	63.5		3275		86%
	61.5 *		3196		83%
RL15	59.5	MAX.	3393		81%
	57.5		3331		78%
	55.5 *		3255		75%
IMR 4350	67.0 *	MAX.	3428		92%
	65.0		3270		90%
	63.0		3156		87%
Hunter Most Accurate Powder Tested	71.0	MAX.	3437		97%
	69.0		3364		95%
	67.0 *		3267		92%
IMR 4831	69.5 *	MAX.	3444		96%
	67.5		3344		94%
	65.5		3219		91%
Viht N560	70.5	MAX.	3475	**	102%
	68.5 *		3359		99%
	66.5		3249		96%
RL19	72.0 *	MAX.	3516	**	102%
	70.0		3400		99%
	68.0		3298		96%

BC=Ballistic Coefficient SD=Sectional Density
*Most Accurate Load Tested **Compressed Load

Use Maximum Loads with Caution
Refer to page 73 for additional safety information

7mm WSM - 140 grain		MAXIMUM S.A.A.M.I. O.A.C.L.		2.860"
		TESTED O.A.C.L.	B.C.	S.D.
AccuBond®	140gr. Spitzer	2.850"	0.485	0.248
Ballistic Tip®	140gr. Spitzer	2.850"	0.485	0.248
CT® Ballistic Silvertip®	140gr. Spitzer	2.850"	0.485	0.248
E-Tip®	140gr. Spitzer	2.820"	0.489	0.248
Due to internal construction differences, always begin with starting loads when using E-Tip® products.				
Partition®	140gr. Spitzer	2.850"	0.434	0.248

CASE TYPE:	Winchester		PRIMER TYPE	WLRM
CASE HOLDS:	74.5	Gr. WATER	BARREL Length/Make	24" Wiseman
			BARREL Twist	1-9 1/2"

POWDER TYPE	POWDER CHG. GRS.		MUZZLE VEL. F.P.S.		LOAD DENSITY (VOLUME)
RL15	54.5	MAX.	3087		76%
	52.5		3020		73%
	50.5 *		2943		71%
Varget	55.0	MAX.	3107		80%
	53.0 *		3018		77%
	51.0		2909		74%
H414	62.0	MAX.	3128		87%
	60.0 *		3027		84%
	58.0		2948		81%
Hunter	66.0	MAX.	3130		93%
	64.0 *		3069		90%
	62.0		2962		88%
IMR 4350 Most Accurate Powder Tested	64.0 *	MAX.	3197		91%
	62.0		3088		88%
	60.0		3001		85%
H4831SC	67.5	MAX.	3218		94%
	65.5 *		3152		92%
	63.5		3071		89%
IMR 4831	65.5 *	MAX.	3224		94%
	63.5		3141		91%
	61.5		3037		88%
Viht N560	67.5 *	MAX.	3262	**	101%
	65.5		3154		98%
	63.5		3050		95%
RL19	68.5	MAX.	3295		100%
	66.5		3211		97%
	64.5 *		3115		94%

BC=Ballistic Coefficient SD=Sectional Density
*Most Accurate Load Tested **Compressed Load

Use Maximum Loads with Caution
Refer to page 73 for additional safety information

7mm WSM - 150 grain

		MAXIMUM S.A.A.M.I. O.A.C.L.		2.860"
		TESTED O.A.C.L.	B.C.	S.D.
Ballistic Tip®	150gr. Spitzer	2.850"	0.493	0.266
CT® Ballistic Silvertip®	150gr. Spitzer	2.850"	0.493	0.266
E-Tip®	150gr. Spitzer	2.820"	0.498	0.266
Due to internal construction differences, always begin with starting loads when using E-Tip® products.				
Partition®	150gr. Spitzer	2.850"	0.456	0.266

CASE TYPE:	Winchester		PRIMER TYPE	WLRM
CASE HOLDS:	73.5	Gr. WATER	BARREL Length/Make	24" Wiseman
			BARREL Twist	1-9 1/2"

POWDER TYPE	POWDER CHG. GRS.		MUZZLE VEL. F.P.S.	LOAD DENSITY (VOLUME)
Varget	52.0	MAX.	2949	76%
	50.0 *		2830	74%
	48.0		2742	71%
H4831SC	65.0	MAX.	3041	92%
	63.0 *		2960	89%
	61.0		2902	86%
IMR 4350	60.5	MAX.	3045	87%
	58.5		2970	84%
	56.5 *		2883	81%
H414	61.0 *	MAX.	3061	86%
	59.0		2983	84%
	57.0		2895	81%
Hunter	65.5	MAX.	3064	94%
	63.5		2979	91%
	61.5 *		2888	88%
IMR 4831	62.5	MAX.	3068	90%
	60.5		3015	88%
	58.5 *		2922	85%
Viht N560	64.0	MAX.	3103	97%
	62.0 *		3009	94%
	60.0		2920	91%
RL19	64.5	MAX.	3129	95%
	62.5		3045	92%
	60.5 *		2959	89%
MAGPRO	73.0	MAX.	3136	** 102%
Most Accurate	71.0 *		3088	100%
Powder Tested	69.0		3052	97%

BC=Ballistic Coefficient SD=Sectional Density
*Most Accurate Load Tested **Compressed Load

Use Maximum Loads with Caution
Refer to page 73 for additional safety information

7mm WSM - 160 grain		MAXIMUM S.A.A.M.I. O.A.C.L.		2.860"
		TESTED O.A.C.L.	B.C.	S.D.
AccuBond®	160gr. Spitzer	2.850"	0.531	0.283
Partition®	160gr. Spitzer	2.850"	0.475	0.283

CASE TYPE:	Winchester		PRIMER TYPE	WLRM
CASE HOLDS:	74.3	Gr. WATER	BARREL Length/Make	24" Wiseman
			BARREL Twist	1-9 1/2"

POWDER TYPE	POWDER CHG. GRS.		MUZZLE VEL. F.P.S.	LOAD DENSITY (VOLUME)
Varget	49.0	MAX.	2742	71%
	47.0 *		2655	68%
	45.0		2560	65%
H4831SC	61.5	MAX.	2881	86%
	59.5		2795	83%
	57.5 *		2710	81%
IMR 4350	57.5 *	MAX.	2894	82%
	55.5		2812	79%
	53.5		2730	76%
H414	57.0	MAX.	2917	80%
	55.0 *		2827	77%
	53.0		2711	74%
IMR 4831 Most Accurate Powder Tested	58.5 *	MAX.	2919	84%
	56.5		2830	81%
	54.5		2756	78%
IMR 7828	67.0	MAX.	2923	96%
	65.0		2802	93%
	63.0 *		2743	90%
Viht N560	59.5	MAX.	2927	89%
	57.5 *		2839	86%
	55.5		2762	83%
Hunter	63.5	MAX.	2954	90%
	61.5		2864	87%
	59.5 *		2807	84%
MAGPRO	70.0	MAX.	3065	97%
	68.0		2971	94%
	66.0 *		2892	92%

BC=Ballistic Coefficient SD=Sectional Density
*Most Accurate Load Tested **Compressed Load

Use Maximum Loads with Caution
Refer to page 73 for additional safety information

7mm WSM - 175 grain		MAXIMUM S.A.A.M.I. O.A.C.L.		2.860"
		TESTED O.A.C.L.	B.C.	S.D.
Partition®	175gr. Spitzer	2.850"	0.519	0.310

CASE TYPE:	Winchester		PRIMER TYPE	WLRM
CASE HOLDS:	72.1	Gr. WATER	BARREL Length/Make	24" Wiseman
			BARREL Twist	1-9 1/2"

POWDER TYPE	POWDER CHG. GRS.		MUZZLE VEL. F.P.S.	LOAD DENSITY (VOLUME)
Varget	47.0	MAX.	2616	70%
	45.0		2536	67%
	43.0 *		2424	64%
H414	54.0	MAX.	2765	78%
	52.0 *		2682	75%
	50.0		2581	72%
Hunter	62.0	MAX.	2779	91%
	60.0		2719	88%
	58.0 *		2667	85%
IMR 4350	56.5	MAX.	2804	83%
	54.5		2718	80%
	52.5 *		2631	77%
IMR 4831	56.5	MAX.	2816	83%
	54.5		2735	80%
	52.5 *		2655	77%
H4831SC	60.0	MAX.	2819	87%
	58.0 *		2741	84%
	56.0		2664	81%
IMR 7828	66.5	MAX.	2841	98%
	64.5 *		2749	95%
	62.5		2640	92%
Viht N560	58.5	MAX.	2855	90%
Most Accurate	56.5		2788	87%
Powder Tested	54.5 *		2680	84%
MAGPRO	71.5 *	MAX.	2951 **	102%
	69.5		2913	99%
	67.5		2851	97%

BC=Ballistic Coefficient SD=Sectional Density
*Most Accurate Load Tested **Compressed Load

Use Maximum Loads with Caution
Refer to page 73 for additional safety information

Diana B. Rupp

7MM REMINGTON MAGNUM

Most of us don't get many chances in our lifetimes to hunt sheep, so when the moment of truth comes, it's crucial that your rifle of choice be up to whatever shot might be presented. When I headed for the mountains of northern British Columbia to try for a Stone sheep, I knew I had the right rifle along: A Remington Model 700 in 7mm Remington Magnum. After five challenging days of hunting, I got a steep downhill shot at 280 yards, and the 7mm Rem Mag never let me down, tipping the ram over in its tracks.

It's been half a century since Remington standardized this cartridge, introducing it in conjunction with their Model 700 bolt-action rifle in 1962, and in that time the 7mm Rem Mag has become one of the most popular cartridges among North America's big-game hunters—and for good reason. It's one of the most versatile and effective hunting cartridges on the market. Factory loadings are plentiful and readily available, and its dies consistently rank in the top ten purchased by handloaders. It's a solid choice for any North American big game short of brown bears. In addition to the Stone sheep, I've used the 7mm Rem Mag to bring down moose, mule deer, pronghorn, and Sitka black-tailed deer.

When it was introduced, it wasn't the first 7mm magnum caliber—Roy Weatherby had developed his 7mm Weatherby Magnum nearly twenty years before—but it was the one that immediately caught on. It quickly displaced the then-popular 264 Winchester Magnum, one of the belted magnum calibers upon which it was based.

Slightly shorter than full-length magnums like the H&Hs and Weatherbys, the 7mm Rem Mag can be chambered in standard-length actions, but ballistically, the 7mm Rem Mag leaves standard cartridges like the 30-06 and 270 Winchester in the dust. Lighter bullets leave the muzzle around 3,200 fps, which means the 7mm Rem Mag has all the reach you need for longer shots at game on open plains or high mountains. For game species requiring a little more punch, it will shoot 175-grain bullets in the 2,850 fps range and 160-grainers at nearly 3,000. The 7mm Rem Mag excels at shooting flat and hitting hard-no wonder it's been one of the most popular hunting cartridges for half a century.

Diana B. Rupp

Diana B. Rupp is Editor in Chief of Sports Afield magazine.

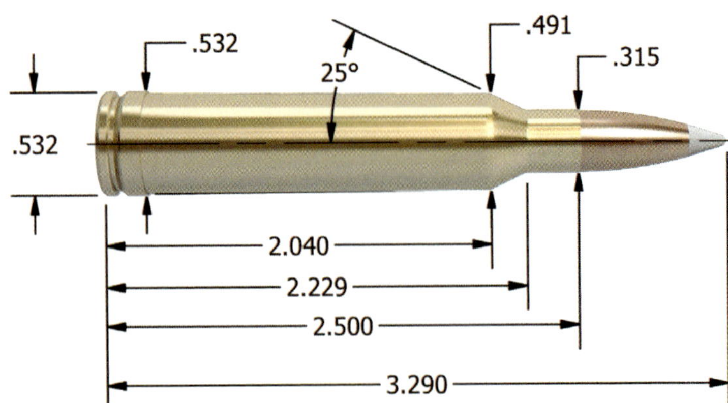

Maximum S.A.A.M.I. Overall Cartridge Length: 3.290"

BULLET CHOICES FOR THE 7MM REMINGTON MAGNUM

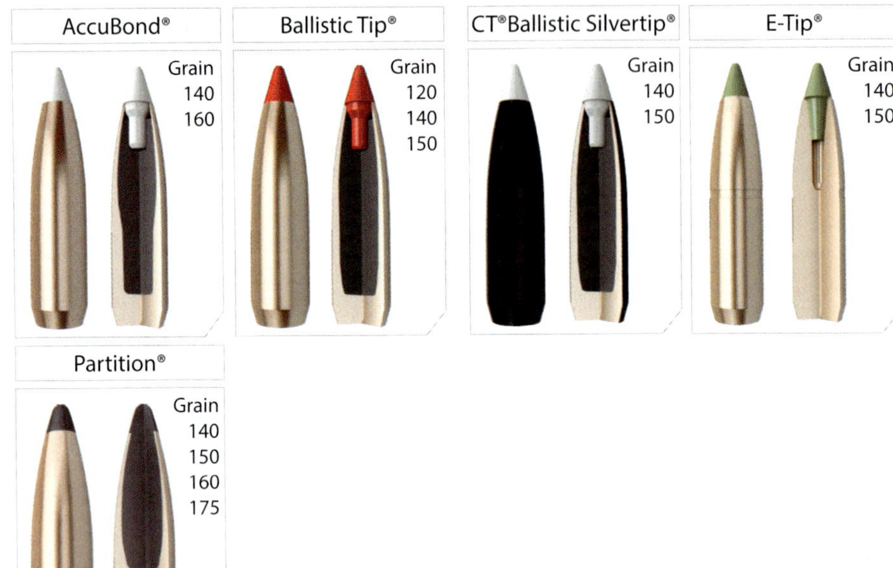

Partition®

Grain
140
150
160
175

7mm Remington Mag - 120 grain		MAXIMUM S.A.A.M.I. O.A.C.L.		3.290"
		TESTED O.A.C.L.	B.C.	S.D.
Ballistic Tip®	120gr. Spitzer	3.290"	0.417	0.213

CASE TYPE:	Nosler		PRIMER TYPE	Fed 215
CASE HOLDS:	81.4	Gr. WATER	BARREL Length/Make	24" Wiseman
			BARREL Twist	1-9"

POWDER TYPE	POWDER CHG. GRS.		MUZZLE VEL. F.P.S.	LOAD DENSITY (VOLUME)
H414	62.0	MAX.	3260	79%
	60.0		3194	77%
	58.0 *		3127	74%
IMR 4350 Most Accurate Powder Tested	66.0 *	MAX.	3348	86%
	64.0		3243	83%
	62.0		3138	81%
IMR 4895	56.0	MAX.	3358	76%
	54.0		3263	73%
	52.0 *		3168	70%
IMR 4831	68.0	MAX.	3370	89%
	66.0		3250	86%
	64.0 *		3130	84%
Viht N160	66.0 *	MAX.	3405	90%
	64.0		3330	87%
	62.0		3255	85%
Viht N560	69.0	MAX.	3497	94%
	67.0		3437	91%
	65.0 *		3369	89%
MAGPRO	81.0	MAX.	3550	** 103%
	79.0		3447	100%
	77.0 *		3339	98%
RL22	71.0 *	MAX.	3562	95%
	69.0		3467	92%
	67.0		3372	89%
RL19	70.0 *	MAX.	3570	93%
	68.0		3470	91%
	66.0		3370	88%

BC=Ballistic Coefficient SD=Sectional Density
*Most Accurate Load Tested **Compressed Load

Use Maximum Loads with Caution
Refer to page 73 for additional safety information

7mm Remington Mag - 140 grain		MAXIMUM S.A.A.M.I. O.A.C.L.		3.290"
		TESTED O.A.C.L.	B.C.	S.D.
AccuBond®	140gr. Spitzer	3.290"	0.485	0.248
Ballistic Tip®	140gr. Spitzer	3.290"	0.485	0.248
CT® Ballistic Silvertip®	140gr. Spitzer	3.290"	0.485	0.248
E-Tip®	140gr. Spitzer	3.260"	0.489	0.248
Due to internal construction differences, always begin with starting loads when using E-Tip® products.				
Partition®	140gr. Spitzer	3.290"	0.434	0.248

CASE TYPE:	Nosler		PRIMER TYPE	Fed 215
CASE HOLDS:	77.8	Gr. WATER	BARREL Length/Make	24" Wiseman
			BARREL Twist	1-9"

POWDER TYPE	POWDER CHG. GRS.		MUZZLE VEL. F.P.S.	LOAD DENSITY (VOLUME)
Viht N165	66.0	MAX.	3061	94%
Most Accurate	64.0		2982	91%
Powder Tested	62.0 *		2903	89%
IMR 4064	56.0	MAX.	3088	79%
	54.0		2983	76%
	52.0 *		2878	73%
IMR 4320	57.0	MAX.	3110	79%
	55.0		3010	76%
	53.0 *		2910	73%
IMR 4895	54.0	MAX.	3128	76%
	52.0		3023	73%
	50.0 *		2918	71%
Viht N160	65.5	MAX.	3137	94%
	63.5 *		3040	91%
	61.5		2952	88%
IMR 4831	66.0	MAX.	3220	90%
	64.0		3120	88%
	62.0 *		3020	85%
MAGPRO	75.5	MAX.	3263	100%
	73.5 *		3171	97%
	71.5		3096	95%
IMR 4350	65.0 *	MAX.	3282	88%
	63.0		3207	86%
	61.0		3132	83%
RL19	65.5 *	MAX.	3318	92%
	63.5		3243	89%
	61.5		3168	86%
RL22	67.5	MAX.	3340	94%
	65.5		3250	92%
	63.5 *		3160	89%

BC=Ballistic Coefficient SD=Sectional Density
*Most Accurate Load Tested **Compressed Load

Use Maximum Loads with Caution
Refer to page 73 for additional safety information

7mm Remington Mag - 150 grain

		MAXIMUM S.A.A.M.I. O.A.C.L.		3.290"
		TESTED O.A.C.L.	B.C.	S.D.
Ballistic Tip®	150gr. Spitzer	3.290"	0.493	0.266
CT® Ballistic Silvertip®	150gr. Spitzer	3.290"	0.493	0.266
E-Tip®	150gr. Spitzer	3.260"	0.498	0.266
Due to internal construction differences, always begin with starting loads when using E-Tip® products.				
Partition®	150gr. Spitzer	3.290"	0.456	0.266

CASE TYPE:	Nosler		PRIMER TYPE	Fed 215
CASE HOLDS:	78.0	Gr. WATER	BARREL Length/Make	24" Wiseman
			BARREL Twist	1-9"

POWDER TYPE	POWDER CHG. GRS.		MUZZLE VEL. F.P.S.	LOAD DENSITY (VOLUME)
IMR 4064	53.0	MAX.	2968	75%
	51.0		2873	72%
	49.0 *		2778	69%
Viht N165	66.0	MAX.	3017	94%
	64.0		2941	91%
	62.0 *		2865	88%
IMR 4320	55.0	MAX.	3048	76%
	53.0		2953	73%
	51.0 *		2858	70%
Viht N160 Most Accurate Powder Tested	64.0 *	MAX.	3048	91%
	62.0		2951	88%
	60.0		2869	85%
H4831SC	65.0	MAX.	3075	87%
	63.0		3023	84%
	61.0 *		2956	81%
MAGPRO	72.5	MAX.	3128	96%
	70.5		3067	93%
	68.5 *		3000	91%
IMR 7828	68.0 *	MAX.	3162	93%
	66.0		3072	90%
	64.0		2947	87%
IMR 4831	65.0	MAX.	3240	89%
	63.0		3130	86%
	61.0 *		3020	83%
IMR 4350	63.0 *	MAX.	3248	85%
	61.0		3138	83%
	59.0		3028	80%

BC=Ballistic Coefficient SD=Sectional Density
*Most Accurate Load Tested **Compressed Load

Use Maximum Loads with Caution
Refer to page 73 for additional safety information

7mm Remington Mag - 160 grain		MAXIMUM S.A.A.M.I. O.A.C.L.		3.290"
		TESTED O.A.C.L.	B.C.	S.D.
AccuBond®	160gr. Spitzer	3.290"	0.531	0.283
Partition®	160gr. Spitzer	3.290"	0.475	0.283

CASE TYPE:	Nosler		PRIMER TYPE	Fed 215
CASE HOLDS:	80.4	Gr. WATER	BARREL Length/Make	24" Wiseman
			BARREL Twist	1-9"

POWDER TYPE	POWDER CHG. GRS.		MUZZLE VEL. F.P.S.	LOAD DENSITY (VOLUME)
Viht N165 Most Accurate Powder Tested	61.0 *	MAX.	2904	84%
	59.0		2846	82%
	57.0		2789	79%
IMR 4064	53.0 *	MAX.	2908	72%
	51.0		2823	70%
	49.0		2738	67%
IMR 4350	60.0 *	MAX.	2998	79%
	58.0		2943	76%
	56.0		2888	74%
IMR 4831	63.0 *	MAX.	3008	83%
	61.0		2933	81%
	59.0		2858	78%
IMR 7828	64.0	MAX.	3015	85%
	62.0		2934	82%
	60.0 *		2859	79%
RL19	61.5 *	MAX.	3046	83%
	59.5		2973	80%
	57.5		2901	78%
RL22	63.0	MAX.	3058	85%
	61.0		2963	82%
	59.0 *		2868	80%
RL25	70.0	MAX.	3066	95%
	68.0 *		2989	92%
	66.0		2931	89%
MAGPRO	72.0	MAX.	3077	92%
	70.0		2989	90%
	68.0 *		2930	87%

BC=Ballistic Coefficient SD=Sectional Density
*Most Accurate Load Tested **Compressed Load

Use Maximum Loads with Caution
Refer to page 73 for additional safety information

7mm Remington Mag - 175 grain

			MAXIMUM S.A.A.M.I. O.A.C.L.		3.290"
			TESTED O.A.C.L.	B.C.	S.D.
Partition®		175gr. Spitzer	3.290"	0.519	0.310

CASE TYPE:	Nosler		PRIMER TYPE	Fed 215
CASE HOLDS:	77.0	Gr. WATER	BARREL Length/Make	24" Wiseman
			BARREL Twist	1-9"

POWDER TYPE	POWDER CHG. GRS.		MUZZLE VEL. F.P.S.	LOAD DENSITY (VOLUME)
IMR 4320	51.0	MAX.	2720	71%
	49.0		2620	68%
	47.0 *		2520	66%
RL19	58.5	MAX.	2820	83%
	56.5		2720	80%
	54.5 *		2619	77%
H1000	65.5	MAX.	2850	89%
	63.5		2800	86%
	61.5 *		2750	84%
IMR 4350	58.0 *	MAX.	2860	80%
	56.0		2760	77%
	54.0		2660	74%
IMR 4831	60.0	MAX.	2870	83%
	58.0		2780	80%
	56.0 *		2690	77%
IMR 7828	63.5	MAX.	2871	88%
	61.5		2806	85%
	59.5 *		2729	82%
Viht N560	60.0 *	MAX.	2878	87%
	58.0		2816	84%
	56.0		2725	81%
MAGPRO	70.0	MAX.	2899	94%
	68.0		2820	91%
	66.0 *		2734	88%
RL22	62.5 *	MAX.	2970	88%
Most Accurate	60.5		2890	85%
Powder Tested	58.5		2810	83%

BC=Ballistic Coefficient SD=Sectional Density
*Most Accurate Load Tested **Compressed Load

Use Maximum Loads with Caution
Refer to page 73 for additional safety information

Jeff H. Johnston

7MM WEATHERBY MAGNUM

The mule deer buck lay bedded in the center of a sage-brush basin. From my vantage 420 yards away, high above on the rimrock, he was all but hidden save for his wide-swept antlers and his nearside shoulder that was exposed thanks to a small pocket in the sage. Normally I'd have crawled closer, but the cliff's edge in front of me did not allow. No matter, the 7mm Weatherby Magnum was rested firmly and zeroed at 200 yards. Oddly enough, there was no wind on that fall day in Wyoming, and frankly, it didn't matter; my 7mm Wby. Mag's 160-grain Nosler Partition would still deliver 2,050 ft.-lbs. of energy a mere 4 inches lower than my point of aim at that distance. I drew a breath, let out half, placed the crosshairs a shade above the point of the buck's shoulder and squeezed the trigger slowly until I felt the wrap of recoil. A millisecond later I heard the stark whack of the Partition as it hit home. The mule deer buck never stood up.

Created in 1944 by the godfather of magnum cartridges, Roy Weatherby, his 7mm Wby. Mag. was designed on a belted 300 H&H case shortened by .301-inch. Case capacity was increased by removing most of the taper via Weatherby's recognizable curved-yet-sharp neck angle. It's longer than the more popular 7mm Remington Magnum by .049-inch, and typically it is loaded to higher pressures, so it averages 175 fps faster when fired with like-weight bullets. Despite its performance advantages, the 7mm Rem. was made available in less expensive rifles and therefore outpaced the Weatherby in sales. For years, the only commercially made rifles chambered for the 7mm Wby. were Weatherbys.

Seven-millimeter projectiles (.284-caliber) are inherently deep penetrating when loaded with heavy-for-caliber bullets due to that caliber's sectional density. For example, a 7mm-160gr bullet has an SD of .283 while a .308-165gr bullet has an SD of .248.

If the legendary hunter Karamoja Bell felled thousands of elephants with his 275 Rigby (similar to the modern 7x57 round), I believe the superior 7mm Wby. Mag., loaded with a premium bullet like Nosler's AccuBond, Partition or E-Tip, could take any animal on this planet with proper shot placement at ranges less than 500 yards.

A life-long hunter and shooter, Jeff H. Johnston began working for the National Rifle Association after graduating from the University of Oklahoma in 2000 with a degree in journalism. In 2009 he was promoted to Managing Editor of American Hunter magazine, where he currently writes articles and edits its "Hardware" gun- and ammo-review section. Johnston is a passionate conservationist and hunter of all game; he has taken several hundred big game animals, from antelope to Cape buffalo.

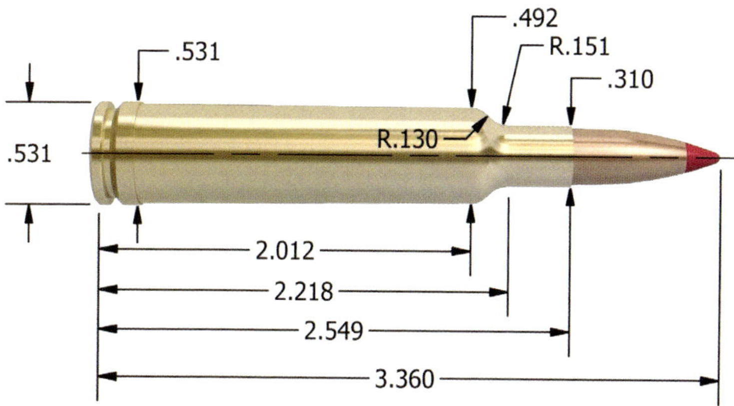

Maximum S.A.A.M.I. Overall Cartridge Length: 3.360"

BULLET CHOICES FOR THE 7MM WEATHERBY MAGNUM

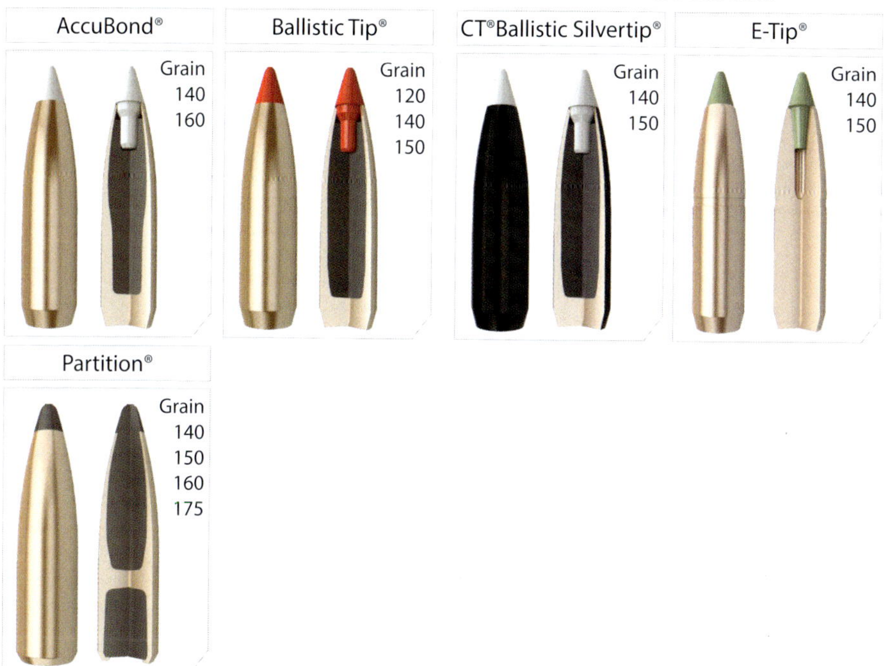

TECHNICAL INFORMATION

As with most Weatherby Magnums, there are a few points to keep in mind when loading for this cartridge.

- The chambers have "freebore", meaning they have a longer throat. For this reason, it is generally not possible to seat a bullet close to, or in contact with the lands (rifling).
- For best accuracy, bullets should be seated as long as the magazine will allow.
- Weatherby rifles of German manufacture (pre-1972) may have barrels with twist rates of 1 turn in 12". These barrels will not reliably stabilize bullets that are heavier than 160 grains.

7mm Weatherby Mag - 120 grain

		MAXIMUM S.A.A.M.I. O.A.C.L.		3.360"
		TESTED O.A.C.L.	B.C.	S.D.
Ballistic Tip®	120gr. Spitzer	3.360"	0.417	0.213

CASE TYPE:	Weatherby		PRIMER TYPE	Rem 9 1/2M
CASE HOLDS:	85.0	Gr. WATER	BARREL Length/Make	24" Lilja
			BARREL Twist	1-9"

POWDER TYPE	POWDER CHG. GRS.		MUZZLE VEL. F.P.S.		LOAD DENSITY (VOLUME)
Viht N165	77.0	MAX.	3422	**	101%
	75.0		3318		98%
	73.0 *		3214		95%
RL22 Most Accurate Powder Tested	75.0 *	MAX.	3467		96%
	73.0		3338		93%
	71.0		3209		91%
IMR 4350	70.0	MAX.	3482		87%
	68.0		3347		85%
	66.0 *		3212		82%
Magnum	87.0	MAX.	3510	**	104%
	85.0 *		3436	**	102%
	83.0		3346		99%
IMR 4831	72.0	MAX.	3540		90%
	70.0		3400		88%
	68.0 *		3260		85%
MAGPRO	86.5	MAX.	3547	**	105%
	84.5		3524	**	102%
	82.5 *		3481		100%
H4831SC	78.0 *	MAX.	3598		96%
	76.0		3511		93%
	74.0		3429		91%
Viht N560	76.5 *	MAX.	3610		100%
	74.5		3505		97%
	72.5		3401		95%
RL25	84.5	MAX.	3653	**	108%
	82.5		3586	**	105%
	80.5 *		3457	**	103%

BC=Ballistic Coefficient SD=Sectional Density
*Most Accurate Load Tested **Compressed Load

Use Maximum Loads with Caution
Refer to page 73 for additional safety information

7mm Weatherby Mag - 140 grain		MAXIMUM S.A.A.M.I. O.A.C.L.		3.360"
		TESTED O.A.C.L.	B.C.	S.D.
AccuBond®	140gr. Spitzer	3.360"	0.485	0.248
Ballistic Tip®	140gr. Spitzer	3.360"	0.485	0.248
CT® Ballistic Silvertip®	140gr. Spitzer	3.360"	0.485	0.248
E-Tip®	140gr. Spitzer	3.330"	0.489	0.248
Due to internal construction differences, always begin with starting loads when using E-Tip® products.				
Partition®	140gr. Spitzer	3.360"	0.434	0.248

CASE TYPE:	Weatherby		PRIMER TYPE	Rem 9 1/2M
CASE HOLDS:	82.0	Gr. WATER	BARREL Length/Make	24" Lilja
			BARREL Twist	1-9"

POWDER TYPE	POWDER CHG. GRS.		MUZZLE VEL. F.P.S.	LOAD DENSITY (VOLUME)
RL25	76.5	MAX.	3277	** 101%
	74.5 *		3181	99%
	72.5		3096	96%
Magnum	82.0	MAX.	3281	** 102%
	80.0		3206	99%
	78.0 *		3132	97%
MAGPRO	79.5	MAX.	3287	100%
	77.5		3259	97%
	75.5 *		3163	95%
IMR 4831	70.0 *	MAX.	3292	91%
	68.0		3237	88%
	66.0		3182	86%
IMR 4350	68.0	MAX.	3320	88%
	66.0		3250	85%
	64.0 *		3180	83%
RL22	73.5 *	MAX.	3336	97%
	71.5		3223	95%
	69.5		3111	92%
H4831SC	74.0 *	MAX.	3353	94%
	72.0		3278	91%
	70.0		3194	89%
H1000	79.0	MAX.	3361	** 101%
Most Accurate	77.0 *		3277	98%
Powder Tested	75.0		3181	96%

BC=Ballistic Coefficient SD=Sectional Density
*Most Accurate Load Tested **Compressed Load

Use Maximum Loads with Caution
Refer to page 73 for additional safety information

7mm Weatherby Mag - 150 grain		MAXIMUM S.A.A.M.I. O.A.C.L.		3.360"
		TESTED O.A.C.L.	B.C.	S.D.
Ballistic Silvertip®	150gr. Spitzer	3.360"	0.493	0.266
Ballistic Tip®	150gr. Spitzer	3.360"	0.493	0.266
E-Tip®	150gr. Spitzer	3.330"	0.498	0.266
Due to internal construction differences, always begin with starting loads when using E-Tip® products.				
Partition®	150gr. Spitzer	3.360"	0.456	0.266

CASE TYPE:	Weatherby		PRIMER TYPE	Rem 9 1/2M
CASE HOLDS:	81.4	Gr. WATER	BARREL Length/Make	24" Lilja
			BARREL Twist	1-9"

POWDER TYPE	POWDER CHG. GRS.		MUZZLE VEL. F.P.S.		LOAD DENSITY (VOLUME)
H4350	67.0	* MAX.	3040		87%
	65.0		2920		85%
	63.0		2800		82%
Viht N165	71.0	* MAX.	3049		97%
	69.0		2959		94%
	67.0		2868		91%
H4831	72.0	* MAX.	3078		92%
	70.0		2973		90%
	68.0		2868		87%
IMR 4350	66.0	* MAX.	3120		86%
	64.0		3060		83%
	62.0		3000		81%
IMR 4831	68.0	* MAX.	3192		89%
	66.0		3097		86%
	64.0		3002		84%
Magnum	81.0	MAX.	3226	**	101%
	79.0		3167		99%
	77.0	*	3053		96%
RL22	72.0	MAX.	3253		96%
	70.0		3147		93%
	68.0	*	3041		91%
H1000	78.0	MAX.	3266		100%
	76.0	*	3184		98%
	74.0		3097		95%
Retumbo	80.0	MAX.	3273	**	103%
Most Accurate	78.0	*	3214	**	101%
Powder Tested	76.0		3169		98%
RL25	77.5	MAX.	3302	**	103%
	75.5		3211	**	101%
	73.5	*	3146		98%

BC=Ballistic Coefficient SD=Sectional Density
*Most Accurate Load Tested **Compressed Load

Use Maximum Loads with Caution
Refer to page 73 for additional safety information

7mm Weatherby Mag - 160 grain		MAXIMUM S.A.A.M.I. O.A.C.L.		3.360"
		TESTED O.A.C.L.	B.C.	S.D.
AccuBond®	160gr. Spitzer	3.360"	0.531	0.283
Partition®	160gr. Spitzer	3.360"	0.475	0.283

CASE TYPE:	Weatherby		PRIMER TYPE	Rem 9 1/2M
CASE HOLDS:	81.0	Gr. WATER	BARREL Length/Make	24" Lilja
			BARREL Twist	1-9"

POWDER TYPE	POWDER CHG. GRS.		MUZZLE VEL. F.P.S.		LOAD DENSITY (VOLUME)
H1000	75.5	MAX.	2922		98%
	73.5		2817		95%
	71.5 *		2712		92%
IMR 4350	64.0	MAX.	3028		84%
	62.0		2943		81%
	60.0 *		2858		78%
IMR 7828	73.0	MAX.	3040		96%
	71.0		2920		93%
	69.0 *		2800		91%
RL22	68.5	MAX.	3050		92%
	66.5		2964		89%
	64.5 *		2878		87%
IMR 4831	66.0 *	MAX.	3080		87%
	64.0		2980		84%
	62.0		2880		81%
Viht N560	69.5	MAX.	3089		95%
	67.5		3022		93%
	65.5 *		2935		90%
H4831SC	70.5	MAX.	3110		91%
	68.5		3035		88%
	66.5 *		2968		86%
Magnum	79.5	MAX.	3111		100%
	77.5 *		3036		97%
	75.5		2970		95%
Retumbo	79.5	MAX.	3189	**	103%
	77.5 *		3131	**	101%
	75.5		3053		98%
RL25 Most Accurate Powder Tested	76.0	MAX.	3197	**	102%
	74.0 *		3117		99%
	72.0		3044		97%

BC=Ballistic Coefficient SD=Sectional Density
*Most Accurate Load Tested **Compressed Load

Use Maximum Loads with Caution
Refer to page 73 for additional safety information

7mm Weatherby Mag - 175 grain		MAXIMUM S.A.A.M.I. O.A.C.L.		3.360"
		TESTED O.A.C.L.	B.C.	S.D.
Partition®	175gr. Spitzer	3.360"	0.519	0.310

CASE TYPE:	Weatherby		PRIMER TYPE	Rem 9 1/2M
CASE HOLDS:	80.0	Gr. WATER	BARREL Length/Make	24" Lilja
			BARREL Twist	1-9"

POWDER TYPE	POWDER CHG. GRS.		MUZZLE VEL. F.P.S.	LOAD DENSITY (VOLUME)
Viht N165	65.0	MAX.	2717	90%
	63.0		2640	88%
	61.0 *		2563	85%
IMR 4350	63.0	MAX.	2882	83%
	61.0		2817	81%
	59.0 *		2752	78%
RL22	67.0 *	MAX.	2921	91%
	65.0		2836	88%
	63.0		2751	86%
IMR 4831	64.0	MAX.	2922	85%
	62.0		2787	82%
	60.0 *		2652	80%
IMR 7828 Most Accurate Powder Tested	72.0	MAX.	2988	96%
	70.0		2883	93%
	68.0 *		2778	90%
RL25	73.0	MAX.	3000	99%
	71.0		2935	96%
	69.0 *		2849	94%
Magnum	77.5 *	MAX.	3003	98%
	75.5		2932	96%
	73.5		2860	93%
H1000	74.0 *	MAX.	3008	97%
	72.0		2936	94%
	70.0		2862	92%
Retumbo	78.0 *	MAX.	3061 **	103%
	76.0		3017	100%
	74.0		2940	97%

BC=Ballistic Coefficient SD=Sectional Density
*Most Accurate Load Tested **Compressed Load

Use Maximum Loads with Caution
Refer to page 73 for additional safety information

7MM SHOOTING TIMES WESTERNER (STW)

There haven't been too many cartridge introductions in the hunting world that have caused as much excitement as when Layne Simpson introduced the 7mm STW (Shooting Times Westerner) to the world. Layne, a good friend and brilliant writer, began work on this cartridge in the early 1980's. His dedication led to some great findings and Layne decided to team up with legendary rifle maker Kenny Jarrett in 1987. The original name for the 7mm STW was actually the 7mm Remington Maximum. Layne changed it to the Shooting Times Westerner to honor his readers and the name stuck when he officially introduced it to the public in 1988. While some people would look at it and say, "It's just an 8mm Remington necked down," I view it as the beginning of true long range ballistics in the modern sense. It was truly revolutionary.

It was after a hunt with Layne that we as a company decided to start loading custom handloaded ammunition in 7mm STW- which subsequently lead us to offer it in the Trophy Grade line-up for retail.

Being out West, we have a passion for long range capabilities. I think this is due to the fact we have mule deer, pronghorn and elk that like to sit just across a canyon. Coupled with a 160-grain AccuBond bullet, this combination of speed, ballistic coefficient, and trajectory make it an absolute dream for anything up to and including moose. I know that Layne himself tested this cartridge on whitetail, black bear, antelope, elk, moose, and even nilgai.

The performance is upwards of 200 fps faster with a 160-grain bullet than the standard 7mm Remington Magnum- and to be brutally honest, the recoil isn't a factor when a 6x6 Rocky Mountain Elk is at the other end of the barrel.

I personally wouldn't hesitate to pack this cartridge along to Africa as a one-size-fits-all plains game gun. The ability to shoot effectively with 140-grain up to 175-grain bullets is a major plus, and paired with "Hodgdon Extreme Powder" (consistent performance regardless of extreme hot or cold), such as Retumbo or H1000, this gem of a cartridge will hammer anything you put in front of it.

John R. Nosler is the VP/GM of Nosler, Inc.

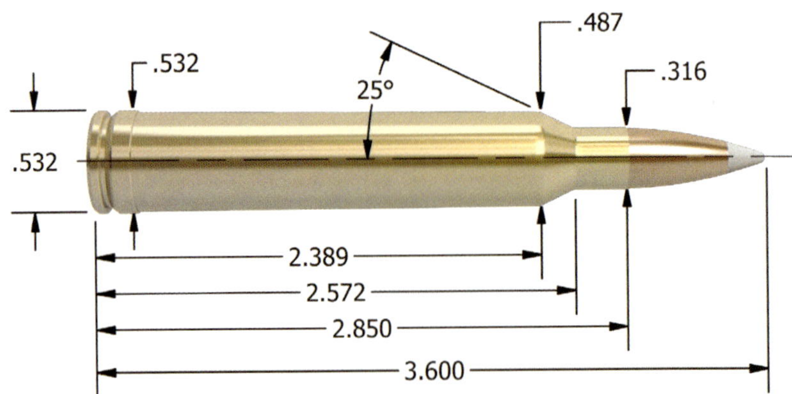

Maximum S.A.A.M.I. Overall Cartridge Length: 3.600″

BULLET CHOICES FOR THE 7MM SHOOTING TIMES WESTERNER

TECHNICAL INFORMATION

This cartridge works well with a wide variety of powders. H1000 seems to be particularly well suited to the heavier bullet weights. While we used Federal 215 primers, any magnum primer will work well with the 7mm STW.

7mm S.T.W. - 120 grain

		MAXIMUM S.A.A.M.I. O.A.C.L.		3.600"
		TESTED O.A.C.L.	B.C.	S.D.
Ballistic Tip®	120gr. Spitzer	3.600"	0.417	0.213

CASE TYPE:	Nosler		PRIMER TYPE	Fed 215
CASE HOLDS:	92.6	Gr. WATER	BARREL Length/Make	26" Lilja
			BARREL Twist	1-9"

POWDER TYPE	POWDER CHG. GRS.		MUZZLE VEL. F.P.S.	LOAD DENSITY (VOLUME)
H4831	78.5	MAX.	3428	88%
	76.5		3324	86%
	74.5	*	3220	84%
Viht N165	80.0	MAX.	3484	96%
	78.0		3389	94%
	76.0	*	3295	91%
RL19	77.5	MAX.	3509	91%
	75.5		3403	89%
	73.5	*	3296	86%
Viht N160	76.5	MAX.	3535	92%
	74.5		3439	89%
	72.5	*	3342	87%
IMR 4350 Most Accurate Powder Tested	76.0	MAX.	3554	87%
	74.0		3453	85%
	72.0	*	3353	82%
IMR 4831	77.5	MAX.	3578	89%
	75.5		3463	87%
	73.5	*	3348	84%
IMR 7828	81.5	MAX.	3580	94%
	79.5		3477	91%
	77.5	*	3374	89%
RL22	80.5	MAX.	3618	94%
	78.5		3511	92%
	76.5	*	3405	90%
Magnum	92.0	MAX.	3624 **	101%
	90.0		3558	99%
	88.0	*	3460	96%

BC=Ballistic Coefficient SD=Sectional Density
*Most Accurate Load Tested **Compressed Load

Use Maximum Loads with Caution
Refer to page 73 for additional safety information

7mm S.T.W. - 140 grain		MAXIMUM S.A.A.M.I. O.A.C.L.		3.600"
		TESTED O.A.C.L.	B.C.	S.D.
AccuBond®	140gr. Spitzer	3.600"	0.485	0.248
Ballistic Tip®	140gr. Spitzer	3.600"	0.485	0.248
CT® Ballistic Silvertip®	140gr. Spitzer	3.600"	0.485	0.248
E-Tip®	140gr. Spitzer	3.570"	0.489	0.248
Due to internal construction differences, always begin with starting loads when using E-Tip® products.				
Partition®	140gr. Spitzer	3.600"	0.434	0.248

CASE TYPE:	Nosler		PRIMER TYPE	Fed 215
CASE HOLDS:	91.3	Gr. WATER	BARREL Length/Make	26" Lilja
			BARREL Twist	1-9"

POWDER TYPE	POWDER CHG. GRS.		MUZZLE VEL. F.P.S.	LOAD DENSITY (VOLUME)
Viht N160	70.0	MAX.	3213	85%
	68.0		3145	83%
	66.0 *		3077	80%
IMR 4350	71.5 *	MAX.	3299	83%
	69.5		3215	81%
	67.5		3130	78%
IMR 4831	73.0 *	MAX.	3319	85%
	71.0		3224	83%
	69.0		3130	80%
H4831 Most Accurate Powder Tested	74.5	MAX.	3319	85%
	72.5		3258	83%
	70.5 *		3198	80%
RL22	75.0	MAX.	3332	89%
	73.0		3241	87%
	71.0 *		3150	85%
IMR 7828	77.5	MAX.	3355	90%
	75.5		3255	88%
	73.5 *		3156	86%
RL19	75.0	MAX.	3362	89%
	73.0		3277	87%
	71.0 *		3192	85%
RL25	81.0	MAX.	3432	96%
	79.0 *		3360	94%
	77.0		3258	92%

BC=Ballistic Coefficient SD=Sectional Density
*Most Accurate Load Tested **Compressed Load

Use Maximum Loads with Caution
Refer to page 73 for additional safety information

7mm S.T.W. - 150 grain		MAXIMUM S.A.A.M.I. O.A.C.L.		3.600"
		TESTED O.A.C.L.	B.C.	S.D.
Ballistic Tip®	150gr. Spitzer	3.600"	0.493	0.266
CT® Ballistic Silvertip®	150gr. Spitzer	3.600"	0.493	0.266
E-Tip®	150gr. Spitzer	3.570"	0.498	0.266
Due to internal construction differences, always begin with starting loads when using E-Tip® products.				
Partition®	150gr. Spitzer	3.600"	0.456	0.266

CASE TYPE:	Nosler		PRIMER TYPE	Fed 215
CASE HOLDS:	90.8	Gr. WATER	BARREL Length/Make	26" Lilja
			BARREL Twist	1-9"

POWDER TYPE	POWDER CHG. GRS.		MUZZLE VEL. F.P.S.	LOAD DENSITY (VOLUME)
Viht N160	69.0	* MAX.	3087	84%
	67.0		3014	82%
	65.0		2940	80%
H4831	72.0	* MAX.	3139	83%
	70.0		3073	80%
	68.0		3007	78%
IMR 4350	69.5	* MAX.	3142	81%
	67.5		3054	79%
	65.5		2967	76%
H1000	79.5	MAX.	3176	92%
	77.5		3090	89%
	75.5	*	3004	87%
RL19 Most Accurate Powder Tested	73.0	* MAX.	3215	87%
	71.0		3125	85%
	69.0		3035	83%
RL22	74.0	* MAX.	3219	89%
	72.0		3134	86%
	70.0		3049	84%
IMR 7828	76.5	* MAX.	3250	90%
	74.5		3158	87%
	72.5		3066	85%
RL25	80.0	* MAX.	3331	96%
	78.0		3254	93%
	76.0		3174	91%

BC=Ballistic Coefficient SD=Sectional Density
*Most Accurate Load Tested **Compressed Load

Use Maximum Loads with Caution
Refer to page 73 for additional safety information

7mm S.T.W. - 160 grain		MAXIMUM S.A.A.M.I. O.A.C.L.		3.600"
		TESTED O.A.C.L.	B.C.	S.D.
AccuBond®	160gr. Spitzer	3.600"	0.531	0.283
Partition®	160gr. Spitzer	3.600"	0.475	0.283

CASE TYPE:	Nosler		PRIMER TYPE	Fed 215
CASE HOLDS:	90.6	Gr. WATER	BARREL Length/Make	26" Lilja
			BARREL Twist	1-9"

POWDER TYPE	POWDER CHG. GRS.		MUZZLE VEL. F.P.S.	LOAD DENSITY (VOLUME)
Viht N165	71.0	MAX.	3006	87%
	69.0		2947	85%
	67.0 *		2888	82%
H4831	69.5	MAX.	3016	80%
	67.5		2947	78%
	65.5 *		2898	75%
IMR 4350	68.0	MAX.	3074	79%
	66.0		3001	77%
	64.0 *		2929	75%
H1000 Most Accurate Powder Tested	78.0	MAX.	3080	90%
	76.0		2997	88%
	74.0 *		2915	86%
RL22	73.0	MAX.	3121	88%
	71.0		3035	85%
	69.0 *		2950	83%
IMR 7828	75.0	MAX.	3123	88%
	73.0		3026	86%
	71.0 *		2929	83%
RL19	70.5	MAX.	3136	85%
	68.5		3096	82%
	66.5 *		3010	80%
Magnum	84.0 *	MAX.	3229	94%
	82.0		3171	92%
	80.0		3118	90%

BC=Ballistic Coefficient SD=Sectional Density
*Most Accurate Load Tested **Compressed Load

Use Maximum Loads with Caution
Refer to page 73 for additional safety information

7mm S.T.W. - 175 grain		MAXIMUM S.A.A.M.I. O.A.C.L.		3.600"
		TESTED O.A.C.L.	B.C.	S.D.
Partition®	175gr. Spitzer	3.600"	0.519	0.310

CASE TYPE:	Nosler		PRIMER TYPE	Fed 215
CASE HOLDS:	89.4	Gr. WATER	BARREL Length/Make	26" Lilja
			BARREL Twist	1-9"

POWDER TYPE	POWDER CHG. GRS.		MUZZLE VEL. F.P.S.	LOAD DENSITY (VOLUME)
Viht N165	68.0 *	MAX.	2829	85%
	66.0		2763	82%
	64.0		2697	80%
IMR 4350	65.0	MAX.	2877	77%
	63.0		2804	75%
	61.0 *		2731	72%
H4831	65.5 *	MAX.	2903	76%
	63.5		2844	74%
	61.5		2784	72%
RL19	67.5	MAX.	2937	82%
	65.5 *		2894	80%
	63.5		2840	77%
H1000 Most Accurate Powder Tested	72.5	MAX.	2956	85%
	70.5		2906	83%
	68.5 *		2856	80%
RL22	69.5	MAX.	2968	85%
	67.5		2888	82%
	65.5 *		2808	80%
IMR 7828	73.5 *	MAX.	3032	87%
	71.5		2949	85%
	69.5		2867	83%
Retumbo	75.0 *	MAX.	3042	88%
	73.0		2976	86%
	71.0		2940	84%

BC=Ballistic Coefficient SD=Sectional Density
*Most Accurate Load Tested **Compressed Load

Use Maximum Loads with Caution
Refer to page 73 for additional safety information

Art Wheaton

7MM REMINGTON ULTRA MAGNUM

My PH said, "Can you make the shot?" "Maybe" I responded, "If I can "bag" my 7mm Remington Ultra Mag Model 700 African Plains rifle."

Chub Eastman, Craig Boddington, Bob Maschmedt, John Dwyer and I were in Tanzania with a camera crew on a mission to score on Cape buffalo, along with selected other species. I had a license for a "Croc" and each day when we returned for lunch at our Safari campsite, my PH would glass the opposite river bank for a nice crocodile. For three days straight, he had disappeared right after lunch for about a half hour, working his way down river, trying to locate a trophy size, seemingly prehistoric monster.

"Crocs" are wise, as the slightest movement from an intruder will trigger their hasty departure into the river depths. It was explained carefully to me that the kill zone was the size of a baseball just at the end of the "smile".
Finally my PH exclaimed, "It's a nice one!" It didn't take long for the camera crew, and the host of observers thinking "I don't want to miss this", filed behind the PH, parade style, through the woods paralleling the bank to a place just opposite the "Croc."

With hand signals, our PH cautioned the "crowd" and beckoned me to sit on the sand behind him. We inched our way ever so slowly to the bank and down the modest incline. Reaching the bottom sandy area, the PH drew with a stick in the sand, the kill zone, on an outline of the "Croc's" head.

A small "grip" or athletic bag with handles was brought along in response to the comment on bagging the rifle. Gently, I slid the bag up and in position, laid the rifle between the handles and adjusted myself into a reasonably comfortable position against the shoulder.
What a chance to choke or be a hero, especially with the audience and camera's rolling. The coach whispered, "Can you see the spot?" Everything felt pretty good as I locked the cross hairs, tightened up on the trigger and the gun went off. The rifle recoiled and the "Croc's" hind legs quivered like a dog scratching in the sand.

Art Wheaton

Art Wheaton is the retired Vice President/General Manager Worldwide Sales of Remington Arms Co. Inc.

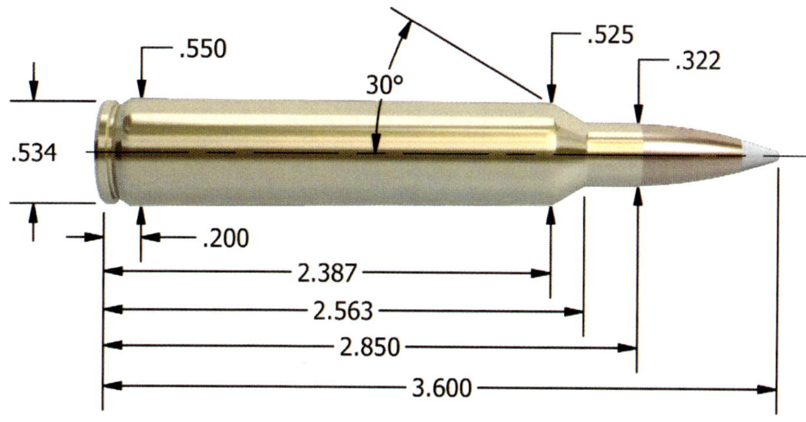

Maximum S.A.A.M.I. Overall Cartridge Length: 3.600"

BULLET CHOICES FOR THE 7MM REMINGTON ULTRA MAGNUM

TECHNICAL INFORMATION

Based on the non-belted, rebated-rim 404 Jeffrey case, the 7mm RUM is without a doubt the highest performing .284 caliber cartridge commercially loaded today. As with any of the "super magnum" sized cartridges that are necked down to smaller calibers, slow powders will work best. Periodic case neck annealing may be necessary to ensure a good seal in the chamber. Signs of a poor seal include excessive blackening beyond the neck and/or case collapsing. IMR 7828 is an excellent powder choice for this cartridge, yielding the most consistent results across all bullet weights.

7mm Rem Ultra Mag - 120 grain		MAXIMUM S.A.A.M.I. O.A.C.L.		3.600"
		TESTED O.A.C.L.	B.C.	S.D.
Ballistic Tip®	120gr. Spitzer	3.570"	0.417	0.213

CASE TYPE:	Nosler		PRIMER TYPE	WLRM
CASE HOLDS:	108.1	Gr. WATER	BARREL Length/Make	26" Wiseman
			BARREL Twist	1-9"

POWDER TYPE	POWDER CHG. GRS.		MUZZLE VEL. F.P.S.	LOAD DENSITY (VOLUME)
H4831	91.0	MAX.	3476	88%
	89.0		3423	86%
	87.0 *		3374	84%
RL22	90.5 *	MAX.	3483	91%
	88.5		3427	89%
	86.5		3373	87%
H1000	97.0	MAX.	3594	94%
	95.0 *		3483	92%
	93.0		3402	90%
IMR 7828	92.5	MAX.	3629	91%
	90.5		3526	89%
	88.5 *		3429	87%
RL25	95.5	MAX.	3639	96%
Most Accurate	93.5		3528	94%
Powder Tested	91.5 *		3445	92%
Retumbo	101.5	MAX.	3704	99%
	99.5		3634	97%
	97.5 *		3560	95%

BC=Ballistic Coefficient SD=Sectional Density
*Most Accurate Load Tested **Compressed Load

Use Maximum Loads with Caution
Refer to page 73 for additional safety information

7mm Rem Ultra Mag - 140 grain		MAXIMUM S.A.A.M.I. O.A.C.L.		3.600"
		TESTED O.A.C.L.	B.C.	S.D.
AccuBond®	140gr. Spitzer	3.570"	0.485	0.248
Ballistic Tip®	140gr. Spitzer	3.570"	0.485	0.248
CT® Ballistic Silvertip®	140gr. Spitzer	3.570"	0.485	0.248
E-Tip®	140gr. Spitzer	3.540"	0.489	0.248
Due to internal construction differences, always begin with starting loads when using E-Tip® products.				
Partition®	140gr. Spitzer	3.570"	0.434	0.248

CASE TYPE:	Nosler		PRIMER TYPE	WLRM
CASE HOLDS:	105.8	Gr. WATER	BARREL Length/Make	26" Wiseman
			BARREL Twist	1-9"

POWDER TYPE	POWDER CHG. GRS.		MUZZLE VEL. F.P.S.		LOAD DENSITY (VOLUME)
H4831	89.0 *	MAX.	3335		88%
	87.0		3289		86%
	85.0		3243		84%
Magnum	94.0	MAX.	3360		90%
	92.0		3326		88%
	90.0 *		3289		86%
Supreme 780	89.0	MAX.	3367		88%
	87.0		3334		86%
	85.0 *		3302		84%
US 869	108.0 *	MAX.	3375	**	111%
	106.0		3305	**	109%
	104.0		3234	**	107%
H1000	95.0 *	MAX.	3421		94%
	93.0		3347		92%
	91.0		3271		90%
RL22	90.0 *	MAX.	3424		92%
	88.0		3368		90%
	86.0		3314		88%
Viht N560	90.0	MAX.	3427		95%
	88.0		3372		92%
	86.0 *		3316		90%
IMR 7828	89.0	MAX.	3459		89%
	87.0		3366		87%
	85.0 *		3278		85%
RL25	91.0	MAX.	3477		99%
Most Accurate	89.0		3407		97%
Powder Tested	87.0 *		3338		95%
Retumbo	96.0	MAX.	3485		96%
	94.0		3372		94%
	92.0 *		3285		92%

BC=Ballistic Coefficient SD=Sectional Density
*Most Accurate Load Tested **Compressed Load

Use Maximum Loads with Caution
Refer to page 73 for additional safety information

7mm Rem Ultra Mag - 150 grain

		MAXIMUM S.A.A.M.I. O.A.C.L.		3.600"
		TESTED O.A.C.L.	B.C.	S.D.
Ballistic Tip®	150gr. Spitzer	3.570"	0.493	0.266
CT® Ballistic Silvertip®	150gr. Spitzer	3.570"	0.493	0.266
E-Tip®	150gr. Spitzer	3.540"	0.498	0.266
Due to internal construction differences, always begin with starting loads when using E-Tip® products.				
Partition®	150gr. Spitzer	3.570"	0.456	0.266

CASE TYPE:	Nosler		PRIMER TYPE	WLRM
CASE HOLDS:	105.4	Gr. WATER	BARREL Length/Make	26" Wiseman
			BARREL Twist	1-9"

POWDER TYPE	POWDER CHG. GRS.		MUZZLE VEL. F.P.S.	LOAD DENSITY (VOLUME)
Viht N165	86.0	MAX.	3180	96%
	84.0		3116	94%
	82.0 *		3047	92%
H4831	85.0	MAX.	3200	84%
	83.0		3157	82%
	81.0 *		3115	80%
Hybrid 100V	76.0	MAX.	3224	74%
	74.0		3166	72%
	72.0 *		3108	70%
Supreme 780	85.5	MAX.	3228	85%
	83.5		3165	83%
	81.5 *		3104	81%
RL22	85.5 *	MAX.	3265	88%
Most Accurate Powder Tested	83.5		3208	86%
	81.5		3153	84%
H1000	92.0	MAX.	3265	91%
	90.0		3212	89%
	88.0 *		3158	87%
US 869	104.5	MAX.	3287	** 108%
	102.5		3209	** 106%
	100.5 *		3129	** 104%
RL25	88.0	MAX.	3336	91%
	86.0 *		3272	89%
	84.0		3210	87%
IMR 7828	87.0	MAX.	3340	88%
	85.0		3243	86%
	83.0 *		3173	84%
Retumbo	94.0	MAX.	3368	94%
	92.0 *		3262	92%
	90.0		3179	90%

BC=Ballistic Coefficient SD=Sectional Density
*Most Accurate Load Tested **Compressed Load

Use Maximum Loads with Caution
Refer to page 73 for additional safety information

7mm Rem Ultra Mag - 160 grain

		MAXIMUM S.A.A.M.I. O.A.C.L.		3.600"
		TESTED O.A.C.L.	B.C.	S.D.
AccuBond®	160gr. Spitzer	3.570"	0.531	0.283
Partition®	160gr. Spitzer	3.570"	0.475	0.283

CASE TYPE:	Nosler		PRIMER TYPE	WLRM
CASE HOLDS:	103.9	Gr. WATER	BARREL Length/Make	26" Wiseman
			BARREL Twist	1-9"

POWDER TYPE	POWDER CHG. GRS.	MUZZLE VEL. F.P.S.	LOAD DENSITY (VOLUME)
H4831	83.0 * MAX.	3115	83%
	81.0	3067	81%
	79.0	3022	79%
RL19	82.0 MAX.	3151	86%
	80.0	3084	84%
	78.0 *	3020	82%
Magnum	90.0 MAX.	3184	88%
	88.0	3107	86%
	86.0 *	3027	84%
Viht N560	85.0 MAX.	3185	91%
	83.0	3121	89%
	81.0 *	3057	87%
RL22	84.0 MAX.	3188	88%
	82.0	3124	86%
	80.0 *	3059	84%
H1000	90.0 * MAX.	3188	91%
	88.0	3137	89%
	86.0	3086	87%
Retumbo	92.0 * MAX.	3229	93%
	90.0	3159	91%
	88.0	3096	89%
IMR 7828 Most Accurate Powder Tested	85.0 MAX.	3231	87%
	83.0	3128	85%
	81.0 *	3061	83%
US 869	103.0 * MAX.	3238	** 108%
	101.0	3158	** 106%
	99.0	3072	** 104%
RL25	88.0 MAX.	3261	92%
	86.0	3194	90%
	84.0 *	3147	88%

BC=Ballistic Coefficient SD=Sectional Density
*Most Accurate Load Tested **Compressed Load

Use Maximum Loads with Caution
Refer to page 73 for additional safety information

7mm Rem Ultra Mag - 175 grain

Partition®	175gr. Spitzer	MAXIMUM S.A.A.M.I. O.A.C.L.		3.600"
		TESTED O.A.C.L.	B.C.	S.D.
		3.570"	0.519	0.310

CASE TYPE:	Nosler		PRIMER TYPE	WLRM
CASE HOLDS:	98.2	Gr. WATER	BARREL Length/Make	26" Wiseman
			BARREL Twist	1-9"

POWDER TYPE	POWDER CHG. GRS.		MUZZLE VEL. F.P.S.		LOAD DENSITY (VOLUME)
H50BMG	96.0	MAX.	2936	**	101%
	94.0		2869		99%
	92.0 *		2801		97%
H4831	81.5	MAX.	3017		86%
	79.5		2967		84%
	77.5 *		2916		82%
RL22	82.0	MAX.	3058		91%
	80.0		2989		89%
	78.0 *		2924		86%
H1000	87.0	MAX.	3065		93%
	85.0		3026		91%
	83.0 *		2984		89%
Magnum	89.0	MAX.	3079		92%
	87.0		3015		90%
	85.0 *		2949		88%
IMR 7828 Most Accurate Powder Tested	83.0	MAX.	3082		90%
	81.0		3014		88%
	79.0 *		2941		86%
Viht N560	84.0	MAX.	3093		95%
	82.0		3022		93%
	80.0 *		2951		91%
RL25	86.0	MAX.	3122		95%
	84.0 *		3052		93%
	82.0		2882		91%
US 869	102.0	MAX.	3154	**	113%
	100.0		3087	**	111%
	98.0 *		3022	**	108%
Retumbo	92.0 *	MAX.	3164		99%
	90.0		3076		96%
	88.0		3022		94%

BC=Ballistic Coefficient SD=Sectional Density
*Most Accurate Load Tested **Compressed Load

Use Maximum Loads with Caution
Refer to page 73 for additional safety information

300 AAC BLACKOUT

The quest to develop a suitable round for the ubiquitous AR-15 that would offer better terminal ballistics and allow effective and reliable suppressed operation has been a long one, fraught with dead ends and shattered dreams. Approved by SAAMI in 2011, the 300 Blackout was the culmination of efforts by Advanced Armament Corp to provide special forces troops with a cartridge that would match the performance of the 7.62x39 soviet round when loaded with a 120gr-class projectile, but could perform just as well when launching a heavier bullet at subsonic velocities. A muzzle velocity of 1050fps is the critical threshold, beyond which the bullet produces a supersonic 'crack', negating one of the chief benefits of a suppressed weapon. Because of this velocity ceiling, the only way to increase muzzle energy is to increase projectile mass, which is why the 300blk is often loaded with the heaviest 30 caliber bullets available.

Based on the 221 Fireball case (itself shortened from the parent 223 Rem) the 300Blk offers the AR-15 owner the option of a round with much better performance at distance than the heaviest 22 caliber options, when compared to its supersonic loadings. It does this while still using a standard bolt and magazine, requiring only the barrel to be changed. When subsonic loads are used, the AR-15 will still function reliably, making it a great tool for harvesting feral hogs where multiple animals can be encountered.

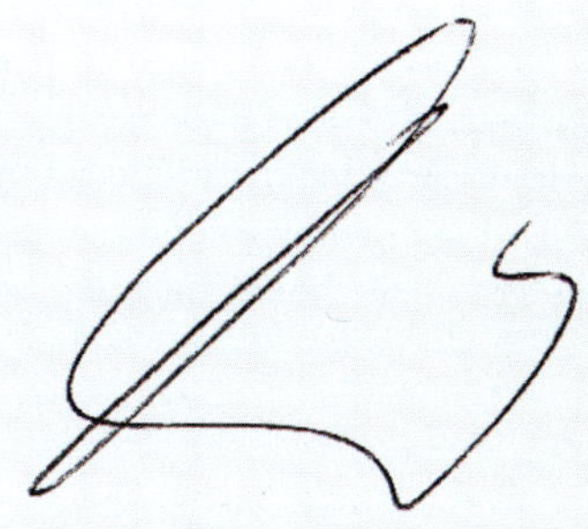

Iain Harrison is the Media Relations Manager for Crimson Trace, TV host and competitive shooter as well as a Nosler Pro-Staff member.

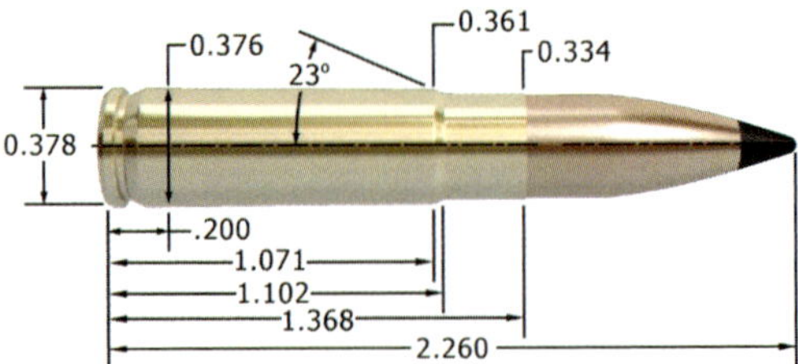

Maximum S.A.A.M.I. Overall Cartridge Length: 2.260"

BULLET CHOICES FOR THE 300 AAC BLACKOUT

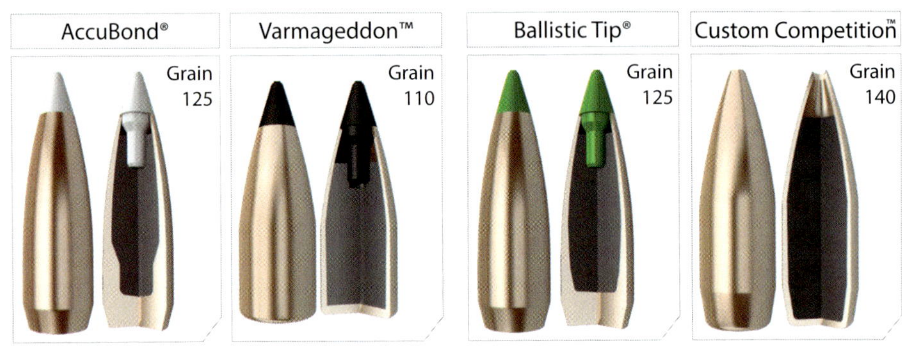

TECHNICAL INFORMATION

The Blackout has proven itself to be an effective medium game cartridge when used at appropriate range. It seems to be particularly useful in feral hog population control efforts. Due to the lower velocities produced by the Blackout, expansion will not be as dramatic as when used in faster 30 caliber cartridges, but our 125 grain AccuBond and Ballistic Tip bullets are excellent choices for hunting with the Blackout.

300 AAC Blackout - 110 grain			MAXIMUM S.A.A.M.I. O.A.C.L.		2.260"
			TESTED O.A.C.L.	B.C.	S.D.
Varmageddon™		110gr. Spitzer	2.025"	0.293	0.166
CASE TYPE:	Hornady	PRIMER TYPE		WSR	
CASE HOLDS:	19.2 Gr. WATER	BARREL Length/Make		16" Pac-Nor	
		BARREL Twist		1-8"	

POWDER TYPE	POWDER CHG. GRS.		MUZZLE VEL. F.P.S.		LOAD DENSITY (VOLUME)
IMR 4227	19.0	MAX.	2138	**	116%
	18.0		2059	**	110%
	17.0 *		1981	**	104%
H110	19.0	MAX.	2249		100%
	18.0		2187		95%
	17.0 *		2121		90%
W296 Most Accurate Powder Tested	19.5 *	MAX.	2291	**	104%
	18.5		2215		98%
	17.5		2141		93%
Lil'Gun	20.0 *	MAX.	2296	**	110%
	19.0		2225	**	104%
	18.0		2155		99%

BC=Ballistic Coefficient SD=Sectional Density
*Most Accurate Load Tested **Compressed Load

Use Maximum Loads with Caution
Refer to page 73 for additional safety information

300 AAC Blackout - 125 grain

		MAXIMUM S.A.A.M.I. O.A.C.L.		2.260"
		TESTED O.A.C.L.	B.C.	S.D.
AccuBond®	125gr. Spitzer	2.060"	0.366	0.188
Ballistic Tip®	125gr. Spitzer	2.060"	0.366	0.188

CASE TYPE:	Hornady		PRIMER TYPE	WSR
CASE HOLDS:	17.5	Gr. WATER	BARREL Length/Make	16" Pac-Nor
			BARREL Twist	1-8"

POWDER TYPE	POWDER CHG. GRS.	MUZZLE VEL. F.P.S.		LOAD DENSITY (VOLUME)
IMR 4227	18.0 * MAX.	2003	**	120%
	17.0	1927	**	114%
	16.0	1845	**	107%
W296 Most Accurate Powder Tested	17.5 * MAX.	2128	**	102%
	16.5	2034		96%
	15.5	1944		90%
Lil'Gun	18.0 * MAX.	2151	**	108%
	17.0	2108	**	102%
	16.0	2067		96%
H110	18.0 * MAX.	2159	**	104%
	17.0	2065		99%
	16.0	1975		93%

BC=Ballistic Coefficient SD=Sectional Density
*Most Accurate Load Tested **Compressed Load

Use Maximum Loads with Caution
Refer to page 73 for additional safety information

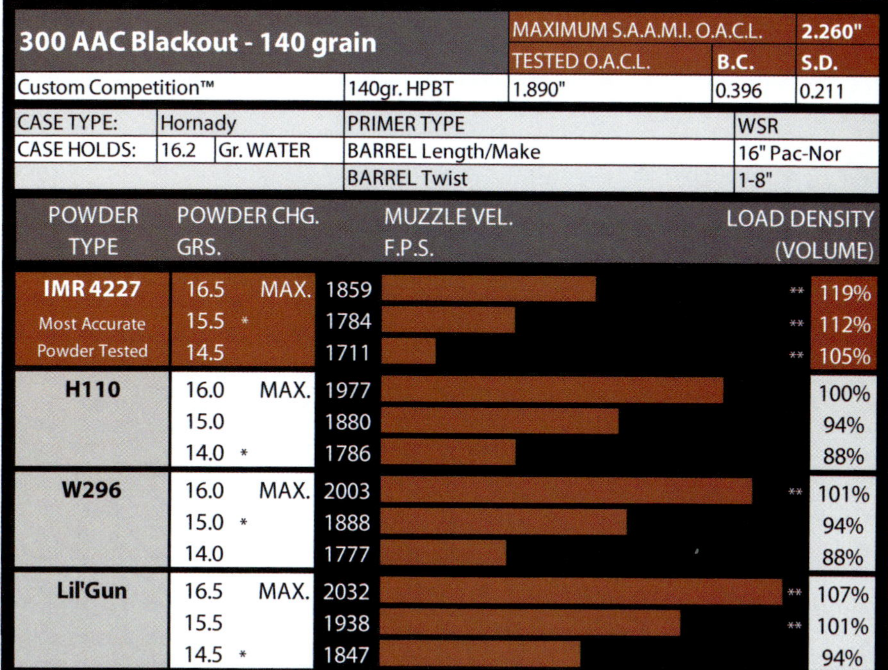

300 AAC Blackout - 140 grain			MAXIMUM S.A.A.M.I. O.A.C.L.		2.260"
			TESTED O.A.C.L.	B.C.	S.D.
Custom Competition™		140gr. HPBT	1.890"	0.396	0.211
CASE TYPE:	Hornady	PRIMER TYPE		WSR	
CASE HOLDS:	16.2 Gr. WATER	BARREL Length/Make		16" Pac-Nor	
		BARREL Twist		1-8"	

POWDER TYPE	POWDER CHG. GRS.		MUZZLE VEL. F.P.S.		LOAD DENSITY (VOLUME)
IMR 4227 Most Accurate Powder Tested	16.5	MAX.	1859	**	119%
	15.5 *		1784	**	112%
	14.5		1711	**	105%
H110	16.0	MAX.	1977		100%
	15.0		1880		94%
	14.0 *		1786		88%
W296	16.0	MAX.	2003	**	101%
	15.0 *		1888		94%
	14.0		1777		88%
Lil'Gun	16.5	MAX.	2032	**	107%
	15.5		1938	**	101%
	14.5 *		1847		94%

BC=Ballistic Coefficient SD=Sectional Density
*Most Accurate Load Tested **Compressed Load

Use Maximum Loads with Caution
Refer to page 73 for additional safety information

Karen Mehall

30-30 WINCHESTER

When my brother, Joe, turned 12, Grandpap Mehall gave him his prized 30-30—a trusty lever-action Winchester Model 1894. I watched this coming-of-age ritual with interest as "Joey" ran his hand along the stock, listening to Grandpap and Dad hail the merits of "the good ole 30-30." I knew what had passed from Joseph Sr. to Joseph III was special, but I would not grasp its significance until I became a hunter.

More than an iconic symbol of the American hunting tradition, the 30-30 is part of its foundation. Little did Winchester know when it unveiled its 30 Winchester Center Fire (WCF) cartridge in 1895 that many would dub it the most popular—and practical—deer cartridge of all time. Chances are, this effective short-to-medium-range round has dropped more deer than any other—thanks to the millions of rifles chambered for it and the availability and comparatively low cost of ammo. The 30 WCF made headlines as the result of Winchester's effort to develop a high-velocity, flat-shooting smokeless powder cartridge. Soon after, Marlin Firearms Company announced its version in conjunction with the Union Metallic Cartridge Company, dubbing it the 30-30. The first .30 designated the caliber in inches and the second noted the smokeless powder charge in grains. Winchester would ultimately adopt Marlin's 30-30 nomenclature.

U.S. President Theodore Roosevelt acquired one of the first "thutty thutty" Winchesters from the factory. But despite impressive velocities, particularly compared to the 45-70 black-powder cartridge of the day, the 30-30 may have faded away had it not been for Winchester's Model 1894. Later called the Model 94, it was chambered for several cartridges but none quite as popular.

Though available in single-shots and bolt-actions, the 30-30 is synonymous with the lever-action. Thousands of 30-30 Winchesters and Marlins were sold in the first 20 years of the 20th century alone. No other caliber competed with it for decades. Factory-loaded Power Point, Core-Lokt and Silvertip cartridges with 150- and 170-grain flat-nose bullets and velocities of 2390 fps and 2200 fps have dropped millions of game, and today's reloader can select bullet weights from 85 to 180 grains.

I was raised in the East where shot opportunities are rarely beyond 75 yards. The 30-30 handily fills that niche. One day, Joe's sons will hoist their great-grandpap's 30-30. It will be reliable, easy to shoot—and the recoil will not rattle their permanent teeth. While new ammo options are inevitable, tradition is slow to wane. Like Grandpap, I respect the 30-30 for epitomizing an era when hunters relied on their woodsmanship skills to put the sneak on their evening meal.

Karen Mehall

Karen Mehall is the Senior Editor of American Hunter

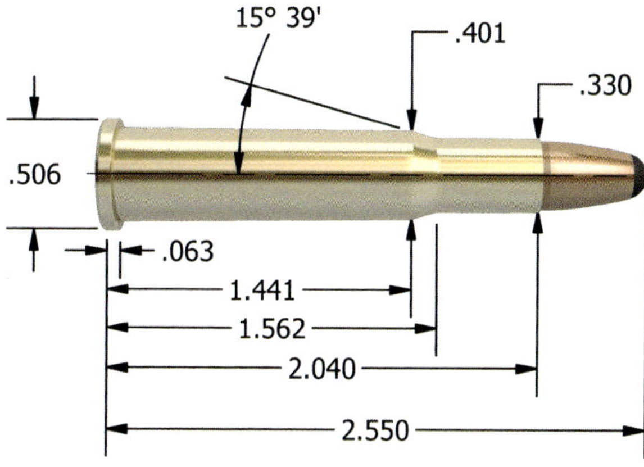

Maximum S.A.A.M.I. Overall Cartridge Length: 2.550"

BULLET CHOICES FOR THE 30-30 WINCHESTER

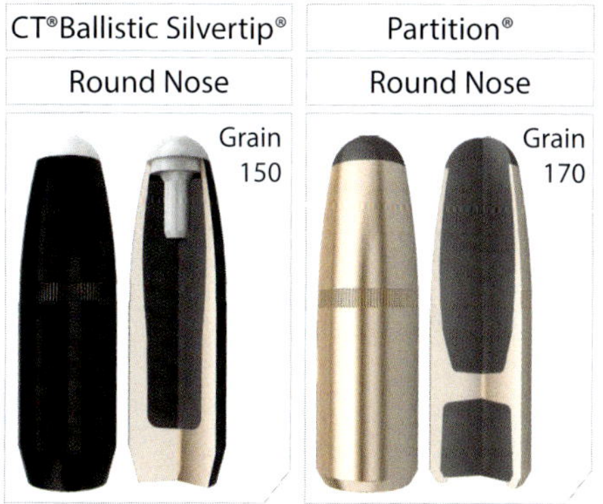

TECHNICAL INFORMATION

Warning: It is not safe to use spitzer style bullets in a tubular magazine. The sharp nose of a spitzer bullet resting on the primer of the next cartridge (in front of it) can cause detonation in the magazine. Our 170gr. Round Nose Partition and 150gr. CT Ballistic Silvertip were designed specifically for the .30-30. They will expand well at .30-30 velocities out to 200 yards, and feature a crimping groove (cannelure). We recommend full-length case resizing and crimping for best results.

30-30 Winchester - 150 grain

30-30 Winchester - 150 grain		MAXIMUM S.A.A.M.I. O.A.C.L.		2.550"
		TESTED O.A.C.L.	B.C.	S.D.
CT® Ballistic Silvertip®	150gr. RN	2.550"	0.232	0.226

CASE TYPE:	Winchester		PRIMER TYPE	WLR
CASE HOLDS:	36.6	Gr. WATER	BARREL Length/Make	20" Wiseman
			BARREL Twist	1-12"

POWDER TYPE	POWDER CHG. GRS.	MUZZLE VEL. F.P.S.	LOAD DENSITY (VOLUME)
W748	35.0 * MAX.	2180	100%
	33.0	2030	94%
	31.0	1890	88%
TAC	31.5 MAX.	2146	88%
	29.5 *	2067	82%
	27.5	1981	77%
IMR 4895 Most Accurate Powder Tested	32.0 MAX.	2247	96%
	30.0	2133	90%
	28.0 *	2020	84%
IMR 3031	31.0 * MAX.	2254	96%
	29.0	2120	90%
	27.0	1980	83%
H4895	33.0 * MAX.	2183	99%
	31.0	2049	93%
	29.0	1914	87%
A-2460	34.0 * MAX.	2193	94%
	32.0	2070	88%
	30.0	1947	83%
Varget	34.5 * MAX.	2206	** 102%
	32.5	2086	96%
	30.5	1967	90%
IMR 4198	26.0 MAX.	2260	81%
	24.0	2110	75%
	22.0 *	1960	68%
IMR 4320	34.0 MAX.	2270	100%
	32.0	2160	94%
	30.0 *	2060	88%

BC=Ballistic Coefficient SD=Sectional Density
*Most Accurate Load Tested **Compressed Load

Use Maximum Loads with Caution
Refer to page 73 for additional safety information

30-30 Winchester - 170 grain		MAXIMUM S.A.A.M.I. O.A.C.L.		2.550"
		TESTED O.A.C.L.	B.C.	S.D.
Partition®	170gr. RN	2.550"	0.252	0.256

CASE TYPE:	Winchester		PRIMER TYPE	WLR
CASE HOLDS:	34.6	Gr. WATER	BARREL Length/Make	20" Wiseman
			BARREL Twist	1-12"

POWDER TYPE	POWDER CHG. GRS.		MUZZLE VEL. F.P.S.		LOAD DENSITY (VOLUME)
RL15	31.5	MAX.	1974		95%
	29.5		1847		89%
	27.5 *		1691		83%
A-2520	32.5	MAX.	1984	**	107%
	30.5 *		1899		100%
	28.5		1784		94%
Benchmark	28.0	MAX.	2000		87%
	26.0 *		1872		81%
	24.0		1750		75%
TAC	30.0	MAX.	2031		88%
	28.0 *		1970		83%
	26.0		1890		77%
H414	37.0 *	MAX.	2062	**	111%
	35.0		1947	**	105%
	33.0		1832		99%
W748	34.0	MAX.	2071	**	102%
	32.0 *		1938		96%
	30.0		1837		90%
IMR 4895	31.0 *	MAX.	2160		98%
	29.0		2030		92%
	27.0		1900		86%
Varget Most Accurate Powder Tested	32.5	MAX.	2162	**	102%
	30.5		1954		95%
	28.5 *		1743		89%
IMR 3031	30.0	MAX.	2168		98%
	28.0		1953		91%
	26.0 *		1738		85%
IMR 4320	33.0	MAX.	2192	**	103%
	31.0		2017		96%
	29.0 *		1842		90%

BC=Ballistic Coefficient SD=Sectional Density
*Most Accurate Load Tested **Compressed Load

Use Maximum Loads with Caution
Refer to page 73 for additional safety information

Wayne van Zwoll

300 SAVAGE

When it lost to the Norwegian-designed Krag-Jorgensen in 1892 government trials, 35-year-old Arthur Savage reconfigured his hammerless No. 1 lever rifle for sportsmen. On April 5, 1894 he formed Savage Repeating Arms Company to build the Model 1895 rifle, first in 32-20, then in 303 Savage.

The 303 Savage hurled 190-grain .308-diameter bullets (not the .311s of the 303 British) as fast as the 30 WCF (30-30) shot 160s. China missionary Harry Caldwell hunted tigers with a Savage rifle in 303. W.T. Hornaday wrote of shooting a "moose and two bull caribou, all killed stone dead in their tracks with one of your incomparable 303 rifles." About 6,000 were shipped between 1895 and 1899, when Savage modified the action. Its enclosed rotary magazine remained.

The subsequent Model 1899 came in several forms and chamberings. In 1913 the 250-3000 replaced the 25-35 – courtesy Charles Newton, who also gave Savage the 22 Hi-Power. These were followed in 1920 by the 300 Savage, when the Model 1899 became the 99. At just 1.871 inch, the 300 Savage hull is shorter than even the 308 Winchester's. Still, it registers 47,000 psi, 5,000 more than a 30-30, which it handily outperforms. Hunters snapped up Savage 99s in .300, to shoot whitetail and black bear in the woods but also for scabbard carry in elk country. The spool magazine permitted use of pointed bullets that shot flat at distance. With 150-grain bullets at 2,670 fps and 180s at 2,370, the .300 put the skids under the .303, dropped after WWII.

Arthur Savage died in 1941 at age 84. Five years later, Savage Arms trimmed the Model 99 stable to three rifles in 250 and 300 Savage. Passing the million-rifle mark in the 1950s, the 99 added newer short-action cartridges, but none would surpass the 300 Savage's lifetime sales record. The last standard 99s were cataloged in 1997 for $650. That year, the firm built 1,000 Model 99-CE (Centennial Edition) rifles in 300 Savage only. Price: $1,660.

The 300 Savage has been chambered in other rifles, notably Remington's 760 pump and 722 bolt guns. A few Winchester 70s were bored to 300 Savage. Standard rifling twist: 1-in-12.

I've used the 300 Savage on whitetail, mule deer and pronghorn, and killed my biggest caribou with it. Adequate for elk and moose, the 300 Savage is still properly paired with an iron-sighted Model 99!

Wayne van Zwoll has published 14 books and several thousand magazine articles on rifles, shooting and big game hunting.

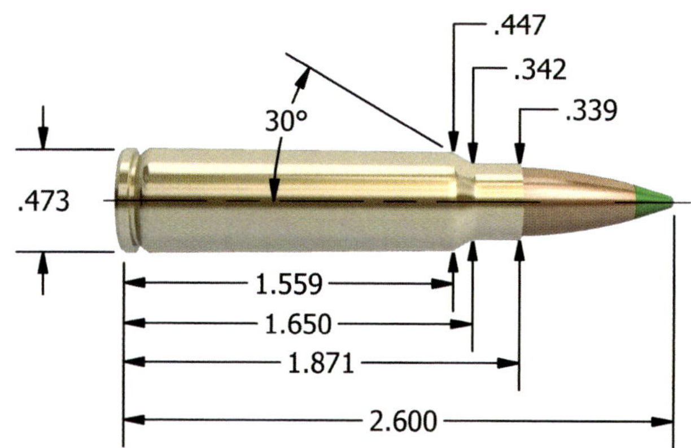

Maximum S.A.A.M.I. Overall Cartridge Length: 2.600"

BULLET CHOICES FOR THE 300 SAVAGE

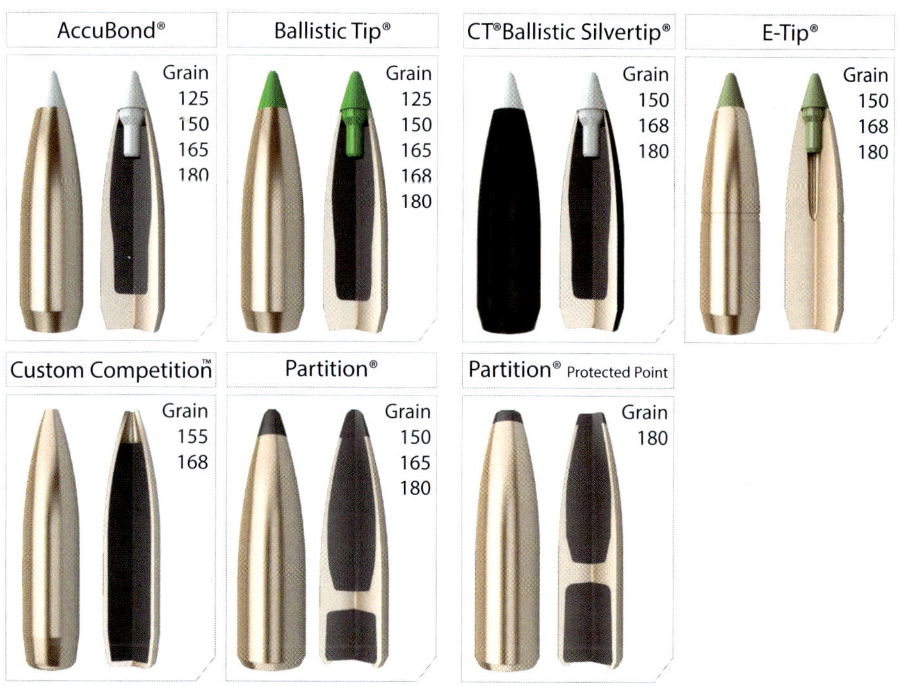

TECHNICAL INFORMATION

When reloading the 300 Savage, we recommend IMR4895 or similar powders. If your rifle is a lever action, pump or semi-automatic, full-length sizing will yield the best results.

300 Savage - 125 grain		MAXIMUM S.A.A.M.I. O.A.C.L.		2.600"
		TESTED O.A.C.L.	B.C.	S.D.
AccuBond®	125gr. Spitzer	2.600"	0.366	0.188
Ballistic Tip®	125gr. Spitzer	2.600"	0.366	0.188

CASE TYPE:	Winchester		PRIMER TYPE	WLR
CASE HOLDS:	47.2	Gr. WATER	BARREL Length/Make	24" Wiseman
			BARREL Twist	1-12"

POWDER TYPE	POWDER CHG. GRS.		MUZZLE VEL. F.P.S.		LOAD DENSITY (VOLUME)
BL-C2	42.5 *	MAX.	2660		91%
	40.5		2530		87%
	38.5		2400		83%
A-2230	41.0 *	MAX.	2768		89%
	39.0		2653		84%
	37.0		2538		80%
H4198	37.0	MAX.	2832		86%
	35.0		2667		81%
	33.0 *		2502		77%
Varget	45.5	MAX.	2947	**	104%
	43.5 *		2830		100%
	41.5		2654		95%
W748 Most Accurate Powder Tested	46.5	MAX.	2965	**	103%
	44.5 *		2893		98%
	42.5		2745		94%
RL7	40.0 *	MAX.	3021		92%
	38.0		2849		88%
	36.0		2676		83%
A-2015	43.5	MAX.	3030	**	105%
	41.5		2873		100%
	39.5 *		2717		95%

BC=Ballistic Coefficient SD=Sectional Density
*Most Accurate Load Tested **Compressed Load

Use Maximum Loads with Caution
Refer to page 73 for additional safety information

300 Savage - 150/155 grain

Bullet	Type	MAXIMUM S.A.A.M.I. O.A.C.L. / TESTED O.A.C.L. (2.600")	B.C.	S.D.
AccuBond®	150gr. Spitzer	2.600"	0.435	0.226
Ballistic Tip®	150gr. Spitzer	2.600"	0.435	0.226
CT® Ballistic Silvertip®	150gr. Spitzer	2.600"	0.435	0.226
E-Tip®	150gr. Spitzer	2.580"	0.469	0.226
Due to internal construction differences, always begin with starting loads when using E-Tip® products.				
Partition®	150gr. Spitzer	2.600"	0.387	0.226
Custom Competition™	155gr. HPBT	2.600"	0.450	0.233

CASE TYPE:	Winchester		PRIMER TYPE	WLR
CASE HOLDS:	44.4	Gr. WATER	BARREL Length/Make	24" Wiseman
			BARREL Twist	1-12"

POWDER TYPE	POWDER CHG. GRS.		MUZZLE VEL. F.P.S.		LOAD DENSITY (VOLUME)
A-2015	37.0	MAX.	2529		95%
	35.0		2397		90%
	33.0 *		2234		84%
IMR 4064	40.0 *	MAX.	2560		99%
	38.0		2460		94%
	36.0		2360		89%
IMR 4895 Most Accurate Powder Tested	40.0	MAX.	2580		99%
	38.0		2450		94%
	36.0 *		2320		89%
IMR 3031	38.0	MAX.	2590		97%
	36.0		2450		92%
	34.0 *		2310		87%
Viht N140	44.0	MAX.	2601	**	115%
	42.0		2486	**	110%
	40.0 *		2371	**	105%
Varget	42.0	MAX.	2662	**	102%
	40.0		2547		97%
	38.0 *		2430		93%
W748	42.0	MAX.	2685		99%
	40.0		2501		94%
	38.0 *		2333		89%

BC=Ballistic Coefficient SD=Sectional Density
*Most Accurate Load Tested **Compressed Load

Use Maximum Loads with Caution
Refer to page 73 for additional safety information

300 Savage - 165/168 grain		MAXIMUM S.A.A.M.I. O.A.C.L.		2.600"
		TESTED O.A.C.L.	B.C.	S.D.
AccuBond®	165gr. Spitzer	2.600"	0.475	0.248
Ballistic Tip®	165gr. Spitzer	2.600"	0.475	0.248
Partition®	165gr. Spitzer	2.600"	0.410	0.248
Ballistic Tip®	168gr. Spitzer	2.600"	0.490	0.253
CT® Ballistic Silvertip®	168gr. Spitzer	2.600"	0.490	0.253
Custom Competition™	168gr. HPBT	2.600"	0.462	0.253
E-Tip®	168gr. Spitzer	2.580"	0.503	0.253

Due to internal construction differences, always begin with starting loads when using E-Tip® products.

CASE TYPE:	Winchester		PRIMER TYPE	WLR
CASE HOLDS:	41.0	Gr. WATER	BARREL Length/Make	24" Wiseman
			BARREL Twist	1-12"

POWDER TYPE	POWDER CHG. GRS.		MUZZLE VEL. F.P.S.		LOAD DENSITY (VOLUME)
IMR 4895	39.0 *	MAX.	2390	**	105%
	37.0		2243		99%
	35.0		2091		94%
IMR 3031	37.0 *	MAX.	2452	**	102%
	35.0		2297		96%
	33.0		2142		91%
IMR 4064	39.0 *	MAX.	2470	**	105%
	37.0		2370		99%
	35.0		2270		94%
Varget	39.5	MAX.	2485	**	104%
	37.5		2396		99%
	35.5 *		2331		94%
A-2230	39.0	MAX.	2488		97%
	37.0		2369		92%
	35.0 *		2251		87%
Viht N140	44.0	MAX.	2492	**	125%
Most Accurate	42.0		2381	**	119%
Powder Tested	40.0 *		2271	**	113%
W748	40.5	MAX.	2540	**	103%
	38.5 *		2428		98%
	36.5		2255		93%

BC=Ballistic Coefficient SD=Sectional Density
*Most Accurate Load Tested **Compressed Load

Use Maximum Loads with Caution
Refer to page 73 for additional safety information

300 Savage - 180 grain		MAXIMUM S.A.A.M.I. O.A.C.L.		2.600"
		TESTED O.A.C.L.	**B.C.**	**S.D.**
AccuBond®	180gr. Spitzer	2.600"	0.507	0.271
Ballistic Tip®	180gr. Spitzer	2.600"	0.507	0.271
CT® Ballistic Silvertip®	180gr. Spitzer	2.600"	0.507	0.271
E-Tip®	180gr. Spitzer	2.580"	0.523	0.271
Due to internal construction differences, always begin with starting loads when using E-Tip® products.				
Partition®	180gr. Spitzer	2.600"	0.474	0.271
Partition®	180gr. PPT	2.600"	0.361	0.271

CASE TYPE:	Winchester		PRIMER TYPE	WLR
CASE HOLDS:	39.7	Gr. WATER	BARREL Length/Make	24" Wiseman
			BARREL Twist	1-12"

POWDER TYPE	POWDER CHG. GRS.		MUZZLE VEL. F.P.S.		LOAD DENSITY (VOLUME)
A-2460	36.0	MAX.	2298		92%
	34.0		2203		87%
	32.0 *		2108		81%
Varget	36.5	MAX.	2347		99%
	34.5		2276		94%
	32.5 *		2228		89%
IMR 4895	38.5 *	MAX.	2350	**	107%
	36.5		2200	**	101%
	34.5		2050		95%
Viht N135	36.5	MAX.	2352	**	107%
	34.5		2234	**	101%
	32.5 *		2125		95%
IMR 3031	36.0 *	MAX.	2360	**	102%
	34.0		2230		97%
	32.0		2100		91%
IMR 4064 Most Accurate Powder Tested	38.0 *	MAX.	2368	**	105%
	36.0		2263		100%
	34.0		2158		94%
A-2015	37.5	MAX.	2502	**	107%
	35.5		2374	**	102%
	33.5 *		2247		96%

BC=Ballistic Coefficient SD=Sectional Density
*Most Accurate Load Tested **Compressed Load

Use Maximum Loads with Caution
Refer to page 73 for additional safety information

Lane Pearce

308 MARLIN EXPRESS

The 308 Marlin Express resulted from a joint venture between Marlin Firearms and Hornady Mfg. Introduced in 2007, Marlin chambered the new round in two, spiffed-up versions of the ever popular Model 336 lever gun. The blued steel 308MX features a 22" barrel and walnut buttstock and forearm while the stainless 308MXLR model has a 24" barrel and laminated furniture.

I participated in a Texas hog hunt soon after the XLR rifle was introduced. It was a first time experience for me trying to spot a sounder of pigs along the edge of a brush-clogged creek from an elevated blind. I finally got a shot right at dusk and plugged a young boar, head-on, just above the snout. The next day, the camp cook directed me to a ground blind a few hundred yards behind the lodge. He allowed that if I'd shoot a young sow, he'd whip up a platter of fresh, pulled pork barbeque for supper.

We ate mighty fine the next evening.

Although the rimmed case 308 ME is similar, it is not simply a reincarnation of the earlier 307 Winchester. The 307 is basically a rimmed 308 Winchester. The 308 ME case is approximately a tenth of an inch shorter to accommodate a longer, sleeker bullet with a higher BC for much improved downrange ballistic performance.

Due to the inherent strength limitations, the 308 ME cannot duplicate the 308 Winchester's ballistics. The .308 ME MAP is ~46,000 psi compared to 52,000 psi for the 308 Win., so it can come close. However, the 308 ME will duplicate (and even improve upon somewhat) the capabilities of the venerable 300 Savage. Marlin's more traditional format features a tubular magazine compared to the original Savage M99's unique rotary mag design.

Factory 308 ME ammo contains a proprietary propellant with improved burn rate characteristics to launch 160-grain bullets with pointed, 'flexible' polymer tips at ~2600 fps. Initially, the flex-tip bullets and special powder were not offered as components. So, we made do with conventional flat-point bullets and several suitable canistered propellants. The lower velocity and ballistic coefficient resulted in reducing the effective 'point-blank' from ~300 to ~250 yards which is still a significant performance boost over the popular 'thutty-thutty' lever gun's capability.

Lane Pearce

Lane Pearce is the Reloading Editor of Shooting Times magazine.

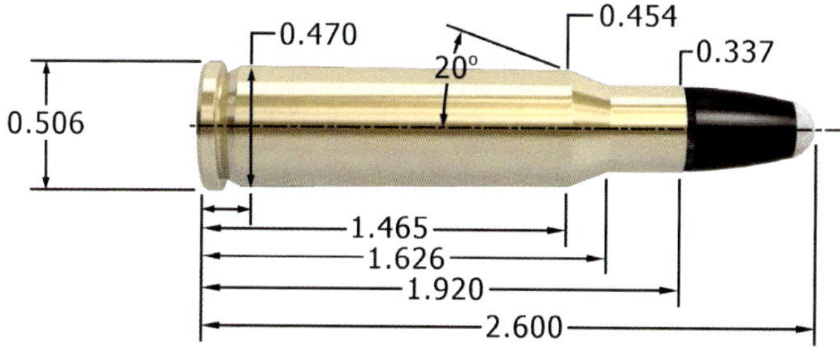

Maximum S.A.A.M.I. Overall Cartridge Length: 2.600"

BULLET CHOICES FOR THE 308 MARLIN EXPRESS

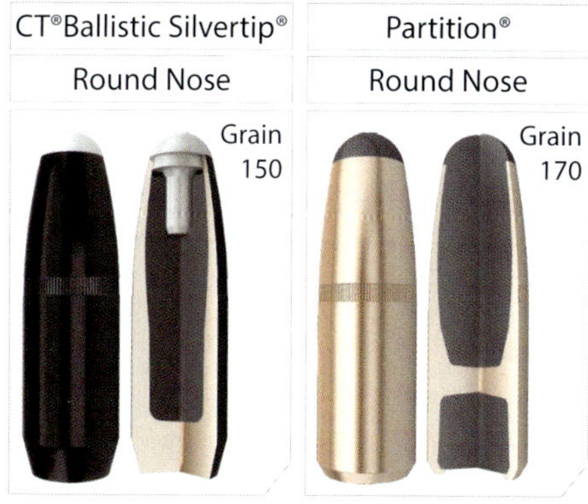

CT® Ballistic Silvertip®	Partition®
Round Nose	Round Nose
Grain 150	Grain 170

TECHNICAL INFORMATION

Designed to approximate 308 Winchester performance in a tube-fed lever-action rifle, the 308 Marlin Express is an excellent short to medium range big game cartridge. Our 170 grain Partition® and 150 grain CT® Ballistic Silvertip® are our only 30 caliber bullets designed for use in rifles with tubular magazines. As such, these are the only bullets we recommend for loading in the 308 Marlin Express.

308 Marlin Express - 150 grain		MAXIMUM S.A.A.M.I. O.A.C.L.		2.600"
		TESTED O.A.C.L.	B.C.	S.D.
CT® Ballistic Silvertip®	150gr. RN	2.400"	0.232	0.226

CASE TYPE:	Hornady		PRIMER TYPE	Fed. 210
CASE HOLDS:	41.7	Gr. WATER	BARREL Length/Make	24" Pac-Nor
			BARREL Twist	1-12"

POWDER TYPE	POWDER CHG. GRS.		MUZZLE VEL. F.P.S.		LOAD DENSITY (VOLUME)
X-Terminator Most Accurate Powder Tested	38.0	MAX.	2583		93%
	36.0		2477		88%
	34.0 *		2366		83%
A-2495	39.0 *	MAX.	2588	**	106%
	37.0		2455	**	101%
	35.0		2309		95%
IMR 4895	41.0	MAX.	2608	**	108%
	39.0		2474	**	103%
	37.0 *		2346		98%
IMR 8208 XBR	39.0	MAX.	2636	**	103%
	37.0 *		2491		98%
	35.0		2347		92%
Varget	42.0	MAX.	2653	**	109%
	40.0		2530	**	104%
	38.0 *		2416		99%
RL15	42.5	MAX.	2659	**	106%
	40.5 *		2536	**	101%
	38.5		2417		96%

BC=Ballistic Coefficient SD=Sectional Density
*Most Accurate Load Tested **Compressed Load

Use Maximum Loads with Caution
Refer to page 73 for additional safety information

308 Marlin Express - 170 grain		MAXIMUM S.A.A.M.I. O.A.C.L.		2.600"
		TESTED O.A.C.L.	B.C.	S.D.
Partition®	170gr. RN	2.380"	0.252	0.256

CASE TYPE:	Hornady		PRIMER TYPE	Fed. 210
CASE HOLDS:	42.2	Gr. WATER	BARREL Length/Make	24" Pac-Nor
			BARREL Twist	1-12"

POWDER TYPE	POWDER CHG. GRS.		MUZZLE VEL. F.P.S.		LOAD DENSITY (VOLUME)
X-Terminator	36.5	MAX.	2430		88%
	34.5		2312		83%
	32.5	*	2187		78%
A-2520	39.5	MAX.	2454		96%
	37.5	*	2359		92%
	35.5		2264		87%
Varget	39.5	MAX.	2458	**	101%
	37.5		2343		96%
	35.5	*	2220		91%
IMR 8208 XBR	38.0	MAX.	2462		99%
Most Accurate	36.0	*	2340		94%
Powder Tested	34.0		2213		89%
IMR 4064	39.5	MAX.	2473	**	103%
	37.5		2346		98%
	35.5	*	2219		92%
RL15	40.5	MAX.	2486		100%
	38.5		2365		95%
	36.5	*	2244		90%

BC=Ballistic Coefficient SD=Sectional Density
*Most Accurate Load Tested **Compressed Load

Use Maximum Loads with Caution
Refer to page 73 for additional safety information

Scott Haugen

30 TC (THOMPSON CENTER)

When Thompson Center introduced the 30 TC, I was hosting a TV show they supported. Initially, I was not quite sure what to expect in terms of what this cartridge had to offer over the 308 Winchester or 30-06. Then I fired it. Instantly, I fell in love with this cartridge, and the moment I saw how it performed on big game, I knew it would be one I'd rely on for years to come.

The 30 TC is a short action cartridge with a case capacity slightly less than that of a 308 Win. The purpose of the 30 TC is to deliver less recoil, yet be able to generate velocities rivaling the popular 30-06.

In recent years, the trend in ammunition has been toward short, fat cartridge designs. This is because gunpowder burns more efficiently in shorter cases of greater diameter, thus optimizing bullet velocities.

The first time I shot this cartridge on the range, its tight groups impressed me. Not only were the groupings of the 150-grain bullets pleasing, but the reduced recoil demanded my attention.

The 30 TC came along at a time when my two sons were young, yet developing an interest in big game hunting. For them, as well as for my small-statured wife, the reduced recoil and knock-down power of this cartridge would be perfect. But before introducing them to this caliber, I wanted to see how it performed on big game.

The first mule deer I shot with the 30 TC dropped the buck on the spot. The first whitetail–a long shot across a big, open canyon–did the same. In the brush-choked blacktail deer country I grew up hunting in, the 30 TC performed flawlessly on those deer. A good friend took a Canadian moose with his 30 TC, motivating me to try it on other big game.

One of the most impressive performances I witnessed with this cartridge was on a massive black bear that my nine year old son, Braxton, took in British Columbia. The shot came inside 100 yards and the bullet hit tight behind the shoulder, taking out both lungs. The 350-pound bear squared just shy of seven feet and sported a skull that just missed the record books. The bear went less than 10 yards before expiring.

Compact with mild recoil and impressive performance, the 30 TC has found a place in the Haugen family.

Scott Haugen

Scott Haugen
TV host, outdoor writer

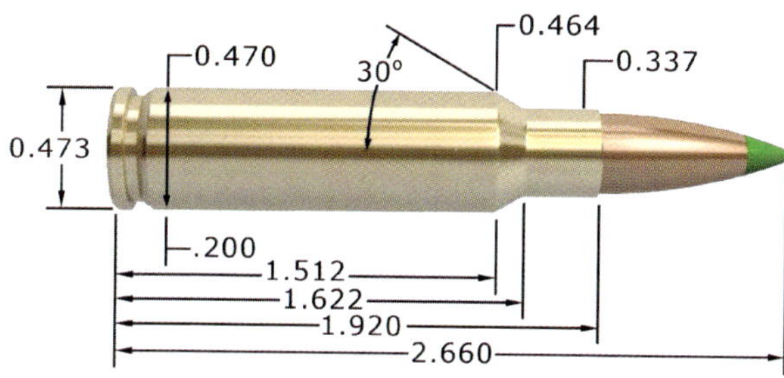

Maximum S.A.A.M.I. Overall Cartridge Length: 2.660"

BULLET CHOICES FOR THE 30 TC (THOMPSON CENTER)

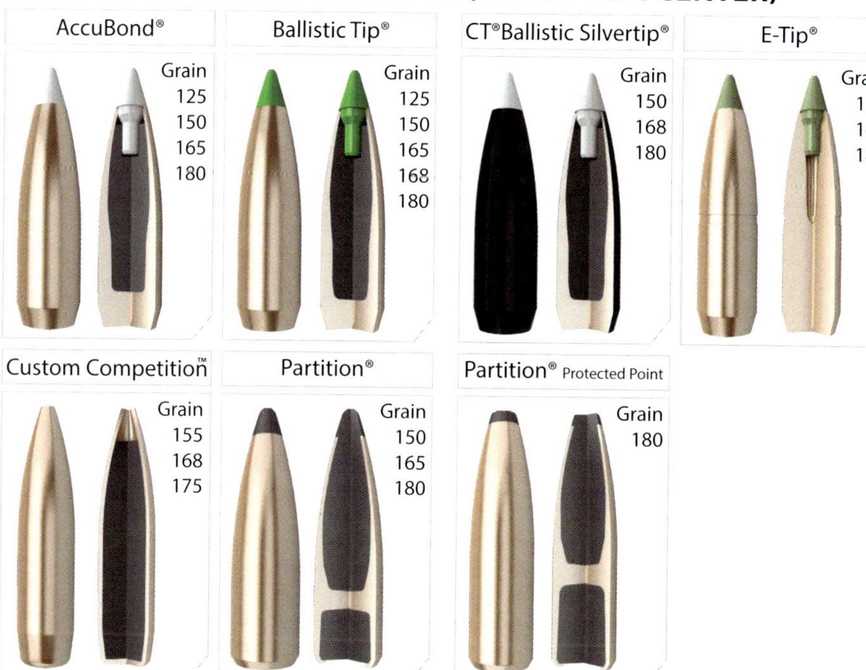

30 Thompson Center - 125 grain

30 Thompson Center - 125 grain		MAXIMUM S.A.A.M.I. O.A.C.L.		2.660"
		TESTED O.A.C.L.	B.C.	S.D.
AccuBond®	125gr. Spitzer	2.640"	0.366	0.188
Ballistic Tip®	125gr. Spitzer	2.640"	0.366	0.188

CASE TYPE:	Hornady		PRIMER TYPE	Fed. 210
CASE HOLDS:	50.0	Gr. WATER	BARREL Length/Make	24" Pac-Nor
			BARREL Twist	1-12"

POWDER TYPE	POWDER CHG. GRS.		MUZZLE VEL. F.P.S.		LOAD DENSITY (VOLUME)
IMR 4320 Most Accurate Powder Tested	49.0	* MAX.	3105	**	105%
	47.0		2984	**	101%
	45.0		2863		97%
RL10x	42.5	* MAX.	3135		96%
	40.5		2999		91%
	38.5		2860		87%
Benchmark	46.5	* MAX.	3187	**	101%
	44.5		3040		96%
	42.5		2897		92%
TAC	48.5	MAX.	3215		99%
	46.5		3116		95%
	44.5	*	3017		91%
A-2460	48.0	MAX.	3247		97%
	46.0		3136		93%
	44.0	*	3032		89%

BC=Ballistic Coefficient SD=Sectional Density
*Most Accurate Load Tested **Compressed Load

Use Maximum Loads with Caution
Refer to page 73 for additional safety information

30 Thompson Center - 150/155 grain		MAXIMUM S.A.A.M.I. O.A.C.L.		2.660"
		TESTED O.A.C.L.	B.C.	S.D.
AccuBond®	150gr. Spitzer	2.640"	0.435	0.226
Ballistic Tip®	150gr. Spitzer	2.640"	0.435	0.226
CT® Ballistic Silvertip®	150gr. Spitzer	2.640"	0.435	0.226
E-Tip®	150gr. Spitzer	2.600"	0.469	0.226
Due to internal construction differences, always begin with starting loads when using E-Tip® products.				
Partition®	150gr. Spitzer	2.590"	0.387	0.226
Custom Competition™	155gr. HPBT	2.640"	0.450	0.233

CASE TYPE:	Hornady		PRIMER TYPE	Fed. 210
CASE HOLDS:	48.5	Gr. WATER	BARREL Length/Make	24" Pac-Nor
			BARREL Twist	1-12"

POWDER TYPE	POWDER CHG. GRS.		MUZZLE VEL. F.P.S.		LOAD DENSITY (VOLUME)
Varget	46.0	MAX.	2855	**	103%
	44.0 *		2734		98%
	42.0		2615		94%
IMR 4320 Most Accurate Powder Tested	46.0	MAX.	2870	**	102%
	44.0		2719		98%
	42.0 *		2568		93%
A-2460	44.0	MAX.	2905		92%
	42.0		2801		87%
	40.0 *		2699		83%
TAC	45.5	MAX.	2920		96%
	43.5		2826		92%
	41.5 *		2734		87%
A-2520	48.0 *	MAX.	2985	**	102%
	46.0		2895		98%
	44.0		2797		94%

BC=Ballistic Coefficient SD=Sectional Density
*Most Accurate Load Tested **Compressed Load

Use Maximum Loads with Caution
Refer to page 73 for additional safety information

30 Thompson Center - 165/168 grain		MAXIMUM S.A.A.M.I. O.A.C.L.		2.660"
		TESTED O.A.C.L.	B.C.	S.D.
AccuBond®	165gr. Spitzer	2.640"	0.475	0.248
Ballistic Tip®	165gr. Spitzer	2.640"	0.475	0.248
Partition®	165gr. Spitzer	2.590"	0.410	0.248
Ballistic Tip®	168gr. Spitzer	2.640"	0.490	0.253
CT® Ballistic Silvertip®	168gr. Spitzer	2.640"	0.490	0.253
Custom Competition™	168gr. HPBT	2.640"	0.462	0.253
E-Tip®	168gr. Spitzer	2.600"	0.503	0.253

Due to internal construction differences, always begin with starting loads when using E-Tip® products.

CASE TYPE:	Hornady		PRIMER TYPE	Fed. 210
CASE HOLDS:	45.7	Gr. WATER	BARREL Length/Make	24" Pac-Nor
			BARREL Twist	1-12"

POWDER TYPE	POWDER CHG. GRS.		MUZZLE VEL. F.P.S.		LOAD DENSITY (VOLUME)
IMR 4320	45.5	MAX.	2756	**	107%
	43.5 *		2638	**	102%
	41.5		2521		98%
Varget	45.5	MAX.	2768	**	108%
	43.5		2667	**	103%
	41.5 *		2563		98%
TAC	43.5 *	MAX.	2770		97%
	41.5		2662		93%
	39.5		2550		88%
IMR 4895 Most Accurate Powder Tested	45.5	MAX.	2791	**	109%
	43.5 *		2681	**	105%
	41.5		2589		100%
A-2520	46.5 *	MAX.	2824	**	105%
	44.5		2752		100%
	42.5		2682		96%

BC=Ballistic Coefficient SD=Sectional Density
*Most Accurate Load Tested **Compressed Load

Use Maximum Loads with Caution
Refer to page 73 for additional safety information

30 Thompson Center - 175/180 grain

30 Thompson Center - 175/180 grain		MAXIMUM S.A.A.M.I. O.A.C.L.		2.660"
		TESTED O.A.C.L.	B.C.	S.D.
Custom Competition™	175gr. HPBT	2.640"	0.505	0.264
AccuBond®	180gr. Spitzer	2.640"	0.507	0.271
Ballistic Tip®	180gr. Spitzer	2.640"	0.507	0.271
CT® Ballistic Silvertip®	180gr. Spitzer	2.640"	0.507	0.271
E-Tip®	180gr. Spitzer	2.600"	0.523	0.271
Due to internal construction differences, always begin with starting loads when using E-Tip® products.				
Partition®	180gr. PPT	2.550"	0.361	0.271
Partition®	180gr. Spitzer	2.590"	0.474	0.271

CASE TYPE:	Hornady		PRIMER TYPE	Fed. 210
CASE HOLDS:	43.5	Gr. WATER	BARREL Length/Make	24" Pac-Nor
			BARREL Twist	1-12"

POWDER TYPE	POWDER CHG. GRS.		MUZZLE VEL. F.P.S.		LOAD DENSITY (VOLUME)
IMR 4320	44.0	MAX.	2648	**	109%
	42.0	*	2568	**	104%
	40.0		2490		99%
TAC	42.5	MAX.	2649		100%
	40.5		2573		95%
	38.5	*	2498		90%
IMR 4895	44.0	MAX.	2666	**	111%
	42.0		2588	**	106%
	40.0	*	2514	**	101%
Varget Most Accurate Powder Tested	44.0	MAX.	2667	**	109%
	42.0		2597	**	104%
	40.0	*	2526		99%
A-2520	44.5	MAX.	2722	**	105%
	42.5	*	2650	**	101%
	40.5		2580		96%

BC=Ballistic Coefficient SD=Sectional Density
*Most Accurate Load Tested **Compressed Load

Use Maximum Loads with Caution
Refer to page 73 for additional safety information

Bob Robb

308 WINCHESTER

In this day and age of high-tech this and that, how is it that the mundane, short-action 308 Winchester -- developed over 60 years ago -- remains one of the biggest sellers among hunters year in and year out?

The .308 is, with microscopic differences, the 7.62x51mm military round that Winchester introduced as a sporting round back in 1952 in a brilliant marketing move. Since shortly after its introduction, the .308 has been in the top five sporting cartridges sold in America, a distinction it retains today.

The reasons are many. First, there isn't a popular rifle maker that doesn't chamber at least one of its bolt guns for the .308. In addition, there are lots of pumps and semiautos, including many built on the AR platform, as well as some lever-actions also chambered for the .308. Also, all the ammunition makers offer a huge array of .308 ammo with a wide range of bullet weights, styles, and designs, too, and you can find it in many a backwoods hardware store should the need arise while traveling. It also doesn't cost an arm and a leg. Handloaders will find .308 brass easy to find, inexpensive, while accurate and efficient loads can be worked up with a wide array of different powders.

The real reason, though, is simply the fact that the .308 is extremely accurate, recoils lightly, and puts an immediate hurt to all but the stoutest of big game animals in North America. I've used mine on everything from deer and pronghorn to black bears and wild hogs and never regretted packing it along on any big game hunting trip. While you can load it with bullets ranging from 110-200 grains, the .308 really shines on big game when using bullets in the 150-168-grain range. For deer-sized game my little .308 seems to prefer the 165-grain bullets, and here I like both the Nosler Ballistic Tip® and Ballistic Silvertip®, both of which group tightly. For bigger stuff game hunting I like the 165- or 180-grain Nosler AccuBond® and Partition®.

Cartridges may come and go, but I am willing to bet that your grandchildren will find the little 308 Winchester as popular when they are your age as it is today. It's that good.

Bob Robb

Bob is the editor of Whitetail Journal and Predator Xtreme magazines.

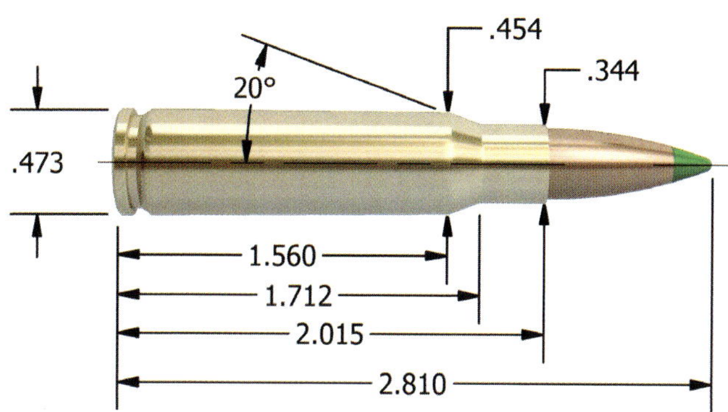

Maximum S.A.A.M.I. Overall Cartridge Length: 2.810"

BULLET CHOICES FOR THE 308 WINCHESTER

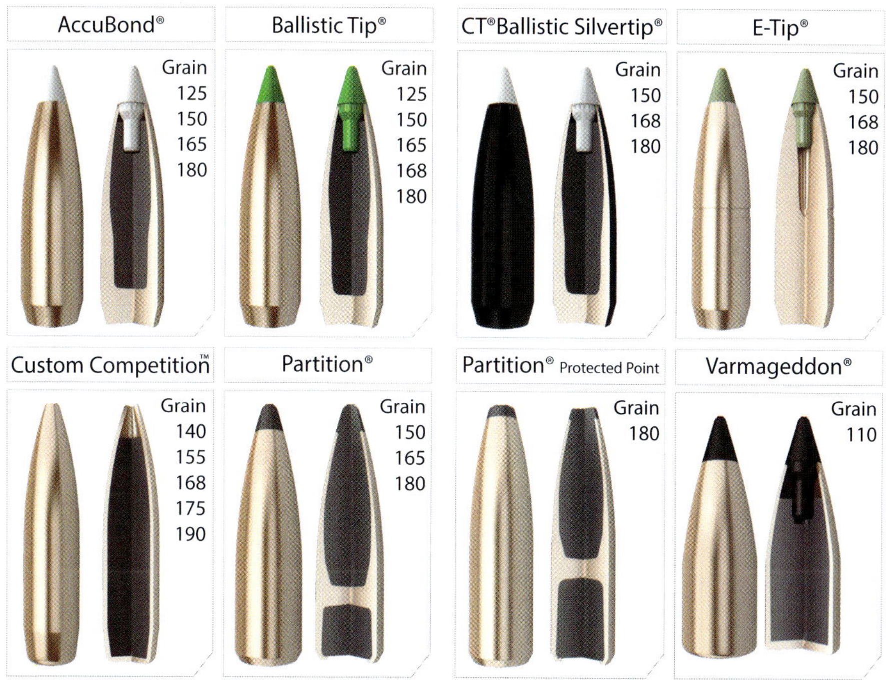

TECHNICAL INFORMATION

The 308 Winchester is an inherently accurate cartridge, and is at its best with medium-to-fast burning powders like Varget and BL-C2. Note that if you are loading heavy bullets with slow powders, you may have difficulty seating the bullet deep enough to clear the rifling or fit in the magazine of your rifle. While most large rifle primers work well in the .308, we had our best results with the Federal 210M. The loads listed here were developed using standard commercial brass. Military brass has less case capacity because of its heavier construction. This results in a smaller combustion chamber, and yields higher pressures. We recommend caution when using military brass, and suggest starting at or below the minimum loads listed. Additionally, if your brass originally had crimped primers, this crimp MUST be removed prior to re-priming. If you will be loading for a semi-auto and find that crimping is necessary, a taper crimp is suggested

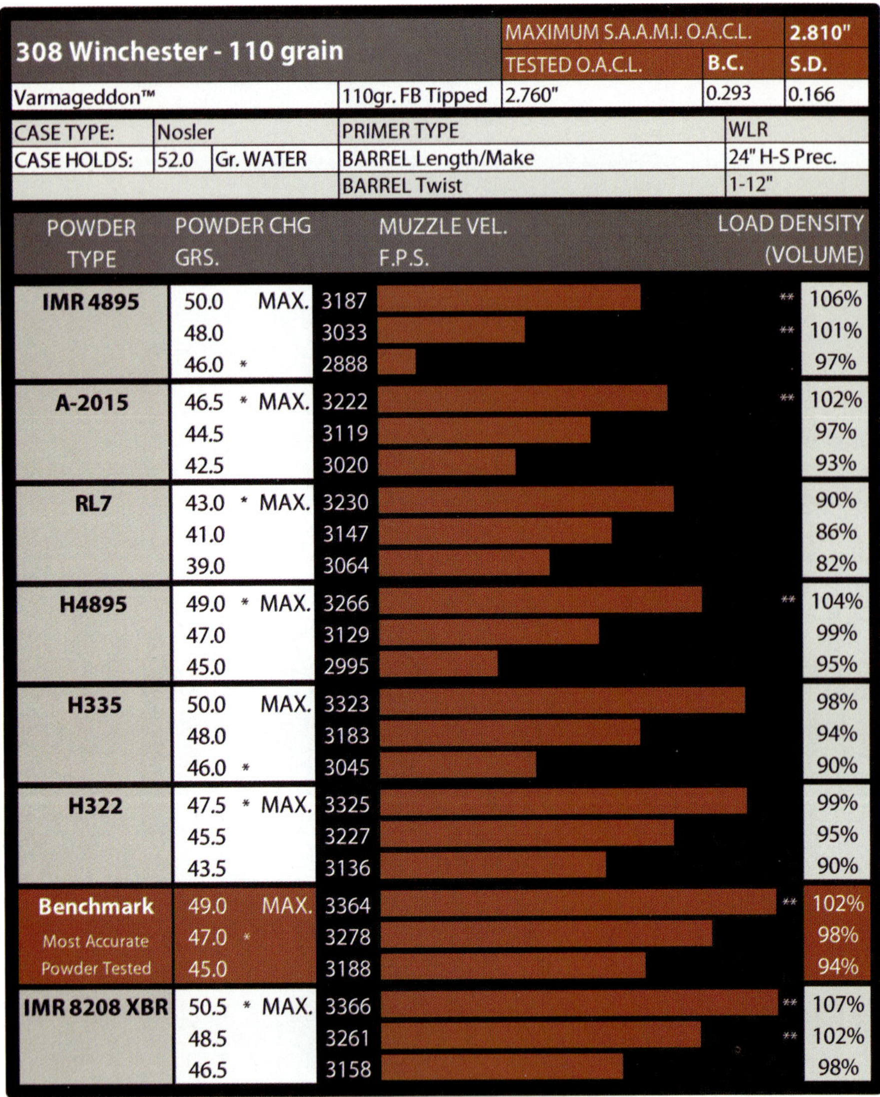

308 Winchester - 110 grain		MAXIMUM S.A.A.M.I. O.A.C.L.		2.810"
		TESTED O.A.C.L.	B.C.	S.D.
Varmageddon™	110gr. FB Tipped	2.760"	0.293	0.166

CASE TYPE:	Nosler		PRIMER TYPE	WLR
CASE HOLDS:	52.0	Gr. WATER	BARREL Length/Make	24" H-S Prec.
			BARREL Twist	1-12"

POWDER TYPE	POWDER CHG GRS.		MUZZLE VEL. F.P.S.		LOAD DENSITY (VOLUME)
IMR 4895	50.0	MAX.	3187	**	106%
	48.0		3033	**	101%
	46.0 *		2888		97%
A-2015	46.5 *	MAX.	3222	**	102%
	44.5		3119		97%
	42.5		3020		93%
RL7	43.0 *	MAX.	3230		90%
	41.0		3147		86%
	39.0		3064		82%
H4895	49.0 *	MAX.	3266	**	104%
	47.0		3129		99%
	45.0		2995		95%
H335	50.0	MAX.	3323		98%
	48.0		3183		94%
	46.0 *		3045		90%
H322	47.5 *	MAX.	3325		99%
	45.5		3227		95%
	43.5		3136		90%
Benchmark Most Accurate Powder Tested	49.0	MAX.	3364	**	102%
	47.0 *		3278		98%
	45.0		3188		94%
IMR 8208 XBR	50.5 *	MAX.	3366	**	107%
	48.5		3261	**	102%
	46.5		3158		98%

BC=Ballistic Coefficient SD=Sectional Density
*Most Accurate Load Tested **Compressed Load

Use Maximum Loads with Caution
Refer to page 73 for additional safety information

308 Winchester - 125 grain

308 Winchester - 125 grain		MAXIMUM S.A.A.M.I. O.A.C.L.		2.810"
		TESTED O.A.C.L.	B.C.	S.D.
AccuBond®	125gr. Spitzer	2.810"	0.366	0.188
Ballistic Tip®	125gr. Spitzer	2.810"	0.366	0.188

CASE TYPE:	Nosler		PRIMER TYPE	Fed 210M
CASE HOLDS:	51.5	Gr. WATER	BARREL Length/Make	24" Lilja
			BARREL Twist	1-10"

POWDER TYPE	POWDER CHG. GRS.		MUZZLE VEL. F.P.S.		LOAD DENSITY (VOLUME)
H335	46.5	* MAX.	2990		92%
	44.5		2870		88%
	42.5		2750		84%
IMR 4895	45.5	* MAX.	3010		97%
	43.5		2840		93%
	41.5		2670		89%
IMR 4320	48.0	* MAX.	3048		100%
	46.0		2883		96%
	44.0		2718		92%
RL7	41.5	* MAX.	3088		88%
	39.5		2953		83%
	37.5		2818		79%
Viht N135	46.5	* MAX.	3088	**	105%
	44.5		2979		100%
	42.5		2870		96%
Varget	50.0	* MAX.	3128	**	105%
	48.0		3039	**	101%
	46.0		2863		97%
A-2460	48.0	MAX.	3150		94%
	46.0		3040		90%
	44.0	*	2930		86%
Viht N140	49.5	* MAX.	3155	**	112%
	47.5		3045	**	107%
	45.5		2934	**	103%
W748 Most Accurate Powder Tested	51.5	* MAX.	3214	**	104%
	49.5		3143		100%
	47.5		3073		96%
TAC	50.0	MAX.	3284		99%
	48.0	*	3173		95%
	46.0		3061		91%

BC=Ballistic Coefficient SD=Sectional Density
*Most Accurate Load Tested **Compressed Load

Use Maximum Loads with Caution
Refer to page 73 for additional safety information

308 Winchester - 140 grain		MAXIMUM S.A.A.M.I. O.A.C.L.		2.810"
		TESTED O.A.C.L.	B.C.	S.D.
Custom Competition™	140gr. HPBT	2.750"	0.396	0.211

CASE TYPE:	Nosler		PRIMER TYPE	WLR
CASE HOLDS:	52.0	Gr. WATER	BARREL Length/Make	24" H-S Prec.
			BARREL Twist	1-12"

POWDER TYPE	POWDER CHG GRS.		MUZZLE VEL. F.P.S.		LOAD DENSITY (VOLUME)
W748	48.5	MAX.	2958		97%
	46.5		2833		93%
	44.5	*	2743		89%
A-2495 Most Accurate Powder Tested	46.5	MAX.	2986	**	102%
	44.5	*	2856		97%
	42.5		2735		93%
TAC	46.5	MAX.	2989		91%
	44.5		2866		87%
	42.5	*	2773		83%
AR-Comp	46.0	MAX.	2994	**	101%
	44.0	*	2923		96%
	42.0		2780		92%
IMR 4895	47.0	MAX.	3005		99%
	45.0	*	2886		95%
	43.0		2753		91%
IMR 8208 XBR	47.0	* MAX.	3012		99%
	45.0		2867		95%
	43.0		2775		91%
RL15	49.0	MAX.	3021		98%
	47.0	*	2909		94%
	45.0		2788		90%
A-2520	48.0	MAX.	3022		95%
	46.0	*	2926		91%
	44.0		2813		87%
Varget	49.0	MAX.	3036	**	102%
	47.0	*	2917		98%
	45.0		2790		94%

BC=Ballistic Coefficient SD=Sectional Density
*Most Accurate Load Tested **Compressed Load

Use Maximum Loads with Caution
Refer to page 73 for additional safety information

308 Winchester - 150/155 grain		MAXIMUM S.A.A.M.I. O.A.C.L.		2.810"
		TESTED O.A.C.L.	B.C.	S.D.
AccuBond®	150gr. Spitzer	2.800"	0.435	0.226
Ballistic Tip®	150gr. Spitzer	2.800"	0.435	0.226
CT® Ballistic Silvertip®	150gr. Spitzer	2.800"	0.435	0.226
E-Tip®	150gr. Spitzer	2.750"	0.469	0.226
Due to internal construction differences, always begin with starting loads when using E-Tip® products.				
Partition®	150gr. Spitzer	2.770"	0.387	0.226
Custom Competition™	155gr. HPBT	2.800"	0.450	0.233

CASE TYPE:	Nosler		PRIMER TYPE	Fed 210M
CASE HOLDS:	49.5	Gr. WATER	BARREL Length/Make	24" Lilja
			BARREL Twist	1-10"

POWDER TYPE	POWDER CHG. GRS.		MUZZLE VEL. F.P.S.		LOAD DENSITY (VOLUME)
IMR 4895	44.5	MAX.	2802		99%
	42.5		2673		94%
	40.5 *		2544		90%
AR-Comp	44.0	MAX.	2830	**	101%
	42.0		2701		96%
	40.0 *		2580		92%
IMR 4320	47.0 *	MAX.	2842	**	102%
	45.0		2727		98%
	43.0		2612		93%
Viht N135	45.0 *	MAX.	2863	**	106%
	43.0		2740	**	101%
	41.0		2617		96%
IMR 3031	45.0	MAX.	2880	**	103%
	43.0		2740		98%
	41.0 *		2600		94%
Viht N140	46.5	MAX.	2887	**	109%
Most Accurate	44.5		2766	**	105%
Powder Tested	42.5 *		2645		100%
Varget	46.5	MAX.	2895	**	102%
	44.5 *		2771		97%
	42.5		2664		93%
IMR 4064	48.0	MAX.	2920	**	107%
	46.0		2830	**	102%
	44.0 *		2740		98%
RL15	46.0	MAX.	2958		97%
	44.0		2843		93%
	42.0 *		2728		88%
TAC	46.0	MAX.	2997		95%
	44.0		2881		91%
	42.0 *		2772		87%

BC=Ballistic Coefficient SD=Sectional Density
*Most Accurate Load Tested **Compressed Load

Use Maximum Loads with Caution
Refer to page 73 for additional safety information

308 Winchester - 165/168 grain		MAXIMUM S.A.A.M.I. O.A.C.L.		2.810"
		TESTED O.A.C.L.	B.C.	S.D.
AccuBond®	165gr. Spitzer	2.800"	0.475	0.248
Ballistic Tip®	165gr. Spitzer	2.800"	0.475	0.248
Partition®	165gr. Spitzer	2.770"	0.410	0.248
Ballistic Tip®	168gr. Spitzer	2.800"	0.490	0.253
CT ® Ballistic Silvertip®	168gr. Spitzer	2.800"	0.490	0.253
Custom Competition™	168gr. HPBT	2.800"	0.462	0.253
E-Tip®	168gr. Spitzer	2.750"	0.503	0.253

Due to internal construction differences, always begin with starting loads when using E-Tip® products.

CASE TYPE:	Nosler		PRIMER TYPE	Fed 210M
CASE HOLDS:	48.3	Gr. WATER	BARREL Length/Make	24" Lilja
			BARREL Twist	1-10"

POWDER TYPE	POWDER CHG. GRS.		MUZZLE VEL. F.P.S.		LOAD DENSITY (VOLUME)
AR-Comp	42.5	MAX.	2682		100%
	40.5 *		2588		95%
	38.5		2421		91%
Viht N140	44.5	MAX.	2695	**	107%
	42.5		2580	**	102%
	40.5 *		2465		98%
BL-C2 Most Accurate Powder Tested	46.5 *	MAX.	2698		98%
	44.5		2583		94%
	42.5		2468		89%
IMR 4064	44.5 *	MAX.	2700	**	101%
	42.5		2630		97%
	40.5		2560		92%
IMR 4895	43.0	MAX.	2708		98%
	41.0		2560		93%
	39.0 *		2412		89%
IMR 3031	43.0	MAX.	2760	**	101%
	41.0		2620		96%
	39.0 *		2480		91%
IMR 4350	50.0	MAX.	2792	**	110%
	48.0		2647	**	105%
	46.0 *		2502	**	101%
RL15	44.0 *	MAX.	2820		95%
	42.0		2740		91%
	40.0		2660		86%
Varget	46.0	MAX.	2820	**	103%
	44.0 *		2758		98%
	42.0		2662		94%

BC=Ballistic Coefficient SD=Sectional Density
*Most Accurate Load Tested **Compressed Load

Use Maximum Loads with Caution
Refer to page 73 for additional safety information

308 Winchester - 175/180 grain

308 Winchester - 175/180 grain		MAXIMUM S.A.A.M.I. O.A.C.L.		2.810"
		TESTED O.A.C.L.	B.C.	S.D.
Custom Competition™	175gr. HPBT	2.800"	0.505	0.264
AccuBond®	180gr. Spitzer	2.800"	0.507	0.271
Ballistic Tip®	180gr. Spitzer	2.800"	0.507	0.271
CT® Ballistic Silvertip®	180gr. Spitzer	2.800"	0.507	0.271
E-Tip®	180gr. Spitzer	2.750"	0.523	0.271
Due to internal construction differences, always begin with starting loads when using E-Tip® products.				
Partition®	180gr. Spitzer	2.770	0.474	0.271
Partition®	180gr. PPT	2.740"	0.361	0.271

CASE TYPE:	Nosler		PRIMER TYPE	WLR
CASE HOLDS:	48.6	Gr. WATER	BARREL Length/Make	24" H-S Prec.
			BARREL Twist	1-12"

POWDER TYPE	POWDER CHG. GRS.		MUZZLE VEL. F.P.S.		LOAD DENSITY (VOLUME)
AR-Comp	41.0	MAX.	2537		96%
	39.0		2417		91%
	37.0 *		2308		87%
IMR 4895 Most Accurate Powder Tested	41.5	MAX.	2559		94%
	39.5 *		2462		89%
	37.5		2311		85%
IMR 8208 XBR	42.0	MAX.	2571		95%
	40.0		2457		90%
	38.0 *		2376		86%
Varget	43.5 *	MAX.	2589		97%
	41.5		2494		92%
	39.5		2341		88%
IMR 4007 SSC	48.0	MAX.	2593	**	106%
	46.0 *		2525	**	101%
	44.0		2384		97%
IMR 4064	43.0	MAX.	2603		97%
	41.0		2433		93%
	39.0 *		2297		88%
A-2520	43.5 *	MAX.	2610		92%
	41.5		2527		88%
	39.5		2435		84%
W748	44.0 *	MAX.	2610		94%
	42.0		2510		90%
	40.0		2386		86%
RL15	44.0 *	MAX.	2617		94%
	42.0		2480		90%
	40.0		2342		86%
W760	48.0	MAX.	2617	**	103%
	46.0 *		2524		99%
	44.0		2412		95%

BC=Ballistic Coefficient SD=Sectional Density
*Most Accurate Load Tested **Compressed Load

Use Maximum Loads with Caution
Refer to page 73 for additional safety information

308 Win - 190 grain		MAXIMUM S.A.A.M.I. O.A.C.L.		2.810"
		TESTED O.A.C.L.	B.C.	S.D.
Custom Competition™	190gr. HPBT	2.800"	0.530	0.286

CASE TYPE:	Nosler		PRIMER TYPE	WLR
CASE HOLDS:	47.0	Gr. WATER	BARREL Length/Make	24" H-S Prec.
			BARREL Twist	1-12"

POWDER TYPE	POWDER CHG. GRS.		MUZZLE VEL. F.P.S.		LOAD DENSITY (VOLUME)
H335	40.0	MAX.	2448		86%
	38.0		2299		82%
	36.0	*	2238		78%
AR-Comp Most Accurate Powder Tested	40.0	MAX.	2491		97%
	38.0		2369		92%
	36.0	*	2256		87%
IMR 4064	42.0	MAX.	2538		98%
	40.0	*	2432		94%
	38.0		2300		89%
A-2520	42.0	MAX.	2551		92%
	40.0	*	2460		88%
	38.0		2360		83%
IMR 4895	42.5	MAX.	2571		99%
	40.5	*	2460		95%
	38.5		2374		90%
Varget	43.0	MAX.	2572		99%
	41.0		2446		94%
	39.0	*	2351		90%
IMR 4007 SSC	47.0	MAX.	2591	**	107%
	45.0	*	2486	**	103%
	43.0		2350		98%
W760	47.0	MAX.	2604	**	105%
	45.0	*	2471		100%
	43.0		2376		96%
2000-MR	46.5	MAX.	2624		100%
	44.5		2520		96%
	42.5	*	2400		91%

BC=Ballistic Coefficient SD=Sectional Density
*Most Accurate Load Tested **Compressed Load

Use Maximum Loads with Caution
Refer to page 73 for additional safety information

30-06 SPRINGFIELD

The 30-06 cartridge was conceived in 1900, when engineers at Springfield Armory began work on a battle rifle to replace the 30-40 Krag-Jorgensen. Their prototype emerged in 1901. Two years later the Model 1903 Springfield appeared. Its 30-caliber rimless cartridge headspaced on the shoulder, like the 8x57 Mauser. A 220-grain bullet at 2,300 fps made the 30-03 a ballistic match for the 8x57 with a 236-grain bullet at 2,125.

A year after the 30-03's debut, Germany switched to a new 154-grain 8mm spitzer at 2,800 fps. The Americans countered with the Ball Cartridge, Caliber .30, Model 1906. It spat a 150-grain bullet at 2,700 fps. Then someone decided to shorten the case .07", to .494." All 30-03 rifles were recalled for rechambering.

The first 30-06 bullets were jacketed with an alloy of 85 percent copper and 15 percent nickel. It did not hold up at 30-06 velocities, and fouling rendered rifles inaccurate. Tin plating reduced fouling, but over time, tin "cold-soldered" to case mouths and could cause pressure spikes. An alloy of zinc and copper, 5-95 or 10-90, solved this problem. It became known as gilding metal.

World War I demonstrated that bullets of high ballistic coefficient increased effective range, so the Army replaced the 150-grain 30-06 bullet with a 173-grain spitzer at 2,646 fps. In 1939, the Army adopted the M-2 load. With a 152-grain spitzer at 2,805 fps, it saw U.S. troops through World War II. At the same time, it established an enviable reputation on the target range. Riflemen began calling it the "all-around cartridge." During its first 25 years, Winchester's Model 70 was sold in 18 chamberings but more than a third of rifles shipped (208,218) were 30-06s! Well-credentialed in all the world's game fields, the 30-06 will handily take any North American animal and all but a few huge beasts abroad. With 180-grain bullets, the 30-06 is a near match, ballistically, for heavy-bullet loads in the 7mm Remington Magnum. Given identical bullets and 200-yard zeros, the '06 hits within 2 vertical inches of the 300 Win. Mag. at 300 yards. It is chambered in more rifles than any other round save, now, the 308 Win. You can buy ought-six ammo in remote outposts, from the Canadian North to sub-Saharan Africa. Such ubiquity endears this cartridge to any traveler who has lost luggage.

Planning a hunt for Alaskan moose and Dall's sheep, I didn't have to look past a converted Springfield in 30-06, with iron sights. It downed both animals cleanly – performance that's distinguished this cartridge for more than a century!

Wayne van Zwoll has published 14 books and several thousand magazine articles on rifles, shooting and big game hunting.

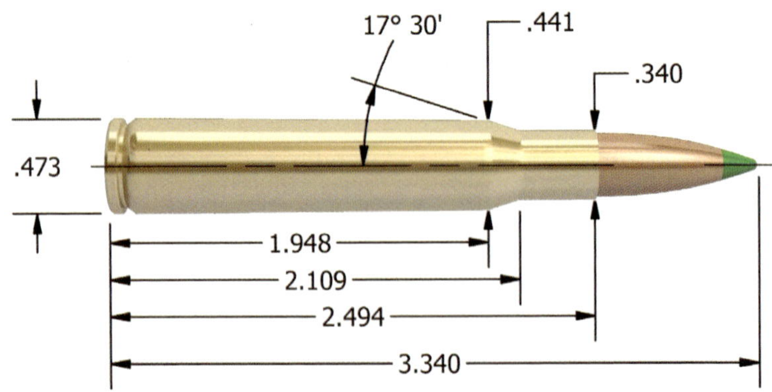

Maximum S.A.A.M.I. Overall Cartridge Length: 3.340"

BULLET CHOICES FOR THE 30-06 SPRINGFIELD

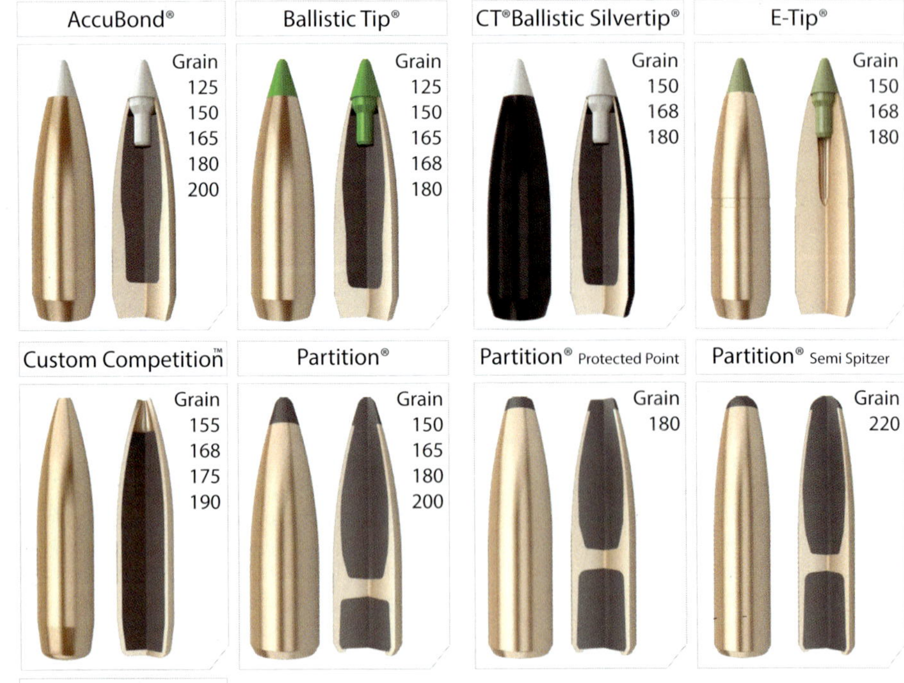

Varmageddon™

Grain
110

TECHNICAL INFORMATION

The loads listed here were developed using standard commercial brass. Military brass has less case capacity because of its heavier construction, resulting in a smaller combustion chamber, and higher pressures. We recommend caution when using military brass, and suggest starting at or below the minimum loads listed. Additionally, if your brass originally had crimped primers, this crimp MUST be removed prior to re-priming.

30-06 Springfield - 110 grain		MAXIMUM S.A.A.M.I. O.A.C.L.		3.340"
		TESTED O.A.C.L.	B.C.	S.D.
Varmageddon™	110gr. Spitzer	3.130"	0.293	0.166

CASE TYPE:	Nosler		PRIMER TYPE	WLR
CASE HOLDS:	65.1	Gr. WATER	BARREL Length/Make	24" H-S Prec.
			BARREL Twist	1-10"

POWDER TYPE	POWDER CHG. GRS.		MUZZLE VEL. F.P.S.	LOAD DENSITY (VOLUME)
Benchmark	54.0	MAX.	3402	90%
	52.0		3333	86%
	50.0 *		3263	83%
RL15 Most Accurate Powder Tested	60.0	MAX.	3419	96%
	58.0 *		3297	93%
	56.0		3180	90%
IMR 4895	58.5	MAX.	3428	99%
	56.5		3295	95%
	54.5 *		3162	92%
Varget	59.0 *	MAX.	3448	98%
	57.0		3363	95%
	55.0		3266	91%

BC=Ballistic Coefficient SD=Sectional Density
*Most Accurate Load Tested **Compressed Load

Use Maximum Loads with Caution
Refer to page 73 for additional safety information

30-06 Springfield - 125 grain

		MAXIMUM S.A.A.M.I. O.A.C.L.		3.340"
		TESTED O.A.C.L.	B.C.	S.D.
AccuBond®	125gr. Spitzer	3.320"	0.366	0.188
Ballistic Tip®	125gr. Spitzer	3.320"	0.366	0.188

CASE TYPE:	Nosler		PRIMER TYPE	WLR
CASE HOLDS:	67.7	Gr. WATER	BARREL Length/Make	24" Lilja
			BARREL Twist	1-10"

POWDER TYPE	POWDER CHG. GRS.		MUZZLE VEL. F.P.S.	LOAD DENSITY (VOLUME)
IMR 4831	61.5 *	MAX.	3112	97%
	59.5		3027	93%
	57.5		2942	90%
H414	60.0	MAX.	3138	92%
	58.0		3033	89%
	56.0 *		2928	86%
IMR 4350	60.5	MAX.	3169	95%
	58.5		3069	91%
	56.5 *		2943	88%
W760	61.0	MAX.	3179	94%
	59.0		3094	91%
	57.0 *		3009	88%
H4895	52.5	MAX.	3200	85%
	50.5		3110	82%
	48.5 *		3020	79%
Viht N140	55.0	MAX.	3246	94%
	53.0		3153	91%
	51.0 *		3069	88%
Big Game	61.0	MAX.	3249	95%
	59.0 *		3198	92%
	57.0		3097	89%
Viht N150	57.0	MAX.	3309	99%
	55.0		3215	96%
	53.0 *		3122	92%
RL15 Most Accurate Powder Tested	56.0	MAX.	3337	86%
	54.0		3217	83%
	52.0 *		3136	80%
Varget	57.5	MAX.	3418	92%
	55.5 *		3267	89%
	53.5		3171	85%

BC=Ballistic Coefficient SD=Sectional Density
*Most Accurate Load Tested **Compressed Load

Use Maximum Loads with Caution
Refer to page 73 for additional safety information

30-06 Springfield - 150/155 grain		MAXIMUM S.A.A.M.I. O.A.C.L.		3.340"
		TESTED O.A.C.L.	B.C.	S.D.
AccuBond®	150gr. Spitzer	3.320"	0.435	0.226
Ballistic Tip®	150gr. Spitzer	3.320"	0.435	0.226
CT® Ballistic Silvertip®	150gr. Spitzer	3.320"	0.435	0.226
E-Tip®	150gr. Spitzer	3.300"	0.469	0.226
Due to internal construction differences, always begin with starting loads when using E-Tip® products.				
Partition®	150gr. Spitzer	3.320"	0.387	0.226
Custom Competition™	155gr. HPBT	3.320"	0.450	0.233

CASE TYPE:	Nosler		PRIMER TYPE	WLR
CASE HOLDS:	62.7	Gr. WATER	BARREL Length/Make	24" Lilja
			BARREL Twist	1-10"

POWDER TYPE	POWDER CHG. GRS.	MUZZLE VEL. F.P.S.	LOAD DENSITY (VOLUME)
IMR 4831	59.0 * MAX.	2862	100%
	57.0	2757	97%
	55.0	2652	93%
IMR 4064	52.0 * MAX.	2908	91%
	50.0	2793	88%
	48.0	2678	84%
W760	57.0 * MAX.	2912	95%
	55.0	2797	92%
	53.0	2682	89%
IMR 4895	51.0 * MAX.	2922	89%
	49.0	2837	86%
	47.0	2752	82%
Viht N150	54.0 MAX.	2939	** 101%
	52.0	2848	98%
	50.0 *	2756	94%
IMR 4320	53.0 MAX.	2940	91%
	51.0	2850	87%
	49.0 *	2760	84%
RL19 Most Accurate Powder Tested	61.5 * MAX.	2982	** 107%
	59.5	2881	** 103%
	57.5	2776	100%
IMR 4350	59.0 * MAX.	3000	100%
	57.0	2880	96%
	55.0	2760	93%
Big Game	58.0 MAX.	3056	97%
	56.0	2979	94%
	54.0 *	2912	91%

BC=Ballistic Coefficient SD=Sectional Density
*Most Accurate Load Tested **Compressed Load

Use Maximum Loads with Caution
Refer to page 73 for additional safety information

30-06 Springfield - 165/168 grain

		MAXIMUM S.A.A.M.I. O.A.C.L.		3.340"
		TESTED O.A.C.L.	B.C.	S.D.
AccuBond®	165gr. Spitzer	3.320"	0.475	0.248
Ballistic Tip®	165gr. Spitzer	3.320"	0.475	0.248
Partition®	165gr. Spitzer	3.320"	0.410	0.248
Ballistic Tip®	168gr. Spitzer	3.320"	0.490	0.253
CT® Ballistic Silvertip®	168gr. Spitzer	3.320"	0.490	0.253
Custom Competition™	168gr. HPBT	3.320"	0.462	0.253
E-Tip®	168gr. Spitzer	3.300"	0.503	0.253

Due to internal construction differences, always begin with starting loads when using E-Tip® products.

CASE TYPE:	Nosler		PRIMER TYPE	WLR
CASE HOLDS:	60.2	Gr. WATER	BARREL Length/Make	24" Lilja
			BARREL Twist	1-10"

POWDER TYPE	POWDER CHG. GRS.		MUZZLE VEL. F.P.S.	LOAD DENSITY (VOLUME)
IMR 4895	48.0	MAX.	2720	88%
	46.0		2640	84%
	44.0 *		2560	80%
IMR 4320	50.0	MAX.	2730	89%
	48.0		2620	86%
	46.0 *		2510	82%
Viht N160	58.0	MAX.	2767	** 107%
	56.0		2664	** 103%
	54.0 *		2560	100%
Viht N150	52.0	MAX.	2802	** 102%
	50.0		2717	98%
	48.0 *		2632	94%
IMR 4831	58.0	MAX.	2820	** 102%
	56.0		2710	99%
	54.0 *		2600	95%
IMR 4350 Most Accurate Powder Tested	57.0 *	MAX.	2832	100%
	55.0		2717	97%
	53.0		2602	93%
Big Game	55.5	MAX.	2859	97%
	53.5		2792	94%
	51.5 *		2725	90%
H4350	57.5 *	MAX.	2872	** 101%
	55.5		2757	98%
	53.5		2642	94%
W760	56.0	MAX.	2882	97%
	54.0		2792	94%
	52.0 *		2720	90%
RL22	63.0	MAX.	3002	** 114%
	61.0		2907	** 110%
	59.0 *		2812	** 107%

BC=Ballistic Coefficient SD=Sectional Density
*Most Accurate Load Tested **Compressed Load

Use Maximum Loads with Caution
Refer to page 73 for additional safety information

30-06 Springfield - 175/180 grain

30-06 Springfield - 175/180 grain		MAXIMUM S.A.A.M.I. O.A.C.L.		3.340"
		TESTED O.A.C.L.	B.C.	S.D.
Custom Competition™	175gr. HPBT	3.330"	0.505	0.264
AccuBond®	180gr. Spitzer	3.330"	0.507	0.271
Ballistic Tip®	180gr. Spitzer	3.330"	0.507	0.271
CT® Ballistic Silvertip®	180gr. Spitzer	3.330"	0.507	0.271
E-Tip®	180gr. Spitzer	3.300"	0.523	0.271
Due to internal construction differences, always begin with starting loads when using E-Tip® products.				
Partition®	180gr. Spitzer	3.310"	0.474	0.271
Partition®	180gr. PPT	3.210"	0.361	0.271

CASE TYPE:	Nosler		PRIMER TYPE	WLR
CASE HOLDS:	63.9	Gr. WATER	BARREL Length/Make	24" H-S Prec.
			BARREL Twist	1-10"

POWDER TYPE	POWDER CHG. GRS.		MUZZLE VEL. F.P.S.		LOAD DENSITY (VOLUME)
IMR 4007 SSC	53.5	MAX.	2700		90%
	51.5		2622		86%
	49.5 *		2497		83%
IMR 4350	56.5 *	MAX.	2734		94%
	54.5		2636		90%
	52.5		2528		87%
RL19	60.0 *	MAX.	2742	**	102%
	58.0		2677		99%
	56.0		2576		95%
4000-MR	57.5	MAX.	2750		95%
	55.5		2665		91%
	53.5 *		2543		88%
H4831	62.0	MAX.	2754	**	101%
	60.0 *		2683		98%
	58.0		2589		95%
Hybrid 100V	56.0 *	MAX.	2767		90%
	54.0		2702		87%
	52.0		2613		84%
H4350	56.5	MAX.	2769		94%
	54.5		2679		90%
	52.5 *		2560		87%
W760	55.5 *	MAX.	2773		91%
	53.5		2686		88%
	51.5		2597		84%
RL22 Most Accurate Powder Tested	62.0 *	MAX.	2812	**	105%
	60.0		2748	**	102%
	58.0		2641		99%
Hunter	59.0	MAX.	2825		97%
	57.0 *		2764		94%
	55.0		2654		91%

BC=Ballistic Coefficient SD=Sectional Density
*Most Accurate Load Tested **Compressed Load

Use Maximum Loads with Caution
Refer to page 73 for additional safety information

30-06 Springfield - 190 grain

		MAXIMUM S.A.A.M.I. O.A.C.L.		3.340"
		TESTED O.A.C.L.	B.C.	S.D.
Custom Competition™	190gr. HPBT	3.330"	0.530	0.286

CASE TYPE:	Nosler		PRIMER TYPE	WLR
CASE HOLDS:	62.7	Gr. WATER	BARREL Length/Make	24" H-S Prec.
			BARREL Twist	1-10"

POWDER TYPE	POWDER CHG. GRS.		MUZZLE VEL. F.P.S.	LOAD DENSITY (VOLUME)
MAGPRO	59.5	MAX.	2560	98%
	57.5		2487	95%
	55.5 *		2393	91%
H4831	57.0	MAX.	2613	95%
	55.0		2494	91%
	53.0 *		2370	88%
IMR 4350 Most Accurate Powder Tested	53.5	MAX.	2620	90%
	51.5		2520	87%
	49.5 *		2394	84%
Hybrid 100V	52.0	MAX.	2620	85%
	50.0 *		2543	82%
	48.0		2445	79%
4000-MR	55.0 *	MAX.	2653	92%
	53.0		2558	89%
	51.0		2433	86%
IMR 7828 SSC	59.5	MAX.	2658	98%
	57.5		2604	95%
	55.5 *		2475	91%
H4350	53.5	MAX.	2661	90%
	51.5 *		2567	87%
	49.5		2468	84%
RL19	56.5	MAX.	2668	98%
	54.5 *		2558	94%
	52.5		2428	91%
W760	53.5	MAX.	2683	89%
	51.5 *		2566	86%
	49.5		2434	83%
Supreme 780	59.5	MAX.	2697	99%
	57.5 *		2589	96%
	55.5		2507	93%

BC=Ballistic Coefficient SD=Sectional Density
*Most Accurate Load Tested **Compressed Load

Use Maximum Loads with Caution
Refer to page 73 for additional safety information

30-06 Springfield - 200 grain

		MAXIMUM S.A.A.M.I. O.A.C.L.		3.340"
		TESTED O.A.C.L.	B.C.	S.D.
AccuBond®	200gr. Spitzer	3.320"	0.588	0.301
Partition®	200gr. Spitzer	3.320"	0.481	0.301

CASE TYPE:	Nosler		PRIMER TYPE	WLR
CASE HOLDS:	59.4	Gr. WATER	BARREL Length/Make	24" Lilja
			BARREL Twist	1-10"

POWDER TYPE	POWDER CHG. GRS.		MUZZLE VEL. F.P.S.		LOAD DENSITY (VOLUME)
IMR 4320	47.0 *	MAX.	2438		85%
	45.0		2343		81%
	43.0		2248		78%
IMR 4350	53.0	MAX.	2528		94%
	51.0		2443		91%
	49.0 *		2358		87%
IMR 4831	55.0	MAX.	2558		99%
	53.0		2463		95%
	51.0 *		2368		91%
H414	53.0 *	MAX.	2560		93%
	51.0		2450		89%
	49.0		2340		86%
W760	52.0 *	MAX.	2568		92%
	50.0		2453		88%
	48.0		2338		85%
Hunter	54.5	MAX.	2591		97%
	52.5 *		2532		93%
	50.5		2462		89%
RL19 Most Accurate Powder Tested	56.0	MAX.	2620	**	102%
	54.0		2520		99%
	52.0 *		2420		95%
Viht N160	57.0	MAX.	2632	**	107%
	55.0		2544	**	103%
	53.0 *		2457		99%
RL22	58.0	MAX.	2688	**	106%
	56.0		2603	**	102%
	54.0 *		2518		99%

BC=Ballistic Coefficient SD=Sectional Density
*Most Accurate Load Tested **Compressed Load

Use Maximum Loads with Caution
Refer to page 73 for additional safety information

30-06 Springfield - 220 grain		MAXIMUM S.A.A.M.I. O.A.C.L.		3.340"
		TESTED O.A.C.L.	B.C.	S.D.
Partition®	220gr. Semi Spitz	3.320"	0.351	0.331

CASE TYPE:	Nosler		PRIMER TYPE	WLR
CASE HOLDS:	61.9	Gr. WATER	BARREL Length/Make	24" Lilja
			BARREL Twist	1-10"

POWDER TYPE	POWDER CHG. GRS.		MUZZLE VEL. F.P.S.	LOAD DENSITY (VOLUME)
W760	48.0	MAX.	2398	81%
	46.0		2296	78%
	44.0 *		2214	74%
Viht N165	56.5	MAX.	2431	** 101%
	54.5		2354	98%
	52.5 *		2275	94%
H4831 Most Accurate Powder Tested	55.0	MAX.	2478	93%
	53.0		2382	89%
	51.0 *		2287	86%
Hunter	54.5	MAX.	2487	93%
	52.5 *		2433	89%
	50.5		2350	86%
IMR 4350	51.5 *	MAX.	2488	88%
	49.5		2400	85%
	47.5		2312	81%
Viht N160	55.5 *	MAX.	2503	100%
	53.5		2415	96%
	51.5		2327	92%
IMR 7828	56.0	MAX.	2507	96%
	54.0		2412	93%
	52.0 *		2316	89%
RL19	55.0	MAX.	2507	97%
	53.0		2416	93%
	51.0 *		2325	90%
RL22	57.0	MAX.	2602	100%
	55.0		2512	97%
	53.0 *		2423	93%

BC=Ballistic Coefficient SD=Sectional Density
*Most Accurate Load Tested **Compressed Load

Use Maximum Loads with Caution
Refer to page 73 for additional safety information

300 HOLLAND & HOLLAND (H&H) MAGNUM

The 300 Holland & Holland was introduced in 1925 by the English firm bearing its name but it did not gain much attention among hunters in America until Ben Comfort used it to win the prestigious 1000-yard Wimbledon Cup at Camp Perry in 1935. Two years later, Winchester offered it in a new rifle called the Model 70. About that same time, Western Cartridge Company started loading the .300 with a 180-grain bullet at 3060 fps which was darned speedy when we consider that in those days the 30-06 was rated at 2700 fps with that same weight bullet.

I became hooked on the 300 H&H quite early in life, but my most memorable hunt with it did not take place until 2006. The Nosler company was nearing its 60th anniversary and since a Winchester Model 70 in 300 H&H Magnum was John Nosler's favorite rifle back when he founded the company, I decided that was what I should use. As his way of icing the cake, a friend who obviously knew people in high places at the Nosler factory managed to round up a few 180-grain Partitions hand-made by John Nosler during the 1940s. One look had convinced me I would be lucky at best, to coax minute-of-moose accuracy from them, but groups on paper measuring less than an inch proved me wrong. During that hunt in the Yukon, I passed on moose and grizzly simply because nothing I saw inside my binocular was better than what I already had back home but the mountain caribou I bagged with a single 60-year-old Partition®, now that's another story entirely.

I don't care what they say about short and fat being the way to go, the 300 H&H has long been, and always will be my favorite magnum cartridge of its caliber. I just counted on one hand the world's big-game animals this grand old cartridge might not be big enough medicine to handle and darned if I didn't have several fingers left over.

Layne Simpson

Layne is Senior Field Editor for Shooting Times, RifleShooter and Guns & Ammo, and Senior Writer for BuckMasters GunHunter. He has also authored books on rifles, shotguns, handguns and hunting.

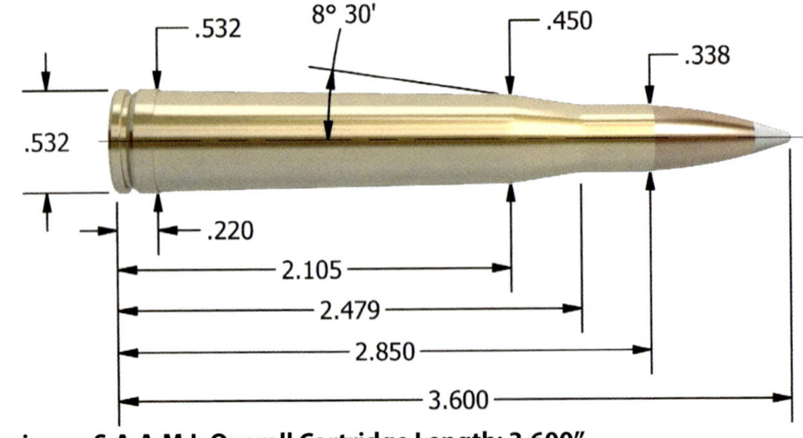

Maximum S.A.A.M.I. Overall Cartridge Length: 3.600"

BULLET CHOICES FOR THE 300 HOLLAND & HOLLAND (H&H) MAGNUM

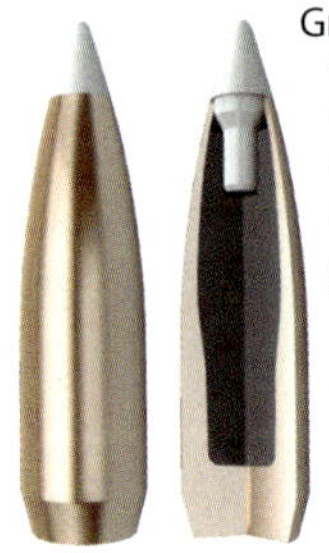

AccuBond®

Grain
125
150
165
180
200

Ballistic Tip®

Grain
125
150
165
168
180

CT® Ballistic Silvertip®

Grain
150
168
180

E-Tip®

Grain
150
168
180

Custom Competition™

Grain
155
168
175
190

Partition®

Grain
150
165
180
200

Partition® Protected Point

Grain
180

Partition® Semi Spitzer

Grain
220

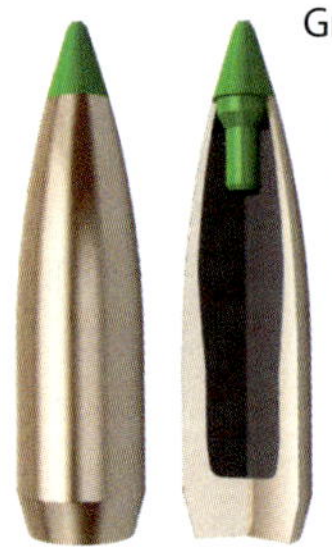

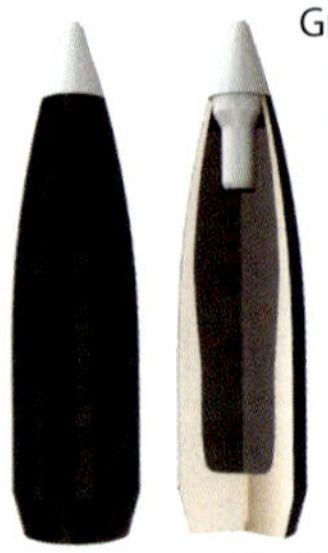

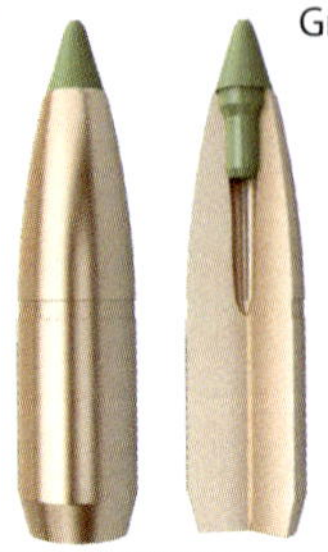

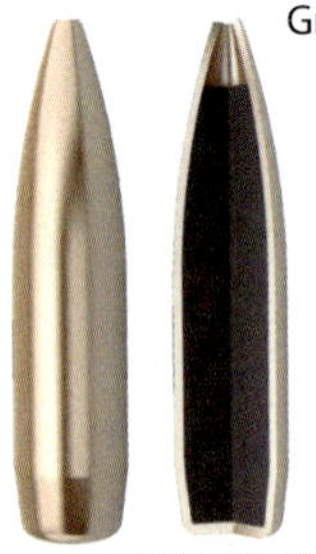

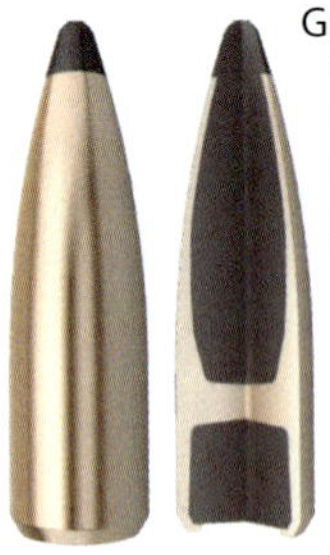

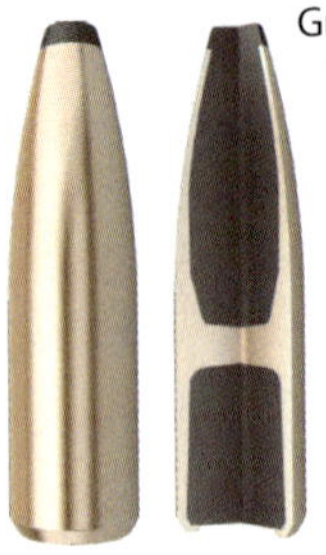

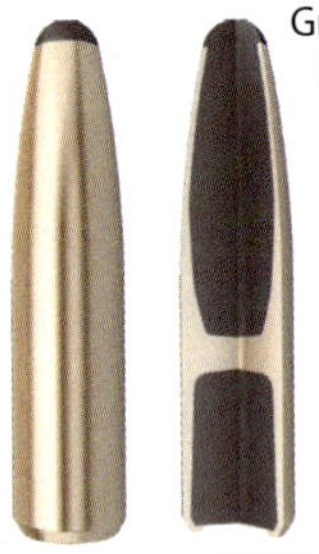

300 H&H Magnum - 125 grain		MAXIMUM S.A.A.M.I. O.A.C.L.		3.600"
		TESTED O.A.C.L.	B.C.	S.D.
AccuBond®	125gr. Spitzer	3.580"	0.366	0.188
Ballistic Tip®	125gr. Spitzer	3.580"	0.366	0.188

CASE TYPE:	Nosler		PRIMER TYPE	CCI 250
CASE HOLDS:	82.6	Gr. WATER	BARREL Length/Make	24" Lilja
			BARREL Twist	1-10"

POWDER TYPE	POWDER CHG. GRS.		MUZZLE VEL. F.P.S.		LOAD DENSITY (VOLUME)
H414	66.0	MAX.	3252		83%
Most Accurate Powder Tested	64.0		3157		81%
	62.0 *		3062		78%
IMR 4831	74.0 *	MAX.	3348		95%
	72.0		3253		93%
	70.0		3158		90%
IMR 4895	63.0	MAX.	3350		84%
	61.0		3240		81%
	59.0 *		3130		78%
IMR 4350	73.0 *	MAX.	3400		94%
	71.0		3270		91%
	69.0		3140		88%
Viht N160	74.5 *	MAX.	3414		100%
	72.5		3329		98%
	70.5		3243		95%
H4350	73.0	MAX.	3438		94%
	71.0		3333		91%
	69.0 *		3228		88%
H4831SC	81.0	MAX.	3451	**	102%
	79.0 *		3383		100%
	77.0		3315		97%
RL19	82.0	MAX.	3520	**	108%
	80.0 *		3446	**	105%
	78.0		3373	**	103%
Viht N550	74.0	MAX.	3590		100%
	72.0		3528		97%
	70.0 *		3396		94%

BC=Ballistic Coefficient SD=Sectional Density
*Most Accurate Load Tested **Compressed Load

Use Maximum Loads with Caution
Refer to page 73 for additional safety information

300 H&H Magnum - 150/155 grain		MAXIMUM S.A.A.M.I. O.A.C.L.		3.600"
		TESTED O.A.C.L.	B.C.	S.D.
AccuBond®	150gr. Spitzer	3.580"	0.435	0.226
Ballistic Tip®	150gr. Spitzer	3.580"	0.435	0.226
CT® Ballistic Silvertip®	150gr. Spitzer	3.580"	0.435	0.226
E-Tip®	150gr. Spitzer	3.550"	0.469	0.226
Due to internal construction differences, always begin with starting loads when using E-Tip® products.				
Partition®	150gr. Spitzer	3.580"	0.387	0.226
Custom Competition™	155gr. HPBT	3.580"	0.450	0.233

CASE TYPE:	Nosler		PRIMER TYPE	WLRM
CASE HOLDS:	81.0	Gr. WATER	BARREL Length/Make	24" Lilja
			BARREL Twist	1-10"

POWDER TYPE	POWDER CHG. GRS.		MUZZLE VEL. F.P.S.		LOAD DENSITY (VOLUME)
RL19	73.0	MAX.	3194		98%
	71.0		3091		95%
	69.0 *		2987		93%
IMR 4064	63.5 *	MAX.	3235		86%
	61.5		3143		83%
	59.5		3058		81%
Viht N160	72.5	MAX.	3255		99%
	70.5		3176		97%
	68.5 *		3096		94%
IMR 4831	71.5	MAX.	3259		94%
	69.5		3167		91%
	67.5 *		3071		89%
H414	65.5	MAX.	3270		84%
	63.5		3170		82%
	61.5 *		3070		79%
H4831SC	77.0 *	MAX.	3290		99%
	75.0		3224		96%
	73.0		3146		94%
IMR 4350 Most Accurate Powder Tested	70.0	MAX.	3300		91%
	68.0		3200		89%
	66.0 *		3100		86%
Viht N560	76.0	MAX.	3301	**	104%
	74.0 *		3202	**	102%
	72.0		3102		99%
RL25	80.5	MAX.	3315	**	108%
	78.5 *		3231	**	105%
	76.5		3154	**	103%

BC=Ballistic Coefficient SD=Sectional Density
*Most Accurate Load Tested **Compressed Load

Use Maximum Loads with Caution
Refer to page 73 for additional safety information

300 H&H Magnum - 165/168 grain

300 H&H Magnum - 165/168 grain		MAXIMUM S.A.A.M.I. O.A.C.L.		3.600"
		TESTED O.A.C.L.	B.C.	S.D.
AccuBond®	165gr. Spitzer	3.580"	0.475	0.248
Ballistic Tip®	165gr. Spitzer	3.580"	0.475	0.248
Partition®	165gr. Spitzer	3.580"	0.410	0.248
Ballistic Tip®	168gr. Spitzer	3.580"	0.490	0.253
CT® Ballistic Silvertip®	168gr. Spitzer	3.580"	0.490	0.253
Custom Competition™	168gr. HPBT	3.580"	0.462	0.253
E-Tip®	168gr. Spitzer	3.550"	0.503	0.253

Due to internal construction differences, always begin with starting loads when using E-Tip® products.

CASE TYPE:	Nosler		PRIMER TYPE	WLRM
CASE HOLDS:	79.6	Gr. WATER	BARREL Length/Make	24" Lilja
			BARREL Twist	1-10"

POWDER TYPE	POWDER CHG. GRS.		MUZZLE VEL. F.P.S.		LOAD DENSITY (VOLUME)
IMR 4320	59.0	MAX.	2940		80%
	57.0		2840		77%
	55.0 *		2740		74%
W760	66.5	MAX.	3072		87%
	64.5 *		3047		85%
	62.5		2969		82%
RL19	71.0	MAX.	3099		97%
	69.0		2989		94%
	67.0 *		2879		91%
H380	67.0 *	MAX.	3100		89%
	65.0		3000		86%
	63.0		2900		83%
IMR 4350	67.0 *	MAX.	3118		89%
	65.0		3033		86%
	63.0		2948		84%
RL22	72.5	MAX.	3133		99%
	70.5		3043		96%
	68.5 *		2953		94%
Viht N560	73.5	MAX.	3135	**	103%
	71.5		3065		100%
	69.5 *		2969		97%
IMR 4831	70.0	MAX.	3148		94%
	68.0		3083		91%
	66.0 *		3018		88%
RL25	76.5 *	MAX.	3150	**	104%
	74.5		3025	**	102%
	72.5		2973		99%
H4350	70.0 *	MAX.	3168		93%
Most Accurate	68.0		3073		90%
Powder Tested	66.0		2978		88%

BC=Ballistic Coefficient SD=Sectional Density
*Most Accurate Load Tested **Compressed Load

Use Maximum Loads with Caution
Refer to page 73 for additional safety information

300 H&H Magnum - 175/180 grain		MAXIMUM S.A.A.M.I. O.A.C.L.		3.600"
		TESTED O.A.C.L.	B.C.	S.D.
Custom Competition™	175gr. HPBT	3.580"	0.505	0.264
AccuBond®	180gr. Spitzer	3.580"	0.507	0.271
Ballistic Tip®	180gr. Spitzer	3.580"	0.507	0.271
CT® Ballistic Silvertip®	180gr. Spitzer	3.580"	0.507	0.271
E-Tip®	180gr. Spitzer	3.550"	0.523	0.271
Due to internal construction differences, always begin with starting loads when using E-Tip® products.				
Partition®	180gr. Spitzer	3.580"	0.474	0.271
Partition®	180gr. PPT	3.550"	0.361	0.271

CASE TYPE:	Nosler		PRIMER TYPE	CCI 250
CASE HOLDS:	77.6	Gr. WATER	BARREL Length/Make	24" Lilja
			BARREL Twist	1-10"

POWDER TYPE	POWDER CHG. GRS.		MUZZLE VEL. F.P.S.	LOAD DENSITY (VOLUME)
H380	62.5	MAX.	2862	85%
	60.5		2767	82%
	58.5 *		2672	79%
H4831SC	69.0	MAX.	2874	93%
	67.0		2798	90%
	65.0 *		2697	87%
Viht N165	70.0 *	MAX.	2918	100%
	68.0		2841	97%
	66.0		2752	95%
IMR 4831 Most Accurate Powder Tested	66.0	MAX.	2940	90%
	64.0		2880	88%
	62.0 *		2820	85%
RL19	69.0 *	MAX.	2958	97%
	67.0		2867	94%
	65.0		2776	91%
Viht N560	69.0	MAX.	2991	99%
	67.0 *		2883	96%
	65.0		2804	93%
RL25	72.5 *	MAX.	3000	** 102%
	70.5		2890	99%
	68.5		2813	96%
IMR 4350	65.0 *	MAX.	3010	89%
	63.0		2950	86%
	61.0		2890	83%
RL22	71.0	MAX.	3023	99%
	69.0		2928	97%
	67.0 *		2834	94%

BC=Ballistic Coefficient SD=Sectional Density
*Most Accurate Load Tested **Compressed Load

Use Maximum Loads with Caution
Refer to page 73 for additional safety information

300 H&H Magnum - 190 grain

		MAXIMUM S.A.A.M.I. O.A.C.L.		3.600"
		TESTED O.A.C.L.	B.C.	S.D.
Custom Competition™	190gr. HPBT	3.580"	0.530	0.286

CASE TYPE:	Nosler		PRIMER TYPE	WLRM
CASE HOLDS:	77.2	Gr. WATER	BARREL Length/Make	24" H-S Prec.
			BARREL Twist	1-10"

POWDER TYPE	POWDER CHG. GRS.	MUZZLE VEL. F.P.S.	LOAD DENSITY (VOLUME)
H4350	63.0 * MAX.	2821	86%
	61.0	2740	84%
	59.0	2654	81%
IMR 4831	65.5 MAX.	2824	90%
	63.5 *	2712	88%
	61.5	2601	85%
Viht N165	68.5 MAX.	2826	99%
	66.5	2730	96%
	64.5 *	2634	93%
Viht N560	68.0 MAX.	2852	98%
	66.0 *	2749	95%
	64.0	2640	92%
RL25	72.0 MAX.	2865	** 101%
	70.0	2788	99%
	68.0 *	2710	96%
RL19	67.0 MAX.	2874	94%
	65.0	2774	92%
	63.0 *	2672	89%
RL22 Most Accurate Powder Tested	68.0 MAX.	2894	96%
	66.0	2790	93%
	64.0 *	2687	90%

BC=Ballistic Coefficient SD=Sectional Density
*Most Accurate Load Tested **Compressed Load

Use Maximum Loads with Caution
Refer to page 73 for additional safety information

300 H&H Magnum - 200 grain		MAXIMUM S.A.A.M.I. O.A.C.L.		3.600"
		TESTED O.A.C.L.	B.C.	S.D.
AccuBond®	200gr. Spitzer	3.580"	0.588	0.301
Partition®	200gr. Spitzer	3.580"	0.481	0.301

CASE TYPE:	Nosler		PRIMER TYPE	CCI 250
CASE HOLDS:	77.0	Gr. WATER	BARREL Length/Make	24" Lilja
			BARREL Twist	1-10"

POWDER TYPE	POWDER CHG. GRS.		MUZZLE VEL. F.P.S.	LOAD DENSITY (VOLUME)
H4350	62.5 *	MAX.	2771	86%
	60.5		2697	83%
	58.5		2624	80%
H4831SC Most Accurate Powder Tested	67.0 *	MAX.	2791	91%
	65.0		2706	88%
	63.0		2665	85%
Viht N550	61.0 *	MAX.	2793	88%
	59.0		2718	85%
	57.0		2607	82%
RL19	65.5 *	MAX.	2796	92%
	63.5		2719	90%
	61.5		2643	87%
Viht N560	66.0	MAX.	2815	95%
	64.0		2718	92%
	62.0 *		2641	89%
IMR 4350	63.0	MAX.	2820	87%
	61.0		2760	84%
	59.0 *		2700	81%
IMR 4831	65.0	MAX.	2822	90%
	63.0		2747	87%
	61.0 *		2672	84%
RL22	67.0 *	MAX.	2827	95%
	65.0		2744	92%
	63.0		2662	89%
RL25	69.5	MAX.	2847	98%
	67.5		2761	95%
	65.5 *		2670	92%

BC=Ballistic Coefficient SD=Sectional Density
*Most Accurate Load Tested **Compressed Load

Use Maximum Loads with Caution
Refer to page 73 for additional safety information

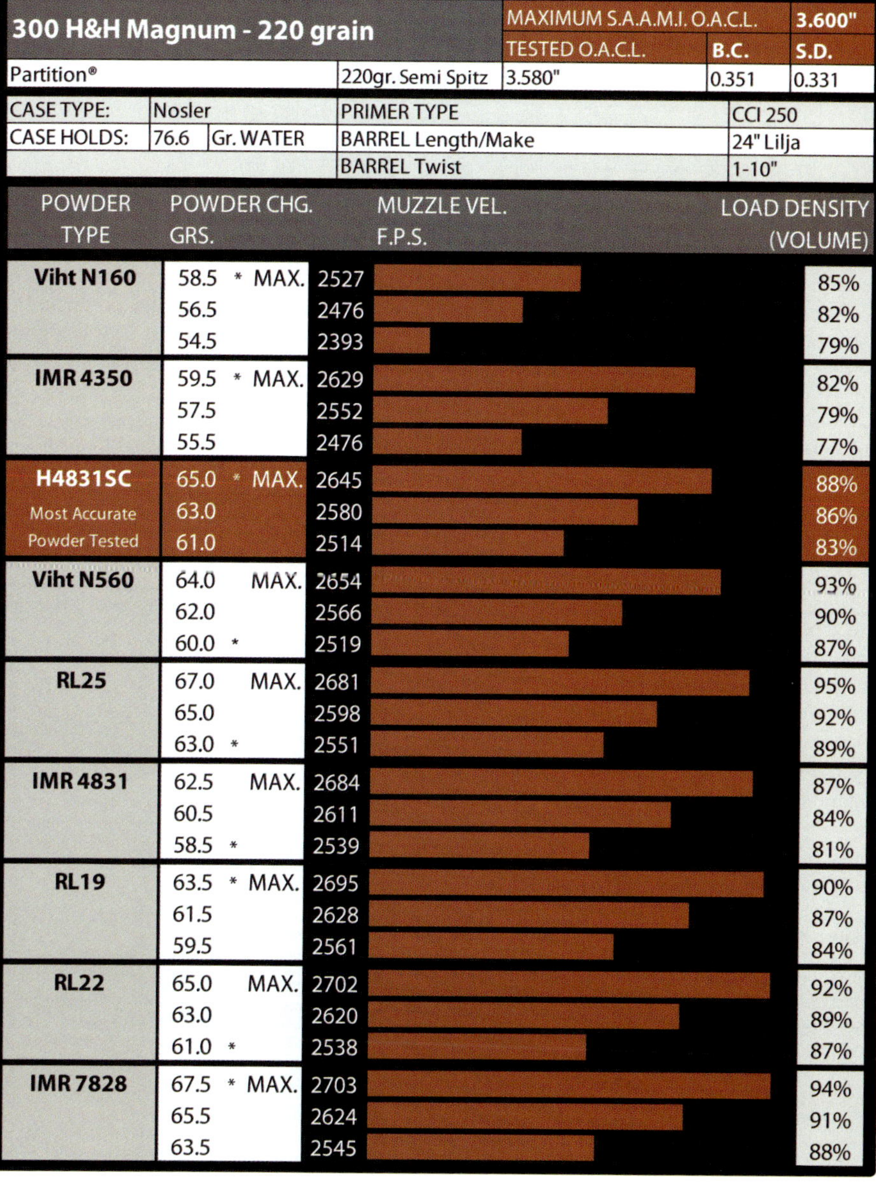

300 H&H Magnum - 220 grain		MAXIMUM S.A.A.M.I. O.A.C.L.		3.600"
		TESTED O.A.C.L.	B.C.	S.D.
Partition®	220gr. Semi Spitz	3.580"	0.351	0.331

CASE TYPE:	Nosler		PRIMER TYPE	CCI 250
CASE HOLDS:	76.6	Gr. WATER	BARREL Length/Make	24" Lilja
			BARREL Twist	1-10"

POWDER TYPE	POWDER CHG. GRS.	MUZZLE VEL. F.P.S.	LOAD DENSITY (VOLUME)
Viht N160	58.5 * MAX.	2527	85%
	56.5	2476	82%
	54.5	2393	79%
IMR 4350	59.5 * MAX.	2629	82%
	57.5	2552	79%
	55.5	2476	77%
H4831SC Most Accurate Powder Tested	65.0 * MAX.	2645	88%
	63.0	2580	86%
	61.0	2514	83%
Viht N560	64.0 MAX.	2654	93%
	62.0	2566	90%
	60.0 *	2519	87%
RL25	67.0 MAX.	2681	95%
	65.0	2598	92%
	63.0 *	2551	89%
IMR 4831	62.5 MAX.	2684	87%
	60.5	2611	84%
	58.5 *	2539	81%
RL19	63.5 * MAX.	2695	90%
	61.5	2628	87%
	59.5	2561	84%
RL22	65.0 MAX.	2702	92%
	63.0	2620	89%
	61.0 *	2538	87%
IMR 7828	67.5 * MAX.	2703	94%
	65.5	2624	91%
	63.5	2545	88%

BC=Ballistic Coefficient SD=Sectional Density
*Most Accurate Load Tested **Compressed Load

Use Maximum Loads with Caution
Refer to page 73 for additional safety information

James Brion

300 WINCHESTER MAGNUM

Any excuse to buy a new gun is a good excuse, so many of us have become collectors of calibers. Each one has its place in our own world of shooting and hunting. However, there is one cartridge that we all shoot the best--the one we shoot the most! Yes, there is truth behind the adage "beware the man who has only one gun….". Since its introduction in 1963, long range competition shooters and sport hunters alike have recognized this and embraced the versatile and accurate 300 Winchester Magnum.

As a teenager in the 1980s I was mentored in the art of reloading for ethical, long-range hunting by a patient old gunsmith who operated out of his garage in the sandhills of Nebraska. Al's first lesson to me was to spend the extra money and put the best bullet up front, the Nosler Partition. From there we developed loads for my 270 Win and later "seven mag" which served me well as a successful open country deer hunter and outfitter.

Back then, I did not know that the day would come when I would need to seek out a new "one gun" for hunting larger quarry around the world. Unlike my favorite deer rifles, this "new" gun would need to possess down range energy rated at the top of the list. I hated the thought of having to reprogram myself to master a new trajectory though. Big game hunting is won or lost in those critical, split second moments when muscle memory takes over and causes you to do the same thing that you have practiced for years. I was delighted to learn that the 300 Win Mag with a 180-grain Partition had similar trajectory to my favorite flat-shooting deer rifles, while delivering down range energy sufficient for any ungulate under say a ton, plus most of the carnivores.

Since then I have taken many of the North American 29 and numerous species on other continents with this cartridge. The 300 Win Mag has modest recoil, yet energy to drop a 1500 pound moose, accuracy to confidently take a mountain sheep and the flat trajectory you need for a pronghorn; all with readily available ammo. There might be a more perfect cartridge for any one of these species, but there isn't a more perfect one for them all.

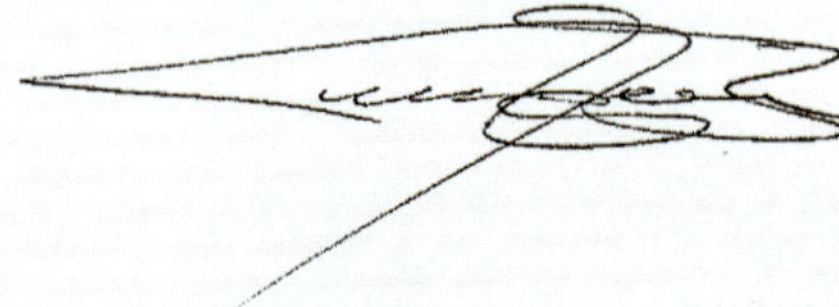

James Brion, Outfitter, Hunting Consultant, Television Producer and Host

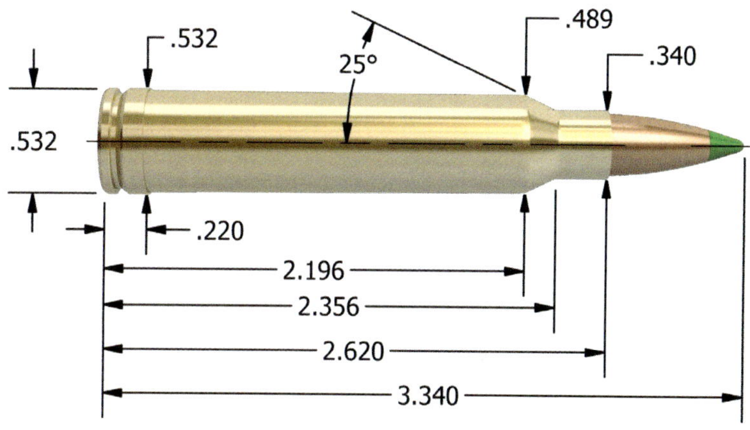

Maximum S.A.A.M.I. Overall Cartridge Length: 3.340"

BULLET CHOICES FOR THE 300 WINCHESTER MAGNUM

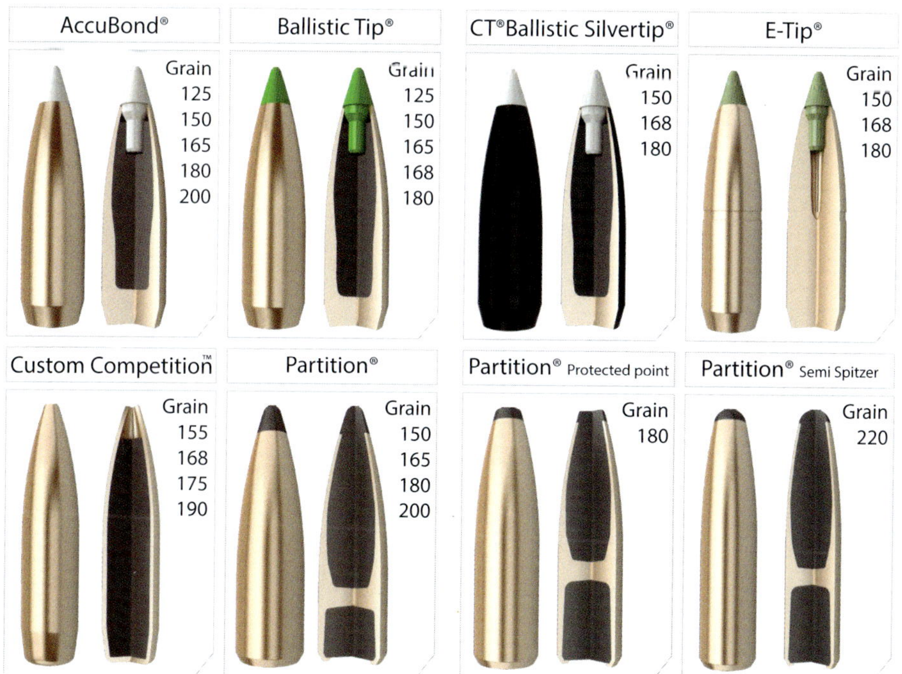

TECHNICAL INFORMATION

The 300 Win Mag's short case neck is not detrimental to accuracy, and it is actually capable of near bench-rest like groups with the proper load in a high quality rifle. If deformation of the lead nose in the magazine box is a problem, our 180-grain Protected Point Partition or AccuBond bullets are good options.

300 Winchester Mag - 125 grain

		MAXIMUM S.A.A.M.I. O.A.C.L.		3.340"
		TESTED O.A.C.L.	B.C.	S.D.
AccuBond®	125gr. Spitzer	3.340"	0.366	0.188
Ballistic Tip®	125gr. Spitzer	3.340"	0.366	0.188

CASE TYPE:	Nosler		PRIMER TYPE	Fed 215
CASE HOLDS:	85.8	Gr. WATER	BARREL Length/Make	24" Lilja
			BARREL Twist	1-10"

POWDER TYPE	POWDER CHG. GRS.		MUZZLE VEL. F.P.S.	LOAD DENSITY (VOLUME)
IMR 4350 Most Accurate Powder Tested	75.0 *	MAX.	3350	93%
	73.0		3220	90%
	71.0		3090	88%
Viht N150	67.5	MAX.	3377	93%
	65.5		3309	90%
	63.5 *		3241	87%
H380	72.5	MAX.	3390	89%
	70.5		3280	86%
	68.5 *		3170	84%
Viht N160	77.5	MAX.	3406	100%
	75.5		3312	98%
	73.5 *		3217	95%
RL15	68.0 *	MAX.	3424	83%
	66.0		3310	80%
	64.0		3196	78%
IMR 4895	67.5	MAX.	3468	86%
	65.5		3363	84%
	63.5 *		3258	81%
H414	74.5 *	MAX.	3472	90%
	72.5		3337	88%
	70.5		3202	86%
RL19	81.0	MAX.	3500	** 103%
	79.0		3405	100%
	77.0 *		3309	98%
H4831SC	84.0 *	MAX.	3521	** 102%
	82.0		3438	100%
	80.0		3362	97%

BC=Ballistic Coefficient SD=Sectional Density
*Most Accurate Load Tested **Compressed Load

Use Maximum Loads with Caution
Refer to page 73 for additional safety information

300 Winchester Mag - 150/155 grain		MAXIMUM S.A.A.M.I. O.A.C.L.		3.340"
		TESTED O.A.C.L.	B.C.	S.D.
AccuBond®	150gr. Spitzer	3.340"	0.435	0.226
Ballistic Tip®	150gr. Spitzer	3.340"	0.435	0.226
CT® Ballistic Silvertip®	150gr. Spitzer	3.340"	0.435	0.226
E-Tip®	150gr. Spitzer	3.310"	0.469	0.226
Due to internal construction differences, always begin with starting loads when using E-Tip® products.				
Partition®	150gr. Spitzer	3.340"	0.387	0.226
Custom Competition™	155gr. HPBT	3.340"	0.450	0.233

CASE TYPE:	Nosler		PRIMER TYPE	Fed 215
CASE HOLDS:	81.7	Gr. WATER	BARREL Length/Make	24" Lilja
			BARREL Twist	1-10"

POWDER TYPE	POWDER CHG. GRS.		MUZZLE VEL. F.P.S.		LOAD DENSITY (VOLUME)
Viht N160	75.5	MAX.	3232	**	103%
	73.5		3142		100%
	71.5 *		3053		97%
H1000	84.0 *	MAX.	3237	**	108%
Most Accurate	82.0		3157	**	105%
Powder Tested	80.0		3076	**	103%
H414	71.0	MAX.	3260		91%
	69.0		3188		88%
	67.0 *		3124		85%
IMR 4320	64.0	MAX.	3260		84%
	62.0		3190		82%
	60.0 *		3120		79%
Viht N165	81.5	MAX.	3266	**	111%
	79.5		3191	**	108%
	77.5 *		3115	**	105%
RL19	76.5 *	MAX.	3272	**	102%
	74.5		3157		99%
	72.5		3042		96%
H4831SC	81.0	MAX.	3318	**	103%
	79.0		3224	**	101%
	77.0 *		3155		98%
IMR 4831	76.0 *	MAX.	3390		99%
	74.0		3320		96%
	72.0		3250		94%
IMR 4350	74.0	MAX.	3420		96%
	72.0		3330		93%
	70.0 *		3240		91%

BC=Ballistic Coefficient SD=Sectional Density
*Most Accurate Load Tested **Compressed Load

Use Maximum Loads with Caution
Refer to page 73 for additional safety information

300 Winchester Mag - 165/168 grain

Bullet	Type	MAXIMUM S.A.A.M.I. O.A.C.L. / TESTED O.A.C.L.	B.C.	S.D.
		MAXIMUM S.A.A.M.I. O.A.C.L.		3.340"
AccuBond®	165gr. Spitzer	3.340"	0.475	0.248
Ballistic Tip®	165gr. Spitzer	3.340"	0.475	0.248
Partition®	165gr. Spitzer	3.340"	0.410	0.248
Ballistic Tip®	168gr. Spitzer	3.340"	0.490	0.253
CT® Ballistic Silvertip®	168gr. Spitzer	3.340"	0.490	0.253
Custom Competition™	168gr. HPBT	3.340"	0.462	0.253
E-Tip®	168gr. Spitzer	3.310"	0.503	0.253

Due to internal construction differences, always begin with starting loads when using E-Tip® products.

CASE TYPE:	Nosler		PRIMER TYPE	Fed 215
CASE HOLDS:	81.1	Gr. WATER	BARREL Length/Make	24" Lilja
			BARREL Twist	1-10"

POWDER TYPE	POWDER CHG. GRS.		MUZZLE VEL. F.P.S.		LOAD DENSITY (VOLUME)
W760	67.0 *	MAX.	3026		87%
	65.0		2935		84%
	63.0		2866		81%
Viht N160	72.5	MAX.	3091		99%
	70.5		3009		97%
	68.5 *		2926		94%
IMR 4320	62.0	MAX.	3118		82%
	60.0		3043		80%
	58.0 *		2968		77%
H414	69.0	MAX.	3140		89%
	67.0		3080		86%
	65.0 *		2990		83%
H4350	73.0	MAX.	3170		95%
	71.0		3070		93%
	69.0 *		2970		90%
RL19	75.0	MAX.	3192	**	101%
Most Accurate Powder Tested	73.0		3087		98%
	71.0 *		2982		95%
IMR 7828	80.0	MAX.	3192	**	105%
	78.0		3107	**	102%
	76.0 *		3022		100%
IMR 4350	72.0	MAX.	3260		94%
	70.0		3170		91%
	68.0 *		3080		89%
IMR 4831	74.0	MAX.	3260		97%
	72.0		3190		94%
	70.0 *		3120		92%
RL22	79.0	MAX.	3290	**	106%
	77.0		3200	**	103%
	75.0 *		3110	**	101%

BC=Ballistic Coefficient SD=Sectional Density
*Most Accurate Load Tested **Compressed Load

Use Maximum Loads with Caution
Refer to page 73 for additional safety information

Nosler®

300 Winchester Mag - 175/180 grain

Bullet	Type	MAXIMUM S.A.A.M.I. O.A.C.L. / TESTED O.A.C.L.	B.C.	3.340" / S.D.
Custom Competition™	175gr. HPBT	3.330"	0.505	0.264
AccuBond®	180gr. Spitzer	3.330"	0.507	0.271
Ballistic Tip®	180gr. Spitzer	3.330"	0.507	0.271
CT® Ballistic Silvertip®	180gr. Spitzer	3.330"	0.507	0.271
E-Tip®	180gr. Spitzer	3.300"	0.523	0.271
Due to internal construction differences, always begin with starting loads when using E-Tip® products.				
Partition®	180gr. Spitzer	3.310"	0.474	0.271
Partition®	180gr. PPT	3.210"	0.361	0.271

CASE TYPE:	Nosler		PRIMER TYPE	WLRM
CASE HOLDS:	82.3	Gr. WATER	BARREL Length/Make	24" H-S Prec.
			BARREL Twist	1-10"

POWDER TYPE	POWDER CHG. GRS.		MUZZLE VEL. F.P.S.		LOAD DENSITY (VOLUME)
H4350	70.0	MAX.	3068		90%
	68.0 *		2980		87%
	66.0		2865		85%
Supreme 780	75.0	MAX.	3076		95%
	73.0 *		2995		93%
	71.0		2919		90%
4000-MR	72.5	MAX.	3080		93%
	70.5 *		2999		90%
	68.5		2925		88%
RL22	75.0	MAX.	3090		99%
	73.0		2996		96%
	71.0 *		2865		94%
IMR 7828 SSC	77.0	MAX.	3100		96%
	75.0 *		2988		94%
	73.0		2905		91%
Hybrid 100V	70.5	MAX.	3109		88%
	68.5 *		3023		86%
	66.5		2956		83%
RL17	70.5	MAX.	3112		89%
	68.5		3011		87%
	66.5 *		2975		84%
H1000 Most Accurate Powder Tested	81.0	MAX.	3123	**	103%
	79.0 *		3080	**	101%
	77.0		2971		98%
IMR 4350	70.0 *	MAX.	3130		90%
	68.0		3060		87%
	66.0		2990		85%
IMR 4831	73.0	MAX.	3160		94%
	71.0		3070		92%
	69.0 *		2980		89%

BC=Ballistic Coefficient SD=Sectional Density
*Most Accurate Load Tested **Compressed Load

Use Maximum Loads with Caution
Refer to page 73 for additional safety information

Nosler®

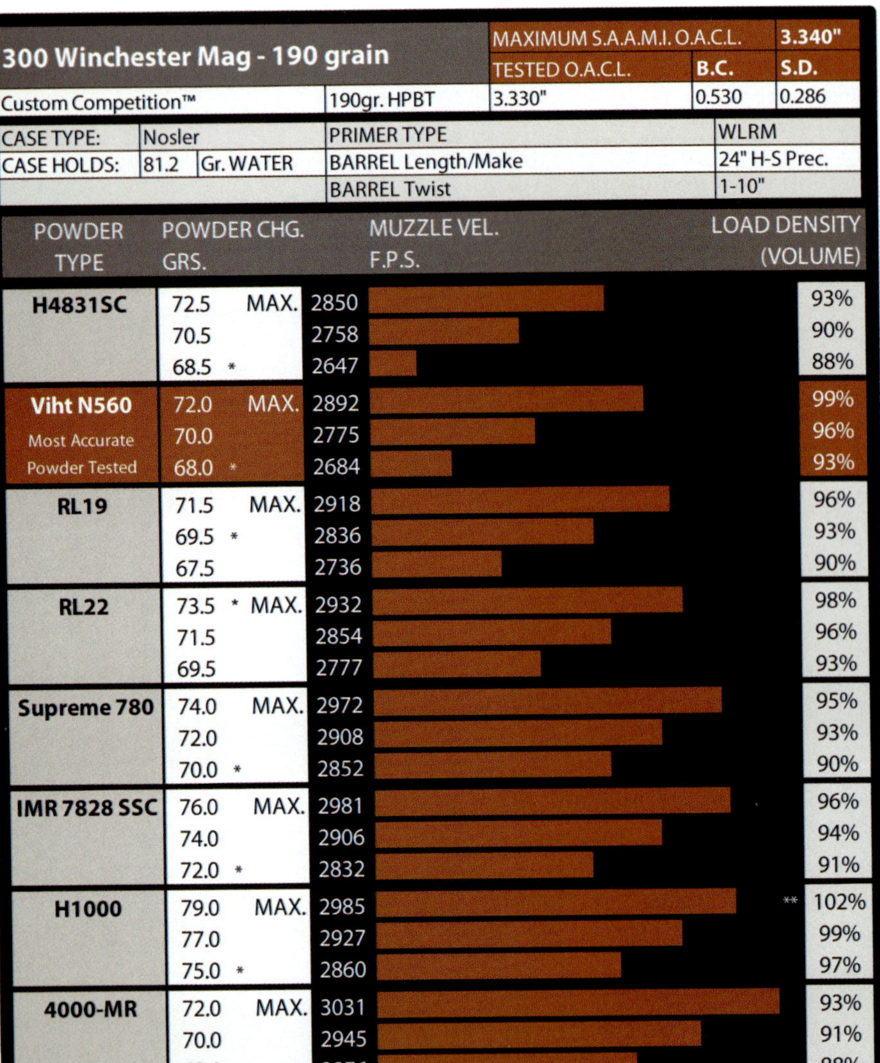

300 Winchester Mag - 190 grain		MAXIMUM S.A.A.M.I. O.A.C.L.		3.340"
		TESTED O.A.C.L.	B.C.	S.D.
Custom Competition™	190gr. HPBT	3.330"	0.530	0.286

CASE TYPE:	Nosler		PRIMER TYPE	WLRM
CASE HOLDS:	81.2	Gr. WATER	BARREL Length/Make	24" H-S Prec.
			BARREL Twist	1-10"

POWDER TYPE	POWDER CHG. GRS.		MUZZLE VEL. F.P.S.	LOAD DENSITY (VOLUME)
H4831SC	72.5	MAX.	2850	93%
	70.5		2758	90%
	68.5 *		2647	88%
Viht N560 Most Accurate Powder Tested	72.0	MAX.	2892	99%
	70.0		2775	96%
	68.0 *		2684	93%
RL19	71.5	MAX.	2918	96%
	69.5 *		2836	93%
	67.5		2736	90%
RL22	73.5 *	MAX.	2932	98%
	71.5		2854	96%
	69.5		2777	93%
Supreme 780	74.0	MAX.	2972	95%
	72.0		2908	93%
	70.0 *		2852	90%
IMR 7828 SSC	76.0	MAX.	2981	96%
	74.0		2906	94%
	72.0 *		2832	91%
H1000	79.0	MAX.	2985	** 102%
	77.0		2927	99%
	75.0 *		2860	97%
4000-MR	72.0	MAX.	3031	93%
	70.0		2945	91%
	68.0 *		2876	88%

BC=Ballistic Coefficient SD=Sectional Density
*Most Accurate Load Tested **Compressed Load

Use Maximum Loads with Caution
Refer to page 73 for additional safety information

300 Winchester Mag - 200 grain		MAXIMUM S.A.A.M.I. O.A.C.L.		3.340"
		TESTED O.A.C.L.	B.C.	S.D.
AccuBond®	200gr. Spitzer	3.340"	0.588	0.301
Partition®	200gr. Spitzer	3.340"	0.481	0.301

CASE TYPE:	Nosler		PRIMER TYPE	Fed 215
CASE HOLDS:	77.4	Gr. WATER	BARREL Length/Make	24" Lilja
			BARREL Twist	1-10"

POWDER TYPE	POWDER CHG. GRS.		MUZZLE VEL. F.P.S.		LOAD DENSITY (VOLUME)
IMR 4320	56.0	MAX.	2648		78%
	54.0		2593		75%
	52.0 *		2538		72%
IMR 4064	58.0	MAX.	2792		82%
	56.0		2737		80%
	54.0 *		2682		77%
Viht N165	71.5	MAX.	2814	**	103%
	69.5		2762		100%
	67.5 *		2710		97%
IMR 7828	74.0	MAX.	2882	**	102%
	72.0		2807		99%
	70.0 *		2732		96%
H4831 Most Accurate Powder Tested	70.5 *	MAX.	2910		95%
	68.5		2810		92%
	66.5		2710		89%
IMR 4350	68.0	MAX.	2950		93%
	66.0		2870		90%
	64.0 *		2790		87%
RL22	72.5 *	MAX.	2960	**	102%
	70.5		2900		99%
	68.5		2840		96%
IMR 4831	71.0	MAX.	2972		98%
	69.0		2917		95%
	67.0 *		2862		92%

BC=Ballistic Coefficient SD=Sectional Density
*Most Accurate Load Tested **Compressed Load

Use Maximum Loads with Caution
Refer to page 73 for additional safety information

300 Winchester Mag - 220 grain		MAXIMUM S.A.A.M.I. O.A.C.L.		3.340"
		TESTED O.A.C.L.	B.C.	S.D.
Partition®	220gr. Semi Spitz	3.340"	0.351	0.331

CASE TYPE:	Nosler		PRIMER TYPE	Fed 215
CASE HOLDS:	79.6	Gr. WATER	BARREL Length/Make	24" Lilja
			BARREL Twist	1-10"

POWDER TYPE	POWDER CHG. GRS.		MUZZLE VEL. F.P.S.	LOAD DENSITY (VOLUME)
A-4350	61.5	MAX.	2575	78%
	59.5		2516	76%
	57.5 *		2457	73%
Viht N165	68.0	MAX.	2598	79%
	66.0		2525	77%
	64.0 *		2450	74%
H1000 Most Accurate Powder Tested	75.5 *	MAX.	2677	74%
	73.5		2593	72%
	71.5		2529	69%
RL19	66.5	MAX.	2699	79%
	64.5		2634	77%
	62.5 *		2570	74%
IMR 7828	68.5	MAX.	2709	79%
	66.5		2655	77%
	64.5 *		2602	74%
IMR 4831	66.5 *	MAX.	2736	82%
	64.5		2676	79%
	62.5		2615	77%
IMR 4350	66.5 *	MAX.	2739	83%
	64.5		2662	81%
	62.5		2585	78%
RL22	68.5	MAX.	2757	83%
	66.5		2706	81%
	64.5 *		2655	78%

BC=Ballistic Coefficient SD=Sectional Density
*Most Accurate Load Tested **Compressed Load

Use Maximum Loads with Caution
Refer to page 73 for additional safety information

300 WINCHESTER SHORT MAGNUM

Every year sees its share of new cartridges, but none in recent memory has been as successful as the 300 Winchester Short Magnum (WSM). Since its introduction in 2001, it has been an instant and long-term success story. The 30 caliber size is very versatile, the compact overall length allows its use in short actions, its accuracy is legendary, and its performance in the field is impressive.

When Bob Nosler and I went to South Africa to test the then-new 180gr AccuBond in 300 WSM, we used it in a variety of circumstances on a broad array of game animals. From springbok to eland, the AccuBond lived up to its reputation, and the results with the 300 WSM were spectacular.

One of my most amusing hunting anecdotes about this cartridge was an unbelievably lucky two-shot offhand volley on a running warthog, across a deep ravine. The first hit him in the spine at 298 yards; the second in the shoulder at 302 yards; both distances measured by a rangefinder. While I like to believe I am a good shot, I am not normally that good. Our PH, Harry Claassens, recognizing the lucky shots for what they were said "I want you to relish those shots, Paul, because I feel certain you will not make two consecutive offhand shots like those for the rest of your life!"

He was right, but the 300 WSM with Nosler up front is one helluva cartridge.

Paul is an active hunter and shooter as well as Nosler's General Counsel.

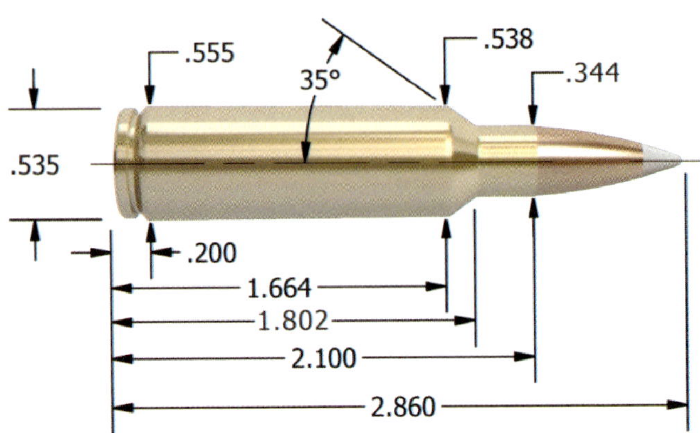

Maximum S.A.A.M.I. Overall Cartridge Length: 2.860"

BULLET CHOICES FOR THE 300 WINCHESTER SHORT MAGNUM

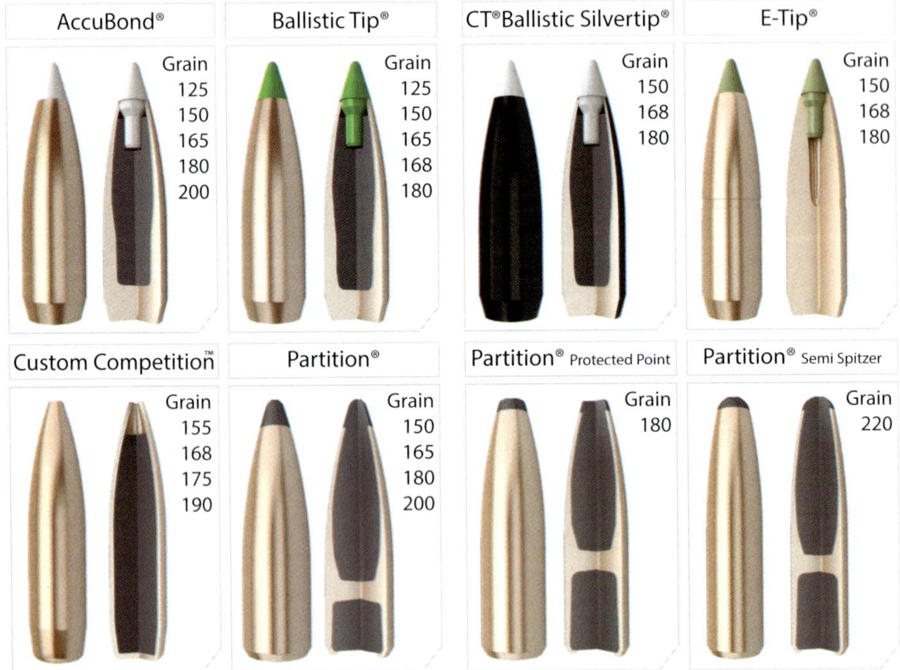

TECHNICAL INFORMATION

While this cartridge is capable of digesting all of the popular 30 caliber weights, best performance will be achieved with 150-180 grain bullets. We had best results with magnum primers and powders in the "medium" burning categories.

300 WSM - 125 grain

		MAXIMUM S.A.A.M.I. O.A.C.L.		2.860"
		TESTED O.A.C.L.	B.C.	S.D.
AccuBond®	125gr. Spitzer	2.835"	0.366	0.188
Ballistic Tip®	125gr. Spitzer	2.835"	0.366	0.188

CASE TYPE:	Nosler		PRIMER TYPE	WLRM
CASE HOLDS:	76.0	Gr. WATER	BARREL Length/Make	26" Wiseman
			BARREL Twist	1-10" 4 groove

POWDER TYPE	POWDER CHG. GRS.		MUZZLE VEL. F.P.S.		LOAD DENSITY (VOLUME)
RL19	75.0	MAX.	3518	**	107%
	73.0 *		3439	**	104%
	71.0		3334	**	102%
Varget	66.0	MAX.	3542		94%
	64.0		3459		91%
	62.0 *		3358		88%
IMR 4350	72.0 *	MAX.	3575		100%
	70.0		3495		97%
	68.0		3387		95%
RL15	66.0	MAX.	3580		90%
	64.0 *		3465		88%
	62.0		3350		85%
H380	72.0	MAX.	3599		100%
	70.0 *		3541		97%
	68.0		3475		94%
IMR 4064	64.5 *	MAX.	3602		93%
	62.5		3469		90%
	60.5		3348		87%
A-2520	65.5	MAX.	3605		89%
	63.5		3562		86%
	61.5 *		3475		83%
H414	72.5 *	MAX.	3634		99%
	70.5		3549		97%
	68.5		3454		94%
Viht N550	70.0	MAX.	3641	**	102%
Most Accurate	68.0		3517		99%
Powder Tested	66.0 *		3390		96%
W760	73.0	MAX.	3659	**	101%
	71.0		3570		98%
	69.0 *		3467		95%

BC=Ballistic Coefficient SD=Sectional Density
*Most Accurate Load Tested **Compressed Load

Use Maximum Loads with Caution
Refer to page 73 for additional safety information

300 WSM - 150/155 grain		MAXIMUM S.A.A.M.I. O.A.C.L.		2.860"
		TESTED O.A.C.L.	B.C.	S.D.
AccuBond®	150gr. Spitzer	2.830"	0.435	0.226
Ballistic Tip®	150gr. Spitzer	2.830"	0.435	0.226
CT® Ballistic Silvertip®	150gr. Spitzer	2.830"	0.435	0.226
E-Tip®	150gr. Spitzer	2.800"	0.469	0.226
Due to internal construction differences, always begin with starting loads when using E-Tip® products.				
Partition®	150gr. Spitzer	2.815"	0.387	0.226
Custom Competition™	155gr. HPBT	2.830"	0.450	0.233

CASE TYPE:	Nosler		PRIMER TYPE	WLRM
CASE HOLDS:	75.1	Gr. WATER	BARREL Length/Make	26" Wiseman
			BARREL Twist	1-10" 4 groove

POWDER TYPE	POWDER CHG. GRS.		MUZZLE VEL. F.P.S.	LOAD DENSITY (VOLUME)
Varget	60.5	MAX.	3184	87%
	58.5		3111	84%
	56.5 *		3027	81%
RL15	61.5	MAX.	3225	85%
	59.5 *		3130	83%
	57.5		3065	80%
RL19	69.5	MAX.	3242	** 101%
	67.5 *		3124	98%
	65.5		3015	95%
IMR 4064 Most Accurate Powder Tested	60.5	MAX.	3242	89%
	58.5 *		3155	86%
	56.5		3063	83%
IMR 4350	67.5	MAX.	3264	95%
	65.5		3171	92%
	63.5 *		3044	89%
H414	67.0	MAX.	3280	93%
	65.0		3185	90%
	63.0 *		3092	87%
IMR 4831	69.0 *	MAX.	3292	98%
	67.0		3190	95%
	65.0		3067	92%
W760	67.5	MAX.	3292	94%
	65.5 *		3214	91%
	63.5		3132	89%
Viht N550	65.5	MAX.	3315	97%
	63.5 *		3193	94%
	61.5		3073	91%

BC=Ballistic Coefficient SD=Sectional Density
*Most Accurate Load Tested **Compressed Load

Use Maximum Loads with Caution
Refer to page 73 for additional safety information

300 WSM - 165/168 grain		MAXIMUM S.A.A.M.I. O.A.C.L.		2.860"
		TESTED O.A.C.L.	B.C.	S.D.
AccuBond®	165gr. Spitzer	2.830"	0.475	0.248
Ballistic Tip®	165gr. Spitzer	2.830"	0.475	0.248
Partition®	165gr. Spitzer	2.815"	0.410	0.248
Ballistic Tip®	168gr. Spitzer	2.830"	0.490	0.253
CT® Ballistic Silvertip®	168gr. Spitzer	2.830"	0.490	0.253
Custom Competition™	168gr. HPBT	2.830"	0.462	0.253
E-Tip®	168gr. Spitzer	2.800"	0.503	0.253

Due to internal construction differences, always begin with starting loads when using E-Tip® products.

CASE TYPE:	Nosler		PRIMER TYPE	WLRM
CASE HOLDS:	74.0	Gr. WATER	BARREL Length/Make	26" Wiseman
			BARREL Twist	1-10" 4 groove

POWDER TYPE	POWDER CHG. GRS.		MUZZLE VEL. F.P.S.	LOAD DENSITY (VOLUME)
H380	64.0	MAX.	3087	91%
Most Accurate	62.0		3017	88%
Powder Tested	60.0 *		2940	85%
RL15	60.0 *	MAX.	3090	84%
	58.0		3003	82%
	56.0		2951	79%
IMR 4320	60.0	MAX.	3104	87%
	58.0 *		3030	84%
	56.0		2951	81%
RL19	68.0	MAX.	3112	100%
	66.0		3001	97%
	64.0 *		2892	94%
IMR 4350	66.0	MAX.	3151	94%
	64.0		3050	92%
	62.0 *		2955	89%
H414	65.5	MAX.	3151	92%
	63.5		3060	89%
	61.5 *		2973	87%
W760	66.5	MAX.	3171	94%
	64.5		3084	91%
	62.5 *		3023	88%
IMR 4831	68.0	MAX.	3174	98%
	66.0		3102	95%
	64.0 *		2978	92%
Viht N550	64.5	MAX.	3181	97%
	62.5		3087	94%
	60.5 *		2997	91%

BC=Ballistic Coefficient SD=Sectional Density
*Most Accurate Load Tested **Compressed Load

Use Maximum Loads with Caution
Refer to page 73 for additional safety information

300 WSM - 175/180 grain

		MAXIMUM S.A.A.M.I. O.A.C.L.		2.860"
		TESTED O.A.C.L.	B.C.	S.D.
Custom Competition™	175gr. HPBT	2.840"	0.505	0.264
AccuBond®	180gr. Spitzer	2.830"	0.507	0.271
Ballistic Tip®	180gr. Spitzer	2.830"	0.507	0.271
CT® Ballistic Silvertip®	180gr. Spitzer	2.830"	0.507	0.271
E-Tip®	180gr. Spitzer	2.800"	0.523	0.271
Due to internal construction differences, always begin with starting loads when using E-Tip® products.				
Partition®	180gr. Spitzer	2.815"	0.474	0.271
Partition®	180gr. PPT	2.715"	0.361	0.271

CASE TYPE:	Nosler		PRIMER TYPE	WLRM
CASE HOLDS:	71.3	Gr. WATER	BARREL Length/Make	24" H-S Prec.
			BARREL Twist	1-10" 4 groove

POWDER TYPE	POWDER CHG. GRS.		MUZZLE VEL. F.P.S.		LOAD DENSITY (VOLUME)
IMR 4350	64.0	* MAX.	2942		95%
	62.0		2857		92%
	60.0		2736		89%
Big Game	61.0	* MAX.	2953		90%
	59.0		2849		87%
	57.0		2798		84%
Hunter	65.5	* MAX.	2966		97%
	63.5		2878		94%
	61.5		2852		91%
H4831SC	69.5	MAX.	2985	**	102%
	67.5		2928		99%
	65.5	*	2851		96%
RL19	67.0	MAX.	3013	**	102%
	65.0		2936		99%
	63.0	*	2836		96%
Hybrid 100V Most Accurate Powder Tested	64.0	* MAX.	3017		93%
	62.0		2892		90%
	60.0		2829		87%
W760	64.0	MAX.	3029		94%
	62.0		2933		91%
	60.0	*	2826		88%
Supreme 780	71.0	* MAX.	3034	**	104%
	69.0		2911	**	101%
	67.0		2812		98%
IMR 7828	70.0	MAX.	3037	**	104%
	68.0	*	2946	**	101%
	66.0		2876		98%
RL22	69.0	MAX.	3057	**	105%
	67.0	*	2972	**	102%
	65.0		2881		99%

BC=Ballistic Coefficient SD=Sectional Density
*Most Accurate Load Tested **Compressed Load

Use Maximum Loads with Caution
Refer to page 73 for additional safety information

300 WSM - 190 grain

		MAXIMUM S.A.A.M.I. O.A.C.L.		2.860"
		TESTED O.A.C.L.	B.C.	S.D.
Custom Competition™	190gr. HPBT	2.840"	0.530	0.286

CASE TYPE:	Nosler		PRIMER TYPE	WLRM
CASE HOLDS:	70.5	Gr. WATER	BARREL Length/Make	24" H-S Prec.
			BARREL Twist	1-10" 4 groove

POWDER TYPE	POWDER CHG. GRS.		MUZZLE VEL. F.P.S.		LOAD DENSITY (VOLUME)
RL19	67.0	MAX.	2881	**	103%
	65.0		2808		100%
	63.0 *		2735		97%
RL22	68.0 *	MAX.	2891	**	105%
	66.0		2832	**	102%
	64.0		2782		99%
H4831SC Most Accurate Powder Tested	69.0	MAX.	2892	**	102%
	67.0 *		2825		99%
	65.0		2750		96%
W760	62.5	MAX.	2901		93%
	60.5 *		2836		90%
	58.5		2777		87%
H414	63.0	MAX.	2908		94%
	61.0		2830		91%
	59.0 *		2748		88%
IMR 7828 SSC	70.0	MAX.	2925	**	102%
	68.0 *		2842		99%
	66.0		2767		97%
Viht N560	69.0	MAX.	2944	**	109%
	67.0		2840	**	106%
	65.0 *		2750	**	102%
Supreme 780	70.0	MAX.	2958	**	104%
	68.0 *		2903	**	101%
	66.0		2856		98%
MAGPRO	75.0	MAX.	2962	**	110%
	73.0 *		2880	**	107%
	71.0		2797	**	104%
Hybrid 100V	63.5	MAX.	2968		93%
	61.5		2875		90%
	59.5 *		2781		87%

BC=Ballistic Coefficient SD=Sectional Density
*Most Accurate Load Tested **Compressed Load

Use Maximum Loads with Caution
Refer to page 73 for additional safety information

300 WSM - 200 grain		MAXIMUM S.A.A.M.I. O.A.C.L.		2.860"
		TESTED O.A.C.L.	B.C.	S.D.
AccuBond®	200gr. Spitzer	2.830"	0.588	0.301
Partition®	200gr. Spitzer	2.815"	0.481	0.301

CASE TYPE:	Nosler		PRIMER TYPE	WLRM
CASE HOLDS:	68.5	Gr. WATER	BARREL Length/Make	26" Wiseman
			BARREL Twist	1-10" 4 groove

POWDER TYPE	POWDER CHG. GRS.		MUZZLE VEL. F.P.S.		LOAD DENSITY (VOLUME)
Viht N550	60.0 *	MAX.	2842		97%
	58.0		2761		94%
	56.0		2682		91%
H414	62.5	MAX.	2851		95%
	60.5 *		2777		92%
	58.5		2696		89%
W760	62.5	MAX.	2853		96%
	60.5 *		2770		92%
	58.5		2680		89%
IMR 4350	62.5	MAX.	2870		97%
	60.5 *		2775		93%
	58.5		2685		90%
IMR 4831	64.5	MAX.	2885		100%
	62.5		2802		97%
	60.5 *		2724		94%
H4831SC	68.0	MAX.	2892	**	103%
	66.0 *		2814		100%
	64.0		2732		97%
IMR 7828 Most Accurate Powder Tested	68.0 *	MAX.	2896	**	106%
	66.0		2824	**	103%
	64.0		2740		99%
RL19	66.0	MAX.	2906	**	105%
	64.0		2833	**	102%
	62.0 *		2745		98%
RL22	67.0	MAX.	2944	**	106%
	65.0 *		2871	**	103%
	63.0		2810		100%

BC=Ballistic Coefficient SD=Sectional Density
*Most Accurate Load Tested **Compressed Load

Use Maximum Loads with Caution
Refer to page 73 for additional safety information

300 WSM - 220 grain

MAXIMUM S.A.A.M.I. O.A.C.L.		2.860"
TESTED O.A.C.L.	B.C.	S.D.
2.815"	0.351	0.331

Partition®	220gr. Semi Spitz

CASE TYPE:	Nosler		PRIMER TYPE	WLRM
CASE HOLDS:	68.7	Gr. WATER	BARREL Length/Make	26" Wiseman
			BARREL Twist	1-10" 4 groove

POWDER TYPE	POWDER CHG. GRS.		MUZZLE VEL. F.P.S.		LOAD DENSITY (VOLUME)
Viht N550	57.0 *	MAX.	2665		92%
	55.0		2596		89%
	53.0		2515		86%
W760	60.0	MAX.	2674		91%
	58.0 *		2607		88%
	56.0		2543		85%
IMR 4350	60.0	MAX.	2677		92%
	58.0 *		2597		89%
	56.0		2519		86%
H414	60.5	MAX.	2682		92%
	58.5 *		2607		89%
	56.5		2548		86%
IMR 4831	62.0	MAX.	2716		96%
Most Accurate	60.0		2638		93%
Powder Tested	58.0 *		2561		90%
H4831SC	65.0	MAX.	2728		99%
	63.0 *		2665		96%
	61.0		2622		92%
RL19	63.5	MAX.	2737		100%
	61.5		2664		97%
	59.5 *		2587		94%
RL22	64.5 *	MAX.	2752	**	102%
	62.5		2689		99%
	60.5		2623		96%
IMR 7828	65.5	MAX.	2762	**	101%
	63.5		2689		98%
	61.5 *		2613		95%

BC=Ballistic Coefficient SD=Sectional Density
*Most Accurate Load Tested **Compressed Load

Use Maximum Loads with Caution
Refer to page 73 for additional safety information

Chub Eastman

300 REMINGTON SHORT ACTION ULTRA MAGNUM

A long anticipated hunting trip to New Zealand had me in a quandary as to what rifle to take. The three critters we would be after were red stag, chamois, and tahr so I wanted a rifle/cartridge combination that would work well on all three, and would be easy to carry as mountains were guaranteed to be on the agenda. I also wanted a rifle that was close to indestructible that could stand up to airline abuse and weather.

A short-barreled Remington Model 7 with a synthetic stock chambered for 300 SAUM (I had ended up with during a horse trade), seemed to be a good choice. Rifle with scope and a magazine full of ammo was no problem to carry, and the 300 SAUM proved to be an excellent choice.

Like most gun nuts, the last few months before the trip were spent at the loading bench and at the range. The goal was to develop one load that would work for all three animals.

Red stag are close to elk size and there was a potential for a long shot. Tahr are about the size of a short-legged mule deer buck and have a reputation of being extremely tough to put on the ground. The chamois is about antelope size.

I lost track of the trips to the range, but the load I settled on was with the 165gr Ballistic Tip®. It shot sub-MOA at nearly 3000 FPS which should have enough horsepower to handle any of the three within a reasonable range.

The red stag was a standing broadside shot at a little over 200 yards. The tahr was a running downhill shot at about 80 yards and the chamois was a quartering away shot at approximately 120 yards. Three shots, three trophies to hang on the wall and memories of a once-in-a-lifetime trip, doesn't get much better than that.

One of the things I remember, besides the successful hunt, was all the rounds that went down range in preparation for the trip. Travel half way around the world and to only fire three shots is an interesting scenario.

Chub Eastman

Chub Eastman is the Rifle Field Editor for Sports Afield.

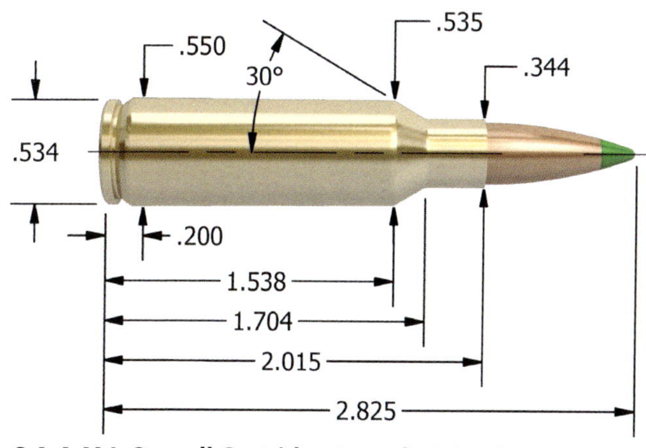

Maximum S.A.A.M.I. Overall Cartridge Length: 2.825"

BULLET CHOICES FOR THE 300 REMINGTON SHORT ACTION ULTRA MAGNUM

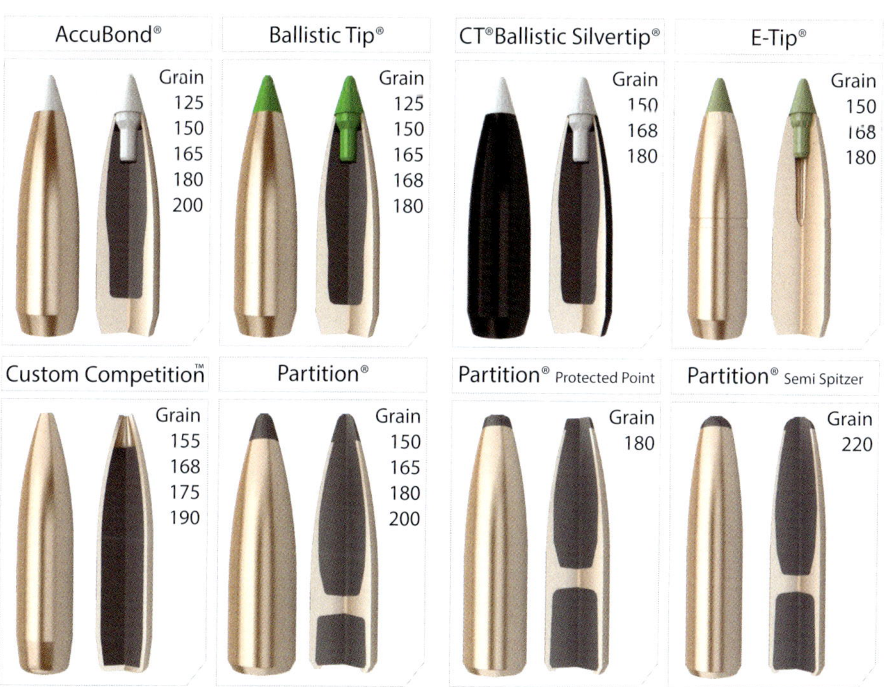

TECHNICAL INFORMATION

Best results were achieved with magnum primers and medium burning powders like Varget and 4350.

300 Rem S.A.U.M. - 125 grain

		MAXIMUM S.A.A.M.I. O.A.C.L.		2.825"
		TESTED O.A.C.L.	B.C.	S.D.
AccuBond®	125gr. Spitzer	2.825"	0.366	0.188
Ballistic Tip®	125gr. Spitzer	2.825"	0.366	0.188

CASE TYPE:	Nosler		PRIMER TYPE	Rem 9 1/2M
CASE HOLDS:	69.4	Gr. WATER	BARREL Length/Make	24" Wiseman
			BARREL Twist	1-10"

POWDER TYPE	POWDER CHG. GRS.		MUZZLE VEL. F.P.S.		LOAD DENSITY (VOLUME)
IMR 4350	68.0	MAX.	3338	**	104%
	66.0		3234	**	101%
	64.0 *		3110		98%
W760	69.0	MAX.	3445	**	104%
	67.0 *		3400	**	101%
	65.0		3316		98%
A-2495 Most Accurate Powder Tested	63.0	MAX.	3505	**	103%
	61.0 *		3384		100%
	59.0		3266		97%
RL15	63.0	MAX.	3523		95%
	61.0		3381		92%
	59.0 *		3267		89%
H4895	61.0	MAX.	3525		97%
	59.0		3409		93%
	57.0 *		3271		90%
Varget	64.0	MAX.	3525		100%
	62.0 *		3424		97%
	60.0		3293		93%
IMR 4064	61.5	MAX.	3537		97%
	59.5		3431		94%
	57.5 *		3292		91%
Viht N540	66.0	MAX.	3610	**	106%
	64.0 *		3485	**	102%
	62.0		3364		99%
A-2700	74.0	MAX.	3632	**	111%
	72.0		3522	**	108%
	70.0 *		3489	**	105%
Viht N550	70.0	MAX.	3634	**	112%
	68.0 *		3546	**	109%
	66.0		3468	**	106%

BC=Ballistic Coefficient SD=Sectional Density
*Most Accurate Load Tested **Compressed Load

Use Maximum Loads with Caution
Refer to page 73 for additional safety information

300 Rem S.A.U.M. - 150/155 grain		MAXIMUM S.A.A.M.I. O.A.C.L.		2.825"
		TESTED O.A.C.L.	B.C.	S.D.
AccuBond®	150gr. Spitzer	2.825"	0.435	0.226
Ballistic Tip®	150gr. Spitzer	2.825"	0.435	0.226
CT® Ballistic Silvertip®	150gr. Spitzer	2.825"	0.435	0.226
E-Tip®	150gr. Spitzer	2.800"	0.469	0.226
Due to internal construction differences, always begin with starting loads when using E-Tip® products.				
Partition®	150gr. Spitzer	2.825"	0.387	0.226
Custom Competition™	155gr. HPBT	2.825"	0.450	0.233

CASE TYPE:	Nosler		PRIMER TYPE	Rem 9 1/2M
CASE HOLDS:	67.5	Gr. WATER	BARREL Length/Make	24" Wiseman
			BARREL Twist	1-10"

POWDER TYPE	POWDER CHG. GRS.		MUZZLE VEL. F.P.S.		LOAD DENSITY (VOLUME)
RL15	59.0 *	MAX.	3203		91%
	57.0		3042		88%
	55.0		2948		85%
IMR 4350	65.0	MAX.	3208	**	102%
	63.0		3116		99%
	61.0 *		3024		96%
Varget Most Accurate Powder Tested	60.0	MAX.	3210		96%
	58.0 *		3079		93%
	56.0		2970		90%
W760	65.0	MAX.	3219	**	101%
	63.0 *		3098		98%
	61.0		2983		95%
IMR 4831	69.0	MAX.	3250	**	109%
	67.0 *		3160	**	106%
	65.0		3096	**	102%
H380	65.5	MAX.	3262	**	102%
	63.5		3150		99%
	61.5 *		3057		96%
H4350	66.5	MAX.	3262	**	104%
	64.5 *		3183	**	101%
	62.5		3095		98%
H414	66.0	MAX.	3270	**	102%
	64.0		3189		99%
	62.0 *		3103		96%
Viht N550	63.5	MAX.	3278	**	105%
	61.5		3162	**	101%
	59.5 *		3050		98%
Big Game	64.0	MAX.	3290		100%
	62.0		3169		97%
	60.0 *		3094		94%

BC=Ballistic Coefficient SD=Sectional Density
*Most Accurate Load Tested **Compressed Load

Use Maximum Loads with Caution
Refer to page 73 for additional safety information

300 Rem S.A.U.M. - 165/168 grain		MAXIMUM S.A.A.M.I. O.A.C.L.		2.825"
		TESTED O.A.C.L.	B.C.	S.D.
AccuBond®	165gr. Spitzer	2.825"	0.475	0.248
Ballistic Tip®	165gr. Spitzer	2.825"	0.475	0.248
Partition®	165gr. Spitzer	2.825"	0.410	0.248
Ballistic Tip®	168gr. Spitzer	2.825"	0.490	0.253
CT® Ballistic Silvertip®	168gr. Spitzer	2.825"	0.498	0.253
Custom Competition™	168gr. HPBT	2.825"	0.462	0.253
E-Tip®	168gr. Spitzer	2.800"	0.503	0.253

Due to internal construction differences, always begin with starting loads when using E-Tip® products.

CASE TYPE:	Nosler		PRIMER TYPE	Rem 9 1/2M
CASE HOLDS:	65.6	Gr. WATER	BARREL Length/Make	24" Wiseman
			BARREL Twist	1-10"

POWDER TYPE	POWDER CHG. GRS.		MUZZLE VEL. F.P.S.		LOAD DENSITY (VOLUME)
Varget	57.5	MAX.	3005		95%
	55.5	*	2908		91%
	53.5		2809		88%
RL15 Most Accurate Powder Tested	57.0	MAX.	3021		91%
	55.0		2889		87%
	53.0	*	2796		84%
Big Game	61.0	MAX.	3030		98%
	59.0		2940		95%
	57.0	*	2841		91%
IMR 4350	64.0	MAX.	3057	**	103%
	62.0		2946		100%
	60.0	*	2861		97%
W760	63.5	MAX.	3082	**	101%
	61.5		2989		98%
	59.5	*	2908		95%
H4350	64.5	MAX.	3095	**	104%
	62.5	*	3025	**	101%
	60.5		2942		98%
Viht N550	61.5	MAX.	3112	**	104%
	59.5		2985	**	101%
	57.5	*	2875		97%
Viht N160	67.0	MAX.	3116	**	113%
	65.0	*	3067	**	110%
	63.0		2955	**	107%
IMR 4831	67.0	MAX.	3118	**	109%
	65.0	*	3017	**	105%
	63.0		2947	**	102%
H414	63.5	MAX.	3129	**	101%
	61.5		3068		98%
	59.5	*	2987		94%

BC=Ballistic Coefficient SD=Sectional Density
*Most Accurate Load Tested **Compressed Load

Use Maximum Loads with Caution
Refer to page 73 for additional safety information

300 Rem S.A.U.M. - 175/180 grain		MAXIMUM S.A.A.M.I. O.A.C.L.		2.825"
		TESTED O.A.C.L.	B.C.	S.D.
Custom Competition™	175gr. HPBT	2.825"	0.505	0.264
AccuBond®	180gr. Spitzer	2.825"	0.507	0.271
Ballistic Tip®	180gr. Spitzer	2.825"	0.507	0.271
CT Ballistic Silvertip®	180gr. Spitzer	2.825"	0.507	0.271
E-Tip®	180gr. Spitzer	2.800"	0.523	0.271
Due to internal construction differences, always begin with starting loads when using E-Tip® products.				
Partition®	180gr. Spitzer	2.825"	0.474	0.271
Partition®	180gr. PPT	2.825"	0.361	0.271

CASE TYPE:	Nosler		PRIMER TYPE	Rem 9 1/2M
CASE HOLDS:	64.3	Gr. WATER	BARREL Length/Make	24" H-S Prec.
			BARREL Twist	1-10"

POWDER TYPE	POWDER CHG. GRS.		MUZZLE VEL. F.P.S.		LOAD DENSITY (VOLUME)
IMR 4007 SSC	58.0 *	MAX.	2905		97%
	56.0		2801		93%
	54.0		2708		90%
Supreme 780	67.0 *	MAX.	2917	**	109%
	65.0		2873	**	106%
	63.0		2789	**	103%
IMR 7828	66.5	MAX.	2944	**	110%
	64.5		2872	**	107%
	62.5 *		2779	**	103%
IMR 4350 Most Accurate Powder Tested	60.5	MAX.	2955		100%
	58.5		2838		96%
	56.5 *		2786		93%
H4350	60.5	MAX.	2955		100%
	58.5 *		2866		96%
	56.5		2772		93%
H414	62.5	MAX.	2963	**	101%
	60.5 *		2882		98%
	58.5		2780		95%
W760	61.0	MAX.	2985		99%
	59.0		2887		96%
	57.0 *		2807		93%
Viht N550	60.0	MAX.	2993	**	104%
	58.0 *		2884		100%
	56.0		2802		97%
RL17	60.0	MAX.	3017	**	101%
	58.0 *		2871		98%
	56.0		2784		95%
IMR 4831	65.5	MAX.	3017	**	108%
	63.5		2937	**	105%
	61.5 *		2865	**	102%

BC=Ballistic Coefficient SD=Sectional Density
*Most Accurate Load Tested **Compressed Load

Use Maximum Loads with Caution
Refer to page 73 for additional safety information

300 Rem S.A.U.M. - 190 grain		MAXIMUM S.A.A.M.I. O.A.C.L.		2.825"
		TESTED O.A.C.L.	B.C.	S.D.
Custom Competition™	190gr. HPBT	2.825"	0.530	0.286

CASE TYPE:	Nosler		PRIMER TYPE	WLRM
CASE HOLDS:	65.0	Gr. WATER	BARREL Length/Make	24" H-S Prec.
			BARREL Twist	1-10"

POWDER TYPE	POWDER CHG. GRS.		MUZZLE VEL. F.P.S.		LOAD DENSITY (VOLUME)
MAGPRO	70.0	MAX.	2849	**	111%
	68.0		2784	**	108%
	66.0	*	2673	**	105%
IMR 7828 SSC	65.5	* MAX.	2852	**	104%
	63.5		2780	**	101%
	61.5		2705		98%
W760	60.0	MAX.	2858		97%
	58.0		2787		93%
	56.0	*	2709		90%
H414	60.0	MAX.	2876		96%
	58.0		2798		93%
	56.0	*	2719		90%
H4831 Most Accurate Powder Tested	66.0	MAX.	2878	**	106%
	64.0	*	2809	**	103%
	62.0		2750		99%
Hybrid 100V	60.0	MAX.	2890		95%
	58.0		2804		92%
	56.0	*	2676		89%
RL19	64.0	MAX.	2898	**	107%
	62.0		2827	**	104%
	60.0	*	2720		100%
Viht N560	65.5	MAX.	2914	**	112%
	63.5	*	2830	**	109%
	61.5		2741	**	105%
RL22	66.0	MAX.	2925	**	110%
	64.0		2853	**	107%
	62.0	*	2729	**	104%

BC=Ballistic Coefficient SD=Sectional Density
*Most Accurate Load Tested **Compressed Load

Use Maximum Loads with Caution
Refer to page 73 for additional safety information

300 Rem S.A.U.M. - 200 grain		MAXIMUM S.A.A.M.I. O.A.C.L.		2.825"
		TESTED O.A.C.L.	B.C.	S.D.
AccuBond®	200gr. Spitzer	2.825"	0.588	0.301
Partition®	200gr. Spitzer	2.825"	0.481	0.301

CASE TYPE:	Nosler		PRIMER TYPE	Rem 9 1/2M
CASE HOLDS:	65.0	Gr. WATER	BARREL Length/Make	24" Wiseman
			BARREL Twist	1-10"

POWDER TYPE	POWDER CHG. GRS.		MUZZLE VEL. F.P.S.		LOAD DENSITY (VOLUME)
Viht N550	57.5	MAX.	2772		98%
	55.5		2687		95%
	53.5 *		2612		91%
W760	60.0	MAX.	2778		97%
	58.0		2674		93%
	56.0 *		2597		90%
IMR 4350	60.5	MAX.	2792		98%
	58.5 *		2708		95%
	56.5		2625		92%
Hunter	62.0	MAX.	2798		100%
	60.0 *		2712		97%
	58.0		2664		94%
H414	61.0	MAX.	2801		98%
	59.0 *		2685		95%
	57.0		2623		91%
H4350	60.0	MAX.	2806		98%
	58.0		2716		94%
	56.0 *		2633		91%
RL19	63.0	MAX.	2810	**	105%
Most Accurate	61.0 *		2710	**	102%
Powder Tested	59.0		2622		99%
Viht N160	62.5 *	MAX.	2830	**	107%
	60.5		2735	**	103%
	58.5		2656		100%
RL22	67.0	MAX.	2870	**	112%
	65.0		2793	**	109%
	63.0 *		2724	**	105%

BC=Ballistic Coefficient SD=Sectional Density
*Most Accurate Load Tested **Compressed Load

Use Maximum Loads with Caution
Refer to page 73 for additional safety information

300 Rem S.A.U.M. - 220 grain		MAXIMUM S.A.A.M.I. O.A.C.L.		2.825"
		TESTED O.A.C.L.	B.C.	S.D.
Partition®	220gr. Semi Spitz	2.825"	0.351	0.331

CASE TYPE:	Nosler		PRIMER TYPE	Rem 9 1/2M
CASE HOLDS:	65.0	Gr. WATER	BARREL Length/Make	24" Wiseman
			BARREL Twist	1-10"

POWDER TYPE	POWDER CHG. GRS.		MUZZLE VEL. F.P.S.		LOAD DENSITY (VOLUME)
H414 Most Accurate Powder Tested	59.0	MAX.	2653		95%
	57.0	*	2569		91%
	55.0		2492		88%
W760	58.0	MAX.	2657		93%
	56.0	*	2563		90%
	54.0		2490		87%
IMR 4350	59.0	MAX.	2662		96%
	57.0	*	2580		93%
	55.0		2484		90%
RL19	61.0	MAX.	2667	**	102%
	59.0	*	2574		99%
	57.0		2489		95%
H4350	58.5	* MAX.	2674		95%
	56.5		2582		92%
	54.5		2495		89%
Viht N160	60.0	MAX.	2681	**	103%
	58.0		2593		99%
	56.0	*	2522		96%
Viht N550	57.5	MAX.	2688		98%
	55.5	*	2596		95%
	53.5		2510		91%
Hunter	61.0	MAX.	2689		99%
	59.0	*	2607		96%
	57.0		2515		92%
RL22	65.5	* MAX.	2754	**	110%
	63.5		2661	**	106%
	61.5		2588	**	103%

BC=Ballistic Coefficient SD=Sectional Density
*Most Accurate Load Tested **Compressed Load

Use Maximum Loads with Caution
Refer to page 73 for additional safety information

300 RUGER COMPACT MAGNUM

I was informed of the new .30-caliber cartridge in 2008, when it was nothing more than a cartridge/firearms concept the folks at Hornady and Ruger were partnering on. But before any preproduction and/or commercial ammo was available, even early indications and reports were encouraging in terms of velocity and energy.

With short-action magnum cartridges all the rage in recent years, there's always a dose of skepticism when another cartridge enters the category, but new cartridge introductions have always intrigued me, and I've seen a fair share in my 35 years in the industry and as editor of Shooting Times and Guns & Ammo magazines, and my 60 years on God's green earth. To mention just a few, rifle cartridges like the 7mm Remington Magnum (1962) and 300 Winchester Magnum (1963) were two of the more significant long-action magnum cartridges from Remington and Winchester, respectively.

My focus here is the .300 RCM, and its deserved position in the short magnum cartridge lineup. Upon learning about the .300 RCM, I was fortunate enough to go with Ruger president and CEO Mike Fifer on a Utah elk hunt, and more recently on an African plains game safari.

I quickly became enamored with the 165-grain bullet weight in the .300 RCM, chambered in the Ruger Hawkeye bolt-action rifle, as it proved to be an outstanding performer on everything from Rocky Mountain elk to plains game ranging from deer-sized game to wildebeest, hartebeest, and kudu. While the ideal bullet weight range is 150- to 180-grains, the velocity and energy combination with the 165-grain bullet is substantial and I believe brings out the best in the cartridge. Muzzle velocity is in the 3,000 fps range and muzzle energy (ft./lbs.) charts at around 3,300. At 100 yards, that translates to approximately 2900 fps and 3100 ft/lbs. The key, for handloaders, is that this compact, short-action cartridge can deliver a payload similar to standard length cartridges/bullet weights while using 10 to 15 percent less propellant.

I've spoken to a number of ballisticians, handloaders and end users and they all view the new short magnum cartridge equally: It's an ideal hunting round. All the gel block

testing and three-shot 100 yard groups over the course of reviewing firearms is noteworthy, but I've always been a firm believer that in the field results trump everything else if you're a hunter. Of course, that means the shooter/hunter has to do his homework, and whether it's firing factory loads or working up handloads, you want to find the right combination of bullet and propellant that shoots best in your particular rifle. The end result is success in the field…and a true indicator of a hunting cartridge's potential.

James W. Bequette

Jim Bequette is Intermedia Outdoors' VP, Group Editorial Director, Shooting

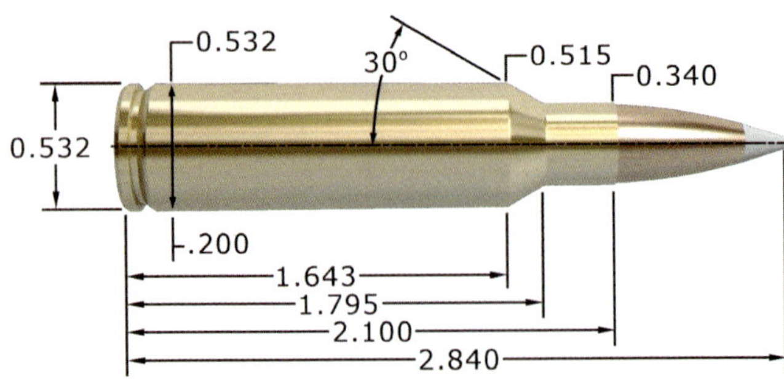

Maximum S.A.A.M.I. Overall Cartridge Length: 2.840"

BULLET CHOICES FOR THE 300 RUGER COMPACT MAGNUM

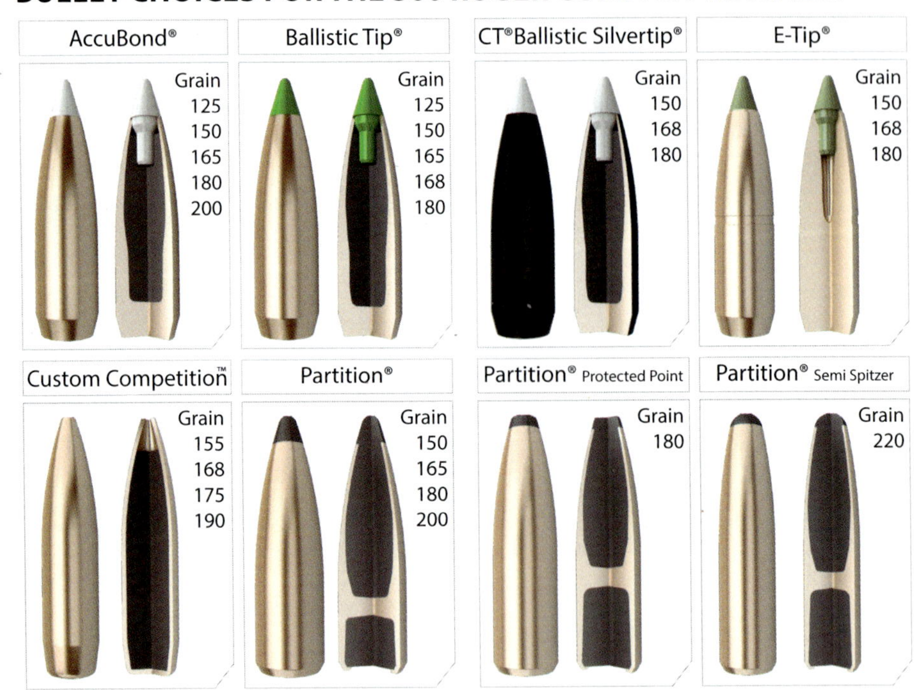

300 RCM- 125 grain			MAXIMUM S.A.A.M.I. O.A.C.L.		2.840"
			TESTED O.A.C.L.	B.C.	S.D.
AccuBond®		125gr. Spitzer	2.800"	0.366	0.188
Ballistic Tip®		125gr. Spitzer	2.800"	0.366	0.188

CASE TYPE:	Hornady		PRIMER TYPE	WLRM
CASE HOLDS:	66.7	Gr. WATER	BARREL Length/Make	24" Pac-Nor
			BARREL Twist	1-10"

POWDER TYPE	POWDER CHG. GRS.		MUZZLE VEL. F.P.S.		LOAD DENSITY (VOLUME)
IMR 4064 Most Accurate Powder Tested	58.0	MAX.	3343		96%
	56.0		3202		92%
	54.0	*	3065		89%
RL15	60.0	MAX.	3354		94%
	58.0		3227		91%
	56.0	*	3099		87%
W760	65.0	MAX.	3394	**	102%
	63.0	*	3270		99%
	61.0		3152		96%
Varget	60.0	MAX.	3406		97%
	58.0		3302		94%
	56.0	*	3198		91%

BC=Ballistic Coefficient SD=Sectional Density
*Most Accurate Load Tested **Compressed Load

Use Maximum Loads with Caution
Refer to page 73 for additional safety information

300 RCM - 150/155 grain		MAXIMUM S.A.A.M.I. O.A.C.L.		2.840"
		TESTED O.A.C.L.	B.C.	S.D.
AccuBond®	150gr. Spitzer	2.800"	0.435	0.226
Ballistic Tip®	150gr. Spitzer	2.800"	0.435	0.226
CT® Ballistic Silvertip®	150gr. Spitzer	2.800"	0.435	0.226
E-Tip®	150gr. Spitzer	2.780"	0.469	0.226
Due to internal construction differences, always begin with starting loads when using E-Tip® products.				
Partition®	150gr. Spitzer	2.820"	0.387	0.226
Custom Competition™	155gr. HPBT	2.760"	0.450	0.233

CASE TYPE:	Hornady		PRIMER TYPE	WLRM
CASE HOLDS:	65.0	Gr. WATER	BARREL Length/Make	24" Pac-Nor
			BARREL Twist	1-10"

POWDER TYPE	POWDER CHG. GRS.		MUZZLE VEL. F.P.S.		LOAD DENSITY (VOLUME)
RL15 Most Accurate Powder Tested	57.5	MAX.	3078		92%
	55.5	*	2967		89%
	53.5		2855		86%
Hybrid 100V	61.5	MAX.	3134		98%
	59.5	*	3022		94%
	57.5		2912		91%
W760	62.0	MAX.	3150		100%
	60.0		3033		97%
	58.0	*	2917		93%
Hunter	67.0	MAX.	3192	**	109%
	65.0		3109	**	105%
	63.0	*	3026	**	102%

BC=Ballistic Coefficient SD=Sectional Density
*Most Accurate Load Tested **Compressed Load

Use Maximum Loads with Caution
Refer to page 73 for additional safety information

300 RCM - 165/168 grain		MAXIMUM S.A.A.M.I. O.A.C.L.		2.840"
		TESTED O.A.C.L.	B.C.	S.D.
AccuBond®	165gr. Spitzer	2.820"	0.475	0.248
Ballistic Tip®	165gr. Spitzer	2.820"	0.475	0.248
Partition®	165gr. Spitzer	2.820"	0.410	0.248
Ballistic Tip®	168gr. Spitzer	2.820"	0.490	0.253
CT® Ballistic Silvertip®	168gr. Spitzer	2.820"	0.490	0.253
Custom Competition™	168gr. HPBT	2.760"	0.462	0.253
E-Tip®	168gr. Spitzer	2.780"	0.503	0.253

Due to internal construction differences, always begin with starting loads when using E-Tip® products.

CASE TYPE:	Hornady		PRIMER TYPE	WLRM
CASE HOLDS:	62.8	Gr. WATER	BARREL Length/Make	24" Pac-Nor
			BARREL Twist	1-10"

POWDER TYPE	POWDER CHG. GRS.		MUZZLE VEL. F.P.S.		LOAD DENSITY (VOLUME)
W760	59.0	MAX.	2901		98%
	57.0	*	2814		95%
	55.0		2735		92%
RL19	63.0	MAX.	2911	**	109%
	61.0		2818	**	106%
	59.0	*	2725	**	102%
Hybrid 100V Most Accurate Powder Tested	59.5	MAX.	2959		98%
	57.5	*	2852		94%
	55.5		2741		91%
Hunter	65.0	MAX.	3020	**	109%
	63.0		2930	**	106%
	61.0	*	2844	**	102%

BC=Ballistic Coefficient SD=Sectional Density
*Most Accurate Load Tested **Compressed Load

Use Maximum Loads with Caution
Refer to page 73 for additional safety information

300 RCM - 175/180 grain		MAXIMUM S.A.A.M.I. O.A.C.L.		2.840"
		TESTED O.A.C.L.	B.C.	S.D.
Custom Competition™	175gr. HPBT	2.760"	0.505	0.264
AccuBond®	180gr. Spitzer	2.820"	0.507	0.271
Ballistic Tip®	180gr. Spitzer	2.820"	0.507	0.271
CT® Ballistic Silvertip®	180gr. Spitzer	2.820"	0.507	0.271
E-Tip®	180gr. Spitzer	2.780"	0.523	0.271
Due to internal construction differences, always begin with starting loads when using E-Tip® products.				
Partition®	180gr. Spitzer	2.820"	0.474	0.271
Partition®	180gr. PPT	2.820"	0.361	0.271

CASE TYPE:	Hornady		PRIMER TYPE	WLRM
CASE HOLDS:	61.4	Gr. WATER	BARREL Length/Make	24" Pac-Nor
			BARREL Twist	1-10"

POWDER TYPE	POWDER CHG. GRS.		MUZZLE VEL. F.P.S.		LOAD DENSITY (VOLUME)
Supreme 780	64.0 *	MAX.	2799	**	109%
	62.0		2728	**	106%
	60.0		2658	**	102%
Hybrid 100V Most Accurate Powder Tested	58.0	MAX.	2867		97%
	56.0		2769		94%
	54.0 *		2665		91%
RL19	63.0	MAX.	2898	**	112%
	61.0 *		2809	**	108%
	59.0		2719	**	104%
Hunter	63.0	MAX.	2906	**	108%
	61.0		2808	**	105%
	59.0 *		2713	**	101%

BC=Ballistic Coefficient SD=Sectional Density
*Most Accurate Load Tested **Compressed Load

Use Maximum Loads with Caution
Refer to page 73 for additional safety information

300 RCM - 190 grain			MAXIMUM S.A.A.M.I. O.A.C.L.		2.840"
			TESTED O.A.C.L.	B.C.	S.D.
Custom Competition™		190gr. HPBT	2.810"	0.530	0.286

CASE TYPE:	Hornady		PRIMER TYPE	WLRM
CASE HOLDS:	61.5	Gr. WATER	BARREL Length/Make	24" Pac-Nor
			BARREL Twist	1-10"

POWDER TYPE	POWDER CHG. GRS.		MUZZLE VEL. F.P.S.		LOAD DENSITY (VOLUME)
IMR 4831	58.0	MAX.	2759		100%
	56.0		2640		97%
	54.0 *		2524		93%
Hybrid 100V	56.0	MAX.	2770		94%
	54.0 *		2674		91%
	52.0		2570		87%
RL19 Most Accurate Powder Tested	61.0	MAX.	2821	**	108%
	59.0		2717	**	104%
	57.0 *		2610	**	101%
Hunter	61.0	MAX.	2824	**	104%
	59.0 *		2753	**	101%
	57.0		2681		98%

BC=Ballistic Coefficient SD=Sectional Density
*Most Accurate Load Tested **Compressed Load

Use Maximum Loads with Caution
Refer to page 73 for additional safety information

300 RCM - 200 grain		MAXIMUM S.A.A.M.I. O.A.C.L.		2.840"
		TESTED O.A.C.L.	B.C.	S.D.
AccuBond®	200gr. Spitzer	2.820"	0.588	0.301
Partition®	200gr. Spitzer	2.780"	0.481	0.301

CASE TYPE:	Hornady		PRIMER TYPE	WLRM
CASE HOLDS:	60.3	Gr. WATER	BARREL Length/Make	24" Pac-Nor
			BARREL Twist	1-10"

POWDER TYPE	POWDER CHG. GRS.		MUZZLE VEL. F.P.S.		LOAD DENSITY (VOLUME)
H4831	61.0 *	MAX.	2653	**	105%
	59.0		2580	**	102%
	57.0		2508		98%
IMR 4350	56.5	MAX.	2692		99%
	54.5		2567		96%
	52.5 *		2491		92%
IMR 4007 SSC	56.0 *	MAX.	2711		100%
	54.0		2613		96%
	52.0		2517		92%
Hunter Most Accurate Powder Tested	60.0	MAX.	2743	**	105%
	58.0		2667	**	101%
	56.0 *		2585		98%

BC=Ballistic Coefficient SD=Sectional Density
*Most Accurate Load Tested **Compressed Load

Use Maximum Loads with Caution
Refer to page 73 for additional safety information

300 RCM - 220 grain		MAXIMUM S.A.A.M.I. O.A.C.L.		2.840"
		TESTED O.A.C.L.	B.C.	S.D.
Partition®	220gr. Semi-Spitz	2.720"	0.351	0.331

CASE TYPE:	Hornady		PRIMER TYPE	WLRM
CASE HOLDS:	58.7	Gr. WATER	BARREL Length/Make	24" Pac-Nor
			BARREL Twist	1-10"

POWDER TYPE	POWDER CHG. GRS.		MUZZLE VEL. F.P.S.		LOAD DENSITY (VOLUME)
H4831	57.0 *	MAX.	2462	**	101%
	55.0		2401		98%
	53.0		2341		94%
H4350	54.0 *	MAX.	2528		97%
	52.0		2452		94%
	50.0		2374		90%
Hunter	57.0	MAX.	2539	**	102%
	55.0		2484		99%
	53.0 *		2421		95%
RL19 Most Accurate Powder Tested	59.0	MAX	2604	**	109%
	57.0		2520	**	106%
	55.0 *		2433	**	102%

BC=Ballistic Coefficient SD=Sectional Density
*Most Accurate Load Tested **Compressed Load

Use Maximum Loads with Caution
Refer to page 73 for additional safety information

Ludo Wurfbain

300 REMINGTON ULTRA MAGNUM

I have liked the .30 caliber magnums for decades. I bought my first one, a custom 300 Weatherby from the old Pachmayr store in downtown Los Angeles when I was in college. From then on, I have owned several .300s until I heard about the 300 Remington Ultra Magnum being announced. I asked Lex Webernick of Rifles, Inc. to build me a gun. It shot very well but it kicked too much for me, however once Lex attached a muzzle brake, all was well—as long as you wore earmuffs.

I took that rifle hunting with me all over, shot it in Asia and had it in Alaska and British Columbia. Strangely enough, it never has seen Africa, although, there is little doubt it would shine there too.

It was a killer and is one of my all-time favorite rifle cartridges. I looked back in my diaries to see where that 300 RUM has followed me and how many shots I had fired through it. Interestingly enough, I did not shoot a lot of game with it but I had it on the range to shoot targets and steel gongs out to 600 yards all the time. In fact, the last time I checked, I had close to a 1000 shots through the barrel, about 98% of them at the range. I suppose this is true for most rifles, still it surprised me. I felled caribou, grizzly, mule deer and a host of other game with it and has proved to be quite deadly in the field. It carries its energy and speed out to well beyond 500 yards in liberal proportions.

If you want a great, flat-shooting cartridge with plenty of down-range energy, look no further than the 300 RUM.

Guide Robb Wiley with my .300 RUM and a very good typical Mule deer from Wyoming

Ludo Wurfbain is the Publisher of Sports Afield and Safari Press

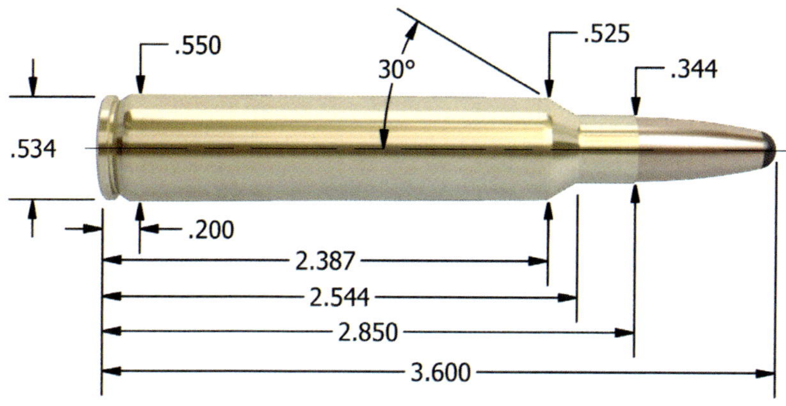

Maximum S.A.A.M.I. Overall Cartridge Length: 3.600"

BULLET CHOICES FOR THE 300 REMINGTON ULTRA MAGNUM

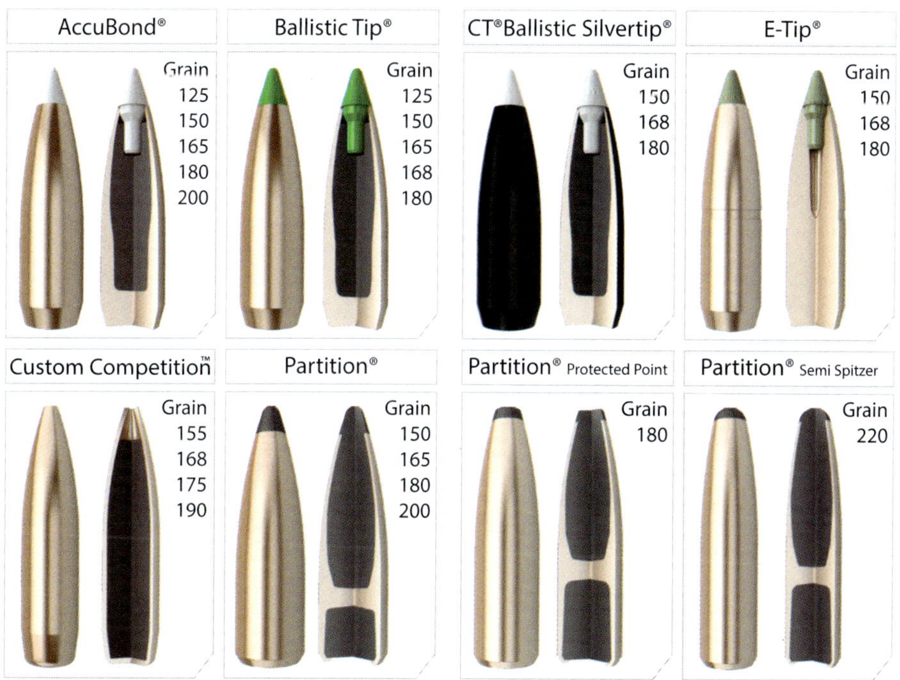

TECHNICAL INFORMATION

Our technicians were surprised to find near bench-rest accuracy was achieved when medium weight bullets were teamed with IMR4350 and a Federal 215 primer.

300 Rem Ultra Mag - 125 grain		MAXIMUM S.A.A.M.I. O.A.C.L.		3.600"
		TESTED O.A.C.L.	B.C.	S.D.
AccuBond®	125gr. Spitzer	3.600"	0.366	0.188
Ballistic Tip®	125gr. Spitzer	3.600"	0.366	0.188

CASE TYPE:	Nosler	PRIMER TYPE	Fed 215
CASE HOLDS:	110.0 Gr. WATER	BARREL Length/Make	26" Wiseman
		BARREL Twist	1-10"

POWDER TYPE	POWDER CHG. GRS.		MUZZLE VEL. F.P.S.		LOAD DENSITY (VOLUME)
H1000	107.0	MAX.	3674	**	102%
	105.0 *		3620		100%
	103.0		3555		98%
Magnum	109.0	MAX.	3781	**	101%
	107.0		3688		99%
	105.0 *		3567		97%
H4831SC	100.0	MAX.	3799		95%
	98.0 *		3746		93%
	96.0		3676		91%
IMR 4350	94.0	MAX.	3813		90%
	92.0		3773		89%
	90.0 *		3673		87%
MAGPRO	105.5	MAX.	3814		99%
	103.5		3751		97%
	101.5 *		3677		95%
IMR 7828	100.0 *	MAX.	3819		97%
	98.0		3733		95%
	96.0		3621		93%
Retumbo	109.0 *	MAX.	3830	**	104%
	107.0		3753	**	102%
	105.0		3659		100%
RL22 Most Accurate Powder Tested	100.5	MAX.	3858		99%
	98.5 *		3730		97%
	96.5		3587		95%
RL25	104.0	MAX.	3909	**	103%
	102.0 *		3804	**	101%
	100.0		3710		99%

BC=Ballistic Coefficient SD=Sectional Density
*Most Accurate Load Tested **Compressed Load

Use Maximum Loads with Caution
Refer to page 73 for additional safety information

300 Rem Ultra Mag - 150/155 grain		MAXIMUM S.A.A.M.I. O.A.C.L.		3.600"
		TESTED O.A.C.L.	B.C.	S.D.
AccuBond®	150gr. Spitzer	3.580"	0.435	0.226
Ballistic Tip®	150gr. Spitzer	3.580"	0.435	0.226
CT® Ballistic Silvertip®	150gr. Spitzer	3.580"	0.435	0.226
E-Tip®	150gr. Spitzer	3.550"	0.469	0.226
Due to internal construction differences, always begin with starting loads when using E-Tip® products.				
Partition®	150gr. Spitzer	3.560"	0.387	0.226
Custom Competition™	155gr. HPBT	3.580"	0.450	0.233

CASE TYPE:	Nosler		PRIMER TYPE	Fed 215
CASE HOLDS:	107.7	Gr. WATER	BARREL Length/Make	26" Wiseman
			BARREL Twist	1-10"

POWDER TYPE	POWDER CHG. GRS.		MUZZLE VEL. F.P.S.	LOAD DENSITY (VOLUME)
H4831SC Most Accurate Powder Tested	92.0	MAX.	3481	89%
	90.0		3409	87%
	88.0 *		3334	85%
IMR 4350	87.0 *	MAX.	3505	85%
	85.0		3435	84%
	83.0		3370	82%
H1000	102.0	MAX.	3513	99%
	100.0 *		3448	97%
	98.0		3380	95%
MAGPRO	98.5	MAX.	3523	94%
	96.5 *		3465	92%
	94.5		3386	90%
IMR 7828	93.0 *	MAX.	3546	92%
	91.0		3466	90%
	89.0		3374	88%
Magnum	104.5	MAX.	3552	99%
	102.5		3412	97%
	100.5 *		3390	95%
Retumbo	102.0	MAX.	3552	100%
	100.0 *		3514	98%
	98.0		3457	96%
RL22	94.5	MAX.	3564	95%
	92.5 *		3497	93%
	90.5		3463	91%
RL25	98.5	MAX.	3630	99%
	96.5		3538	97%
	94.5 *		3450	95%

BC=Ballistic Coefficient SD=Sectional Density
*Most Accurate Load Tested **Compressed Load

Use Maximum Loads with Caution
Refer to page 73 for additional safety information

300 Rem Ultra Mag - 165/168 grain		MAXIMUM S.A.A.M.I. O.A.C.L.		3.600"
		TESTED O.A.C.L.	B.C.	S.D.
AccuBond®	165gr. Spitzer	3.580"	0.475	0.248
Ballistic Tip®	165gr. Spitzer	3.580"	0.475	0.248
Partition®	165gr. Spitzer	3.580"	0.410	0.248
Ballistic Tip®	168gr. Spitzer	3.580"	0.490	0.253
CT® Ballistic Silvertip®	168gr. Spitzer	3.580"	0.490	0.253
Custom Competition™	168gr. HPBT	3.580"	0.462	0.253
E-Tip®	168gr. Spitzer	3.550"	0.503	0.253

Due to internal construction differences, always begin with starting loads when using E-Tip® products.

CASE TYPE:	Nosler		PRIMER TYPE	Fed 215
CASE HOLDS:	104.7	Gr. WATER	BARREL Length/Make	26" Wiseman
			BARREL Twist	1-10"

POWDER TYPE	POWDER CHG. GRS.		MUZZLE VEL. F.P.S.	LOAD DENSITY (VOLUME)
IMR 4350	84.0	MAX.	3317	85%
	82.0		3254	83%
	80.0 *		3156	81%
H4831SC	90.0 *	MAX.	3328	90%
	88.0		3275	88%
	86.0		3209	86%
H1000 Most Accurate Powder Tested	98.0	MAX.	3335	98%
	96.0 *		3270	96%
	94.0		3198	94%
IMR 7828	90.0	MAX.	3340	91%
	88.0 *		3256	89%
	86.0		3170	87%
RL22	90.0	MAX.	3368	93%
	88.0 *		3312	91%
	86.0		3249	89%
MAGPRO	95.0	MAX.	3370	94%
	93.0		3331	92%
	91.0 *		3312	90%
Retumbo	98.0 *	MAX.	3391	99%
	96.0		3330	97%
	94.0		3254	95%
Magnum	102.0	MAX.	3439	99%
	100.0		3366	97%
	98.0 *		3285	95%
RL25	96.5	MAX.	3483	100%
	94.5 *		3370	98%
	92.5		3283	96%

BC=Ballistic Coefficient SD=Sectional Density
*Most Accurate Load Tested **Compressed Load

Use Maximum Loads with Caution
Refer to page 73 for additional safety information

300 Rem Ultra Mag - 175/180 grain

Bullet	Type	MAXIMUM S.A.A.M.I. O.A.C.L. 3.600" / TESTED O.A.C.L.	B.C.	S.D.
Custom Competition™	175gr. HPBT	3.580"	0.505	0.264
AccuBond®	180gr. Spitzer	3.580"	0.507	0.271
Ballistic Tip®	180gr. Spitzer	3.580"	0.507	0.271
CT® Ballistic Silvertip®	180gr. Spitzer	3.580"	0.507	0.271
E-Tip®	180gr. Spitzer	3.550"	0.523	0.271
Due to internal construction differences, always begin with starting loads when using E-Tip® products.				
Partition®	180gr. Spitzer	3.560"	0.474	0.271
Partition®	180gr. PPT	3.460"	0.361	0.271

CASE TYPE:	Nosler		PRIMER TYPE	WLRM
CASE HOLDS:	105.2	Gr. WATER	BARREL Length/Make	24" H-S Prec.
			BARREL Twist	1-10"

POWDER TYPE	POWDER CHG. GRS.		MUZZLE VEL. F.P.S.	LOAD DENSITY (VOLUME)
Viht N560	86.0 *	MAX.	3105	91%
	84.0		3054	89%
	82.0		2983	87%
IMR 4350 Most Accurate Powder Tested	82.0	MAX.	3187	82%
	80.0 *		3125	80%
	78.0		3053	78%
H1000	94.0	MAX.	3200	94%
	92.0 *		3142	92%
	90.0		3068	90%
Retumbo	95.0	MAX.	3201	95%
	93.0 *		3121	93%
	91.0		3032	91%
H4831SC	88.0	MAX.	3205	87%
	86.0 *		3157	85%
	84.0		3089	83%
RL22	87.0 *	MAX.	3220	90%
	85.0		3146	88%
	83.0		3077	86%
4000-MR	84.0	MAX.	3223	84%
	82.0 *		3153	82%
	73.0		3065	73%
IMR 7828	89.0	MAX.	3229	90%
	87.0 *		3137	88%
	85.0		3093	86%
MAGPRO	94.0	MAX.	3243	92%
	92.0		3193	90%
	90.0 *		3126	88%
RL25	95.0	MAX.	3359	98%
	93.0		3271	96%
	91.0 *		3198	94%

BC=Ballistic Coefficient SD=Sectional Density
*Most Accurate Load Tested **Compressed Load

Use Maximum Loads with Caution
Refer to page 73 for additional safety information

300 Rem Ultra Mag - 190 grain		MAXIMUM S.A.A.M.I. O.A.C.L.		3.600"
		TESTED O.A.C.L.	B.C.	S.D.
Custom Competition™	190gr. HPBT	3.580"	0.530	0.286

CASE TYPE:	Nosler		PRIMER TYPE	WLRM
CASE HOLDS:	104.4	Gr. WATER	BARREL Length/Make	24" H-S Prec.
			BARREL Twist	1-10"

POWDER TYPE	POWDER CHG. GRS.		MUZZLE VEL. F.P.S.	LOAD DENSITY (VOLUME)
IMR 7828 SSC	88.0 *	MAX.	3106	87%
	86.0		3061	85%
	84.0		3012	83%
RL22	87.5 *	MAX.	3146	91%
	85.5		3096	89%
	83.5		3033	87%
RL25	92.0 *	MAX.	3159	96%
	90.0		3112	94%
	88.0		3032	92%
Magnum	93.0	MAX.	3173	90%
	91.0		3124	88%
	89.0 *		3071	87%
H1000 Most Accurate Powder Tested	92.5	MAX.	3173	93%
	90.5 *		3103	91%
	88.5		3049	89%
Retumbo	95.0	MAX.	3190	96%
	93.0		3115	94%
	91.0 *		3032	92%
Viht N560	89.0	MAX.	3192	95%
	87.0		3156	93%
	85.0 *		3093	90%

BC=Ballistic Coefficient SD=Sectional Density
*Most Accurate Load Tested **Compressed Load

Use Maximum Loads with Caution
Refer to page 73 for additional safety information

300 Rem Ultra Mag - 200 grain		MAXIMUM S.A.A.M.I. O.A.C.L.		3.600"
		TESTED O.A.C.L.	B.C.	S.D.
AccuBond®	200gr. Spitzer	3.580"	0.588	0.301
Partition®	200gr. Spitzer	3.580"	0.481	0.301

CASE TYPE:	Nosler		PRIMER TYPE	WLRM
CASE HOLDS:	102.8	Gr. WATER	BARREL Length/Make	24" H-S Prec.
			BARREL Twist	1-10"

POWDER TYPE	POWDER CHG. GRS.		MUZZLE VEL. F.P.S.	LOAD DENSITY (VOLUME)
Viht N170	82.0	MAX.	2940	89%
	80.0	*	2885	86%
	78.0		2830	84%
IMR 4350	81.0	MAX.	2995	83%
	79.0		2953	81%
	77.0	*	2916	79%
H1000	92.0	MAX.	3042	94%
	90.0		2988	92%
	88.0	*	2927	90%
H4831SC	86.5	MAX.	3057	88%
	84.5		3013	86%
	82.5	*	2967	84%
Magnum Most Accurate Powder Tested	93.0	* MAX.	3070	92%
	91.0		3005	90%
	89.0		2941	88%
Retumbo	91.5	* MAX.	3070	94%
	89.5		3015	92%
	87.5		2960	90%
IMR 7828	87.5	MAX.	3102	91%
	85.5		3023	88%
	83.5	*	2951	86%
MAGPRO	90.0	MAX.	3177	90%
	88.0	*	3019	88%
	86.0		2963	86%
RL25	91.5	MAX.	3185	97%
	89.5		3123	95%
	87.5	*	3064	93%

BC=Ballistic Coefficient SD=Sectional Density
*Most Accurate Load Tested **Compressed Load

Use Maximum Loads with Caution
Refer to page 73 for additional safety information

300 Rem Ultra Mag - 220 grain		MAXIMUM S.A.A.M.I. O.A.C.L.		3.600"
		TESTED O.A.C.L.	B.C.	S.D.
Partition®	220gr. Semi-Spitz	3.580"	0.351	0.331

CASE TYPE:	Nosler		PRIMER TYPE	WLRM
CASE HOLDS:	102.9	Gr. WATER	BARREL Length/Make	24" H-S Prec.
			BARREL Twist	1-10"

POWDER TYPE	POWDER CHG. GRS.		MUZZLE VEL. F.P.S.	LOAD DENSITY (VOLUME)
Viht N170	80.5	MAX.	2817	87%
	78.5 *		2753	85%
	76.5		2714	83%
IMR 4350 Most Accurate Powder Tested	77.0 *	MAX.	2835	91%
	75.0		2796	89%
	73.0		2741	86%
H4831SC	81.5	MAX.	2870	83%
	79.5 *		2848	80%
	77.5		2788	78%
IMR 7828	83.0 *	MAX.	2884	86%
	81.0		2818	84%
	79.0		2778	82%
H1000	89.0	MAX.	2903	91%
	87.0		2849	89%
	85.0 *		2796	86%
MAGPRO	88.0	MAX.	2904	88%
	86.0 *		2847	86%
	84.0		2819	84%
Magnum	90.0	MAX.	2916	89%
	88.0		2896	87%
	86.0 *		2823	85%
Retumbo	88.0 *	MAX.	2917	90%
	86.0		2864	88%
	84.0		2811	86%
RL25	89.0	MAX.	2973	94%
	87.0 *		2921	92%
	85.0		2872	90%

BC=Ballistic Coefficient SD=Sectional Density
*Most Accurate Load Tested **Compressed Load

Use Maximum Loads with Caution
Refer to page 73 for additional safety information

Joseph von Benedikt

300 WEATHERBY MAGNUM

For the big-game-cartridge addict, the 300 Weatherby Magnum is perhaps the most capable round available. It has the versatility of a Swiss Army knife and the horsepower and panache of a '67 Shelby Mustang. For a cartridge of its extreme performance, it is relatively mild-mannered in recoil and easy to handload. Those who know me, know that the elk-hunting disease has a terminal hold on me, so when I say that the 300 Weatherby is my favorite elk medicine it is not a statement I make lightly.

Though it was spawned back in 1944, I believe it is of better design than the recent afterburner-like .30 magnums that have come to market. It has a long, projectile-gripping neck that aids concentricity and rarely allows a bullet's base to intrude into powder space. The double-radiused shoulder is of debatable value but seems to aid smooth feeding and oozes cool factor. The cartridge is overbored, but not to the obscene extent of the 300 Remington Ultra Mag and the 300 Weatherby's own steroidal brother, the 30-378 Weatherby Magnum.

My handload pushes a 180-grain AccuBond at 3,175 fps, carries a whopping 4,043 ft-lbs of energy at the muzzle, and has an astonishing point-blank range of 498 yards on an elk-size (18 inch) vital zone. Due to the projectile's excellent BC and high starting velocity, it maintains enough speed to carry more than 1,500 ft-lbs of energy and to reliably expand out past 800 yards, which is much farther than I'll shoot at game.

One fall, deep in Montana's Bob Marshall Wilderness, I shot a heavy old bear at 130 yards in the willow-choked bottom of a high alpine basin. As bears will, it dove into the thickest cover available. While giving it time to expire, I spotted a big mule deer on the move, sneaking over a high ridge across the basin. It was 604 yards away, but conditions were perfect, and my rifle was a superbly accurate 300 Weatherby Magnum. I got steady over my daypack, calculated the effect that the steep angle would have on the shot, and rolled that old timberline monarch down the mountain.

Though the distance and angle gave me pause, I never questioned whether or not I had enough cartridge.

Joseph von Benedikt

Joseph von Benedikt is Editor in Chief of Shooting Times magazine

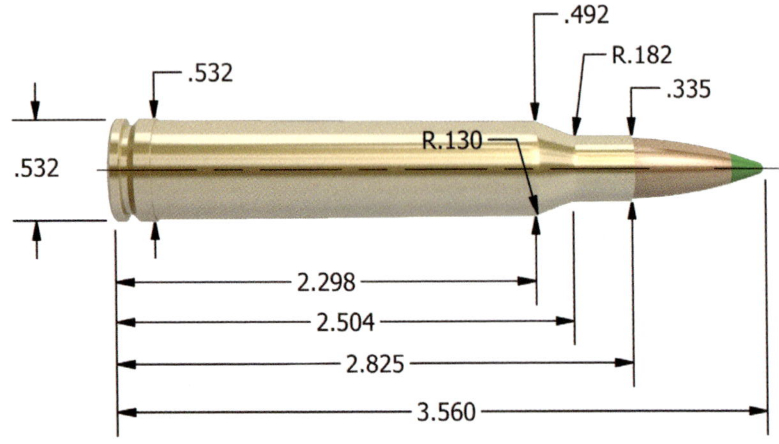

Maximum S.A.A.M.I. Overall Cartridge Length: 3.560"

BULLET CHOICES FOR THE 300 WEATHERBY MAGNUM

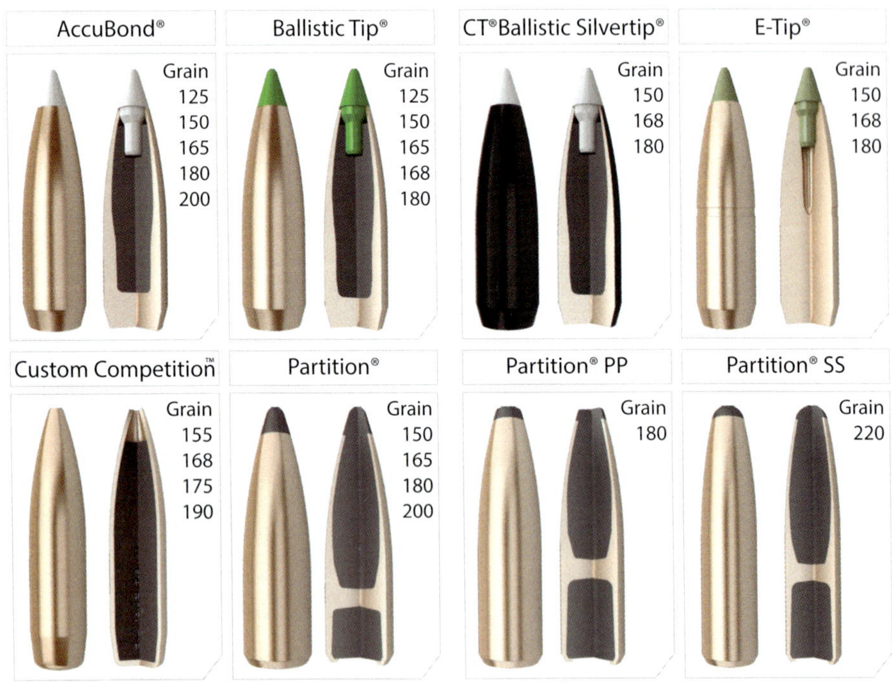

AccuBond®	Ballistic Tip®	CT® Ballistic Silvertip®	E-Tip®
Grain 125, 150, 165, 180, 200	Grain 125, 150, 165, 168, 180	Grain 150, 168, 180	Grain 150, 168, 180

Custom Competition™	Partition®	Partition® PP	Partition® SS
Grain 155, 168, 175, 190	Grain 150, 165, 180, 200	Grain 180	Grain 220

TECHNICAL INFORMATION

As with most Weatherby Magnums, there are a few points to keep in mind when loading for this cartridge:

- The chambers have "freebore", meaning they have a longer throat. For this reason, it is generally not possible to seat a bullet close to, or in contact with the lands (rifling).
- For best accuracy, bullets should be seated as long as the magazine will allow.
- Weatherby rifles of German manufacture (pre-1972) may have barrels with twist rates of 1 turn in 12". These barrels will not reliably stabilize bullets that are heavier than 180 grains.

300 Weatherby Mag - 125 grain

Bullet		MAXIMUM S.A.A.M.I. O.A.C.L.		3.560"
		TESTED O.A.C.L.	B.C.	S.D.
AccuBond®	125gr. Spitzer	3.530"	0.366	0.188
Ballistic Tip®	125gr. Spitzer	3.530"	0.366	0.188

CASE TYPE:	Nosler		PRIMER TYPE	WLRM
CASE HOLDS:	96.0	Gr. WATER	BARREL Length/Make	26" Pac-Nor
			BARREL Twist	1-10"

POWDER TYPE	POWDER CHG. GRS.		MUZZLE VEL. F.P.S.		LOAD DENSITY (VOLUME)
IMR 4350	80.5	* MAX.	3390		89%
	78.5		3270		87%
	76.5		3150		84%
IMR 4831	82.0	* MAX.	3442		91%
	80.0		3317		89%
	78.0		3192		86%
RL19	83.0	* MAX.	3462		94%
	81.0		3347		92%
	79.0		3232		89%
H4350	81.5	* MAX.	3472		90%
	79.5		3357		88%
	77.5		3242		85%
Viht N165 Most Accurate Powder Tested	88.0	* MAX.	3541	**	102%
	86.0		3462		100%
	84.0		3384		97%
Viht N160	83.5	MAX.	3561		97%
	81.5		3451		94%
	79.5	*	3341		92%
RL22	89.0	* MAX.	3682	**	101%
	87.0		3577		99%
	85.0		3472		96%

BC=Ballistic Coefficient SD=Sectional Density
*Most Accurate Load Tested **Compressed Load

Use Maximum Loads with Caution
Refer to page 73 for additional safety information

300 Weatherby Mag - 150/155 grain		MAXIMUM S.A.A.M.I. O.A.C.L.		3.560"
		TESTED O.A.C.L.	B.C.	S.D.
AccuBond®	150gr. Spitzer	3.550"	0.435	0.226
Ballistic Tip®	150gr. Spitzer	3.550"	0.435	0.226
CT® Ballistic Silvertip®	150gr. Spitzer	3.550"	0.435	0.226
E-Tip®	150gr. Spitzer	3.500"	0.469	0.226
Due to internal construction differences, always begin with starting loads when using E-Tip® products.				
Partition®	150gr. Spitzer	3.520"	0.386	0.226
Custom Competition™	155gr. HPBT	3.540"	0.450	0.233

CASE TYPE:	Nosler		PRIMER TYPE	WLRM
CASE HOLDS:	92.2	Gr. WATER	BARREL Length/Make	26" Pac-Nor
			BARREL Twist	1-10"

POWDER TYPE	POWDER CHG. GRS.		MUZZLE VEL. F.P.S.		LOAD DENSITY (VOLUME)
H4831 Most Accurate Powder Tested	85.0	MAX.	3232		96%
	83.0		3127		94%
	81.0	*	3022		92%
Hybrid 100V	77.0	MAX.	3259		86%
	75.0	*	3188		84%
	73.0		3137		82%
RL19	81.0	MAX.	3327		95%
	79.0		3206		93%
	77.0	*	3085		91%
Viht N160	81.0	MAX.	3333		98%
	79.0		3236		95%
	77.0	*	3139		93%
IMR 4831	81.0	MAX.	3352		93%
	79.0		3257		91%
	77.0	*	3162		89%
Supreme 780	85.5	MAX.	3363		97%
	83.5	*	3290		95%
	81.5		3204		93%
Norma MRP	86.0	MAX.	3387	**	104%
	84.0	*	3265	**	101%
	82.0		3166		99%
IMR 7828	89.0	* MAX.	3408	**	103%
	87.0		3313		100%
	85.0		3218		98%
IMR 4350	80.0	MAX.	3412		92%
	78.0		3277		90%
	76.0	*	3142		87%

BC=Ballistic Coefficient SD=Sectional Density
*Most Accurate Load Tested **Compressed Load

Use Maximum Loads with Caution
Refer to page 73 for additional safety information

300 Weatherby Mag - 165/168 grain

		MAXIMUM S.A.A.M.I. O.A.C.L.		3.560"
		TESTED O.A.C.L.	B.C.	S.D.
AccuBond®	165gr. Spitzer	3.550"	0.475	0.248
Ballistic Tip®	165gr. Spitzer	3.550"	0.475	0.248
Partition®	165gr. Spitzer	3.520	0.410	0.248
Ballistic Tip®	168gr. Spitzer	3.550"	0.490	0.253
CT® Ballistic Silvertip®	168gr. Spitzer	3.550"	0.490	0.253
Custom Competition™	168gr. HPBT	3.540"	0.462	0.253
E-Tip®	168gr. Spitzer	3.500"	0.503	0.253

Due to internal construction differences, always begin with starting loads when using E-Tip® products.

CASE TYPE:	Nosler		PRIMER TYPE	WLRM
CASE HOLDS:	88.8	Gr. WATER	BARREL Length/Make	26" Pac-Nor
			BARREL Twist	1-10"

POWDER TYPE	POWDER CHG. GRS.		MUZZLE VEL. F.P.S.		LOAD DENSITY (VOLUME)
RL19	80.0	MAX.	3113		98%
	78.0 *		3013		95%
	76.0		2969		93%
H4831	82.0	MAX.	3144		96%
	80.0		3046		94%
	78.0 *		2960		91%
IMR 4350	76.0	MAX.	3153		91%
	74.0 *		3060		88%
	72.0		2986		86%
IMR 4831 Most Accurate Powder Tested	79.0	MAX.	3158		95%
	77.0		3103		92%
	75.0 *		3048		90%
MAGPRO	88.0	MAX.	3171	**	102%
	86.0		3115		100%
	84.0 *		3063		98%
Magnum	91.0	MAX.	3217	**	104%
	89.0 *		3176	**	102%
	87.0		3130		99%
IMR 7828 SSC	83.5 *	MAX.	3224		97%
	81.5		3137		95%
	79.5		3068		92%
Hybrid 100V	78.5 *	MAX.	3230		91%
	76.5		3171		89%
	74.5		3095		86%
H1000	90.0 *	MAX.	3248	**	106%
	88.0		3188	**	104%
	86.0		3132	**	101%
Supreme 780	85.5	MAX.	3281	**	101%
	83.5		3219		98%
	81.5 *		3124		96%

BC=Ballistic Coefficient SD=Sectional Density
*Most Accurate Load Tested **Compressed Load

Use Maximum Loads with Caution
Refer to page 73 for additional safety information

300 Weatherby Mag - 175/180 grain		MAXIMUM S.A.A.M.I. O.A.C.L.		3.560"
		TESTED O.A.C.L.	B.C.	S.D.
Custom Competition™	175gr. HPBT	3.550"	0.505	0.264
AccuBond®	180gr. Spitzer	3.550"	0.507	0.271
Ballistic Tip®	180gr. Spitzer	3.550"	0.507	0.271
CT® Ballistic Silvertip®	180gr. Spitzer	3.550"	0.507	0.271
E-Tip®	180gr. Spitzer	3.500"	0.523	0.271
Due to internal construction differences, always begin with starting loads when using E-Tip® products.				
Partition®	180gr. Spitzer	3.520"	0.474	0.271
Partition®	180gr. PPT	3.420"	0.361	0.271

CASE TYPE:	Nosler		PRIMER TYPE	WLRM
CASE HOLDS:	91.8	Gr. WATER	BARREL Length/Make	26" Pac-Nor
			BARREL Twist	1-10"

POWDER TYPE	POWDER CHG. GRS.		MUZZLE VEL. F.P.S.		LOAD DENSITY (VOLUME)
IMR 4350	74.5	MAX.	3024		86%
	72.5 *		2939		84%
	70.5		2855		81%
IMR 4831	77.0	MAX.	3056		89%
	75.0		3001		87%
	73.0 *		2946		85%
RL22	80.0 *	MAX.	3132		95%
	78.0		3039		92%
	76.0		2945		90%
Supreme 780	83.0	MAX.	3136		95%
	81.0		3100		92%
	79.0 *		3041		90%
H1000	89.0	MAX.	3146	**	102%
	87.0		3071		99%
	85.0 *		3012		97%
RL25	86.5	MAX.	3152	**	102%
	84.5 *		3071		100%
	82.5		3002		98%
Hybrid 100V	78.0	MAX.	3162		88%
	76.0 *		3096		85%
	74.0		2993		83%
IMR 7828 Most Accurate Powder Tested	84.5 *	MAX.	3185		98%
	82.5		3091		96%
	80.5		2998		93%
RL19	79.0	MAX.	3198		94%
	77.0 *		3088		91%
	75.0		2992		89%

BC=Ballistic Coefficient SD=Sectional Density
*Most Accurate Load Tested **Compressed Load

Use Maximum Loads with Caution
Refer to page 73 for additional safety information

300 Weatherby Mag - 190 grain

		MAXIMUM S.A.A.M.I. O.A.C.L.			3.560"
		TESTED O.A.C.L.		B.C.	S.D.
Custom Competition™	190gr. HPBT	3.550"		0.530	0.286
CASE TYPE:	Nosler	PRIMER TYPE		WLRM	
CASE HOLDS:	92.8 Gr. WATER	BARREL Length/Make		26" Pac-Nor	
		BARREL Twist		1-10"	

POWDER TYPE	POWDER CHG. GRS.		MUZZLE VEL. F.P.S.	LOAD DENSITY (VOLUME)
IMR 4831	75.0	MAX.	2957	86%
	73.0		2868	84%
	71.0 *		2783	81%
Viht N165 Most Accurate Powder Tested	80.5	MAX.	2996	96%
	78.5		2924	94%
	76.5 *		2850	92%
Supreme 780	80.0 *	MAX.	3031	90%
	78.0		2974	88%
	76.0		2913	86%
RL22	79.0	MAX.	3043	93%
	77.0		2962	90%
	75.0 *		2880	88%
IMR 4350	75.0	MAX.	3082	86%
	73.0		3024	83%
	71.0 *		2962	81%
Hybrid 100V	75.0	MAX.	3109	83%
	73.0		3037	81%
	71.0 *		2966	79%
IMR 7828 SSC	83.0 *	MAX.	3135	92%
	81.0		3043	90%
	79.0		2955	88%
H1000	88.0	MAX.	3153	99%
	86.0		3083	97%
	84.0 *		3015	95%

BC=Ballistic Coefficient SD=Sectional Density
*Most Accurate Load Tested **Compressed Load

Use Maximum Loads with Caution
Refer to page 73 for additional safety information

300 Weatherby Mag - 200 grain		MAXIMUM S.A.A.M.I. O.A.C.L.		3.560"
		TESTED O.A.C.L.	B.C.	S.D.
AccuBond®	200gr. Spitzer	3.550"	0.588	0.301
Partition®	200gr. Spitzer	3.510"	0.481	0.301

CASE TYPE:	Nosler		PRIMER TYPE	WLRM
CASE HOLDS:	87.6	Gr. WATER	BARREL Length/Make	26" Pac-Nor
			BARREL Twist	1-10"

POWDER TYPE	POWDER CHG. GRS.		MUZZLE VEL. F.P.S.		LOAD DENSITY (VOLUME)
IMR 4831	72.0	MAX.	2818		87%
	70.0		2763		85%
	68.0	*	2708		83%
Viht N165	78.0	MAX.	2894		99%
	76.0		2830		96%
	74.0	*	2767		94%
Hybrid 100V	74.0	MAX.	2924		87%
	72.0		2872		85%
	70.0	*	2789		82%
Retumbo	86.0	MAX.	2928	**	103%
	84.0		2866	**	101%
	82.0	*	2807		99%
IMR 4350	74.0	MAX.	2960		89%
	72.0		2880		87%
	70.0	*	2800		85%
RL22	78.0	MAX.	2982		97%
Most Accurate	76.0		2897		94%
Powder Tested	74.0	*	2812		92%
RL25	85.0	MAX.	3008	**	105%
	83.0		2916	**	103%
	81.0	*	2871	**	101%
IMR 7828	83.0	MAX.	3028	**	101%
	81.0		2943		98%
	79.0	*	2858		96%
H1000	87.0	MAX.	3039	**	104%
	85.0		2972	**	102%
	83.0	*	2892		99%

BC=Ballistic Coefficient SD=Sectional Density
*Most Accurate Load Tested **Compressed Load

Use Maximum Loads with Caution
Refer to page 73 for additional safety information

300 Weatherby Mag - 220 grain

300 Weatherby Mag - 220 grain		MAXIMUM S.A.A.M.I. O.A.C.L.		3.560"
		TESTED O.A.C.L.	B.C.	S.D.
Partition®	220gr. Semi-Spitz	3.550"	0.351	0.331

CASE TYPE:	Nosler		PRIMER TYPE	WLRM
CASE HOLDS:	90.2	Gr. WATER	BARREL Length/Make	26" Pac-Nor
			BARREL Twist	1-10"

POWDER TYPE	POWDER CHG. GRS.		MUZZLE VEL. F.P.S.	LOAD DENSITY (VOLUME)
A-4350 Most Accurate Powder Tested	68.5	* MAX.	2643	85%
	66.5		2569	83%
	64.5		2496	80%
Viht N165	75.5	* MAX.	2722	93%
	73.5		2654	91%
	71.5		2586	88%
H1000	81.0	MAX.	2724	94%
	79.0		2661	92%
	77.0	*	2598	89%
RL19	72.0	MAX.	2745	87%
	70.0		2671	84%
	68.0	*	2597	82%
IMR 4350	70.5	* MAX.	2770	83%
	68.5		2697	80%
	66.5		2623	78%
Magnum	83.0	MAX.	2772	93%
	81.0	*	2702	91%
	79.0		2604	89%
RL22	75.0	* MAX.	2805	90%
	73.0		2726	88%
	71.0		2647	86%
IMR 7828	77.0	MAX.	2827	91%
	75.0		2746	88%
	73.0	*	2664	86%
H4831SC	76.0	* MAX.	2840	88%
	74.0		2755	85%
	72.0		2710	83%
Norma MRP	80.0	MAX.	2841	99%
	78.0		2774	96%
	76.0	*	2703	94%

BC=Ballistic Coefficient SD=Sectional Density
*Most Accurate Load Tested **Compressed Load

Use Maximum Loads with Caution
Refer to page 73 for additional safety information

Aaron Carter

30-378 WEATHERBY MAGNUM

Developed by Roy Weatherby at the behest of the U.S. Army in 1959, the 30-378 Wby. Mag. is essentially a "necked-down" 378 Wby. Mag., thereby taking advantage of the parent case's voluminous interior to propel bullets to considerable velocities; it's about as much performance-wise that can be realized from a .30-cal. cartridge while exhibiting a semblance of efficiency.

The 30-378 Wby. Mag. can utilize Nosler's full lineup of .30-cal. projectiles, ranging from the 125-gr. Ballistic Tip—posting a top velocity nearing 3,850 f.p.s.—to the 220-gr. Semi-Spitzer Partition. However, the cartridge's large case makes it especially susceptible to the law of diminishing returns. When compared to its elder sibling, the 300 Wby. Mag., immediately noticeable is the 30-378's inefficiency when lightweight projectiles and traditional-type propellants are used; substantial increases in propellant usage are necessary for relatively small velocity gains. The cartridge, however, truly "comes into its element" with bullets weighing 180-grains or heavier.

Heavy, aerodynamic .30-cal. bullets, like those of many other calibers, have lengths that force them to be seated deeply in the cases of most cartridges to meet maximum cartridge overall lengths (COL), resulting in the displacement of much propellant. The capacity of the 30-378 Wby. Mag.'s case is such that even when lengthy, spitzer projectiles—especially those with secant ogives—are used, there is still sufficient space for large charges of slow-burning propellants. As such, the 30-378 Wby. Mag. can safely attain velocities unapproachable by any other non-proprietary factory .30-cal. chambering. This both flattens trajectories and reduces time to target, the latter resulting in less wind deflection—hence the reason long-range competitors and hunters have taken to the cartridge.

Because of the extreme velocities at which the 30-378 Wby. Mag. propels bullets, those of controlled expansion are a necessity. To me, Nosler's standout is the 200-gr. AccuBond; it'll retain upward of 70 percent of its original weight, open at velocities as low as 1,800 f.p.s (a concern at longer ranges) and has an incredible ballistic coefficient of .588, which translates to remarkable downrange ballistics. For example, when zeroed at 250 yds., the AccuBond—propelled to 3200 f.p.s.—drops 5.4" at 300 yds. and 15.5" at 400 yds., where it still has 3,634 ft.-lbs. of energy remaining—ample for nearly all of the world's game.

As for shooting, a 30-378 Wby. Mag.-chambered rifle, as long as it is equipped with a muzzle brake, to me felt recoil is comparable to that of a standard-weight hunting rifle in 270 Win., if not lighter; without a brake, not so much. Perhaps the worst attribute is the attendant muzzle blast; however, if hearing protection is worn, as should be done for all shooting, this should pose no problem. It's simply one of the costs associated with harnessing the power of one of the most potent sporting cartridges in existence. For those who are unafraid, they'll be rewarded handsomely in the field.

Aaron Carter

Aaron Carter, Managing Editor, American Rifleman

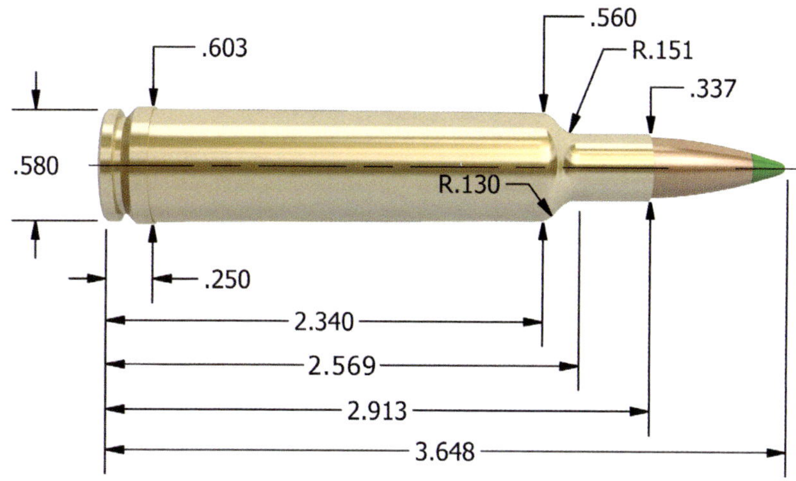

Maximum S.A.A.M.I. Overall Cartridge Length: 3.648"

BULLET CHOICES FOR THE 30-378 WEATHERBY MAGNUM

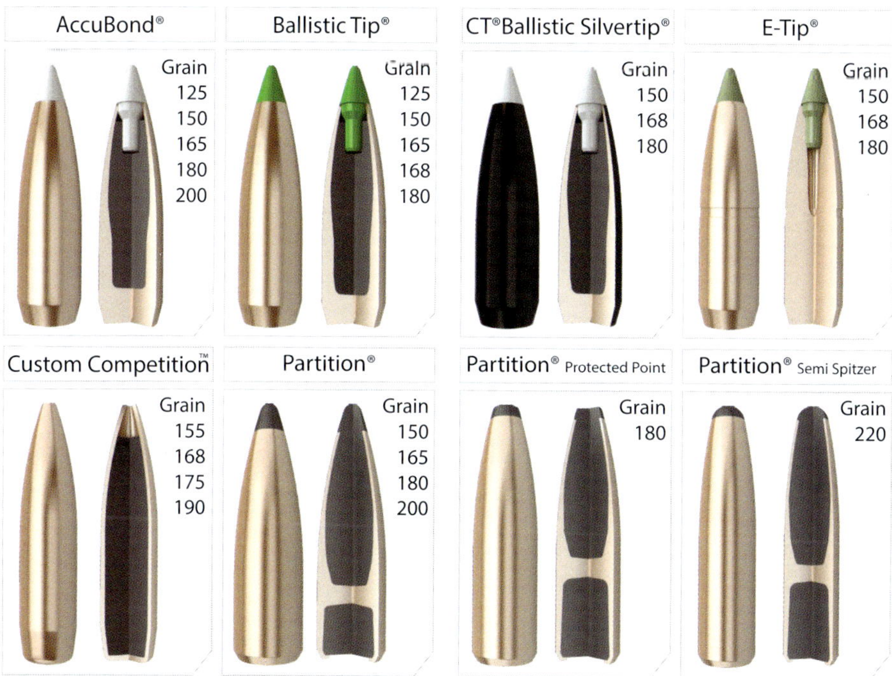

TECHNICAL INFORMATION

Best accuracy with this cartridge will generally be achieved by seating the bullet as long as the magazine will allow. There are a couple of points to keep in mind with this cartridge when establishing a seating depth:

- Always check to ensure that you can eject a loaded cartridge, as some Weatherby rifles have magazines that are longer than the ejection port in the receiver.
- Factory chambers have "freebore", meaning they have a longer throat. For this reason, it is generally not possible to seat a bullet close to, or in contact with the lands (rifling).

30-378 Weatherby Mag - 125 grain		MAXIMUM S.A.A.M.I. O.A.C.L.		3.648"
		TESTED O.A.C.L.	B.C.	S.D.
AccuBond®	125gr. Spitzer	3.630"	0.366	0.188
Ballistic Tip®	125gr. Spitzer	3.630"	0.366	0.188

CASE TYPE:	Nosler		PRIMER TYPE	Fed 215
CASE HOLDS:	128	Gr. WATER	BARREL Length/Make	26" Lilja
			BARREL Twist	1-10"

POWDER TYPE	POWDER CHG. GRS.	MUZZLE VEL. F.P.S.	LOAD DENSITY (VOLUME)
H4831SC Most Accurate Powder Tested	109.0 MAX.	3787	89%
	107.0	3745	87%
	105.0 *	3642	86%
H1000	120.0 MAX.	3800	98%
	118.0	3768	97%
	116.0 *	3694	95%
IMR 7828	109.0 MAX.	3812	91%
	107.0 *	3774	89%
	105.0	3669	87%
Viht N165	113.0 MAX.	3835	98%
	111.0 *	3728	97%
	109.0	3661	95%

BC=Ballistic Coefficient SD=Sectional Density
*Most Accurate Load Tested **Compressed Load

Use Maximum Loads with Caution
Refer to page 73 for additional safety information

30-378 Wby Mag - 150/155 grain		MAXIMUM S.A.A.M.I. O.A.C.L.		3.648"
		TESTED O.A.C.L.	B.C.	S.D.
AccuBond®	150gr. Spitzer	3.630"	0.435	0.226
Ballistic Tip®	150gr. Spitzer	3.630"	0.435	0.226
CT® Ballistic Silvertip®	150gr. Spitzer	3.630"	0.435	0.226
E-Tip®	150gr. Spitzer	3.600"	0.469	0.226
Due to internal construction differences, always begin with starting loads when using E-Tip® products.				
Partition®	150gr. Spitzer	3.630"	0.387	0.226
Custom Competition™	155gr. HPBT	3.630"	0.450	0.233

CASE TYPE:	Nosler		PRIMER TYPE	Fed 215
CASE HOLDS:	124.0	Gr. WATER	BARREL Length/Make	26" Lilja
			BARREL Twist	1-10"

POWDER TYPE	POWDER CHG. GRS.		MUZZLE VEL. F.P.S.	LOAD DENSITY (VOLUME)
H1000	105.0	MAX.	3394	89%
	103.0		3327	87%
	101.0 *		3260	85%
RL22	97.0 *	MAX.	3446	85%
	95.0		3382	83%
	93.0		3318	82%
H50BMG Most Accurate Powder Tested	120.0	MAX.	3449	100%
	118.0		3397	99%
	116.0 *		3345	97%
IMR 7828	102.0	MAX.	3498	88%
	100.0		3425	86%
	98.0 *		3351	84%

BC=Ballistic Coefficient SD=Sectional Density
*Most Accurate Load Tested **Compressed Load

Use Maximum Loads with Caution
Refer to page 73 for additional safety information

30-378 Wby Mag - 165/168 grain		MAXIMUM S.A.A.M.I. O.A.C.L.		3.648"
		TESTED O.A.C.L.	B.C.	S.D.
AccuBond®	165gr. Spitzer	3.630"	0.475	0.248
Ballistic Tip®	165gr. Spitzer	3.630"	0.475	0.248
Partition®	165gr. Spitzer	3.630"	0.410	0.248
Ballistic Tip®	168gr. Spitzer	3.630"	0.490	0.253
CT® Ballistic Silvertip®	168gr. Spitzer	3.630"	0.490	0.253
Custom Competition™	168gr. HPBT	3.630"	0.462	0.253
E-Tip®	168gr. Spitzer	3.600"	0.503	0.253

Due to internal construction differences, always begin with starting loads when using E-Tip® products.

CASE TYPE:	Nosler		PRIMER TYPE	Fed 215
CASE HOLDS:	122.0	Gr. WATER	BARREL Length/Make	26" Lilja
			BARREL Twist	1-10"

POWDER TYPE	POWDER CHG. GRS.		MUZZLE VEL. F.P.S.	LOAD DENSITY (VOLUME)
H1000	99.5 *	MAX.	3257	85%
	97.5		3203	84%
	95.5		3150	82%
RL22	94.0 *	MAX.	3336	84%
	92.0		3280	82%
	90.0		3224	80%
H50BMG Most Accurate Powder Tested	116.0	MAX.	3337	99%
	114.0		3289	97%
	112.0 *		3242	95%
IMR 7828	99.0	MAX.	3378	86%
	97.0		3307	85%
	95.0 *		3237	83%

BC=Ballistic Coefficient SD=Sectional Density
*Most Accurate Load Tested **Compressed Load

Use Maximum Loads with Caution
Refer to page 73 for additional safety information

30-378 Wby Mag - 175/180 grain		MAXIMUM S.A.A.M.I. O.A.C.L.		3.648"
		TESTED O.A.C.L.	B.C.	S.D.
Custom Competition™	175gr. HPBT	3.630"	0.505	0.264
AccuBond®	180gr. Spitzer	3.630"	0.507	0.271
Ballistic Tip®	180gr. Spitzer	3.630"	0.507	0.271
CT® Ballistic Silvertip®	180gr. Spitzer	3.630"	0.507	0.271
E-Tip®	180gr. Spitzer	3.600"	0.523	0.271
Due to internal construction differences, always begin with starting loads when using E-Tip® products.				
Partition®	180gr. Spitzer	3.630"	0.474	0.271
Partition®	180gr. PPT	3.600"	0.361	0.271

CASE TYPE:	Nosler		PRIMER TYPE	Fed 215
CASE HOLDS:	121.0	Gr. WATER	BARREL Length/Make	26" Lilja
			BARREL Twist	1-10"

POWDER TYPE	POWDER CHG. GRS.		MUZZLE VEL. F.P.S.	LOAD DENSITY (VOLUME)
RL22	89.5	MAX.	3146	80%
	87.5		3085	79%
	85.5	*	3024	77%
H1000	100.0	MAX.	3187	87%
	98.0		3128	85%
	96.0	*	3070	83%
H50BMG Most Accurate Powder Tested	114.0	MAX.	3248	98%
	112.0		3203	96%
	110.0	*	3158	94%
IMR 7828	97.0	MAX.	3264	85%
	95.0		3190	84%
	93.0	*	3115	82%

BC=Ballistic Coefficient SD=Sectional Density
*Most Accurate Load Tested **Compressed Load

Use Maximum Loads with Caution
Refer to page 73 for additional safety information

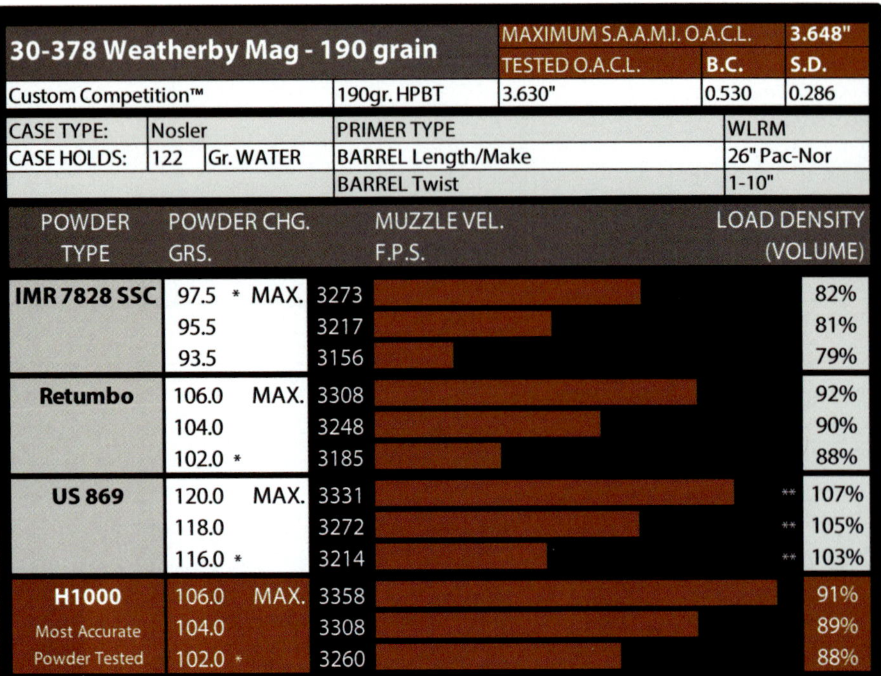

30-378 Weatherby Mag - 190 grain		MAXIMUM S.A.A.M.I. O.A.C.L.		3.648"
		TESTED O.A.C.L.	B.C.	S.D.
Custom Competition™	190gr. HPBT	3.630"	0.530	0.286

CASE TYPE:	Nosler		PRIMER TYPE	WLRM
CASE HOLDS:	122	Gr. WATER	BARREL Length/Make	26" Pac-Nor
			BARREL Twist	1-10"

POWDER TYPE	POWDER CHG. GRS.		MUZZLE VEL. F.P.S.		LOAD DENSITY (VOLUME)
IMR 7828 SSC	97.5 *	MAX.	3273		82%
	95.5		3217		81%
	93.5		3156		79%
Retumbo	106.0	MAX.	3308		92%
	104.0		3248		90%
	102.0 *		3185		88%
US 869	120.0	MAX.	3331	**	107%
	118.0		3272	**	105%
	116.0 *		3214	**	103%
H1000 Most Accurate Powder Tested	106.0	MAX.	3358		91%
	104.0		3308		89%
	102.0 *		3260		88%

BC=Ballistic Coefficient SD=Sectional Density
*Most Accurate Load Tested **Compressed Load

Use Maximum Loads with Caution
Refer to page 73 for additional safety information

30-378 Weatherby Mag - 200 grain		MAXIMUM S.A.A.M.I. O.A.C.L.		3.648"
		TESTED O.A.C.L.	B.C.	S.D.
AccuBond®	200gr. Spitzer	3.630"	0.588	0.301
Partition®	200gr. Spitzer	3.630"	0.481	0.301

CASE TYPE:	Nosler		PRIMER TYPE	Fed 215
CASE HOLDS:	121	Gr. WATER	BARREL Length/Make	26" Lilja
			BARREL Twist	1-10"

POWDER TYPE	POWDER CHG. GRS.		MUZZLE VEL. F.P.S.	LOAD DENSITY (VOLUME)
RL22	85.5 *	MAX.	2966	77%
	83.5		2908	75%
	81.5		2855	73%
H1000	99.0	MAX.	3068	86%
	97.0		3014	84%
	95.0 *		2959	82%
H50BMG Most Accurate Powder Tested	112.0	MAX.	3068	96%
	110.0		3017	94%
	108.0 *		2965	93%
IMR 7828	94.0	MAX.	3102	83%
	92.0		3047	81%
	90.0 *		2993	79%

BC=Ballistic Coefficient SD=Sectional Density
*Most Accurate Load Tested **Compressed Load

Use Maximum Loads with Caution
Refer to page 73 for additional safety information

30-378 Weatherby Mag - 220 grain		MAXIMUM S.A.A.M.I. O.A.C.L.		3.648"
		TESTED O.A.C.L.	B.C.	S.D.
Partition®	220gr. Semi-Spitz	3.630"	0.351	0.331

CASE TYPE:	Nosler		PRIMER TYPE	Fed 215
CASE HOLDS:	121	Gr. WATER	BARREL Length/Make	26" Lilja
			BARREL Twist	1-10"

POWDER TYPE	POWDER CHG. GRS.		MUZZLE VEL. F.P.S.	LOAD DENSITY (VOLUME)
IMR 7828 Most Accurate Powder Tested	91.0 *	MAX.	2879	80%
	89.0		2839	78%
	87.0		2782	77%
H1000	98.5 *	MAX.	2923	85%
	96.5		2865	84%
	94.5		2843	82%
H50BMG	110.0	MAX.	2980	94%
	108.0		2936	93%
	106.0 *		2888	91%
Viht N165	93.0	MAX.	2992	86%
	91.0		2927	84%
	89.0 *		2870	82%

BC=Ballistic Coefficient SD=Sectional Density
*Most Accurate Load Tested **Compressed Load

Use Maximum Loads with Caution
Refer to page 73 for additional safety information

8X57 JS MAUSER

A number of conundrums and ironies are attached to the 8x57mm Mauser. One irony is that Mauser had nothing to do with developing this cartridge and did not chamber any of their bolt actions for it until the Model 1898. Yet in America at least, 8x57mm has become the cartridge most identified with Mauser rifles and most researchers feel that of over 100 million Mauser rifles produced, the majority were indeed chambered for 8x57mm.

A conundrum about the 8x57mm is that such is the designation used here but in Europe it is primarily known as 7.9x57mm or 7.92x57mm. Here's an interesting bit of trivia. German K98k rifles from circa 1935 until the early 1940s, were stamped with the exact barrel bore diameter atop the chamber. I have one K98k stamped "7.9" and another stamped "7.91." Also it is ironic that many people think the 8x57mm Mauser was particular to Germany's military forces. It was not. Among others Poland, Czechoslovakia, Romania, Yugoslavia, and even China adopted this cartridge for their armies at one time or the other.

Then there is the matter of the "J" and "JS" designations. Actually the "J" is a mistranslation for "I." Regardless "J" (or "I" in Europe) on a box of ammunition means its bullets are .318 inch. If a box of ammunition is labeled "JS" ("IS" in Europe) then its bullets are .323 inch. Generally speaking, the above named military organizations had loads ranging from 154 grain bullets at a nominal 2,880 fps to 198 grain bullets at a nominal 2,540 fps.

Here's a final puzzle. Even though the 8x57mm was capable of more muzzle energy with flatter trajectory than the 7x57mm it never developed a sizeable following among sportsmen in North America. Regardless, an 8x57mm rifle loaded with Nosler's 200 grain Accubond bullet with a muzzle velocity of about 2,500 fps would be suitable for any non-dangerous game animal.

Mike Venturino at the shooting bench with one of his World War II vintage German sniper rifles.

Michael L. Venturino

Mike Venturino is a Contributing Editor to Rifle, Handloader, Guns, and American Handgunner magazines.

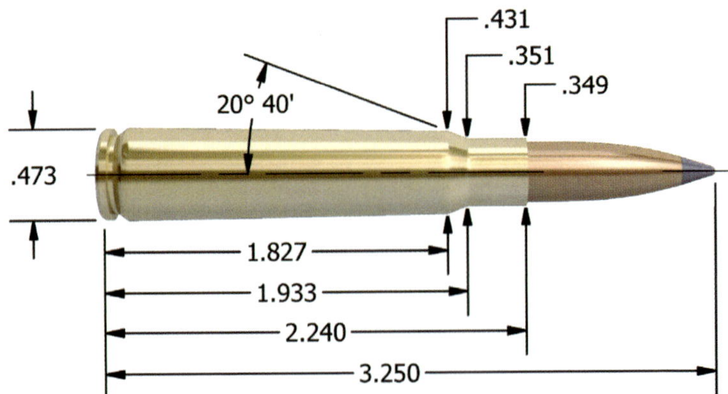

Maximum S.A.A.M.I. Overall Cartridge Length: 3.250"

BULLET CHOICES FOR THE 8X57 JS MAUSER

TECHNICAL INFORMATION

There are two versions of the 8x57mm Mauser cartridge. The original version of this cartridge, the 8x57mm (J), was designed for use with .318" diameter bullets. Later the cartridge was redesigned to utilize a bullet of .323" diameter and was designated the 8x57mm (JS). This is an important distinction to make, as firing a .323" bullet through a firearm barreled for the 8x57mm (J) may cause dangerous chamber pressures. The data listed is intended for use with firearms chambered for the 8x57mm (JS), and .323" diameter bullets only! These loads are intended for use only with new firearms in good condition. If you have an old military rifle, or are unsure about the bore diameter, it must be thoroughly inspected by a competent gunsmith prior to use.

8x57mm JS Mauser - 180 grain		MAXIMUM S.A.A.M.I. O.A.C.L.		3.250"
		TESTED O.A.C.L.	B.C.	S.D.
Ballistic Tip®	180gr. Spitzer	3.030"	0.394	0.246
CT® Ballistic Silvertip®	180gr. Spitzer	3.030"	0.394	0.246
E-Tip®	180gr. Spitzer	3.030"	0.427	0.246

Due to internal construction differences, always begin with starting loads when using E-Tip® products.

CASE TYPE:	Federal		PRIMER TYPE	WLR
CASE HOLDS:	52.4	Gr. WATER	BARREL Length/Make	24" Lilja
			BARREL Twist	1-10"

POWDER TYPE	POWDER CHG. GRS.		MUZZLE VEL. F.P.S.		LOAD DENSITY (VOLUME)
Viht N140	46.0	MAX.	2541	**	102%
	44.0 *		2444		98%
	42.0		2334		93%
IMR 4350	52.0	MAX.	2545	**	105%
	50.0		2494	**	101%
	48.0 *		2403		97%
IMR 4064 Most Accurate Powder Tested	47.0	MAX.	2636		99%
	45.0		2532		94%
	43.0 *		2410		90%
RL15	48.5	MAX.	2667		96%
	46.5		2576		92%
	44.5 *		2462		88%
Varget	49.0	MAX.	2669	**	101%
	47.0		2567		97%
	45.0 *		2464		93%

BC=Ballistic Coefficient SD=Sectional Density
*Most Accurate Load Tested **Compressed Load

Use Maximum Loads with Caution
Refer to page 73 for additional safety information

8x57mm JS Mauser - 200 grain		MAXIMUM S.A.A.M.I. O.A.C.L.		3.250"
		TESTED O.A.C.L.	B.C.	S.D.
AccuBond®	200gr. Spitzer	3.030"	0.450	0.274
Custom Competition™	200gr. HPBT	3.030"	0.520	0.274
Partition®	200gr. Spitzer	3.030"	0.426	0.274

CASE TYPE:	Federal		PRIMER TYPE	WLR
CASE HOLDS:	51.5	Gr. WATER	BARREL Length/Make	24" Lilja
			BARREL Twist	1-10"

POWDER TYPE	POWDER CHG. GRS.		MUZZLE VEL. F.P.S.		LOAD DENSITY (VOLUME)
Varget Most Accurate Powder Tested	47.5	MAX.	2530		100%
	45.5		2451		96%
	43.5 *		2341		91%
Viht N160	54.0	MAX.	2547	**	117%
	52.0		2458	**	112%
	50.0 *		2370	**	108%
H380	51.5	MAX.	2560	**	105%
	49.5		2480	**	101%
	47.5 *		2400		97%
Viht N150	49.5 *	MAX.	2635	**	113%
	47.5		2550	**	109%
	45.5		2465	**	104%
IMR 4064	46.0 *	MAX.	2638		98%
	44.0		2543		94%
	42.0		2448		90%
IMR 4350	52.0 *	MAX.	2698	**	107%
	50.0		2583	**	103%
	48.0		2468		99%

BC=Ballistic Coefficient SD=Sectional Density
*Most Accurate Load Tested **Compressed Load

Use Maximum Loads with Caution
Refer to page 73 for additional safety information

325 WINCHESTER SHORT MAGNUM

The 325 WSM was introduced in 2005, and is the fourth in the WSM series behind the 300, 7mm, and 270. It is an excellent cartridge, and appears to be gaining popularity on its 270 and 7mm brethren. The cartridge cognoscenti at the time of its debut were sure that the next in the series would be a .338-caliber, but Winchester's research showed that the desired ballistics were better achieved with an 8mm bullet, so the 325 WSM was born. The 325 WSM is actually a cleverly disguised 8mm cartridge, as the bullet diameter is .323 inch. Note that it is not called the "8mm WSM" or even the "323 WSM." Metric calibers have seldom been popular in America, so this was probably a smart marketing move.

The 325 WSM is a very efficient cartridge, and its ballistics tread closely on the heels of the much larger 8mm Remington Magnum. Nosler offers four excellent 8mm bullets that make the 325 a versatile performer. For deer and antelope, the 180-grain Ballistic Tip is a good fit, and a good all-around bullet is the 200-grain AccuBond. For just about any size game, at any reasonable range, the famous Partition bullet in 200-grain weight is always a good choice. Lastly, for those denizens who need a lead-free bullet, the new 180-grain E-Tip is the obvious answer. Reliable expansion and deep penetration are assured with any of these bullets, their trajectories approximate that of the 300 magnums, and they deliver a lot of down-range energy.

My experience with the 325 WSM is limited to hundreds of rounds of test loads, and a couple of plump Missouri whitetail bucks taken on my acreage. My deer load used the 200-grain AccuBond, and while the ranges were not long, one shot apiece was all it took. Both now gaze eternally at me as I write these lines.

The 325 WSM is available in several light and handy rifles, and stoked with Nosler bullets, it makes a great all around rifle for a wide array of big game. It combines power and efficiency in a modern 8mm cartridge, and is a superb addition to the American hunter's repertoire.

Steve is a Contributing Editor of Guns & Ammo, and writes for numerous firearms publications from his Missouri farm.

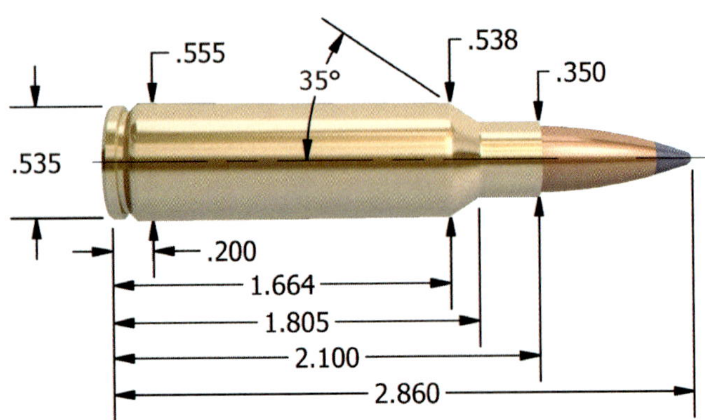

Maximum S.A.A.M.I. Overall Cartridge Length: 2.860"

BULLET CHOICES FOR THE 325 WINCHESTER SHORT MAGNUM

TECHNICAL INFORMATION

The 325 WSM is an excellent cartridge for all North American big game. It has less recoil than the 300 Win Mag but nearly the same energy of the 338 Win Mag with a 225 grain load. All of the WSM cartridges are based on the 404 Jeffery non-belted magnum cartridge which is shortened to fit a short action.

325 WSM - 180 grain		MAXIMUM S.A.A.M.I. O.A.C.L.		2.860"
		TESTED O.A.C.L.	B.C.	S.D.
Ballistic Tip®	180gr. Spitzer	2.840"	0.394	0.246
E-Tip®	180gr. Spitzer	2.800"	0.427	0.246

Due to internal construction differences, always begin with starting loads when using E-Tip® products.

CASE TYPE:	Winchester		PRIMER TYPE	WLRM
CASE HOLDS:	71.7	Gr. WATER	BARREL Length/Make	26" Wiseman
			BARREL Twist	1-10"

POWDER TYPE	POWDER CHG. GRS.		MUZZLE VEL. F.P.S.		LOAD DENSITY (VOLUME)
IMR 4350	67.0	MAX.	3055		99%
	65.0 *		2987		96%
	63.0		2899		93%
H414	66.0	MAX.	3056		96%
	64.0 *		2980		93%
	62.0		2890		90%
W760 Most Accurate Powder Tested	66.0	MAX.	3060		95%
	64.0		2980		92%
	62.0 *		2884		89%
H4350	66.0	MAX.	3071		97%
	64.0 *		2960		94%
	62.0		2892		92%
IMR 4831	69.0	MAX.	3082	**	102%
	67.0 *		3014		99%
	65.0		2946		96%
Viht N550	64.0	MAX.	3083		99%
	62.0		2997		96%
	60.0 *		2900		93%
Big Game	66.0	MAX.	3095		97%
	64.0		3012		94%
	62.0 *		2954		91%
Hunter	70.5 *	MAX.	3125	**	104%
	68.5		3060	**	101%
	66.5		2986		98%
RL19	73.0 *	MAX.	3153	**	111%
	71.0		3093	**	108%
	69.0		3032	**	105%

BC=Ballistic Coefficient SD=Sectional Density
*Most Accurate Load Tested **Compressed Load

Use Maximum Loads with Caution
Refer to page 73 for additional safety information

325 WSM - 200 grain		MAXIMUM S.A.A.M.I. O.A.C.L.		2.860"
		TESTED O.A.C.L.	B.C.	S.D.
AccuBond®	200gr. Spitzer	2.860"	0.450	0.274
Custom Competition™	200gr. HPBT	2.860"	0.520	0.274
Partition®	200gr. Spitzer	2.860"	0.426	0.274

CASE TYPE:	Winchester		PRIMER TYPE	WLRM
CASE HOLDS:	68.9	Gr. WATER	BARREL Length/Make	26" Wiseman
			BARREL Twist	1-10"

POWDER TYPE	POWDER CHG. GRS.		MUZZLE VEL. F.P.S.		LOAD DENSITY (VOLUME)
W760	63.0	* MAX.	2891		94%
	61.0		2810		91%
	59.0		2741		88%
IMR 4350	65.0	* MAX.	2899		100%
	63.0		2846		97%
	61.0		2744		94%
H414	64.0	MAX.	2916		97%
	62.0	*	2867		94%
	60.0		2780		91%
Big Game	63.0	* MAX.	2922		96%
	61.0		2853		93%
	59.0		2775		90%
H4350 Most Accurate Powder Tested	64.5	MAX.	2925		99%
	62.5	*	2872		96%
	60.5		2809		93%
IMR 4831	67.0	MAX.	2928	**	103%
	65.0		2864		100%
	63.0	*	2790		97%
Viht N560	70.0	MAX.	2947	**	113%
	68.0		2868	**	110%
	66.0	*	2770	**	106%
Hunter	67.0	MAX.	2951	**	102%
	65.0		2860		99%
	63.0	*	2785		96%
RL19	69.0	MAX.	2964	**	109%
	67.0	*	2883	**	106%
	65.0		2802	**	103%

BC=Ballistic Coefficient SD=Sectional Density
*Most Accurate Load Tested **Compressed Load

Use Maximum Loads with Caution
Refer to page 73 for additional safety information

8MM REMINGTON MAGNUM

American riflemen have long thirsted for velocity and power…although, seemingly, we're not always prepared to pay their price in recoil. In the late 1950s and on through the '60s the style ran toward belted magnums based on the 375 H&H case shortened to a maximum of 30-06-length. Even the majority of Roy Weatherby's cartridges were shortened to fit in standard-length actions. Introduced in 1978 at the tail end of this "magnum craze," the 8mm Remington Magnum was a departure. It used the full-length 375 H&H, blown out with a fairly short neck, for massive case capacity. This was a daring move, made more so by the fact that it was the first American 8mm, with limited bullet selection in a relatively unfamiliar diameter.

Despite this, initial indications were that Remington's "Big Eight," truly a big brother to their wildly popular 7mm Remington Magnum, was finding favor with American elk, moose, and bear hunters. I used the cartridge when it was new, and it was indeed a real thumper for elk, but initial success quickly waned. The cartridge required a full-length (375 H&H) action, was regarded as a heavy kicker (which it is!) and, inexplicably, factory loads have always been much milder than the case capacity and strong modern actions should allow.

Today the 8mm Remington Magnum barely hangs on as a factory offering, but with good handloads it comes alive, and remains an under-rated and often overlooked choice for larger game. In the mid-90s I returned to the "Big Eight" in a long-barreled rifle with an amazingly accurate barrel, and I have used it quite a bit ever since. Bullet selection is not robust, but is more than adequate for the cartridge's purposes. I have used it for a lot of elk hunting, a wide array of Africa's larger plains game, and although my rifle is heavy, I've hauled it up some number of sheep mountains—just because it works so well!

My opinion is that it carries far better at longer ranges than the 338 Winchester Magnum, and although recoil is stout, it is considerably less than the faster .33s, while essentially as effective. I am also convinced that, with the .323-inch bullet's greater frontal diameter, it hits noticeably harder than any 30-caliber. Although my long (28-inch!) barrel achieves considerably higher velocity, a 24-inch bar-

rel easily exceeds 3000 fps with either the 200-grain Partition or AccuBond, both superb choices for elk in big country. For lighter game the 180-grain Ballistic Tip can exceed 3250 fps, making it a flat-shooting wonder. Slow-burning powders are essential; historically I've had the best luck with good old IMR 4831, but there are a number of good propellants that achieve both velocity and accuracy. I think it's too late for the 8mm Remington to achieve significant popularity, and I'm not one of the guys who gets his kicks from championing lost causes…but I believe the cartridge deserves much more credit than it currently receives, and I will continue to use it.

Craig Boddington

Author and Television personality Craig Boddington is Executive Field Editor, Intermedia Outdoor Group as well as a Contributing Editor to Sports Afield.

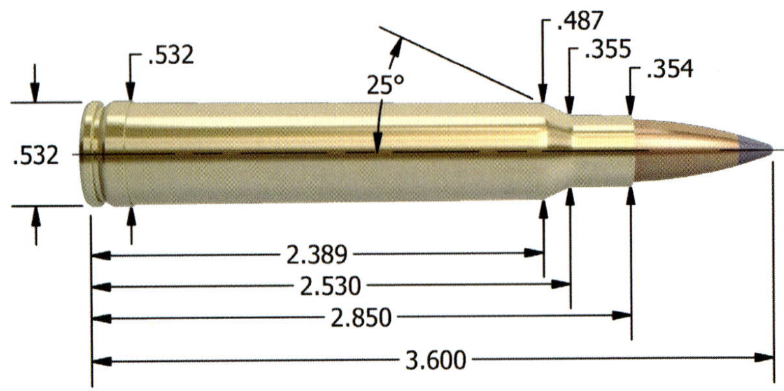

Maximum S.A.A.M.I. Overall Cartridge Length: 3.600"

BULLET CHOICES FOR THE 8MM REMINGTON MAGNUM

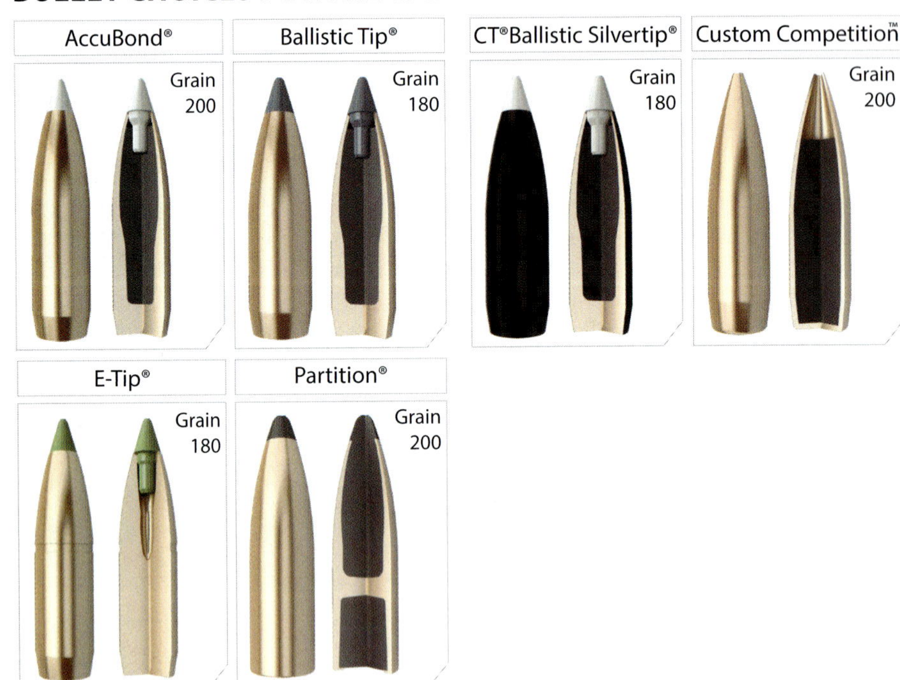

TECHNICAL INFORMATION

This cartridge performs best with magnum primers and slow burning powders. As with any magnum cartridge, we recommend three-shot groups when working up loads.

8mm Remington Mag - 180 grain

8mm Remington Mag - 180 grain		MAXIMUM S.A.A.M.I. O.A.C.L.		3.600"
		TESTED O.A.C.L.	B.C.	S.D.
Ballistic Tip®	180gr. Spitzer	3.580"	0.394	0.246
CT® Ballistic Silvertip®	180gr. Spitzer	3.580"	0.394	0.246
E-Tip®	180gr. Spitzer	3.530"	0.427	0.246

Due to internal construction differences, always begin with starting loads when using E-Tip® products.

CASE TYPE:	Remington		PRIMER TYPE	Fed 215
CASE HOLDS:	90.4	Gr. WATER	BARREL Length/Make	24" Wiseman
			BARREL Twist	1-10"

POWDER TYPE	POWDER CHG. GRS.		MUZZLE VEL. F.P.S.		LOAD DENSITY (VOLUME)
H4831SC Most Accurate Powder Tested	85.0	MAX.	3175		98%
	83.0		3112		96%
	81.0 *		3054		93%
IMR 4350	79.0	MAX.	3190		92%
	77.0		3121		90%
	75.0 *		3062		88%
RL19	83.0	MAX.	3215		100%
	81.0		3164		97%
	79.0 *		3096		95%
Viht N165	87.5 *	MAX.	3231	**	108%
	85.5		3207	**	105%
	83.5		3153	**	103%
IMR 7828	88.0	MAX.	3260	**	104%
	86.0 *		3199	**	101%
	84.0		3110		99%
RL22	88.0 *	MAX.	3283	**	106%
	86.0		3242	**	103%
	84.0		3181	**	101%

BC=Ballistic Coefficient SD=Sectional Density
*Most Accurate Load Tested **Compressed Load

Use Maximum Loads with Caution
Refer to page 73 for additional safety information

8mm Remington Mag - 200 grain		MAXIMUM S.A.A.M.I. O.A.C.L.		3.600"
		TESTED O.A.C.L.	B.C.	S.D.
AccuBond®	200gr. Spitzer	3.580"	0.450	0.274
Custom Competition™	200gr. HPBT	3.600"	0.520	0.274
Partition®	200gr. Spitzer	3.550"	0.426	0.274

CASE TYPE:	Remington		PRIMER TYPE	Fed 215
CASE HOLDS:	91.1	Gr. WATER	BARREL Length/Make	24" Wiseman
			BARREL Twist	1-10"

POWDER TYPE	POWDER CHG. GRS.		MUZZLE VEL. F.P.S.	LOAD DENSITY (VOLUME)
IMR 7828	83.5	MAX.	2930	98%
	81.5		2850	95%
	79.5 *		2770	93%
RL22 Most Accurate Powder Tested	80.5 *	MAX.	2950	96%
	78.5		2850	94%
	76.5		2750	91%
Viht N165	79.0 *	MAX.	2980	96%
	77.0		2924	94%
	75.0		2867	91%
IMR 4350	74.0 *	MAX.	3030	86%
	72.0		2960	84%
	70.0		2890	81%
IMR 4831	78.0	MAX.	3050	91%
	76.0		2980	89%
	74.0 *		2910	86%

BC=Ballistic Coefficient SD=Sectional Density
*Most Accurate Load Tested **Compressed Load

Use Maximum Loads with Caution
Refer to page 73 for additional safety information

Jack Mitchell

338 FEDERAL

What I thought I heard, apparently wasn't what they said, which has been a common theme during my gun writing career. I thought Trijicon invited me on a writers' hunt to Alaska for a combination fishing excursion and black bear hunt. In an attempt to add a dash of flair to the article for Safari Club International and with just about everyone in the world of "gundom" building AR-style rifles, I ordered a DPMS Panther in their new 338 Federal offering. I knew little of this cartridge and with the various components not arriving until the last minute, there was no chance to put the rifle together for some range time. But hey, what could go wrong?

I grabbed a copy of "Nosler's Reloading Guide No. 6" to research my new 338 Federal DPMS AR. Drew Goodlin, Manager of New Product Development for Federal Cartridge Corporation had been a huge fan of the .308 caliber and Nosler Partition bullets. On a moose hunt in Alaska, he had a few scary moments hearing a grizzly while packing out the meat and realizing he was undergunned with his 308 Winchester. Realizing Nosler made .33 caliber bullets, he began work to develop a more powerful caliber using the .308 case.

After much experimentation, Goodlin came up with the .338-308 and is now dubbed the 338 Federal. Using 210-grain Nosler Partition bullets he found that with every animal he shot, the bullet went through both sides. The new caliber provides similar energy to the 7mm Remington Magnum anointing it a big game caliber. I now had little doubt the rifle and cartridge would have any problem with a small or large black bear.

After our arrival in Alaska, I had just enough time to zero my rifle before three of us were flown into some wilderness area with no name. One day later, my buddy had a nice black bear. That's when our guide told me we also had a grizzly permit. Naturally, the next day a very large, suicidal grizzly began his march—and then a sprint right into our camp.

My first shot hit him squarely and he went down. My follow up shot ended the event. The second 210-grain Nosler Partition hit him high on the left side and exited just below his beltline. Talk about penetration. Thank you Drew Goodlin, wherever you are.

Jack is Special Projects Editor for Safari Magazine.

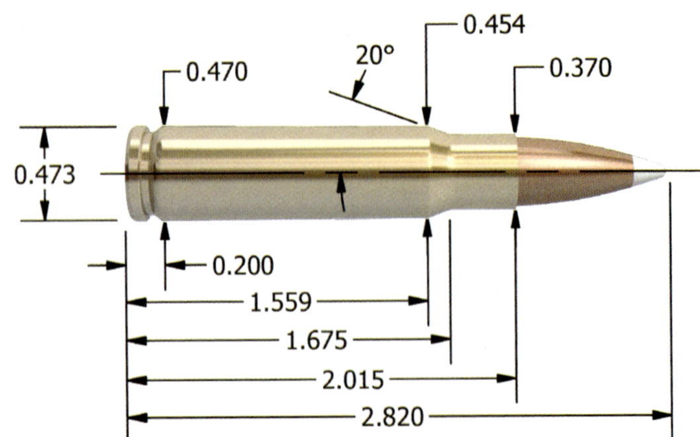

Maximum S.A.A.M.I. Overall Cartridge Length: 2.820"

BULLET CHOICES FOR THE 338 FEDERAL

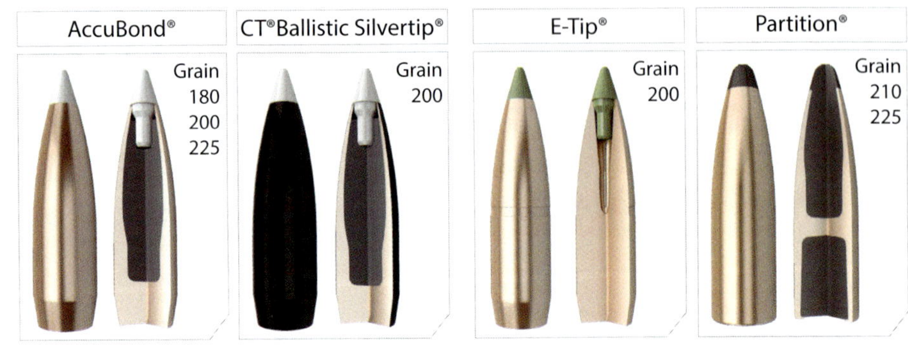

TECHNICAL INFORMATION

The 338 Federal is a rifle cartridge that provides the energy of a 7mm Remington Magnum in a necked-up 308 Winchester case. It is a big game caliber designed with reasonable recoil for today's lightweight rifles. Load development work in the Nosler lab showed it to be a very accurate and well-mannered cartridge.

338 Federal - 180 grain

338 Federal - 180 grain		MAXIMUM S.A.A.M.I. O.A.C.L.		2.820"
		TESTED O.A.C.L.	B.C.	S.D.
AccuBond®	180gr. Spitzer	2.800"	0.372	0.225

CASE TYPE:	Federal		PRIMER TYPE	Fed 210
CASE HOLDS:	45.2	Gr. WATER	BARREL Length/Make	24" Pac-Nor
			BARREL Twist	1-10"

POWDER TYPE	POWDER CHG. GRS.		MUZZLE VEL. F.P.S.		LOAD DENSITY (VOLUME)
RL15	48.0	MAX.	2648	**	111%
	46.0 *		2587	**	106%
	44.0		2486	**	101%
IMR 4895	47.0	MAX.	2657	**	114%
	45.0 *		2575	**	109%
	43.0		2488	**	105%
IMR 3031	43.5	MAX.	2667	**	109%
	41.5 *		2611	**	104%
	39.5		2503		99%
W748	49.0	MAX.	2678	**	113%
	47.0		2611	**	108%
	45.0 *		2555	**	104%
A-2495 Most Accurate Powder Tested	45.0	MAX.	2679	**	113%
	43.0 *		2601	**	108%
	41.0		2523	**	103%
Viht N140	47.5	MAX.	2686	**	122%
	45.5 *		2613	**	117%
	43.5		2528	**	112%
H335	46.0	MAX.	2687	**	103%
	44.0		2606		99%
	42.0 *		2499		94%
Benchmark	45.0 *	MAX.	2697	**	108%
	43.0		2621	**	103%
	41.0		2485		98%
Varget	48.0	MAX.	2706	**	115%
	46.0 *		2604	**	110%
	44.0		2495	**	105%
TAC	47.0	MAX.	2728	**	106%
	45.0		2646	**	102%
	43.0 *		2526		97%

BC=Ballistic Coefficient SD=Sectional Density
*Most Accurate Load Tested **Compressed Load

Use Maximum Loads with Caution
Refer to page 73 for additional safety information

338 Federal - 200/210 grain		MAXIMUM S.A.A.M.I. O.A.C.L.		2.820"
		TESTED O.A.C.L.	B.C.	S.D.
AccuBond®	200gr. Spitzer	2.800"	0.414	0.250
CT® Ballistic Silvertip®	200gr. Spitzer	2.800"	0.414	0.250
E-Tip®	200gr. Spitzer	2.750"	0.425	0.250
Due to internal construction differences, always begin with starting loads when using E-Tip® products.				
Partition®	210gr. Spitzer	2.760"	0.400	0.263
CASE TYPE: Federal		PRIMER TYPE		Fed 210
CASE HOLDS: 44.5 Gr. WATER		BARREL Length/Make		24" Pac-Nor
		BARREL Twist		1-10"

POWDER TYPE	POWDER CHG. GRS.		MUZZLE VEL. F.P.S.		LOAD DENSITY (VOLUME)
A-2495	41.5 *	MAX.	2413	**	106%
	39.5		2291	**	101%
	37.5		2185		96%
H335	42.5	MAX.	2431		97%
	40.5 *		2326		92%
	38.5		2129		88%
IMR 3031	40.5	MAX.	2441	**	103%
	38.5		2329		98%
	36.5 *		2191		93%
TAC	44.0	MAX.	2447	**	101%
	42.0 *		2383		96%
	40.0		2303		92%
Viht N140	44.0 *	MAX.	2450	**	115%
	42.0		2352	**	110%
	40.0		2251	**	105%
Benchmark	42.0	MAX.	2452	**	102%
	40.0		2345		97%
	38.0 *		2254		92%
IMR 4895	44.5	MAX.	2467	**	110%
	42.5 *		2390	**	105%
	40.5		2239		100%
Varget	45.0 *	MAX.	2483	**	109%
	43.0		2404	**	104%
	41.0		2281		100%
RL15	46.0	MAX.	2513	**	108%
	44.0 *		2426	**	103%
	42.0		2344		98%
W748 Most Accurate Powder Tested	47.0	MAX.	2529	**	110%
	45.0 *		2466	**	105%
	43.0		2399	**	101%

BC=Ballistic Coefficient SD=Sectional Density
*Most Accurate Load Tested **Compressed Load

Use Maximum Loads with Caution
Refer to page 73 for additional safety information

338 Federal - 225 grain		MAXIMUM S.A.A.M.I. O.A.C.L.		2.820"
		TESTED O.A.C.L.	B.C.	S.D.
AccuBond®	225gr. Spitzer	2.800"	0.550	0.281
Partition®	225gr. Spitzer	2.760"	0.454	0.281

CASE TYPE:	Federal		PRIMER TYPE	Fed 210
CASE HOLDS:	42.3	Gr. WATER	BARREL Length/Make	24" Pac-Nor
			BARREL Twist	1-10"

POWDER TYPE	POWDER CHG. GRS.		MUZZLE VEL. F.P.S.		LOAD DENSITY (VOLUME)
IMR 3031	40.0	MAX.	2363	**	107%
	38.0 *		2298	**	102%
	36.0		2190		96%
A-2495	42.0	MAX.	2372	**	113%
	40.0 *		2274	**	107%
	38.0		2181	**	102%
H335	41.5 *	MAX.	2374		100%
	39.5		2285		95%
	37.5		2223		90%
TAC	42.5	MAX.	2382	**	103%
	40.5 *		2287		98%
	38.5		2200		93%
Benchmark	41.0 *	MAX.	2386	**	105%
	39.0		2282		100%
	37.0		2192		95%
Viht N135	42.0	MAX.	2394	**	115%
	40.0		2318	**	110%
	38.0 *		2222	**	104%
IMR 4895	43.5	MAX.	2398	**	113%
	41.5		2304	**	108%
	39.5 *		2217	**	103%
RL15 Most Accurate Powder Tested	45.0	MAX.	2408	**	111%
	43.0		2312	**	106%
	41.0 *		2230	**	101%
Varget	44.0	MAX.	2424	**	112%
	42.0 *		2349	**	107%
	40.0		2249	**	102%
W748	46.0	MAX.	2428	**	113%
	44.0 *		2342	**	108%
	42.0		2261	**	103%

BC=Ballistic Coefficient SD=Sectional Density
*Most Accurate Load Tested **Compressed Load

Use Maximum Loads with Caution
Refer to page 73 for additional safety information

Steve Comus

338-06 A-SQUARE

The 338-06 is an amazing, yet under-appreciated cartridge. For reasons known only to mystic forces in the cosmos, it just hasn't captured the imaginations of the masses. And so it is with many good cartridges.

Tis a pity, because the 338-06 affords a very efficient use of the case. When one can throw fatter, heavier bullets at valid velocities, it's good.

I have been shooting and loading the 338-06 for decades, and have been amazed with the way it kills animals big and small. There is a quality to the kill that puts this cartridge in a different category than its parent. It simply puts animals down – with authority. That's nice.

Typically, the 338-06 enters the discussion relative to bigger beasts like bears, moose or elk. That's fine, because with those animals, it shines and takes a back seat to none of the belted magnums. Truth is, that it also works amazingly well for deer and wild pigs. In fact, it is an ultra-hogger because it hits really hard and anchors the boars in ways that only the larger diameter bullets can do.

Within normal hunting distances, there is no handicap with the 338-06 when compared to its parent or similar cartridges. Within 300 yards, it is great. Within 100 yards, it is awesome.

Loading for the 338-06 is easy and enjoyable. A single pass through a sizing die with a tapered expander turns a 30-06 Springfield into a 338-06, or one can start with 338-06 brass. A variety of standard powders fuel this cartridge well. No need for niche powders.

Another advantage of the 338-06 compared to the belted magnums is that magazines can hold one more round. This is something to consider when the target is big and bad. Although I've never had to shoot any animal repeatedly with this cartridge. A single well-placed bullet seems to do the trick in a hurry.

Historically, the concept of a 338-06 goes back to the 318 Westley Richards (.330" bullet diameter) in 1910. After WWII, Elmer Keith, Charles O'Neil and Don Hopkins came up with the 333 OKH (.333" diameter bullet). In the late 1950s when good .338" bullets became available, the wildcat 338-06 evolved. It was standardized in the 1990s, but has not experienced a significant following since.

Because of the cartridge dimensions, it is relatively easy to make a really handy, accurate hunting rifle for this round.

Steve Comus

Steve has been a professional photojournalist for more than 50 years, and has written/edited/published articles and books about shooting and hunting for the past 30 years. He has been a hunter since 1949, and has hunted a wide variety of game on numerous continents around the world.

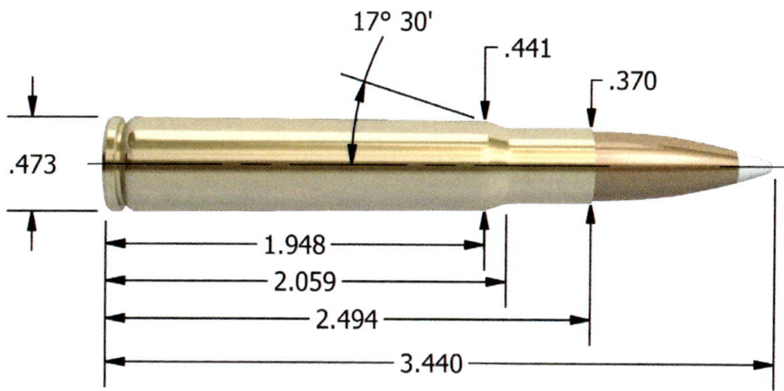

Maximum S.A.A.M.I. Overall Cartridge Length: 3.440"

BULLET CHOICES FOR THE 338-06 A-SQUARE

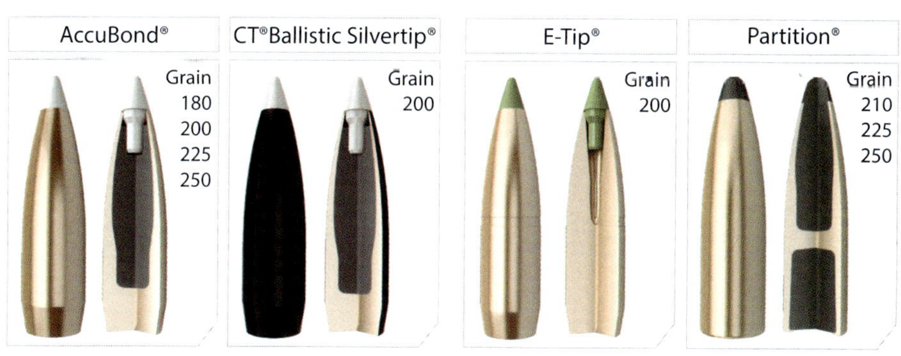

TECHNICAL INFORMATION

Case forming for the 338-06 A-Square consists simply of sizing a standard .30-06 case over a tapered expander ball. No fire forming is required. Either the .338-caliber AccuBond or Partition bullets will handle everything up through the moose category. As a general rule, any powder that works well in the .30-06 will certainly work in the 338-06 A-Square. Our favorites are IMR4064, IMR4320 and RL15.

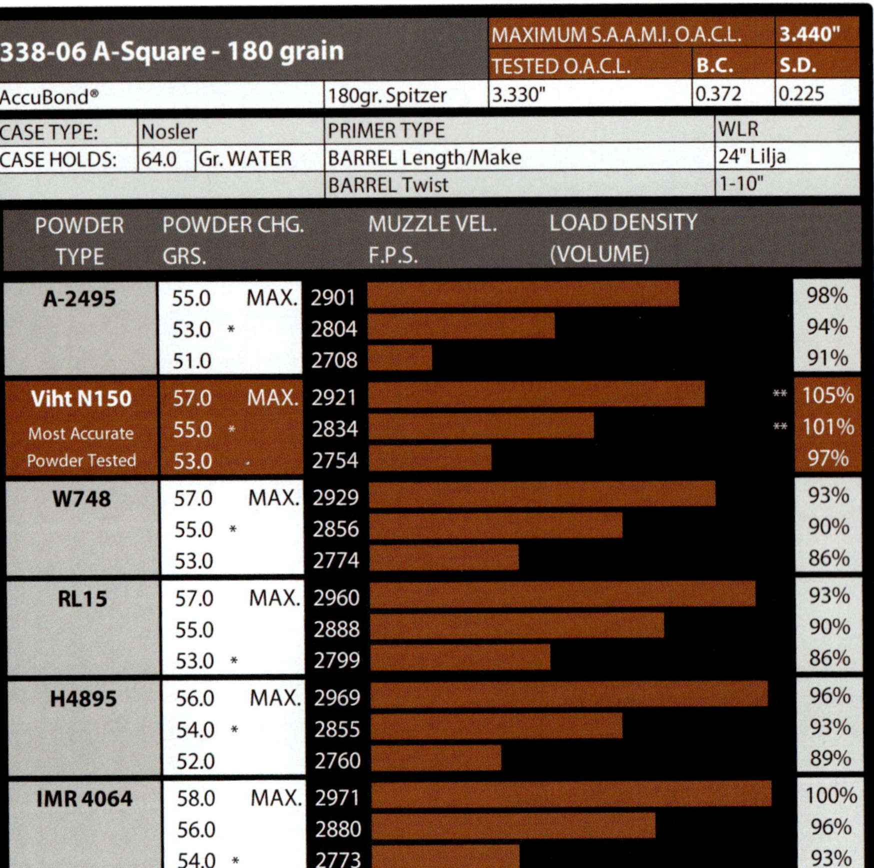

338-06 A-Square - 180 grain

AccuBond®	180gr. Spitzer	MAXIMUM S.A.A.M.I. O.A.C.L.	3.440"	
		TESTED O.A.C.L.	B.C.	S.D.
		3.330"	0.372	0.225

CASE TYPE:	Nosler	PRIMER TYPE	WLR
CASE HOLDS:	64.0 Gr. WATER	BARREL Length/Make	24" Lilja
		BARREL Twist	1-10"

POWDER TYPE	POWDER CHG. GRS.		MUZZLE VEL. F.P.S.	LOAD DENSITY (VOLUME)
A-2495	55.0	MAX.	2901	98%
	53.0 *		2804	94%
	51.0		2708	91%
Viht N150 Most Accurate Powder Tested	57.0	MAX.	2921	** 105%
	55.0 *		2834	** 101%
	53.0		2754	97%
W748	57.0	MAX.	2929	93%
	55.0 *		2856	90%
	53.0		2774	86%
RL15	57.0	MAX.	2960	93%
	55.0		2888	90%
	53.0 *		2799	86%
H4895	56.0	MAX.	2969	96%
	54.0 *		2855	93%
	52.0		2760	89%
IMR 4064	58.0	MAX.	2971	100%
	56.0		2880	96%
	54.0 *		2773	93%

BC=Ballistic Coefficient SD=Sectional Density
*Most Accurate Load Tested **Compressed Load

Use Maximum Loads with Caution
Refer to page 73 for additional safety information

338-06 A-Square - 200/210 grain		MAXIMUM S.A.A.M.I. O.A.C.L.		3.440"
		TESTED O.A.C.L.	B.C.	S.D.
AccuBond®	200gr. Spitzer	3.330"	0.414	0.250
CT® Ballistic Silvertip®	200gr. Spitzer	3.330"	0.414	0.250
E-Tip®	200gr. Spitzer	3.330"	0.425	0.250
Due to internal construction differences, always begin with starting loads when using E-Tip® products.				
Partition®	210gr. Spitzer	3.330"	0.400	0.263

CASE TYPE:	Nosler		PRIMER TYPE	WLR
CASE HOLDS:	62.6	Gr. WATER	BARREL Length/Make	24" Lilja
			BARREL Twist	1-10"

POWDER TYPE	POWDER CHG. GRS.		MUZZLE VEL. F.P.S.		LOAD DENSITY (VOLUME)
IMR 4350	60.0 *	MAX.	2610	**	101%
	58.0		2532		98%
	56.0		2455		95%
H380	58.5	MAX.	2630		98%
	56.5		2550		95%
	54.5 *		2470		92%
Viht N150	54.0	MAX.	2642	**	101%
	52.0		2567		98%
	50.0 *		2492		94%
RL15	53.5	MAX.	2668		89%
	51.5		2563		86%
	49.5 *		2458		82%
IMR 4320 Most Accurate Powder Tested	52.0 *	MAX.	2690		89%
	50.0		2590		86%
	48.0		2490		82%

BC=Ballistic Coefficient SD=Sectional Density
*Most Accurate Load Tested **Compressed Load

Use Maximum Loads with Caution
Refer to page 73 for additional safety information

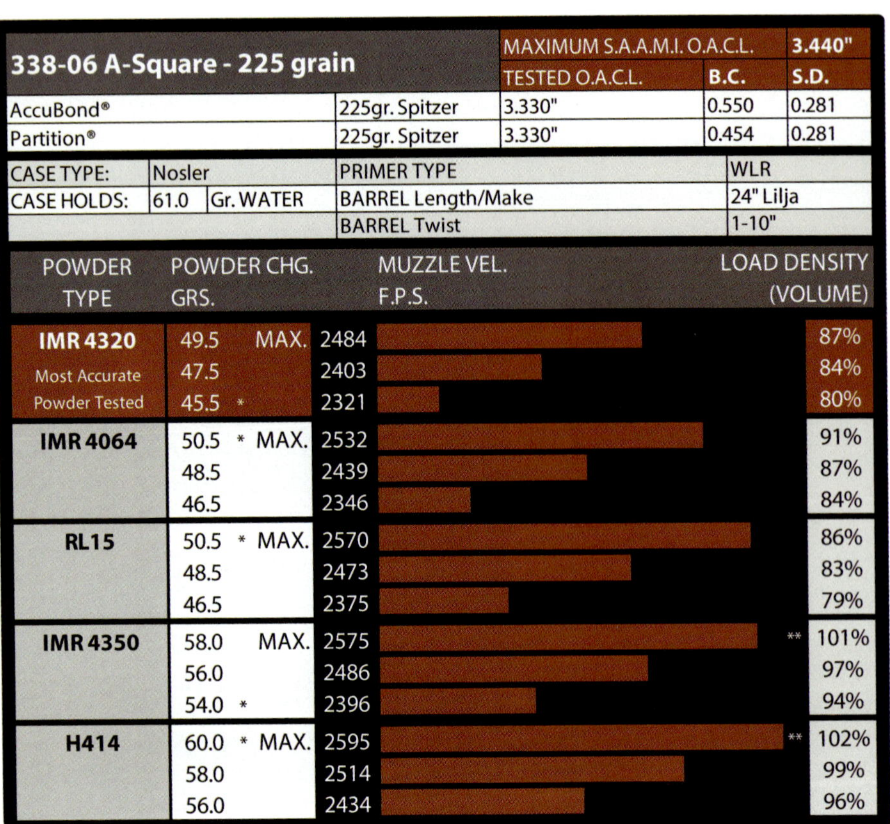

338-06 A-Square - 225 grain		MAXIMUM S.A.A.M.I. O.A.C.L.		3.440"
		TESTED O.A.C.L.	B.C.	S.D.
AccuBond®	225gr. Spitzer	3.330"	0.550	0.281
Partition®	225gr. Spitzer	3.330"	0.454	0.281

CASE TYPE:	Nosler		PRIMER TYPE	WLR
CASE HOLDS:	61.0	Gr. WATER	BARREL Length/Make	24" Lilja
			BARREL Twist	1-10"

POWDER TYPE	POWDER CHG. GRS.		MUZZLE VEL. F.P.S.		LOAD DENSITY (VOLUME)
IMR 4320 Most Accurate Powder Tested	49.5	MAX.	2484		87%
	47.5		2403		84%
	45.5 *		2321		80%
IMR 4064	50.5 *	MAX.	2532		91%
	48.5		2439		87%
	46.5		2346		84%
RL15	50.5 *	MAX.	2570		86%
	48.5		2473		83%
	46.5		2375		79%
IMR 4350	58.0	MAX.	2575	**	101%
	56.0		2486		97%
	54.0 *		2396		94%
H414	60.0 *	MAX.	2595	**	102%
	58.0		2514		99%
	56.0		2434		96%

BC=Ballistic Coefficient SD=Sectional Density
*Most Accurate Load Tested **Compressed Load

Use Maximum Loads with Caution
Refer to page 73 for additional safety information

338-06 A-Square - 250 grain		MAXIMUM S.A.A.M.I. O.A.C.L.		3.440"
		TESTED O.A.C.L.	B.C.	S.D.
AccuBond®	250gr. Spitzer	3.330"	0.575	0.313
Partition®	250gr. Spitzer	3.330"	0.473	0.313

CASE TYPE:	Nosler		PRIMER TYPE	WLR
CASE HOLDS:	57.6	Gr. WATER	BARREL Length/Make	24" Lilja
			BARREL Twist	1-10"

POWDER TYPE	POWDER CHG. GRS.		MUZZLE VEL. F.P.S.		LOAD DENSITY (VOLUME)
IMR 4320	48.5 *	MAX.	2372		91%
	46.5		2257		87%
	44.5		2142		83%
IMR 4350	56.5	MAX.	2410	**	104%
	54.5		2320		100%
	52.5 *		2230		96%
Viht N150	50.5 *	MAX.	2413	**	103%
	48.5		2341		99%
	46.5		2268		95%
H380 Most Accurate Powder Tested	55.5	MAX.	2418	**	101%
	53.5		2343		98%
	51.5 *		2268		94%
RL15	50.0 *	MAX.	2420		90%
	48.0		2320		87%
	46.0		2220		83%
W760	53.5	MAX.	2424		97%
	51.5		2345		94%
	49.5 *		2267		90%

BC=Ballistic Coefficient SD=Sectional Density
*Most Accurate Load Tested **Compressed Load

Use Maximum Loads with Caution
Refer to page 73 for additional safety information

Dave Petzal

338 WINCHESTER MAGNUM

The late Finn Aagaard said that it was probably the most useful belted magnum cartridge to be introduced since World War II. To that I'll add that it's the best elk cartridge, period, and that it's as versatile a round as you can take to Africa for everything short of elephant and Cape buffalo. I'm referring to that unique mixture of finesse and brute force called the 338 Winchester Magnum.

Finesse, because you can load it with 200- and 210-grain bullets at 2900 fps, and hit "way out there," and the bullets will arrive with enough force to put anything down, or you can stuff it with 250-grain Partitions, and flatten big critters at close or intermediate range.

I don't ascribe magical qualities to cartridges, but it does seem that the 338 Win. Mag. pounds stuff extra hard. In New Zealand a couple of years ago, I shot a red stag (which is about the size of a small elk) and my guide said, "He went down so hard it looked like someone jerked the earth out from under him." This seems to be the case very often. If you do your part, what you shoot either goes down in its tracks or you find it within 50 yards; no matter how big, no matter how tough.

For years, I believed devoutly that the 250-grain Partition was the only bullet to use in a .338. I loaded them to 2,550-2,600 fps and never had a complaint. However, I now think that the 225-grain is the better way to go. Loaded to 2,750, it shoots flatter, kicks somewhat less, and will do anything the heavier slug will do.

Yes, the .338 does kick. You will not think you're shooting a 270 Winchester.

But, it's not all that bad, and with a little practice just about anyone can master it. What you get in return is a rifle that you can hunt with anywhere and never wish you had brought a bigger gun.

Dave Petzal

David E. Petzal is the Rifles Field Editor for Field and Stream.

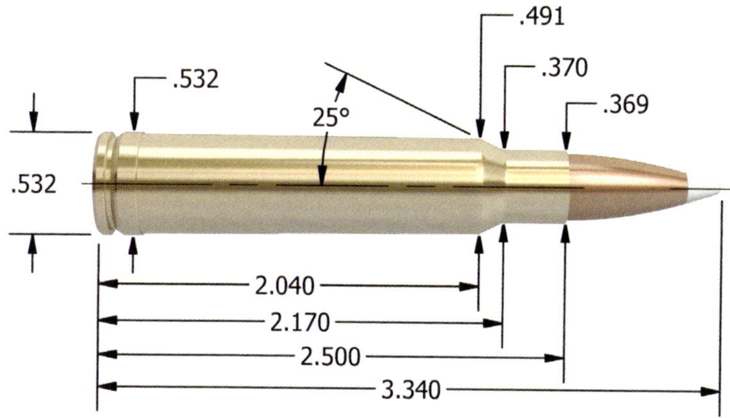

Maximum S.A.A.M.I. Overall Cartridge Length: 3.340"

BULLET CHOICES FOR THE 338 WINCHESTER MAGNUM

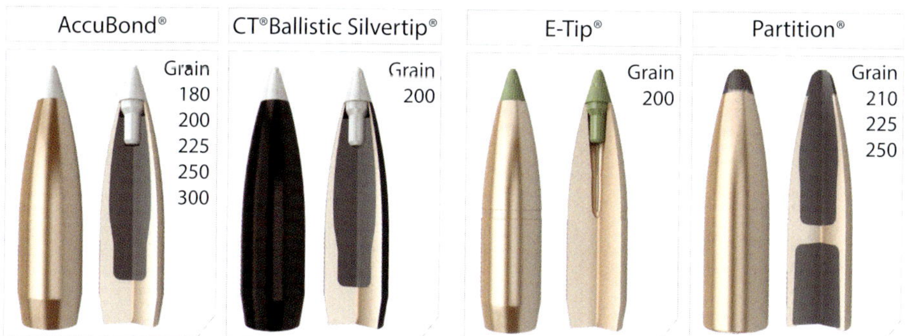

TECHNICAL INFORMATION

The 338 Winchester Magnum is one of the most accurate of the big magnum cartridges. In the lab, we have seen excellent results with medium-burning powders such as 4350, 4831, and RL19.

338 Winchester Mag - 180 grain		MAXIMUM S.A.A.M.I. O.A.C.L.		3.340"
		TESTED O.A.C.L.	B.C.	S.D.
AccuBond®	180gr. Spitzer	3.330"	0.372	0.225

CASE TYPE:	Nosler		PRIMER TYPE	Fed. 215
CASE HOLDS:	78.6	Gr. WATER	BARREL Length/Make	24" Wiseman
			BARREL Twist	1-10"

POWDER TYPE	POWDER CHG. GRS.		MUZZLE VEL. F.P.S.		LOAD DENSITY (VOLUME)
H4831SC	78.0	* MAX.	3048	**	103%
	76.0		2984	**	101%
	74.0		2891		98%
RL19	78.0	MAX.	3120	**	108%
	76.0	*	3057	**	105%
	74.0		2954	**	102%
Viht N150 Most Accurate Powder Tested	67.0	* MAX.	3149		100%
	65.0		3076		97%
	63.0		3002		94%
H414	74.0	* MAX.	3163		98%
	72.0		3080		95%
	70.0		3007		93%
W760	73.0	* MAX.	3177		97%
	71.0		3108		95%
	69.0		3030		92%
RL15	68.0	* MAX.	3189		90%
	66.0		3114		87%
	64.0		3017		85%
IMR 4350	75.0	* MAX.	3260	**	101%
	73.0		3168		98%
	71.0		3075		96%
H4350	76.0	MAX.	3281	**	102%
	74.0	*	3193		100%
	72.0		3112		97%

BC=Ballistic Coefficient SD=Sectional Density
*Most Accurate Load Tested **Compressed Load

Use Maximum Loads with Caution
Refer to page 73 for additional safety information

338 Winchester Mag - 200/210 grain		MAXIMUM S.A.A.M.I. O.A.C.L.		3.340"
		TESTED O.A.C.L.	B.C.	S.D.
AccuBond®	200gr. Spitzer	3.330"	0.414	0.250
CT® Ballistic Silvertip®	200gr. Spitzer	3.330"	0.414	0.250
E-Tip®	200gr. Spitzer	3.280"	0.425	0.250
Due to internal construction differences, always begin with starting loads when using E-Tip® products.				
Partition®	210gr. Spitzer	3.330"	0.400	0.263

CASE TYPE:	Nosler		PRIMER TYPE	Fed. 215
CASE HOLDS:	76.8	Gr. WATER	BARREL Length/Make	24" Wiseman
			BARREL Twist	1-10"

POWDER TYPE	POWDER CHG. GRS.		MUZZLE VEL. F.P.S.		LOAD DENSITY (VOLUME)
IMR 4831	75.0	MAX.	2872	**	104%
	73.0		2777	**	101%
	71.0 *		2682		98%
IMR 4350	73.0	MAX.	2902	**	101%
	71.0		2827		98%
	69.0 *		2752		95%
H414	71.0 *	MAX.	2905		96%
	69.0		2829		94%
	67.0		2753		91%
Viht N160	74.0 *	MAX.	2915	**	107%
	72.0		2831	**	104%
	70.0		2748	**	101%
RL15	65.5	MAX.	2940		89%
	63.5		2851		86%
	61.5 *		2728		83%
W760	71.5	MAX.	2985		97%
	69.5		2902		95%
	67.5 *		2819		92%
RL19	76.0	MAX.	3020	**	108%
Most Accurate	74.0		2940	**	105%
Powder Tested	72.0 *		2860	**	102%

BC=Ballistic Coefficient SD=Sectional Density
*Most Accurate Load Tested **Compressed Load

Use Maximum Loads with Caution
Refer to page 73 for additional safety information

338 Winchester Mag - 225 grain		MAXIMUM S.A.A.M.I. O.A.C.L.		3.340"
		TESTED O.A.C.L.	B.C.	S.D.
AccuBond®	225gr. Spitzer	3.330"	0.550	0.281
Partition®	225gr. Spitzer	3.300"	0.454	0.281

CASE TYPE:	Nosler		PRIMER TYPE	Fed. 215
CASE HOLDS:	76.4	Gr. WATER	BARREL Length/Make	24" Wiseman
			BARREL Twist	1-10"

POWDER TYPE	POWDER CHG. GRS.		MUZZLE VEL. F.P.S.		LOAD DENSITY (VOLUME)
H414	67.5	MAX.	2677		92%
	65.5		2594		89%
	63.5 *		2506		87%
A-4350	67.5	MAX.	2716		99%
	65.5		2614		96%
	63.5 *		2511		93%
RL19	74.0 *	MAX.	2782	**	105%
Most Accurate	72.0		2715	**	102%
Powder Tested	70.0		2639		100%
IMR 4831	71.0 *	MAX.	2807		99%
	69.0		2725		96%
	67.0		2643		93%
H4831SC	75.0	MAX.	2810	**	102%
	73.0 *		2772		100%
	71.0		2693		97%
Viht N160	73.0	MAX.	2833	**	106%
	71.0		2766	**	103%
	69.0 *		2699		100%
IMR 4350	71.5 *	MAX.	2882		99%
	69.5		2808		96%
	67.5		2733		93%

BC=Ballistic Coefficient SD=Sectional Density
*Most Accurate Load Tested **Compressed Load

Use Maximum Loads with Caution
Refer to page 73 for additional safety information

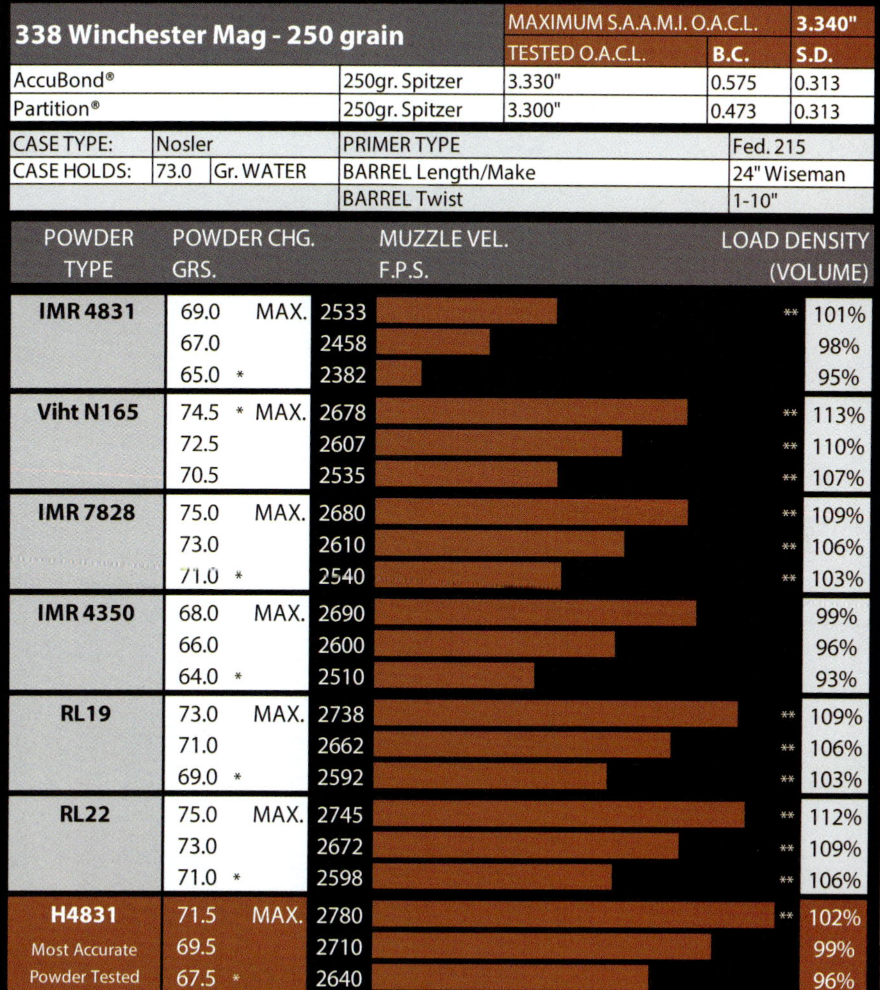

338 Winchester Mag - 250 grain		MAXIMUM S.A.A.M.I. O.A.C.L.		3.340"
		TESTED O.A.C.L.	B.C.	S.D.
AccuBond®	250gr. Spitzer	3.330"	0.575	0.313
Partition®	250gr. Spitzer	3.300"	0.473	0.313

CASE TYPE:	Nosler	PRIMER TYPE	Fed. 215
CASE HOLDS:	73.0 Gr. WATER	BARREL Length/Make	24" Wiseman
		BARREL Twist	1-10"

POWDER TYPE	POWDER CHG. GRS.		MUZZLE VEL. F.P.S.		LOAD DENSITY (VOLUME)
IMR 4831	69.0	MAX.	2533	**	101%
	67.0		2458		98%
	65.0 *		2382		95%
Viht N165	74.5 *	MAX.	2678	**	113%
	72.5		2607	**	110%
	70.5		2535	**	107%
IMR 7828	75.0	MAX.	2680	**	109%
	73.0		2610	**	106%
	71.0 *		2540	**	103%
IMR 4350	68.0	MAX.	2690		99%
	66.0		2600		96%
	64.0 *		2510		93%
RL19	73.0	MAX.	2738	**	109%
	71.0		2662	**	106%
	69.0 *		2592	**	103%
RL22	75.0	MAX.	2745	**	112%
	73.0		2672	**	109%
	71.0 *		2598	**	106%
H4831 Most Accurate Powder Tested	71.5	MAX.	2780	**	102%
	69.5		2710		99%
	67.5 *		2640		96%

BC=Ballistic Coefficient SD=Sectional Density
*Most Accurate Load Tested **Compressed Load

Use Maximum Loads with Caution
Refer to page 73 for additional safety information

338 Winchester Mag - 300 grain

338 Winchester Mag - 300 grain		MAXIMUM S.A.A.M.I. O.A.C.L.		3.340"
		TESTED O.A.C.L.	B.C.	S.D.
AccuBond®	300gr. Spitzer	3.330"	0.720	0.375

CASE TYPE:	Nosler		PRIMER TYPE	WLRM
CASE HOLDS:	70.3	Gr. WATER	BARREL Length/Make	24" Lilja
			BARREL Twist	1-10"

POWDER TYPE	POWDER CHG. GRS.		MUZZLE VEL. F.P.S.		LOAD DENSITY (VOLUME)
Hybrid 100V	60.5	MAX.	2339		89%
	58.5 *		2279		86%
	56.5		2188		83%
IMR 4350	61.0	MAX.	2341		92%
	59.0 *		2286		89%
	57.0		2199		86%
IMR 7828 SSC Most Accurate Powder Tested	68.0	MAX.	2341		100%
	66.0		2299		97%
	64.0 *		2254		94%
RL25	71.0	MAX.	2346	**	110%
	69.0 *		2262	**	107%
	67.0		2193	**	104%
H1000	70.0	MAX.	2372	**	104%
	68.0 *		2321	**	101%
	66.0		2271		98%
H4831	67.0	MAX.	2393		99%
	65.0 *		2341		96%
	63.0		2290		93%
Viht N560	68.5	MAX.	2399	**	108%
	66.5		2331	**	105%
	64.5 *		2296	**	102%
Supreme 780	71.0	MAX.	2432	**	106%
	69.0		2376	**	103%
	67.0 *		2326		100%

BC=Ballistic Coefficient SD=Sectional Density
*Most Accurate Load Tested **Compressed Load

Use Maximum Loads with Caution
Refer to page 73 for additional safety information

340 WEATHERBY MAGNUM

The 340 Weatherby: the thought of having to shoot one for a customer in 1977 had me flinching before I got to the range. Having experienced the recoil of Weatherby's 7mm and 300, this was not going to be a good day.

First shot – must have been a light load. Second shot – same thing, a lot of push but not the sharp bite of the 7mm and 300 Wby. The next day, my 300 Mark V was headed to Weatherby for rebarreling. I've never regretted that decision. I shortened the barrel to 24 inches for better handling in the dark timber, and had MagnaPort vent the barrel. It's now possible to watch animals go down in the scope while shooting off hand. Over the years, I have sold many 340s to friends and customers, all were very skeptical until they shot one. Give them a rifle and a box of ammo and all I ever got back was a box of empties.

With 180- and 200-grain AccuBonds at 3200-3300 fps, you have a very flat shooting, long-range cartridge for the largest plains game. At the other end of the spectrum, a 250-grain Partition at 2800 fps is most adequate for any soft-skinned animal that walks.

So what do I like? A 225-grain Partition or AccuBond at 3000 fps has covered most of my hunting for the last 30 years: antelope, deer and most importantly, elk. Zeroed 3 inches high at 100 yards, I'm dead on at 275 yards and down only 30 inches at 500; which is further than I will shoot at a game animal.

Mike is a long-time handloader, hunter and Customer Service Director at Nosler, Inc.

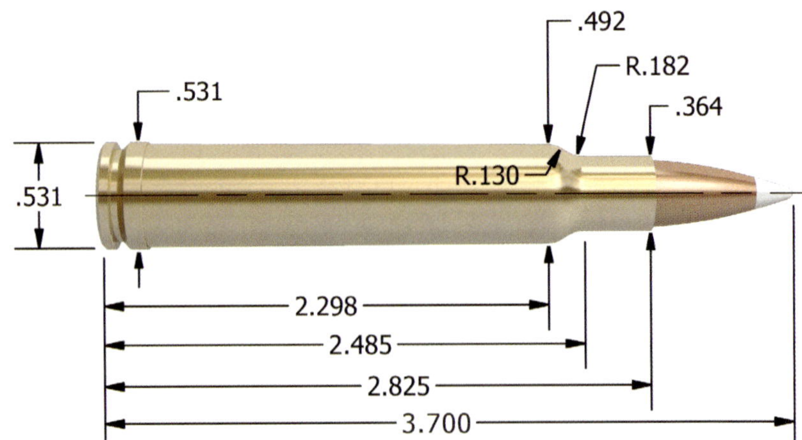

Maximum S.A.A.M.I. Overall Cartridge Length: 3.700"

BULLET CHOICES FOR THE 340 WEATHERBY MAGNUM

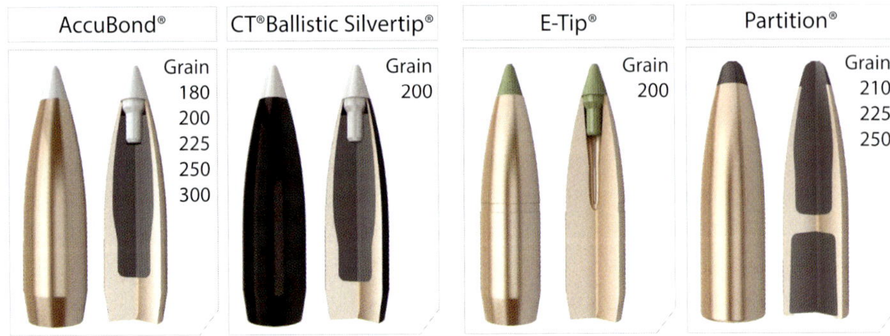

TECHNICAL INFORMATION

As with the rest of the Weatherby offerings, there are a few points to keep in mind when loading for this cartridge:

• Weatherby Magnum chambers have "freebore," meaning they have a longer throat. For this reason, it is generally not possible to seat a bullet close to, or in contact with the lands (rifling).
• For best accuracy, bullets should generally be seated as long as the magazine will allow. As with all belted magnums, range work is best limited to three-shot groups.

340 Weatherby Mag - 180 grain		MAXIMUM S.A.A.M.I. O.A.C.L.		3.700"
		TESTED O.A.C.L.	B.C.	S.D.
AccuBond®	180gr. Spitzer	3.575"	0.372	0.225

CASE TYPE:	Weatherby		PRIMER TYPE	Fed 215
CASE HOLDS:	94.6	Gr. WATER	BARREL Length/Make	26" Lilja
			BARREL Twist	1-10"

POWDER TYPE	POWDER CHG. GRS.		MUZZLE VEL. F.P.S.		LOAD DENSITY (VOLUME)
Viht N160	87.0	* MAX.	3265	**	102%
	85.0		3177		100%
	83.0		3085		97%
RL22	90.0	* MAX.	3266	**	103%
	88.0		3172	**	101%
	86.0		3075		99%
IMR 7828	93.0	MAX.	3267	**	105%
	91.0	*	3236	**	102%
	89.0		3203		100%
RL19	88.0	* MAX.	3269	**	101%
	86.0		3179		99%
	84.0		3069		97%
IMR 4350 Most Accurate Powder Tested	87.5	MAX.	3359		98%
	85.5	*	3322		96%
	83.5		3278		93%
IMR 4831	88.0	* MAX.	3364		99%
	86.0		3290		97%
	84.0		3218		94%

BC=Ballistic Coefficient SD=Sectional Density
*Most Accurate Load Tested **Compressed Load

Use Maximum Loads with Caution
Refer to page 73 for additional safety information

340 Weatherby Mag - 200/210 grain		MAXIMUM S.A.A.M.I. O.A.C.L.		3.700"
		TESTED O.A.C.L.	B.C.	S.D.
AccuBond®	200gr. Spitzer	3.580"	0.414	0.250
CT® Ballistic Silvertip®	200gr. Spitzer	3.580"	0.414	0.250
E-Tip®	200gr. Spitzer	3.550"	0.425	0.250
Due to internal construction differences, always begin with starting loads when using E-Tip® products.				
Partition®	210gr. Spitzer	3.580"	0.400	0.263
CASE TYPE:	Weatherby	PRIMER TYPE	Fed 215	
CASE HOLDS:	94.0 Gr. WATER	BARREL Length/Make	26" Lilja	
		BARREL Twist	1-10"	

POWDER TYPE	POWDER CHG. GRS.		MUZZLE VEL. F.P.S.	LOAD DENSITY (VOLUME)
IMR 4320	69.0	MAX.	2912	79%
	67.0		2817	77%
	65.0 *		2722	74%
Viht N165	87.0 *	MAX.	2977	** 103%
	85.0		2905	100%
	83.0		2834	98%
H4831	87.0	MAX.	3036	96%
	85.0		2973	94%
	83.0 *		2909	92%
RL19 Most Accurate Powder Tested	85.5	MAX.	3063	99%
	83.5		2983	97%
	81.5 *		2904	94%
IMR 4831	85.5	MAX.	3140	97%
	83.5		3050	94%
	81.5 *		2960	92%
IMR 4350	83.5	MAX.	3223	94%
	81.5		3109	92%
	79.5 *		2995	89%

BC=Ballistic Coefficient SD=Sectional Density
*Most Accurate Load Tested **Compressed Load

Use Maximum Loads with Caution
Refer to page 73 for additional safety information

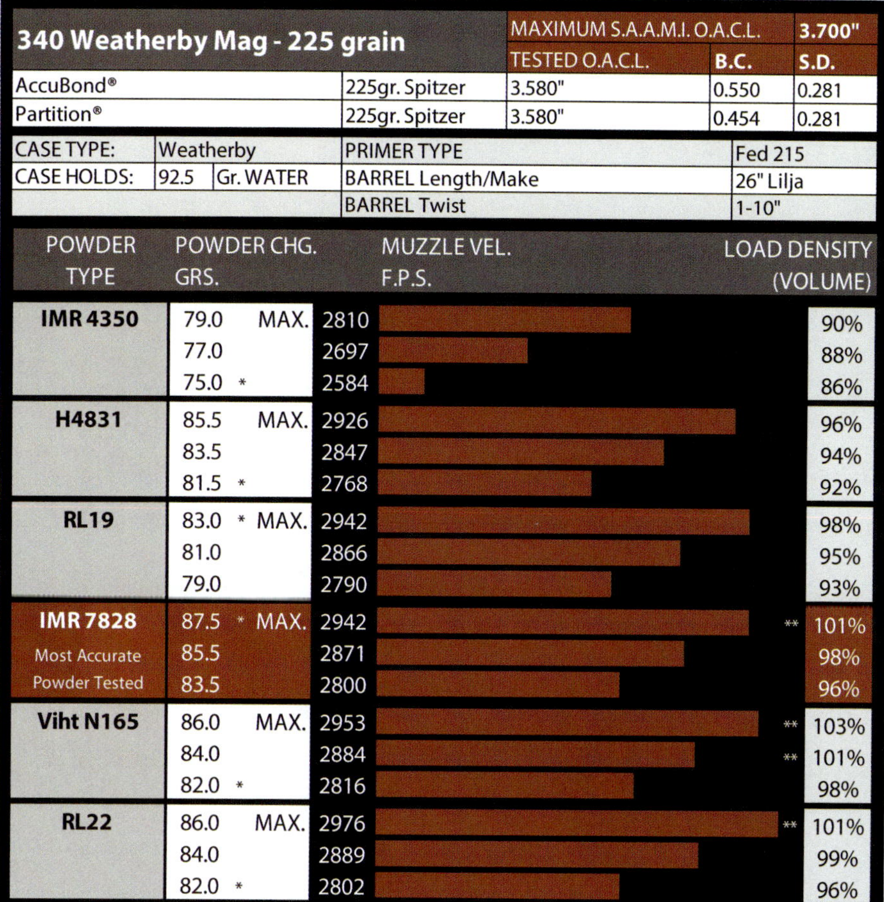

340 Weatherby Mag - 225 grain		MAXIMUM S.A.A.M.I. O.A.C.L.		3.700"
		TESTED O.A.C.L.	B.C.	S.D.
AccuBond®	225gr. Spitzer	3.580"	0.550	0.281
Partition®	225gr. Spitzer	3.580"	0.454	0.281

CASE TYPE:	Weatherby		PRIMER TYPE	Fed 215
CASE HOLDS:	92.5	Gr. WATER	BARREL Length/Make	26" Lilja
			BARREL Twist	1-10"

POWDER TYPE	POWDER CHG. GRS.		MUZZLE VEL. F.P.S.		LOAD DENSITY (VOLUME)
IMR 4350	79.0	MAX.	2810		90%
	77.0		2697		88%
	75.0 *		2584		86%
H4831	85.5	MAX.	2926		96%
	83.5		2847		94%
	81.5 *		2768		92%
RL19	83.0 *	MAX.	2942		98%
	81.0		2866		95%
	79.0		2790		93%
IMR 7828 Most Accurate Powder Tested	87.5 *	MAX.	2942	**	101%
	85.5		2871		98%
	83.5		2800		96%
Viht N165	86.0	MAX.	2953	**	103%
	84.0		2884	**	101%
	82.0 *		2816		98%
RL22	86.0	MAX.	2976	**	101%
	84.0		2889		99%
	82.0 *		2802		96%

BC=Ballistic Coefficient SD=Sectional Density
*Most Accurate Load Tested **Compressed Load

Use Maximum Loads with Caution
Refer to page 73 for additional safety information

340 Weatherby Mag - 250 grain		MAXIMUM S.A.A.M.I. O.A.C.L.		3.700"
		TESTED O.A.C.L.	B.C.	S.D.
AccuBond®	250gr. Spitzer	3.580"	0.575	0.313
Partition®	250gr. Spitzer	3.580"	0.473	0.313

CASE TYPE:	Weatherby		PRIMER TYPE	Fed 215
CASE HOLDS:	89.0	Gr. WATER	BARREL Length/Make	26" Lilja
			BARREL Twist	1-10"

POWDER TYPE	POWDER CHG. GRS.		MUZZLE VEL. F.P.S.		LOAD DENSITY (VOLUME)
IMR 4320	66.0	MAX.	2680		80%
	64.0		2590		77%
	62.0 *		2500		75%
IMR 4831	77.5	MAX.	2700		93%
	75.5		2620		90%
	73.5 *		2540		88%
IMR 4350	76.0	MAX.	2732		90%
	74.0		2647		88%
	72.0 *		2562		86%
Viht N165	85.0 *	MAX.	2773	**	106%
	83.0		2705	**	104%
	81.0		2637	**	101%
RL19	81.0	MAX.	2801		99%
	79.0		2736		96%
	77.0 *		2670		94%
IMR 7828 Most Accurate Powder Tested	86.0 *	MAX.	2818	**	103%
	84.0		2733		100%
	82.0		2648		98%
RL22	83.0 *	MAX.	2829	**	101%
	81.0		2765		99%
	79.0		2700		96%

BC=Ballistic Coefficient SD=Sectional Density
*Most Accurate Load Tested **Compressed Load

Use Maximum Loads with Caution
Refer to page 73 for additional safety information

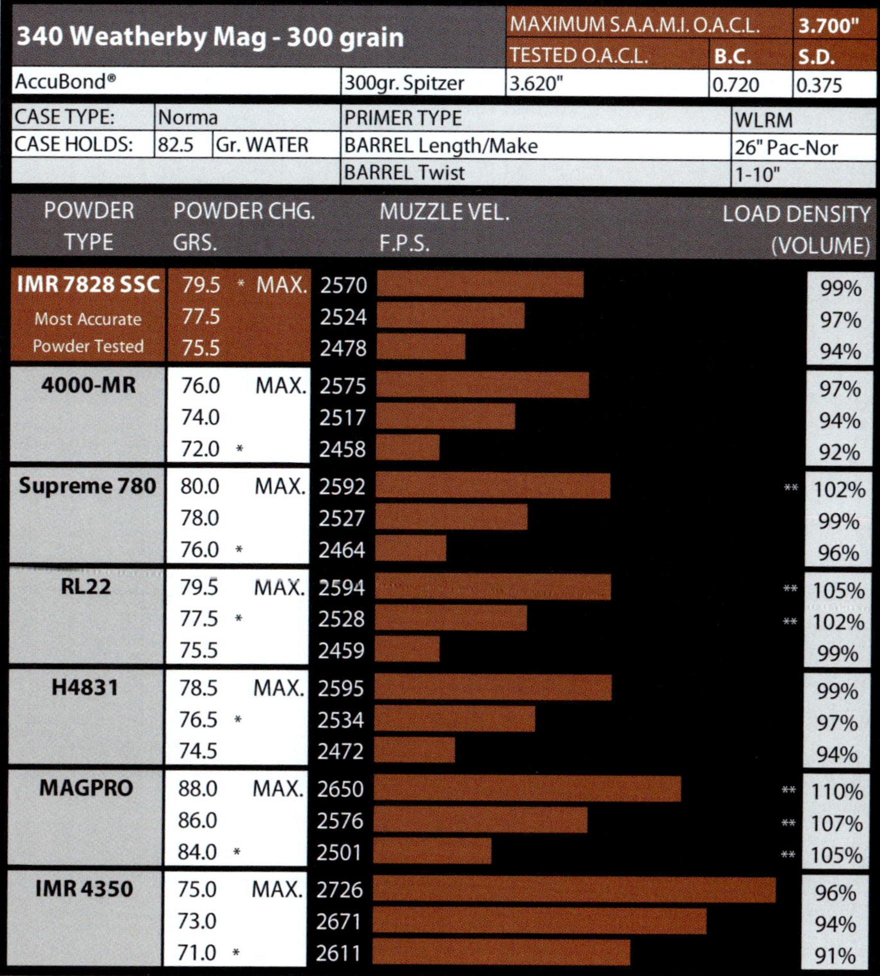

340 Weatherby Mag - 300 grain		MAXIMUM S.A.A.M.I. O.A.C.L.		3.700"
		TESTED O.A.C.L.	B.C.	S.D.
AccuBond®	300gr. Spitzer	3.620"	0.720	0.375

CASE TYPE:	Norma		PRIMER TYPE	WLRM
CASE HOLDS:	82.5	Gr. WATER	BARREL Length/Make	26" Pac-Nor
			BARREL Twist	1-10"

POWDER TYPE	POWDER CHG. GRS.		MUZZLE VEL. F.P.S.		LOAD DENSITY (VOLUME)
IMR 7828 SSC Most Accurate Powder Tested	79.5 *	MAX.	2570		99%
	77.5		2524		97%
	75.5		2478		94%
4000-MR	76.0	MAX.	2575		97%
	74.0		2517		94%
	72.0 *		2458		92%
Supreme 780	80.0	MAX.	2592	**	102%
	78.0		2527		99%
	76.0 *		2464		96%
RL22	79.5	MAX.	2594	**	105%
	77.5 *		2528	**	102%
	75.5		2459		99%
H4831	78.5	MAX.	2595		99%
	76.5 *		2534		97%
	74.5		2472		94%
MAGPRO	88.0	MAX.	2650	**	110%
	86.0		2576	**	107%
	84.0 *		2501	**	105%
IMR 4350	75.0	MAX.	2726		96%
	73.0		2671		94%
	71.0 *		2611		91%

BC=Ballistic Coefficient SD=Sectional Density
*Most Accurate Load Tested **Compressed Load

Use Maximum Loads with Caution
Refer to page 73 for additional safety information

Mark A. Keefe IV

338 REMINGTON ULTRA MAGNUM

Remington changed the cartridge landscape with the introduction of a 300 Ultra Mag in 1999, a magnum that fit in long-action rifles but could deliver more than 3,400 feet per second with a 150-grain bullet. Remington engineers did this by starting with the 404 Jeffery case, but with a fat case body measuring 0.550", a 0.527" rebated rim (so it would work with bolt faces of belted magnum sizes), a relatively short neck and a 30-degree shoulder. The new result was a case 3.60" long with around 20 percent more powder capacity than a 300 Winchester Magnum.

If upgrading power of the .300 magnums was good, necking the Ultra Mag case up to .338" was even better, which is exactly what Remington did when it introduced the 338 Ultra Mag in 2002. With the 338 Ultra Mag, you can drive a 250-grain bullet up to the 2900 feet per second mark, while you are really stretching safe limits to get to 2700 feet per second with the 338 Winchester Magnum. Only the 338-378 Weatherby Magnum and the 338 Lapua Magnum have the edge on the 338 Ultra Magnum when it comes to powder capacity and velocity, and that edge comes at the price of larger actions and rifles.

I have found the 338 Ultra Mag to be the most accurate in the family. It is very responsive to handloading, and it doesn't burn through barrels every few hundred rounds. The recoil is stout, but manageable, and I have hunted Nilgai, elk, hogs and javelina with this cartridge. The elk was a 400 yarder that crumbled the old bull on the spot, literally last day, last light of a hunt that, frankly, had not gone well. The beauty of this cartridge is that it shoots relatively heavy bullets at high velocity so bullet drop isn't nearly as big a factor as my old friend the 338 Winchester Magnum. I have hunted with or shot all the Ultra Mags save the .416, and the 338 Ultra Magnum is likely the best cartridge to come from the Ultra Mag family. And you can download it to 338 Federal levels for practice, whitetails and other tasks that do not demand the full potential of the cartridge. The only gun with the words "Ultra Mag" on the barrel in my gun room has the numbers ".338" ahead of it.

Mark A. Keefe, IV

Mark A. Keefe, IV, is Editor In Chief of American Rifleman magazine.

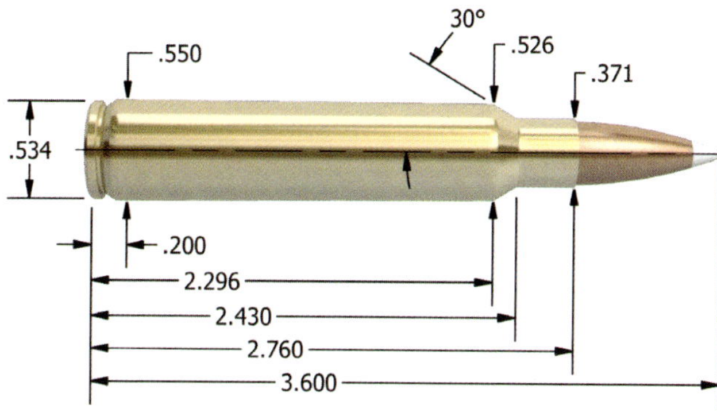

Maximum S.A.A.M.I. Overall Cartridge Length: 3.600"

BULLET CHOICES FOR THE 338 REMINGTON ULTRA MAGNUM

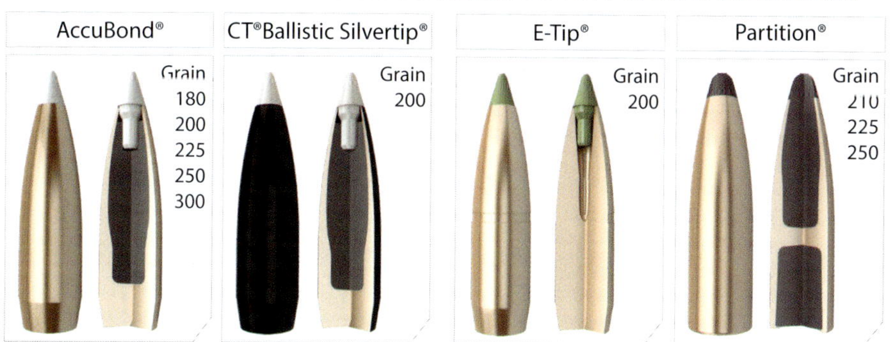

TECHNICAL INFORMATION

The 338 RUM delivers surprising accuracy for a large magnum cartridge. It has become the standard .338 caliber test round for the Nosler Ballistics Lab. While nearly any medium to slow burning powder will work well in the big RUM, IMR 7828 and H1000 are particularly good choices.

338 Rem Ultra Mag - 180 grain		MAXIMUM S.A.A.M.I. O.A.C.L.		3.600"
		TESTED O.A.C.L.	B.C.	S.D.
AccuBond®	180gr. Spitzer	3.575"	0.372	0.225

CASE TYPE:	Remington		PRIMER TYPE	Fed. 215
CASE HOLDS:	103.2	Gr. WATER	BARREL Length/Make	26" Wiseman
			BARREL Twist	1-10"

POWDER TYPE	POWDER CHG. GRS.		MUZZLE VEL. F.P.S.		LOAD DENSITY (VOLUME)
H4831SC	99.0	MAX.	3443		100%
	97.0		3365		98%
	95.0 *		3290		96%
IMR 4350	93.0 *	MAX.	3514		95%
	91.0		3417		93%
	89.0		3292		91%
RL22	101.0 *	MAX.	3522	**	106%
	99.0		3457	**	104%
	97.0		3362	**	102%
IMR 4831 Most Accurate Powder Tested	97.0	MAX.	3529		100%
	95.0 *		3436		98%
	93.0		3366		96%
IMR 7828	102.0 *	MAX.	3535	**	105%
	100.0		3454	**	103%
	98.0		3375	**	101%

BC=Ballistic Coefficient SD=Sectional Density
*Most Accurate Load Tested **Compressed Load

Use Maximum Loads with Caution
Refer to page 73 for additional safety information

338 Rem Ultra Mag - 200/210 grain		MAXIMUM S.A.A.M.I. O.A.C.L.		3.600"
		TESTED O.A.C.L.	B.C.	S.D.
AccuBond®	200gr. Spitzer	3.575"	0.414	0.250
CT® Ballistic Silvertip®	200gr. Spitzer	3.575"	0.414	0.250
E-Tip®	200gr. Spitzer	3.520"	0.425	0.250
Due to internal construction differences, always begin with starting loads when using E-Tip® products.				
Partition®	210gr. Spitzer	3.550"	0.400	0.263

CASE TYPE:	Remington	PRIMER TYPE	Fed. 215
CASE HOLDS:	102.9 Gr. WATER	BARREL Length/Make	26" Wiseman
		BARREL Twist	1-10"

POWDER TYPE	POWDER CHG. GRS.		MUZZLE VEL. F.P.S.	LOAD DENSITY (VOLUME)
H4831SC Most Accurate Powder Tested	91.0	MAX.	3170	92%
	89.0 *		3065	90%
	87.0		2940	88%
IMR 4831	89.0	MAX.	3210	92%
	87.0 *		3145	90%
	85.0		3065	88%
RL22	93.0 *	MAX.	3230	98%
	91.0		3165	96%
	89.0		3090	94%
H1000	102.0 *	MAX.	3230	** 104%
	100.0		3146	** 102%
	98.0		3091	100%
IMR 7828	95.0 *	MAX.	3270	98%
	93.0		3190	96%
	91.0		3105	94%
IMR 4350	89.0 *	MAX.	3291	92%
	87.0		3186	89%
	85.0		3087	87%

BC=Ballistic Coefficient SD=Sectional Density
*Most Accurate Load Tested **Compressed Load

Use Maximum Loads with Caution
Refer to page 73 for additional safety information

338 Rem Ultra Mag - 225 grain		MAXIMUM S.A.A.M.I. O.A.C.L.		3.600"
		TESTED O.A.C.L.	B.C.	S.D.
AccuBond®	225gr. Spitzer	3.575"	0.550	0.281
Partition®	225gr. Spitzer	3.550"	0.454	0.281

CASE TYPE:	Remington	PRIMER TYPE	Fed. 215
CASE HOLDS:	101.0 Gr. WATER	BARREL Length/Make	26" Wiseman
		BARREL Twist	1-10"

POWDER TYPE	POWDER CHG. GRS.		MUZZLE VEL. F.P.S.	LOAD DENSITY (VOLUME)
H4831SC	90.5	MAX.	3095	93%
	88.5		3050	91%
	86.5 *		3010	89%
IMR 4350	86.0	MAX.	3123	90%
	84.0		3059	88%
	82.0 *		3010	86%
H1000	100.0	MAX.	3125	** 101%
	98.0		3070	99%
	96.0 *		3020	96%
IMR 7828	93.0 *	MAX.	3147	98%
	91.0		3086	96%
	89.0		3008	94%
IMR 4831	89.0 *	MAX.	3153	94%
	87.0		3084	92%
	85.0		2992	90%
RL22 Most Accurate Powder Tested	91.0 *	MAX.	3160	98%
	89.0		3115	96%
	87.0		3070	94%
RL25	98.0	MAX.	3227	** 105%
	96.0 *		3178	** 103%
	94.0		3131	** 101%

BC=Ballistic Coefficient SD=Sectional Density
*Most Accurate Load Tested **Compressed Load

Use Maximum Loads with Caution
Refer to page 73 for additional safety information

338 Rem Ultra Mag - 250 grain		MAXIMUM S.A.A.M.I. O.A.C.L.		3.600"
		TESTED O.A.C.L.	B.C.	S.D.
AccuBond®	250gr. Spitzer	3.575"	0.575	0.313
Partition®	250gr. Spitzer	3.550"	0.473	0.313

CASE TYPE:	Nosler		PRIMER TYPE	Fed. 215
CASE HOLDS:	99.1	Gr. WATER	BARREL Length/Make	26" Wiseman
			BARREL Twist	1-10"

POWDER TYPE	POWDER CHG. GRS.		MUZZLE VEL. F.P.S.	LOAD DENSITY (VOLUME)
IMR 4350	79.0 *	MAX.	2876	84%
	77.0		2811	82%
	75.0		2775	80%
RL22	84.0	MAX.	2920	92%
	82.0 *		2875	90%
	80.0		2840	88%
IMR 4831	84.5	MAX.	2938	91%
	82.5 *		2907	89%
	80.5		2860	86%
Viht N560	84.5	MAX.	2947	95%
	82.5		2906	92%
	80.5 *		2857	90%
IMR 7828	87.0	MAX.	2955	93%
	85.0		2898	91%
	83.0 *		2855	89%
RL25 Most Accurate Powder Tested	91.0 *	MAX.	2990	100%
	89.0		2918	98%
	87.0		2872	95%

BC=Ballistic Coefficient SD=Sectional Density
*Most Accurate Load Tested **Compressed Load

Use Maximum Loads with Caution
Refer to page 73 for additional safety information

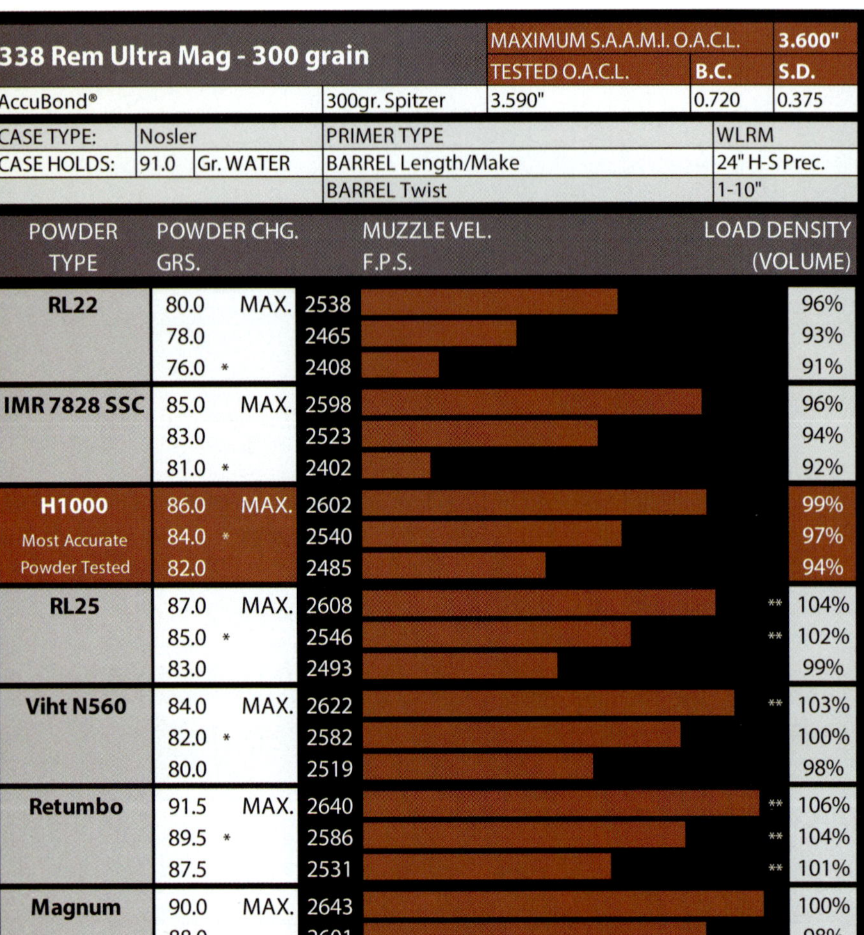

338 Rem Ultra Mag - 300 grain		MAXIMUM S.A.A.M.I. O.A.C.L.		3.600"
		TESTED O.A.C.L.	B.C.	S.D.
AccuBond®	300gr. Spitzer	3.590"	0.720	0.375

CASE TYPE:	Nosler		PRIMER TYPE	WLRM
CASE HOLDS:	91.0	Gr. WATER	BARREL Length/Make	24" H-S Prec.
			BARREL Twist	1-10"

POWDER TYPE	POWDER CHG. GRS.		MUZZLE VEL. F.P.S.		LOAD DENSITY (VOLUME)
RL22	80.0	MAX.	2538		96%
	78.0		2465		93%
	76.0 *		2408		91%
IMR 7828 SSC	85.0	MAX.	2598		96%
	83.0		2523		94%
	81.0 *		2402		92%
H1000	86.0	MAX.	2602		99%
Most Accurate	84.0 *		2540		97%
Powder Tested	82.0		2485		94%
RL25	87.0	MAX.	2608	**	104%
	85.0 *		2546	**	102%
	83.0		2493		99%
Viht N560	84.0	MAX.	2622	**	103%
	82.0 *		2582		100%
	80.0		2519		98%
Retumbo	91.5	MAX.	2640	**	106%
	89.5 *		2586	**	104%
	87.5		2531	**	101%
Magnum	90.0	MAX.	2643		100%
	88.0		2601		98%
	86.0 *		2544		96%

BC=Ballistic Coefficient SD=Sectional Density
*Most Accurate Load Tested **Compressed Load

Use Maximum Loads with Caution
Refer to page 73 for additional safety information

Adam Heggenstaller

338 LAPUA MAGNUM

Even though I had the Nightforce scope cranked up to 15X, the silhouette target standing on the Utah hillside still looked small. Nearly a mile of snow-covered sagebrush separated me from that rectangular piece of steel, making it practically untouchable with all but a handful of cartridges. Nevertheless, I nudged the reticle into the wind, pressed the trigger of my Desert Tactical Arms Stealth Recon Scout rifle and waited to see what would happen 1,600 long yards downrange.

A couple seconds after the suppressor-muffled report, I heard my spotter's voice over my right shoulder. "Hit," he reported, without so much as a hint of surprise in his tone. A military sniper with multiple deployments to Iraq and Afghanistan, he knew exactly what to expect from the .338 Lapua Magnum. I, on the other hand, am still a bit shocked I made that shot.

The .338 Lapua Magnum received CIP recognition about 25 years ago, and its performance at ultra-long ranges, particularly in combat, has since become legendary. Case in point: In November 2009, British Army sniper Craig Harrison used a .338 Lapua Magnum L115A3 rifle, manufactured by Accuracy International, to take out a pair of Taliban combatants in Afghanistan's Helmand Province with two consecutive shots from a verified range of 2,707 yards. In case you don't have your calculator handy, that's more than 1.5 miles.

The cartridge began as a .416 Rigby necked down for a 250-grain, .338 bullet in 1983 at the request of the U.S. Navy. However, testing revealed the parent case was too weak to handle the pressure required to meet the velocity goal of 3,000 fps, and the project was canceled. Lapua later redesigned the .338/416 case by shortening it by almost 2 mm, making the web and a portion of the sidewall thicker—and thus stronger—and hardening the brass at the case head while keeping it comparatively soft at the mouth. The Finnish company named the revamped cartridge the .338 Lapua Magnum and put it into production, with Accuracy International becoming the primary manufacturer of rifles chambered for the round. Other precision-rifle makers soon followed, and today companies including Barrett, Remington and Savage chamber bolt guns for the round.

With a healthy dose of slow-burning powder, the .338 Lapua Magnum is capable of pushing a 250-grain bullet in the neighborhood of 2,900 fps. Under most conditions, that bullet will remain supersonic to 1,200 yards or more, making the .338 Lapua Magnum a favorite of long-range target shooters and hunters who enjoy shooting big game at extended ranges. In military circles, it fills the void between the .300 Winchester Magnum and the .50 BMG. In civilian terms, the .338 Lapua Magnum simply hits stuff a long, long way off.

Adam is the Executive Editor of the National Rifle Association's Shooting Illustrated magazine.

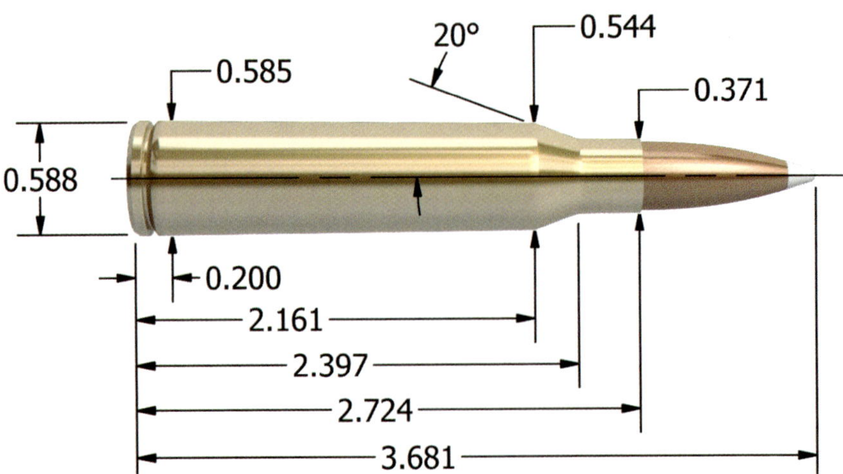

Maximum S.A.A.M.I. Overall Cartridge Length: 3.681"

BULLET CHOICES FOR THE 338 LAPUA MAGNUM

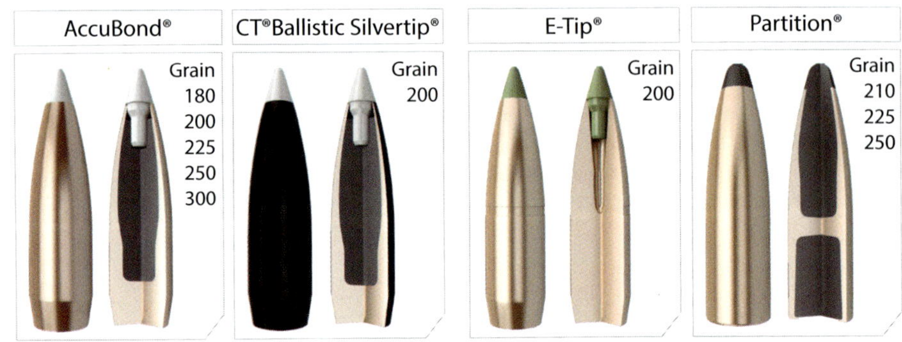

TECHNICAL INFORMATION

The 338 Lapua (8.6 x 70 mm or 8.58 x 70 mm) is a specialized, rimless, bottlenecked cartridge developed for military sniping duty. Not only is it a combat-proven round, it is increasingly used by big game hunters and civilian long-range shooting enthusiasts. The big Lapua is suitable for hunting any game animal on the planet.

338 Lapua Magnum - 180 grain		MAXIMUM S.A.A.M.I. O.A.C.L.		3.681"
		TESTED O.A.C.L.	B.C.	S.D.
AccuBond®	180gr. Spitzer	3.580"	0.372	0.225

CASE TYPE:	Norma	PRIMER TYPE	Fed. 215
CASE HOLDS:	107.0 Gr. WATER	BARREL Length/Make	26" Pac-Nor
		BARREL Twist	1-10"

POWDER TYPE	POWDER CHG. GRS.		MUZZLE VEL. F.P.S.		LOAD DENSITY (VOLUME)
H414 Most Accurate Powder Tested	84.5	MAX.	3337		82%
	82.5 *		3261		80%
	80.5		3220		78%
Magnum	109.0	MAX.	3349	**	103%
	107.0		3292	**	102%
	105.0 *		3234		100%
H4831SC	99.0	MAX.	3400		96%
	97.0		3348		94%
	95.0 *		3273		92%
H1000	104.0 *	MAX.	3401	**	102%
	102.0		3384		100%
	100.0		3321		98%
Retumbo	108.0	MAX.	3408	**	106%
	106.0		3360	**	104%
	104.0 *		3284	**	102%
H4350	90.5	MAX.	3421		90%
	88.5 *		3331		88%
	86.5		3280		86%
RL19	95.5	MAX.	3455		97%
	93.5 *		3355		95%
	91.5		3294		93%
IMR 4831	94.5	MAX.	3455		94%
	92.5 *		3359		92%
	90.5		3282		90%
IMR 7828	100.0	MAX.	3456		99%
	98.0 *		3381		97%
	96.0		3274		95%

BC=Ballistic Coefficient SD=Sectional Density
*Most Accurate Load Tested **Compressed Load

Use Maximum Loads with Caution
Refer to page 73 for additional safety information

338 Lapua Magnum - 200/210 grain		MAXIMUM S.A.A.M.I. O.A.C.L.		3.681"
		TESTED O.A.C.L.	B.C.	S.D.
AccuBond®	200gr. Spitzer	3.580"	0.414	0.250
CT® Ballistic Silvertip®	200gr. Spitzer	3.580"	0.414	0.250
E-Tip®	200gr. Spitzer	3.580"	0.425	0.250
Due to internal construction differences, always begin with starting loads when using E-Tip® products.				
Partition®	210gr. Spitzer	3.540"	0.400	0.263

CASE TYPE:	Norma		PRIMER TYPE	Fed. 215
CASE HOLDS:	106.0	Gr. WATER	BARREL Length/Make	26" Pac-Nor
			BARREL Twist	1-10"

POWDER TYPE	POWDER CHG. GRS.		MUZZLE VEL. F.P.S.	LOAD DENSITY (VOLUME)
Magnum	100.5 *	MAX.	3065	96%
	98.5		3021	94%
	96.5		2950	92%
H4831SC	90.0	MAX.	3114	88%
	88.0		3091	86%
	86.0 *		3019	85%
IMR 4350	84.0	MAX.	3126	84%
	82.0 *		3095	82%
	80.0		3055	80%
H1000	95.5 *	MAX.	3128	94%
	93.5		3098	92%
	91.5		3035	90%
Viht N560	88.5	MAX.	3138	93%
	86.5		3087	91%
	84.5 *		2999	89%
IMR 7828	90.5	MAX.	3143	91%
	88.5 *		3077	89%
	86.5		2998	87%
RL22	90.0	MAX.	3173	92%
	88.0		3110	90%
	86.0 *		3016	88%
RL19 Most Accurate Powder Tested	88.0	MAX.	3182	90%
	86.0 *		3137	88%
	84.0		3052	86%
MAGPRO	95.0	MAX.	3188	92%
	93.0 *		3142	90%
	91.0		3073	89%

BC=Ballistic Coefficient SD=Sectional Density
*Most Accurate Load Tested **Compressed Load

Use Maximum Loads with Caution
Refer to page 73 for additional safety information

338 Lapua Magnum - 225 grain

338 Lapua Magnum - 225 grain		MAXIMUM S.A.A.M.I. O.A.C.L.		3.681"
		TESTED O.A.C.L.	B.C.	S.D.
AccuBond®	225gr. Spitzer	3.580"	0.550	0.281
Partition®	225gr. Spitzer	3.540"	0.454	0.281

CASE TYPE:	Norma		PRIMER TYPE	Fed. 215
CASE HOLDS:	105.0	Gr. WATER	BARREL Length/Make	26" Pac-Nor
			BARREL Twist	1-10"

POWDER TYPE	POWDER CHG. GRS.		MUZZLE VEL. F.P.S.	LOAD DENSITY (VOLUME)
H4831SC	86.0	MAX.	2983	85%
	84.0	*	2939	83%
	82.0		2882	81%
H1000	93.0	MAX.	2989	93%
	91.0		2937	91%
	89.0	*	2874	89%
Viht N165	88.0	MAX.	2999	93%
	86.0	*	2935	91%
	84.0		2892	89%
Viht N560 Most Accurate Powder Tested	86.0	MAX.	3028	91%
	84.0	*	2951	89%
	82.0		2899	87%
Magnum	100.0	MAX.	3037	97%
	98.0		3004	95%
	96.0	*	2919	93%
RL19	86.0	MAX.	3044	89%
	84.0		2973	87%
	82.0	*	2936	85%
IMR 7828	89.0	MAX.	3048	90%
	87.0		2958	88%
	85.0	*	2904	86%
Retumbo	97.0	* MAX.	3080	97%
	95.0		3013	95%
	93.0		2953	93%
Viht N170	98.5	* MAX.	3104	** 104%
	96.5		3028	** 102%
	94.5		2989	100%

BC=Ballistic Coefficient SD=Sectional Density
*Most Accurate Load Tested **Compressed Load

Use Maximum Loads with Caution
Refer to page 73 for additional safety information

338 Lapua Magnum - 250 grain		MAXIMUM S.A.A.M.I. O.A.C.L.		3.681"
		TESTED O.A.C.L.	B.C.	S.D.
AccuBond®	250gr. Spitzer	3.580"	0.575	0.313
Partition®	250gr. Spitzer	3.540"	0.473	0.313

CASE TYPE:	Norma		PRIMER TYPE	Fed. 215
CASE HOLDS:	101.7	Gr. WATER	BARREL Length/Make	26" Pac-Nor
			BARREL Twist	1-10"

POWDER TYPE	POWDER CHG. GRS.		MUZZLE VEL. F.P.S.	LOAD DENSITY (VOLUME)
Viht N165	87.0	MAX.	2873	95%
	85.0 *		2826	93%
	83.0		2759	91%
H4831SC	86.5	MAX.	2877	89%
	84.5		2836	87%
	82.5 *		2772	85%
H1000	94.0 *	MAX.	2898	97%
	92.0		2864	95%
	90.0		2803	93%
IMR 7828	87.0	MAX.	2899	91%
	85.0		2829	89%
	83.0 *		2762	87%
Viht N560 Most Accurate Powder Tested	85.0	MAX.	2902	93%
	83.0 *		2829	91%
	81.0		2772	88%
MAGPRO	90.0	MAX.	2920	91%
	88.0 *		2874	89%
	86.0		2834	87%
RL19	85.0	MAX.	2926	91%
	83.0 *		2868	89%
	81.0		2827	87%
Retumbo	95.0 *	MAX.	2934	98%
	93.0		2881	96%
	91.0		2829	94%
RL22	86.0	MAX.	2941	92%
	84.0 *		2869	90%
	82.0		2813	88%

BC=Ballistic Coefficient SD=Sectional Density
*Most Accurate Load Tested **Compressed Load

Use Maximum Loads with Caution
Refer to page 73 for additional safety information

338 Lapua Magnum - 300 grain		MAXIMUM S.A.A.M.I. O.A.C.L.		3.681"
		TESTED O.A.C.L.	B.C.	S.D.
AccuBond®	300gr. Spitzer	3.650"	0.720	0.375

CASE TYPE:	Nosler		PRIMER TYPE	WLRM
CASE HOLDS:	98.0	Gr. WATER	BARREL Length/Make	24" Pac-Nor
			BARREL Twist	1-10"

POWDER TYPE	POWDER CHG. GRS.		MUZZLE VEL. F.P.S.		LOAD DENSITY (VOLUME)
IMR 7828 SSC	85.0	MAX.	2595		89%
	83.0		2544		87%
	81.0 *		2471		85%
US869	104.0	MAX.	2640	**	115%
	102.0		2590	**	113%
	100.0 *		2525	**	111%
Viht N560	84.0	MAX.	2641		95%
	82.0		2589		93%
	80.0 *		2530		91%
Supreme 780 Most Accurate Powder Tested	84.0	MAX.	2647		90%
	82.0 *		2581		88%
	80.0		2528		85%
RL25	88.0	MAX.	2667		98%
	86.0		2602		95%
	84.0 *		2535		93%
Magnum	91.0	MAX.	2677		94%
	89.0 *		2641		92%
	87.0		2583		90%
H1000	91.0	MAX.	2698		97%
	89.0 *		2634		95%
	87.0		2586		93%
Retumbo	94.0	MAX.	2732	**	101%
	92.0		2663		99%
	90.0 *		2611		97%

BC=Ballistic Coefficient SD=Sectional Density
*Most Accurate Load Tested **Compressed Load

Use Maximum Loads with Caution
Refer to page 73 for additional safety information

Joe Busalacchi

338-378 WEATHERBY MAGNUM

From the development of the 333OKH by Charles O'Neil, Elmer Keith and Don Hopkins in 1945, the era of big caliber 338's had begun. Bullets to handle the faster velocities and freight-train like impact of the big 33's were very limited. One such bullet that was able to hold together with deep penetration, were those produced by Fred Barnes, founder of Barnes Bullets. However, soon John Nosler created his famous Partition Bullet in 1946 and as they say; the rest is history.

Nosler set the bar to new heights and the birth of "premium bullets" was born. Hunters now had a bullet that was both very accurate and reliable for deep penetration, deep into the vitals of the toughest game. When Roy Weatherby came out with his 378 Magnum cartridge in 1953, Keith and good friend R.W. Thomson, necked the case down to .338 caliber. After almost 30 years as a Wildcat, in 1998 Weatherby re-formed the case to create the 338-378 Weatherby Magnum.

On several occasions, I had the pleasure of talking with Big John Nosler and in 1968 he sent me some of his .338 Partition bullets that he made on his home lathe, for testing. Those prototype bullets predated Nosler's use of the impact extrusion method and I considered them to be some of the greatest bullets ever produced.

Today, with newer developments in the Nosler AccuBond and E-Tip Bullets, new levels of performance and controlled expansion are being reached. The 338-378 Weatherby Magnum is a powerful cartridge capable of taking any dangerous game of the world and the cartridge has become popular with long range target shooters as well.

On a recent prairie dog shoot, I had taken a PD at over 700 yards using my 204 Ruger. So back home I decided, why not take a varmint at long range with the 338-378! I modified the tips of some of my supply of old style 200-grain Ballistic Tips. Out in the field I spotted a ground squirrel, ranged at 483 yards. I made the correct MOA adjustment on my Leupold Tactical scope and I let fly. As I drove out to the spot, not being too picturesque, all that was left was the bottom quarter of the ground squirrel that included the tail and some hide. The point I am making is that the big 338-378 is not only a powerful cartridge for taking big game but also one that is very accurate at long range shooting too.

Joe Busalacchi

Joe Busalacchi is an avid hunter, reloader, and freelance writer.

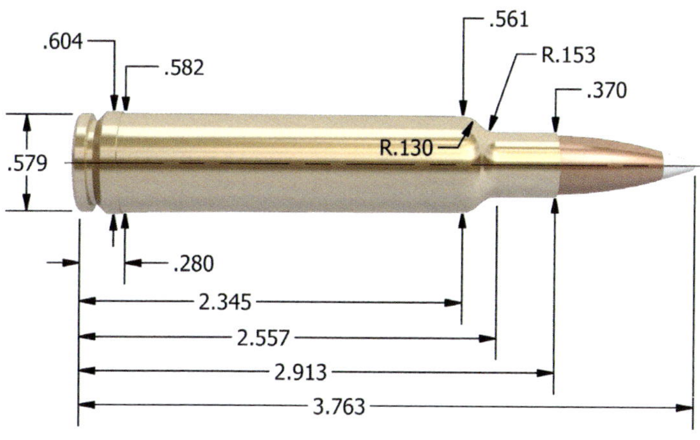

Maximum S.A.A.M.I. Overall Cartridge Length: 3.763"

BULLET CHOICES FOR THE 338-378 WEATHERBY MAGNUM

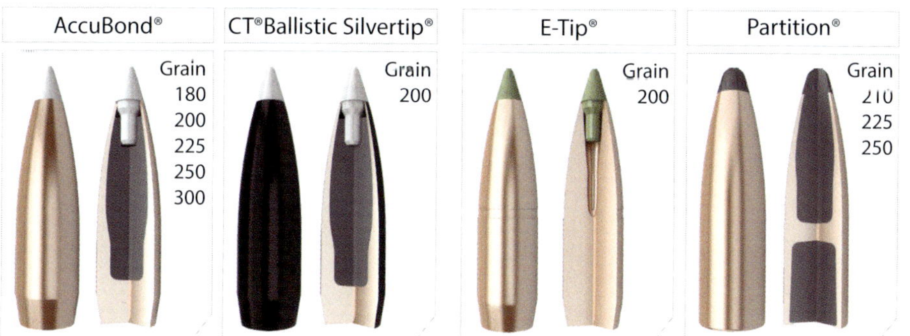

AccuBond®	CT® Ballistic Silvertip®	E-Tip®	Partition®
Grain	Grain	Grain	Grain
180	200	200	210
200			225
225			250
250			
300			

TECHNICAL INFORMATION

We achieved best results with slow powders like H-1000 and Viht N165.

338-378 Weatherby Mag - 180 grain

		MAXIMUM S.A.A.M.I. O.A.C.L.		3.763"
		TESTED O.A.C.L.	B.C.	S.D.
AccuBond®	180gr. Spitzer	3.700"	0.372	0.225

CASE TYPE:	Weatherby		PRIMER TYPE	WLRM
CASE HOLDS:	123.0	Gr. WATER	BARREL Length/Make	26" Wiseman
			BARREL Twist	1-10"

POWDER TYPE	POWDER CHG. GRS.		MUZZLE VEL. F.P.S.	LOAD DENSITY (VOLUME)
H1000 Most Accurate Powder Tested	118.0	MAX.	3496	100%
	116.0		3434	99%
	114.0 *		3372	97%
Viht N165	112.0	MAX.	3575	** 101%
	110.0 *		3482	99%
	108.0		3416	98%
IMR 7828	110.5 *	MAX.	3612	96%
	108.5		3525	94%
	106.5		3406	92%

BC=Ballistic Coefficient SD=Sectional Density
*Most Accurate Load Tested **Compressed Load

Use Maximum Loads with Caution
Refer to page 73 for additional safety information

338-378 Wby Mag - 200/210 grain		MAXIMUM S.A.A.M.I. O.A.C.L.		3.763"
		TESTED O.A.C.L.	B.C.	S.D.
AccuBond®	200gr. Spitzer	3.700	0.414	0.250
CT® Ballistic Silvertip®	200gr. Spitzer	3.700	0.414	0.250
E-Tip®	200gr. Spitzer	3.700"	0.425	0.250
Due to internal construction differences, always begin with starting loads when using E-Tip® products.				
Partition®	210gr. Spitzer	3.700	0.400	0.263

CASE TYPE:	Weatherby	PRIMER TYPE	WLRM
CASE HOLDS:	121.5 Gr. WATER	BARREL Length/Make	26" Wiseman
		BARREL Twist	1-10"

POWDER TYPE	POWDER CHG. GRS.	MUZZLE VEL. F.P.S.	LOAD DENSITY (VOLUME)
H1000 Most Accurate Powder Tested	113.0 MAX.	3276	97%
	111.0 *	3201	96%
	109.0	3160	94%
Viht N165	107.0 MAX.	3294	98%
	105.0 *	3231	96%
	103.0	3163	94%
IMR 7828	107.0 * MAX.	3365	94%
	105.0	3314	92%
	103.0	3241	90%

BC=Ballistic Coefficient SD=Sectional Density
*Most Accurate Load Tested **Compressed Load

Use Maximum Loads with Caution
Refer to page 73 for additional safety information

338-378 Weatherby Mag - 225 grain		MAXIMUM S.A.A.M.I. O.A.C.L.		3.763"
		TESTED O.A.C.L.	B.C.	S.D.
AccuBond®	225gr. Spitzer	3.700"	0.550	0.281
Partition®	225gr. Spitzer	3.700"	0.454	0.281

CASE TYPE:	Weatherby		PRIMER TYPE	WLRM
CASE HOLDS:	121.4	Gr. WATER	BARREL Length/Make	26" Wiseman
			BARREL Twist	1-10"

POWDER TYPE	POWDER CHG. GRS.		MUZZLE VEL. F.P.S.	LOAD DENSITY (VOLUME)
H4831SC	102.0	MAX.	3147	88%
	100.0		3089	86%
	98.0 *		3030	84%
Viht N165 Most Accurate Powder Tested	104.5	MAX.	3152	96%
	102.5 *		3082	94%
	100.5		3028	92%
H1000	111.0	MAX.	3175	96%
	109.0 *		3125	94%
	107.0		3082	92%
IMR 7828	104.0	MAX.	3220	91%
	102.0 *		3151	89%
	100.0		3091	88%

BC=Ballistic Coefficient SD=Sectional Density
*Most Accurate Load Tested **Compressed Load

Use Maximum Loads with Caution
Refer to page 73 for additional safety information

338-378 Weatherby Mag - 250 grain		MAXIMUM S.A.A.M.I. O.A.C.L.		3.763"
		TESTED O.A.C.L.	B.C.	S.D.
AccuBond®	250gr. Spitzer	3.700"	0.575	0.313
Partition®	250gr. Spitzer	3.700"	0.473	0.313

CASE TYPE:	Weatherby		PRIMER TYPE	WLRM
CASE HOLDS:	119.0	Gr. WATER	BARREL Length/Make	26" Wiseman
			BARREL Twist	1-10"

POWDER TYPE	POWDER CHG. GRS.		MUZZLE VEL. F.P.S.	LOAD DENSITY (VOLUME)
Viht N165 Most Accurate Powder Tested	98.0 *	MAX.	2905	92%
	96.0		2832	90%
	94.0		2775	88%
H4831SC	99.0 *	MAX.	2966	87%
	97.0		2928	85%
	95.0		2878	83%
H1000	107.0 *	MAX.	2998	94%
	105.0		2952	92%
	103.0		2891	91%
IMR 7828	101.0	MAX.	3034	90%
	99.0		2976	89%
	97.0 *		2915	87%

BC=Ballistic Coefficient SD=Sectional Density
*Most Accurate Load Tested **Compressed Load

Use Maximum Loads with Caution
Refer to page 73 for additional safety information

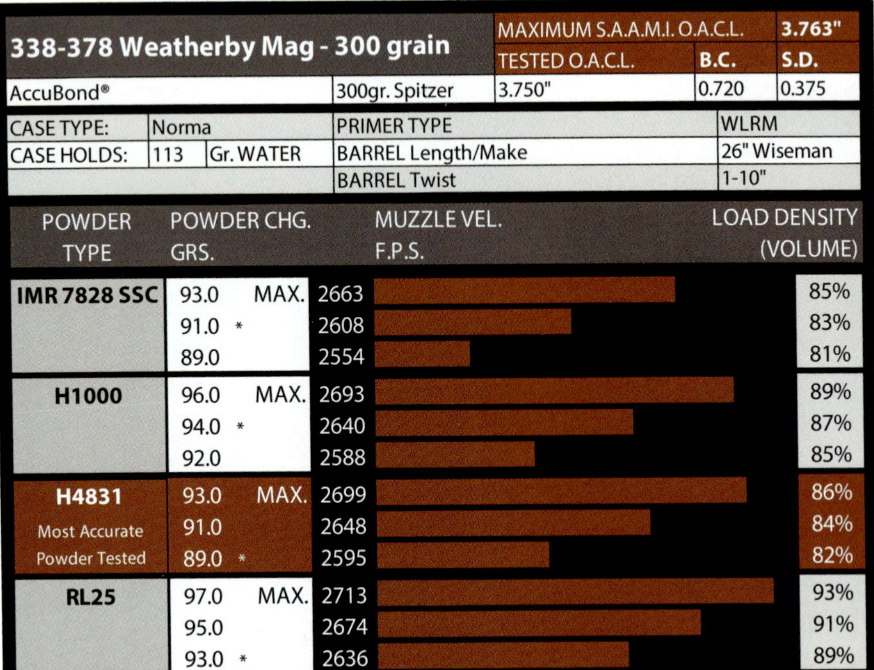

338-378 Weatherby Mag - 300 grain		MAXIMUM S.A.A.M.I. O.A.C.L.		3.763"
		TESTED O.A.C.L.	B.C.	S.D.
AccuBond®	300gr. Spitzer	3.750"	0.720	0.375

CASE TYPE:	Norma		PRIMER TYPE	WLRM
CASE HOLDS:	113	Gr. WATER	BARREL Length/Make	26" Wiseman
			BARREL Twist	1-10"

POWDER TYPE	POWDER CHG. GRS.		MUZZLE VEL. F.P.S.	LOAD DENSITY (VOLUME)
IMR 7828 SSC	93.0	MAX.	2663	85%
	91.0 *		2608	83%
	89.0		2554	81%
H1000	96.0	MAX.	2693	89%
	94.0 *		2640	87%
	92.0		2588	85%
H4831	93.0	MAX.	2699	86%
Most Accurate	91.0		2648	84%
Powder Tested	89.0 *		2595	82%
RL25	97.0	MAX.	2713	93%
	95.0		2674	91%
	93.0 *		2636	89%

BC=Ballistic Coefficient SD=Sectional Density
*Most Accurate Load Tested **Compressed Load

Use Maximum Loads with Caution
Refer to page 73 for additional safety information

35 WHELEN

I own rifles chambered for every commercial .358 rifle cartridge on the market, as well as several obsolete cartridges. I have even designed a couple of 35-caliber wildcat cartridges. It is my favorite caliber for hunting big game and this love affair started with the 35 Whelen.

I first discovered Colonel Townsend Whelen as a teenager when I read an article he wrote for the December 1906 issue of Outdoor Life, titled "Red Letter Days in British Columbia." I have probably read a few million more words about hunting in the years since, but nothing moved me more than that article about exploring in the Canadian Wilderness. Whelen lived out his last years in Vermont, close to my home, and as a fan I longed for a rifle chambered for his most famous cartridge, the 35 Whelen.

A neighbor had a gunsmith re-barrel an old Remington Model 721 to 35 Whelen and once I witnessed its effectiveness on big game I knew I had to have one. Somehow, thirty years later, that "one" has evolved into a half dozen. I have used the 35 Whelen to take game from moose to rabbits and it has never let me down. It was not by accident that the only hunting cartridge mentioned by name in my action novel "The 14th Reinstated" is the 35 Whelen.

There is some controversy about the origins of this cartridge. Some claim it was designed by James Howe and named for his friend Townsend Whelen, but it's reported that in a 1923 issue of American Rifleman Whelen called the 35 Whelen, "the first cartridge that I designed." I would think that would settle the argument.

The 35 Whelen is simply a 30-06 case necked up to take a .358 diameter bullet. But, that simple change makes a world of difference in performance. With today's powders, the Whelen can push a 225-grain bullet to 2,800 fps, which is the same as a handloaded 30-06 pushes a 180-grain bullet. This results in a 25% increase in muzzle energy. Using AccuBond bullets and a 200-yard zero, the difference in point-of-impact between the two cartridges at 300-yards is only .42-inch. At 400-yards the difference is 1.47-inches. Yet the Whelen has more energy at any hunting range and hits the critter with a bigger, heavier bullet. If you are a serious big game hunter, why compromise?

Bryce M. Towsley

Bryce M. Towsley is a Field Editor for American Rifleman

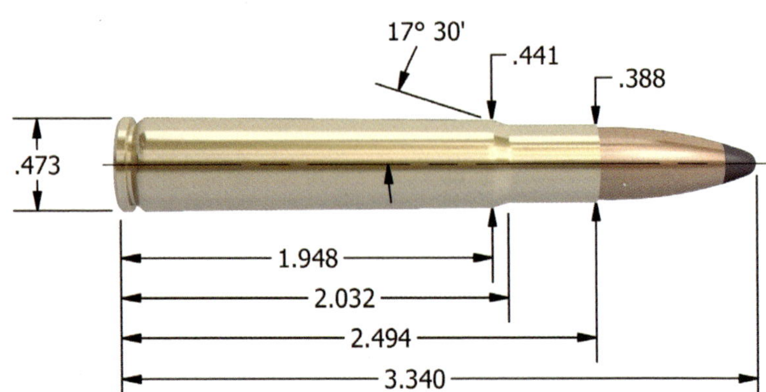

Maximum S.A.A.M.I. Overall Cartridge Length: 3.340"

BULLET CHOICES FOR THE 35 WHELEN

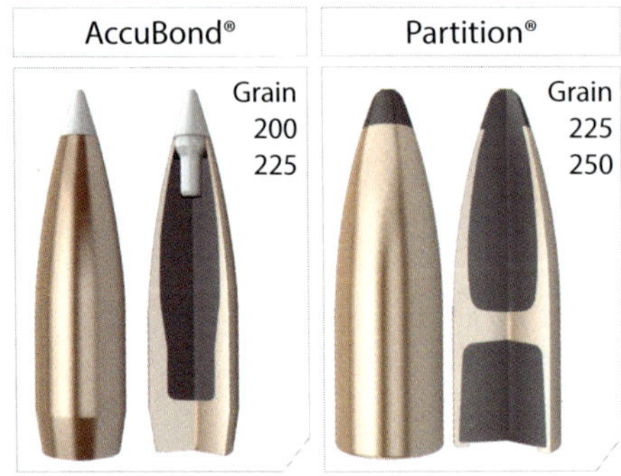

TECHNICAL INFORMATION

The 35 Whelen is a well-behaved cartridge, and is at its best with medium burning powders like IMR4350. The 200 and 225-grain AccuBond® are excellent bullet choices for this cartridge, and is suitable for any North American big game.

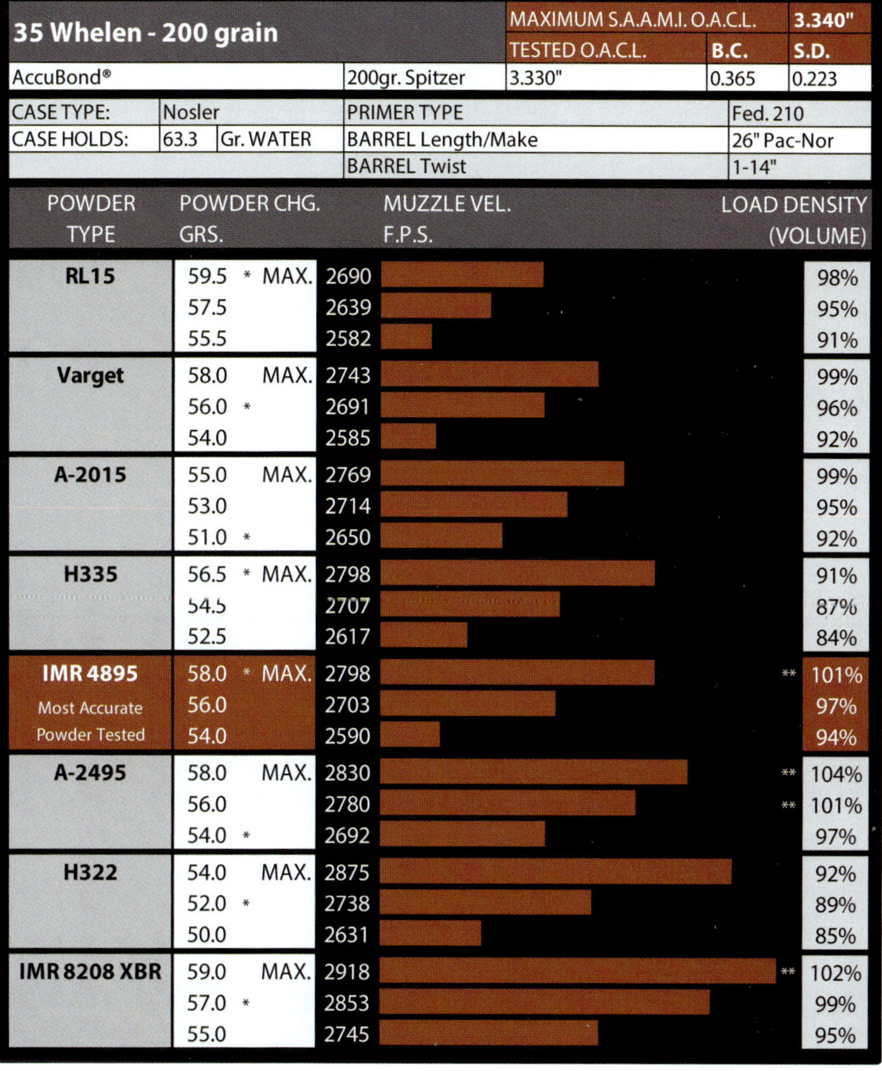

35 Whelen - 200 grain		MAXIMUM S.A.A.M.I. O.A.C.L.		3.340"
		TESTED O.A.C.L.	B.C.	S.D.
AccuBond®	200gr. Spitzer	3.330"	0.365	0.223

CASE TYPE:	Nosler	PRIMER TYPE	Fed. 210
CASE HOLDS:	63.3 Gr. WATER	BARREL Length/Make	26" Pac-Nor
		BARREL Twist	1-14"

POWDER TYPE	POWDER CHG. GRS.	MUZZLE VEL. F.P.S.	LOAD DENSITY (VOLUME)
RL15	59.5 * MAX.	2690	98%
	57.5	2639	95%
	55.5	2582	91%
Varget	58.0 MAX.	2743	99%
	56.0 *	2691	96%
	54.0	2585	92%
A-2015	55.0 MAX.	2769	99%
	53.0	2714	95%
	51.0 *	2650	92%
H335	56.5 * MAX.	2798	91%
	54.5	2707	87%
	52.5	2617	84%
IMR 4895 Most Accurate Powder Tested	58.0 * MAX.	2798	** 101%
	56.0	2703	97%
	54.0	2590	94%
A-2495	58.0 MAX.	2830	** 104%
	56.0	2780	** 101%
	54.0 *	2692	97%
H322	54.0 MAX.	2875	92%
	52.0 *	2738	89%
	50.0	2631	85%
IMR 8208 XBR	59.0 MAX.	2918	** 102%
	57.0 *	2853	99%
	55.0	2745	95%

BC=Ballistic Coefficient SD=Sectional Density
*Most Accurate Load Tested **Compressed Load

Use Maximum Loads with Caution
Refer to page 73 for additional safety information

35 Whelen - 225 grain		MAXIMUM S.A.A.M.I. O.A.C.L.		3.340"
		TESTED O.A.C.L.	B.C.	S.D.
AccuBond®	225gr. Spitzer	3.330"	0.430	0.251
Partition®	225gr. Spitzer	3.300"	0.421	0.251

CASE TYPE:	Nosler		PRIMER TYPE	Fed. 210
CASE HOLDS:	66.4	Gr. WATER	BARREL Length/Make	26" Pac-Nor
			BARREL Twist	1-14"

POWDER TYPE	POWDER CHG. GRS.		MUZZLE VEL. F.P.S.	LOAD DENSITY (VOLUME)
A-2015	49.0	MAX.	2462	84%
	47.0		2390	80%
	45.0 *		2318	77%
A-2495 Most Accurate Powder Tested	53.0 *	MAX.	2574	91%
	51.0		2499	87%
	49.0		2425	84%
H380	61.0	MAX.	2621	97%
	59.0		2573	94%
	57.0 *		2527	90%
IMR 8208 XBR	55.0	MAX.	2662	91%
	53.0 *		2591	88%
	51.0		2521	84%
RL10x	50.0	MAX.	2663	85%
	48.0		2562	81%
	46.0 *		2460	78%
H4895	57.0 *	MAX.	2752	94%
	55.0		2673	91%
	53.0		2591	88%
W748	61.0 *	MAX.	2773	96%
	59.0		2721	93%
	57.0		2662	89%
RL15	59.0 *	MAX.	2789	93%
	57.0		2734	89%
	55.0		2698	86%
Varget	60.5 *	MAX.	2800	99%
	58.5		2742	95%
	56.5		2685	92%
IMR 4064	58.0 *	MAX.	2805	96%
	56.0		2738	93%
	54.0		2668	89%

BC=Ballistic Coefficient SD=Sectional Density
*Most Accurate Load Tested **Compressed Load

Use Maximum Loads with Caution
Refer to page 73 for additional safety information

35 Whelen - 250 grain

Partition®	250gr. Spitzer	MAXIMUM S.A.A.M.I. O.A.C.L.		3.340"
		TESTED O.A.C.L.	B.C.	S.D.
		3.330"	0.446	0.279

CASE TYPE:	Nosler		PRIMER TYPE	Fed. 210
CASE HOLDS:	62.2	Gr. WATER	BARREL Length/Make	26" Pac-Nor
			BARREL Twist	1-14"

POWDER TYPE	POWDER CHG. GRS.		MUZZLE VEL. F.P.S.	LOAD DENSITY (VOLUME)
A-2015	48.0	MAX.	2399	88%
	46.0		2314	84%
	44.0 *		2226	80%
A-2495	51.0	MAX.	2452	93%
	49.0		2382	90%
	47.0 *		2309	86%
A-2520	52.0	MAX.	2472	86%
	50.0		2393	83%
	48.0 *		2314	80%
IMR 4320	53.0	MAX.	2497	92%
	51.0		2411	88%
	49.0 *		2326	85%
RL15	53.0	MAX.	2506	89%
	51.0		2416	85%
	49.0 *		2327	82%
H380	59.0	MAX.	2545	100%
	57.0		2474	96%
	55.0 *		2399	93%
Varget Most Accurate Powder Tested	55.0 *	MAX.	2591	96%
	53.0		2507	92%
	51.0		2421	89%
IMR 4064	55.0	MAX.	2598	97%
	53.0 *		2503	94%
	51.0		2409	90%
H4895	55.0 *	MAX.	2614	97%
	53.0		2536	94%
	51.0		2462	90%
W748	58.0 *	MAX.	2637	97%
	56.0		2581	94%
	54.0		2490	90%

BC=Ballistic Coefficient SD=Sectional Density
*Most Accurate Load Tested **Compressed Load

Use Maximum Loads with Caution
Refer to page 73 for additional safety information

Buck Pope

350 REMINGTON MAGNUM

The 350 Remington Magnum was introduced in 1965 by Remington as the most powerful short action caliber in the world. It was also introduced in the new Remington Model 600 carbine which had an 18 ½ inch length barrel. The new short action Remington carbine with a weight of less than six pounds, offered a muzzle blast and recoil that was very noticeable. This cartridge/ rifle combination had a rather short life mainly due to the wicked muzzle blast and recoil generated by this very light and short barreled rifle. In 1968, Remington introduced a new carbine, Model 660 which had a 20 inch barrel which was an improvement.

The 350 Remington Magnum cartridge is basically a 375 H & H Magnum case shortened, necked down to .358" and with 25 degree shoulder. The cartridge has a .532" diameter belted case with a 2.800" O.A.L. and is currently available in commercial loads of 200, 225 and 250 grain bullets.

Nosler Custom Ammunition offers a 225-grain Nosler Partition bullet rated at 2550 fps and with an energy at the muzzle of 3248 ft.-lbs.

I have hunted elk with this cartridge and become rather fond of it. I had a 22-inch rifle that weighed around eight pounds and this made the cartridge very easy to shoot. I handloaded and hunted with the 225-grain Nosler Partition bullet. I used IMR 4320 and got good velocity and excellent accuracy. Even today in Alaska, you will still see this cartridge and often in the little Remington carbine, which makes for a light, short type packing rifle. The ballistics of the 350 Remington Magnum are very similar to the 35 Whelen.

This cartridge is ideal for regions such as Alaska where you can have a short action large bore magnum cartridge that offers heavier bullet weights for game such as moose and bear. This cartridge has a reputation for hitting hard and putting game on the ground in very short order.

Buck Pope

Buck Pope is a freelance gun and hunting writer.

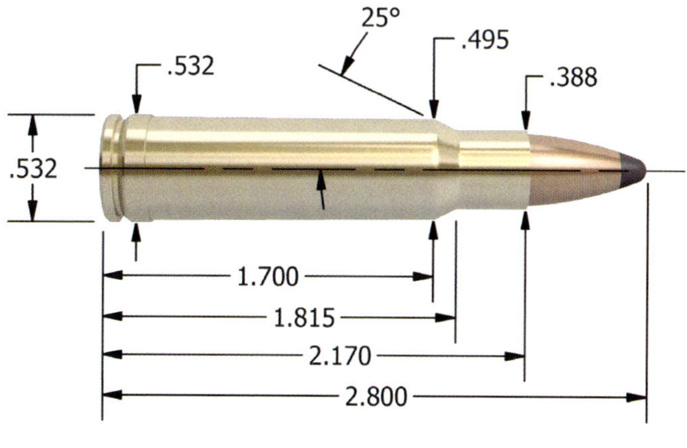

Maximum S.A.A.M.I. Overall Cartridge Length: 2.800"

BULLET CHOICES FOR THE 350 REMINGTON MAGNUM

Partition®

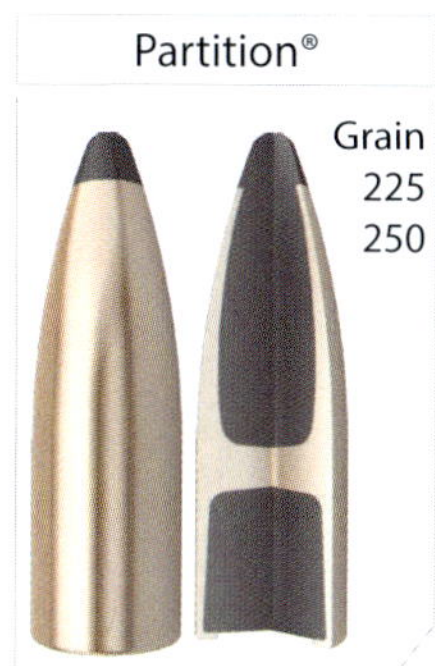

Grain
225
250

TECHNICAL INFORMATION

At 2.800" C.O.A.L., the 350 Remington Magnum is a true, short action magnum cartridge. Because of its sleek design, our 200 and 225-grain AccuBond® will not work in rifles with standard magazines. If your rifle is a single-shot, or will accept longer cartridge lengths, this bullet is an excellent choice for the 350 Remington Magnum. For our Partition bullets, we recommend medium-to-fast powders like IMR4064, IMR4320, and RL15.

350 Remington Magnum 225 grain		MAXIMUM S.A.A.M.I. O.A.C.L.		2.800"
		TESTED O.A.C.L.	B.C.	S.D.
Partition®	225gr. Spitzer	2.780"	0.430	0.251

CASE TYPE:	Remington		PRIMER TYPE	CCI 250
CASE HOLDS:	62.6	Gr. WATER	BARREL Length/Make	22" Lilja
			BARREL Twist	1-16"

POWDER TYPE	POWDER CHG. GRS.		MUZZLE VEL. F.P.S.		LOAD DENSITY (VOLUME)
H380	60.5	MAX.	2528	**	102%
	58.5		2443		98%
	56.5 *		2358		95%
RL7	47.0	MAX.	2570		82%
	45.0		2460		78%
	43.0 *		2350		75%
H4895	56.0 *	MAX.	2592		98%
	54.0		2497		95%
	52.0		2402		91%
A-2015	53.0 *	MAX.	2621		96%
	51.0		2508		93%
	49.0		2394		89%
IMR 4320 Most Accurate Powder Tested	60.0	MAX.	2640	**	103%
	58.0		2550		100%
	56.0 *		2460		96%
W748	59.0 *	MAX.	2700		98%
	57.0		2618		95%
	55.0		2535		92%

BC=Ballistic Coefficient SD=Sectional Density
*Most Accurate Load Tested **Compressed Load

Use Maximum Loads with Caution
Refer to page 73 for additional safety information

350 Remington Magnum 250 grain			MAXIMUM S.A.A.M.I. O.A.C.L.		2.800"
			TESTED O.A.C.L.	B.C.	S.D.
Partition®		250gr. Spitzer	2.780"	0.446	0.279

CASE TYPE:	Remington		PRIMER TYPE	CCI 250
CASE HOLDS:	58.6	Gr. WATER	BARREL Length/Make	22" Lilja
			BARREL Twist	1-16"

POWDER TYPE	POWDER CHG. GRS.		MUZZLE VEL. F.P.S.	LOAD DENSITY (VOLUME)
IMR 4320	53.0	MAX.	2380	97%
	51.0		2290	94%
	49.0 *		2201	90%
RL15 Most Accurate Powder Tested	55.0	MAX.	2422	98%
	53.0		2322	94%
	51.0 *		2223	91%
A-2460	52.0 *	MAX.	2454	90%
	50.0		2382	86%
	48.0		2309	83%
A-2015	51.0	MAX.	2471	99%
	49.0		2385	95%
	47.0 *		2299	91%
A-2520	55.0 *	MAX.	2507	97%
	53.0		2432	93%
	51.0		2357	90%
W748	58.0	MAX.	2571 **	103%
	56.0		2472	100%
	54.0 *		2374	96%

BC=Ballistic Coefficient SD=Sectional Density
*Most Accurate Load Tested **Compressed Load

Use Maximum Loads with Caution
Refer to page 73 for additional safety information

Gary Lewis

9.3X62 MAUSER

When the rifle I had planned to shoot was not ready to go, I had to borrow one. Never had I shot it before, but I listened as Chub Eastman recounted the ballistics and the benefits of the 9.3x62mm Mauser. He handed me the ammunition and I pushed it into the loops on my belt.

Out there in the trees, I might encounter anything from impala to elephant, lion, leopard or Cape buffalo.

Now these animals were created from cardboard and the trees were juniper, not mopane. There was nothing more at stake than bragging rights at dinner that night, and maybe a prize or two. But the adrenaline was real.

In the meager shade, a lion stood, yellow eyes focused on the object of its hate. A rope ran through a pulley to the cardboard lion. Tied to the back of the all-electric ATV, the four-wheeled cart with the two-dimensional lion reached top speed in about two jumps.

With no sound save the bump of tires over tufts of bunch grass, the target hurtled toward me from a distance of 40 yards. I fired, picked up the target again over the top of the front sight and squeezed the trigger again when the lion was 15 feet off the barrel in a cloud of dust.

Sunlight streamed through two holes, one centered between the eyes, the other just out.

At the end of it, that open-sighted Swedish Husqvarna 9.3x62 bolt-action earned me first place in my division. It also earned my respect – a cartridge that has stood the test of time since it was introduced by Otto Bock in 1905 and still holds the ground between the 35 Whelen and the 375 H&H.

It is highly regarded in Europe and Africa for its mild recoil and effectiveness on game from roe deer to eland. On North American soil, it is gaining favor as the right choice for bear, elk, moose and any other creature that is best dispatched with one round, placed in just the right spot.

Gary A. Lewis

Gary, with John Nosler, is the author of John Nosler – Going Ballistic. Lewis makes his home in Bend, Oregon

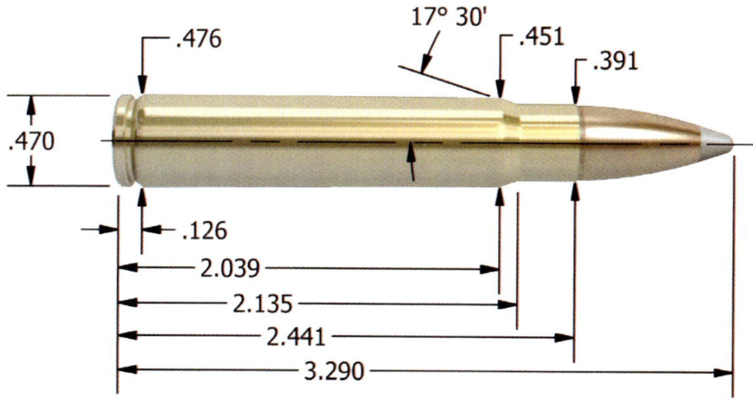

Maximum S.A.A.M.I. Overall Cartridge Length: 3.290"

BULLET CHOICES FOR THE 9.3X62 MAUSER

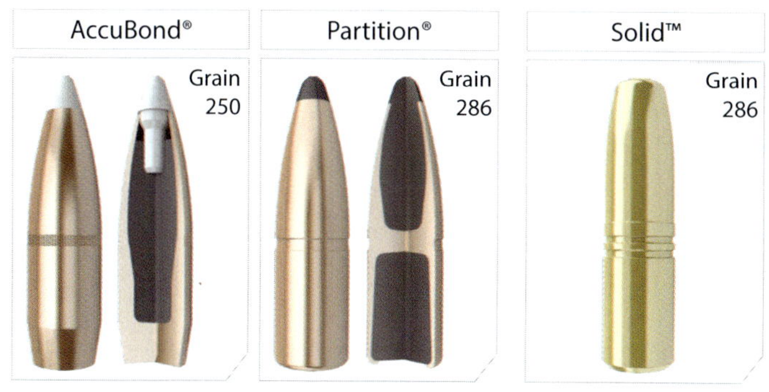

TECHNICAL INFORMATION

The 9.3x62mm Mauser has become our primary cartridge for 9.3mm production testing. We have seen great results with Varget and Viht N135.

9.3x62mm Mauser - 250 grain		MAXIMUM CIP O.A.C.L.		3.290"
		TESTED O.A.C.L.	B.C.	S.D.
AccuBond®	250gr. Spitzer	3.230"	0.494	0.267

CASE TYPE:	Norma		PRIMER TYPE	WLR
CASE HOLDS:	62.5	Gr. WATER	BARREL Length/Make	26" Lo. Walther
			BARREL Twist	1-10"

POWDER TYPE	POWDER CHG. GRS.		MUZZLE VEL. F.P.S.		LOAD DENSITY (VOLUME)
A-2015	51.5	MAX.	2476		94%
	49.5 *		2365		90%
	47.5		2307		86%
Viht N135	53.5	MAX.	2487		100%
	51.5		2429		96%
	49.5 *		2326		92%
IMR 3031	53.5	MAX.	2514		97%
	51.5		2438		93%
	49.5 *		2345		89%
Viht N140	57.0 *	MAX.	2524	**	106%
	55.0		2482	**	102%
	53.0		2392		99%
Varget Most Accurate Powder Tested	56.5	MAX.	2525		98%
	54.5		2452		94%
	52.5 *		2383		91%
IMR 4895	56.0	MAX.	2551		98%
	54.0 *		2502		95%
	52.0		2398		91%
IMR 4320	59.5 *	MAX.	2569	**	102%
	57.5		2515		99%
	55.5		2415		95%
IMR 4064	56.0	MAX.	2582		98%
	54.0		2467		95%
	52.0 *		2341		91%

BC=Ballistic Coefficient SD=Sectional Density
*Most Accurate Load Tested **Compressed Load

Use Maximum Loads with Caution
Refer to page 73 for additional safety information

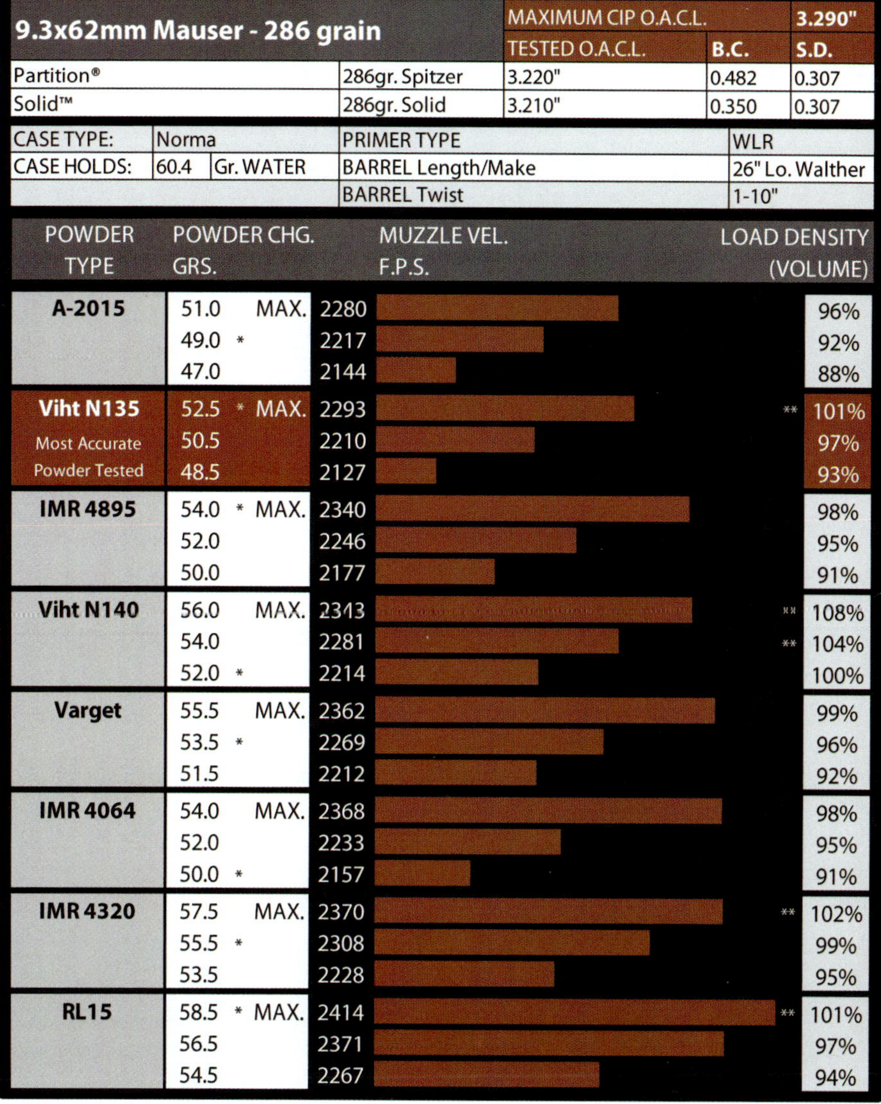

9.3x62mm Mauser - 286 grain		MAXIMUM CIP O.A.C.L.		3.290"
		TESTED O.A.C.L.	B.C.	S.D.
Partition®	286gr. Spitzer	3.220"	0.482	0.307
Solid™	286gr. Solid	3.210"	0.350	0.307

CASE TYPE:	Norma		PRIMER TYPE	WLR
CASE HOLDS:	60.4	Gr. WATER	BARREL Length/Make	26" Lo. Walther
			BARREL Twist	1-10"

POWDER TYPE	POWDER CHG. GRS.		MUZZLE VEL. F.P.S.		LOAD DENSITY (VOLUME)
A-2015	51.0	MAX.	2280		96%
	49.0 *		2217		92%
	47.0		2144		88%
Viht N135 Most Accurate Powder Tested	52.5 *	MAX.	2293	**	101%
	50.5		2210		97%
	48.5		2127		93%
IMR 4895	54.0 *	MAX.	2340		98%
	52.0		2246		95%
	50.0		2177		91%
Viht N140	56.0	MAX.	2343	**	108%
	54.0		2281	**	104%
	52.0 *		2214		100%
Varget	55.5	MAX.	2362		99%
	53.5 *		2269		96%
	51.5		2212		92%
IMR 4064	54.0	MAX.	2368		98%
	52.0		2233		95%
	50.0 *		2157		91%
IMR 4320	57.5	MAX.	2370	**	102%
	55.5 *		2308		99%
	53.5		2228		95%
RL15	58.5 *	MAX.	2414	**	101%
	56.5		2371		97%
	54.5		2267		94%

BC=Ballistic Coefficient SD=Sectional Density
*Most Accurate Load Tested **Compressed Load

Use Maximum Loads with Caution
Refer to page 73 for additional safety information

Roland Zeitler

9.3X74R

The 9.3x74R came onto the market in 1905, the same year in which the 9.3x62 was introduced. The 9.3x74R is the rimmed sister-cartridge to the 9.3x62 with nearly the same power. Practically speaking, both cartridges are equal, and very similar to the 375 H&H Flanged Magnum. Being designed for use in double rifles, the case of the 9.3x74R has a rim. The 9.3x74R case is long and has lots of internal volume. Throughout Europe, it is the most popular cartridge for double rifles. It is not a cartridge suited to use on dangerous African game like lion, Cape buffalo or elephant. It is also not legal for dangerous game in most African countries. In Africa, it is best suited as a plains game cartridge for species such as greater kudu, sable, wildebeest or roan. The cartridge is most famous in the northern hemisphere for hunting big game throughout Europe. It works well on the big bears (polar, grizzly, brown bear), muskox, bison, moose and elk. In Europe, it is a perfect cartridge for red deer, moose and wild boar. It is best used at short and medium ranges, and many hunters prefer it for hunting thick woods. The most common use of the cartridge is for driven big game hunts. I've used the 9.3x74R mostly in my Krieghoff double rifle, which is a very well made rifle that will place bullets from both barrels to the same point of impact at 100 meters (109.36 yards). The rifle is well balanced and I can place accurate shots very quickly. In my handloads for this rifle, I use the 286gr Nosler Partition and the 250gr Nosler AccuBond. With either bullet, the regulation of my rifle's barrels holds true, striking the same point of impact whether loaded with the AccuBond or Partition. The accuracy is outstanding. Throughout the years, I've taken many a wild boar on driven hunts. I've also taken red deer, fallow deer and roe deer with this rifle. Every time, the Nosler bullets worked perfectly. Terminal performance was perfect and most game dropped at the shot. I like the cartridge, and I like Nosler bullets, because I can rely on them to perform every time. That's all I need.

Roland Zeitler

Roland Zeitler is a noted European freelance writer.

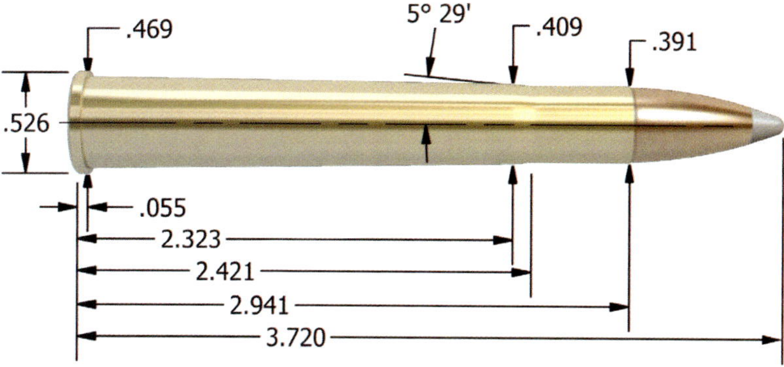

Maximum S.A.A.M.I. Overall Cartridge Length: 3.720"

BULLET CHOICES FOR THE 9.3X74R

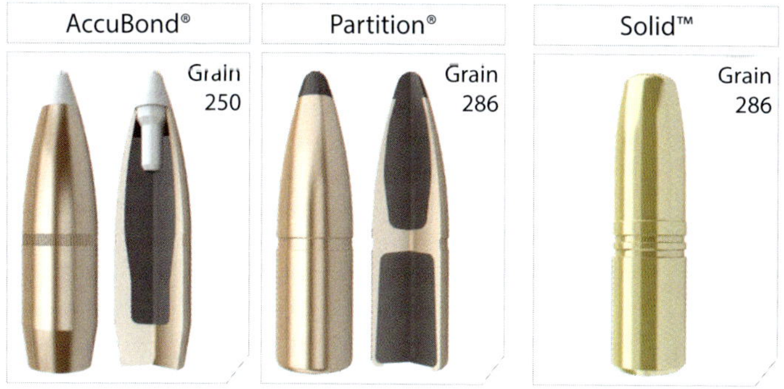

TECHNICAL INFORMATION

Because of its 3.720" overall cartridge length, the 9.3x74R is most commonly seen in break action single-shot rifles. With no magazine length constraints, best accuracy will be realized by adjusting the bullet seating depth in relation to the rifling.

9.3x74R - 250 grain		MAXIMUM CIP O.A.C.L.		3.720"
		TESTED O.A.C.L.	B.C.	S.D.
AccuBond®	250gr. Spitzer	3.710"	0.494	0.267

CASE TYPE:	Norma		PRIMER TYPE	Fed. 210
CASE HOLDS:	68.5	Gr. WATER	BARREL Length/Make	26" Lilja
			BARREL Twist	1-14"

POWDER TYPE	POWDER CHG. GRS.		MUZZLE VEL. F.P.S.		LOAD DENSITY (VOLUME)
Viht N135 Most Accurate Powder Tested	55.5	MAX.	2482		94%
	53.5		2401		91%
	51.5 *		2336		87%
Varget	57.5	MAX.	2490		91%
	55.5 *		2396		88%
	53.5		2312		84%
Viht N140	58.5	MAX.	2515		99%
	56.5 *		2447		96%
	54.5		2365		93%
H380	62.5	MAX.	2540		96%
	60.5 *		2474		93%
	58.5		2405		90%
IMR 4350	68.0 *	MAX.	2553	**	105%
	66.0		2473	**	102%
	64.0		2411		99%
IMR 4064	59.0	MAX.	2558		95%
	57.0 *		2483		91%
	55.0		2410		88%
RL15	61.0	MAX.	2579		93%
	59.0 *		2499		90%
	57.0		2430		87%
Viht N550	67.0	MAX.	2653	**	109%
	65.0		2566	**	105%
	63.0 *		2490	**	102%

BC=Ballistic Coefficient SD=Sectional Density
*Most Accurate Load Tested **Compressed Load

Use Maximum Loads with Caution
Refer to page 73 for additional safety information

9.3x74R - 286 grain		MAXIMUM CIP O.A.C.L.		3.720"
		TESTED O.A.C.L.	B.C.	S.D.
Partition®	286gr. Spitzer	3.710"	0.482	0.307
Solid™	286gr. Solid	3.700"	0.350	0.307

CASE TYPE:	Norma		PRIMER TYPE	Fed. 210
CASE HOLDS:	66.7	Gr. WATER	BARREL Length/Make	24" Lilja
			BARREL Twist	1-14"

POWDER TYPE	POWDER CHG. GRS.		MUZZLE VEL. F.P.S.	LOAD DENSITY (VOLUME)
Viht N140 Most Accurate Powder Tested	55.5	MAX.	2284	97%
	53.5 *		2212	93%
	51.5		2132	90%
Varget	54.5	MAX.	2324	88%
	52.5 *		2238	85%
	50.5		2165	82%
Viht N150	56.5	MAX.	2328	100%
	54.5		2260	96%
	52.5 *		2193	93%
RL15	58.0	MAX.	2350	91%
	56.0		2271	87%
	54.0 *		2204	84%
IMR 4831	64.5 *	MAX.	2360	** 103%
	62.5		2304	100%
	60.5		2226	96%
IMR 4064	56.5 *	MAX.	2373	93%
	54.5		2301	90%
	52.5		2211	86%
IMR 4350	64.5	MAX.	2398	** 102%
	62.5		2322	99%
	60.5 *		2253	96%

BC=Ballistic Coefficient SD=Sectional Density
*Most Accurate Load Tested **Compressed Load

Use Maximum Loads with Caution
Refer to page 73 for additional safety information

J. Scott Olmsted

375 HOLLAND & HOLLAND MAGNUM

If dangerous game is part of the plan, get a "three-seven-five" and learn to shoot it well. You'll discover it expels enough power—on average 4,000 foot-pounds of energy—to penetrate and kill the world's toughest game. You'll feel its recoil is stout but manageable. You'll appreciate the fact that a 9 1/2- or 10-pound rifle so chambered is not a burden in your hands. Most importantly, you'll wield a rifle that fires a bullet you can place confidently rather than one that scares you.

The 375 H&H Magnum was developed in 1912 by the British firm Holland & Holland for its magnum Mauser rifles built for Africa. It's enjoyed a loyal following ever since.

Perhaps due to its age, many hunters think it archaic. It's neither the best choice for long range nor the best as a stopper at close range. It delivers more power than needed for all North American game except the big bears, and in less than ideal situations it's a bit shy on bullet diameter for the biggest African game. Its long, belted, rimless case tapers slowly to a slight shoulder—an outmoded look compared to all the fat cartridges, with sharp shoulders these days, that produce plenty of velocity and power from shorter cases. Many smaller calibers shoot flatter and hit almost as hard; many larger calibers shoot just as flat and hit harder. Since 1988, many other .375's have come on-line that do one thing or another better than the hundred-year-old original—whether they fly faster or flatter, or hit harder (often on both ends).

But the original three-seven-five remains the most useful cartridge ever developed. It is not loaded hot enough to bind actions—a considerable concern in warmer climates. That long, slender case makes for butter-smooth feeding. Its trajectory loaded with 260-grain bullets nearly duplicates that of the 30-06 with 180-grain bullets—a good thing to know if one encounters an antelope at 300 yards. Its most useful load, a 300-grainer, delivers enough energy for everything from brown bears to Cape buffalo—and even elephant, under normal conditions.

A rifle chambered in 375 H&H Mag. is the best rig a hunter could carry in Africa, and set up in a synthetic stock and stainless steel, it would be excellent in brown bear country. If one cartridge is meant to do it all, this is it.

J. Scott Olmsted is the editor in chief of the NRA's American Hunter magazine; an NRA-certified firearm instructor; a former Marine infantry squad leader; and a graduate of the University of Missouri School of Journalism.

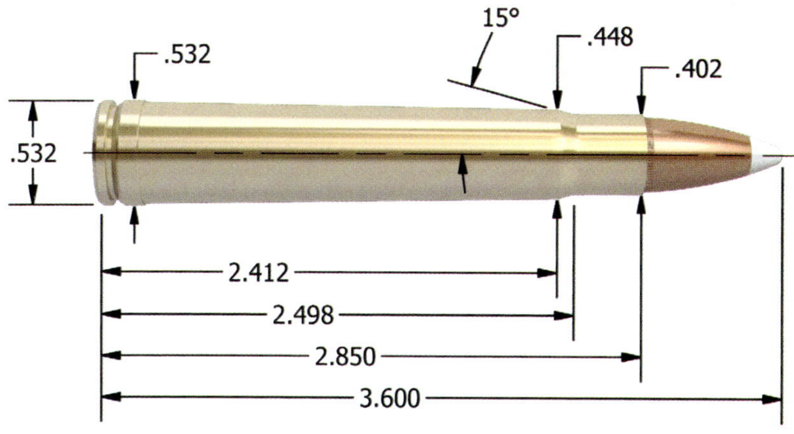

Maximum S.A.A.M.I. Overall Cartridge Length: 3.600"

BULLET CHOICES FOR THE 375 HOLLAND & HOLLAND (H&H) MAGNUM

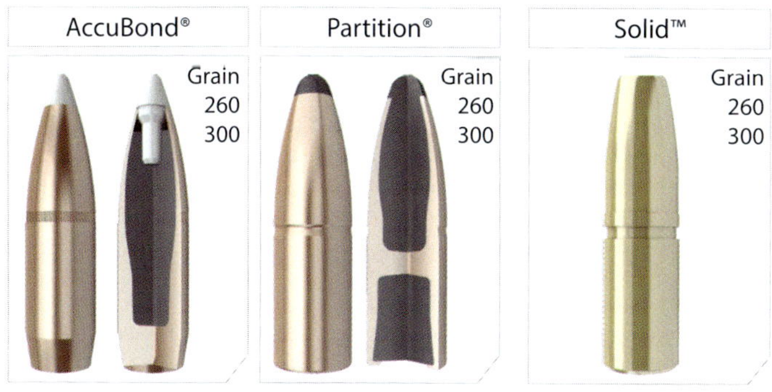

TECHNICAL INFORMATION

Our 260 grain AccuBond or Partition bullets are excellent choices for North American hunting. For dangerous game, we recommend our 300 grain bullet designs.

375 H&H Magnum - 260 grain		MAXIMUM S.A.A.M.I. O.A.C.L.		3.600"
		TESTED O.A.C.L.	B.C.	S.D.
AccuBond®	260gr. Spitzer	3.575"	0.473	0.264
Partition®	260gr. Spitzer	3.495"	0.314	0.264
Solid™	260gr. Solid	3.590"	0.254	0.264

CASE TYPE:	Nosler		PRIMER TYPE	Rem. 9 1/2M
CASE HOLDS:	83.4	Gr. WATER	BARREL Length/Make	24" Lilja
			BARREL Twist	1-12"

POWDER TYPE	POWDER CHG. GRS.		MUZZLE VEL. F.P.S.		LOAD DENSITY (VOLUME)
RL19	80.0 *	MAX.	2604	**	104%
	78.0		2537	**	102%
	76.0		2470		99%
Viht N160	81.0 *	MAX.	2681	**	108%
	79.0		2621	**	105%
	77.0		2561	**	103%
IMR 4350	78.5	MAX.	2707		100%
	76.5		2639		97%
	74.5 *		2570		95%
IMR 4064	70.0	MAX.	2721		92%
	68.0		2617		90%
	66.0 *		2512		87%
IMR 4831	80.5	MAX.	2734	**	103%
	78.5		2669		100%
	76.5 *		2605		98%
W760	78.5	MAX.	2747		99%
	76.5		2690		96%
	74.5 *		2633		94%
RL15 Most Accurate Powder Tested	73.0	MAX.	2793		91%
	71.0		2721		89%
	69.0 *		2648		86%

BC=Ballistic Coefficient SD=Sectional Density
*Most Accurate Load Tested **Compressed Load

Use Maximum Loads with Caution
Refer to page 73 for additional safety information

375 H&H Magnum - 300 grain		MAXIMUM S.A.A.M.I. O.A.C.L.		3.600"
		TESTED O.A.C.L.	B.C.	S.D.
AccuBond®	300gr. Spitzer	3.575"	0.485	0.305
Partition®	300gr. Spitzer	3.550"	0.398	0.305
Solid™	300gr. Solid	3.570"	0.300	0.305

CASE TYPE:	Nosler		PRIMER TYPE	Rem. 9 1/2M
CASE HOLDS:	78.5	Gr. WATER	BARREL Length/Make	24" Lilja
			BARREL Twist	1-12"

POWDER TYPE	POWDER CHG. GRS.		MUZZLE VEL. F.P.S.		LOAD DENSITY (VOLUME)
IMR 4350	75.0	MAX.	2452	**	101%
	73.0		2367		98%
	71.0 *		2282		96%
Viht N160	78.5 *	MAX.	2462	**	111%
	76.5		2406	**	108%
	74.5		2350	**	105%
RL19 Most Accurate Powder Tested	79.0 *	MAX.	2470	**	109%
	77.0		2410	**	107%
	75.0		2350	**	104%
RL15	68.0	MAX.	2490		90%
	66.0		2425		88%
	64.0 *		2361		85%
IMR 4831	78.0	MAX.	2513	**	106%
	76.0		2402	**	103%
	74.0 *		2300		100%
H4895	67.5	MAX.	2532		94%
	65.5		2466		92%
	63.5 *		2392		89%
W760	78.0 *	MAX.	2600	**	104%
	76.0		2564	**	101%
	74.0		2517		99%

BC=Ballistic Coefficient SD=Sectional Density
*Most Accurate Load Tested **Compressed Load

Use Maximum Loads with Caution
Refer to page 73 for additional safety information

Phil Shoemaker

375 RUGER

Hailed as the "Queen of calibers," the 375 H&H has an enviable track record on big game from whitetail deer to elephants. The main drawback of this centenarian cartridge however is that it was intended for an archaic, temperature sensitive cordite powder. It's excessively long, highly tapered, belted design requires a heavy, magnum length action.

Those faults were corrected with the introduction of the 375 Ruger. It is a modern, compact cartridge specifically designed to fit in standard length actions. With a slightly greater powder capacity, it reaches equal velocities in shortened barrels or, in the tropical heat of Africa, achieves equal velocity with lower chamber pressures.

But the appeal of the 375 is not isolated to the Dark Continent. In the vast wilderness of Alaska, where hunters travel on foot, few things are more appreciated than short, handy, powerful rifles. The acceptance of the 375 Ruger in Alaska has been phenomenal.

It is a highly versatile, flat shooting, accurate cartridge. With 260 gr AccuBond bullets, it works fantastic for distant blacktail deer and goats on Kodiak Island, as well as for Dall sheep and caribou in the Brooks range. It is also as fine an elk and moose round as you can find.

Among Alaska's bear hunters, the 375 has long been the standard by which all other calibers are judged. It combines the critical bullet mass and bone smashing raw power, required for close encounters with the biggest of brown bears, with a flat trajectory to reach across expanses of open tundra.

The 375 Ruger has become the default rifle for Alaskan guides who specialize in massive bruins because, when you combine the proven performance of the 375, with the equally legendary 300 grain Partition, you have an unbeatable combination.

Phil Shoemaker is an Alaskan Master Guide and occasional outdoor writer

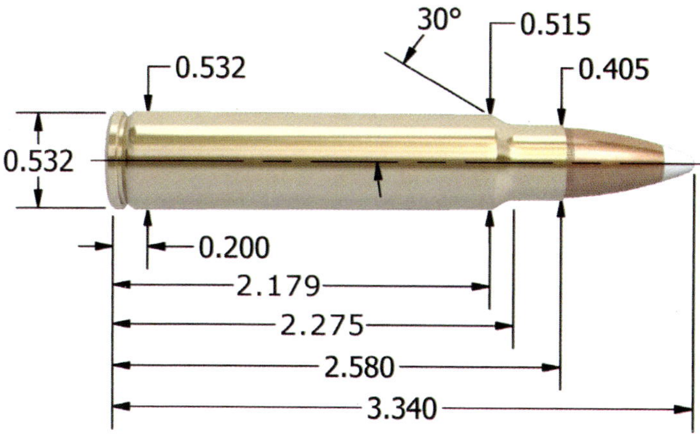

Maximum S.A.A.M.I. Overall Cartridge Length: 3.340″

BULLET CHOICES FOR THE 375 RUGER

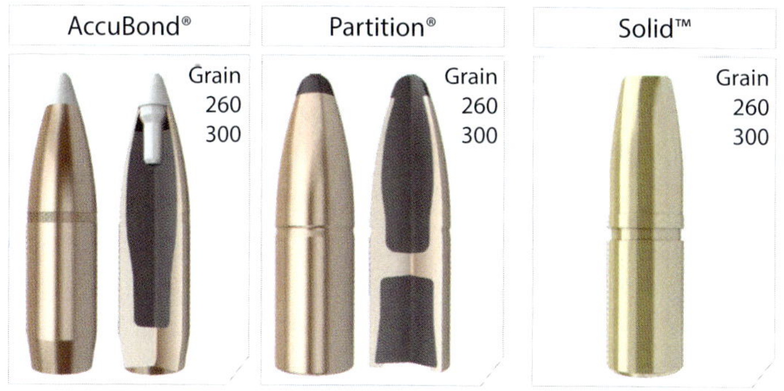

TECHNICAL INFORMATION

The 375 Ruger is capable of exceeding 375 H&H performance levels with a standard length action. This is due to its rimless, beltless case (the same head diameter as "belted cases") having greater capacity from the increased body diameter. *NOTE – Due to its ogive curve, our 260-grain Partition must be seated deeper than the crimping groove to avoid interference with the rifling. Heavy interference with the rifling will cause elevated pressures.*

375 Ruger - 260 grain		MAXIMUM S.A.A.M.I. O.A.C.L.		3.340"
		TESTED O.A.C.L.	B.C.	S.D.
AccuBond®	260gr. Spitzer	3.320"	0.473	0.264
Partition®	260gr. Spitzer	3.240"	0.314	0.264
Solid™	260gr. Solid	3.335"	0.254	0.264

CASE TYPE:	Hornady		PRIMER TYPE	WLRM
CASE HOLDS:	85.5	Gr. WATER	BARREL Length/Make	26" Pac-Nor
			BARREL Twist	1-12"

POWDER TYPE	POWDER CHG. GRS.		MUZZLE VEL. F.P.S.		LOAD DENSITY (VOLUME)
H380	81.0 *	MAX.	2854		100%
	79.0		2798		97%
	77.0		2697		95%
IMR 4007 SSC	79.0	MAX.	2872		99%
	77.0		2799		97%
	75.0 *		2730		94%
W760	79.0 *	MAX.	2876		97%
	77.0		2824		94%
	75.0		2771		92%
RL15	76.0	MAX.	2883		93%
	74.0		2803		90%
	72.0 *		2742		88%
Hybrid 100V Most Accurate Powder Tested	81.0 *	MAX.	2885		98%
	79.0		2818		95%
	77.0		2751		93%
RL17	81.5	MAX.	2927		99%
	79.5		2846		96%
	77.5 *		2767		94%
Big Game	82.5	MAX.	2931	**	102%
	80.5 *		2849		99%
	78.5		2790		97%
Hunter	88.0	MAX.	2941	**	108%
	86.0		2870	**	106%
	84.0 *		2801	**	103%

BC=Ballistic Coefficient SD=Sectional Density
*Most Accurate Load Tested **Compressed Load

Use Maximum Loads with Caution
Refer to page 73 for additional safety information

375 Ruger - 300 grain

375 Ruger - 300 grain		MAXIMUM S.A.A.M.I. O.A.C.L.		3.340"
		TESTED O.A.C.L.	B.C.	S.D.
AccuBond®	300gr. Spitzer	3.325"	0.485	0.305
Partition®	300gr. Spitzer	3.300"	0.398	0.305
Solid™	300gr. Solid	3.320"	0.300	0.305

CASE TYPE:	Hornady		PRIMER TYPE	WLRM
CASE HOLDS:	80.7	Gr. WATER	BARREL Length/Make	26" Pac-Nor
			BARREL Twist	1-12"

POWDER TYPE	POWDER CHG. GRS.		MUZZLE VEL. F.P.S.		LOAD DENSITY (VOLUME)
W760	73.0 *	MAX.	2613		95%
	71.0		2551		92%
	69.0		2488		90%
Varget	71.0	MAX.	2615		95%
	69.0 *		2554		92%
	67.0		2494		90%
RL15 Most Accurate Powder Tested	72.0	MAX.	2619		93%
	70.0 *		2579		90%
	68.0		2507		88%
IMR 4007SSC	77.5	MAX.	2667	**	103%
	75.5		2588		100%
	73.5 *		2509		98%
RL17	78.5	MAX.	2672	**	101%
	76.5		2613		98%
	74.5 *		2557		96%
Hybrid 100V	78.0 *	MAX.	2680		100%
	76.0		2607		97%
	74.0		2537		95%
Hunter	85.0 *	MAX.	2706	**	111%
	83.0		2653	**	108%
	81.0		2597	**	106%
IMR 4350	83.0	MAX.	2715	**	109%
	81.0 *		2655	**	106%
	79.0		2595	**	104%

BC=Ballistic Coefficient SD=Sectional Density
*Most Accurate Load Tested **Compressed Load

Use Maximum Loads with Caution
Refer to page 73 for additional safety information

Art Wheaton

375 REMINGTON ULTRA MAGNUM

Our Professional Hunter glassed the herd, seemingly being lazy in the trees about 250 yards away. We decided on an immediate stalk. A clump of trees between us made it easier to close some of the distance undetected, then, we began on our hands and knees. Not really fun on the wrists but we gained the decided advantage of a small tree covered hill, then got to our feet putting every little depression and tree trunk between us and the buffs.

He led the way with the sticks in hand until we had managed to make it to a little less than 100 yards, the PH glassing closely for a fine specimen with nice boss and horn. Whispering to me, he pointed out a nice bull. After looking through the scope, I was able to identify the target.

A couple more steps out from behind a large tree trunk he planted the shooting sticks and identified the bull again to ensure we were together on the right one. Planting the rifle on the sticks and quickly adjusting their lengths for height, the Leupold 1 3/4X6 cranked to maximum, my Professional Hunter coached me to break the right shoulder, knowing full well the herd would be off in a cloud of dust and a well-placed shot would immobilize the trophy animal. Upon impact, and as if choreographed, the wounded bull stumbled but remained upright while his pals took off. It should be noted that this procedure on a very dangerous buffalo is important as many "dead" buffalo have killed a PH or pursuer alike. The left shoulder was exposed as the bull turned and was facing to my left. It was down at the second shot and as we carefully approached, I was instructed to send a solid to the exposed spine making for an insurance policy that the animal would not get up to take revenge on us…standard procedure again. The 260gr. bullets had done their job, the solid for good measure, and now his glaring eyes look down upon me from a high wall. A vivid reminder of my first of two buffalo taken with the Remington 375 Ultra Mag and Remington African Big Game rifle on their maiden voyage in Tanzania, Africa.

Art Wheaton is retired Vice President General Manager Worldwide Sales of Remington Arms Co. Inc.

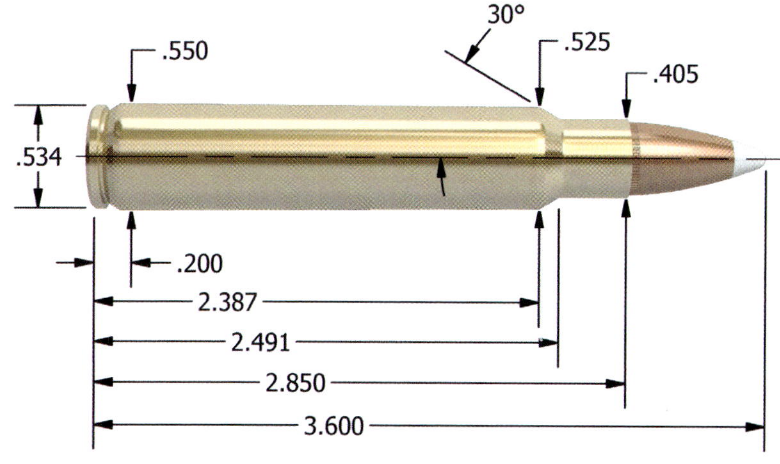

Maximum S.A.A.M.I. Overall Cartridge Length: 3.600"

BULLET CHOICES FOR THE 375 REMINGTON ULTRA MAGNUM

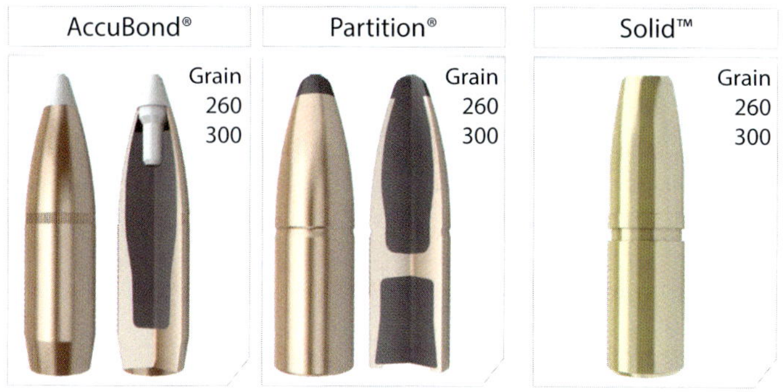

TECHNICAL INFORMATION

The most accurate of the 375's, the RUM has edged-out the old H&H as our standard 375 caliber test vehicle. H4831SC and H1000 will generally provide best results, especially when teamed with magnum primers.

375 Rem Ultra Mag - 260 grain

		MAXIMUM S.A.A.M.I. O.A.C.L.		3.600"
		TESTED O.A.C.L.	B.C.	S.D.
AccuBond®	260gr. Spitzer	3.575"	0.473	0.264
Partition®	260gr. Spitzer	3.495"	0.314	0.264
Solid™	260gr. Solid	3.590"	0.254	0.264

CASE TYPE:	Remington	PRIMER TYPE	Rem. 9 1/2M
CASE HOLDS:	105.2 Gr. WATER	BARREL Length/Make	26" Lilja
		BARREL Twist	1-12"

POWDER TYPE	POWDER CHG. GRS.	MUZZLE VEL. F.P.S.	LOAD DENSITY (VOLUME)
H1000	107.0 * MAX.	2935	** 107%
	105.0	2905	** 105%
	103.0	2835	** 103%
H4831SC Most Accurate Powder Tested	101.0 MAX.	3017	100%
	99.0	2961	98%
	97.0 *	2907	96%
IMR 7828	102.0 * MAX.	3018	** 103%
	100.0	2968	** 101%
	98.0	2928	99%
IMR 4350	96.0 * MAX.	3026	97%
	94.0	2976	95%
	92.0	2913	93%

BC=Ballistic Coefficient SD=Sectional Density
*Most Accurate Load Tested **Compressed Load

Use Maximum Loads with Caution
Refer to page 73 for additional safety information

375 Rem Ultra Mag - 300 grain		MAXIMUM S.A.A.M.I. O.A.C.L.		3.600"
		TESTED O.A.C.L.	B.C.	S.D.
AccuBond®	300gr. Spitzer	3.575"	0.485	0.305
Partition®	300gr. Spitzer	3.550"	0.398	0.305
Solid™	300gr. Solid	3.570"	0.300	0.305

CASE TYPE:	Remington		PRIMER TYPE	Rem. 9 1/2M
CASE HOLDS:	100.8	Gr. WATER	BARREL Length/Make	26" Lilja
			BARREL Twist	1-12"

POWDER TYPE	POWDER CHG. GRS.		MUZZLE VEL. F.P.S.		LOAD DENSITY (VOLUME)
H1000 Most Accurate Powder Tested	105.0 *	MAX.	2793	**	109%
	103.0		2745	**	107%
	101.0		2702	**	105%
H4831SC	97.0 *	MAX.	2820		100%
	95.0		2768		98%
	93.0		2705		96%
IMR 4350	94.0	MAX.	2849		99%
	92.0 *		2807		97%
	90.0		2760		94%
IMR 7828	100.0	MAX.	2859	**	106%
	98.0 *		2800	**	103%
	96.0		2734	**	101%

BC=Ballistic Coefficient SD=Sectional Density
*Most Accurate Load Tested **Compressed Load

Use Maximum Loads with Caution
Refer to page 73 for additional safety information

John R. Nosler

416 REMINGTON MAGNUM

My personal history with the 416 Remington Magnum is one of necessity. It has become my go-to rifle in Africa-out of sheer happenstance. On three separate occasions, (2 buffalo and 1 hippo), I have had one of those "uh-oh" moments. The kind of moment where I began to think I will be the latest Darwin Award Winner- "annually given to individuals who contribute to human evolution by self-selecting themselves out of the gene pool through putting themselves in life-threatening situations."

The most memorable safari was in the Selous of Tanzania. I was carrying a 458 Lott at the time and we were after hippo or "river horse." Getting close enough to make a shot, I squeezed the trigger and he dove into the water. Our PH calmly stated, "He's done." We waited for five minutes then a large shadow began to appear in the river about 10 yards further from our last known sighting. The hippo surfaced- unscathed. I immediately fired again and without question knew I had missed. He dove under the water-we backed up nervously, awaiting the impending charge. Five more minutes passed and behold the hippo appeared again, just out of range. He was swimming towards a pod of 8-10 hippos. The bull of that particular group decided to let his presence be known. The two bulls fought until the bull we had been hunting turned and began swimming full steam-towards us. An entertaining sight for sure, but in the back of my mind, I was questioning my shooting ability. I quickly checked my scope to ensure it was on low power- when the entire tube moved in the rings. The nightmare just got worse. Realizing I had seconds before nearly 4 tons of anger reached us- I looked back at my hunting partner and Nosler CFO, Mark Roberts, "Mark, give me your rifle" I exclaimed. I have never questioned Mark's integrity or personal loyalty, but for one moment- I would swear there was hesitation, and to be honest, I can't blame him. He handed me his 416 Remington, I placed it on the shooting sticks, and at about 30 yards, a 400gr Partition went right thru the massive hippo head. Once again, the 416 Remington Magnum to the rescue.

John R. Nosler is the VP/GM of Nosler, Inc.

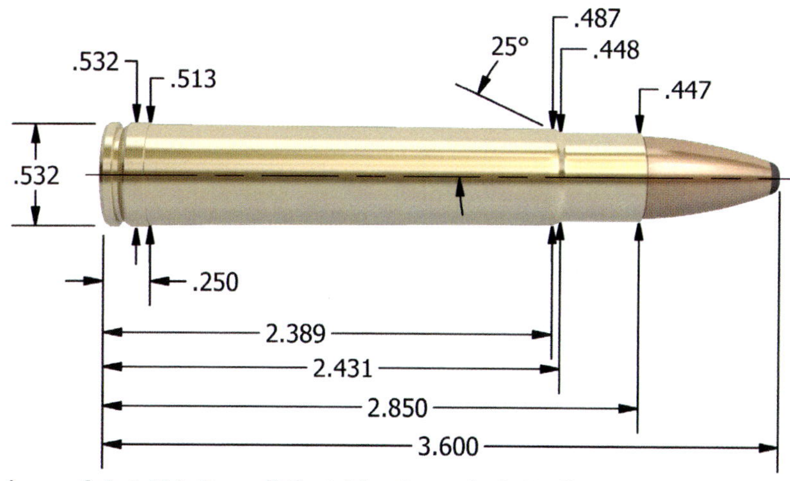

Maximum S.A.A.M.I. Overall Cartridge Length: 3.600″

BULLET CHOICES FOR THE 416 REMINGTON MAGNUM

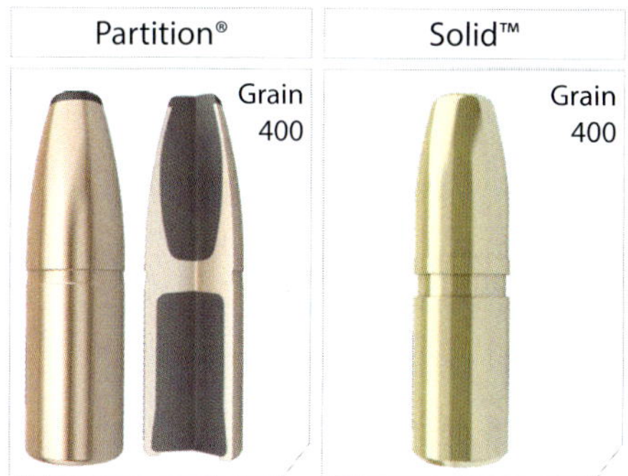

TECHNICAL INFORMATION

The 416 Remington offers the same performance level as the Rigby, but has the advantage of readily available rifles and ammunition based on a standard belted magnum case. Primarily used for African game, this cartridge certainly fits the category of "overkill" for most North American animals with the exception of Alaskan bear. With a nearly straight-wall case, very slow-burning powders are not very efficient in the 416 Rem. Our best results were seen with Varget and RL-15.

416 Remington Magnum - 400 grain

416 Remington Magnum - 400 grain		MAXIMUM S.A.A.M.I. O.A.C.L.		3.600"
		TESTED O.A.C.L.	B.C.	S.D.
Partition®	400gr. Spitzer	3.590"	0.390	0.330
Solid®	400gr. Solid	3.600"	0.289	0.330

CASE TYPE:	Norma		PRIMER TYPE	WLRM
CASE HOLDS:	80.0	Gr. WATER	BARREL Length/Make	24" H-S Prec.
			BARREL Twist	1-14"

POWDER TYPE	POWDER CHG. GRS.		MUZZLE VEL. F.P.S.		LOAD DENSITY (VOLUME)
A-2015	71.0	MAX.	2345	**	101%
	69.0	*	2305		98%
	67.0		2257		95%
IMR 8208 XBR	72.0	MAX.	2365		99%
	70.0	*	2306		96%
	68.0		2246		93%
W748	80.0	* MAX.	2392	**	104%
	78.0		2320	**	102%
	76.0		2266		99%
Big Game	83.0	* MAX.	2395	**	109%
	81.0		2356	**	107%
	79.0		2322	**	104%
IMR 4007 SSC	84.0	* MAX.	2415	**	113%
	82.0		2403	**	110%
	80.0		2321	**	107%
IMR 4064 Most Accurate Powder Tested	77.5	* MAX.	2438	**	106%
	75.5		2378	**	104%
	73.5		2332	**	101%
Varget	79.0	* MAX.	2468	**	107%
	77.0		2417	**	104%
	75.0		2365	**	101%
RL15	81.0	MAX.	2503	**	105%
	79.0	*	2453	**	103%
	77.0		2404		100%

BC=Ballistic Coefficient SD=Sectional Density
*Most Accurate Load Tested **Compressed Load

Use Maximum Loads with Caution
Refer to page 73 for additional safety information

Mark A. Keefe IV

416 RIGBY

When it comes to heavy bore, bolt-action rifle cartridges, it was the 416 Rigby, a proprietary cartridge introduced by John Rigby & Co., that first gave the English double rifle a run for its money. The 416 wasn't the first of the big British magazine rifle cartridges—the 404 Rimless Nitro Express or 404 Jeffrey appeared earlier, as did the 425 Westley Richards Magnum—but it is undoubtedly the best known and most used today. Intended as an alternative to rimmed Nitro Express cartridges, the .416 Rigby initially propelled (via Cordite) a 410-grain roundnose bullet at 2,370 feet per second, delivering more than 5,000 foot pounds of energy. This gave the Rigby a performance advantage of nearly 1,000 foot-pounds of energy over the .450/500 Nitro Express 3 ¼" with a 400-grain bullet at 2,150 feet per second.

Now more than a century old, the 416 Rigby's virtues remain the same as they did when announced by the Irish (and London) gunmaker in 1911.

The beltless, rimless 416 Rigby has voluminous capacity, fairly low pressures and a gentle (by today's standards) shoulder. Recoil is stout but manageable for an experienced shooter. The 416 Rigby was designed first and foremost to be reliable, to feed and extract without fault. It also shoots fairly flat, so can be used on other game if only one rifle is to be taken on safari. I used the 416 Rigby on my first Cape buffalo hunt in Zimbabwe's Selous, where I took a 60-yard broadside shot on a dugga boy. The 400-grain bullet passed completely through the top of the bull's heart and exited the other side. As expected, the Magnum Mauser performed flawlessly, digesting handloads and factory loads without difficulty.

The only drawback to this cartridge, in my view, is its length of 3.750" which requires a true, magnum length action. Some long actions (not magnum length) were done by Rigby with a bullet-nose cut on the rear receiver ring and the magazine well opened up to allow for the .416" cartridge. Likely the most famous such rifle was used by professional hunter Harry Selby. After his double rifle was run over by a "safari car" he bought a 416 Rigby magazine rifle in Nairobi until he could get around to getting another double. He never did get around to it, and it served Selby well for more than 50 hunting seasons. It is my undisputed favorite in a large-bore magazine rifle.

Mark A. Keefe, IV

Mark A. Keefe, IV, is Editor In Chief of American Rifleman magazine.

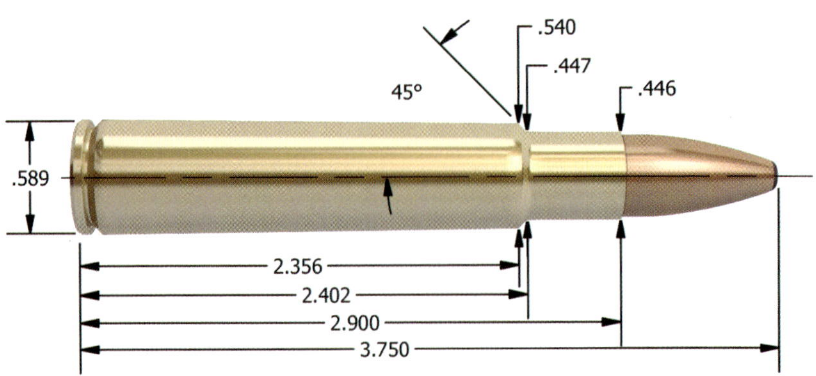

Maximum S.A.A.M.I. Overall Cartridge Length: 3.750"

BULLET CHOICES FOR THE 416 RIGBY

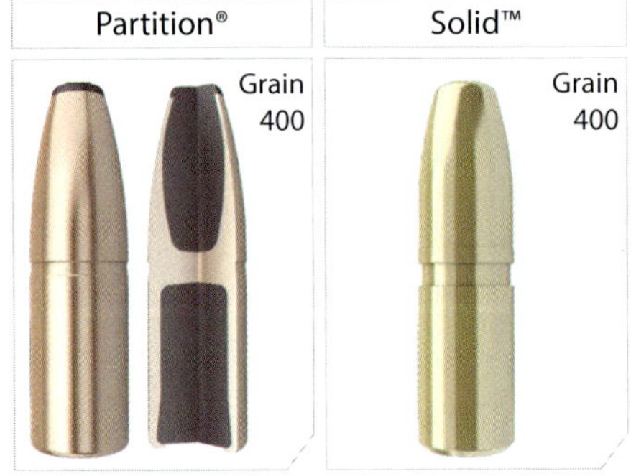

TECHNICAL INFORMATION

Until the late 20th century, the term "Rigby" inspired visions of African safaris and charging lions to the few who even recognized the term. With the availability of factory ammunition and readily available rifles in 416 Rigby, this cartridge has now become much more commonplace in North America. Loaded with the 400-grain Partition, the Rigby will stop anything that walks.

416 Rigby - 400 grain		MAXIMUM S.A.A.M.I. O.A.C.L.		3.750"
		TESTED O.A.C.L.	B.C.	S.D.
Partition®	400gr. Spitzer	3.630"	0.390	0.330
Solid®	400gr. Solid	3.655"	0.289	0.330

CASE TYPE:	Norma	PRIMER TYPE	WLRM
CASE HOLDS:	106.0 Gr. WATER	BARREL Length/Make	24" H-S Prec.
		BARREL Twist	1-14"

POWDER TYPE	POWDER CHG. GRS.		MUZZLE VEL. F.P.S.		LOAD DENSITY (VOLUME)
IMR 4350	92.0	MAX.	2427		92%
Most Accurate	90.0		2378		90%
Powder Tested	88.0	*	2324		88%
IMR 4007 SSC	90.0	MAX.	2430		91%
	88.0	*	2384		89%
	86.0		2360		87%
IMR 7828 SSC	102.0	* MAX.	2450		99%
	100.0		2398		97%
	98.0		2347		95%
H4831SC	99.0	* MAX.	2462		97%
	97.0		2427		95%
	95.0		2392		93%
RL19	101.0	MAX.	2479	**	104%
	99.0	*	2426	**	102%
	97.0		2400		99%
Hybrid 100 V	92.0	MAX.	2490		89%
	90.0	*	2428		88%
	88.0		2378		86%
RL22	104.0	MAX.	2515	**	107%
	102.0		2472	**	105%
	100.0	*	2421	**	103%
Supreme 780	103.0	MAX.	2538	**	102%
	101.0		2491		100%
	99.0	*	2461		98%

BC=Ballistic Coefficient SD=Sectional Density
*Most Accurate Load Tested **Compressed Load

Use Maximum Loads with Caution
Refer to page 73 for additional safety information

Buck Pope

416 RUGER

The 416 Ruger cartridge was introduced in 2008 by the joint venture of Ruger and Hornady. The 416 Ruger was made available to consumers in 2009, led off with the Ruger Hawkeye Compact rifle in the 416 Ruger cartridge and referred to as the Ruger African Rifle.

The 416 Ruger is basically the 375 Ruger cartridge necked up to .416." This is a beltless cartridge with a length of 3.340 inches and is designed to function in medium-length magnum actions. Cartridge design is such that it is slightly larger in diameter than your case diameter based off the 404 Jeffrey case. With the use of the latest powder technology, Hornady was able to design a cartridge that is noticeably shorter than the other 416 calibers and yet offers excellent ballistics. Even with this shorter cartridge case, the 416 Ruger provides ballistics that almost match the well established veteran 416 Rigby and does match the more recently introduced 416 Remington Magnum cartridge. The ballistics show the 400gr bullet is rated at 2400 fps and offers 5115 ft.-lbs. of energy at the muzzle. It was designed to handle the most dangerous big game animals to be found in the world.

I decided several years back, I had to have this cartridge for a possible cape buffalo hunt in the near future. I had Danny Pedersen of Classic Barrel & Gunworks in Prescott, Arizona build me such a rifle. The rifle is built off a Winchester Model 70 Stainless Classic action and fitted with one of Pedersen's custom integral quarter-rib, cut-rifling barrels. In addition, stock maker Robert Szweda of RMS Custom in Prescott Valley, Arizona did the stock work.

This rifle, which I call "The Cape Rifle," firing my own handloaded 400-grain Nosler Partition bullets, assembles sub-inch groups at 100 yards. This caliber in addition to being an ideal African dangerous game cartridge for animals such as buffalo, hippo and elephant, is also an excellent selection for the brown bears found in Alaska and Russia. The 416 Ruger is rapidly gaining popularity as an ideal .416 caliber cartridge.

Buck Pope

Buck Pope is a freelance gun and hunting writer.

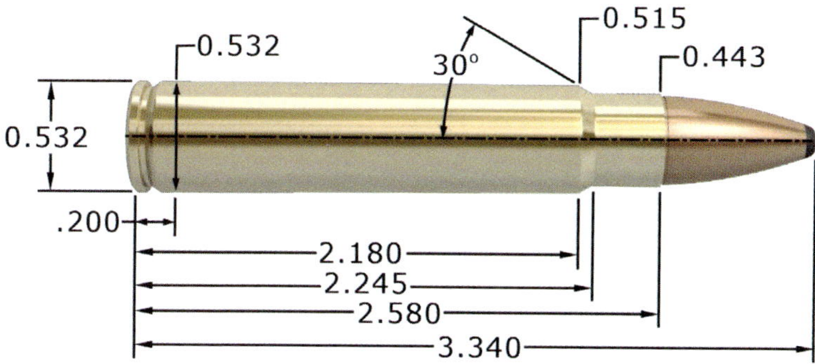

Maximum S.A.A.M.I. Overall Cartridge Length: 3.340"

BULLET CHOICES FOR THE 416 RUGER

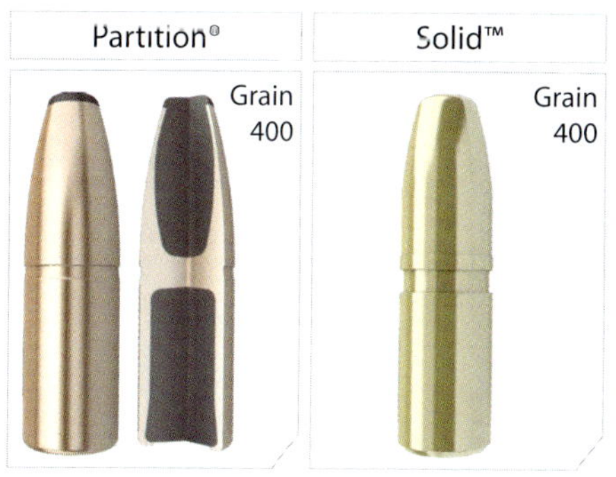

416 Ruger - 400 grain

		MAXIMUM S.A.A.M.I. O.A.C.L.		3.340"
		TESTED O.A.C.L.	B.C.	S.D.
Partition®	400gr. Spitzer	3.310"	0.390	0.330
Solid™	400gr. Solid	3.320"	0.289	0.330

CASE TYPE:	Hornady		PRIMER TYPE	WLRM
CASE HOLDS:	75.8	Gr. WATER	BARREL Length/Make	24" Pac-Nor
			BARREL Twist	1-14"

POWDER TYPE	POWDER CHG. GRS.		MUZZLE VEL. F.P.S.		LOAD DENSITY (VOLUME)
IMR 4320	69.0 *	MAX.	2227		98%
	67.0		2163		95%
	65.0		2101		92%
Varget Most Accurate Powder Tested	70.0	MAX.	2279		100%
	68.0		2225		97%
	66.0 *		2171		94%
H335	67.5 *	MAX.	2279		90%
	65.5		2164		88%
	63.5		2108		85%
Big Game	77.0	MAX.	2345	**	107%
	75.0		2304	**	104%
	73.0 *		2260	**	101%
RL15	74.5	MAX.	2381	**	102%
	72.5 *		2294		100%
	70.5		2212		97%
W760	82.0	MAX.	2392	**	113%
	80.0		2328	**	111%
	78.0 *		2258	**	108%
2000-MR	80.0	MAX.	2424	**	107%
	78.0		2357	**	104%
	76.0 *		2291	**	101%

BC=Ballistic Coefficient SD=Sectional Density
*Most Accurate Load Tested **Compressed Load

Use Maximum Loads with Caution
Refer to page 73 for additional safety information

458 WINCHESTER MAGNUM

The "four five eight" as the 458 Winchester Magnum is referred to by serious dangerous game hunters is one of the finest stopping calibers available. It was introduced by Winchester in 1956 in their standard length Model 70 bolt action rifle to compete in power with big bore doubles. Chambered by several manufacturers today, it is a wonderful life insurance policy for those who risk death or injury hunting the world's most dangerous critters. It is the most used fight stopping cartridge by African professional hunters. 500-grain solid bullets are required for elephants, buffaloes, rhinos, hippos and the like. 500-grain soft points will drop a lion in his tracks. I see .458 expanding bullets used in Alaska by guides as stopping rifles for big bears.

500-grain bullets from factory and hand loads top out around 2250 fps with about 5500 ft-lbs of muzzle energy. Arguments about bolt guns versus doubles will continue forever, but anyone who has staked his life on the 458 will agree that it has the horsepower to put nasty, dangerous animals down quickly. 458's, 500/465's, 470's and the like should be considered equally effective. Two fast shots are wonderful, but I love having four! The 458 is an almost straight-walled, belted case and easy to load. I suggest once fired cases for dangerous game hunting loads since they have already proven their integrity.

I killed my last elephant bull two years ago in very thick bush with visibility of only a few yards at most. We were able to get within about ten steps of the bull we'd followed for several hours. He was facing away at an angle and I slipped a 500-grain Nosler Solid just behind his ribcage toward the off-shoulder… a kill shot for sure. The bull ran straight ahead and I broke his right hip with shot number two, and he immediately went down (elephants can't go on three legs). The first bullet was recovered under the skin of the off shoulder after breaking ribs and destroying both lungs. The second bullet broke his hip, traveled almost the full length of his body, and was recovered in his left lung. That is penetration plus! Those bullets are not deformed in any way except for rifling engraving and could be loaded and shot again.

458 recoil is hefty, but controllable and quick for successive shots in a ten or more pound rifle. It is NOT a long range cartridge nor will you ever shoot it very far. Intelligent guides and hunters get close enough to the quarry to burn hair with the muzzle flash. Much less margin for shooting error exists when you are

up close and personal. Sighted in dead on at 100 yards is about the same at 50 yards. When I go back to Zimbabwe in August for another elephant bull, my 458 will be on my shoulder doing what it does best.

Roger Roberts

Roger G. Roberts founded Roberts Sales Company, Inc. and has worked with Nosler, Inc. for 33 years. He has taken 16 African safaris, hunted in four Canadian Provinces, Australia, many of the central and western United States, and Alaska. Virtually every trophy has fallen to his own Nosler bullet handloads.

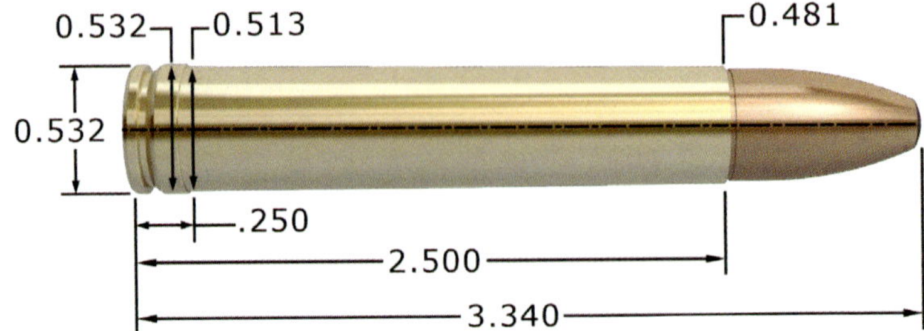

Maximum S.A.A.M.I. Overall Cartridge Length: 3.340"

BULLET CHOICES FOR THE 458 WINCHESTER MAGNUM

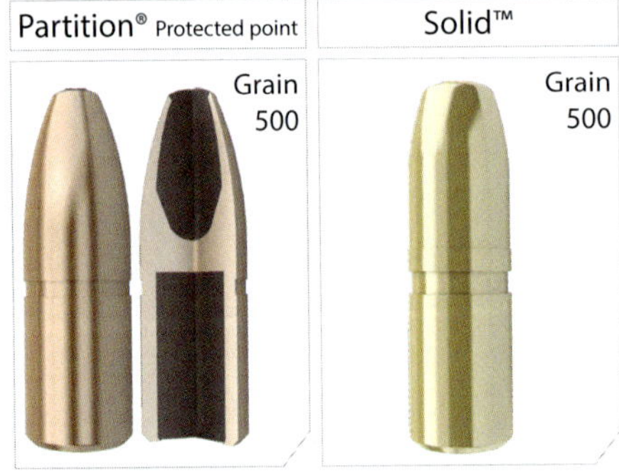

458 Winchester Magnum - 500 grain

		MAXIMUM S.A.A.M.I. O.A.C.L.		3.340"
		TESTED O.A.C.L.	B.C.	S.D.
Partition®	500gr. PPT	3.320"	0.389	0.341
Solid™	500gr. Solid	3.320"	0.246	0.341

CASE TYPE:	Winchester		PRIMER TYPE	WLRM
CASE HOLDS:	64.2	Gr. WATER	BARREL Length/Make	24" Pac-Nor
			BARREL Twist	1-14"

POWDER TYPE	POWDER CHG. GRS.		MUZZLE VEL. F.P.S.		LOAD DENSITY (VOLUME)
RL7	58.0	MAX.	1989		98%
	56.0		1915		95%
	54.0 *		1856		91%
W748	71.0	MAX.	2029	**	115%
	69.0		1978	**	112%
	67.0 *		1936	**	109%
H335 Most Accurate Powder Tested	68.0 *	MAX.	2092	**	108%
	66.0		2007	**	104%
	64.0		1950	**	101%

BC=Ballistic Coefficient SD=Sectional Density
*Most Accurate Load Tested **Compressed Load

Use Maximum Loads with Caution
Refer to page 73 for additional safety information

Bob Nosler

458 LOTT

When my son, John, and I planned our dream trip to Tanzania for elephant, buffalo, and lion, we wanted to take a rifle that offered "serious firepower", yet was versatile enough to use on larger plains game. Since this trip was also the final field test for our new line of Nosler Safari Ammunition, we decided on a pair of rifles in 458 Lott.

From the first day in camp, our PH had been telling us about a nice male lion that was on the verge of becoming a problem for the local Maasai people. Since we knew this lion was in the area, we began a daily routine of checking our hanging baits for signs of feeding.

John was first to try the 458 Lott by bagging a mature Cape buffalo and, as expected, the 500 grain Nosler Partition made quick work of the massive animal. Since it was clear that there were only a few legal elephants in the area, we decided to stake out our hanging baits in hopes of bagging the lion.

While Africa has many sights and sounds that can make the hunter wonder if he or she brought "enough gun", a sound that will stick in my memory is the deep throated roar of an adult male lion! Again, in the capable hands of my son John, the 458 Lott loaded with 500 grain Nosler Partitions proved to be more than enough gun for a one-shot kill on the big cat.

While I wasn't able to cross paths with a trophy elephant, I didn't leave Tanzania empty-handed. Just three hundred yards from where John killed the lion was a small group of "Ninja Bulls". These are buffalo that are not part of the main herds, and they seem to be very territorial.

When folks hear that I used a 458 Lott, the first question is usually something like "How bad does it kick?" At the range, or off the bench, the 458 Lott kicks like the proverbial mule—but, from personal experience, when you pull the trigger on a Cape buffalo, you hardly notice the recoil!

Bob and John Nosler celebrate their successful stalk on Bob's cape buffalo.

Bob Nosler

Bob Nosler is the CEO/President of Nosler, Inc.

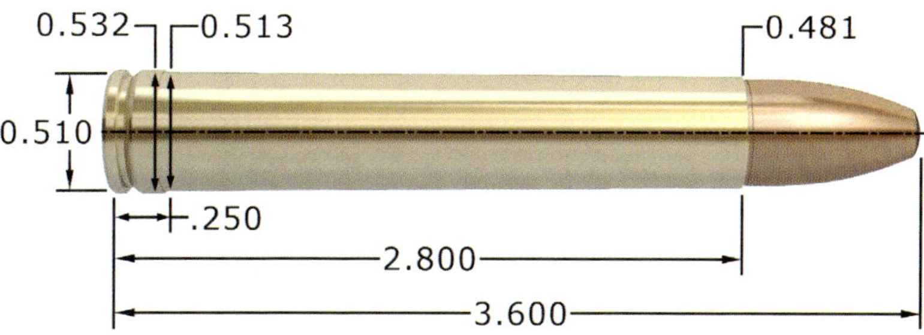

Maximum S.A.A.M.I. Overall Cartridge Length: 3.600"

BULLET CHOICES FOR THE 458 LOTT

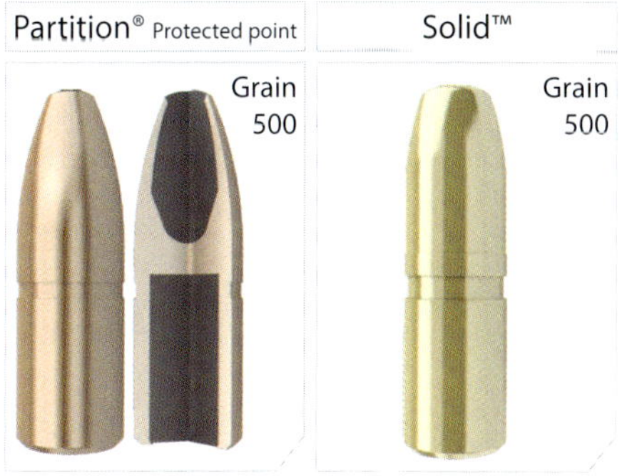

458 Lott - 500 grain

458 Lott - 500 grain		MAXIMUM S.A.A.M.I. O.A.C.L.		3.600"
		TESTED O.A.C.L.	B.C.	S.D.
Partition®	500gr. PPT	3.590"	0.389	0.341
Solid®	500gr. Solid	3.590"	0.246	0.341

CASE TYPE:	Norma		PRIMER TYPE	WLRM
CASE HOLDS:	75.8	Gr. WATER	BARREL Length/Make	24" Pac-Nor
			BARREL Twist	1-14"

POWDER TYPE	POWDER CHG. GRS.		MUZZLE VEL. F.P.S.		LOAD DENSITY (VOLUME)
Varget	77.0 *	MAX.	2142	**	110%
	75.0		2108	**	107%
	73.0		2075	**	104%
H4895	74.0	MAX.	2149	**	107%
	72.0 *		2098	**	104%
	70.0		2075	**	101%
IMR 4895 Most Accurate Powder Tested	76.0	MAX.	2153	**	110%
	74.0 *		2120	**	107%
	72.0		2105	**	104%
TAC	74.0 *	MAX.	2159		100%
	72.0		2097		97%
	70.0		2047		94%
RL15	79.0	MAX.	2207	**	109%
	77.0		2152	**	106%
	75.0 *		2105	**	103%

BC=Ballistic Coefficient SD=Sectional Density
*Most Accurate Load Tested **Compressed Load

Use Maximum Loads with Caution
Refer to page 73 for additional safety information

44 REMINGTON MAGNUM (RIFLE)

The 44 Remington Magnum cartridge was developed in the mid-1950s as a powerful handgun cartridge. Rifles chambered for the 44 Magnum appeared shortly thereafter, with single-shot, bolt-action, and lever-action carbines and rifles currently in production. Lever-action carbines, weighing between 5 and 8.5 pounds with 18- to 20-inch barrels and tubular magazines holding up to 10 cartridges, are a popular choice in today's marketplace.

Most commercial ammunition for the 44 Magnum is loaded with flat-nose or flexible-tip jacketed bullets of 180- to 275- grains. In carbines, muzzle velocities approximate 1,400 to 2,000 feet per second with muzzle energies of 1,000 to 1,900 foot pounds, making the 44 Magnum an excellent cartridge for deer and black bear at distances out to approximately 100 yards. Cartridges with bullets of 300 grains and heavier are also available. In some lever-action models, however, the rifling twist may be too slow to stabilize heavy bullets, and cartridge overall length and bullet configuration (e.g. some lead bullets) may preclude their use. For those who prefer a mild cartridge, lead-bullet loads employed by cowboy-action shooters are available. Also, 44 Special rounds can be used in the 44 Magnum.

The 44 Magnum case is easy to load and handloads can be tailored for the task at hand. A moderate powder charge behind a 200-grain bullet provides a load with light recoil that is suitable for target use out to 100 yards. A full-power load with 240-grain bullets will knock over metallic silhouette rams at plus 200 yards. If lead bullets are used in barrels with shallow rifling, velocities may need to be kept under 1,500 fps to achieve acceptable accuracy.

A talented female artist who lives in northwest Montana with bears and wolves for neighbors, sums up the virtues of the 44 Magnum. "A lever-action carbine in 44 Magnum suits my needs because it is compact, lightweight, holds enough rounds, and I can sling it and take it anywhere I walk, or tuck it in a scabbard if on horseback. The cartridge packs enough punch for game or personal defense. I recently had to grab my carbine, walk out on the porch, and take a lead on a coyote chasing a fawn across my yard. The shot tumbled the 'yote end over end. A carbine chambered for the 44 Magnum cartridge is versatile, handy, and easy to use."

Roy Welch

Roy is a freelance writer on rifle and shooting topics.

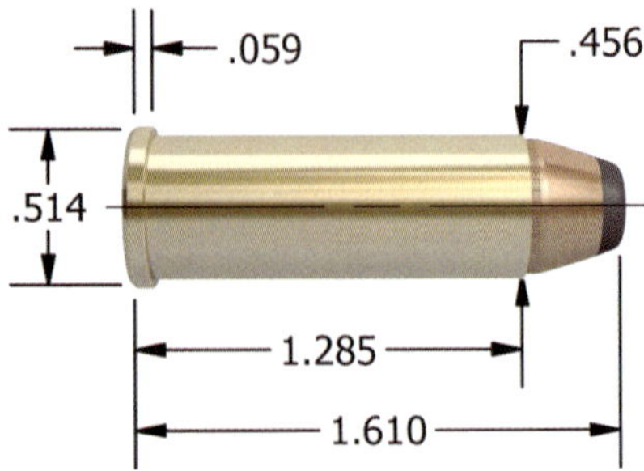

Maximum S.A.A.M.I. Overall Cartridge Length: 1.610"

BULLET CHOICES FOR THE 44 REMINGTON MAGNUM (RIFLE)

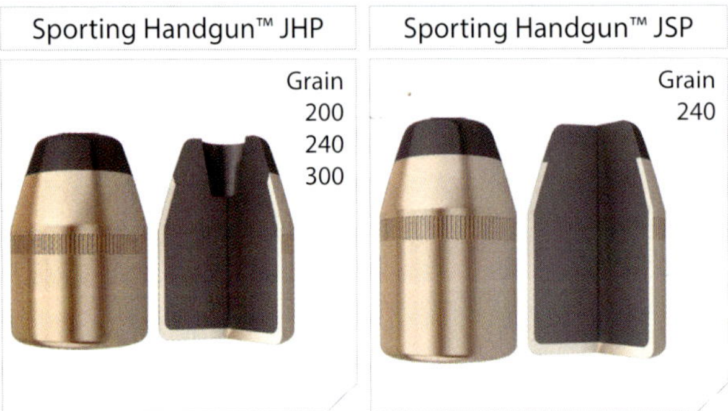

TECHNICAL INFORMATION

Because of its superior accuracy with all bullet weights, H110 has become a favorite of the technicians in our Ballistics Lab. For consistent performance, we recommend a good, heavy roll crimp.

44 Rem Mag (Rifle) - 200 grain		MAXIMUM S.A.A.M.I. O.A.C.L.		1.610"
		TESTED O.A.C.L.	B.C.	S.D.
Sporting Handgun™	200gr JHP	1.580"	0.151	0.155

CASE TYPE:	Winchester		PRIMER TYPE	WLP
CASE HOLDS:	24.0	Gr. WATER	BARREL Length/Make	18.5" Ruger
			BARREL Twist	1-38"

POWDER TYPE	POWDER CHG. GRS.		MUZZLE VEL. F.P.S.		LOAD DENSITY (VOLUME)
SR4756	13.7 *	MAX.	1579		96%
	13.2		1544		92%
	12.7		1508		89%
HS7	17.8	MAX.	1635		82%
	17.3		1582		80%
	16.8 *		1331		78%
A-No.7 Most Accurate Powder Tested	18.4	MAX.	1693		78%
	17.9		1628		76%
	17.4 *		1563		74%
2400	21.8 *	MAX.	1761	**	106%
	21.3		1735	**	103%
	20.8		1710	**	101%
H4227	24.9 *	MAX.	1797	**	119%
	24.4		1767	**	117%
	23.9		1733	**	114%
A #9	23.5	MAX.	1845	**	105%
	23.0		1830	**	102%
	22.5 *		1814		100%

BC=Ballistic Coefficient SD=Sectional Density
*Most Accurate Load Tested **Compressed Load

Use Maximum Loads with Caution
Refer to page 73 for additional safety information

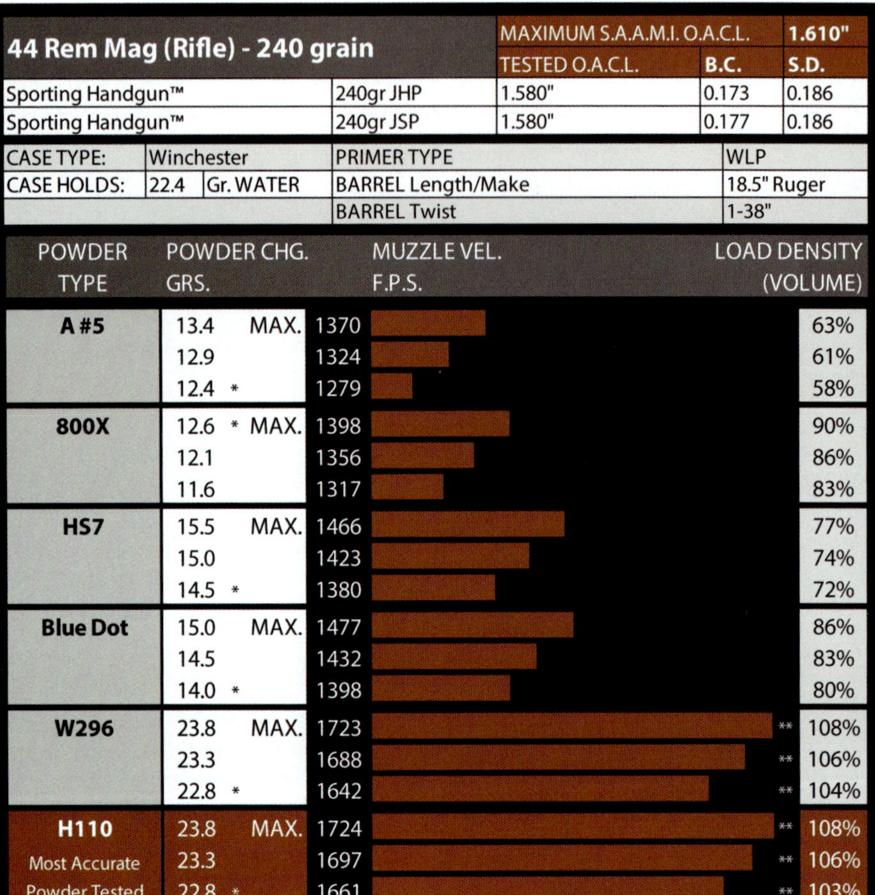

44 Rem Mag (Rifle) - 240 grain		MAXIMUM S.A.A.M.I. O.A.C.L.		1.610"
		TESTED O.A.C.L.	B.C.	S.D.
Sporting Handgun™	240gr JHP	1.580"	0.173	0.186
Sporting Handgun™	240gr JSP	1.580"	0.177	0.186

CASE TYPE:	Winchester		PRIMER TYPE	WLP
CASE HOLDS:	22.4	Gr. WATER	BARREL Length/Make	18.5" Ruger
			BARREL Twist	1-38"

POWDER TYPE	POWDER CHG. GRS.		MUZZLE VEL. F.P.S.		LOAD DENSITY (VOLUME)
A #5	13.4	MAX.	1370		63%
	12.9		1324		61%
	12.4 *		1279		58%
800X	12.6 *	MAX.	1398		90%
	12.1		1356		86%
	11.6		1317		83%
HS7	15.5	MAX.	1466		77%
	15.0		1423		74%
	14.5 *		1380		72%
Blue Dot	15.0	MAX.	1477		86%
	14.5		1432		83%
	14.0 *		1398		80%
W296	23.8	MAX.	1723	**	108%
	23.3		1688	**	106%
	22.8 *		1642	**	104%
H110 Most Accurate Powder Tested	23.8	MAX.	1724	**	108%
	23.3		1697	**	106%
	22.8 *		1661	**	103%

BC=Ballistic Coefficient SD=Sectional Density
*Most Accurate Load Tested **Compressed Load

Use Maximum Loads with Caution
Refer to page 73 for additional safety information

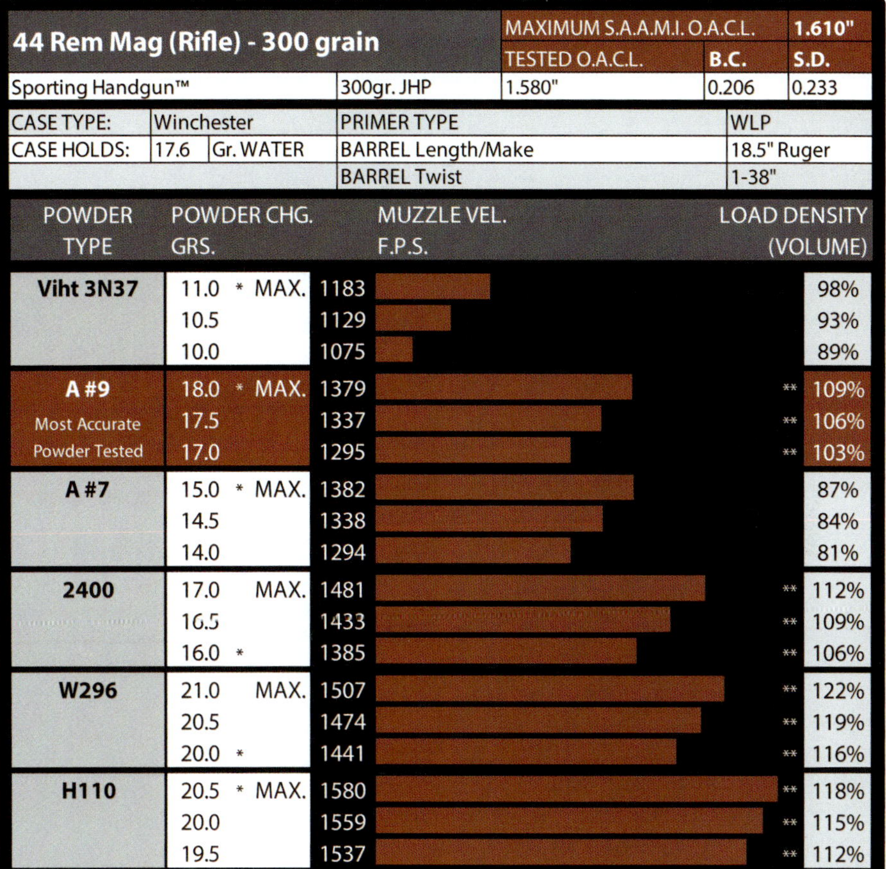

44 Rem Mag (Rifle) - 300 grain		MAXIMUM S.A.A.M.I. O.A.C.L.		1.610"
		TESTED O.A.C.L.	B.C.	S.D.
Sporting Handgun™	300gr. JHP	1.580"	0.206	0.233

CASE TYPE:	Winchester		PRIMER TYPE	WLP
CASE HOLDS:	17.6	Gr. WATER	BARREL Length/Make	18.5" Ruger
			BARREL Twist	1-38"

POWDER TYPE	POWDER CHG. GRS.		MUZZLE VEL. F.P.S.		LOAD DENSITY (VOLUME)
Viht 3N37	11.0 *	MAX.	1183		98%
	10.5		1129		93%
	10.0		1075		89%
A #9 Most Accurate Powder Tested	18.0 *	MAX.	1379	**	109%
	17.5		1337	**	106%
	17.0		1295	**	103%
A #7	15.0 *	MAX.	1382		87%
	14.5		1338		84%
	14.0		1294		81%
2400	17.0	MAX.	1481	**	112%
	16.5		1433	**	109%
	16.0 *		1385	**	106%
W296	21.0	MAX.	1507	**	122%
	20.5		1474	**	119%
	20.0 *		1441	**	116%
H110	20.5 *	MAX.	1580	**	118%
	20.0		1559	**	115%
	19.5		1537	**	112%

BC=Ballistic Coefficient SD=Sectional Density
*Most Accurate Load Tested **Compressed Load

Use Maximum Loads with Caution
Refer to page 73 for additional safety information

Richard Mann

45-70 GOVERNMENT (STRONG ACTION)

The 45-70 Government has been with us for 140 years. It will likely be the last cartridge pried from the cold, dead hands of American gun owners. Originally a military cartridge, loaded with black powder, it was not powerful by modern standards. That did not stop it from becoming popular. It launched a big bullet, punched a big hole, and the Trapdoor Springfield made history on the buffalo plains and the Indian wars.

Since then, the 45-70 has been chambered in countless rifles. Many of those actions, like the old "trapdoor" are what we today would consider weak. Because of this, factory 45-70 ammunition is held to a maximum pressure of not more than 28,000 psi. In most cases, actual chamber pressures are much less. Still, the 45-70 hangs on and is the darling of many big game hunters, especially those who are fond of Marlin's model 1895 lever gun.

The savviest of the shooters understand that rifles like the 1895 Marlin, Browning 1886, Ruger No. 1 and modern bolt actions, this old war horse can take on a different personality if handloaded. These actions can withstand much more pressure and even outperform the belted 450 Marlin cartridge which was designed to be the 45-70 of the future.

With Nosler's 300-grain Ballistic Tip® and Ballistic Silvertip®, you can take your slick handling Marlin 1895 or Ruger No. 1 and tackle any big game animal in North America. No, the 45-70 is not a long range cartridge but hunting has never been about long range. The real beauty of this cartridge, when fired from rifles with strong actions, is that it lets you decide how much power you need, based on the game you're hunting; light factory loads for deer and hogs or handloads on steroids for elk or griz. You cannot argue with the versatility of the 45-70.

When folks asked me what rifle and cartridge I was taking to Mozambique for buffalo, many were shocked when I replied that I was going "cowboy" with a Marlin lever gun and the 45-70. A lesson in ballistics followed and I could see the wheels in their heads begin to turn as those hunters were contemplating how a .45-70 might fit into their arsenal. Don't underestimate the 45-70; it's been around for almost 150 years for good reason.

Richard is a freelance gun and outdoor writer from West Virginia.

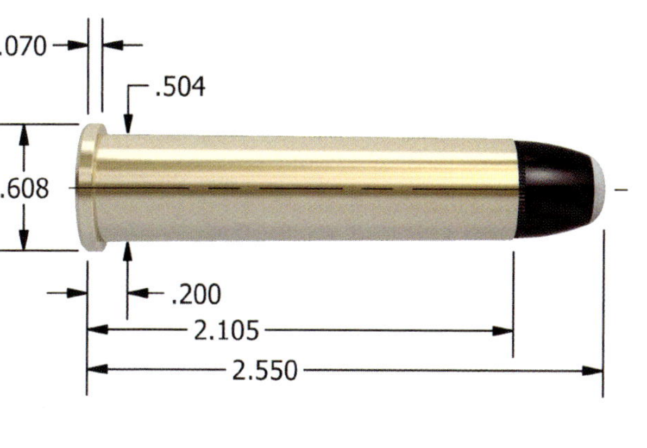

Maximum S.A.A.M.I. Overall Cartridge Length: 2.550"

BULLET CHOICE FOR THE 45-70 GOVERNMENT (STRONG ACTION)

CT® Ballistic Silvertip®

Round Nose

Grain
300

TECHNICAL INFORMATION

WARNING: This data is only for use in the new breed of "Strong Action" rifles such as those listed below:

- Newly manufactured Marlin Model 1895 rifles
- Browning Model 1886 rifles
- Replica (not original) Sharps Model 1874 single shots
- Original Winchester Model 1886 lever action and Model 1885 single-shot rifles known to be in good condition that have been thoroughly checked out by a competent gunsmith.

Additionally, this data may be safely used with any rifle classified as "stronger" than those listed above, such as:

- Ruger No. 1 and No. 3 single-shots
- Browning model 1885 single shot
- Mauser M98 bolt-actions properly converted to fire 45-70 ammunition.

Under no circumstances should this data be used in Original Trapdoor Springfields, modern trapdoor replicas, original Sharps Model 1874 rifles, any rolling block actions, or other original (old) 45-70 rifles. If you are not SURE as to the strength of your particular rifle, contact the manufacturer or Customer Service.

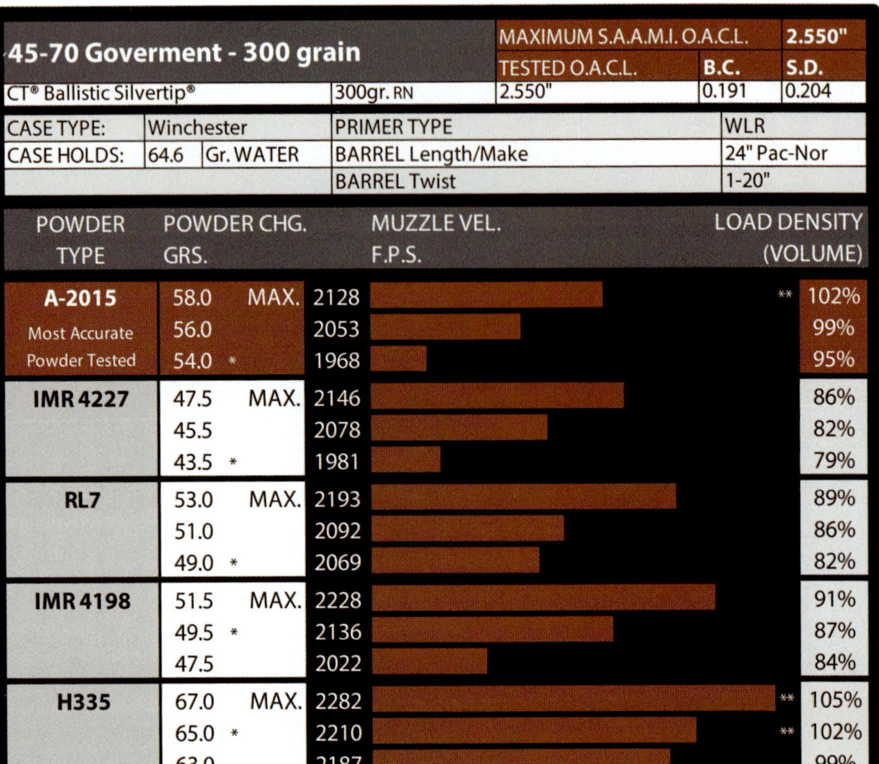

45-70 Goverment - 300 grain		MAXIMUM S.A.A.M.I. O.A.C.L.		2.550"
		TESTED O.A.C.L.	B.C.	S.D.
CT® Ballistic Silvertip®	300gr. RN	2.550"	0.191	0.204

CASE TYPE:	Winchester		PRIMER TYPE	WLR
CASE HOLDS:	64.6	Gr. WATER	BARREL Length/Make	24" Pac-Nor
			BARREL Twist	1-20"

POWDER TYPE	POWDER CHG. GRS.		MUZZLE VEL. F.P.S.		LOAD DENSITY (VOLUME)
A-2015 Most Accurate Powder Tested	58.0	MAX.	2128	**	102%
	56.0		2053		99%
	54.0	*	1968		95%
IMR 4227	47.5	MAX.	2146		86%
	45.5		2078		82%
	43.5	*	1981		79%
RL7	53.0	MAX.	2193		89%
	51.0		2092		86%
	49.0	*	2069		82%
IMR 4198	51.5	MAX.	2228		91%
	49.5	*	2136		87%
	47.5		2022		84%
H335	67.0	MAX.	2282	**	105%
	65.0	*	2210	**	102%
	63.0		2187		99%

BC=Ballistic Coefficient SD=Sectional Density
*Most Accurate Load Tested **Compressed Load

Use Maximum Loads with Caution
Refer to page 73 for additional safety information

HANDGUN CARTRIDGE RELOADING DATA

Dave Workman

380 AUTO (ACP)

Introduced back in 1908 as a step up from the 32 ACP for which Colt had chambered its classic flat Pocket Automatic, the 380 ACP has carved for itself a notch in history over the past century-plus, not only as a defensive round but also for the guns in which it has been chambered.

Simultaneously unveiled in Europe as the 9mm Browning Short, it's also been called the 9mm Kurz, 9mm Corto, 9x17 and a few other things as well by some of my pals who have reloaded it over the years, only because they've found that fumbling with small cases on a loading press requires a bit more dexterity than it takes to handle a 30-06 empty.

But the cartridge fascinates me because it is better than I believe some folks give it credit for. And let's face it, that Model 1908 Colt pistol just has to be one of the best little handguns ever built for the 380 Auto (and it looked great in those old gangster movies).

Topped with a .355-caliber bullet, the 380 ACP, thanks in large part to modern powder development and bullet design, delivers a bit more punch.

Today, one can brew up loads with hollow-point bullets that are far more effective than the original "ball" ammo. In modern handguns such as Ruger's LCP, the newest incarnation of Colt's classic Mustang Pocketlite, the Sig Sauer P238, Kel-Tec's P-3AT, the Kahr P380 and, naturally, the current edition of the famed Walther PPK, the 380 ACP really delivers.

These pistols would not be around if the cartridge did not enjoy considerable popularity among concealed carry advocates and for home protection. Nosler's 115-grain JHP is one of the heavier projectiles available for this cartridge. Loaded over 2.6 grains of Hodgdon's HP-38, one of my favorite pistol powders, that bullet leaves the muzzle at 720 fps, but with 3.3 grains of Unique, current data reveals it will scoot out the barrel at more than 800 fps.

A real sizzler load for this cartridge is 3.1 grains of SR 7625, clocking just over 830 fps, and that definitely wrings everything possible out of Nosler's JHP bullet.

Call me crazy, but the 380 ACP in a pinch can put a rabbit in the pot if someone is a relatively decent shot, and I would not be reluctant to plug a grouse with one of today's pistols if the opportunity arose.

Dave Workman with the Ruger LCP in 380 ACP.

Dave Workman

Dave Workman is an author, senior editor at TheGunMag.com, and the communications director for the Citizens Committee for the Right to Keep and Bear Arms.

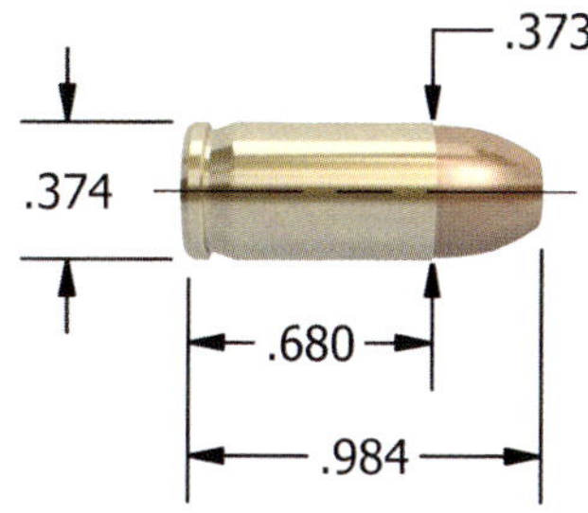

Maximum S.A.A.M.I. Overall Cartridge Length: .984" / Minimum .940"

BULLET CHOICES FOR THE 380 AUTO (ACP)

Sporting Handgun™ JHP

Grain
115

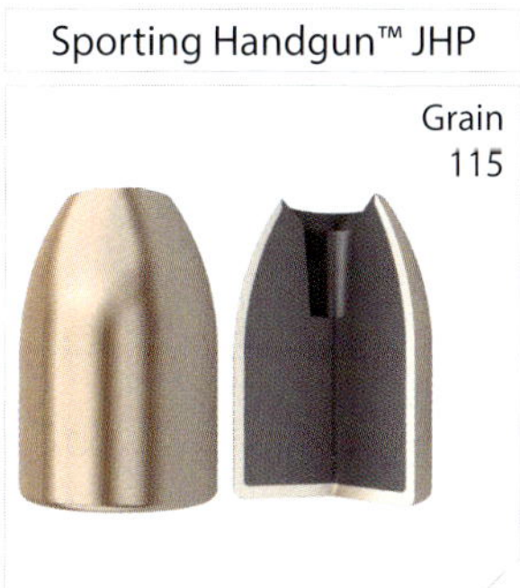

TECHNICAL INFORMATION

The .380 Automatic (ACP) has grown in popularity over the last several years due to the vast quantity and variety of .380 pocket pistols being sold. Like most pistol cartridges, it headspaces on the case mouth. When loading for this cartridge, bell the case mouth just enough to reliably guide the bullet into position and then secure the bullet with a taper crimp. Using this seating and crimping technique will help ensure proper headspacing.

The S.A.A.M.I. overall cartridge lengths for this cartridge are .940" min. and .984" max. Seating on the higher end of this range is best so as not to encroach on the .380's already limited case capacity, provided that the longer rounds will function well in your particular firearm.

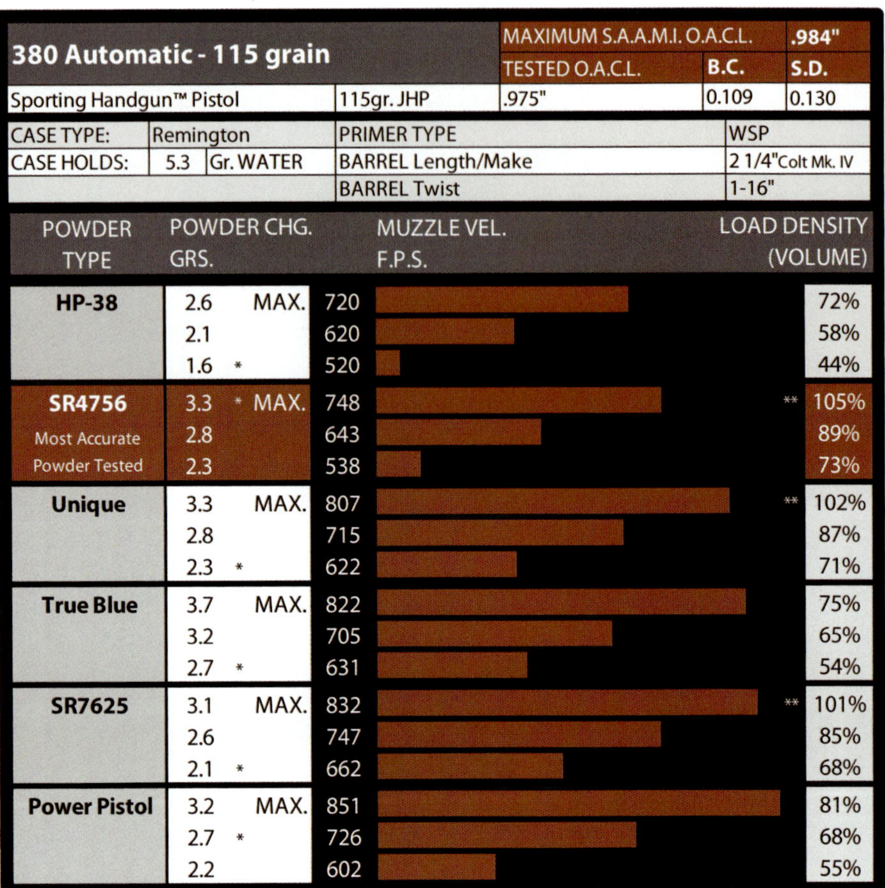

380 Automatic - 115 grain				MAXIMUM S.A.A.M.I. O.A.C.L.		.984"
				TESTED O.A.C.L.	B.C.	S.D.
Sporting Handgun™ Pistol		115gr. JHP		.975"	0.109	0.130
CASE TYPE:	Remington	PRIMER TYPE				WSP
CASE HOLDS:	5.3 Gr. WATER	BARREL Length/Make				2 1/4" Colt Mk. IV
		BARREL Twist				1-16"

POWDER TYPE	POWDER CHG. GRS.		MUZZLE VEL. F.P.S.		LOAD DENSITY (VOLUME)
HP-38	2.6	MAX.	720		72%
	2.1		620		58%
	1.6	*	520		44%
SR4756 Most Accurate Powder Tested	3.3	* MAX.	748	**	105%
	2.8		643		89%
	2.3		538		73%
Unique	3.3	MAX.	807	**	102%
	2.8		715		87%
	2.3	*	622		71%
True Blue	3.7	MAX.	822		75%
	3.2		705		65%
	2.7	*	631		54%
SR7625	3.1	MAX.	832	**	101%
	2.6		747		85%
	2.1	*	662		68%
Power Pistol	3.2	MAX.	851		81%
	2.7	*	726		68%
	2.2		602		55%

BC=Ballistic Coefficient SD=Sectional Density
*Most Accurate Load Tested **Compressed Load

Use Maximum Loads with Caution
Refer to page 73 for additional safety information

9MM LUGER (PARABELLUM)

The 9mm Parabellum is 111 years young. Happy birthday! Being asked to write a few words on the most popular handgun cartridge ever designed is a somewhat intimidating task, as there's been enough words expended on Georg Luger's creation to fill a Manhattan phone book, but it's easy to appreciate why this round has been so popular for so long. Economical and versatile with moderate recoil, it can be loaded with a wide range of bullet weights and powders and fill a variety of roles, ranging from the deadly serious task of self-defense, to casual plinking or introducing a new shooter to the world of centerfire handguns. There are several very good reasons why it can be found in the magazines of action pistol shooters, as it easily makes minor power factor and can be incredibly accurate. Many Open Division USPSA shooters have also discovered that, properly loaded and launched from a custom pistol that locks up like a bank vault, the round can be driven hard enough to make major PF too, though those loads are both beyond SAAMI specs and the scope of this book.

Designed from the outset as a high pressure cartridge, the 9mm's limited case capacity makes it a little trickier to reload than say, the 45ACP. A small decrease in OAL can produce a large increase in pressure, so stick to the recommended minimum lengths in your reloading manual. It's generally OK to load longer than book values, up to the point where the bullet's ogive approaches the lands; for certain loads in certain guns a longer round will produce better accuracy. Which ones? Therein lies the science and mystery of reloading – you'll have to do your own experiments to find out.

As long as there are still firearms being built, you can bet a large quantity of them will be chambered in 9mm Para. Here's to another century of life for this exceptional cartridge.

Iain Harrison is the Media Relations Manager for Crimson Trace, TV host and competitive shooter as well as a Nosler Pro-Staff member.

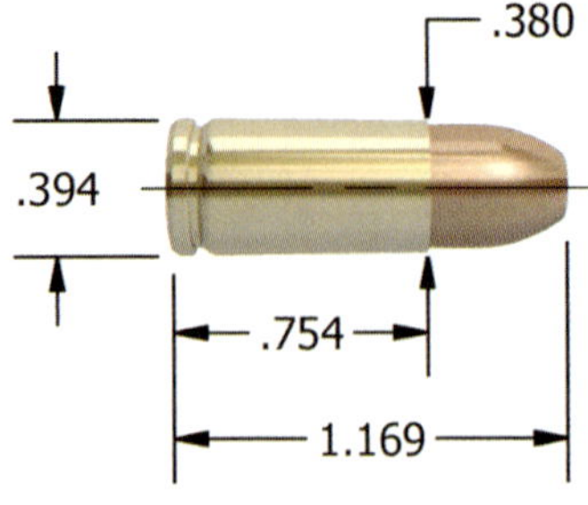

Maximum S.A.A.M.I. Overall Cartridge Length: 1.169"

BULLET CHOICES FOR THE 9MM LUGER (PARABELLUM)

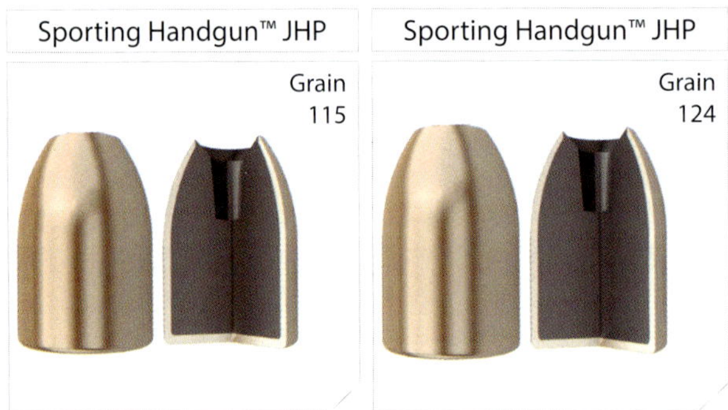

TECHNICAL INFORMATION

Like most pistol cartridges, the 9mm headspaces on the case mouth. When loading for this cartridge, bell the case mouth just enough to reliably guide the bullet into position and then secure the bullet with a taper crimp. Using this seating and crimping technique will help ensure proper headspacing.

The S.A.A.M.I. overall cartridge lengths for this cartridge are 1.000" min. and 1.169" max. We suggest seating on the higher end of this range, provided that the longer rounds will function well in your particular firearm.

This load data is not +P rated and does not exceed the parameters for standard 9mm Luger pressures.

9mm Luger - 115 grain

9mm Luger - 115 grain		MAXIMUM S.A.A.M.I. O.A.C.L.		1.169"
		TESTED O.A.C.L.	B.C.	S.D.
Sporting Handgun™ Pistol	115gr. JHP	1.100"	0.109	0.130

CASE TYPE:	Winchester		PRIMER TYPE	Rem 1 1/2
CASE HOLDS:	10.0	Gr. WATER	BARREL Length/Make	4" Douglas
			BARREL Twist	1-16"

POWDER TYPE	POWDER CHG. GRS.		MUZZLE VEL. F.P.S.		LOAD DENSITY (VOLUME)
HS-6	7.1	* MAX.	990		79%
	6.6		880		73%
	6.1		770		68%
A-No.7	8.5	* MAX.	998		86%
	8.0		933		81%
	7.5		868		76%
HP-38	5.2	MAX.	1042		76%
	4.7		937		69%
	4.2	*	832		61%
A-No.5	6.7	MAX.	1091		71%
	6.2	*	1009		65%
	5.7		927		60%
Unique	6.1	MAX.	1120		100%
	5.6		1060		92%
	5.1	*	1000		84%
Blue Dot	8.5	MAX.	1130	**	109%
	8.0		1080	**	103%
	7.5	*	1030		96%
Auto-Comp	5.5	* MAX.	1139		64%
	5.0		1055		58%
	4.5		979		52%
W231	5.0	* MAX.	1140		73%
	4.5		1033		66%
	4.0		932		58%
SR4756	6.5	* MAX.	1178	**	109%
Most Accurate	6.0		1083	**	101%
Powder Tested	5.5		988		92%
Power Pistol	6.6	MAX.	1198		88%
	6.1	*	1107		81%
	5.6		1031		75%

BC=Ballistic Coefficient SD=Sectional Density
*Most Accurate Load Tested **Compressed Load

Use Maximum Loads with Caution
Refer to page 73 for additional safety information

9mm Luger - 124 grain		MAXIMUM S.A.A.M.I. O.A.C.L.		1.169"
		TESTED O.A.C.L.	B.C.	S.D.
Sporting Handgun™ Pistol	124gr. JHP	1.120"	0.118	0.141

CASE TYPE:	Winchester		PRIMER TYPE	Rem 1 1/2
CASE HOLDS:	10.0	Gr. WATER	BARREL Length/Make	4" Douglas
			BARREL Twist	1-16"

POWDER TYPE	POWDER CHG. GRS.		MUZZLE VEL. F.P.S.	LOAD DENSITY (VOLUME)
A-No.5	5.3 *	MAX.	1022	56%
	4.8		937	51%
	4.3		852	45%
Bullseye	4.4	MAX.	1046	71%
	3.9 *		953	63%
	3.4		859	55%
W231	4.6 *	MAX.	1048	67%
	4.1		957	60%
	3.6		861	53%
True Blue Most Accurate Powder Tested	5.8	MAX.	1099	62%
	5.3 *		1004	57%
	4.8		909	51%
Universal	4.8	MAX.	1104	76%
	4.3		1032	68%
	3.8 *		959	60%
Power Pistol	6.1	MAX.	1111	81%
	5.6 *		1034	75%
	5.1		957	68%
Titegroup	4.3	MAX.	1114	57%
	3.8 *		1002	51%
	3.3		890	44%
A-No.7	7.9	MAX.	1140	80%
	7.4		1076	75%
	6.9 *		1011	70%
Auto-Comp	5.1 *	MAX.	1144	59%
	4.6		1061	53%
	4.1		978	48%
Unique	5.6	MAX.	1147	92%
	5.1 *		1046	84%
	4.6		981	75%

BC=Ballistic Coefficient SD=Sectional Density
*Most Accurate Load Tested **Compressed Load

Use Maximum Loads with Caution
Refer to page 73 for additional safety information

Dave Workman

357 SIG

When Federal Cartridge and Sig collaborated back in 1994 to produce a semi-auto cartridge capable of delivering the punch of a 357 Magnum with lighter bullets, the end product was a bottleneck cartridge dubbed the 357 SIG.

Duplicating the performance of the 357 Magnum was not the only hurdle. The developers also had to make the cartridge fit in a semi-auto platform that could accommodate a cartridge in the range of a 9mm. On top of that, they wanted this new round to deliver about 500 foot-pounds of muzzle energy. That's exactly what they ended up with, and the round has already proven itself a real fight-stopper.

Contrary to popular belief, this is not simply a necked-down 40 S&W case with a .355-caliber bullet. The cartridge is based on that case, of course, but the 357 SIG case is slightly longer.

First introduced in the SIG P229, the cartridge now finds a home in several different pistols, including the popular P250, a gun I had the chance to put through its paces in this caliber, and it is a winner. Other manufacturers are chambering pistol models for the 357 SIG, and you're likely to find this trend continuing.

Using Nosler's 115-grain JHP, a handloader can turn out some terrific ammunition, producing screaming velocities upwards of 1,400 fps with 9.1 grains of Vihtavouri N350, and many loads jumping above 1,300 fps using Unique, Blue Dot or HS6 powders.

Because of its necked-down design, the 357 SIG ramps reliably and consistently. During my evaluations, I did not experience a single malfunction.

The 357 SIG actually uses a .355-caliber bullet, the same projectile as the 380 ACP and 9mm. This round truly gets the best performance out of that projectile.

In the field, I've found the 357 SIG to be a flat shooter, and recoil is remarkably easy to handle in the pistols chambered for the round. Indeed, this is a rather pleasant cartridge to shoot, and reloading the brass — because it is necked down — is pretty easy.

Popular powders for this cartridge include Unique, Universal, Blue Dot and HS6.

Workman put the 357 SIG through its paces in a Sig Sauer P250.

Dave Workman

Dave Workman is an author, senior editor at TheGunMag.com, and the communications director for the Citizens Committee for the Right to Keep and Bear Arms.

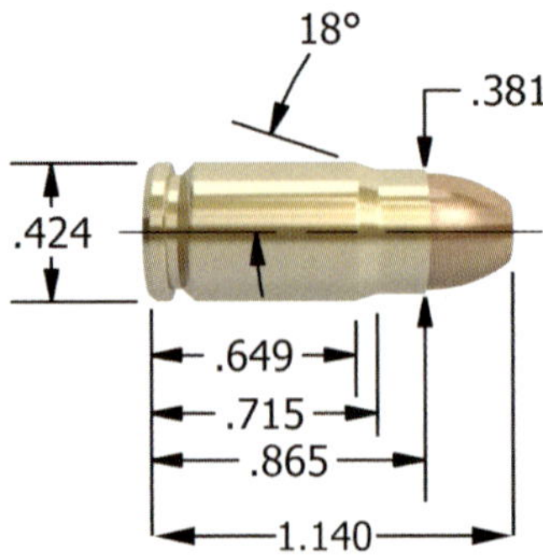

Maximum S.A.A.M.I. Overall Cartridge Length: 1.140"

BULLET CHOICES FOR THE 357 SIG

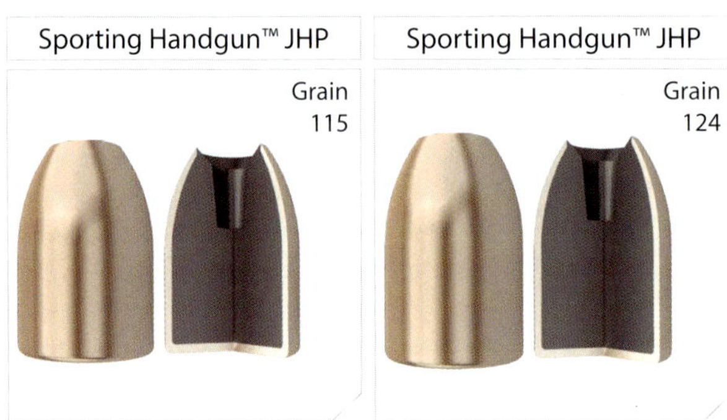

TECHNICAL INFORMATION

Contrary to the cartridge designation, the 357 Sig uses 9mm (.355") bullets. Both our 115 and 124 grain 9mm Sporting Handgun Pistol bullets are suitable for loading in the 357 Sig.

357 Sig - 115 grain

		MAXIMUM S.A.A.M.I. O.A.C.L.		1.140"
		TESTED O.A.C.L.	B.C.	S.D.
Sporting Handgun™ Pistol	115gr. JHP	1.140"	0.109	0.130

CASE TYPE:	Federal		PRIMER TYPE	WSP
CASE HOLDS:	13.8	Gr. WATER	BARREL Length/Make	3.9" Sig
			BARREL Twist	1-16"

POWDER TYPE	POWDER CHG. GRS.			MUZZLE VEL. F.P.S.		LOAD DENSITY (VOLUME)
A-No.5	9.4		MAX.	1308		72%
	8.9	*		1236		68%
	8.4			1164		64%
Unique	7.7		MAX.	1326		91%
	7.2	*		1272		86%
	6.7			1218		80%
Power Pistol	9.0		MAX.	1328		87%
	8.5			1257		82%
	8.0	*		1185		77%
HS-6	9.9		MAX.	1355		80%
	9.4			1274		76%
	8.9	*		1198		72%
Universal	7.2		MAX.	1362		83%
	6.7	*		1298		77%
	6.2			1234		71%
Blue Dot	10.8	*	MAX.	1368		100%
	10.3			1311		96%
	9.8			1262		91%
Viht N350	9.1		MAX.	1425	**	116%
	8.6	*		1366	**	109%
	8.1			1312	**	103%
Viht 3N37	8.9	*	MAX.	1438	**	101%
	8.4			1349		95%
	7.9			1276		89%
A-No.7 Most Accurate Powder Tested	12.1		MAX.	1442		89%
	11.6	*		1391		85%
	11.1			1322		82%
A-No.9	14.3		MAX.	1454	**	111%
	13.8			1399	**	107%
	13.3	*		1347	**	103%

BC=Ballistic Coefficient SD=Sectional Density
*Most Accurate Load Tested **Compressed Load

Use Maximum Loads with Caution
Refer to page 73 for additional safety information

357 Sig - 124 grain		MAXIMUM S.A.A.M.I. O.A.C.L.		1.140"
		TESTED O.A.C.L.	B.C.	S.D.
Sporting Handgun™ Pistol	124gr. JHP	1.120"	0.118	0.141

CASE TYPE:	Federal		PRIMER TYPE	WSP
CASE HOLDS:	13.8	Gr. WATER	BARREL Length/Make	3.9" Sig
			BARREL Twist	1-16"

POWDER TYPE	POWDER CHG. GRS.		MUZZLE VEL. F.P.S.		LOAD DENSITY (VOLUME)
True Blue	8.5	MAX.	1250		66%
	8.0	*	1178		62%
	7.5		1106		58%
HS-6	9.1	MAX.	1276		73%
	8.6		1189		69%
	8.1	*	1102		65%
SR4756	8.7	MAX.	1284	**	106%
	8.2	*	1237		100%
	7.7		1181		94%
Unique	7.6	* MAX.	1288		90%
	7.1		1216		84%
	6.6		1145		78%
Power Pistol	8.5	MAX.	1314		82%
Most Accurate	8.0	*	1230		77%
Powder Tested	7.5		1146		72%
Viht N350	8.2	MAX.	1318	**	104%
	7.7		1256		98%
	7.2	*	1198		92%
A-No.5	9.2	MAX.	1328		70%
	8.7	*	1240		66%
	8.2		1155		63%
A-No.7	11.1	* MAX.	1336		82%
	10.6		1257		78%
	10.1		1165		74%
Blue Dot	10.1	MAX.	1345		94%
	9.6	*	1268		89%
	9.1		1191		85%
A-No.9	13.0	MAX.	1376	**	101%
	12.5		1312		97%
	12.0	*	1248		93%

BC=Ballistic Coefficient SD=Sectional Density
*Most Accurate Load Tested **Compressed Load

Use Maximum Loads with Caution
Refer to page 73 for additional safety information

Roy Huntington

38 SPECIAL

Not only is the grand 38 Special one of the classic cartridges of all time, but at about 115 years old, it has aged gracefully, adapting and re-inventing itself as the decades have passed. As one of the rare few cartridges making the transition from black powder to today's cutting-edge, high performance loadings, the 38 Special is, simply put — Special.

Based on the 38 Long Colt, the S&W 38 Special was first chambered in the S&W Model of 1899, with the bullet weight jumping from 150 to 158 grains. The Model of 1899 was also the beginning of the equally classic S&W K-frame revolver. That marriage of a stout cartridge, well-suited for self defense or law enforcement, with the nearly perfect size of the K-frame, created an instant success, and remains so today.

While many of today's handgunners are more comfortable with a semi-auto, legions of today's handgunners — including me — cut their teeth on that classic 158-grain round-nosed lead 38 Special load. Eventually, due to demand from law enforcement and the consumer market, more effective loads were developed, ultimately resulting in today's high-performance defensive factory loads. But even many of those early loads remain effective. During the early years of my police career, a 6" S&W Model 10, loaded with 158-grain LSWC bullets, rode in my duty holster. Even now, I wouldn't feel under-gunned with that set-up!

When I was 16, and a certifiable gun-crank, I invested about 20 hard-earned dollars in a Lee loader, a can of Unique, a 100-pack of primers and a hundred 158 grain LSWC bullets. After getting things sorted out, I loaded my first centerfire cartridge — a 38 Special. As time passed, I joined the police department, becoming a member of the pistol team competing in that timeless tradition of the PPC match. From 1975 to 1979, I fired upwards of 25,000 rounds of 38 Special 148-grain wadcutter loads yearly, honing my skills.

That background of precision target shooting has paid dividends to this day, making me a better all-around marksman with rifles or handguns. And it's all because of the 38 Special and its ability to effortlessly manage self-defense, accuracy, ease of re-loading and economy. I hope 100 years from now, reloading manuals will still be touting the "classic" 38 Special!

Roy Huntington is the editor of American Handgunner Magazine

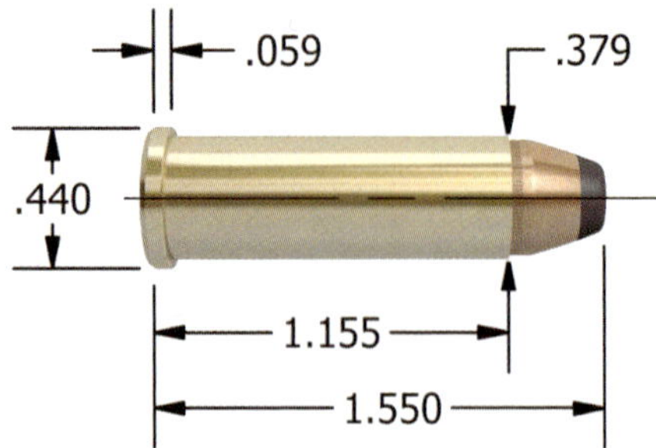

Maximum S.A.A.M.I. Overall Cartridge Length: 1.550"

BULLET CHOICE FOR THE 38 SPECIAL

Sporting Handgun™ JHP

TECHNICAL INFORMATION

Loading the .38 Special with slower burning powders like A-No.5 or HS-6 will produce the best results in barrels four inches or longer. In short barreled revolvers, faster burning powders like A-No.2 or WST are more efficient.

When loading for the 38 Special, as with any revolver cartridge, seat the bullet so that the top of the cannelure is just visible above the case mouth and then give it a good roll crimp. The crimp should be such that you can just see the case mouth beginning to deform the ridges in the cannelure.

38 Special - 158 grain			MAXIMUM S.A.A.M.I. O.A.C.L.		1.550"
			TESTED O.A.C.L.	B.C.	S.D.
Sporting Handgun™ Revolver		158gr. JHP	1.440"	0.182	0.177
CASE TYPE:	Federal		PRIMER TYPE		WSP
CASE HOLDS:	15.6	Gr. WATER	BARREL Length/Make		7.75"H-S Prec.
			BARREL Twist		1-16"

POWDER TYPE	POWDER CHG. GRS.		MUZZLE VEL. F.P.S.	LOAD DENSITY (VOLUME)
W231	4.0	MAX.	672	37%
	3.6		567	33%
	3.0 *		462	28%
WST	3.6 *	MAX.	692	44%
	3.1		577	38%
	2.6		462	32%
Power Pistol	5.1 *	MAX.	701	44%
	4.6		618	39%
	4.1		549	35%
Unique	4.8 *	MAX.	746	50%
	4.3		663	45%
	3.8		581	40%
A-No.2	4.0	MAX.	760	39%
	3.5		680	35%
	3.0 *		600	30%
Universal	4.4	MAX.	782	45%
	3.9 *		657	40%
	3.4		532	35%
HS-6 Most Accurate Powder Tested	6.5	MAX.	810	46%
	6.0		723	43%
	5.5 *		644	39%
Bullseye	4.1 *	MAX.	830	42%
	3.6		730	37%
	3.1		630	32%
A-No.5	6.2 *	MAX.	840	42%
	5.7		750	38%
	5.2		660	35%
SR4756	5.2 *	MAX.	865	56%
	4.7		751	51%
	4.2		637	45%

BC=Ballistic Coefficient SD=Sectional Density
*Most Accurate Load Tested **Compressed Load

Use Maximum Loads with Caution
Refer to page 73 for additional safety information

Brian Pearce

357 MAGNUM

After the introduction of the N-Frame Smith & Wesson 38/44 Outdoorsman and Heavy Duty revolvers around 1930, experimenters such as Phil Sharpe, Doug Wesson, Elmer Keith and others began stoking the .38 Special to significantly higher pressures (and velocities) than factory loads. Ultimately, the decision was made to lengthen the case and call it "Magnum," the first of its kind. This joint effort between Smith & Wesson and Winchester resulted in the 357 Magnum cartridge appearing in 1935. The S&W revolvers were marvelous, high-grade guns with each being custom-built for the owner, along with a certificate of registration specifying features, accuracy and for whom it was built. It was by far the most expensive gun in the company line. It was The Great Depression and sales were expected to be limited, however, it sold remarkably well. This gun and cartridge significantly changed the future of handgunning.

In the post-World War II era, the popularity of the 357 Magnum grew to amazing status. Many guns have been thus chambered and have included small-frame revolvers, many sizes of double and single action six-guns, single shots and a variety of rifles. It has served well for hunting, defense, precision target work and law enforcement applications. Frankly, few revolver cartridges can boast of such extreme versatility, especially when handloaded.

Being primarily an outdoorsman and shooter, the 357 Magnum has served me well. It has taken black bear, deer, antelope and considerable lesser game. I know men that it has saved from both man and beast.

While hunting varmints with a friend who works for a major Eastern gun company and feels that handguns are not effective beyond 20 yards, we stumbled onto a big jackrabbit that jumped and bounded his way to a distant hillside then stopped. I reached for the Smith & Wesson N-frame 357 Magnum with a 5-inch barrel resting on my hip. He rolled his eyes in total disbelief and mumbled something about a steak dinner. In a sitting position, with my back up against the pickup tire and elbows resting between my knees, the sights were carefully aligned. The rabbit jumped then landed belly up, which was followed by the sound of the hollow-point bullet hitting home. My associate took more than 100 long paces to retrieve him. More importantly, he picked up the dinner tab. When kept within its proper limits, it is truly an outstanding cartridge.

Brian Pearce

Brian Pearce is a contributing editor for Rifle and Handloader magazines.

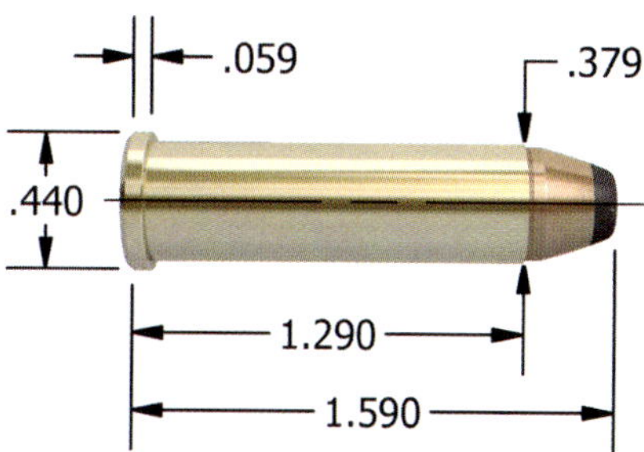

Maximum S.A.A.M.I. Overall Cartridge Length: 1.590"

BULLET CHOICE FOR THE 357 MAGNUM

Sporting Handgun™ JHP

TECHNICAL INFORMATION

The 357 Magnum was, at the time of its introduction, the most powerful revolver cartridge in existence. Today it falls well short of that mark, but is still a relatively powerful round.

38 Special ammunition can be safely and accurately fired through a 357 Magnum revolver, and is excellent for practice and plinking. When loading full-power .357 rounds, we have always had the best luck with slower powders like 2400, H110, and W296. When using these slow powders, a small pistol magnum primer and a heavy roll crimp are necessary to help achieve consistent ignition. We have had great results with the following procedure:

- Seat the bullet to where you can just see the top edge of the cannelure.
- Adjust your crimp so that the case mouth is deforming the ridges in the cannelure and biting all the way to the bottom of the groove.

357 Magnum - 158 grain		MAXIMUM S.A.A.M.I. O.A.C.L.		1.590"
		TESTED O.A.C.L.	B.C.	S.D.
Sporting Handgun™ Revolver	158gr. JHP	1.585"	0.182	0.177

CASE TYPE:	Federal		PRIMER TYPE	WSPM
CASE HOLDS:	17.6	Gr. WATER	BARREL Length/Make	8.3" H-S Prec.
			BARREL Twist	1-16"

POWDER TYPE	POWDER CHG. GRS.		MUZZLE VEL. F.P.S.	LOAD DENSITY (VOLUME)
IMR 4227	15.0	* MAX.	1095	100%
	14.5		1064	96%
	14.0		1033	93%
Blue Dot	10.4	MAX.	1211	76%
	9.9	*	1153	72%
	9.4		1095	68%
Enforcer	13.7	MAX.	1298	83%
	13.2	*	1260	80%
	12.7		1222	77%
SR4756	7.9	MAX.	1340	75%
	7.4		1310	71%
	6.9	*	1280	66%
A-No.7 Most Accurate Powder Tested	11.2	* MAX.	1420	65%
	10.7		1340	62%
	10.2		1260	59%
A-No.5	9.6	MAX.	1428	57%
	9.1		1343	54%
	8.6	*	1258	51%
800-X	8.7	* MAX.	1448	79%
	8.2		1373	75%
	7.7		1298	70%
H110	15.9	* MAX.	1490	92%
	15.4		1470	89%
	14.9		1460	86%
2400	12.3	* MAX.	1520	81%
	11.8		1500	78%
	11.3		1480	75%
W296	14.8	* MAX.	1540	86%
	14.3		1530	83%
	13.8		1520	80%

BC=Ballistic Coefficient SD=Sectional Density
*Most Accurate Load Tested **Compressed Load

Use Maximum Loads with Caution
Refer to page 73 for additional safety information

357 REMINGTON MAXIMUM

The year 1983 marked the introduction of a handgun and a cartridge that caught the imagination of a generation of revolver shooters. Here was a load that could push a 158-grain handgun bullet at close to 2,000 feet per second. For those that needed more energy on target, the 357 Maximum achieved 1,500 feet per second with a 180-grain bullet. For the hunter that fancied revolver shooting, this was a true deer gun, especially when using a longer barrel.

When single-shot handguns were chambered for the cartridge, accuracy potential was extended beyond 100 yards.

When I read that Ruger ceased production of the 357 Maximum, my interest was piqued. When the opportunity came along to buy one, it was gently used and still reclined in the box that the original owner had taken it home in. I snapped it up.

No holster I owned would fit the New Model Super Blackhawk, so that was the next purchase as well as a box of 38s and another of 357 Magnum to go with the supply of 357 Maximum ammunition from the original owner.

After a few trips to the range, the 357 Maximum became a regular companion on hunts for feral sheep and pigs, which was a suitable use for a cartridge that has gained acclaim for its prowess on the silhouette range.

Often, I wore the gun/holster rig over the shoulder with the gun in the cross-draw position, which made getting that 10-inch barrel out of the holster an easy proposition.

The versatility of a handgun chambered for this cartridge allows even the most timid shooter to put rounds downrange. And no one is under-gunned when the cylinder is stoked with 357 Maximum. It is a great choice for the silhouette shooter, the handgun deer hunter or the sportsman following the hounds on a bear or cougar hunt.

Gary A. Lewis

Gary, with John Nosler, is the author of John Nosler – Going Ballistic. Lewis makes his home in Bend, Oregon

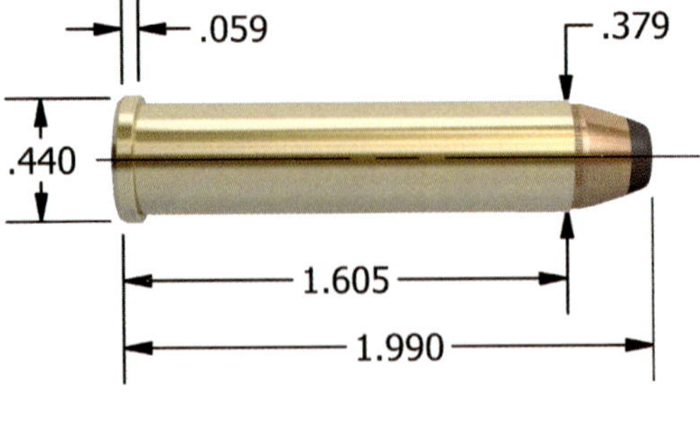

Maximum S.A.A.M.I. Overall Cartridge Length: 1.990"

BULLET CHOICES FOR THE 357 REMINGTON MAXIMUM

Sporting Handgun™ JHP

TECHNICAL INFORMATION

The 357 Maximum was developed primarily as a silhouette round. Because of the velocities that can be achieved with the Maximum, our 158 grain Sporting Handgun Revolver bullet is a great choice for hunting. It will expand well on deer and antelope-sized game out to 150 yards.

This cartridge performs best with very slow burning powders and a heavy roll crimp to help achieve consistent ignition. When we load for the Maximum, we use the following steps:

- Seat the bullet to where you can just see the top edge of the cannelure.
- Adjust your crimp so that the case mouth is deforming the ridges in the cannelure and biting all the way to the bottom of the groove.

The Maximum actually uses small rifle primers rather than small pistol magnum primers. This is because of its long powder column and higher working pressures.

357 Remington Max. - 158 grain				MAXIMUM O.A.C.L.		1.990"
					B.C.	S.D.
Sporting Handgun™ Revolver			158gr. JHP		0.182	0.177
CASE TYPE:	Federal		PRIMER TYPE			Fed 205
CASE HOLDS:	24.2	Gr. WATER	BARREL Length/Make			10.5" H-S Prec.
			BARREL Twist			1-18.75"

POWDER TYPE	POWDER CHG. GRS.		MUZZLE VEL. F.P.S.		LOAD DENSITY (VOLUME)
RL7 Most Accurate Powder Tested	26.5	MAX.	1839	**	119%
	26.0		1803	**	117%
	25.5 *		1768	**	115%
H110	23.5	MAX.	1940		99%
	23.0		1900		96%
	22.5 *		1860		94%
A-1680	26.6	MAX.	1992	**	116%
	26.1		1957	**	114%
	25.6 *		1921	**	111%

BC=Ballistic Coefficient SD=Sectional Density
*Most Accurate Load Tested **Compressed Load

Use Maximum Loads with Caution
Refer to page 73 for additional safety information

Massad Ayoob

40 S&W

In the year 1990, the US police transition to auto pistols was in full swing. Sentiment was split between the 16-shot 9mms and 8-shot .45s of the time, and the FBI's 10mm had not caught on. It was a perfect time for Smith & Wesson and Winchester to introduce a compromise, the 40 S&W. It emerged concurrent with S&W's 4006 pistol, with a 12-shot capacity that exactly split the difference. Glock's swift introduction of their 16-shot .40, the G22, made the caliber all the more desirable. Soon the 40 S&W was the issue cartridge for more US law enforcement agencies than any other.

The combined efforts of S&W's Tom Campbell, Paul Liebenberg, and outside consultants Whit Collins and Ed Hobbe had created a round powerful enough to satisfy large-bore advocates, with enough magazine capacity to satisfy those who sought more firepower. The 40 S&W had its roots in the wildcat Centimeter cartridge Liebenberg had created earlier to fit 9mm pistol platforms, while meeting major power factor in IPSC matches, where the .40 found success.

The result was a remarkably versatile cartridge. The first generation load (still the most popular) is a 180 grain JHP at the subsonic velocity of around 1000 feet per second, with muzzle energy of 400 ft-lbs. This was most often compared to the old .38-40 (180 grains at 975 fps and 380 fpe), but was also in the ballpark with the most popular police 45 ACP hollow point of the time, the 185-grain at a nominal 1000 fps/411 fpe.

Some departments, such as CHP, have been happy with the 180-grain load from 1990 to present. New loads soon emerged that gave more options. A 155-grain .40 JHP at roughly 1200 fps/500 fpe proved remarkably effective for the US Border Patrol, though they eventually went to milder rounds, reportedly to save gun wear and reduce recoil. More departments chose the 165-grain JHP at up to 1150 fps and 485 foot-pounds, which proved dramatically successful for such departments as Nashville Metro and Tulsa PD. For comparison, a 158-grain 357 Magnum hollow point is spec'd for 1235 fps and 640 fpe.

The .40 S&W remains an excellent compromise cartridge in police service weapons, and also in the subcompact pistols of lawfully armed citizens who want large-bore performance in a 9mm-size concealment gun.

Massad Ayoob is the handgun editor of GUNS magazine and law enforcement editor of AMERICAN HANDGUNNER.

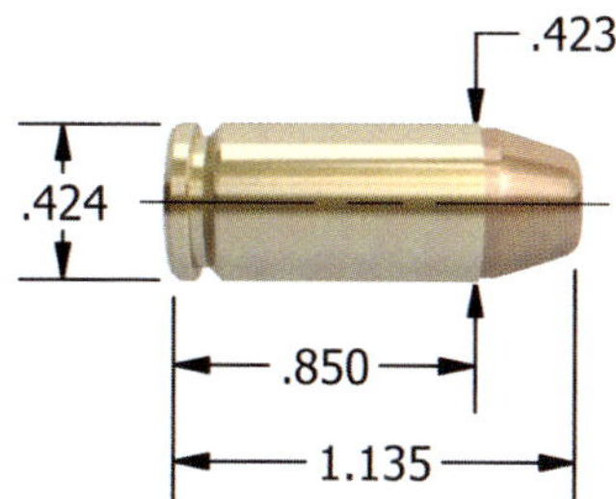

Maximum S.A.A.M.I. Overall Cartridge Length: 1.135"

BULLET CHOICES FOR THE 40 S&W

Sporting Handgun™ JHP

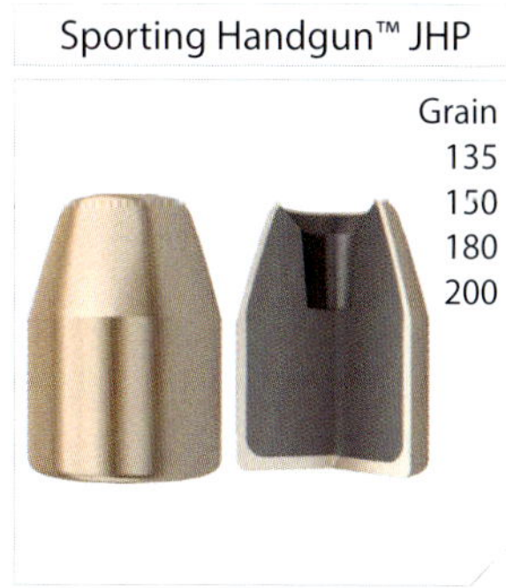

Grain
135
150
180
200

TECHNICAL INFORMATION

Like most pistol cartridges, the 40 S&W headspaces on the case mouth. When loading for this cartridge, bell the case mouth just enough to reliably guide the bullet into position and then secure the bullet with a taper crimp. Using this seating and crimping technique will help ensure proper headspacing.

The favorite all-around 40 S&W load in our Ballistics Lab consists of our 135 grain bullet and Blue Dot. All of the standard small pistol primers available give good results with this combination.

The S.A.A.M.I. overall cartridge lengths for this cartridge are 1.085" min. and 1.135" max. Seating to anywhere in this range is acceptable, provided that the rounds will function well in your particular firearm.

Nosler

40 S&W - 135 grain		MAXIMUM S.A.A.M.I. O.A.C.L.		1.135"
		TESTED O.A.C.L.	B.C.	S.D.
Sporting Handgun™ Pistol	135gr. JHP	1.125"	0.093	0.121

CASE TYPE:	Winchester		PRIMER TYPE	WSP
CASE HOLDS:	12.7	Gr. WATER	BARREL Length/Make	4" S&W
			BARREL Twist	1-16"

POWDER TYPE	POWDER CHG. GRS.		MUZZLE VEL. F.P.S.	LOAD DENSITY (VOLUME)
WST	6.8	* MAX.	1108	** 102%
	6.3		1027	94%
	5.8		945	87%
A-No.5	9.0	* MAX.	1117	75%
	8.5		1091	70%
	8.0		1065	66%
Unique	7.2	* MAX.	1125	93%
	6.7		1042	86%
	6.2		958	80%
Bullseye	6.4	* MAX.	1160	81%
	5.9		1071	75%
	5.4		981	69%
HS-7	10.5	* MAX.	1162	92%
	10.0		1106	87%
	9.5		1050	83%
Green Dot	5.8	MAX.	1166	88%
	5.3		1047	80%
	4.8	*	929	73%
A-No.7	11.9	MAX.	1224	95%
	11.4		1190	91%
	10.9	*	1156	87%
A-No.2	6.6	* MAX.	1232	80%
	6.1		1197	74%
	5.6		1162	68%
WSF	7.8	* MAX.	1242	82%
	7.3		1157	77%
	6.8		1072	71%
Blue Dot	12.0	* MAX.	1320	** 121%
Most Accurate	11.5		1265	** 116%
Powder Tested	11.0		1210	** 111%

BC=Ballistic Coefficient SD=Sectional Density
*Most Accurate Load Tested **Compressed Load

Use Maximum Loads with Caution
Refer to page 73 for additional safety information

40 S&W - 150 grain			MAXIMUM S.A.A.M.I. O.A.C.L.		1.135"
			TESTED O.A.C.L.	B.C.	S.D.
Sporting Handgun™ Pistol		150gr. JHP	1.125"	0.106	0.134

CASE TYPE:	Winchester		PRIMER TYPE	WSP
CASE HOLDS:	11.3	Gr. WATER	BARREL Length/Make	4" S&W
			BARREL Twist	1-16"

POWDER TYPE	POWDER CHG. GRS.		MUZZLE VEL. F.P.S.	LOAD DENSITY (VOLUME)
A-No.7	10.5	MAX.	1083	94%
	10.0		1019	90%
	9.5 *		955	85%
HP-38	6.6 *	MAX.	1107	85%
	6.1		1022	79%
	5.6		937	72%
A-No.2	6.4	MAX.	1118	87%
	5.9		1070	80%
	5.4 *		1022	74%
A-No.5 Most Accurate Powder Tested	8.7 *	MAX.	1133	81%
	8.2		1031	76%
	7.7		988	72%
HS-6	9.2 *	MAX.	1137	90%
	8.7		1075	86%
	8.2		1013	81%
W231	6.5 *	MAX.	1143	84%
	6.0		1081	78%
	5.5		1018	71%
Power Pistol	8.0	MAX.	1170	94%
	7.5		1102	88%
	7.0 *		1035	83%

BC=Ballistic Coefficient SD=Sectional Density
*Most Accurate Load Tested **Compressed Load

Use Maximum Loads with Caution
Refer to page 73 for additional safety information

40 S&W - 180 grain		MAXIMUM S.A.A.M.I. O.A.C.L.		1.135"
		TESTED O.A.C.L.	B.C.	S.D.
Sporting Handgun™ Pistol	180gr. JHP	1.125"	0.147	0.161

CASE TYPE:	Winchester		PRIMER TYPE	WSP
CASE HOLDS:	9.9	Gr. WATER	BARREL Length/Make	4" S&W
			BARREL Twist	1-16"

POWDER TYPE	POWDER CHG. GRS.		MUZZLE VEL. F.P.S.	LOAD DENSITY (VOLUME)
WST	4.8	* MAX.	860	92%
	4.3		768	83%
	3.8		675	73%
Bullseye	4.8	* MAX.	901	78%
	4.3		797	70%
	3.8		694	62%
HS-7	7.9	MAX.	918	89%
	7.4		861	83%
	6.9	*	804	77%
Green Dot	4.9	* MAX.	922	95%
	4.4		825	85%
	3.9		727	76%
A-No.7	8.8	* MAX.	944	90%
	8.3		914	85%
	7.8		884	80%
A-No.2	5.3	MAX.	955	82%
	4.8		898	75%
	4.3	*	842	67%
A-No.5	7.2	* MAX.	971	77%
	6.7		936	71%
	6.2		901	66%
WSF	5.9	* MAX.	981	79%
Most Accurate	5.4		897	73%
Powder Tested	4.9		813	66%

BC=Ballistic Coefficient SD=Sectional Density
*Most Accurate Load Tested **Compressed Load

Use Maximum Loads with Caution
Refer to page 73 for additional safety information

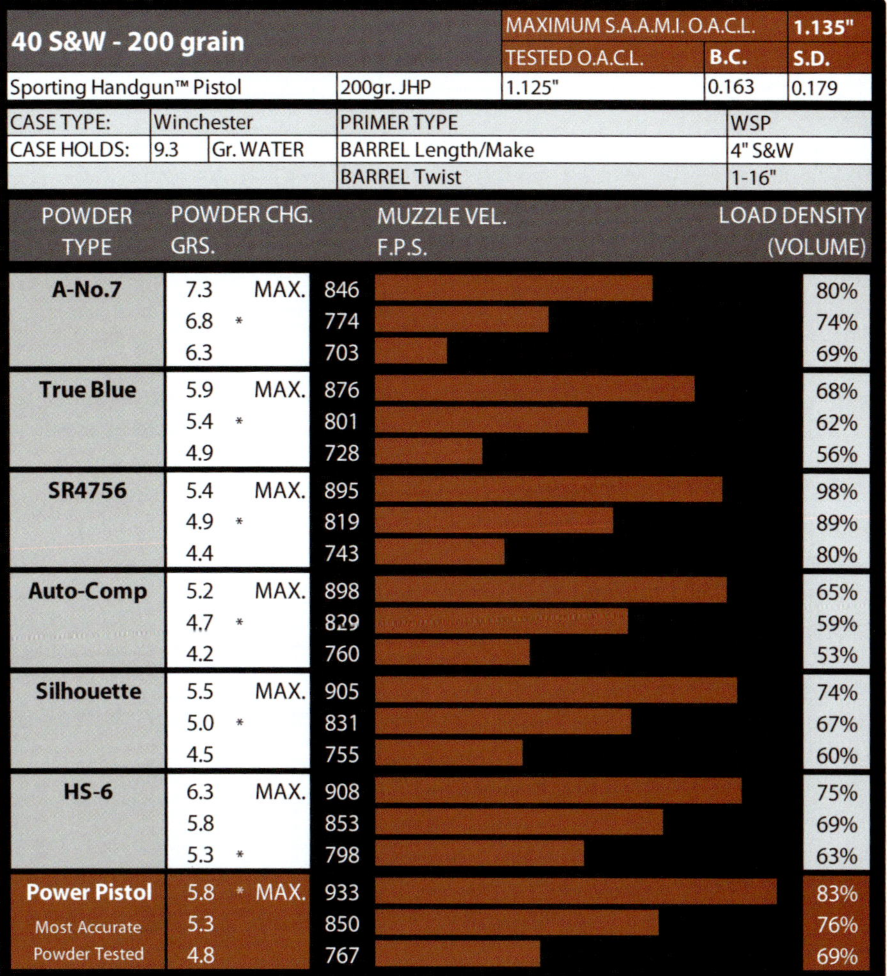

40 S&W - 200 grain

		MAXIMUM S.A.A.M.I. O.A.C.L.		1.135"
		TESTED O.A.C.L.	B.C.	S.D.
Sporting Handgun™ Pistol	200gr. JHP	1.125"	0.163	0.179

CASE TYPE:	Winchester		PRIMER TYPE	WSP
CASE HOLDS:	9.3	Gr. WATER	BARREL Length/Make	4" S&W
			BARREL Twist	1-16"

POWDER TYPE	POWDER CHG. GRS.		MUZZLE VEL. F.P.S.	LOAD DENSITY (VOLUME)
A-No.7	7.3	MAX.	846	80%
	6.8 *		774	74%
	6.3		703	69%
True Blue	5.9	MAX.	876	68%
	5.4 *		801	62%
	4.9		728	56%
SR4756	5.4	MAX.	895	98%
	4.9 *		819	89%
	4.4		743	80%
Auto-Comp	5.2	MAX.	898	65%
	4.7 *		829	59%
	4.2		760	53%
Silhouette	5.5	MAX.	905	74%
	5.0 *		831	67%
	4.5		755	60%
HS-6	6.3	MAX.	908	75%
	5.8		853	69%
	5.3 *		798	63%
Power Pistol	5.8 *	MAX.	933	83%
Most Accurate Powder Tested	5.3		850	76%
	4.8		767	69%

BC=Ballistic Coefficient SD=Sectional Density
*Most Accurate Load Tested **Compressed Load

Use Maximum Loads with Caution
Refer to page 73 for additional safety information

Mike McNett

10MM AUTO

The 10mm may just be the most versatile handgun round ever developed. It possesses more speed and energy than the 9mm, 40S&W, 45ACP, and even the 357 Magnum. Loaded with Nosler's excellent 135gr, 150gr, and 180gr JHP bullets, the 10mm is a very powerful and accurate defensive tool. Also, when loaded properly with Nosler's 180gr or 200gr JHP, it can and has cleanly taken deer, hogs, mountain lion, wolf, antelope, and even elk! The beauty of this cartridge lies in its great versatility. If recoil is an issue, it can be downloaded to match 40S&W power, making it very soft to shoot. Because it is capable of such a wide spectrum of uses, the 10mm is what rides my hip most of the time.

The 10mm cartridge was brought to life by Dornaus & Dixon, Jeff Cooper, and Norma in 1982-1983 while trying to develop a better cartridge option for our military sidearm. The original loading launched a 200gr flat point bullet at 1200fps! The idea was that a soldier could accurately engage enemies out to 100yds without excessive bullet drop, with enough power and penetration to get the job done. Loaded with a 190gr JHP, it was adopted by the FBI in 1988 as their official sidearm and used as such until the 40S&W supplanted it in 1992 because of its softer recoil.

If you want a handgun round that is effective for carry and personal defense as well as very capable of hunting medium sized game, the 10mm is as good as it gets. Teaming this excellent cartridge with Nosler's JHP bullets creates the ideal blend of power, accuracy, and terminal performance all in one cartridge. I have owned and shot every model of 10mm that I know of and cannot remember one that I didn't like! Take my word for it, once you try the 10mm you will be hooked!

Mike McNett is the owner and founder of McNett's Doubletap Ammunition.

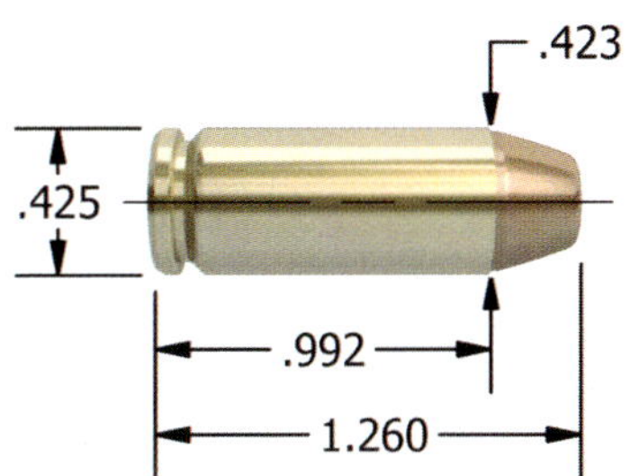

Maximum S.A.A.M.I. Overall Cartridge Length: 1.260"

BULLET CHOICES FOR THE 10MM AUTO

Sporting Handgun™ JHP

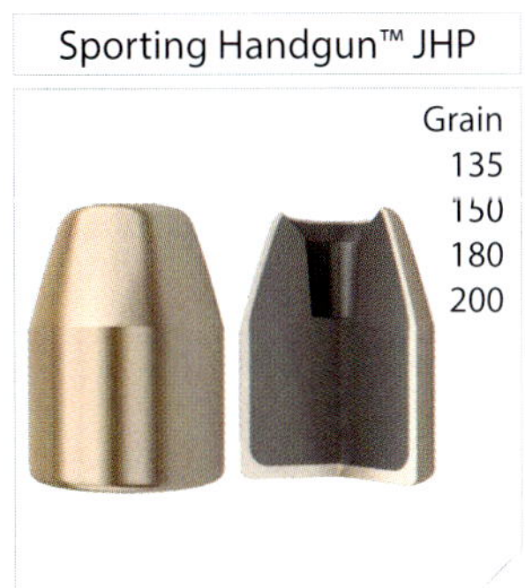

TECHNICAL INFORMATION

The 10mm Auto is one of the more powerful pistol cartridges on the market today. For this reason, it is one of the few pistol cartridges that is capable of efficiently taking big-game animals, with proper shot placement of course. If you choose to use the 10mm for hunting applications, we would recommend using either our 180 or 200 grain JHP Sporting Handgun Pistol bullets.

Like most pistol cartridges, the 10mm headspaces on the case mouth. When loading for this cartridge, bell the case mouth just enough to reliably guide the bullet into position and then secure the bullet with a taper crimp. Using this seating and crimping technique will help ensure proper headspacing.

The S.A.A.M.I. overall cartridge lengths for this cartridge range from 1.240" min. to 1.260" max. Seating to anywhere in this range is acceptable, but there may be a depth at which your particular firearm performs the best.

10mm Automatic - 135 grain		MAXIMUM S.A.A.M.I. O.A.C.L.		1.260"
		TESTED O.A.C.L.	B.C.	S.D.
Sporting Handgun™ Pistol	135gr. JHP	1.260"	0.093	0.121

CASE TYPE:	Winchester		PRIMER TYPE	WLP
CASE HOLDS:	17.3	Gr. WATER	BARREL Length/Make	6" Omega
			BARREL Twist	1-16"

POWDER TYPE	POWDER CHG. GRS.		MUZZLE VEL. F.P.S.	LOAD DENSITY (VOLUME)
W231	7.5	MAX.	1289	63%
	7.0		1224	59%
	6.5	*	1160	55%
Unique	8.5	MAX.	1340	81%
	8.0		1263	76%
	7.5	*	1185	71%
HS-6	11.0	MAX.	1356	71%
	10.5		1280	67%
	10.0	*	1205	64%
A-No.7	13.2	MAX.	1405	77%
	12.7		1341	75%
	12.2	*	1287	72%
A-No.5	11.0	MAX.	1417	67%
	10.5		1362	64%
	10.0	*	1307	61%
Blue Dot	12.5	MAX.	1459	93%
Most Accurate	12.0		1383	89%
Powder Tested	11.5	*	1308	85%

BC=Ballistic Coefficient SD=Sectional Density
*Most Accurate Load Tested **Compressed Load

Use Maximum Loads with Caution
Refer to page 73 for additional safety information

10mm Automatic - 150 grain		MAXIMUM S.A.A.M.I. O.A.C.L.		1.260"
		TESTED O.A.C.L.	B.C.	S.D.
Sporting Handgun™ Pistol	150gr. JHP	1.260"	0.106	0.134

CASE TYPE:	Winchester		PRIMER TYPE	WLP
CASE HOLDS:	15.1	Gr. WATER	BARREL Length/Make	6" Omega
			BARREL Twist	1-16"

POWDER TYPE	POWDER CHG. GRS.		MUZZLE VEL. F.P.S.	LOAD DENSITY (VOLUME)
W231	7.0	MAX.	1197	68%
	6.5		1132	63%
	6.0	*	1068	58%
Unique Most Accurate Powder Tested	8.0	MAX.	1239	87%
	7.5		1198	81%
	7.0	*	1157	76%
HS-6	10.5	MAX.	1290	77%
	10.0		1223	74%
	9.5	*	1156	70%
Blue Dot	11.2	MAX.	1292	95%
	10.7		1220	91%
	10.2	*	1148	87%
A-No.5	10.1	MAX.	1296	70%
	9.6		1240	67%
	9.1	*	1184	63%
A-No.7	12.7	MAX.	1358	85%
	12.2		1306	82%
	11.7	*	1253	79%

BC=Ballistic Coefficient SD=Sectional Density
*Most Accurate Load Tested **Compressed Load

Use Maximum Loads with Caution
Refer to page 73 for additional safety information

10mm Automatic - 180 grain		MAXIMUM S.A.A.M.I. O.A.C.L.		1.260"
		TESTED O.A.C.L.	B.C.	S.D.
Sporting Handgun™ Pistol	180gr. JHP	1.260"	0.147	0.161

CASE TYPE:	Winchester		PRIMER TYPE	WLP
CASE HOLDS:	14.1	Gr. WATER	BARREL Length/Make	6" Omega
			BARREL Twist	1-16"

POWDER TYPE	POWDER CHG. GRS.		MUZZLE VEL. F.P.S.	LOAD DENSITY (VOLUME)
W231 Most Accurate Powder Tested	5.8	MAX.	1019	60%
	5.3		961	55%
	4.8 *		904	50%
Unique	6.8	MAX.	1092	79%
	6.3		1051	73%
	5.8 *		993	67%
HS-6	8.6 *	MAX.	1122	68%
	8.1		1057	64%
	7.6		992	60%
A-No.5	8.7	MAX.	1142	65%
	8.2		1079	61%
	7.7 *		1017	57%
Blue Dot	10.2	MAX.	1190	93%
	9.7		1130	88%
	9.2 *		1070	84%
A-No. 7	11.0	MAX.	1190	79%
	10.5		1143	76%
	10.0 *		1096	72%

BC=Ballistic Coefficient SD=Sectional Density
*Most Accurate Load Tested **Compressed Load

Use Maximum Loads with Caution
Refer to page 73 for additional safety information

10mm Automatic - 200 grain		MAXIMUM S.A.A.M.I. O.A.C.L.		1.260"
		TESTED O.A.C.L.	B.C.	S.D.
Sporting Handgun™ Pistol	200gr. JHP	1.260"	0.163	0.179

CASE TYPE:	Winchester		PRIMER TYPE	WLP
CASE HOLDS:	13.5	Gr. WATER	BARREL Length/Make	6" Omega
			BARREL Twist	1-16"

POWDER TYPE	POWDER CHG. GRS.		MUZZLE VEL. F.P.S.	LOAD DENSITY (VOLUME)
Universal	5.8	* MAX.	1023	68%
	5.3		955	62%
	4.8		887	56%
HS-6	8.4	MAX.	1039	69%
	7.9	*	976	65%
	7.4		913	61%
Unique	6.1	MAX.	1058	74%
	5.6	*	966	68%
	5.1		873	62%
Power Pistol	7.1	MAX.	1085	70%
	6.6		1028	65%
	6.1	*	971	60%
A-No.7	9.8	MAX.	1088	74%
	9.3		1036	70%
	7.8	*	985	59%
800-X	7.7	MAX.	1124	91%
	7.2	*	1077	85%
	6.7		1019	79%
Blue Dot	9.3	* MAX.	1137	91%
Most Accurate	8.8		1094	86%
Powder Tested	8.3		1053	81%
A-No.9	12.4	MAX.	1148	98%
	11.9	*	1090	94%
	11.4		1032	90%

BC=Ballistic Coefficient SD=Sectional Density
*Most Accurate Load Tested **Compressed Load

Use Maximum Loads with Caution
Refer to page 73 for additional safety information

Tom Gresham

41 REMINGTON MAGNUM

One of the best ways to start a fight among a bunch of serious handgunners is to offer the suggestion that the 41 Remington Magnum is just as good as the 44 Remington Magnum. Then, just sit back and watch the fireworks.

A decidedly divisive cartridge, the 41 Remington Magnum (41 Mag) continues to support a small group of avid fans as well as a larger group which holds that there's not much need for this revolver cartridge. Truth is, the .41 has a place, and the handloader is in the best position to get the most out of it.

The 44 Mag. may be Elmer Keith's brainchild, but the short man with the big hat teamed up with the tall Border Patrolman, Bill Jordan, to create what they thought would be a great round for law enforcement. It might have been, had Remington heeded the advice of this learned duo, who called for two loads. One was a full-power recipe pushing a 210-grain bullet close to 1,400 fps. That would be for hunting. For police, a lead bullet going in the neighborhood of 900 fps would be controllable, but would be a good stopper of bad guys.

Remington did offer two loads, but they cranked up the lighter load a good 200fps faster than Keith's recommendation. Perhaps that was the reason, or there were others, but the big .41 never caught on with the police.

Hunters, however, realized that the .41 would do pretty much anything the .44 would do. (Yes, those are fighting words.) Handgun makers turned out single action and double action revolves, single shots, semi-auto pistols, and lever-action rifles for the .41.

Push a good 210-grain bullet at 1,300 to 1,400 fps, and the .41 is within reach of the ballistics of the 44 Mag, in energy and trajectory. After all, there's only 0.019 inches difference in the diameter of the two bullets (.410 versus .429). Load it down, and you have a comfortable revolver for an afternoon at the range.

Handgunners also can choose 41 Mag revolvers for self defense. With a short barrel, that original Keith-Jordan idea of a "police load" comes back into play.

I've loved the 41 Magnum ever since I fired one of the original S&W Model 57 revolvers at age 14. I still have that revolver, 47 years later. It's that good.

Tom Gresham

Tom Gresham is host of "Tom Gresham's Gun Talk," a nationally-syndicated talk radio program, as well as co-host of Gun Talk and Guns and Gear television series. He's also the co-author of "Weatherby: The Man, The Gun, The Legend"

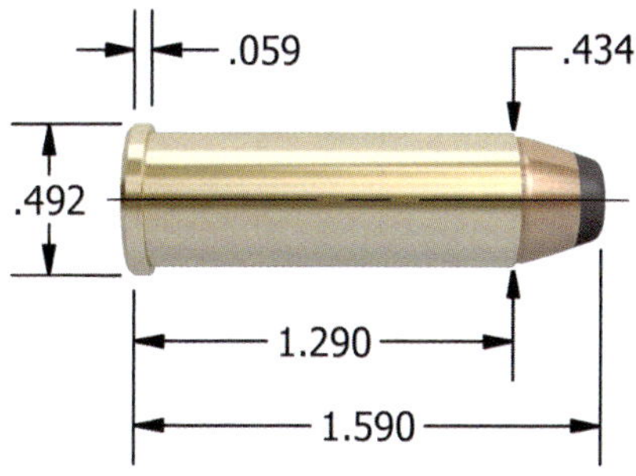

Maximum S.A.A.M.I. Overall Cartridge Length: 1.590"

BULLET CHOICES FOR THE 41 REMINGTON MAGNUM

Sporting Handgun™ JHP

TECHNICAL INFORMATION

The 41 Magnum, like the 44 Magnum, is best loaded with slower burning powders like A-No.9, H110, and W296. All of these powders require a heavy roll crimp to help achieve consistent ignition.

We have had great results with the following procedure:

• Seat the bullet to where you can just see the top edge of the cannelure.

• Adjust your crimp so that the case mouth is deforming the ridges in the cannelure and biting all the way to the bottom of the groove.

41 Remington Magnum - 210 grain		MAXIMUM S.A.A.M.I. O.A.C.L.		1.590"
		TESTED O.A.C.L.	B.C.	S.D.
Sporting Handgun™ Revolver	210gr. JHP	1.585"	0.170	0.178

CASE TYPE:	Federal		PRIMER TYPE	Fed 150
CASE HOLDS:	20.2	Gr. WATER	BARREL Length/Make	10" Douglas
			BARREL Twist	1-14"

POWDER TYPE	POWDER CHG. GRS.		MUZZLE VEL. F.P.S.		LOAD DENSITY (VOLUME)
W231	8.6	MAX.	1253		62%
	8.1		1198		59%
	7.6 *		1143		55%
SR4756	11.3	MAX.	1450		94%
	10.8		1430		90%
	10.3 *		1410		86%
A-No.7	15.1	MAX.	1477		76%
	14.6		1431		73%
	14.1 *		1385		71%
2400	16.7	MAX.	1500		96%
	16.2		1460		93%
	15.7 *		1420		90%
800X	12.1 *	MAX.	1510		96%
	11.6		1490		92%
	11.1		1470		88%
A-No.9	17.6 *	MAX.	1521		93%
	17.1		1485		91%
	16.6		1450		88%
H4227	19.9	MAX.	1530	**	113%
	19.4		1500	**	110%
	18.9 *		1470	**	108%
W296 Most Accurate Powder Tested	20.5	MAX.	1599	**	104%
	20.0		1563	**	101%
	19.5 *		1528		99%
H110	20.9	MAX.	1650	**	105%
	20.4		1620	**	103%
	19.9 *		1590		100%

BC=Ballistic Coefficient SD=Sectional Density
*Most Accurate Load Tested **Compressed Load

Use Maximum Loads with Caution
Refer to page 73 for additional safety information

44 SPECIAL

In late 1907 Smith & Wesson introduced a new sixgun and a new cartridge. The sixgun was the now almost mythical Triple-Lock and the cartridge was the 44 Special. Smith & Wesson had a very strong revolver in this first swing-out cylindered N-frame and a new cartridge case, which was simply a lengthened 44 Russian. For whatever the reason the new combination was not capitalized on, but rather the 44 Special was loaded to the same ballistics as the 44 Russian. It would be up to experimenters such as Elmer Keith and the ".44 Associates" to load the 44 Special to its full potential. The Triple-Lock evolved into the 44 Special Model 1926 which was used as the platform for the .38/44 Heavy Duty and the 357 Magnum. The Model 1926 became the Model 1950 Target and it was then used as the platform for the 44 Magnum. So, it is easy to see the far-reaching effect of the 44 Special sixgun.

The 44 Special, probably because of its lack of factory loadings, has almost always been a Connoisseur's Cartridge as it can only reach its full potential through handloading. The 44 Special cartridge is particularly appealing, at least I think so, because of the great sixguns it has been chambered in such as the Smith & Wesson .44 Hand Ejectors and the Colt New Service. Today, it is found in such soul-stirring sixguns as the Colt, USFA, and Freedom Arms Single Actions, the Colt New Frontier, and several versions of the Ruger New Model Flat-Top Blackhawk.

The 44 Special is not a 44 Magnum and that is one of the things that make it really special. For those who want extreme power the answer is still the 44 Magnum, however for those who want an everyday, easy-packing, great-shooting sixgun, the 44 Special will do just about anything we need or want to do with a sixgun. At the end of a long day with great time spent in the desert, foothills, forest, or mountains, the lighter 44 Special sixgun which has been carried all day pays huge dividends.

Good Shootin' and God Bless,

John A. Taffin

John Taffin is a contributing editor to Guns and American Handgunner magazines.

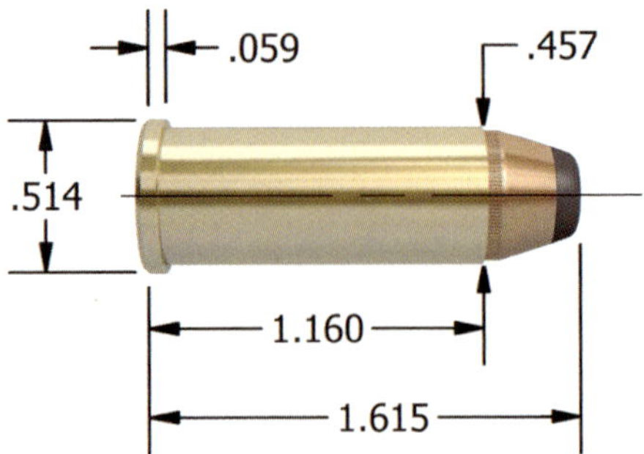

Maximum S.A.A.M.I. Overall Cartridge Length: 1.615″

BULLET CHOICES FOR THE 44 SPECIAL

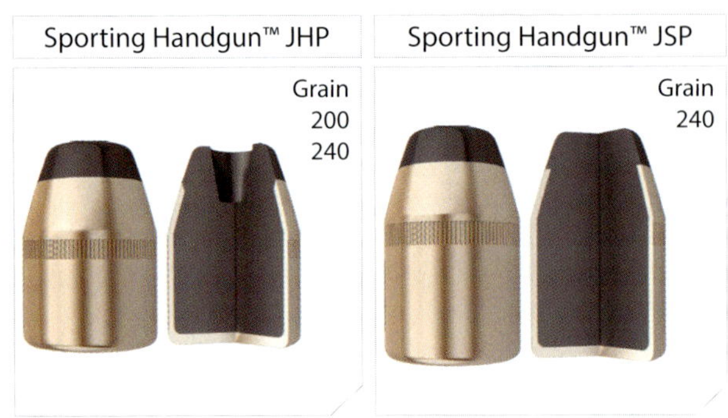

TECHNICAL INFORMATION

When loading for the 44 Special, as with any revolver cartridge, bell the case mouth just enough to reliably guide the bullet into position, seat the bullet so that the top of the cannelure is just visible above the case mouth and then give it a good roll crimp. The crimp should be such that you can just see the case mouth beginning to deform the ridges in the cannelure.

44 Special - 200 grain

		MAXIMUM S.A.A.M.I. O.A.C.L.		1.615"
		TESTED O.A.C.L.	B.C.	S.D.
Sporting Handgun™ Revolver	200gr. JHP	1.455"	0.151	0.155

CASE TYPE:	Remington		PRIMER TYPE	Rem 2 1/2
CASE HOLDS:	22.5	Gr. WATER	BARREL Length/Make	8.2" Douglas
			BARREL Twist	1-20"

POWDER TYPE	POWDER CHG. GRS.		MUZZLE VEL. F.P.S.	LOAD DENSITY (VOLUME)
Bullseye	6.0	MAX.	893	43%
	5.5 *		845	39%
	5.0		797	36%
A-No.5	8.6	MAX.	900	40%
	8.1		840	38%
	7.6 *		780	36%
Herco	8.0 *	MAX.	940	62%
	7.5		870	58%
	7.0		800	55%
Unique	8.0 *	MAX.	948	58%
	7.5		883	55%
	7.0		818	51%
Power Pistol	8.5	MAX.	968	50%
	8.0		917	47%
	7.5 *		865	44%
W231	7.3	MAX.	970	47%
Most Accurate	6.8		909	44%
Powder Tested	6.3 *		849	41%
IMR 4227	16.4	MAX.	972	85%
	15.9 *		897	83%
	15.4		822	80%
Universal	6.9 *	MAX.	977	49%
	6.4		894	45%
	5.9		811	42%

BC=Ballistic Coefficient SD=Sectional Density
*Most Accurate Load Tested **Compressed Load

Use Maximum Loads with Caution
Refer to page 73 for additional safety information

44 Special - 240 grain		MAXIMUM S.A.A.M.I. O.A.C.L.		1.615"
		TESTED O.A.C.L.	B.C.	S.D.
Sporting Handgun™ Revolver	240gr. JHP	1.455"	0.173	0.186
Sporting Handgun™ Revolver	240gr. JSP	1.455"	0.177	0.186

CASE TYPE:	Remington		PRIMER TYPE	Rem 2 1/2
CASE HOLDS:	21.8	Gr. WATER	BARREL Length/Make	8.2" Douglas
			BARREL Twist	1-20"

POWDER TYPE	POWDER CHG. GRS.		MUZZLE VEL. F.P.S.	LOAD DENSITY (VOLUME)
SR7625	6.1	* MAX.	680	48%
	5.6		600	44%
	5.1		520	40%
A-No.5	7.2	* MAX.	730	35%
	6.7		650	32%
	6.2		570	30%
Blue Dot	9.0	* MAX.	758	53%
Most Accurate	8.5		693	50%
Powder Tested	8.0		628	47%
Unique	7.5	MAX.	823	56%
	7.0	*	776	53%
	6.5		729	49%

BC=Ballistic Coefficient SD=Sectional Density
*Most Accurate Load Tested **Compressed Load

Use Maximum Loads with Caution
Refer to page 73 for additional safety information

44 REMINGTON MAGNUM

Since its introduction back in 1955-1956, the 44 Remington Magnum has become one of the most popular handgun cartridges in modern times. Thanks to the efforts of Elmer Keith, along with Remington, Smith & Wesson, Ruger, and a few others, the 44 Magnum has a stream of followers to this day. Based on a lengthened 44 Special case, the 44 Magnum makes an ideal hunting cartridge for close range opportunities on any game in North America. The "double four" is the proverbial workhorse for many handgun hunters. Originally intended for revolvers, the cartridge has also been chambered in several light, compact semi-automatic carbines and lever actions.

In my early days of handgun hunting, I took black bear, mountain lion, whitetail, antelope, and a bunch of big mean hogs. Ruger's Super Blackhawk or a Smith & Wesson Model 29 were always on my side. Later on, T/C's Contender got in on the action. The 44 Magnum never failed me and consistently delivered a clean, quick, humane kill on any intended quarry. Even in Africa and Australia the cartridge performed flawlessly, taking several head of big game. My wife and I both shot dandy bushpigs on our first safari in Zimbabwe back in 1983 with Ruger's Super Blackhawk 44 Magnum. Simply put, the cartridge is very effective on a wide variety of game.

The 44 Magnum is extremely flexible especially for handloaders. Nosler's 240-grain JSP and JHP make a wise choice for medium-sized game like whitetail deer. Their 300-grain JHP is ideal for larger game such as black bear, boar, or even elk. Inside the realm of effective revolver range, the 44 Magnum is a good choice for hunting in the woods or timber. This cartridge has taken every huntable species of big game in North America and a multitude of African game including Cape buffalo and elephant.

While the 44 Magnum has been surpassed in power by larger bores with intimidating muzzle blast, many shooters find their honest tolerance level to recoil maxes out with the double four. According to die sales from a leading manufacturer, the 44 Magnum far exceeds those larger, rhino clobbering rounds. Bigger doesn't always equate to better. The 44 Magnum has definitely been satisfying shooters, punching tags, and filling freezers for over fifty years. Even Dirty Harry can't top that!

Mark Hampton

Mark Hampton has written numerous articles and published two books related to hunting. He has hunted on six continents in thirty countries and taken over 160 different species of big game with a handgun alone. Mark is considered one of the world's top handgun hunters.

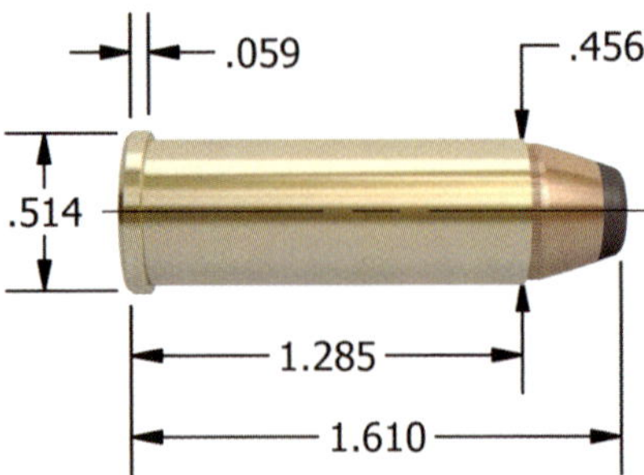

Maximum S.A.A.M.I. Overall Cartridge Length: 1.610"

BULLET CHOICES FOR THE 44 REMINGTON MAGNUM

TECHNICAL INFORMATION

Because of its manageable power and excellent accuracy, the 44 Magnum is one of the best handgun hunting cartridges out there. For most big game hunting applications, either of our 240 grain Sporting Handgun Revolver bullets are an excellent choice. Our 300 grain Hollow Point will offer a bit more penetration for larger game animals.

W296 is our favorite powder for the 44 Magnum, and H110 is another good choice. With any powder, and especially the two that we mention here, a heavy roll crimp is a requisite to help achieve consistent ignition.

- Seat the bullet to where you can just see the top edge of the cannelure.
- Adjust your crimp so that the case mouth is deforming the ridges in the cannelure and biting all the way to the bottom of the groove.

For practicing and plinking, loading with a faster powder and our 200 grain bullet will produce a lower level of recoil. 44 Special ammunition can be safely fired from a 44 Magnum revolver, and is another good low-recoil option.

44 Remington Magnum - 200 grain		MAXIMUM S.A.A.M.I. O.A.C.L.		1.610"
		TESTED O.A.C.L.	B.C.	S.D.
Sporting Handgun™ Revolver	200gr. JHP	1.580"	0.151	0.155

CASE TYPE:	Winchester		PRIMER TYPE	WLP
CASE HOLDS:	24.0	Gr. WATER	BARREL Length/Make	8.25" Douglas
			BARREL Twist	1-20"

POWDER TYPE	POWDER CHG. GRS.		MUZZLE VEL. F.P.S.		LOAD DENSITY (VOLUME)
Unique	12.5	MAX.	1406		85%
	12.0 *		1368		82%
	11.5		1330		79%
SR4756	13.7 *	MAX.	1482		96%
	13.2		1447		92%
	12.7		1411		89%
True Blue	15.2	MAX.	1492		68%
	14.7		1439		66%
	14.2 *		1386		63%
Blue Dot	17.0 *	MAX.	1524		91%
	16.5		1495		88%
	16.0		1466		85%
A-No.7 Most Accurate Powder Tested	18.4	MAX.	1580		78%
	17.9		1515		76%
	17.4 *		1450		74%
2400	21.8 *	MAX.	1619	**	106%
	21.3		1593	**	103%
	20.8		1568	**	101%
A-No.9	23.5	MAX.	1662	**	105%
	23.0		1647	**	102%
	22.5 *		1631		100%
IMR 4227	27.0	MAX.	1672	**	132%
	26.5		1639	**	129%
	26.0 *		1606	**	127%
H110	28.0	MAX.	1798	**	118%
	27.5 *		1749	**	116%
	27.0		1699	**	114%

BC=Ballistic Coefficient SD=Sectional Density
*Most Accurate Load Tested **Compressed Load

Use Maximum Loads with Caution
Refer to page 73 for additional safety information

44 Remington Magnum - 240 grain		MAXIMUM S.A.A.M.I. O.A.C.L.		1.610"
		TESTED O.A.C.L.	B.C.	S.D.
Sporting Handgun™ Revolver	240gr. JHP	1.580"	0.173	0.186
Sporting Handgun™ Revolver	240gr. JSP	1.580"	0.177	0.186

CASE TYPE:	Winchester		PRIMER TYPE	WLP
CASE HOLDS:	22.4	Gr. WATER	BARREL Length/Make	8.25" Douglas
			BARREL Twist	1-20"

POWDER TYPE	POWDER CHG. GRS.		MUZZLE VEL. F.P.S.		LOAD DENSITY (VOLUME)
A-No.5	13.4	MAX.	1329		63%
	12.9		1283		61%
	12.4 *		1238		58%
800-X	12.6 *	MAX.	1360		90%
	12.1		1320		86%
	11.6		1280		83%
Enforcer	21.5	MAX.	1384	**	102%
	21.0		1342		100%
	20.5 *		1310		97%
Blue Dot	15.0	MAX.	1390		86%
	14.5		1350		83%
	14.0 *		1310		80%
A-No.9	21.0	MAX.	1396		100%
	20.5 *		1351		98%
	20.0		1306		95%
IMR 4227	23.0	MAX.	1419	**	120%
	22.5 *		1380	**	117%
	22.0		1341	**	115%
2400	21.0 *	MAX.	1429	**	109%
	20.5		1388	**	106%
	20.0		1337	**	104%
W296	23.8	MAX.	1450	**	108%
	23.3		1410	**	106%
	22.8 *		1370	**	104%
H110 Most Accurate Powder Tested	23.8	MAX.	1472	**	108%
	23.3		1442	**	106%
	22.8 *		1412	**	103%
Lil'Gun	23.2	MAX.	1576	**	109%
	22.7 *		1547	**	107%
	22.2		1515	**	104%

BC=Ballistic Coefficient SD=Sectional Density
*Most Accurate Load Tested **Compressed Load

Use Maximum Loads with Caution
Refer to page 73 for additional safety information

44 Remington Magnum - 300 grain		MAXIMUM S.A.A.M.I. O.A.C.L.		1.610"
		TESTED O.A.C.L.	B.C.	S.D.
Sporting Handgun™ Revolver	300gr. JHP	1.580"	0.206	0.233

CASE TYPE:	Winchester		PRIMER TYPE	WLP
CASE HOLDS:	17.6	Gr. WATER	BARREL Length/Make	8.25" Douglas
			BARREL Twist	1-20"

POWDER TYPE	POWDER CHG. GRS.		MUZZLE VEL. F.P.S.		LOAD DENSITY (VOLUME)
Viht 3N37	11.0 *	MAX.	1078		98%
	10.5		1024		93%
	10.0		970		89%
Blue Dot	12.0	MAX.	1123		87%
	11.5		1082		84%
	11.0 *		1039		80%
Enforcer	17.0	MAX.	1164	**	103%
	16.5 *		1116		100%
	16.0		1057		97%
A-No.7	15.0 *	MAX.	1187		87%
	14.5		1143		84%
	14.0		1099		81%
Lil'Gun	17.0	MAX.	1245	**	102%
	16.5 *		1221		99%
	16.0		1192		96%
A-No.9 Most Accurate Powder Tested	18.0	MAX.	1249	**	109%
	17.5 *		1207	**	106%
	17.0		1165	**	103%
W296	21.0	MAX.	1344	**	122%
	20.5		1311	**	119%
	20.0 *		1278	**	116%
IMR 4227	20.0	MAX.	1345	**	133%
	19.5 *		1310	**	130%
	19.0		1279	**	126%
2400	17.0	MAX.	1365	**	112%
	16.5		1313	**	109%
	16.0 *		1217	**	106%
H110	20.5 *	MAX.	1423	**	118%
	20.0		1402	**	115%
	19.5		1380	**	112%

BC=Ballistic Coefficient SD=Sectional Density
*Most Accurate Load Tested **Compressed Load

Use Maximum Loads with Caution
Refer to page 73 for additional safety information

Clint Smith

45 AUTO

It is my privilege to introduce one of the most respected handgun cartridges of all time, the 45 ACP. Spanning military history from the time of horse mounted cavalry, to Tomahawk missiles, this cartridge has served in all United States military actions since its adoption in the year 1911. Its enviable reputation for effectiveness was earned during two World Wars, and in actions from Vera Cruz to Vietnam as well as all of the War on Terror's desert/ mountain conflicts to date.

This cartridge, although considered archaic by some standards, is useable in both semi-automatic pistols and revolvers and is still used in modest numbers today by American law enforcement officers, and in even larger numbers by civilian shooters for defense and sport.

The standard cartridge is a 230-grain full metal-jacketed projectile at a nominal 850 feet per second. Though its effectiveness is questioned by some, most of the naysayers haven't been in a tight spot with a handgun. Although not necessarily the best handgun cartridge made for hunting purposes, the quantum leaps forward in bullet design and performance over the last two decades make it a rock solid choice for personal defense…and it will work for hunting if you need it too.

I started carrying a 1911 pistol in 1968 and I don't recall ever feeling "uncomfortable" about only having the 45 ACP for defense.
During my tenure serving as operations officer at the American Pistol Institute/Gunsite with Colonel Jeff Cooper, many of the positive attributes of the 45 ACP cartridge were reinforced by my experiences working, teaching and shooting it in quantity. Having used the 45 ACP for hunting a bit, it is my opinion, that with the correct projectile, placed well, at an effective reasonable range, the cartridge is very functional on deer-sized animals.

Having taught in my own business since 1983 with I.T.C. and Thunder Ranch, Inc. and while teaching literally tens of thousands of students over that thirty year span, the single most prolific handgun cartridge I have seen in attendance to the school is the 45 ACP.

If I taught for thirty more years, I believe that trend would continue. It will be some time far in the future before the .45 caliber will not be present on the firing line, whether at paper targets or in real life combat. The 45 ACP has a place of honor in firearms history…and it is well deserved.

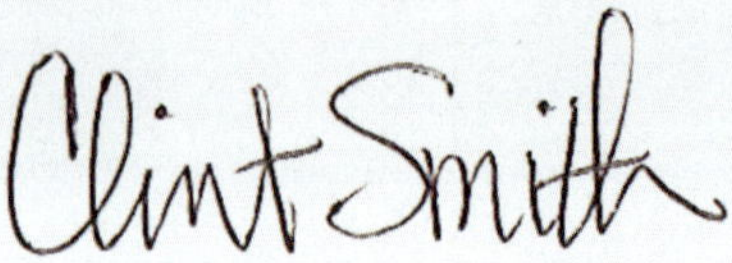

Clint Smith

Clint Smith is the President and Director of Thunder Ranch, as well as a contributing editor for both Guns and American Handgunner magazines.

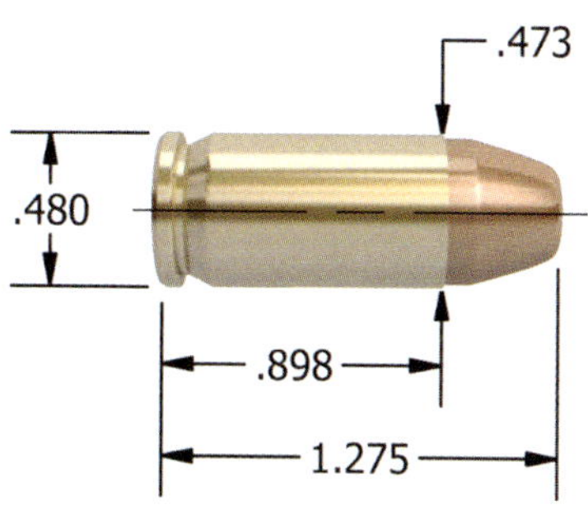

Maximum S.A.A.M.I. Overall Cartridge Length: 1.275″ Maximum, 1.190″ Minimum

BULLET CHOICES FOR THE 45 AUTO

TECHNICAL INFORMATION

The 45 ACP is a great choice for competitive shooters and plinkers alike. A-No.2, W231, and A-No.5 are the powders that we use most commonly in the lab.

Like most pistol cartridges, the 45 ACP headspaces on the case mouth. When loading for this cartridge, bell the case mouth just enough to reliably guide the bullet into position and then secure the bullet with a taper crimp. Using this seating and crimping technique will help ensure proper headspacing.

45 Automatic (ACP) - 185 grain

45 Automatic (ACP) - 185 grain		MAXIMUM S.A.A.M.I. O.A.C.L.		1.275"
		TESTED O.A.C.L.	B.C.	S.D.
Custom Competition™	185gr. JHP	1.200"	0.142	0.130

CASE TYPE:	Winchester		PRIMER TYPE	Fed. 150
CASE HOLDS:	19.2	Gr. WATER	BARREL Length/Make	5" Douglas
			BARREL Twist	1-16"

POWDER TYPE	POWDER CHG. GRS.		MUZZLE VEL. F.P.S.	LOAD DENSITY (VOLUME)
W231	6.1 *	MAX.	812	46%
	5.6		747	43%
	5.1		682	39%
Bullseye	5.4	MAX.	840	45%
	4.9		780	41%
	4.4 *		720	37%
HP-38	6.2	MAX.	850	47%
	5.7		790	43%
	5.2 *		730	40%
SR7625	6.9 *	MAX.	888	62%
	6.4		823	57%
	5.9		758	53%
A-No.2	6.3	MAX.	888	50%
	5.8		833	46%
	5.3 *		778	42%
Green Dot	6.4 *	MAX.	908	64%
	5.9		843	59%
	5.4		778	54%
A-No.5 Most Accurate Powder Tested	9.6	MAX.	924	53%
	9.1		864	50%
	8.6 *		804	47%
Viht N320	6.2 *	MAX.	938	65%
	5.7		813	59%
	5.2		688	54%
Power Pistol	8.5	MAX.	963	59%
	8.0		921	56%
	7.5 *		879	52%
Titegroup	5.7	MAX.	980	40%
	5.2 *		916	36%
	4.7		852	33%

BC=Ballistic Coefficient SD=Sectional Density
*Most Accurate Load Tested **Compressed Load

Use Maximum Loads with Caution
Refer to page 73 for additional safety information

45 Automatic (ACP) - 230 grain		MAXIMUM S.A.A.M.I. O.A.C.L.		1.275"
		TESTED O.A.C.L.	B.C.	S.D.
Sporting Handgun™ Pistol	230gr. FMJ	1.200"	0.183	0.162
Sporting Handgun™ Pistol	230gr. JHP	1.200"	0.175	0.162

CASE TYPE:	Winchester		PRIMER TYPE	Fed. 150
CASE HOLDS:	17.7	Gr. WATER	BARREL Length/Make	5" Douglas
			BARREL Twist	1-16"

POWDER TYPE	POWDER CHG. GRS.		MUZZLE VEL. F.P.S.	LOAD DENSITY (VOLUME)
HP-38	5.5	* MAX.	760	45%
	5.0		700	41%
	4.5		640	37%
WST	4.8	* MAX.	769	52%
	4.3		723	46%
	3.8		678	41%
W231 Most Accurate Powder Tested	5.8	* MAX.	790	48%
	5.3		730	44%
	4.8		670	40%
Bullseye	5.1	* MAX.	790	46%
	4.6		730	42%
	4.1		670	37%
700-X	5.0	MAX.	800	56%
	4.5		735	51%
	4.0	*	670	45%
Titegroup	4.7	MAX.	802	38%
	4.2	*	708	34%
	3.7		616	30%
Unique	6.5	MAX.	808	60%
	6.0		753	56%
	5.5	*	698	51%
HS-6	8.1	MAX.	811	51%
	7.6	*	773	48%
	7.1		739	45%
A-No.5	8.5	* MAX.	840	51%
	8.0		790	48%
	7.5		740	45%
Longshot	6.9	MAX.	896	50%
	6.4		839	46%
	5.9	*	771	44%

BC=Ballistic Coefficient SD=Sectional Density
*Most Accurate Load Tested **Compressed Load

Use Maximum Loads with Caution
Refer to page 73 for additional safety information

Charles E. Petty

45 COLT (SINGLE ACTION ARMY AND REPLICAS)

If ever there was a "venerable" cartridge, it is the 45 Colt. Common (but wrong) usage aside, it is neither short nor long, it is just Colt. It made the transition from black to smokeless powder gracefully, and from 1873 to this very day, is an effective, and potent cartridge.

But, it is also a victim of a built-in disparity between bore and chamber mouth dimensions that probably didn't matter with black powder or when all bullets were lead with hollow bases, but gave early single-actions a reputation for questionable accuracy with modern components.

When I was a kid, original Colt single-action revolvers were fairly common but far beyond my meager means. I had to make do with replicas from Hawes or Hy Hunter. The word "clone" was far in the future. Later on, I was trained as a bullseye pistol gunsmith and shooter, and my world revolved around 1911s.

Fast forward now to Y2K. I was offered a first generation 45 SAA made in 1909. The price was right and even though the condition wasn't great, that was trumped by the fact that it wore a set of sights made by King Gun Sight Co. back in the 1930s.

Over a couple of years, the gun was painstakingly restored, and when I was finally able to shoot it, I was pleasantly surprised with the accuracy. Measurements revealed cylinder dimensions with modern values which lead me to think the cylinder started out as a smaller caliber and was rechambered. No harm at all.

Loading the 45 Colt is straightforward, but my greatest concern is that somebody will do damage to the gun or themselves with overly enthusiastic handloads in old guns. Colts with serial numbers below 192000 are black powder guns, and most authorities advise against shooting them with smokeless powder loads. Colts made after that point or any of the new reproductions are fine.

The other thing that bothers me is the possibility that a hotrod load meant for- and perfectly safe in- a Ruger single-action be fired in an old Colt. I know I'm a curmudgeon, but I have been known to tell people that if they want a magnum they should have bought one.

Charlie Petty

Charles is a contributing editor for Handloader and The American Rifleman.

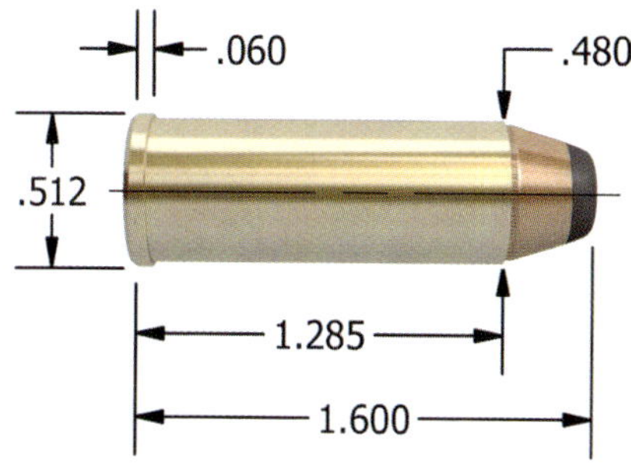

Maximum S.A.A.M.I. Overall Cartridge Length: 1.600″

BULLET CHOICES FOR THE 45 COLT (SINGLE ACTION ARMY AND REPLICAS)

Sporting Handgun™ JHP

Grain
250

TECHNICAL INFORMATION

This data is intended for original Single Action Army, Ruger New Model Vaquero, and SAA replica revolvers. For "Strong Action" handguns please refer to "45 Colt (Ruger and TC Contender)".

45 Colt (SAA & Replicas) - 250 grain		MAXIMUM S.A.A.M.I. O.A.C.L.		1.600"
		TESTED O.A.C.L.	B.C.	S.D.
Sporting Handgun™ Revolver	250gr. JHP	1.590"	0.177	0.176

CASE TYPE:	Winchester		PRIMER TYPE	Rem 2 1/2
CASE HOLDS:	24.3	Gr. WATER	BARREL Length/Make	7.25" Douglas
			BARREL Twist	1-16"

POWDER TYPE	POWDER CHG. GRS.		MUZZLE VEL. F.P.S.	LOAD DENSITY (VOLUME)
HP-38	6.7	MAX.	712	40%
	6.2		647	37%
	5.7 *		582	34%
A-No.5	9.8 *	MAX.	740	42%
	9.3		680	40%
	8.8		620	38%
HS-6	9.8 *	MAX.	742	45%
	9.3		677	43%
	8.8		612	40%
W231 Most Accurate Powder Tested	6.9 *	MAX.	748	41%
	6.4		663	38%
	5.9		578	35%
800-X	7.4 *	MAX.	794	49%
	6.9		743	45%
	6.4		692	42%
Blue Dot	11.4	MAX.	800	60%
	10.9		750	58%
	10.4 *		700	55%
Herco	8.1	MAX.	802	58%
	7.6		737	55%
	7.1 *		672	51%
IMR 4227	17.5 *	MAX.	810	84%
	17.0		780	82%
	16.5		750	79%
SR4756	8.6	MAX.	839	59%
	8.1 *		782	56%
	7.6		725	53%
HS-7	12.1	MAX.	840	55%
	11.6 *		800	53%
	11.1		760	51%

BC=Ballistic Coefficient SD=Sectional Density
*Most Accurate Load Tested **Compressed Load

Use Maximum Loads with Caution
Refer to page 73 for additional safety information

45 COLT (RUGER, T/C CONTENDER & ENCORE)

The 45 Colt was originally designed with a .455-inch, 250-grain roundnose lead bullet seated over a modest dose of black powder in a 1.265-inch copper case. Other than a few improvements in case and bullet design, and a switch to smokeless powder, it remained essentially unchanged until the late 1960s when the introduction of .451- to .452-inch jacketed bullets, coupled with increasing use of relatively heavy doses of slower-burning pistol powder, threatened to burst the old Colt Single Action Army at the seams.

There were other problems as well. Prior to the 1980s, reloading dies were still designed around .454-inch lead bullets and failed to reduce the case neck enough to provide a grip on .451/.452-inch jacketed bullets, and chamber throats in Colt cylinders were grossly oversized, measuring up to .457 inch.

Difficulties associated with oversized case necks were eventually solved by modifications to steel and carbide reloading dies, to produce an inside neck diameter of .448 to .449 inch to ensure sufficient bullet pull and proper ignition with slower-burning pistol powders, but bullets were still designed around impact velocities and pressures associated with Colt SAAs. Introduction of the Sturm, Ruger 45 Colt Blackhawk revolver in the early 1970s changed the whole ball game.

Nosler offers one jacketed bullet design that is satisfactory for higher pressures and velocities developed for the Ruger and T/C 45 Colt. The Sporting Handgun Jacketed Hollow Point (JHP) has a broad application, ranging from loads encroaching on 1,000 fps (7.5-inch barrel) at 21,000 psi (the same as 45 ACP) and somewhat beyond 25,000 psi where muzzle velocity sneaks past 1,100 fps, depending on barrel length.

Dave Scovill is the editor in chief for Wolfe Publications

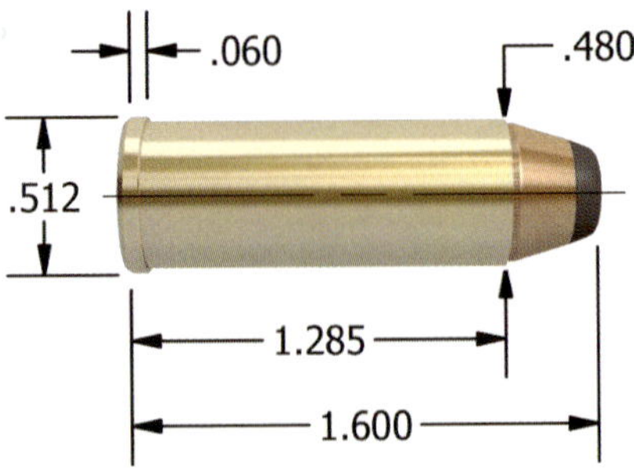

Maximum S.A.A.M.I. Overall Cartridge Length: 1.600"

BULLET CHOICES FOR THE 45 COLT (RUGER, T/C CONTENDER & ENCORE)

Sporting Handgun™ JHP

Grain
250

TECHNICAL INFORMATION

This data is intended for "Strong Action" handguns. For Single Action Army and replica revolvers, please refer to "45 Colt (SAA and replicas)". Note: This data should NOT be used in the Ruger New Model Vaquero (Manufactured in 2005 or newer). When loading for the New Model Vaquero, only the data listed for the Colt SAA and replicas should be used.

45 Colt (Ruger/TC) - 250 grain		MAXIMUM S.A.A.M.I. O.A.C.L.		1.600"
		TESTED O.A.C.L.	B.C.	S.D.
Sporting Handgun™ Revolver	250gr. JHP	1.590"	0.177	0.176

CASE TYPE:	Winchester		PRIMER TYPE	WLP
CASE HOLDS:	27.0	Gr. WATER	BARREL Length/Make	12" Contender
			BARREL Twist	1-16"

POWDER TYPE	POWDER CHG. GRS.	MUZZLE VEL. F.P.S.	LOAD DENSITY (VOLUME)
A-No.5	12.0 * MAX.	933	47%
	11.5	876	45%
	11.0	820	43%
A-No.7	15.0 MAX.	968	56%
	14.5	897	55%
	14.0 *	827	53%
W231	10.0 MAX.	1078	54%
	9.5	1030	51%
	9.0 *	981	49%
2400 Most Accurate Powder Tested	20.0 * MAX.	1103	86%
	19.5	1075	84%
	19.0	1043	82%
IMR 4227	25.0 * MAX.	1119	** 108%
	24.5	1087	** 106%
	24.0	1056	** 104%
A-No.9	18.0 MAX.	1139	71%
	17.5	1082	69%
	17.0 *	1025	67%
Unique	11.3 MAX.	1159	69%
	10.8	1116	66%
	10.3 *	1072	63%
800-X	13.8 MAX.	1273	82%
	13.3	1224	79%
	12.8 *	1176	76%
Lil'Gun	25.0 MAX.	1401	97%
	24.5 *	1378	96%
	24.0	1352	94%
W296	25.0 MAX.	1420	94%
	24.5 *	1382	93%
	24.0	1346	91%

BC=Ballistic Coefficient SD=Sectional Density
*Most Accurate Load Tested **Compressed Load

Use Maximum Loads with Caution
Refer to page 73 for additional safety information

APPENDIX

The following data is for classic cartridges, wildcats, and cartridges that are no longer in current production.

Ashland, Oregon circa 1950.
John A. Nosler making bullets one at a time.

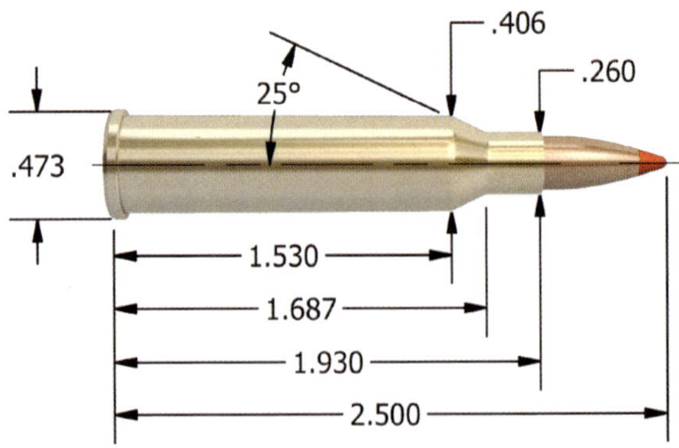

Maximum S.A.A.M.I. Overall Cartridge Length: 2.500"

BULLET CHOICES FOR THE 225 WINCHESTER

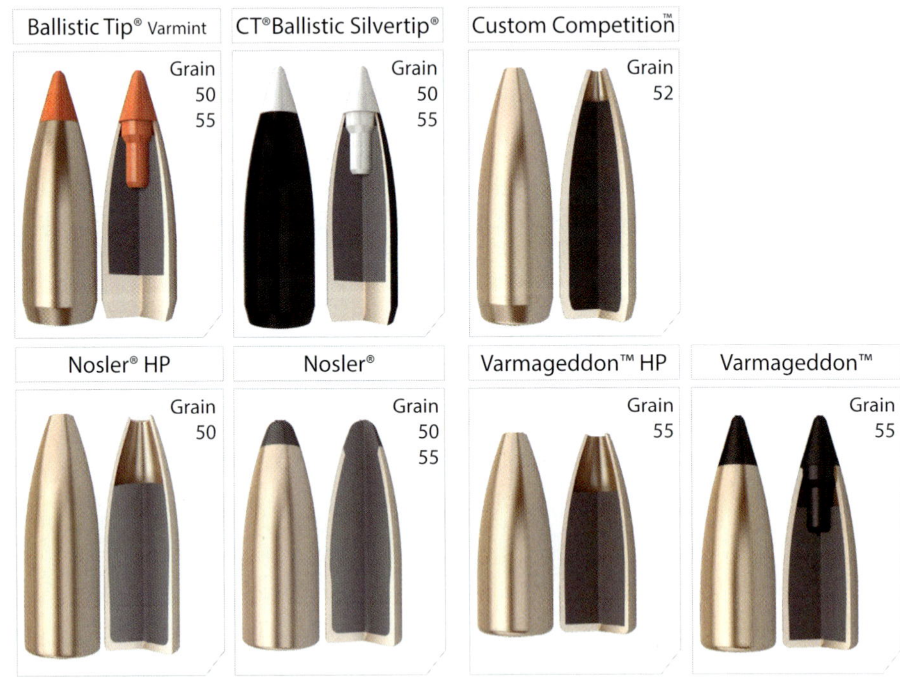

TECHNICAL INFORMATION

While the 225 Winchester may be called "obsolete" or "old school" by some, the performance of this cartridge is anything but out of date; with performance rivaling the 22-250 Remington. Even though this is a semi-rimmed cartridge, it still headspaces on the shoulder. As such, optimum accuracy and brass life can be achieved by neck sizing when you are reloading cases fired in the same gun for which you are reloading. Fortunately, 225 Winchester brass is not too difficult to find, as the process to make it from 30-30 Winchester cases is no small task.

225 Winchester - 50/52 grain		MAXIMUM S.A.A.M.I. O.A.C.L.		2.500"
		TESTED O.A.C.L.	B.C.	S.D.
Ballistic Tip®	50gr. Spitzer	2.500"	0.238	0.142
CT® Ballistic Silvertip®	50gr. Spitzer	2.500"	0.238	0.142
Nosler®	50gr. FBHP	2.500"	0.192	0.142
Nosler®	50gr. FBSP	2.500"	0.195	0.142
Custom Competition™	52gr. HPBT	2.500"	0.220	0.148

CASE TYPE:	Winchester		PRIMER TYPE	WLR
CASE HOLDS:	40.0	Gr. WATER	BARREL Length/Make	24" Lilja
			BARREL Twist	1-14"

POWDER TYPE	POWDER CHG. GRS.	MUZZLE VEL. F.P.S.	LOAD DENSITY (VOLUME)
IMR 4350	37.0 * MAX.	3420	98%
	35.0	3170	93%
	33.0	2920	87%
IMR 3031	32.0 MAX.	3440	90%
	30.0	3190	85%
	28.0 *	2940	79%
H380	37.0 MAX.	3472	97%
	35.0	3297	92%
	33.0 *	3122	87%
IMR 4895	32.5 MAX.	3482	89%
	30.5	3237	84%
	28.5 *	2992	78%
IMR 4064 Most Accurate Powder Tested	34.0 * MAX.	3650	93%
	32.0	3440	88%
	30.0	3230	82%

BC=Ballistic Coefficient SD=Sectional Density
*Most Accurate Load Tested **Compressed Load

Use Maximum Loads with Caution
Refer to page 73 for additional safety information

225 Winchester - 55 grain		MAXIMUM S.A.A.M.I. O.A.C.L.		2.500"
		TESTED O.A.C.L.	B.C.	S.D.
Ballistic Tip®	55gr. Spitzer	2.500"	0.267	0.157
CT® Ballistic Silvertip®	55gr. Spitzer	2.500"	0.267	0.157
Nosler®	55gr. FBSP	2.500"	0.218	0.157
Varmageddon™	55gr. FBHP	2.500"	0.210	0.157
Varmageddon™	55gr. FB Tipped	2.500"	0.255	0.157

CASE TYPE:	Winchester		PRIMER TYPE	WLR
CASE HOLDS:	40.0	Gr. WATER	BARREL Length/Make	24" Lilja
			BARREL Twist	1-14"

POWDER TYPE	POWDER CHG. GRS.	MUZZLE VEL. F.P.S.	LOAD DENSITY (VOLUME)
IMR 3031	31.0 MAX.	3338	88%
	29.0	3103	82%
	27.0 *	2868	76%
IMR 4350	37.0 * MAX.	3342	98%
	35.0	3117	93%
	33.0	2892	87%
H380	37.0 * MAX.	3450	97%
	35.0	3270	92%
	33.0	3090	87%
IMR 4895	32.0 * MAX.	3508	88%
	30.0	3203	82%
	28.0	2898	77%
IMR 4064	33.5 * MAX.	3520	92%
Most Accurate	31.5	3280	87%
Powder Tested	29.5	3040	81%

BC=Ballistic Coefficient SD=Sectional Density
*Most Accurate Load Tested **Compressed Load

Use Maximum Loads with Caution
Refer to page 73 for additional safety information

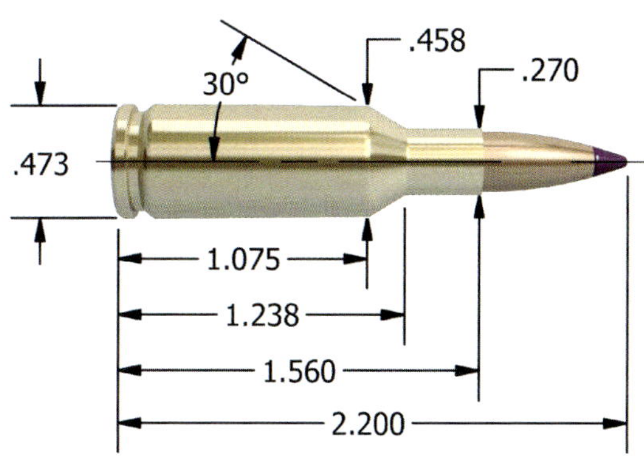

Maximum Overall Cartridge Length: 2.200"

BULLET CHOICES FOR THE 6MM BR REMINGTON

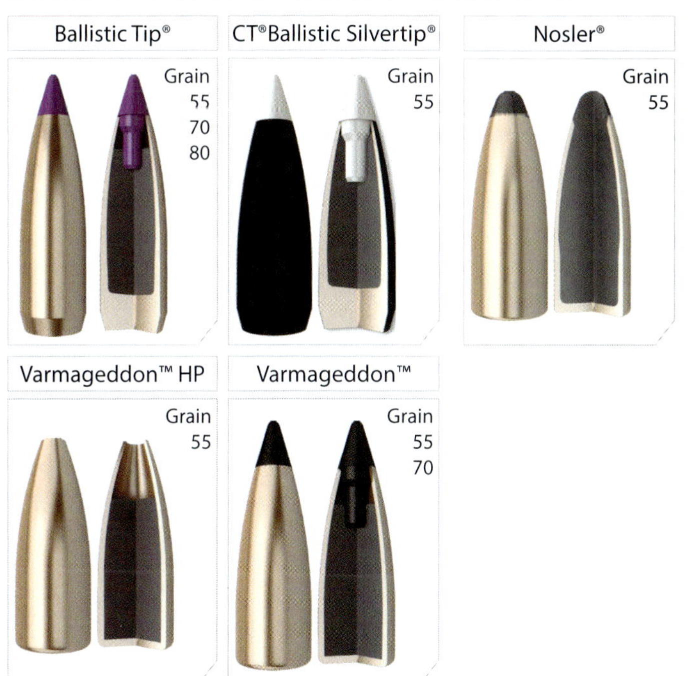

TECHNICAL INFORMATION

The 6mm BR is capable of exceptional accuracy and, when teamed with our 6mm 55-grain Ballistic Tip® will easily work in the 3700 FPS range. Since the standard twist for the 6mm BR is 1 turn in 14", bullets heavier than 80 grains will not stabilize properly, and accuracy may suffer.

6mm BR Remington - 55 grain		MAXIMUM O.A.C.L.		2.200"
		TESTED O.A.C.L.	B.C.	S.D.
Ballistic Tip®	55gr. Spitzer	2.150"	0.276	0.133
CT® Ballistic Silvertip®	55gr. Spitzer	2.150"	0.276	0.133
Nosler®	55gr. FBSP	2.150"	0.203	0.133
Varmageddon™	55gr. FBHP	2.150"	0.192	0.133
Varmageddon™	55gr. FB Tipped	2.150"	0.252	0.133

CASE TYPE:	Remington		PRIMER TYPE	CCI 450
CASE HOLDS:	35.8	Gr. WATER	BARREL Length/Make	24" Lilja
			BARREL Twist	1-14"

POWDER TYPE	POWDER CHG. GRS.		MUZZLE VEL. F.P.S.		LOAD DENSITY (VOLUME)
Viht N120	28.5	MAX.	3626		97%
	26.5		3392		90%
	24.5 *		3158		83%
IMR 4227	27.0 *	MAX.	3630		88%
	25.0		3401		82%
	23.0		3171		75%
A-2015	32.0 *	MAX.	3649	**	102%
	30.0		3412		95%
	28.0		3175		89%
W748	36.0	MAX.	3667	**	105%
	34.0		3415		99%
	32.0 *		3163		93%
H4895	35.0 *	MAX.	3674	**	107%
	33.0		3437	**	101%
	31.0		3200		95%
H322	32.5 *	MAX.	3682		98%
	30.5		3439		92%
	28.5		3196		86%
A-1680	29.5 *	MAX.	3700		87%
	27.5		3433		81%
	25.5		3167		75%
IMR 4198	29.5 *	MAX.	3708		94%
	27.5		3425		87%
	25.5		3141		81%
RL7	31.0	MAX.	3720		94%
Most Accurate	29.0		3416		88%
Powder Tested	27.0 *		3112		82%
A-2230	35.0 *	MAX.	3810		100%
	33.0		3543		94%
	31.0		3276		88%

BC=Ballistic Coefficient SD=Sectional Density
*Most Accurate Load Tested **Compressed Load

Use Maximum Loads with Caution
Refer to page 73 for additional safety information

6mm BR Remington - 70 grain

		MAXIMUM O.A.C.L.		2.200"
		TESTED O.A.C.L.	B.C.	S.D.
Ballistic Tip®	70gr. Spitzer	2.200"	0.310	0.169
Varmageddon™	70gr. FB Tipped	2.200"	0.334	0.169

CASE TYPE:	Remington		PRIMER TYPE	CCI 450
CASE HOLDS:	35.8	Gr. WATER	BARREL Length/Make	24" Lilja
			BARREL Twist	1-14"

POWDER TYPE	POWDER CHG. GRS.		MUZZLE VEL. F.P.S.		LOAD DENSITY (VOLUME)
RL7	27.5 *	MAX.	3211		83%
	25.5		3023		77%
	23.5		2834		71%
H335	32.5	MAX.	3313		92%
	30.5		3114		86%
	28.5 *		2932		81%
H4895 Most Accurate Powder Tested	32.0	MAX.	3321		98%
	30.0 *		3117		92%
	28.0		2919		86%
H322	30.5	MAX.	3354		92%
	28.5		3163		86%
	26.5 *		2973		80%
A-2015	30.5	MAX.	3374		88%
	28.5		3168		82%
	26.5 *		2962		76%
IMR 3031	31.5 *	MAX.	3377		99%
	29.5		3133		93%
	27.5		2890		87%
Benchmark	31.5	MAX.	3382		95%
	29.5		3198		89%
	27.5 *		3017		83%
A-2460	32.0	MAX.	3384		90%
	30.0		3190		85%
	28.0 *		2997		79%
Viht N135	32.5 *	MAX.	3384	**	106%
	30.5		3203		99%
	28.5		3021		93%
Varget	34.0	MAX.	3450	**	103%
	32.0		3273		97%
	30.0 *		3095		91%

BC=Ballistic Coefficient SD=Sectional Density
*Most Accurate Load Tested **Compressed Load

Use Maximum Loads with Caution
Refer to page 73 for additional safety information

6mm BR Remington - 80 grain		MAXIMUM O.A.C.L.		2.200"
		TESTED O.A.C.L.	B.C.	S.D.
Ballistic Tip®	80gr. Spitzer	2.200"	0.329	0.194

CASE TYPE:	Remington		PRIMER TYPE	CCI 450
CASE HOLDS:	33.7	Gr. WATER	BARREL Length/Make	24" Lilja
			BARREL Twist	1-14"

POWDER TYPE	POWDER CHG. GRS.		MUZZLE VEL. F.P.S.		LOAD DENSITY (VOLUME)
RL7	26.5	MAX.	3018		85%
	24.5 *		2803		79%
	22.5		2621		73%
A-2015	29.0	MAX.	3039		98%
	27.0 *		2873		91%
	25.0		2710		84%
Viht N135	30.0	MAX.	3079	**	104%
	28.0		2896		97%
	26.0 *		2728		90%
H322	29.0 *	MAX.	3144		93%
	27.0		3007		87%
	25.0		2837		80%
A-2460	31.0	MAX.	3159		93%
	29.0		2974		87%
	27.0 *		2744		81%
Varget	32.5	MAX.	3163	**	104%
	30.5		2992		98%
	28.5 *		2841		91%
H4895	31.5	MAX.	3175	**	103%
	29.5 *		2986		96%
	27.5		2794		90%
IMR 3031 Most Accurate Powder Tested	30.5	MAX.	3185	**	102%
	28.5 *		2985		96%
	26.5		2773		89%
Benchmark	30.5	MAX.	3192		98%
	28.5 *		3021		91%
	26.5		2877		85%
A-2520	33.0	MAX.	3260	**	101%
	31.0		3058		95%
	29.0 *		2844		89%

BC=Ballistic Coefficient SD=Sectional Density
*Most Accurate Load Tested **Compressed Load

Use Maximum Loads with Caution
Refer to page 73 for additional safety information

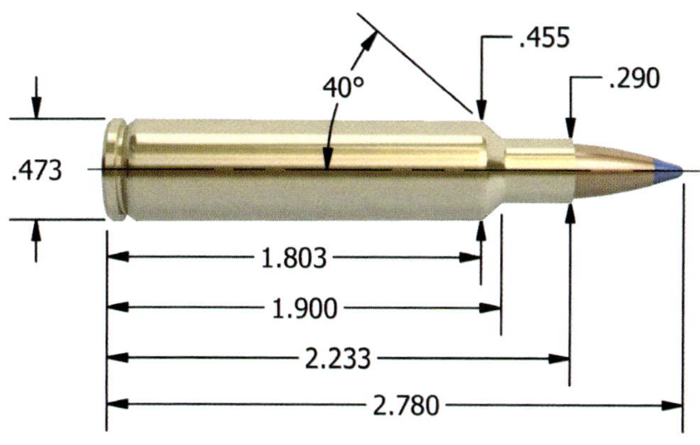

Maximum Overall Cartridge Length: 2.780"

BULLET CHOICES FOR THE 257 ROBERTS ACKLEY IMPROVED

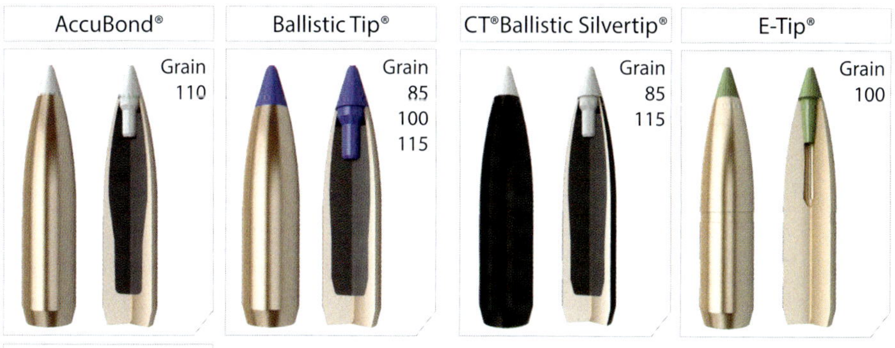

Partition®

Grain
100
115
120

TECHNICAL INFORMATION

While there are several versions of .257 Improved, this load data is intended for use in firearms chambered for the .257 Roberts Ackley Improved (40° shoulder angle). The other prominent version is the .257 RCBS Improved, which features a 28° shoulder. If you are not certain about the chamber design in your rifle, we suggest consulting your gunsmith.

To fire-form brass for the .257 Ackley Improved, we suggest the following procedure:

- Start with a standard .257 Roberts cases
- Select a light load of relatively fast powder (IMR4895 is great) from the standard .257 Roberts data.
- Seat the bullet well out into the rifling, so that it makes good contact with the lands when the cartridge is chambered. This holds the case head against the bolt face and eliminates case stretching in the web area, a cause of case head separation.
- After fire-forming, follow load data for the .257 Roberts Ackley Improved.

257 Roberts A.I. - 85 grain		MAXIMUM O.A.C.L.		2.780"
		TESTED O.A.C.L.	B.C.	S.D.
Ballistic Tip®	85gr. Spitzer	2.780"	0.329	0.183
CT® Ballistic Silvertip®	85gr. Spitzer	2.780"	0.329	0.183

CASE TYPE:	Winchester		PRIMER TYPE	Fed 210
CASE HOLDS:	58.5	Gr. WATER	BARREL Length/Make	24" Wiseman
			BARREL Twist	1-10"

POWDER TYPE	POWDER CHG. GRS.		MUZZLE VEL. F.P.S.	LOAD DENSITY (VOLUME)
Big Game Most Accurate Powder Tested	48.5	MAX.	3388	87%
	46.5		3235	84%
	44.5	*	3136	80%
H380	49.5	* MAX.	3400	89%
	47.5		3280	85%
	45.5		3160	82%
IMR 4895	45.0	* MAX.	3402	85%
	43.0		3237	81%
	41.0		3072	77%
Varget	45.5	MAX.	3407	84%
	43.5	*	3323	80%
	41.5		3153	77%
H4831	56.0	MAX.	3482	100%
	54.0		3327	96%
	52.0	*	3172	93%
Viht N150	47.5	MAX.	3487	96%
	45.5		3363	92%
	43.5	*	3239	87%
IMR 4350	52.5	* MAX.	3488	95%
	50.5		3353	91%
	48.5		3218	88%
RL19	53.5	* MAX.	3508	99%
	51.5		3303	96%
	49.5		3098	92%
W760	51.5	* MAX.	3518	92%
	49.5		3403	89%
	47.5		3288	85%

BC=Ballistic Coefficient SD=Sectional Density
*Most Accurate Load Tested **Compressed Load

Use Maximum Loads with Caution
Refer to page 73 for additional safety information

257 Roberts A.I. - 100 grain		MAXIMUM O.A.C.L.		2.780"
		TESTED O.A.C.L.	B.C.	S.D.
Ballistic Tip®	100gr. Spitzer	2.780"	0.393	0.216
E-Tip®	100gr. Spitzer	2.780"	0.409	0.216
Due to internal construction differences, always begin with starting loads when using E-Tip® products.				
Partition®	100gr. Spitzer	2.780"	0.377	0.216

CASE TYPE:	Winchester		PRIMER TYPE	Fed 210
CASE HOLDS:	56.6	Gr. WATER	BARREL Length/Make	24" Wiseman
			BARREL Twist	1-10"

POWDER TYPE	POWDER CHG. GRS.		MUZZLE VEL. F.P.S.	LOAD DENSITY (VOLUME)
H380	45.5	* MAX.	3028	85%
	43.5		2913	81%
	41.5		2798	77%
RL19	49.5	* MAX.	3088	95%
	47.5		2953	91%
	45.5		2818	87%
RL15	43.0	* MAX.	3132	79%
	41.0		3007	75%
	39.0		2882	72%
W760	47.0	MAX.	3139	87%
	45.0	*	3012	83%
	43.0		2916	80%
H4831SC	53.5	MAX.	3165	98%
	51.5	*	3095	95%
	49.5		3001	91%
Viht N160	50.0	* MAX.	3173	98%
	48.0		3064	94%
	46.0		2955	90%
IMR 4350 Most Accurate Powder Tested	50.0	* MAX.	3208	93%
	48.0		3093	90%
	46.0		2978	86%
IMR 4895	44.5	* MAX.	3222	86%
	42.5		3067	83%
	40.5		2912	79%
Hunter	50.5	MAX.	3226	94%
	48.5		3150	90%
	46.5	*	3025	86%

BC=Ballistic Coefficient SD=Sectional Density
*Most Accurate Load Tested **Compressed Load

Use Maximum Loads with Caution
Refer to page 73 for additional safety information

257 Roberts A.I. - 110 grain		MAXIMUM O.A.C.L.		2.780"
		TESTED O.A.C.L.	B.C.	S.D.
AccuBond®	110gr. Spitzer	2.780"	0.418	0.238

CASE TYPE:	Winchester		PRIMER TYPE	Fed 210
CASE HOLDS:	56.0	Gr. WATER	BARREL Length/Make	24" Wiseman
			BARREL Twist	1-10"

POWDER TYPE	POWDER CHG. GRS.		MUZZLE VEL. F.P.S.	LOAD DENSITY (VOLUME)
Big Game	44.0	MAX.	2960	83%
	42.0		2860	79%
	40.0 *		2746	75%
W760	45.5	MAX.	2981	85%
	43.5 *		2890	81%
	41.5		2808	78%
IMR 4831	48.0	MAX.	3001	91%
	46.0		2880	87%
	44.0 *		2767	84%
IMR 4350	47.5	MAX.	3002	90%
	45.5 *		2927	86%
	43.5		2787	82%
Viht N160	47.5	MAX.	3010	94%
	45.5 *		2895	90%
	43.5		2759	86%
H4831SC Most Accurate Powder Tested	51.5	MAX.	3036	96%
	49.5 *		2968	92%
	47.5		2856	88%
Hunter	48.5 *	MAX.	3038	91%
	46.5		2930	87%
	44.5		2862	84%
RL19	49.5 *	MAX.	3044	96%
	47.5		2934	92%
	45.5		2822	88%
RL22	51.0	MAX.	3055	99%
	49.0 *		2982	95%
	47.0		2860	91%

BC=Ballistic Coefficient SD=Sectional Density
*Most Accurate Load Tested **Compressed Load

Use Maximum Loads with Caution
Refer to page 73 for additional safety information

257 Roberts A.I. 115 grain

		MAXIMUM O.A.C.L.		2.780"
		TESTED O.A.C.L.	B.C.	S.D.
Ballistic Tip®	115gr. Spitzer	2.780"	0.453	0.249
CT® Ballistic Silvertip®	115gr. Spitzer	2.780"	0.453	0.249
Partition®	115gr. Spitzer	2.780"	0.389	0.249

CASE TYPE:	Winchester		PRIMER TYPE	Fed 210
CASE HOLDS:	56.0	Gr. WATER	BARREL Length/Make	24" Wiseman
			BARREL Twist	1-10"

POWDER TYPE	POWDER CHG. GRS.		MUZZLE VEL. F.P.S.	LOAD DENSITY (VOLUME)
A-4350	44.5	* MAX.	2835	89%
	42.5		2694	85%
	40.5		2552	81%
IMR 4831	45.5	* MAX.	2869	86%
	43.5		2724	83%
	41.5		2579	79%
IMR 4350	45.5	* MAX.	2931	86%
	43.5		2792	82%
	41.5		2654	78%
RL15	41.0	MAX.	2950	76%
	39.0		2833	73%
	37.0	*	2715	69%
Hunter	47.5	MAX.	2958	89%
	45.5		2909	86%
	43.5	*	2810	82%
RL19 Most Accurate Powder Tested	47.5	MAX.	2960	92%
	45.5		2816	88%
	43.5	*	2672	84%
Viht N160	47.5	MAX.	2963	94%
	45.5		2851	90%
	43.5	*	2739	86%
W760	45.0	MAX.	3010	84%
	43.0		2856	80%
	41.0	*	2735	77%
H4831SC	50.5	MAX.	3012	94%
	48.5		2895	90%
	46.5	*	2778	86%

BC=Ballistic Coefficient SD=Sectional Density
*Most Accurate Load Tested **Compressed Load

Use Maximum Loads with Caution
Refer to page 73 for additional safety information

257 Roberts A.I. - 120 grain

		MAXIMUM O.A.C.L.		2.780"
		TESTED O.A.C.L.	B.C.	S.D.
Partition®	120gr. Spitzer	2.780"	0.391	0.260

CASE TYPE:	Winchester		PRIMER TYPE	Fed 210
CASE HOLDS:	56.6	Gr. WATER	BARREL Length/Make	24" Wiseman
			BARREL Twist	1-10"

POWDER TYPE	POWDER CHG. GRS.		MUZZLE VEL. F.P.S.	LOAD DENSITY (VOLUME)
H380	43.0	MAX.	2798	80%
	41.0		2673	76%
	39.0 *		2548	73%
Viht N160	46.5	MAX.	2841	91%
	44.5		2729	87%
	42.5 *		2612	83%
RL22	47.5 *	MAX.	2847	91%
	45.5		2732	87%
	43.5		2617	84%
IMR 7828	50.5 *	MAX.	2858	95%
	48.5		2723	91%
	46.5		2588	87%
IMR 4350	45.5	MAX.	2858	85%
	43.5		2743	81%
	41.5 *		2628	78%
Varget	41.0	MAX.	2875	78%
	39.0		2755	74%
	37.0 *		2672	71%
Hunter	47.5	MAX.	2882	88%
	45.5 *		2788	85%
	43.5		2694	81%
RL19 Most Accurate Powder Tested	47.5 *	MAX.	2898	91%
	45.5		2773	87%
	43.5		2648	84%
H4831SC	50.5	MAX.	2972	93%
	48.5		2855	89%
	46.5 *		2738	86%

BC=Ballistic Coefficient SD=Sectional Density
*Most Accurate Load Tested **Compressed Load

Use Maximum Loads with Caution
Refer to page 73 for additional safety information

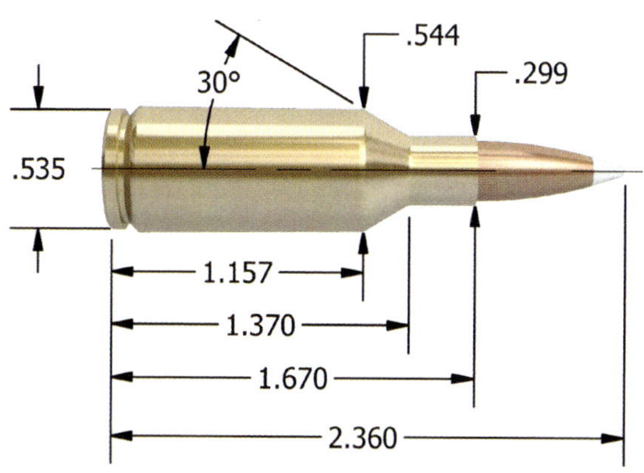

Maximum S.A.A.M.I. Overall Cartridge Length: 2.360″

BULLET CHOICES FOR THE 25 WINCHESTER SUPER SHORT MAGNUM

TECHNICAL INFORMATION

The most recent and largest caliber member of the WSSM family, the 25 WSSM epitomizes the "short mag" and "super short" philosophy. With an equivalent bullet loaded to SAAMI maximum overall cartridge length, the 25 WSSM case has over 20% less capacity than the 25-06 Remington. For this reason, equivalent powders loaded with charge weights around 10% less than the 25-06 will generate comparable velocities in the 25 WSSM. The 25-06 also requires a standard 30-06 length action, whereas the WSSM case (which is almost a full inch shorter), uses a special "super short" action. The resulting rifle is smaller, lighter, more compact, and quicker handling.

25 WSSM - 85 grain		MAXIMUM S.A.A.M.I. O.A.C.L.		2.360"
		TESTED O.A.C.L.	B.C.	S.D.
Ballistic Tip®	85gr. Spitzer	2.350"	0.329	0.183
CT® Ballistic Silvertip®	85gr. Spitzer	2.350"	0.329	0.183

CASE TYPE:	Winchester		PRIMER TYPE	WLR
CASE HOLDS:	51.8	Gr. WATER	BARREL Length/Make	24" Wiseman
			BARREL Twist	1-10"

POWDER TYPE	POWDER CHG. GRS.		MUZZLE VEL. F.P.S.	LOAD DENSITY (VOLUME)
IMR 4895	42.0 *	MAX.	3479	89%
	40.0		3351	85%
	38.0		3229	81%
IMR 3031	41.0	MAX.	3485	89%
	39.0		3375	85%
	37.0 *		3240	81%
Varget	43.5	MAX.	3488	91%
	41.5		3379	87%
	39.5 *		3269	82%
IMR 4320	45.0	MAX.	3505	93%
	43.0 *		3381	89%
	41.0		3261	85%
H4895	42.5	MAX.	3507	90%
	40.5 *		3396	86%
	38.5		3276	82%
IMR 4064	44.0 *	MAX.	3528	93%
	42.0		3419	89%
	40.0		3287	85%
IMR 4350 Most Accurate Powder Tested	50.0	MAX.	3528	** 102%
	48.0 *		3452	98%
	46.0		3310	94%
H380	49.0	MAX.	3544	100%
	47.0 *		3444	96%
	45.0		3348	91%
Big Game	49.0	MAX.	3548	100%
	47.0 *		3457	96%
	45.0		3334	91%
BL-C2	47.0		3563	92%
	45.0 *		3429	88%
	43.0		3342	84%

BC=Ballistic Coefficient SD=Sectional Density
*Most Accurate Load Tested **Compressed Load

Use Maximum Loads with Caution
Refer to page 73 for additional safety information

25 WSSM - 100 grain		MAXIMUM S.A.A.M.I. O.A.C.L.		2.360"
		TESTED O.A.C.L.	B.C.	S.D.
Ballistic Tip®	100gr. Spitzer	2.350"	0.393	0.216
E-Tip®	100gr. Spitzer	2.320"	0.409	0.216
Due to internal construction differences, always begin with starting loads when using E-Tip® products.				
Partition®	100gr. Spitzer	2.350"	0.377	0.216

CASE TYPE:	Winchester		PRIMER TYPE	WLR
CASE HOLDS:	50.2	Gr. WATER	BARREL Length/Make	24" Wiseman
			BARREL Twist	1-10"

POWDER TYPE	POWDER CHG. GRS.		MUZZLE VEL. F.P.S.	LOAD DENSITY (VOLUME)
IMR 4895	39.5	MAX.	3161	86%
	37.5		3035	82%
	35.5 *		2913	78%
Varget Most Accurate Powder Tested	41.0 *	MAX.	3184	88%
	39.0		3075	84%
	37.0		2957	80%
H380	45.0	MAX.	3216	94%
	43.0 *		3103	90%
	41.0		2981	86%
IMR 4064	41.5 *	MAX.	3219	91%
	39.5		3091	86%
	37.5		2960	82%
H4350	47.0	MAX.	3252	99%
	45.0		3144	95%
	43.0 *		3028	91%
IMR 4350	46.5 *	MAX.	3255	98%
	44.5		3136	94%
	42.5		3012	90%
IMR 4831	48.5	MAX.	3268	** 103%
	46.5		3153	99%
	44.5 *		3019	94%
Big Game	46.0	MAX.	3269	96%
	44.0		3175	92%
	42.0 *		3050	88%
H414	47.5	MAX.	3276	99%
	45.5		3159	94%
	43.5 *		3043	90%
W760	47.5	MAX.	3291	99%
	45.5		3180	95%
	43.5 *		3053	91%

BC=Ballistic Coefficient SD=Sectional Density
*Most Accurate Load Tested **Compressed Load

Use Maximum Loads with Caution
Refer to page 73 for additional safety information

25 WSSM - 110 grain		MAXIMUM S.A.A.M.I. O.A.C.L.		2.360"
		TESTED O.A.C.L.	B.C.	S.D.
AccuBond®	110gr. Spitzer	2.350"	0.418	0.238

CASE TYPE:	Winchester		PRIMER TYPE	WLR
CASE HOLDS:	49.2	Gr. WATER	BARREL Length/Make	24" Wiseman
			BARREL Twist	1-10"

POWDER TYPE	POWDER CHG. GRS.		MUZZLE VEL. F.P.S.		LOAD DENSITY (VOLUME)
IMR 4064	40.0	MAX.	3010		89%
	38.0 *		2885		85%
	36.0		2756		80%
H4831SC	48.0	MAX.	3064	**	102%
	46.0 *		2950		97%
	44.0		2852		93%
H380	44.0	MAX.	3072		94%
	42.0 *		2984		90%
	40.0		2868		86%
IMR 4350	44.5 *	MAX.	3077		96%
	42.5		2952		91%
	40.5		2818		87%
H4350	45.0	MAX.	3096		97%
	43.0 *		2982		92%
	41.0		2868		88%
H414	46.0 *	MAX.	3105		97%
	44.0		3027		93%
	42.0		2927		89%
Big Game	45.5	MAX.	3130		97%
	43.5		3042		93%
	41.5 *		2922		89%
W760	46.0	MAX.	3135		98%
Most Accurate	44.0		3038		94%
Powder Tested	42.0 *		2928		89%
IMR 4831	48.0	MAX.	3140	**	104%
	46.0		3012		99%
	44.0 *		2887		95%
Hunter	49.5	MAX.	3198	**	106%
	47.5 *		3111	**	102%
	45.5		3043		97%

BC=Ballistic Coefficient SD=Sectional Density
*Most Accurate Load Tested **Compressed Load

Use Maximum Loads with Caution
Refer to page 73 for additional safety information

25 WSSM - 115 grain

25 WSSM - 115 grain		MAXIMUM S.A.A.M.I. O.A.C.L.		2.360"
		TESTED O.A.C.L.	B.C.	S.D.
Ballistic Tip®	115gr. Spitzer	2.350"	0.453	0.249
CT® Ballistic Silvertip®	115gr. Spitzer	2.350"	0.453	0.249
Partition®	115gr. Spitzer	2.350"	0.389	0.249

CASE TYPE:	Winchester		PRIMER TYPE	WLR
CASE HOLDS:	49.0	Gr. WATER	BARREL Length/Make	24" Wiseman
			BARREL Twist	1-10"

POWDER TYPE	POWDER CHG. GRS.		MUZZLE VEL. F.P.S.		LOAD DENSITY (VOLUME)
IMR 4064	39.5	MAX.	2937		89%
Most Accurate	37.5		2820		84%
Powder Tested	35.5 *		2683		80%
Big Game	43.0	MAX.	2970		92%
	41.0		2854		88%
	39.0 *		2748		84%
H4831SC	48.0	MAX.	2974	**	102%
	46.0		2876		98%
	44.0 *		2771		94%
IMR 4350	44.5	MAX.	2983		96%
	42.5 *		2910		92%
	40.5		2782		87%
H380	44.0	MAX.	3000		95%
	42.0		2901		90%
	40.0 *		2790		86%
IMR 4831	46.5	MAX.	3018	**	101%
	44.5 *		2888		97%
	42.5		2740		92%
H4350	45.0	MAX.	3026		97%
	43.0 *		2927		93%
	41.0		2815		89%
H414	45.5	MAX.	3049		97%
	43.5		2939		92%
	41.5 *		2824		88%
W760	45.5 *	MAX.	3052		97%
	43.5		2938		93%
	41.5		2829		89%
Hunter	48.0	MAX.	3083	**	103%
	46.0 *		3010		99%
	44.0		2942		95%

BC=Ballistic Coefficient SD=Sectional Density
*Most Accurate Load Tested **Compressed Load

Use Maximum Loads with Caution
Refer to page 73 for additional safety information

25 WSSM - 120 grain

		MAXIMUM S.A.A.M.I. O.A.C.L.		2.360"
		TESTED O.A.C.L.	B.C.	S.D.
Partition®	120gr. Spitzer	2.350"	0.391	0.260

CASE TYPE:	Winchester		PRIMER TYPE	WLR
CASE HOLDS:	48.8	Gr. WATER	BARREL Length/Make	24" Wiseman
			BARREL Twist	1-10"

POWDER TYPE	POWDER CHG. GRS.		MUZZLE VEL. F.P.S.		LOAD DENSITY (VOLUME)
IMR 4064	38.0	MAX.	2873		86%
	36.0	*	2734		81%
	34.0		2618		77%
H380	43.0	MAX.	2951		93%
	41.0		2851		88%
	39.0	*	2740		84%
Big Game	43.0	MAX.	2958		93%
	41.0	*	2860		88%
	39.0		2738		84%
H4831SC	48.0	MAX.	2962	**	102%
	46.0		2878		98%
	44.0	*	2775		94%
IMR 4350 Most Accurate Powder Tested	44.0	MAX.	2974		95%
	42.0		2860		91%
	40.0	*	2755		87%
H414	45.0	MAX.	2999		96%
	43.0		2907		92%
	41.0	*	2782		88%
IMR 7828	49.0	MAX.	3005	**	107%
	47.0	*	2864	**	102%
	45.0		2757		98%
W760	45.0	MAX.	3012		97%
	43.0	*	2893		92%
	41.0		2790		88%
H4350	45.0	MAX.	3019		98%
	43.0		2922		93%
	41.0	*	2808		89%
Hunter	48.0	MAX.	3024	**	104%
	46.0	*	2936		99%
	44.0		2819		95%

BC=Ballistic Coefficient SD=Sectional Density
*Most Accurate Load Tested **Compressed Load

Use Maximum Loads with Caution
Refer to page 73 for additional safety information

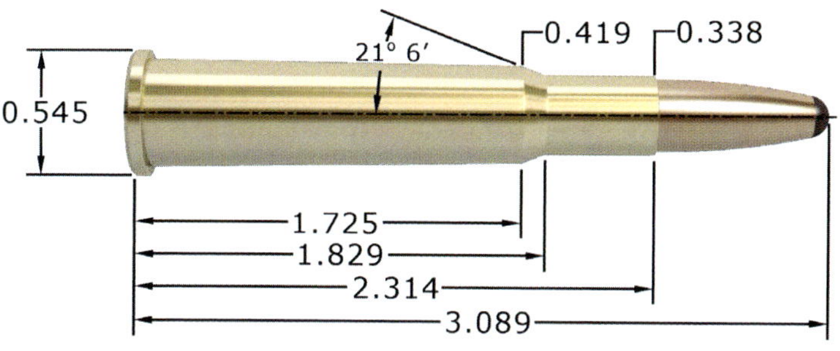

Maximum S.A.A.M.I. Overall Cartridge Length: 3.089″

BULLET CHOICES FOR THE 30/40 KRAG

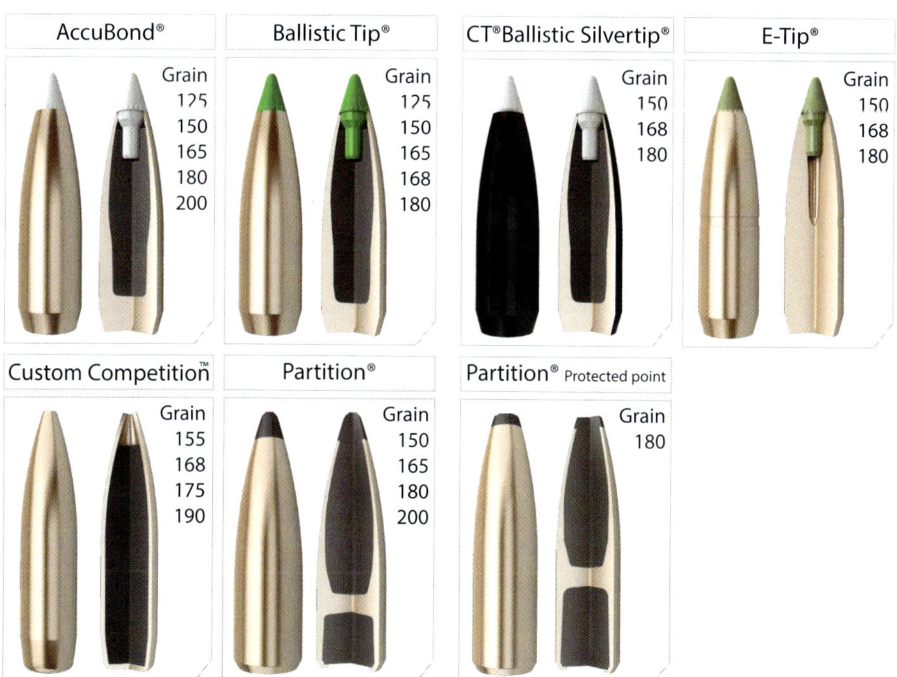

TECHNICAL INFORMATION

Adopted by the US Army in 1892 to replace the 45-70, the "Krag" has served both military and civilian needs for over a century. Most military chambers had long throats to accommodate the long, 220-grain round-nose service bullet. Because of this, modern bullets lighter than 180 grains may not be as accurate as heavier choices. Additionally, some of the older military rifles can have large groove diameter barrels. If you are unsure of the size of your particular bore, have it checked by a competent gunsmith.

30-40 Krag - 150/155 grain		MAXIMUM S.A.A.M.I. O.A.C.L.		3.089"
			B.C.	S.D.
AccuBond®	150gr. Spitzer		0.435	0.226
Ballistic Tip®	150gr. Spitzer		0.435	0.226
CT® Ballistic Silvertip®	150gr. Spitzer		0.435	0.226
E-Tip®	150gr. Spitzer		0.469	0.226
Due to internal construction differences, always begin with starting loads when using E-Tip® products.				
Partition®	150gr. Spitzer		0.387	0.226
Custom Competition™	155gr. HPBT		0.450	0.226

CASE TYPE:	Remington	PRIMER TYPE	CCI 200
		BARREL Length/Make	30" Spfld M1898
		BARREL Twist	1-10"

POWDER TYPE	POWDER CHG. GRS.		MUZZLE VEL. F.P.S.		LOAD DENSITY (VOLUME)
IMR 4895	40.0	MAX.	2400		80%
Most Accurate	38.0		2300		76%
Powder Tested	36.0 *		2200		72%
IMR 3031	40.0	MAX.	2520		80%
	38.0		2410		76%
	36.0 *		2300		72%
IMR 4831	51.0 *	MAX.	2580	**	102%
	49.0		2450		98%
	47.0		2320		94%
IMR 4350	50.0 *	MAX.	2710		100%
	48.0		2580		96%
	46.0		2450		92%

BC=Ballistic Coefficient SD=Sectional Density
*Most Accurate Load Tested **Compressed Load

Use Maximum Loads with Caution
Refer to page 73 for additional safety information

30-40 Krag - 165/168 grain		MAXIMUM S.A.A.M.I. O.A.C.L.	3.089"
		B.C.	S.D.
AccuBond®	165gr. Spitzer	0.475	0.248
Ballistic Tip®	165gr. Spitzer	0.475	0.248
Partition®	165gr. Spitzer	0.410	0.248
Ballistic Tip®	168gr. Spitzer	0.490	0.253
CT® Ballistic Silvertip®	168gr. Spitzer	0.490	0.253
Custom Competition™	168gr. HPBT	0.462	0.253
E-Tip®	168gr. Spitzer	0.503	0.253

Due to internal construction differences, always begin with starting loads when using E-Tip® products.

CASE TYPE:	Remington	PRIMER TYPE	CCI 200
		BARREL Length/Make	30" Spfld M1898
		BARREL Twist	1-10"

POWDER TYPE	POWDER CHG. GRS.	MUZZLE VEL. F.P.S.	LOAD DENSITY (VOLUME)
IMR 4895 Most Accurate Powder Tested	38.0 * MAX.	2258	79%
	36.0	2163	75%
	34.0	2068	71%
IMR 3031	38.0 MAX.	2350	79%
	36.0	2240	75%
	34.0 *	2130	71%
IMR 4831	49.0 * MAX.	2430	100%
	47.0	2300	96%
	45.0	2170	91%
IMR 4350	48.0 MAX.	2570 **	102%
	46.0	2430	98%
	44.0 *	2290	94%

BC=Ballistic Coefficient SD=Sectional Density
*Most Accurate Load Tested **Compressed Load

Use Maximum Loads with Caution
Refer to page 73 for additional safety information

30-40 Krag - 175/180 grain		MAXIMUM S.A.A.M.I. O.A.C.L.		3.089"
			B.C.	S.D.
Custom Competition™	175gr. HPBT		0.505	0.264
AccuBond®	180gr. Spitzer		0.507	0.271
Ballistic Tip®	180gr. Spitzer		0.507	0.271
CT® Ballistic Silvertip®	180gr. Spitzer		0.507	0.271
E-Tip®	180gr. Spitzer		0.523	0.271
Due to internal construction differences, always begin with starting loads when using E-Tip® products.				
Partition®	180gr. Spitzer		0.474	0.271
Partition®	180gr. PPT		0.361	0.271

CASE TYPE:	Remington	PRIMER TYPE	CCI 200
		BARREL Length/Make	30" Spfld M1898
		BARREL Twist	1-10"

POWDER TYPE	POWDER CHG. GRS.		MUZZLE VEL. F.P.S.	LOAD DENSITY (VOLUME)
IMR 3031	36.0	MAX.	2200	76%
	34.0		2100	72%
	32.0 *		2000	68%
IMR 4895	37.0 *	MAX.	2200	78%
	35.0		2090	74%
	33.0		1980	70%
IMR 4831	47.0	MAX.	2370	99%
	45.0		2230	95%
	43.0 *		2090	91%
IMR 4350 Most Accurate Powder Tested	46.0	MAX.	2460	97%
	44.0		2330	93%
	42.0 *		2190	89%

BC=Ballistic Coefficient SD=Sectional Density
*Most Accurate Load Tested **Compressed Load

Use Maximum Loads with Caution
Refer to page 73 for additional safety information

30-40 Krag - 200 grain

		MAXIMUM S.A.A.M.I. O.A.C.L.	3.089
		B.C.	S.D.
AccuBond®	200gr. Spitzer	0.588	0.301
Partition®	200gr. Spitzer	0.481	0.301

CASE TYPE:	Remington	PRIMER TYPE	CCI
		BARREL Length/Make	30" Spfld M1898
		BARREL Twist	1-10"

POWDER TYPE	POWDER CHG. GRS.		MUZZLE VEL. F.P.S.	LOAD DENSITY (VOLUME)
IMR 3031	34.0	MAX.	2080	72%
	32.0		2000	68%
	30.0 *		1920	64%
IMR 4895	36.0	MAX.	2152	76%
	34.0		2037	72%
	32.0 *		1922	68%
IMR 4831	45.0 *	MAX.	2238	95%
	43.0		2123	91%
	41.0		2008	87%
IMR 4350 Most Accurate Powder Tested	44.0 *	MAX.	2348	93%
	42.0		2233	89%
	40.0		2118	86%

BC=Ballistic Coefficient SD=Sectional Density
*Most Accurate Load Tested **Compressed Load

Use Maximum Loads with Caution
Refer to page 73 for additional safety information

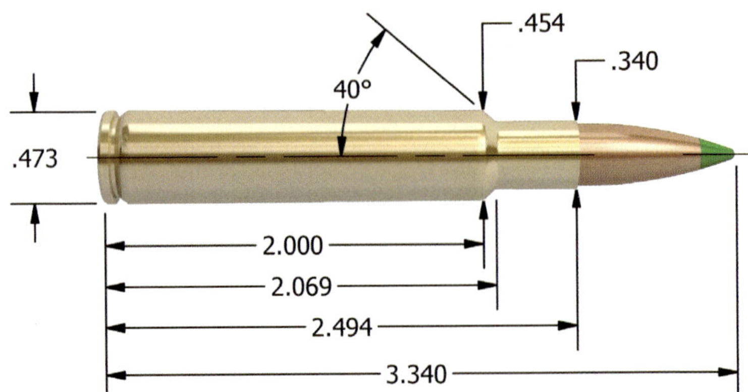

Maximum Overall Cartridge Length: 3.340"

BULLET CHOICES FOR THE 30-06 SPRINGFIELD ACKLEY IMPROVED

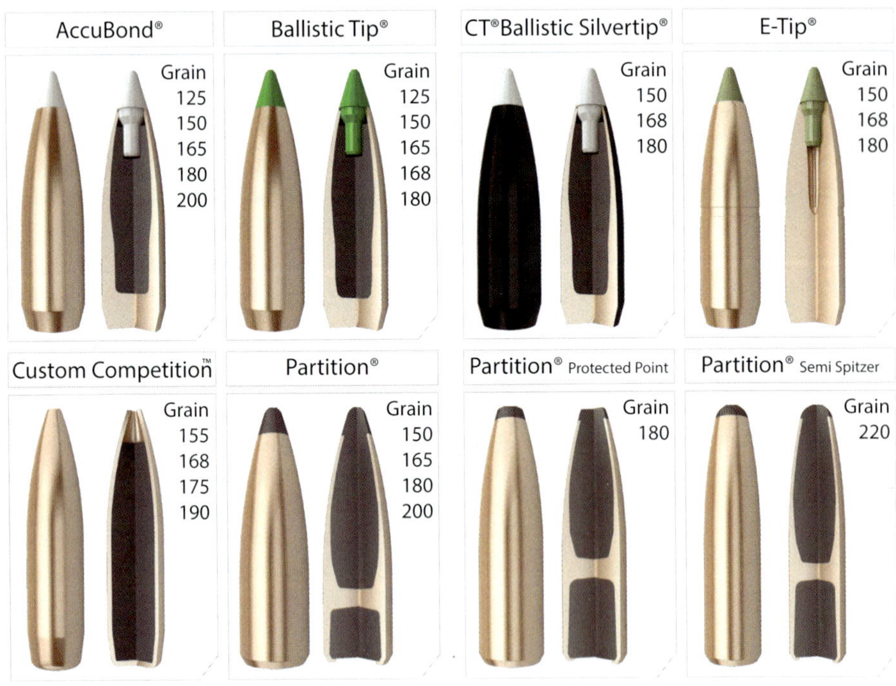

TECHNICAL INFORMATION

This load data is intended for use only with firearms chambered for the .30-06 Springfield Ackley Improved (40-degree shoulder angle).

30-06 Springfield A.I. - 125 grain

30-06 Springfield A.I. - 125 grain		MAXIMUM O.A.C.L.		3.340"
			B.C.	S.D.
AccuBond®	125gr. Spitzer		0.366	0.188
Ballistic Tip®	125gr. Spitzer		0.366	0.188

CASE TYPE:	Nosler		PRIMER TYPE	WLR
CASE HOLDS:	69.1	Gr. WATER	BARREL Length/Make	24" Lilja
			BARREL Twist	1-10"

POWDER TYPE	POWDER CHG. GRS.		MUZZLE VEL. F.P.S.		LOAD DENSITY (VOLUME)
H414	62.5	MAX.	3215		94%
	60.5		3127		91%
	58.5 *		3038		88%
IMR 4350	64.0	MAX.	3279		98%
	62.0		3161		95%
	60.0 *		3042		92%
RL15	55.5	MAX.	3286		84%
	53.5		3179		81%
	51.5 *		3072		78%
W760 Most Accurate Powder Tested	63.0 *	MAX.	3309		95%
	61.0		3198		92%
	59.0		3087		89%
RL19	67.0	MAX.	3316	**	105%
	65.0		3219	**	102%
	63.0 *		3121		99%
Viht N150	58.0	MAX.	3316		99%
	56.0		3229		95%
	54.0 *		3141		92%

BC=Ballistic Coefficient SD=Sectional Density
*Most Accurate Load Tested **Compressed Load

Use Maximum Loads with Caution
Refer to page 73 for additional safety information

30-06 Springfield A.I. - 150/155 grain		MAXIMUM O.A.C.L.	3.340"
		B.C.	S.D.
AccuBond®	150gr. Spitzer	0.435	0.226
Ballistic Tip®	150gr. Spitzer	0.435	0.226
CT® Ballistic Silvertip®	150gr. Spitzer	0.435	0.226
E-Tip®	150gr. Spitzer	0.469	0.226
Due to internal construction differences, always begin with starting loads when using E-Tip® products.			
Partition®	150gr. Spitzer	0.387	0.226
Custom Competition™	155gr. HPBT	0.450	0.233

CASE TYPE:	Nosler		PRIMER TYPE	WLR
CASE HOLDS:	66.0	Gr. WATER	BARREL Length/Make	24" Lilja
			BARREL Twist	1-10"

POWDER TYPE	POWDER CHG. GRS.	MUZZLE VEL. F.P.S.	LOAD DENSITY (VOLUME)
H414	60.0 * MAX.	3001	95%
	58.0	2905	92%
	56.0	2810	88%
RL15	53.5 MAX.	3036	84%
	51.5	2934	81%
	49.5 *	2832	78%
Viht N150	55.0 MAX.	3038	98%
	53.0	2920	94%
	51.0 *	2802	91%
RL19 Most Accurate Powder Tested	63.0 * MAX.	3058	** 104%
	61.0	2947	100%
	59.0	2835	97%
IMR 4350	61.5 * MAX.	3071	99%
	59.5	2963	95%
	57.5	2854	92%
W760	60.0 * MAX.	3075	95%
	58.0	2968	92%
	56.0	2861	89%

BC=Ballistic Coefficient SD=Sectional Density
*Most Accurate Load Tested **Compressed Load

Use Maximum Loads with Caution
Refer to page 73 for additional safety information

30-06 Springfield A.I. - 165/168 grain		MAXIMUM O.A.C.L.		3.340"
			B.C.	S.D.
AccuBond®	165gr. Spitzer		0.475	0.248
Ballistic Tip®	165gr. Spitzer		0.475	0.248
Partition®	165gr. Spitzer		0.410	0.248
Ballistic Tip®	168gr. Spitzer		0.490	0.253
CT® Ballistic Silvertip®	168gr. Spitzer		0.490	0.253
Custom Competition™	168gr. HPBT		0.462	0.253
E-Tip®	168gr. Spitzer		0.503	0.253

Due to internal construction differences, always begin with starting loads when using E-Tip® products.

CASE TYPE:	Nosler		PRIMER TYPE	WLR
CASE HOLDS:	65.4	Gr. WATER	BARREL Length/Make	24" Lilja
			BARREL Twist	1-10"

POWDER TYPE	POWDER CHG. GRS.		MUZZLE VEL. F.P.S.		LOAD DENSITY (VOLUME)
H414	58.0 *	MAX.	2881		92%
	56.0		2796		89%
	54.0		2711		86%
Viht N160	60.0	MAX.	2912	**	102%
	58.0		2817		99%
	56.0 *		2722		95%
H4831 Most Accurate Powder Tested	62.5 *	MAX.	2918		100%
	60.5		2824		96%
	58.5		2730		93%
RL19	60.5	MAX.	2936	**	101%
	58.5		2823		97%
	56.5 *		2710		94%
IMR 4350	59.0	MAX.	2940		95%
	57.0		2838		92%
	55.0 *		2736		89%

BC=Ballistic Coefficient SD=Sectional Density
*Most Accurate Load Tested **Compressed Load

Use Maximum Loads with Caution
Refer to page 73 for additional safety information

30-06 Springfield A.I. - 175/180 grain		MAXIMUM O.A.C.L.	3.340"
		B.C.	S.D.
Custom Competition™	175gr. HPBT	0.505	0.264
AccuBond®	180gr. Spitzer	0.507	0.271
Ballistic Tip®	180gr. Spitzer	0.507	0.271
CT® Ballistic Silvertip®	180gr. Spitzer	0.507	0.271
E-Tip®	180gr. Spitzer	0.523	0.271
Due to internal construction differences, always begin with starting loads when using E-Tip® products.			
Partition®	180gr. Spitzer	0.474	0.271
Partition®	180gr. PPT	0.361	0.271

CASE TYPE:	Nosler		PRIMER TYPE	WLR
CASE HOLDS:	64.1	Gr. WATER	BARREL Length/Make	24" Lilja
			BARREL Twist	1-10"

POWDER TYPE	POWDER CHG. GRS.		MUZZLE VEL. F.P.S.	LOAD DENSITY (VOLUME)
H414	57.5	MAX.	2787	93%
	55.5		2705	90%
	53.5 *		2622	87%
H4831	60.0 *	MAX.	2798	98%
	58.0		2708	94%
	56.0		2618	91%
Viht N165	63.0	MAX.	2812	** 109%
	61.0		2723	** 106%
	59.0 *		2633	** 102%
IMR 4350	56.5	MAX.	2835	93%
	54.5		2719	90%
	52.5 *		2602	87%
RL22	62.0	MAX.	2985	** 105%
Most Accurate Powder Tested	60.0		2881	** 102%
	58.0 *		2779	98%

BC=Ballistic Coefficient SD=Sectional Density
*Most Accurate Load Tested **Compressed Load

Use Maximum Loads with Caution
Refer to page 73 for additional safety information

30-06 Springfield A.I. - 200 grain		MAXIMUM O.A.C.L.		3.340"
			B.C.	S.D.
AccuBond®	200gr. Spitzer		0.588	0.301
Partition®	200gr. Spitzer		0.481	0.301

CASE TYPE:	Nosler		PRIMER TYPE	WLR
CASE HOLDS:	63.0	Gr. WATER	BARREL Length/Make	24" Lilja
			BARREL Twist	1-10"

POWDER TYPE	POWDER CHG. GRS.		MUZZLE VEL. F.P.S.		LOAD DENSITY (VOLUME)
IMR 4350	55.5	* MAX.	2651		93%
	53.5		2550		90%
	51.5		2448		87%
H4831	59.0	* MAX.	2656		98%
	57.0		2571		94%
	55.0		2487		91%
Viht N165	60.5	* MAX.	2673	**	107%
	58.5		2591	**	103%
	56.5		2509		100%
RL19 Most Accurate Powder Tested	57.5	MAX.	2737		99%
	55.5		2649		96%
	53.5	*	2562		92%
RL22	59.5	* MAX.	2768	**	103%
	57.5		2670		99%
	55.5		2572		96%

BC=Ballistic Coefficient SD=Sectional Density
*Most Accurate Load Tested **Compressed Load

Use Maximum Loads with Caution
Refer to page 73 for additional safety information

30-06 Springfield A.I. - 220 grain		MAXIMUM O.A.C.L.	3.340"
		B.C.	S.D.
Partition®	220gr. Semi Spitz	0.351	0.331

CASE TYPE:	Nosler		PRIMER TYPE	WLR
CASE HOLDS:	62.6	Gr. WATER	BARREL Length/Make	24" Lilja
			BARREL Twist	1-10"

POWDER TYPE	POWDER CHG. GRS.	MUZZLE VEL. F.P.S.	LOAD DENSITY (VOLUME)
Viht N165	58.0 * MAX.	2509	** 103%
	56.0	2428	99%
	54.0	2347	96%
H4831 Most Accurate Powder Tested	57.5 * MAX.	2541	96%
	55.5	2464	92%
	53.5	2380	89%
IMR 4350	53.0 MAX.	2575	90%
	51.0	2481	86%
	49.0 *	2388	83%
RL22	58.5 * MAX.	2670	** 102%
	56.5	2598	98%
	54.5	2528	95%

BC=Ballistic Coefficient SD=Sectional Density
*Most Accurate Load Tested **Compressed Load

Use Maximum Loads with Caution
Refer to page 73 for additional safety information

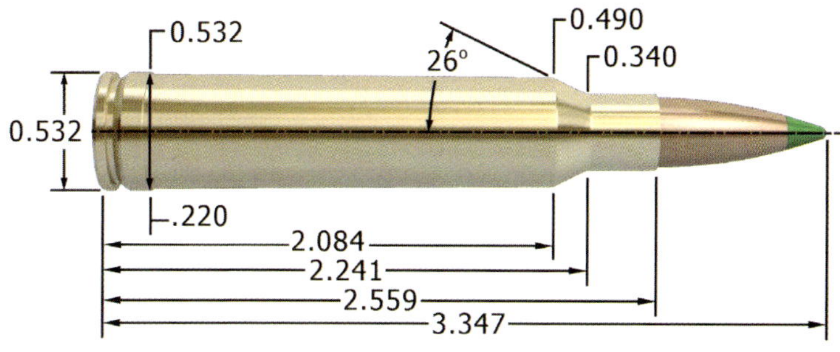

Maximum Overall Cartridge Length: 3.347"

BULLET CHOICES FOR THE 308 NORMA MAGNUM

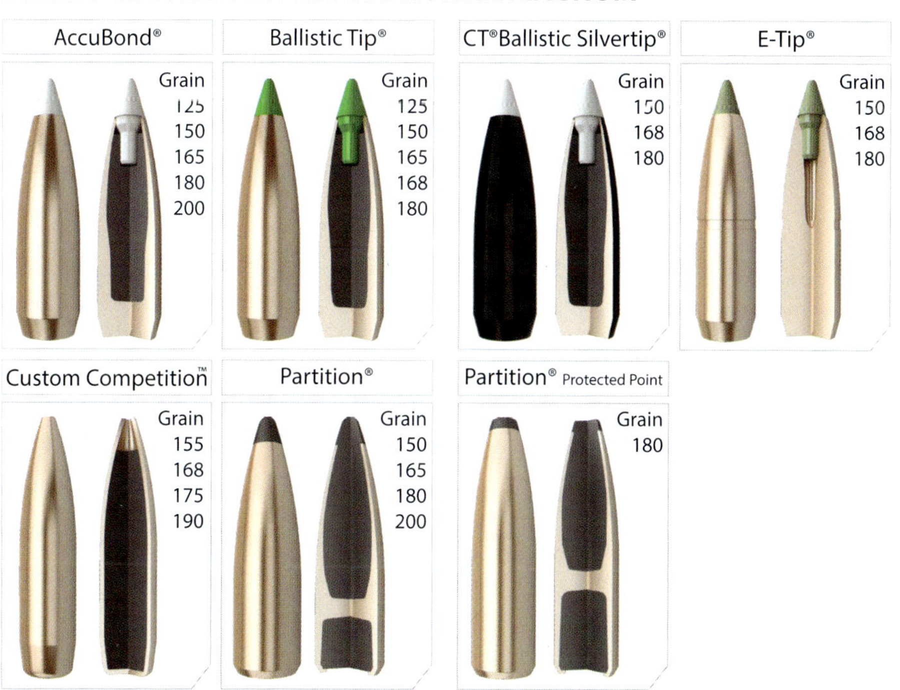

308 Norma Magnum - 125 grain		MAXIMUM O.A.C.L.		3.347"
			B.C.	S.D.
AccuBond®	125gr. Spitzer		0.366	0.188
Ballistic Tip®	125gr. Spitzer		0.366	0.188

CASE TYPE:	Norma		PRIMER TYPE	Fed 215
CASE HOLDS:	84.0	Gr. WATER	BARREL Length/Make	24" Wiseman
				1-10"

POWDER TYPE	POWDER CHG. GRS.		MUZZLE VEL. F.P.S.	LOAD DENSITY (VOLUME)
Viht N150	64.0 *	MAX.	3335	76%
	62.0		3264	74%
	60.0		3193	71%
IMR 4320	66.0	MAX.	3350	79%
	64.0		3220	76%
	62.0 *		3090	74%
IMR 4350	75.0	MAX.	3352	89%
	73.0		3247	87%
	71.0 *		3142	85%
Viht N160	75.5	MAX.	3426	90%
Most Accurate	73.5		3331	88%
Powder Tested	71.5 *		3237	85%
H414	71.5 *	MAX.	3430	85%
	69.5		3340	83%
	67.5		3250	80%
RL15	65.5	MAX.	3457	78%
	63.5		3376	76%
	61.5 *		3295	73%
H4831	79.0	MAX.	3532	94%
	77.0		3447	92%
	75.0 *		3362	89%

BC=Ballistic Coefficient SD=Sectional Density
*Most Accurate Load Tested **Compressed Load

Use Maximum Loads with Caution
Refer to page 73 for additional safety information

308 Norma Magnum - 150/155 grain		MAXIMUM O.A.C.L.		3.347"
			B.C.	S.D.
AccuBond®	150gr. Spitzer		0.435	0.226
Ballistic Tip®	150gr. Spitzer		0.435	0.226
CT® Ballistic Silvertip®	150gr. Spitzer		0.435	0.226
E-Tip®	150gr. Spitzer		0.469	0.226
Due to internal construction differences, always begin with starting loads when using E-Tip® products.				
Partition®	150gr. Spitzer		0.387	0.226
Custom Competition™	155gr. HPBT		0.450	0.233

CASE TYPE:	Norma		PRIMER TYPE	Fed 215
CASE HOLDS:	81.0	Gr. WATER	BARREL Length/Make	24" Wiseman
				1-10"

POWDER TYPE	POWDER CHG. GRS.		MUZZLE VEL. F.P.S.	LOAD DENSITY (VOLUME)
H414	69.0	MAX.	3040	85%
	67.0		2950	83%
	65.0 *		2860	80%
Viht N150	62.0	MAX.	3119	77%
	60.0		3049	74%
	58.0 *		2987	72%
IMR 4895	60.0	MAX.	3170	74%
	58.0		3020	72%
	56.0 *		2870	69%
RL15	63.0 *	MAX.	3206	78%
	61.0		3127	75%
	59.0		3048	73%
IMR 4831	74.0 *	MAX.	3208	91%
	72.0		3103	89%
	70.0		2998	86%
IMR 4064	62.0	MAX.	3210	77%
	60.0		3090	74%
	58.0 *		2970	72%
IMR 4350	72.0 *	MAX.	3232	89%
Most Accurate	70.0		3107	86%
Powder Tested	68.0		2982	84%

BC=Ballistic Coefficient SD=Sectional Density
*Most Accurate Load Tested **Compressed Load

Use Maximum Loads with Caution
Refer to page 73 for additional safety information

308 Norma Magnum - 165/168 grain		MAXIMUM O.A.C.L.		3.347"
			B.C.	S.D.
AccuBond®	165gr. Spitzer		0.475	0.248
Ballistic Tip®	165gr. Spitzer		0.475	0.248
Partition®	165gr. Spitzer		0.410	0.248
Ballistic Tip®	168gr. Spitzer		0.490	0.253
CT® Ballistic Silvertip®	168gr. Spitzer		0.490	0.253
Custom Competition™	168gr. HPBT		0.462	0.253
E-Tip®	168gr. Spitzer		0.503	0.253

Due to internal construction differences, always begin with starting loads when using E-Tip® products.

CASE TYPE:	Norma		PRIMER TYPE	Fed 215
CASE HOLDS:	73.0	Gr. WATER	BARREL Length/Make	24" Wiseman
				1-10"

POWDER TYPE	POWDER CHG. GRS.		MUZZLE VEL. F.P.S.	LOAD DENSITY (VOLUME)
IMR 4320	60.0 *	MAX.	2920	82%
	58.0		2870	79%
	56.0		2820	77%
H4831 Most Accurate Powder Tested	73.5 *	MAX.	3042	** 101%
	71.5		2947	98%
	69.5		2852	95%
IMR 4831	72.0 *	MAX.	3080	99%
	70.0		3000	96%
	68.0		2920	93%
IMR 4064	61.0	MAX.	3082	84%
	59.0		2987	81%
	57.0 *		2892	78%
IMR 4350	70.0 *	MAX.	3088	96%
	68.0		2963	93%
	66.0		2838	90%
Viht N160	70.0 *	MAX.	3089	96%
	68.0		3008	93%
	66.0		2928	90%
RL19	72.0	MAX.	3153	99%
	70.0		3056	96%
	68.0 *		2960	93%

BC=Ballistic Coefficient SD=Sectional Density
*Most Accurate Load Tested **Compressed Load

Use Maximum Loads with Caution
Refer to page 73 for additional safety information

308 Norma Magnum - 180 grain		MAXIMUM O.A.C.L.		3.347"
			B.C.	S.D.
AccuBond®	180gr. Spitzer		0.507	0.271
Ballistic Tip®	180gr. Spitzer		0.507	0.271
CT® Ballistic Silvertip®	180gr. Spitzer		0.507	0.271
E-Tip®	180gr. Spitzer		0.523	0.271
Due to internal construction differences, always begin with starting loads when using E-Tip® products.				
Partition®	180gr. Spitzer		0.474	0.271
Partition®	180gr. PPT		0.361	0.271

CASE TYPE:	Norma		PRIMER TYPE	Fed 215
CASE HOLDS:	69.4	Gr. WATER	BARREL Length/Make	24" Wiseman
				1-10"

POWDER TYPE	POWDER CHG. GRS.		MUZZLE VEL. F.P.S.		LOAD DENSITY (VOLUME)
IMR 7828	72.5	MAX.	2860	**	104%
Most Accurate	70.5		2770	**	102%
Powder Tested	68.5 *		2680		99%
IMR 4350	68.0	MAX.	2900		98%
	66.0		2820		95%
	64.0 *		2740		92%
Viht N160	67.0	MAX.	2906		97%
	65.0		2830		94%
	63.0 *		2753		91%
IMR 4895	58.0	MAX.	2932		84%
	56.0		2787		81%
	54.0 *		2642		78%
RL19	69.0 *	MAX.	2956		99%
	67.0		2865		97%
	65.0		2774		94%
IMR 4831	71.0 *	MAX.	2980	**	102%
	69.0		2880		99%
	67.0		2780		97%

BC=Ballistic Coefficient SD=Sectional Density
*Most Accurate Load Tested **Compressed Load

Use Maximum Loads with Caution
Refer to page 73 for additional safety information

308 Norma Magnum - 200 grain		MAXIMUM O.A.C.L.		3.347"
			B.C.	S.D.
AccuBond®	200gr. Spitzer		0.588	0.301
Partition®	200gr. Spitzer		0.481	0.301

CASE TYPE:	Norma		PRIMER TYPE	Fed 215
CASE HOLDS:	66.7	Gr. WATER	BARREL Length/Make	24" Wiseman
				1-10"

POWDER TYPE	POWDER CHG. GRS.		MUZZLE VEL. F.P.S.	LOAD DENSITY (VOLUME)
IMR 4895	54.0 *	MAX.	2610	81%
	52.0		2550	78%
	50.0		2490	75%
IMR 4320	56.0	MAX.	2610	84%
	54.0		2550	81%
	52.0 *		2490	78%
IMR 4350	66.0 *	MAX.	2740	99%
	64.0		2640	96%
	62.0		2540	93%
Viht N165 Most Accurate Powder Tested	67.0	MAX.	2747	100%
	65.0		2678	97%
	63.0 *		2608	94%
IMR 4831	69.0 *	MAX.	2760	** 103%
	67.0		2690	100%
	65.0		2620	97%
H4831	68.0 *	MAX.	2820	** 102%
	66.0		2760	99%
	64.0		2690	96%
RL22	68.5	MAX.	2850	** 103%
	66.5		2768	100%
	64.5 *		2685	97%

BC=Ballistic Coefficient SD=Sectional Density
*Most Accurate Load Tested **Compressed Load

Use Maximum Loads with Caution
Refer to page 73 for additional safety information

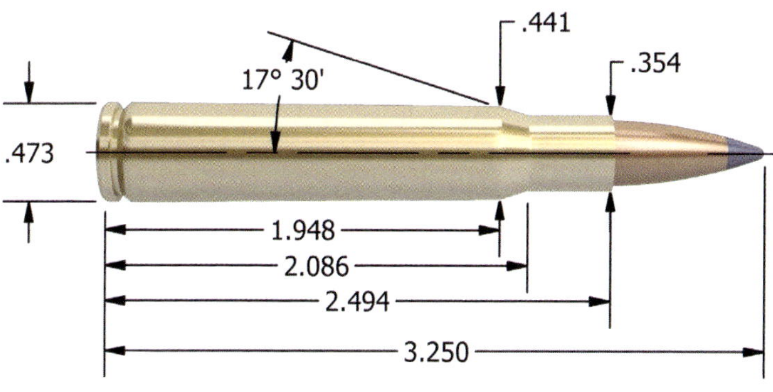

Maximum Overall Cartridge Length: 3.250"

BULLET CHOICES FOR THE 8MM-06

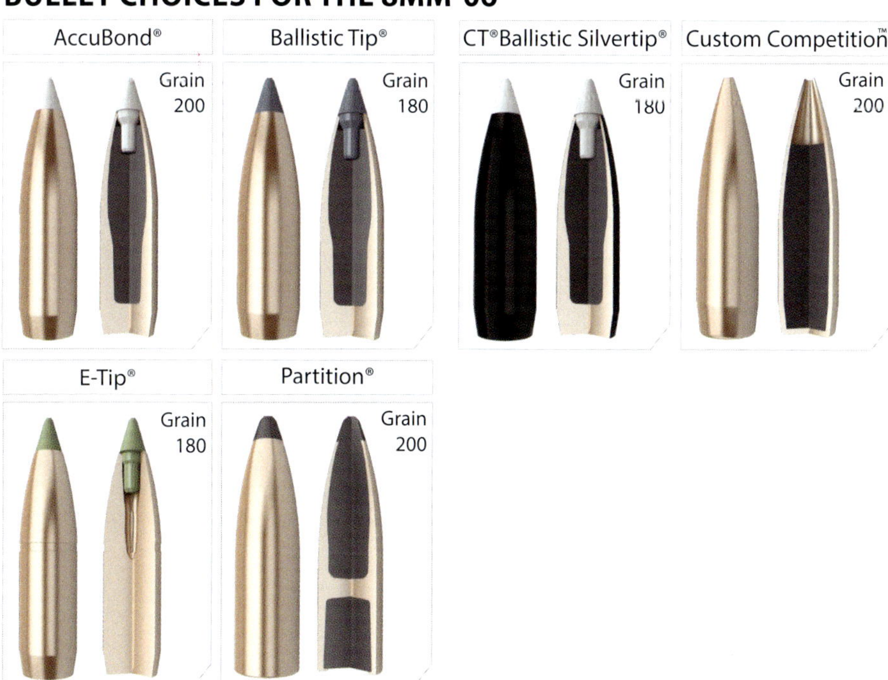

TECHNICAL INFORMATION

Our technicians found exceptional performance with the 180 grain Ballistic Tip and IMR 4350. Sizing standard .30-06 brass over a tapered expander ball easily forms cases for the 8mm-06. No fire-forming is required.

8mm-06 - 180 grain		MAXIMUM O.A.C.L.		3.250"
			B.C.	S.D.
Ballistic Tip®	180gr. Spitzer		0.394	0.246
CT® Ballistic Silvertip®	180gr. Spitzer		0.394	0.246
E-Tip®	180gr. Spitzer		0.427	0.246

Due to internal construction differences, always begin with starting loads when using E-Tip® products.

CASE TYPE:	Winchester (30-06)		PRIMER TYPE	Fed 210M
CASE HOLDS:	60.1	Gr. WATER	BARREL Length/Make	24" Wiseman
			BARREL Twist	1-10"

POWDER TYPE	POWDER CHG. GRS.		MUZZLE VEL. F.P.S.	LOAD DENSITY (VOLUME)
RL15	48.0 *	MAX.	2684	83%
	46.0		2575	80%
	44.0		2468	76%
IMR 4350 Most Accurate Powder Tested	56.0	MAX.	2704	99%
	54.0		2636	95%
	52.0 *		2525	92%
H380	56.0	MAX.	2763	98%
	54.0		2698	95%
	52.0 *		2608	91%

BC=Ballistic Coefficient SD=Sectional Density
*Most Accurate Load Tested **Compressed Load

Use Maximum Loads with Caution
Refer to page 73 for additional safety information

8mm-06 - 200 grain		MAXIMUM O.A.C.L.	3.250"	
			B.C.	S.D.
AccuBond®	200gr. Spitzer		0.450	0.274
Custom Competition™	200gr. HPBT		0.520	0.274
Partition®	200gr. Spitzer		0.426	0.274

CASE TYPE:	Winchester (30-06)		PRIMER TYPE	Fed 210M
CASE HOLDS:	60.7	Gr. WATER	BARREL Length/Make	24" Wiseman
			BARREL Twist	1-10"

POWDER TYPE	POWDER CHG. GRS.		MUZZLE VEL. F.P.S.		LOAD DENSITY (VOLUME)
RL19 Most Accurate Powder Tested	56.5	* MAX.	2571	**	101%
	54.5		2469		98%
	52.5		2364		94%
H4831SC	59.0	* MAX.	2602	**	101%
	57.0		2518		98%
	55.0		2414		94%
IMR 4350	55.5	MAX.	2628		97%
	53.5	*	2502		93%
	51.5		2426		90%

BC=Ballistic Coefficient SD=Sectional Density
*Most Accurate Load Tested **Compressed Load

Use Maximum Loads with Caution
Refer to page 73 for additional safety information

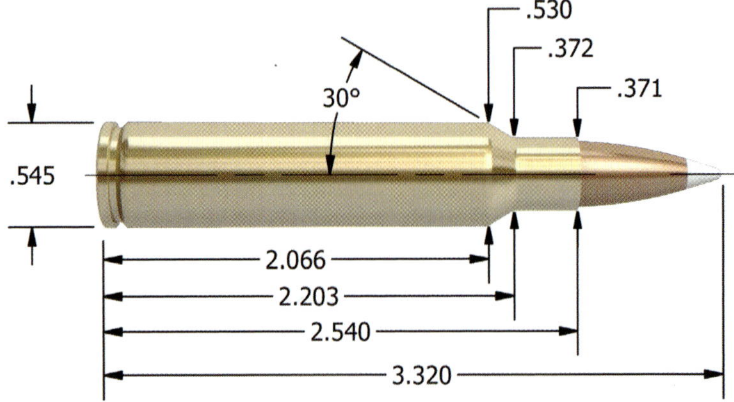

Maximum Overall Cartridge Length: 3.320"

BULLET CHOICES FOR THE 330 DAKOTA

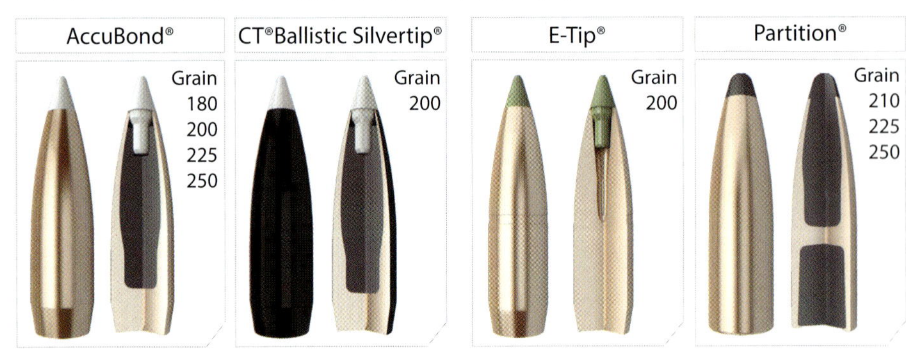

TECHNICAL INFORMATION

We observed best performance with standard powders like 4350 and 4831.

330 Dakota - 180 grain				MAXIMUM O.A.C.L.		3.320"
					B.C.	S.D.
AccuBond®			180gr. Spitzer		0.372	0.225
CASE TYPE:	Dakota		PRIMER TYPE		Fed 215	
CASE HOLDS:	84.8	Gr. WATER	BARREL Length/Make		26" Lilja	
			BARREL Twist		1-10"	

POWDER TYPE	POWDER CHG. GRS.		MUZZLE VEL. F.P.S.		LOAD DENSITY (VOLUME)
RL19	85.5	MAX.	3168	**	110%
	83.5		3110	**	107%
	81.5 *		3026	**	104%
Varget	73.5 *	MAX.	3181		94%
	71.5		3072		91%
	69.5		2992		89%
H4831SC Most Accurate Powder Tested	86.0	MAX.	3192		106%
	84.0 *		3131	**	103%
	82.0		3037	**	101%
IMR 4350	81.5	MAX.	3205		102%
	79.5 *		3165		99%
	77.5		3030		97%

BC=Ballistic Coefficient SD=Sectional Density
*Most Accurate Load Tested **Compressed Load

Use Maximum Loads with Caution
Refer to page 73 for additional safety information

330 Dakota - 200/210 grain		MAXIMUM O.A.C.L.		3.320"
			B.C.	S.D.
AccuBond®	200gr. Spitzer		0.414	0.250
CT® Ballistic Silvertip®	200gr. Spitzer		0.414	0.250
E-Tip®	200gr. Spitzer		0.425	0.250
Due to internal construction differences, always begin with starting loads when using E-Tip® products.				
Partition®	210gr. Spitzer		0.400	0.263

CASE TYPE:	Dakota		PRIMER TYPE	Fed 215
CASE HOLDS:	86.0	Gr. WATER	BARREL Length/Make	26" Lilja
			BARREL Twist	1-10"

POWDER TYPE	POWDER CHG. GRS.		MUZZLE VEL. F.P.S.		LOAD DENSITY (VOLUME)
RL19	83.0	MAX.	2948	**	105%
	81.0		2891	**	102%
	79.0 *		2786		100%
IMR 4350	78.5 *	MAX.	2975		97%
	76.5		2874		94%
	74.5		2803		92%
Viht N160 Most Accurate Powder Tested	81.0	MAX.	2989		105%
	79.0 *		2914	**	102%
	77.0		2832		99%
H4831SC	84.0	MAX.	2989		102%
	82.0		2959		99%
	80.0 *		2901		97%

BC=Ballistic Coefficient SD=Sectional Density
*Most Accurate Load Tested **Compressed Load

Use Maximum Loads with Caution
Refer to page 73 for additional safety information

330 Dakota - 225 grain		MAXIMUM O.A.C.L.		3.320"
			B.C.	S.D.
AccuBond®	225gr. Spitzer		0.550	0.281
Partition®	225gr. Spitzer		0.454	0.281

CASE TYPE:	Dakota		PRIMER TYPE	Fed 215
CASE HOLDS:	85.7	Gr. WATER	BARREL Length/Make	26" Lilja
			BARREL Twist	1-10"

POWDER TYPE	POWDER CHG. GRS.		MUZZLE VEL. F.P.S.		LOAD DENSITY (VOLUME)
H4831SC	79.0	MAX.	2842		96%
	77.0 *		2786		94%
	75.0		2715		91%
Viht N160	77.5	MAX.	2862		100%
	75.5 *		2798		98%
	73.5		2725		95%
IMR 4350 Most Accurate Powder Tested	76.5 *	MAX.	2887		94%
	74.5		2832		92%
	72.5		2769		90%
RL19	82.0	MAX.	2894	**	104%
	80.0		2875	**	101%
	78.0 *		2817		99%

BC=Ballistic Coefficient SD=Sectional Density
*Most Accurate Load Tested **Compressed Load

Use Maximum Loads with Caution
Refer to page 73 for additional safety information

330 Dakota - 250 grain		MAXIMUM O.A.C.L.		3.320"
			B.C.	S.D.
AccuBond®	250gr. Spitzer		0.575	0.313
Partition®	250gr. Spitzer		0.473	0.313

CASE TYPE:	Dakota		PRIMER TYPE	Fed 215
CASE HOLDS:	84.6	Gr. WATER	BARREL Length/Make	26" Lilja
			BARREL Twist	1-10"

POWDER TYPE	POWDER CHG. GRS.		MUZZLE VEL. F.P.S.		LOAD DENSITY (VOLUME)
IMR 4350	72.5	MAX.	2685		91%
Most Accurate Powder Tested	70.5		2625		88%
	68.5 *		2555		86%
Viht N160	75.0 *	MAX.	2688		99%
	73.0		2644		96%
	71.0		2576		93%
RL19	78.5 *	MAX.	2730	**	101%
	76.5		2682		98%
	74.5		2605		96%
H4831SC	78.0	MAX.	2736		96%
	76.0 *		2687		94%
	74.0		2619		91%

BC=Ballistic Coefficient SD=Sectional Density
*Most Accurate Load Tested **Compressed Load

Use Maximum Loads with Caution
Refer to page 73 for additional safety information

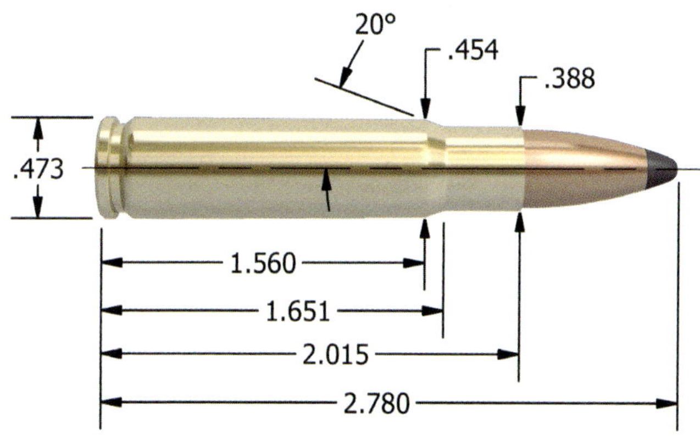

Maximum S.A.A.M.I. Overall Cartridge Length: 2.780"

BULLET CHOICES FOR THE 358 WINCHESTER

Partition®

TECHNICAL INFORMATION

Because of their sleek design, AccuBond bullets will not work in rifles with standard-length magazines. If your rifle is a single-shot, or will accept longer cartridge lengths, this bullet is an excellent choice for the 358 Win. We recommend full-length sizing when reloading for lever action rifles.

358 Winchester - 225 grain		MAXIMUM S.A.A.M.I. O.A.C.L.		2.780"
		TESTED O.A.C.L.	B.C.	S.D.
Partition®	225gr. Spitzer	2.780"	0.430	0.251

CASE TYPE:	Winchester		PRIMER TYPE	WLR
CASE HOLDS:	48.5	Gr. WATER	BARREL Length/Make	23" Lilja
			BARREL Twist	1-12"

POWDER TYPE	POWDER CHG. GRS.		MUZZLE VEL. F.P.S.		LOAD DENSITY (VOLUME)
H380	50.0	MAX.	2192	**	109%
	48.0		2107	**	104%
	46.0 *		2022		100%
RL7 Most Accurate Powder Tested	36.5 *	MAX.	2220		82%
	34.5		2100		77%
	32.5		1980		73%
IMR 4320	47.0	MAX.	2390	**	104%
	45.0		2290		100%
	43.0 *		2190		95%
Viht N135	45.0	MAX.	2401	**	108%
	43.0		2332	**	103%
	41.0 *		2262		98%
W748	48.5 *	MAX.	2426	**	104%
	46.5		2359		100%
	44.5		2292		96%
IMR 4895	49.5	MAX.	2528	**	112%
	47.5		2423	**	108%
	45.5 *		2318	**	103%

BC=Ballistic Coefficient SD=Sectional Density
*Most Accurate Load Tested **Compressed Load

Use Maximum Loads with Caution
Refer to page 73 for additional safety information

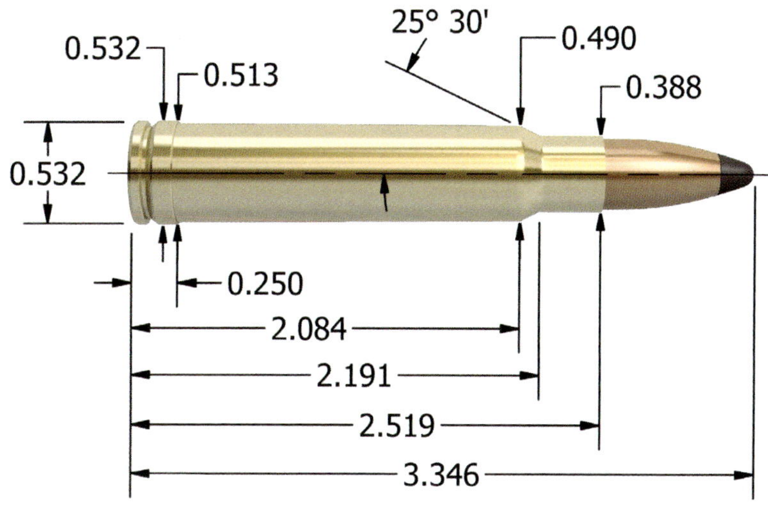

Maximum CIP Overall Cartridge Length: 3.346"

BULLET CHOICES FOR THE 358 NORMA MAGNUM

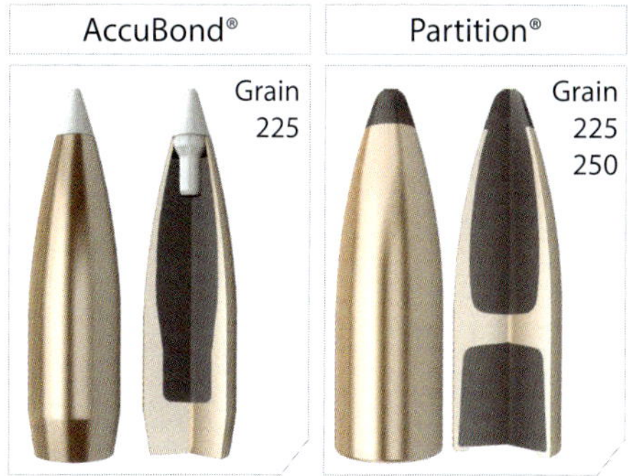

TECHNICAL INFORMATION

Falling just below the 375 H&H in performance, the 358 Norma Mag is an excellent choice for the larger species of North American game. Brass can be made from 338 Win Mag cases by necking them up to 35 caliber with a tapered expander ball. As with all belted magnums, brass life can be extended by neck sizing.

358 Norma Magnum - 225 grain		MAXIMUM CIP O.A.C.L.		3.346"
			B.C.	S.D.
AccuBond®	225gr. Spitzer		0.421	0.251
Partition®	225gr. Spitzer		0.430	0.251

CASE TYPE:	Norma		PRIMER TYPE	Fed 215
CASE HOLDS:	80.0	Gr. WATER	BARREL Length/Make	24" Lilja
			BARREL Twist	1-12"

POWDER TYPE	POWDER CHG. GRS.		MUZZLE VEL. F.P.S.	LOAD DENSITY (VOLUME)
A-2700	72.5	MAX.	2756	91%
	70.5		2689	88%
	68.5 *		2623	86%
H380	67.0	MAX.	2760	84%
	65.0		2680	81%
	63.0 *		2600	79%
IMR 4320	68.0	MAX.	2790	85%
	66.0		2710	83%
	64.0 *		2630	80%
IMR 4895	67.5 *	MAX.	2828	84%
	65.5		2743	82%
	63.5		2658	79%
Viht N160	79.5 *	MAX.	2857	99%
	77.5		2794	97%
	75.5		2729	94%
IMR 4350	75.0	MAX.	2878	94%
Most Accurate	73.0		2803	91%
Powder Tested	71.0 *		2728	89%

BC=Ballistic Coefficient SD=Sectional Density
*Most Accurate Load Tested **Compressed Load

Use Maximum Loads with Caution
Refer to page 73 for additional safety information

358 Norma Magnum - 250 grain

358 Norma Magnum - 250 grain		MAXIMUM CIP O.A.C.L.	3.346"
		B.C.	S.D.
Partition®	250gr. Spitzer	0.446	0.279

CASE TYPE:	Norma		PRIMER TYPE	Fed 215
CASE HOLDS:	78.0	Gr. WATER	BARREL Length/Make	24" Lilja
			BARREL Twist	1-12"

POWDER TYPE	POWDER CHG. GRS.		MUZZLE VEL. F.P.S.	LOAD DENSITY (VOLUME)
RL19	71.5	MAX.	2536	92%
	69.5		2464	89%
	67.5 *		2391	87%
IMR 4320	62.0	MAX.	2552	79%
	60.0		2474	77%
	58.0 *		2394	74%
H414	72.0 *	MAX.	2565	92%
	70.0		2510	90%
	68.0		2455	87%
RL15 Most Accurate Powder Tested	62.0 *	MAX.	2568	79%
	60.0		2486	77%
	58.0		2403	74%
IMR 4350	72.0	MAX.	2628	92%
	70.0		2556	90%
	68.0 *		2485	87%

BC=Ballistic Coefficient SD=Sectional Density
*Most Accurate Load Tested **Compressed Load

Use Maximum Loads with Caution
Refer to page 73 for additional safety information

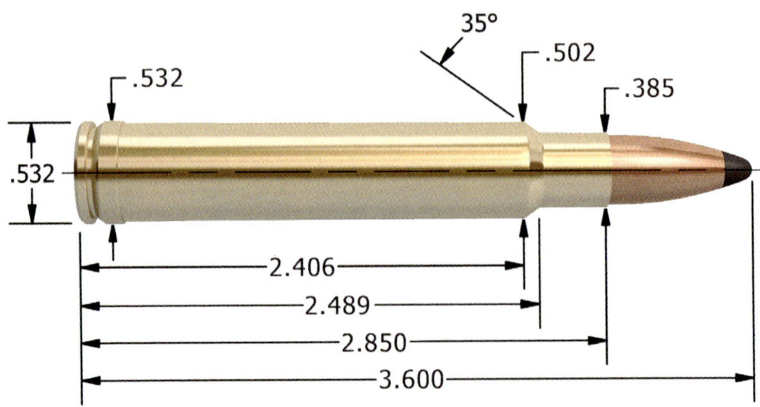

Maximum S.A.A.M.I. Overall Cartridge Length: 3.600"

BULLET CHOICES FOR THE 358 SHOOTING TIMES ALASKAN

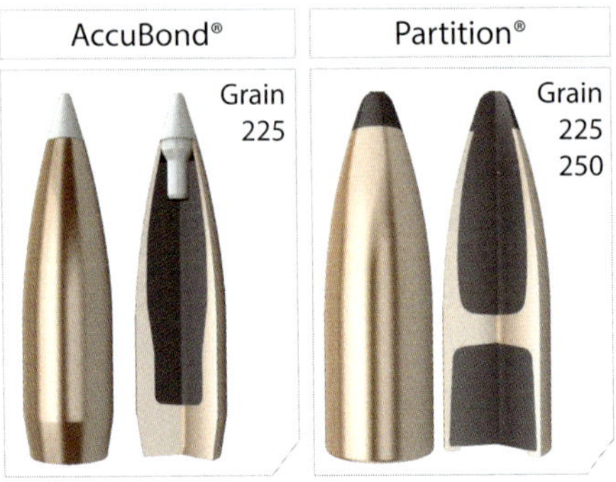

TECHNICAL INFORMATION

Brass for the "STA" is easily formed by necking-up 8mm Remington Magnum cases in a sizing die equipped with a tapered expander ball. To fire-form, we recommend reducing our starting load of IMR 4350 by 2.0 grains, and seating the bullet to contact the lands of the barrel. Fire forming in this manner will help prevent the possibility of premature case head separation. For subsequent firings, seat the bullet deep enough to avoid contact with lands (rifling).

358 STA - 225 grain		MAXIMUM S.A.A.M.I. O.A.C.L.		3.650"
			B.C.	S.D.
AccuBond®	225gr. Spitzer		0.421	0.251
Partition®	225gr. Spitzer		0.430	0.251

CASE TYPE:	Remington (8mm Ma		PRIMER TYPE	Fed 215
CASE HOLDS:	95.0	Gr. WATER	BARREL Length/Make	26" Wiseman
			BARREL Twist	1-12"

POWDER TYPE	POWDER CHG. GRS.		MUZZLE VEL. F.P.S.	LOAD DENSITY (VOLUME)
Viht N160 Most Accurate Powder Tested	85.5	MAX.	2950	100%
	83.5		2870	98%
	81.5 *		2792	95%
IMR 4350	85.5 *	MAX.	3008	95%
	83.5		2931	93%
	81.5		2852	91%
H4831	91.0 *	MAX.	3021	100%
	89.0		2955	98%
	87.0		2889	95%
RL19	89.5 *	MAX.	3081 **	102%
	87.5		2993	100%
	85.5		2905	98%

BC=Ballistic Coefficient SD=Sectional Density
*Most Accurate Load Tested **Compressed Load

Use Maximum Loads with Caution
Refer to page 73 for additional safety information

358 STA - 250 grain		MAXIMUM S.A.A.M.I. O.A.C.L.		3.650"
			B.C.	S.D.
Partition®	250gr. Spitzer		0.446	0.279

CASE TYPE:	Remington (8mm Ma	PRIMER TYPE	Fed 215
CASE HOLDS:	91.4 Gr. WATER	BARREL Length/Make	26" Wiseman
		BARREL Twist	1-12"

POWDER TYPE	POWDER CHG. GRS.		MUZZLE VEL. F.P.S.		LOAD DENSITY (VOLUME)
Viht N160	81.5 *	MAX.	2781		99%
	79.5		2704		97%
	77.5		2628		94%
IMR 7828 Most Accurate Powder Tested	88.5	MAX.	2823	**	103%
	86.5		2755	**	101%
	84.5 *		2688		98%
IMR 4350	81.5 *	MAX.	2827		94%
	79.5		2757		92%
	77.5		2686		90%
H4831	88.5	MAX.	2878	**	101%
	86.5		2820		99%
	84.5 *		2763		96%
RL22	87.0 *	MAX.	2898	**	103%
	85.0		2824	**	101%
	83.0		2751		99%

BC=Ballistic Coefficient SD=Sectional Density
*Most Accurate Load Tested **Compressed Load

Use Maximum Loads with Caution
Refer to page 73 for additional safety information

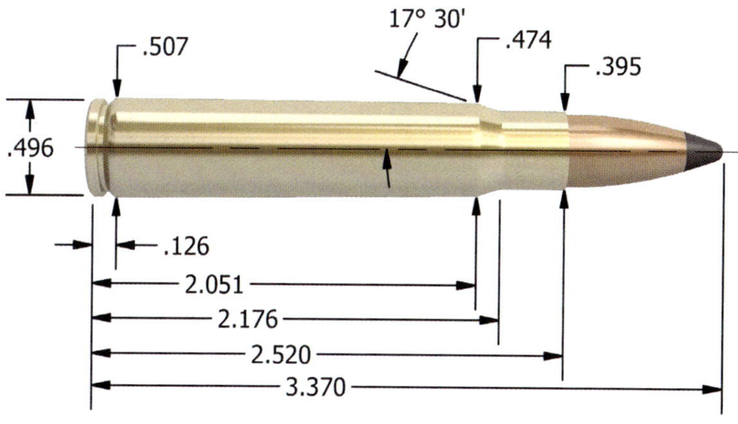

Maximum CIP Overall Cartridge Length: 3.370"

BULLET CHOICES FOR THE 9.3X64 BRENNEKE

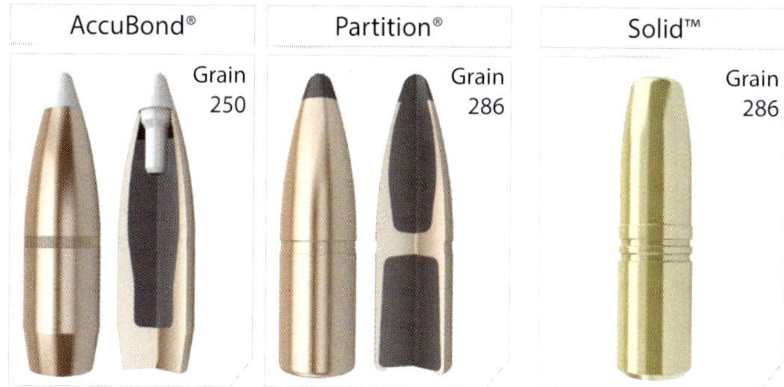

TECHNICAL INFORMATION

With a CIP maximum pressure level of almost 64,000 psi, the 9.3x64mm Brenneke is the fastest and most powerful of the modern 9.3mm (0.366") cartridges. Unfortunately for the enthusiasts of this cartridge, cases can be difficult to find in the US. Loaded with either Partition or AccuBond bullets, the 9.3x64mm is more than suitable for any North American game animal.

9.3x64 Brenneke - 250 grain

		MAXIMUM CIP O.A.C.L.		3.370"
		TESTED O.A.C.L.	B.C.	S.D.
AccuBond®	250gr. Spitzer	3.345"	0.494	0.267

CASE TYPE:	RWS		PRIMER TYPE	Rem.9 1/2M
CASE HOLDS:	73.0	Gr. WATER	BARREL Length/Make	25" Lothar Walt.
			BARREL Twist	1-14"

POWDER TYPE	POWDER CHG. GRS.		MUZZLE VEL. F.P.S.		LOAD DENSITY (VOLUME)
Viht N140	65.0	MAX.	2725	**	104%
	63.0 *		2653		100%
	61.0		2572		97%
IMR 4895	64.0	MAX.	2754		96%
	62.0		2683		93%
	60.0 *		2611		90%
IMR 4350	76.0 *	MAX.	2784	**	110%
	74.0		2703	**	107%
	72.0		2621	**	104%
RL15 Most Accurate Powder Tested	67.0	MAX.	2802		96%
	65.0		2730		93%
	63.0 *		2640		90%

BC=Ballistic Coefficient SD=Sectional Density
*Most Accurate Load Tested **Compressed Load

Use Maximum Loads with Caution
Refer to page 73 for additional safety information

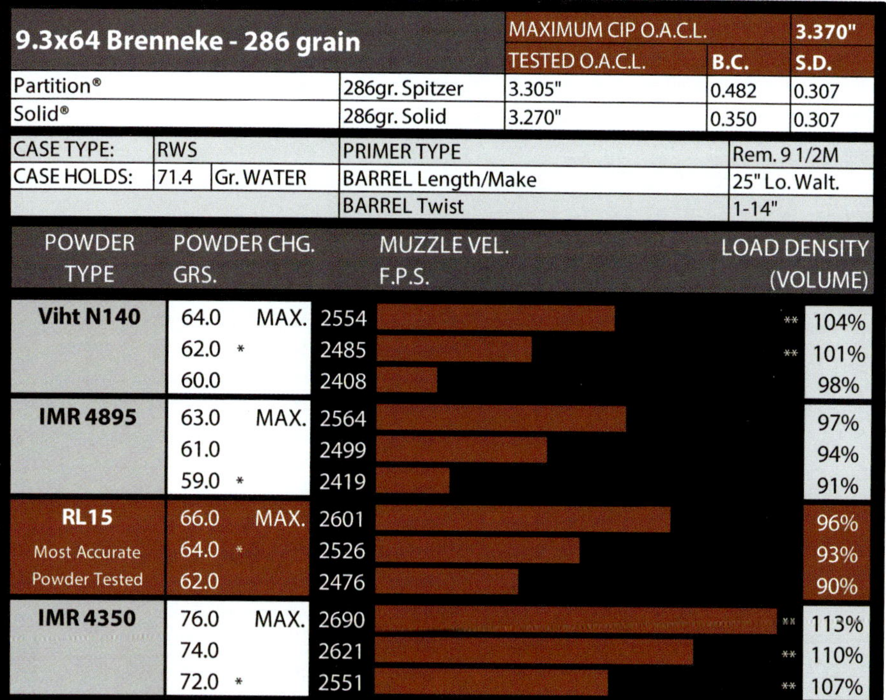

9.3x64 Brenneke - 286 grain		MAXIMUM CIP O.A.C.L.		3.370"
		TESTED O.A.C.L.	B.C.	S.D.
Partition®	286gr. Spitzer	3.305"	0.482	0.307
Solid®	286gr. Solid	3.270"	0.350	0.307

CASE TYPE:	RWS		PRIMER TYPE	Rem. 9 1/2M
CASE HOLDS:	71.4	Gr. WATER	BARREL Length/Make	25" Lo. Walt.
			BARREL Twist	1-14"

POWDER TYPE	POWDER CHG. GRS.		MUZZLE VEL. F.P.S.		LOAD DENSITY (VOLUME)
Viht N140	64.0	MAX.	2554	**	104%
	62.0	*	2485	**	101%
	60.0		2408		98%
IMR 4895	63.0	MAX.	2564		97%
	61.0		2499		94%
	59.0	*	2419		91%
RL15 Most Accurate Powder Tested	66.0	MAX.	2601		96%
	64.0	*	2526		93%
	62.0		2476		90%
IMR 4350	76.0	MAX.	2690	**	113%
	74.0		2621	**	110%
	72.0	*	2551	**	107%

BC=Ballistic Coefficient SD=Sectional Density
*Most Accurate Load Tested **Compressed Load

Use Maximum Loads with Caution
Refer to page 73 for additional safety information

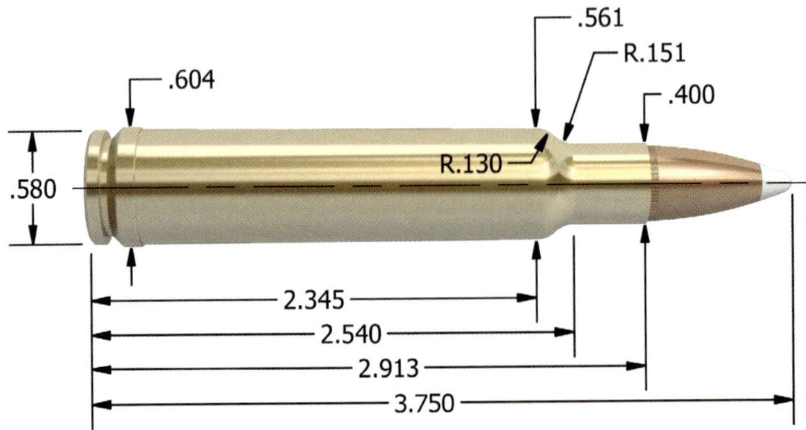

Maximum S.A.A.M.I. Overall Cartridge Length: 3.750"

BULLET CHOICES FOR THE 378 WEATHERBY MAGNUM

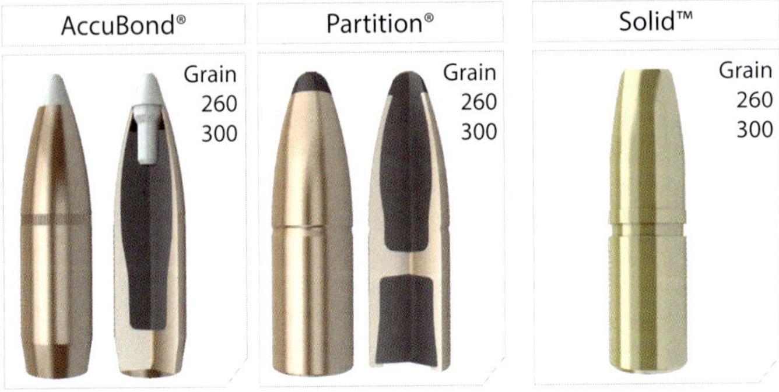

TECHNICAL INFORMATION

Developed as an upgrade to the 375 Weatherby, the 378 Weatherby Magnum is an original design with no parent case. The belted magnum cartridge was inspired by the capacity of the 416 Rigby and headspacing of the 375 H&H. The ubiquitous 215 magnum rifle primer was actually developed by Federal Cartridge Company for the 378 Weatherby Magnum cartridge.

As with most Weatherby Magnums, there are a few points to keep in mind when loading for this cartridge:

- The chambers have "freebore", meaning they have a longer throat.
- For best accuracy, bullets should be seated as long as the magazine will allow.
- We recommend three-shot groups when working-up loads.

378 Weatherby Magnum - 260 grain

		MAXIMUM S.A.A.M.I. O.A.C.L.		3.750"
		TESTED O.A.C.L.	B.C.	S.D.
AccuBond®	260gr. Spitzer	3.650"	0.473	0.264
Partition®	260gr. Spitzer	3.570"	0.314	0.264
Solid®	260gr. Solid	3.675"	0.254	0.264

CASE TYPE:	Weatherby		PRIMER TYPE	Rem. 9 1/2M
CASE HOLDS:	122.0	Gr. WATER	BARREL Length/Make	26" Shilen
			BARREL Twist	1-12"

POWDER TYPE	POWDER CHG. GRS.		MUZZLE VEL. F.P.S.	LOAD DENSITY (VOLUME)
RL22	107.0	MAX.	2902	95%
	105.0		2815	94%
	103.0 *		2732	92%
RL19	105.5 *	MAX.	2909	94%
	103.5		2826	92%
	101.5		2742	90%
IMR 4350	103.0	MAX.	2970	89%
	101.0		2883	88%
	99.0 *		2793	86%
Viht N160	105.5	MAX.	3002	96%
	103.5		2939	94%
	101.5 *		2875	92%
IMR 7828	113.0	MAX.	3045	99%
Most Accurate	111.0		2961	97%
Powder Tested	109.0 *		2877	95%

BC=Ballistic Coefficient SD=Sectional Density
*Most Accurate Load Tested **Compressed Load

Use Maximum Loads with Caution
Refer to page 73 for additional safety information

378 Weatherby Magnum - 300 grain		MAXIMUM S.A.A.M.I. O.A.C.L.		3.750"
		TESTED O.A.C.L.	B.C.	S.D.
AccuBond®	300gr. Spitzer	3.650"	0.485	0.305
Partition®	300gr. Spitzer	3.625"	0.398	0.305
Solid®	300gr. Solid	3.645"	0.300	0.305

CASE TYPE:	Weatherby	PRIMER TYPE	Rem. 9 1/2M
CASE HOLDS:	116.0 Gr. WATER	BARREL Length/Make	26" Shilen
		BARREL Twist	1-12"

POWDER TYPE	POWDER CHG. GRS.	MUZZLE VEL. F.P.S.		LOAD DENSITY (VOLUME)
H1000	115.0 * MAX.	2650	**	104%
	113.0	2620	**	102%
	111.0	2590		100%
RL22	104.5 * MAX.	2789		98%
	102.5	2678		96%
	100.5	2590		94%
IMR 4320	88.0 MAX.	2790		82%
	86.0	2740		80%
	84.0 *	2690		78%
H4831 Most Accurate Powder Tested	113.0 * MAX.	2838	**	101%
	111.0	2763		100%
	109.0	2688		98%
IMR 4350	101.0 MAX.	2850		92%
	99.0	2790		90%
	97.0 *	2730		88%
IMR 7828	111.0 * MAX.	2893	**	102%
	109.0	2848		100%
	107.0	2802		98%

BC=Ballistic Coefficient SD=Sectional Density
*Most Accurate Load Tested **Compressed Load

Use Maximum Loads with Caution
Refer to page 73 for additional safety information

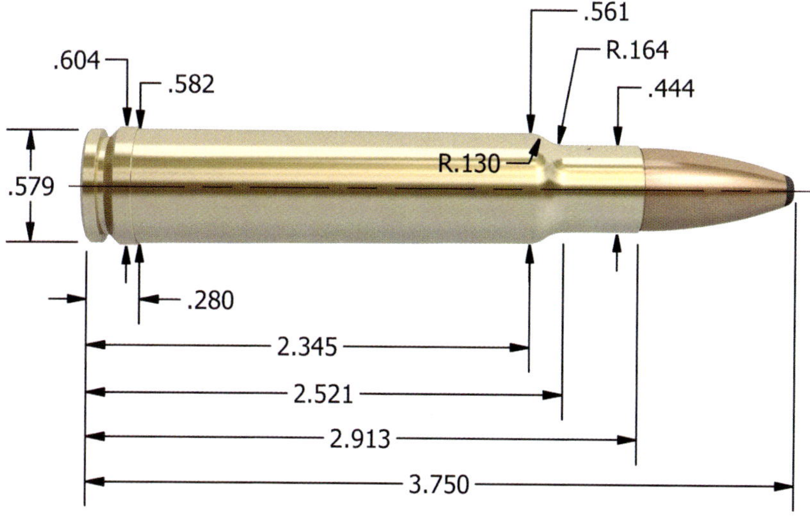

Maximum S.A.A.M.I. Overall Cartridge Length: 3.750"

BULLET CHOICES FOR THE 416 WEATHERBY MAGNUM

Partition®	Solid™
Grain 400	Grain 400

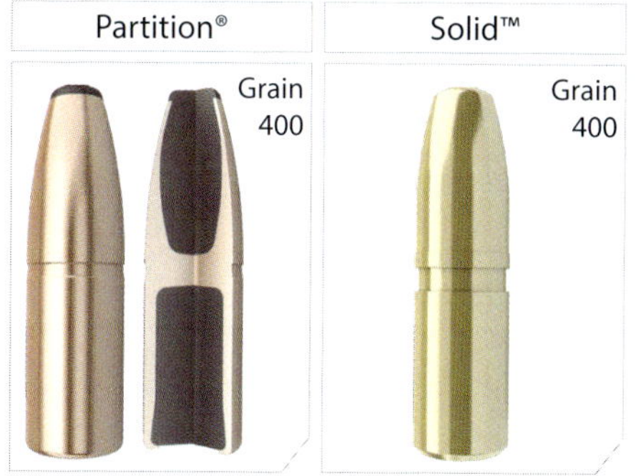

416 Weatherby Magnum - 400 grain		MAXIMUM S.A.A.M.I. O.A.C.L.		3.750"
		TESTED O.A.C.L.	B.C.	S.D.
Partition®	400gr. Spitzer	3.700"	0.390	0.330
Solid®	400gr. Solid	3.700"	0.289	0.330

CASE TYPE:	Weatherby	PRIMER TYPE	Fed 215
CASE HOLDS:	114.5 Gr. WATER	BARREL Length/Make	26" Weatherby
		BARREL Twist	1-14"

POWDER TYPE	POWDER CHG. GRS.	MUZZLE VEL. F.P.S.	LOAD DENSITY (VOLUME)
H4831SC	113.0 * MAX.	2667	** 103%
	111.0	2630	** 101%
	109.0	2599	99%
RL19	115.0 MAX.	2696	** 109%
	113.0 *	2641	** 107%
	111.0	2601	** 105%
IMR 7828 Most Accurate Powder Tested	117.0 MAX.	2704	** 109%
	115.0 *	2667	** 107%
	113.0	2640	** 105%
RL22	117.0 MAX.	2725	** 111%
	115.0	2679	** 109%
	113.0 *	2607	** 107%

BC=Ballistic Coefficient SD=Sectional Density
*Most Accurate Load Tested **Compressed Load

Use Maximum Loads with Caution
Refer to page 73 for additional safety information

UNIVERSAL DROP TABLE FOR RIFLE BULLETS

Universal Table for Rifle Bullets with Ballistic Coefficients of .110-.119

Sight Above Bore (in.) +1.5

Range (yds)	Velocity (fps)	Bullet Path Zero (100 yds)	Zero (200 yds)	Zero (300 yds)	Zero (400 yds)	Zero (500 yds)
0	4200	-1.5	-1.5	-1.5	-1.5	-1.5
100	3283	0.0	0.9	2.9	6.7	12.4
200	2546	-1.8	0.0	3.9	11.5	23.1
300	1907	-8.6	-5.9	0.0	11.4	28.7
400	1393	-26.6	-23.1	-15.2	0.0	23.1
500	1038	-62.1	-57.7	-47.9	-28.8	0.0
600	911	-128.7	-123.3	-111.6	-88.7	-54.1
0	4100	-1.5	-1.5	-1.5	-1.5	-1.5
100	3199	0.0	1.0	3.3	7.1	13.1
200	2478	-1.9	0.0	4.7	12.2	24.3
300	1854	-9.9	-7.0	0.0	11.3	29.4
400	1347	-28.3	-24.4	-15.1	0.0	24.1
500	1008	-65.6	-60.7	-49.0	-30.2	0.0
600	898	-135.5	-129.7	-115.6	-93.0	-56.8
0	4000	-1.5	-1.5	-1.5	-1.5	-1.5
100	3115	0.0	1.0	3.6	7.5	13.9
200	2409	-2.1	0.0	5.0	13.0	25.7
300	1795	-10.7	-7.5	0.0	11.9	31.0
400	1300	-30.1	-25.9	-15.9	0.0	25.4
500	977	-69.4	-64.1	-51.6	-31.7	0.0
600	891	-144.1	-137.8	-122.7	-98.9	-60.8
0	3900	-1.5	-1.5	-1.5	-1.5	-1.5
100	3047	0.0	1.0	3.7	7.9	14.5
200	2339	-2.0	0.0	5.3	13.8	27.0
300	1736	-11.1	-8.0	0.0	12.6	32.5
400	1253	-31.6	-27.5	-16.8	0.0	26.5
500	947	-72.6	-67.5	-54.2	-33.1	0.0
600	879	-151.1	-145.0	-129.0	-103.8	-64.0
0	3800	-1.5	-1.5	-1.5	-1.5	-1.5
100	2964	0.0	1.3	3.9	8.4	16.1
200	2272	-2.6	0.0	5.3	14.2	29.5
300	1675	-11.8	-7.9	0.0	13.3	36.3
400	1206	-33.5	-28.3	-17.7	0.0	30.7
500	971	-80.3	-73.8	-60.5	-38.4	0.0
600	871	-162.1	-154.3	-138.4	-111.8	-65.7
0	3700	-1.5	-1.5	-1.5	-1.5	-1.5
100	2881	0.0	1.4	4.2	8.9	17.0
200	2202	-2.8	0.0	5.6	15.0	31.1
300	1614	-12.7	-8.4	0.0	14.0	38.3
400	1158	-35.6	-29.9	-18.7	0.0	32.3
500	949	-84.9	-77.8	-63.8	-40.4	0.0
600	860	-171.0	-162.4	-145.6	-117.5	-69.1
0	3600	-1.5	-1.5	-1.5	-1.5	-1.5
100	2797	0.0	1.6	4.5	9.4	18.0
200	2130	-3.1	0.0	5.9	15.8	32.9
300	1553	-13.5	-8.8	0.0	14.8	40.6
400	1110	-37.8	-31.6	-19.8	0.0	34.3
500	926	-90.2	-82.4	-67.7	-42.9	0.0
600	849	-180.4	-171.1	-153.4	-123.7	-72.2
0	3500	-1.5	-1.5	-1.5	-1.5	-1.5
100	2712	0.0	1.7	4.8	10.0	20.0
200	2058	-3.4	0.0	6.2	16.6	36.6
300	1490	-14.4	-9.3	0.0	15.6	45.6
400	1062	-40.1	-33.3	-20.8	0.0	39.9
500	920	-100.0	-91.5	-76.0	-49.9	0.0
600	838	-190.5	-180.3	-161.6	-130.3	-70.4

Universal Table for Rifle Bullets with Ballistic Coefficients of .120-.129

Sight Above Bore (in.) +1.5

		Bullet Path				
Range (yds)	Velocity (fps)	Zero (100 yds)	Zero (200 yds)	Zero (300 yds)	Zero (400 yds)	Zero (500 yds)
0	4200	-1.5	-1.5	-1.5	-1.5	-1.5
100	3276	0.0	0.9	2.9	6.7	12.6
200	2536	-1.8	0.0	3.9	11.7	23.4
300	1894	-8.6	-5.9	0.0	11.6	29.1
400	1379	-26.9	-23.3	-15.4	0.0	23.4
500	1026	-62.9	-58.4	-48.5	-29.2	0.0
600	905	-130.3	-125.0	-113.1	-90.0	-54.9
0	4100	-1.5	-1.5	-1.5	-1.5	-1.5
100	3193	0.0	1.0	3.3	7.1	13.3
200	2468	-1.9	0.0	4.7	12.3	24.6
300	1841	-10.0	-7.1	0.0	11.4	29.8
400	1333	-28.6	-24.7	-15.2	0.0	24.5
500	996	-66.4	-61.5	-49.7	-30.6	0.0
600	898	-138.1	-132.3	-118.1	-95.3	-58.5
0	4000	-1.5	-1.5	-1.5	-1.5	-1.5
100	3109	0.0	1.1	3.6	7.6	14.0
200	2399	-2.1	0.0	5.1	13.1	25.9
300	1782	-10.7	-7.6	0.0	12.1	31.3
400	1286	-30.4	-26.2	-16.1	0.0	25.6
500	966	-70.0	-64.8	-52.1	-32.0	0.0
600	886	-145.6	-139.2	-124.1	-99.9	-61.5
0	3900	-1.5	-1.5	-1.5	-1.5	-1.5
100	3041	0.0	1.0	3.7	8.0	14.7
200	2329	-2.0	0.0	5.4	13.9	27.3
300	1723	-11.1	-8.1	0.0	12.7	32.9
400	1240	-31.8	-27.7	-17.0	0.0	26.8
500	936	-73.3	-68.2	-54.8	-33.6	0.0
600	874	-152.7	-146.6	-130.5	-105.0	-64.7
0	3800	-1.5	-1.5	-1.5	-1.5	-1.5
100	2958	0.0	1.3	4.0	8.4	16.2
200	2262	-2.6	0.0	5.3	14.3	29.8
300	1663	-11.9	-8.0	0.0	13.4	36.8
400	1193	-33.8	-28.5	-17.9	0.0	31.1
500	963	-81.1	-74.6	-61.3	-38.9	0.0
600	867	-163.3	-155.4	-139.5	-112.7	-65.9
0	3700	-1.5	-1.5	-1.5	-1.5	-1.5
100	2875	0.0	1.4	4.2	9.0	17.2
200	2192	-2.9	0.0	5.6	15.1	31.5
300	1602	-12.7	-8.4	0.0	14.2	38.8
400	1145	-35.9	-30.1	-18.9	0.0	32.8
500	941	-85.8	-78.7	-64.6	-41.0	0.0
600	856	-172.3	-163.7	-146.8	-118.5	-69.3
0	3600	-1.5	-1.5	-1.5	-1.5	-1.5
100	2791	0.0	1.6	4.5	9.5	18.2
200	2121	-3.1	0.0	5.9	15.9	33.2
300	1540	-13.6	-8.9	0.0	15.0	40.9
400	1097	-38.1	-31.8	-19.9	0.0	34.6
500	919	-90.8	-83.0	-68.2	-43.3	0.0
600	845	-181.8	-172.5	-154.7	-124.7	-72.8
0	3500	-1.5	-1.5	-1.5	-1.5	-1.5
100	2707	0.0	1.7	4.8	10.1	19.9
200	2049	-3.4	0.0	6.3	16.8	36.3
300	1478	-14.5	-9.4	0.0	15.8	45.1
400	1049	-40.4	-33.6	-21.0	0.0	39.1
500	924	-99.4	-90.8	-75.2	-48.9	0.0
600	834	-192.0	-181.8	-163.0	-131.5	-72.8

Universal Table for Rifle Bullets with Ballistic Coefficients of .130-.139

Sight Above Bore (in.) +1.5

		Bullet Path				
Range (yds)	Velocity (fps)	Zero (100 yds)	Zero (200 yds)	Zero (300 yds)	Zero (400 yds)	Zero (500 yds)
0	4200	-1.5	-1.5	-1.5	-1.5	-1.5
100	3339	0.0	0.8	2.7	5.7	10.5
200	2640	-1.7	0.0	3.7	9.7	19.4
300	2030	-8.2	-5.6	0.0	9.0	23.4
400	1501	-22.8	-19.4	-11.9	0.0	19.3
500	1098	-52.6	-48.4	-39.0	-24.1	0.0
600	934	-111.9	-106.8	-95.6	-77.7	-48.8
0	4100	-1.5	-1.5	-1.5	-1.5	-1.5
100	3255	0.0	0.9	2.9	6.5	11.9
200	2570	-1.9	0.0	3.9	11.2	22.0
300	1968	-8.7	-5.9	0.0	10.9	27.0
400	1473	-26.1	-22.4	-14.5	0.0	21.6
500	1113	-59.5	-54.9	-45.1	-27.0	0.0
600	939	-120.7	-115.2	-103.4	-81.7	-49.3
0	4000	-1.5	-1.5	-1.5	-1.5	-1.5
100	3170	0.0	1.0	3.1	7.0	12.6
200	2500	-2.0	0.0	4.1	11.9	23.3
300	1907	-9.2	-6.2	0.0	11.7	28.7
400	1424	-27.8	-23.8	-15.6	0.0	22.7
500	1080	-63.2	-58.2	-47.9	-28.4	0.0
600	927	-129.3	-123.3	-110.9	-87.6	-53.5
0	3900	-1.5	-1.5	-1.5	-1.5	-1.5
100	3099	0.0	1.0	3.5	7.3	13.3
200	2429	-2.0	0.0	5.0	12.7	24.6
300	1851	-10.4	-7.5	0.0	11.5	29.5
400	1375	-29.3	-25.3	-15.3	0.0	24.0
500	1047	-66.5	-61.6	-49.2	-30.0	0.0
600	919	-134.0	-128.1	-113.2	-90.2	-54.2
0	3800	-1.5	-1.5	-1.5	-1.5	-1.5
100	3016	0.0	1.1	3.7	7.8	14.1
200	2357	-2.1	0.0	5.3	13.5	26.1
300	1789	-11.2	-8.0	0.0	12.2	31.2
400	1325	-31.2	-27.0	-16.3	0.0	25.2
500	1014	-70.6	-65.3	-51.9	-31.5	0.0
600	909	-142.9	-136.5	-120.5	-96.1	-58.2
0	3700	-1.5	-1.5	-1.5	-1.5	-1.5
100	2931	0.0	1.4	4.0	8.3	15.0
200	2288	-2.7	0.0	5.3	13.9	27.2
300	1727	-12.0	-7.9	0.0	12.9	32.9
400	1276	-33.2	-27.8	-17.3	0.0	26.6
500	981	-74.9	-68.1	-54.9	-33.3	0.0
600	900	-152.7	-144.6	-128.7	-102.8	-62.8
0	3600	-1.5	-1.5	-1.5	-1.5	-1.5
100	2847	0.0	1.5	4.3	8.9	15.9
200	2215	-3.0	0.0	5.6	14.7	28.8
300	1663	-12.9	-8.4	0.0	13.7	34.8
400	1226	-35.5	-29.5	-18.3	0.0	28.1
500	948	-79.4	-72.0	-57.9	-35.1	0.0
600	887	-161.5	-152.5	-135.7	-108.3	-66.2
0	3500	-1.5	-1.5	-1.5	-1.5	-1.5
100	2761	0.0	1.6	4.6	9.4	17.6
200	2142	-3.3	0.0	5.9	15.6	31.9
300	1600	-13.8	-8.9	0.0	14.5	39.0
400	1176	-37.8	-31.2	-19.4	0.0	32.7
500	971	-88.0	-79.9	-65.1	-40.8	0.0
600	878	-173.4	-163.6	-145.8	-116.7	-67.7

Universal Table for Rifle Bullets with Ballistic Coefficients of .150-.159

Sight Above Bore (in.) +1.5

Range (yds)	Velocity (fps)	Bullet Path Zero (100 yds)	Zero (200 yds)	Zero (300 yds)	Zero (400 yds)	Zero (500 yds)
0	4200	-1.5	-1.5	-1.5	-1.5	-1.5
100	3442	0.0	0.8	2.5	5.1	9.0
200	2812	-1.5	0.0	3.4	8.6	16.4
300	2257	-7.4	-5.1	0.0	7.8	19.5
400	1768	-20.2	-17.2	-10.4	0.0	15.7
500	1364	-44.9	-41.1	-32.6	-19.6	0.0
600	1077	-88.8	-84.2	-74.0	-58.5	-34.9
0	4100	-1.5	-1.5	-1.5	-1.5	-1.5
100	3356	0.0	0.8	2.6	5.4	9.5
200	2740	-1.7	0.0	3.6	9.1	17.4
300	2193	-7.9	-5.4	0.0	8.2	20.6
400	1711	-21.5	-18.1	-10.9	0.0	16.6
500	1318	-47.6	-43.4	-34.4	-20.7	0.0
600	1046	-93.8	-88.7	-77.9	-61.5	-36.7
0	4000	-1.5	-1.5	-1.5	-1.5	-1.5
100	3270	0.0	0.9	2.8	5.7	10.1
200	2667	-1.9	0.0	3.8	9.5	18.3
300	2128	-8.5	-5.7	0.0	8.6	21.8
400	1654	-22.8	-19.1	-11.5	0.0	17.5
500	1272	-50.4	-45.8	-36.3	-21.9	0.0
600	1014	-99.3	-93.7	-82.3	-65.1	-38.8
0	3900	-1.5	-1.5	-1.5	-1.5	-1.5
100	3195	0.0	0.9	2.9	6.0	10.6
200	2594	-1.8	0.0	4.0	10.1	19.4
300	2063	-8.8	-6.0	0.0	9.1	23.0
400	1596	-23.8	-20.1	-12.1	0.0	18.6
500	1225	-53.0	-48.4	-38.4	-23.3	0.0
600	983	-104.3	-98.8	-86.8	-68.6	-40.7
0	3800	-1.5	-1.5	-1.5	-1.5	-1.5
100	3110	0.0	1.0	3.1	6.3	11.2
200	2519	-2.0	0.0	4.2	10.6	20.4
300	1997	-9.3	-6.3	0.0	9.6	24.4
400	1538	-25.2	-21.2	-12.7	0.0	19.7
500	1178	-56.1	-51.1	-40.6	-24.7	0.0
600	998	-114.5	-108.4	-95.8	-76.7	-47.1
0	3700	-1.5	-1.5	-1.5	-1.5	-1.5
100	3024	0.0	1.1	3.6	7.2	11.9
200	2444	-2.2	0.0	5.0	12.3	21.6
300	1935	-10.8	-7.5	0.0	10.8	24.8
400	1501	-28.9	-24.5	-14.5	0.0	18.6
500	1131	-59.4	-53.9	-41.4	-23.3	0.0
600	974	-121.6	-115.1	-100.0	-78.3	-50.4
0	3600	-1.5	-1.5	-1.5	-1.5	-1.5
100	2937	0.0	1.4	3.9	7.8	13.5
200	2372	-2.7	0.0	5.0	12.8	24.3
300	1869	-11.7	-7.6	0.0	11.6	28.9
400	1446	-31.0	-25.6	-15.5	0.0	23.1
500	1134	-67.6	-60.8	-48.2	-28.8	0.0
600	963	-133.1	-124.9	-109.8	-86.5	-51.9
0	3500	-1.5	-1.5	-1.5	-1.5	-1.5
100	2850	0.0	1.5	4.2	8.3	14.5
200	2296	-3.0	0.0	5.4	13.6	26.0
300	1803	-12.6	-8.0	0.0	12.4	30.9
400	1392	-33.3	-27.3	-16.6	0.0	24.6
500	1096	-72.4	-64.9	-51.5	-30.8	0.0
600	954	-139.6	-130.5	-114.4	-89.5	-52.6

Universal Table for Rifle Bullets with Ballistic Coefficients of .160-.169

Sight Above Bore (in.) +1.5

		Bullet Path				
Range (yds)	Velocity (fps)	Zero (100 yds)	Zero (200 yds)	Zero (300 yds)	Zero (400 yds)	Zero (500 yds)
0	3800	-1.5	-1.5	-1.5	-1.5	-1.5
100	3201	0.0	0.9	2.9	5.6	9.6
200	2680	-1.9	0.0	3.8	9.4	17.4
300	2214	-8.6	-5.8	0.0	8.4	20.3
400	1795	-22.6	-18.8	-11.2	0.0	15.9
500	1441	-48.1	-43.4	-33.8	-19.8	0.0
600	1169	-91.2	-85.6	-74.1	-57.4	-33.6
0	3700	-1.5	-1.5	-1.5	-1.5	-1.5
100	3114	0.0	1.0	3.1	6.0	10.3
200	2603	-2.0	0.0	4.1	10.0	18.5
300	2144	-9.2	-6.1	0.0	8.9	21.6
400	1733	-24.1	-20.0	-11.8	0.0	17.0
500	1389	-51.3	-46.2	-36.0	-21.3	0.0
600	1132	-97.3	-91.2	-78.9	-61.2	-35.7
0	3600	-1.5	-1.5	-1.5	-1.5	-1.5
100	3026	0.0	1.1	3.3	6.4	11.0
200	2525	-2.2	0.0	4.3	10.6	19.7
300	2074	-9.8	-6.5	0.0	9.4	23.0
400	1670	-25.6	-21.2	-12.5	0.0	18.2
500	1337	-54.8	-49.2	-38.4	-22.7	0.0
600	1095	-103.8	-97.2	-84.2	-65.3	-38.1
0	3500	-1.5	-1.5	-1.5	-1.5	-1.5
100	2938	0.0	1.4	3.8	6.8	11.7
200	2449	-2.8	0.0	4.8	10.9	20.7
300	2006	-11.4	-7.2	0.0	9.2	23.8
400	1606	-27.4	-21.8	-12.2	0.0	19.5
500	1284	-58.6	-51.7	-39.7	-24.4	0.0
600	1058	-110.9	-102.6	-88.2	-69.8	-40.5
0	3400	-1.5	-1.5	-1.5	-1.5	-1.5
100	2854	0.0	1.4	4.0	7.7	12.3
200	2370	-2.8	0.0	5.1	12.6	21.9
300	1936	-11.9	-7.7	0.0	11.1	25.2
400	1558	-30.7	-25.1	-14.9	0.0	18.7
500	1232	-61.7	-54.7	-41.9	-23.3	0.0
600	1022	-117.2	-108.8	-93.4	-71.1	-43.1
0	3300	-1.5	-1.5	-1.5	-1.5	-1.5
100	2766	0.0	1.5	4.3	8.3	13.9
200	2291	-3.1	0.0	5.5	13.5	24.8
300	1865	-12.9	-8.3	0.0	12.1	28.9
400	1498	-33.2	-27.1	-16.1	0.0	22.5
500	1211	-69.6	-62.0	-48.2	-28.1	0.0
600	1024	-129.4	-120.3	-103.7	-79.6	-45.9
0	3200	-1.5	-1.5	-1.5	-1.5	-1.5
100	2677	0.0	1.7	4.6	9.0	15.0
200	2211	-3.3	0.0	5.9	14.6	26.7
300	1793	-13.9	-8.9	0.0	13.0	31.1
400	1439	-35.8	-29.1	-17.3	0.0	24.2
500	1168	-75.0	-66.7	-51.8	-30.3	0.0
600	996	-138.8	-128.7	-110.9	-85.0	-48.7
0	3100	-1.5	-1.5	-1.5	-1.5	-1.5
100	2588	0.0	1.8	5.0	9.7	16.2
200	2130	-3.7	0.0	6.4	15.7	28.7
300	1721	-15.0	-9.5	0.0	14.0	33.6
400	1379	-38.7	-31.4	-18.7	0.0	26.1
500	1125	-81.0	-71.8	-56.0	-32.6	0.0
600	980	-152.6	-141.6	-122.6	-94.5	-55.4

Universal Table for Rifle Bullets with Ballistic Coefficients of .170-.179

Sight Above Bore (in.) +1.5

Range (yds)	Velocity (fps)	Bullet Path Zero (100 yds)	Zero (200 yds)	Zero (300 yds)	Zero (400 yds)	Zero (500 yds)
0	3800	-1.5	-1.5	-1.5	-1.5	-1.5
100	3234	0.0	0.9	2.8	5.4	9.1
200	2737	-1.8	0.0	3.7	9.0	16.3
300	2291	-8.3	-5.6	0.0	7.9	18.9
400	1888	-21.6	-18.0	-10.5	0.0	14.7
500	1537	-45.4	-40.8	-31.5	-18.3	0.0
600	1231	-81.3	-75.8	-64.6	-48.8	-26.8
0	3700	-1.5	-1.5	-1.5	-1.5	-1.5
100	3145	0.0	1.0	3.0	5.8	9.7
200	2659	-2.0	0.0	4.0	9.6	17.4
300	2220	-8.9	-5.9	0.0	8.4	20.1
400	1824	-23.1	-19.1	-11.2	0.0	15.6
500	1484	-48.4	-43.5	-33.6	-19.5	0.0
600	1214	-90.5	-84.6	-72.7	-55.9	-32.4
0	3600	-1.5	-1.5	-1.5	-1.5	-1.5
100	3057	0.0	1.1	3.2	6.2	10.4
200	2580	-2.2	0.0	4.2	10.2	18.6
300	2148	-9.6	-6.3	0.0	9.0	21.5
400	1759	-24.7	-20.4	-12.0	0.0	16.7
500	1430	-51.9	-46.4	-35.9	-20.9	0.0
600	1175	-96.8	-90.3	-77.6	-59.7	-34.5
0	3500	-1.5	-1.5	-1.5	-1.5	-1.5
100	2968	0.0	1.3	3.4	6.6	11.1
200	2502	-2.7	0.0	4.2	10.5	19.6
300	2076	-10.3	-6.3	0.0	9.6	23.1
400	1694	-26.5	-21.1	-12.7	0.0	18.1
500	1375	-55.7	-48.9	-38.5	-22.6	0.0
600	1136	-103.6	-95.5	-83.0	-63.9	-36.8
0	3400	-1.5	-1.5	-1.5	-1.5	-1.5
100	2884	0.0	1.4	3.8	6.9	11.8
200	2423	-2.7	0.0	4.9	11.2	20.8
300	2007	-11.5	-7.4	0.0	9.3	23.8
400	1628	-27.8	-22.3	-12.4	0.0	19.4
500	1321	-58.9	-52.1	-39.7	-24.2	0.0
600	1097	-110.2	-102.0	-87.2	-68.6	-39.5
0	3300	-1.5	-1.5	-1.5	-1.5	-1.5
100	2795	0.0	1.5	4.1	7.9	12.6
200	2343	-3.0	0.0	5.3	12.8	22.2
300	1935	-12.4	-8.0	0.0	11.3	25.3
400	1577	-31.6	-25.6	-15.0	0.0	18.8
500	1266	-63.0	-55.5	-42.2	-23.5	0.0
600	1058	-117.9	-109.0	-93.1	-70.5	-42.4
0	3200	-1.5	-1.5	-1.5	-1.5	-1.5
100	2705	0.0	1.6	4.5	8.5	13.4
200	2262	-3.3	0.0	5.7	13.8	23.6
300	1862	-13.5	-8.6	0.0	12.1	26.9
400	1516	-34.2	-27.6	-16.2	0.0	19.6
500	1211	-67.2	-59.1	-44.8	-24.6	0.0
600	1051	-129.8	-120.0	-102.9	-78.6	-49.1
0	3100	-1.5	-1.5	-1.5	-1.5	-1.5
100	2616	0.0	1.8	4.9	9.3	15.3
200	2181	-3.6	0.0	6.2	15.0	27.0
300	1788	-14.6	-9.3	0.0	13.2	31.2
400	1454	-37.1	-29.9	-17.6	0.0	24.0
500	1193	-76.4	-67.5	-52.0	-30.1	0.0
600	1021	-139.8	-129.1	-110.5	-84.2	-48.1

Universal Table for Rifle Bullets with Ballistic Coefficients of .180-.189						
					Sight Above Bore (in.)	+1.5
		Bullet Path				
Range (yds)	Velocity (fps)	Zero (100 yds)	Zero (200 yds)	Zero (300 yds)	Zero (400 yds)	Zero (500 yds)
0	4200	-1.5	-1.5	-1.5	-1.5	-1.5
100	3610	0.0	0.6	2.0	3.9	6.6
200	3097	-1.1	0.0	2.8	6.7	12.0
300	2640	-5.9	-4.2	0.0	5.9	13.8
400	2226	-15.7	-13.4	-7.8	0.0	10.6
500	1850	-32.8	-30.0	-23.0	-13.2	0.0
600	1523	-60.9	-57.5	-49.1	-37.4	-21.5
0	4100	-1.5	-1.5	-1.5	-1.5	-1.5
100	3521	0.0	0.6	2.1	4.2	7.0
200	3019	-1.2	0.0	2.9	7.1	12.7
300	2571	-6.3	-4.4	0.0	6.2	14.6
400	2162	-16.7	-14.2	-8.3	0.0	11.2
500	1792	-34.8	-31.7	-24.3	-14.0	0.0
600	1474	-64.7	-61.0	-52.1	-39.7	-23.0
0	4000	-1.5	-1.5	-1.5	-1.5	-1.5
100	3432	0.0	0.8	2.4	4.4	7.4
200	2945	-1.6	0.0	3.2	7.3	13.2
300	2502	-7.2	-4.8	0.0	6.1	15.0
400	2098	-17.7	-14.5	-8.1	0.0	11.8
500	1734	-36.9	-32.9	-24.9	-14.8	0.0
600	1425	-68.8	-64.0	-54.4	-42.3	-24.5
0	3900	-1.5	-1.5	-1.5	-1.5	-1.5
100	3351	0.0	0.8	2.5	4.9	7.7
200	2867	-1.6	0.0	3.4	8.1	13.9
300	2432	-7.5	-5.1	0.0	7.1	15.7
400	2036	-19.4	-16.2	-9.4	0.0	11.5
500	1675	-38.7	-34.7	-26.2	-14.4	0.0
600	1376	-72.6	-67.7	-57.6	-43.4	-26.1
0	3800	-1.5	-1.5	-1.5	-1.5	-1.5
100	3262	0.0	0.9	2.7	5.2	8.6
200	2788	-1.8	0.0	3.6	8.6	15.4
300	2360	-8.1	-5.4	0.0	7.5	17.6
400	1972	-20.7	-17.2	-10.0	0.0	13.5
500	1627	-42.8	-38.4	-29.4	-16.9	0.0
600	1326	-77.1	-71.8	-61.0	-46.0	-25.7
0	3900	-1.5	-1.5	-1.5	-1.5	-1.5
100	3351	0.0	0.8	2.5	4.9	7.7
200	2867	-1.6	0.0	3.4	8.1	13.9
300	2432	-7.5	-5.1	0.0	7.1	15.7
400	2036	-19.4	-16.2	-9.4	0.0	11.5
500	1675	-38.7	-34.7	-26.2	-14.4	0.0
600	1376	-72.6	-67.7	-57.6	-43.4	-26.1
0	3800	-1.5	-1.5	-1.5	-1.5	-1.5
100	3262	0.0	0.9	2.7	5.2	8.6
200	2788	-1.8	0.0	3.6	8.6	15.4
300	2360	-8.1	-5.4	0.0	7.5	17.6
400	1972	-20.7	-17.2	-10.0	0.0	13.5
500	1627	-42.8	-38.4	-29.4	-16.9	0.0
600	1326	-77.1	-71.8	-61.0	-46.0	-25.7
0	3700	-1.5	-1.5	-1.5	-1.5	-1.5
100	3174	0.0	1.0	2.9	5.6	9.2
200	2709	-2.0	0.0	3.8	9.2	16.4
300	2288	-8.7	-5.8	0.0	8.0	18.9
400	1907	-22.2	-18.3	-10.6	0.0	14.5
500	1571	-45.9	-41.1	-31.4	-18.1	0.0
600	1277	-81.8	-75.9	-64.4	-48.4	-26.7

Universal Table for Rifle Bullets with Ballistic Coefficients of .190-.199

Sight Above Bore (in.) +1.5

Range (yds)	Velocity (fps)	Bullet Path Zero (100 yds)	Zero (200 yds)	Zero (300 yds)	Zero (400 yds)	Zero (500 yds)
0	2300	-1.5	-1.5	-1.5	-1.5	-1.5
100	1889	0.0	4.3	9.9	19.4	30.3
200	1532	-8.6	0.0	11.3	30.2	52.3
300	1221	-29.7	-16.9	0.0	28.4	59.5
400	1054	-77.5	-60.4	-37.9	0.0	44.8
500	949	-151.3	-130.6	-99.1	-55.9	0.0
600	874	-267.3	-241.6	-196.2	-144.6	-78.7
0	2200	-1.5	-1.5	-1.5	-1.5	-1.5
100	1798	0.0	4.8	11.9	21.3	33.5
200	1456	-9.6	0.0	14.2	33.0	57.5
300	1190	-35.7	-21.3	0.0	28.1	65.0
400	1017	-85.1	-65.9	-37.5	0.0	49.1
500	931	-167.7	-143.8	-108.3	-61.3	0.0
600	858	-288.4	-259.7	-210.1	-153.8	-87.1
0	2100	-1.5	-1.5	-1.5	-1.5	-1.5
100	1707	0.0	5.4	13.2	24.2	36.5
200	1380	-10.8	0.0	15.7	37.6	62.2
300	1135	-39.6	-23.5	0.0	32.9	69.8
400	991	-96.7	-75.2	-40.9	0.0	49.2
500	911	-182.4	-155.5	-116.4	-61.5	0.0
600	843	-309.4	-272.1	-224.1	-164.3	-90.5
0	2000	-1.5	-1.5	-1.5	-1.5	-1.5
100	1626	0.0	5.5	14.1	26.1	39.9
200	1304	-10.9	0.0	17.3	41.2	68.8
300	1081	-42.3	-25.9	0.0	36.0	77.3
400	970	-104.3	-82.5	-44.0	0.0	55.1
500	894	-199.3	-172.0	-122.9	-68.9	0.0
600	828	-334.4	-290.7	-239.9	-177.9	-95.3
0	1900	-1.5	-1.5	-1.5	-1.5	-1.5
100	1541	0.0	5.8	15.3	28.0	43.2
200	1229	-11.6	0.0	18.9	42.9	74.9
300	1029	-45.8	-28.3	0.0	36.1	83.9
400	951	-111.9	-85.7	-48.1	0.0	56.9
500	876	-211.1	-178.1	-131.9	-71.2	0.0
600	813	-350.9	-311.0	-255.7	-183.0	-100.8
0	1800	-1.5	-1.5	-1.5	-1.5	-1.5
100	1455	0.0	7.5	17.7	31.4	46.9
200	1190	-15.0	0.0	20.5	47.9	78.9
300	1017	-53.2	-30.8	0.0	41.0	87.6
400	930	-125.7	-95.8	-54.7	0.0	62.1
500	858	-234.7	-192.3	-140.9	-77.6	0.0
600	798	-378.4	-334.0	-271.4	-193.4	-106.3
0	1700	-1.5	-1.5	-1.5	-1.5	-1.5
100	1368	0.0	8.4	20.7	34.7	50.5
200	1127	-16.9	0.0	24.5	52.6	84.2
300	986	-62.1	-36.8	0.0	42.1	89.4
400	910	-138.9	-101.2	-56.1	0.0	63.2
500	841	-252.5	-207.4	-149.1	-79.0	0.0
600	783	-407.4	-357.9	-293.3	-209.1	-114.4
0	1600	-1.5	-1.5	-1.5	-1.5	-1.5
100	1281	0.0	9.5	22.7	38.6	56.1
200	1065	-19.0	0.0	26.4	58.2	93.3
300	966	-68.0	-37.7	0.0	47.7	100.4
400	888	-148.3	-110.4	-63.6	0.0	70.3
500	823	-271.7	-223.4	-167.4	-87.9	0.0
600	767	-440.3	-382.7	-317.3	-221.9	-116.4

Universal Table for Rifle Bullets with Ballistic Coefficients of .190-.199

Sight Above Bore (in.) +1.5

Range (yds)	Velocity (fps)	Bullet Path Zero (100 yds)	Zero (200 yds)	Zero (300 yds)	Zero (400 yds)	Zero (500 yds)
0	1500	-1.5	-1.5	-1.5	-1.5	-1.5
100	1222	0.0	10.7	24.8	41.3	59.8
200	1037	-21.3	0.0	28.3	61.4	98.3
300	942	-74.4	-42.5	0.0	49.6	104.9
400	867	-165.3	-119.7	-66.1	0.0	73.8
500	806	-293.9	-240.7	-174.9	-92.2	0.0
600	752	-474.2	-409.3	-331.4	-232.2	-121.5
0	1400	-1.5	-1.5	-1.5	-1.5	-1.5
100	1146	0.0	11.6	27.2	44.9	65.5
200	989	-23.2	0.0	31.1	66.6	107.7
300	915	-81.5	-46.7	0.0	53.2	110.0
400	846	-179.6	-129.3	-71.0	0.0	77.0
500	787	-319.3	-258.3	-183.4	-97.7	0.0
600	735	-512.5	-440.9	-349.5	-246.0	-129.7
0	1300	-1.5	-1.5	-1.5	-1.5	-1.5
100	1066	0.0	14.0	31.2	50.1	71.0
200	967	-27.9	0.0	34.4	72.3	114.1
300	889	-90.5	-51.6	0.0	56.9	119.6
400	824	-194.5	-140.6	-75.9	0.0	83.6
500	768	-347.1	-279.2	-199.3	-104.5	0.0
600	718	-555.6	-474.1	-375.9	-263.1	-139.8
0	1200	-1.5	-1.5	-1.5	-1.5	-1.5
100	1021	0.0	15.6	34.0	54.5	77.3
200	933	-31.2	0.0	36.8	77.9	123.4
300	860	-99.6	-55.2	0.0	61.6	129.8
400	800	-214.1	-155.7	-82.1	0.0	91.0
500	746	-379.8	-308.4	-216.4	-113.8	0.0
600	698	-605.1	-513.6	-405.1	-284.1	-149.5
0	1100	-1.5	-1.5	-1.5	-1.5	-1.5
100	970	0.0	17.4	37.6	60.3	84.6
200	894	-34.7	0.0	40.5	85.9	134.5
300	828	-112.9	-60.7	0.0	68.1	141.1
400	771	-241.2	-171.7	-90.7	0.0	99.4
500	721	-421.4	-334.4	-234.9	-123.7	0.0
600	675	-667.2	-563.1	-443.0	-310.5	-161.5
0	1000	-1.5	-1.5	-1.5	-1.5	-1.5
100	910	0.0	20.4	44.0	68.0	95.8
200	842	-40.8	0.0	47.3	96.0	151.5
300	784	-130.1	-70.9	0.0	76.0	159.0
400	732	-272.6	-192.0	-101.5	0.0	110.6
500	686	-478.8	-378.9	-264.8	-138.1	0.0
600	642	-757.6	-637.1	-500.4	-348.6	-182.7
0	900	-1.5	-1.5	-1.5	-1.5	-1.5
100	832	0.0	25.5	51.0	80.9	115.9
200	775	-51.0	0.0	54.0	113.7	180.7
300	725	-153.9	-81.3	0.0	88.8	187.1
400	679	-323.6	-227.3	-118.4	0.0	131.0
500	636	-568.2	-451.8	-311.7	-163.7	0.0
600	596	-895.9	-759.8	-587.9	-410.3	-217.6

Universal Table for Rifle Bullets with Ballistic Coefficients of .200-.209

Sight Above Bore (in.) +1.5

Range (yds)	Velocity (fps)	Bullet Path Zero (100 yds)	Zero (200 yds)	Zero (300 yds)	Zero (400 yds)	Zero (500 yds)
0	4400	-1.5	-1.5	-1.5	-1.5	-1.5
100	3773	0.0	0.5	1.7	3.5	6.0
200	3228	-0.9	0.0	2.5	6.1	11.0
300	2747	-5.2	-3.8	0.0	5.4	12.7
400	2313	-14.2	-12.3	-7.2	0.0	9.7
500	1921	-29.8	-27.5	-21.1	-12.1	0.0
600	1576	-55.6	-52.8	-45.2	-34.4	-19.8
0	4300	-1.5	-1.5	-1.5	-1.5	-1.5
100	3685	0.0	0.5	1.9	3.8	6.3
200	3151	-1.0	0.0	2.7	6.5	11.6
300	2679	-5.6	-4.0	0.0	5.7	13.4
400	2251	-15.0	-12.9	-7.6	0.0	10.2
500	1865	-31.6	-29.0	-22.3	-12.8	0.0
600	1528	-59.0	-55.9	-47.9	-36.5	-21.1
0	4200	-1.5	-1.5	-1.5	-1.5	-1.5
100	3597	0.0	0.6	2.0	4.0	6.7
200	3074	-1.1	0.0	2.8	6.8	12.2
300	2610	-6.0	-4.2	0.0	6.0	14.1
400	2189	-15.9	-13.6	-8.0	0.0	10.8
500	1808	-33.4	-30.6	-23.5	-13.6	0.0
600	1480	-62.6	-59.2	-50.7	-38.7	-22.5
0	4100	-1.5	-1.5	-1.5	-1.5	-1.5
100	3509	0.0	0.6	2.1	4.2	7.1
200	2996	-1.3	0.0	3.0	7.2	12.9
300	2541	-6.4	-4.4	0.0	6.3	14.9
400	2126	-16.9	-14.3	-8.4	0.0	11.5
500	1751	-35.4	-32.2	-24.8	-14.3	0.0
600	1432	-66.4	-62.6	-53.7	-41.1	-23.9
0	4000	-1.5	-1.5	-1.5	-1.5	-1.5
100	3420	0.0	0.8	2.4	4.7	7.5
200	2923	-1.6	0.0	3.2	7.7	13.4
300	2473	-7.3	-4.8	0.0	6.8	15.3
400	2065	-18.7	-15.5	-9.0	0.0	11.3
500	1693	-37.5	-33.5	-25.4	-14.1	0.0
600	1384	-70.5	-65.7	-56.0	-42.4	-25.5
0	3900	-1.5	-1.5	-1.5	-1.5	-1.5
100	3339	0.0	0.8	2.5	4.9	7.9
200	2846	-1.6	0.0	3.4	8.2	14.1
300	2403	-7.6	-5.2	0.0	7.2	16.0
400	2002	-19.7	-16.5	-9.6	0.0	11.7
500	1635	-39.3	-35.2	-26.7	-14.7	0.0
600	1336	-74.3	-69.4	-59.1	-44.7	-27.1
0	3800	-1.5	-1.5	-1.5	-1.5	-1.5
100	3251	0.0	0.9	2.7	5.3	8.8
200	2767	-1.8	0.0	3.6	8.8	15.7
300	2332	-8.2	-5.5	0.0	7.7	18.1
400	1937	-21.1	-17.5	-10.2	0.0	13.9
500	1590	-43.9	-39.3	-30.2	-17.4	0.0
600	1287	-78.8	-73.4	-62.5	-47.1	-27.8
0	3700	-1.5	-1.5	-1.5	-1.5	-1.5
100	3162	0.0	1.0	2.9	5.6	9.4
200	2689	-2.0	0.0	3.9	9.3	16.8
300	2260	-8.8	-5.9	0.0	8.1	19.4
400	1873	-22.6	-18.7	-10.9	0.0	15.0
500	1535	-47.0	-42.1	-32.3	-18.7	0.0
600	1238	-83.6	-77.7	-66.0	-49.7	-28.9

Universal Table for Rifle Bullets with Ballistic Coefficients of .200-.209

Sight Above Bore (in.) +1.5

		Bullet Path				
Range (yds)	Velocity (fps)	Zero (100 yds)	Zero (200 yds)	Zero (300 yds)	Zero (400 yds)	Zero (500 yds)
0	3600	-1.5	-1.5	-1.5	-1.5	-1.5
100	3073	0.0	1.1	3.1	6.0	10.1
200	2609	-2.2	0.0	4.1	9.9	18.0
300	2188	-9.4	-6.2	0.0	8.7	20.8
400	1808	-24.2	-19.9	-11.6	0.0	16.1
500	1480	-50.3	-45.0	-34.6	-20.1	0.0
600	1219	-93.2	-86.7	-74.3	-56.9	-32.8
0	3500	-1.5	-1.5	-1.5	-1.5	-1.5
100	2984	0.0	1.3	3.4	6.5	10.8
200	2531	-2.6	0.0	4.1	10.3	19.0
300	2116	-10.1	-6.2	0.0	9.3	22.3
400	1742	-25.9	-20.7	-12.4	0.0	17.3
500	1425	-54.0	-47.4	-37.1	-21.6	0.0
600	1179	-99.8	-91.9	-79.6	-60.9	-35.0
0	3400	-1.5	-1.5	-1.5	-1.5	-1.5
100	2899	0.0	1.3	3.8	6.8	11.5
200	2451	-2.7	0.0	4.8	10.9	20.3
300	2045	-11.3	-7.2	0.0	9.2	23.2
400	1675	-27.2	-21.9	-12.2	0.0	18.6
500	1369	-57.3	-50.7	-38.6	-23.3	0.0
600	1138	-106.5	-98.5	-84.0	-65.7	-37.7
0	3300	-1.5	-1.5	-1.5	-1.5	-1.5
100	2810	0.0	1.5	4.1	7.7	12.3
200	2370	-3.0	0.0	5.2	12.4	21.6
300	1972	-12.2	-7.8	0.0	10.8	24.6
400	1620	-30.7	-24.8	-14.4	0.0	18.3
500	1314	-61.3	-53.9	-40.9	-22.9	0.0
600	1099	-113.8	-105.0	-89.4	-67.7	-40.3
0	3200	-1.5	-1.5	-1.5	-1.5	-1.5
100	2720	0.0	1.6	4.4	8.3	13.1
200	2289	-3.2	0.0	5.6	13.4	23.0
300	1899	-13.2	-8.4	0.0	11.8	26.1
400	1557	-33.3	-26.9	-15.7	0.0	19.1
500	1258	-65.6	-57.5	-43.6	-23.9	0.0
600	1059	-122.1	-112.4	-95.6	-72.1	-43.3
0	3100	-1.5	-1.5	-1.5	-1.5	-1.5
100	2630	0.0	1.8	4.8	9.0	14.8
200	2207	-3.5	0.0	6.0	14.5	26.1
300	1825	-14.4	-9.0	0.0	12.7	30.1
400	1495	-36.1	-29.0	-17.0	0.0	23.1
500	1230	-74.1	-65.2	-50.2	-28.9	0.0
600	1051	-134.8	-124.1	-106.0	-80.6	-45.8
0	3000	-1.5	-1.5	-1.5	-1.5	-1.5
100	2540	0.0	1.9	5.2	9.8	16.1
200	2125	-3.9	0.0	6.5	15.8	28.3
300	1750	-15.6	-9.8	0.0	13.9	32.6
400	1432	-39.3	-31.6	-18.5	0.0	25.0
500	1184	-80.4	-70.7	-54.3	-31.2	0.0
600	1020	-145.7	-134.0	-114.4	-86.7	-49.2
0	2900	-1.5	-1.5	-1.5	-1.5	-1.5
100	2451	0.0	2.2	5.4	10.5	17.3
200	2045	-4.4	0.0	6.5	16.6	30.1
300	1675	-16.3	-9.7	0.0	15.1	35.5
400	1369	-41.9	-33.1	-20.2	0.0	27.1
500	1138	-86.3	-75.3	-59.1	-33.9	0.0
600	1000	-159.0	-145.9	-126.4	-96.2	-55.5

Universal Table for Rifle Bullets with Ballistic Coefficients of .200-.209

Sight Above Bore (in.) +1.5

Range (yds)	Velocity (fps)	Bullet Path Zero (100 yds)	Zero (200 yds)	Zero (300 yds)	Zero (400 yds)	Zero (500 yds)
0	2800	-1.5	-1.5	-1.5	-1.5	-1.5
100	2361	0.0	2.4	6.3	11.3	18.6
200	1963	-4.9	0.0	7.8	17.7	32.4
300	1613	-18.9	-11.6	0.0	15.0	36.9
400	1307	-45.2	-35.5	-20.0	0.0	29.2
500	1094	-93.1	-80.9	-61.5	-36.6	0.0
600	982	-170.1	-155.5	-132.2	-102.3	-58.4
0	2700	-1.5	-1.5	-1.5	-1.5	-1.5
100	2270	0.0	2.7	7.0	12.2	20.1
200	1881	-5.4	0.0	8.6	19.1	34.8
300	1542	-20.9	-12.9	0.0	15.7	39.4
400	1244	-48.9	-38.1	-21.0	0.0	31.5
500	1050	-100.5	-87.1	-65.6	-39.4	0.0
600	963	-185.0	-168.9	-143.1	-111.7	-64.4
0	2600	-1.5	-1.5	-1.5	-1.5	-1.5
100	2178	0.0	3.0	7.7	14.1	22.6
200	1798	-6.0	0.0	9.4	22.3	39.2
300	1472	-23.1	-14.2	0.0	19.2	44.6
400	1214	-56.4	-44.5	-25.6	0.0	33.9
500	1040	-112.9	-98.0	71.4	-42.4	0.0
600	948	-201.8	-183.9	-155.6	-117.2	-66.3
0	2500	-1.5	-1.5	-1.5	-1.5	-1.5
100	2087	0.0	3.3	8.5	15.5	24.5
200	1715	-6.6	0.0	10.4	24.4	42.5
300	1403	-25.5	-15.5	0.0	21.1	48.2
400	1163	-62.0	-48.8	-28.1	0.0	36.1
500	1007	-122.7	-106.2	-80.3	-45.2	0.0
600	929	-218.4	-198.6	-167.5	-125.4	-71.2
0	2400	-1.5	-1.5	-1.5	-1.5	-1.5
100	1999	0.0	3.7	8.9	16.6	26.9
200	1643	-7.4	0.0	10.5	25.9	46.5
300	1334	-26.8	-15.8	0.0	23.1	54.0
400	1113	-66.5	-51.8	-30.8	0.0	41.2
500	990	-134.7	-116.2	-90.0	-51.5	0.0
600	912	-237.3	-215.1	-183.6	-137.4	-75.6
0	2300	-1.5	-1.5	-1.5	-1.5	-1.5
100	1908	0.0	4.2	9.7	18.1	29.5
200	1565	-8.3	0.0	11.1	27.9	50.6
300	1265	-29.1	-16.7	0.0	25.2	59.3
400	1064	-72.4	-55.8	-33.6	0.0	45.4
500	970	-147.3	-126.5	-98.8	-56.8	0.0
600	895	-256.4	-231.4	-198.1	-147.7	-79.6
0	2200	-1.5	-1.5	-1.5	-1.5	-1.5
100	1817	0.0	4.7	11.5	20.6	31.4
200	1488	-9.4	0.0	13.7	31.9	53.9
300	1225	-34.6	-20.6	0.0	27.3	60.8
400	1047	-82.6	-63.9	-36.4	0.0	45.1
500	950	-157.2	-134.7	-101.4	-56.4	0.0
600	878	-277.8	-249.7	-208.5	-153.9	-79.9
0	2100	-1.5	-1.5	-1.5	-1.5	-1.5
100	1725	0.0	5.3	12.9	22.7	35.4
200	1411	-10.5	0.0	15.2	34.8	60.2
300	1168	-38.6	-22.9	0.0	29.3	67.5
400	1010	-90.6	-69.6	-39.1	0.0	50.8
500	931	-176.8	-150.5	-112.4	-63.5	0.0
600	862	-299.3	-267.8	-222.0	-163.3	-87.1

Universal Table for Rifle Bullets with Ballistic Coefficients of .210-.219

Sight Above Bore (in.) +1.5

Range (yds)	Velocity (fps)	Bullet Path Zero (100 yds)	Zero (200 yds)	Zero (300 yds)	Zero (400 yds)	Zero (500 yds)
0	4400	-1.5	-1.5	-1.5	-1.5	-1.5
100	3800	0.0	0.5	1.7	3.4	5.7
200	3276	-0.9	0.0	2.5	5.9	10.5
300	2811	-5.1	-3.7	0.0	5.2	12.1
400	2391	-13.7	-11.9	-6.9	0.0	9.2
500	2009	-28.6	-26.3	-20.1	-11.5	0.0
600	1658	-50.9	-48.2	-40.8	-30.4	-16.6
0	4300	-1.5	-1.5	-1.5	-1.5	-1.5
100	3712	0.0	0.5	1.8	3.6	6.1
200	3198	-1.0	0.0	2.6	6.3	11.1
300	2742	-5.4	-3.9	0.0	5.5	12.7
400	2328	-14.6	-12.5	-7.3	0.0	9.7
500	1952	-30.3	-27.8	-21.2	-12.1	0.0
600	1618	-55.7	-52.7	-44.8	-33.9	-19.4
0	4200	-1.5	-1.5	-1.5	-1.5	-1.5
100	3624	0.0	0.6	1.9	3.9	6.4
200	3121	-1.1	0.0	2.8	6.6	11.7
300	2672	-5.8	-4.1	0.0	5.8	13.5
400	2265	-15.5	-13.2	-7.7	0.0	10.3
500	1894	-32.2	-29.4	-22.4	-12.8	0.0
600	1568	-59.2	-55.9	-47.6	-36.0	-20.6
0	4100	-1.5	-1.5	-1.5	-1.5	-1.5
100	3535	0.0	0.6	2.1	4.1	6.8
200	3042	-1.2	0.0	2.9	7.0	12.4
300	2602	-6.2	-4.4	0.0	6.1	14.2
400	2200	-16.5	-14.0	-8.1	0.0	10.8
500	1836	-34.1	-31.0	-23.7	-13.5	0.0
600	1518	-63.0	-59.3	-50.5	-38.3	-22.1
0	4000	-1.5	-1.5	-1.5	-1.5	-1.5
100	3445	0.0	0.8	2.3	4.4	7.2
200	2967	-1.6	0.0	3.1	7.2	12.9
300	2533	-7.0	-4.7	0.0	6.0	14.7
400	2136	-17.4	-14.3	-8.0	0.0	11.5
500	1777	-36.2	-32.3	-24.5	-14.4	0.0
600	1468	-67.0	-62.3	-52.9	-40.9	-23.6
0	3900	-1.5	-1.5	-1.5	-1.5	-1.5
100	3363	0.0	0.8	2.5	4.8	7.6
200	2889	-1.6	0.0	3.3	7.9	13.6
300	2462	-7.4	-5.0	0.0	6.9	15.4
400	2073	-19.0	-15.9	-9.2	0.0	11.3
500	1718	-38.0	-34.0	-25.7	-14.2	0.0
600	1419	-70.7	-65.9	-55.9	-42.1	-25.1
0	3800	-1.5	-1.5	-1.5	-1.5	-1.5
100	3275	0.0	0.9	2.7	5.1	8.1
200	2810	-1.8	0.0	3.6	8.4	14.4
300	2390	-8.0	-5.3	0.0	7.3	16.2
400	2008	-20.4	-16.9	-9.8	0.0	11.8
500	1657	-40.3	-35.9	-27.0	-14.8	0.0
600	1368	-75.3	-70.0	-59.3	-44.7	-26.9
0	3700	-1.5	-1.5	-1.5	-1.5	-1.5
100	3186	0.0	1.0	2.9	5.5	9.0
200	2731	-1.9	0.0	3.8	9.0	16.0
300	2318	-8.6	-5.7	0.0	7.8	18.3
400	1942	-21.9	-18.0	-10.5	0.0	13.9
500	1610	-44.8	-40.0	-30.5	-17.4	0.0
600	1318	-80.0	-74.2	-62.8	-47.1	-26.2

Universal Table for Rifle Bullets with Ballistic Coefficients of .210-.219

Sight Above Bore (in.) +1.5

Range (yds)	Velocity (fps)	Bullet Path Zero (100 yds)	Zero (200 yds)	Zero (300 yds)	Zero (400 yds)	Zero (500 yds)
0	3600	-1.5	-1.5	-1.5	-1.5	-1.5
100	3097	0.0	1.1	3.1	5.9	9.6
200	2651	-2.1	0.0	4.0	9.6	17.1
300	2245	-9.2	-6.1	0.0	8.4	19.6
400	1876	-23.5	-19.3	-11.2	0.0	15.0
500	1553	-48.1	-42.8	-32.7	-18.8	0.0
600	1267	-85.1	-78.7	-66.6	-49.8	-27.3
0	3500	-1.5	-1.5	-1.5	-1.5	-1.5
100	3007	0.0	1.2	3.3	6.3	10.4
200	2570	-2.3	0.0	4.3	10.3	18.4
300	2172	-9.9	-6.4	0.0	9.0	21.1
400	1809	-25.2	-20.6	-12.0	0.0	16.2
500	1496	-51.8	-46.0	-35.2	-20.3	0.0
600	1242	-94.6	-87.6	-74.7	-56.7	-32.4
0	3400	-1.5	-1.5	-1.5	-1.5	-1.5
100	2921	0.0	1.3	3.6	6.6	11.0
200	2491	-2.6	0.0	4.7	10.6	19.4
300	2100	-10.9	-7.0	0.0	9.0	22.1
400	1742	-26.5	-21.3	-12.0	0.0	17.5
500	1439	-55.0	-48.5	-36.9	-21.9	0.0
600	1199	-101.1	-93.2	-79.3	-61.3	-35.0
0	3300	-1.5	-1.5	-1.5	-1.5	-1.5
100	2832	0.0	1.4	4.0	7.1	11.8
200	2410	-2.9	0.0	5.0	11.3	20.7
300	2026	-11.9	-7.5	0.0	9.4	23.6
400	1674	-28.3	-22.5	-12.5	0.0	18.9
500	1382	-59.1	-51.9	-39.3	-23.7	0.0
600	1157	-108.6	-100.0	-84.9	-66.2	-37.7
0	3200	-1.5	-1.5	-1.5	-1.5	-1.5
100	2742	0.0	1.6	4.3	8.0	12.7
200	2328	-3.2	0.0	5.4	12.9	22.2
300	1951	-12.9	-8.2	0.0	11.2	25.1
400	1617	-32.1	-25.8	-14.9	0.0	18.6
500	1325	-63.3	-55.4	-41.8	-23.2	0.0
600	1116	-116.5	-107.0	-90.7	-68.4	-40.5
0	3100	-1.5	-1.5	-1.5	-1.5	-1.5
100	2651	0.0	1.7	4.7	8.7	13.6
200	2245	-3.5	0.0	5.9	14.0	23.7
300	1876	-14.1	-8.8	0.0	12.1	26.8
400	1553	-34.9	-27.9	-16.2	0.0	19.5
500	1267	-68.0	-59.3	-44.6	-24.4	0.0
600	1075	-125.2	-114.7	-97.0	-72.8	-43.5
0	3000	-1.5	-1.5	-1.5	-1.5	-1.5
100	2561	0.0	1.9	5.1	9.5	15.4
200	2162	-3.9	0.0	6.4	15.2	26.9
300	1801	-15.3	-9.5	0.0	13.2	30.8
400	1489	-38.0	-30.3	-17.6	0.0	23.4
500	1237	-76.9	-67.2	-51.3	-29.3	0.0
600	1062	-138.7	-127.1	-108.1	-81.6	-46.5
0	2900	-1.5	-1.5	-1.5	-1.5	-1.5
100	2471	0.0	2.1	5.3	10.1	16.5
200	2082	-4.3	0.0	6.4	16.0	28.8
300	1725	-16.0	-9.6	0.0	14.4	33.6
400	1425	-40.6	-32.1	-19.2	0.0	25.5
500	1189	-82.6	-72.0	-56.0	-31.9	0.0
600	1031	-148.3	-135.5	-116.3	-87.4	-49.1

Universal Table for Rifle Bullets with Ballistic Coefficients of .210-.219

Sight Above Bore (in.) +1.5

Range (yds)	Velocity (fps)	Bullet Path Zero (100 yds)	Zero (200 yds)	Zero (300 yds)	Zero (400 yds)	Zero (500 yds)
0	2800	-1.5	-1.5	-1.5	-1.5	-1.5
100	2380	0.0	2.4	6.1	11.0	17.9
200	1999	-4.7	0.0	7.5	17.2	31.1
300	1658	-18.4	-11.3	0.0	14.6	35.4
400	1361	-44.0	-34.5	-19.4	0.0	27.8
500	1142	-89.7	-77.9	-59.0	-34.7	0.0
600	1000	-160.0	-145.8	-123.2	-94.0	-52.3
0	2700	-1.5	-1.5	-1.5	-1.5	-1.5
100	2289	0.0	2.6	6.8	11.9	19.5
200	1916	-5.3	0.0	8.3	18.6	33.6
300	1587	-20.3	-12.4	0.0	15.4	38.1
400	1297	-47.6	-37.1	-20.6	0.0	30.2
500	1096	-97.3	-84.1	-63.5	-37.7	0.0
600	989	-175.3	-159.5	-134.8	-103.9	-58.6
0	2600	-1.5	-1.5	-1.5	-1.5	-1.5
100	2197	0.0	2.9	7.5	12.8	21.1
200	1832	-5.9	0.0	9.1	19.8	36.2
300	1516	-22.4	-13.6	0.0	16.1	40.8
400	1234	-51.4	-39.7	-21.5	0.0	32.8
500	1051	-105.3	-90.6	-67.9	-41.0	0.0
600	969	-191.0	-173.4	-146.1	-113.9	-64.6
0	2500	-1.5	-1.5	-1.5	-1.5	-1.5
100	2118	0.0	3.2	8.1	14.6	23.1
200	1772	-6.4	0.0	9.8	22.8	39.7
300	1475	-24.4	-14.7	0.0	19.5	44.9
400	1234	-58.5	-45.6	-26.0	0.0	33.8
500	1065	-115.4	-99.3	-74.8	-42.3	0.0
600	968	-202.1	-182.8	-153.3	-114.4	-63.6
0	2400	-1.5	-1.5	-1.5	-1.5	-1.5
100	2016	0.0	3.2	8.7	16.2	25.5
200	1665	-6.4	0.0	11.1	25.9	44.6
300	1375	-26.2	-16.7	0.0	22.2	50.2
400	1152	-64.6	-51.8	-29.6	0.0	37.4
500	1006	-127.5	-111.5	-83.7	-46.7	0.0
600	932	-225.9	-206.8	-173.5	-129.0	-73.0
0	2300	-1.5	-1.5	-1.5	-1.5	-1.5
100	1925	0.0	4.0	9.5	17.6	28.3
200	1595	-8.1	0.0	11.0	27.1	48.4
300	1305	-28.6	-16.4	0.0	24.2	56.2
400	1102	-70.4	-54.2	-32.3	0.0	42.6
500	991	-141.3	-121.0	-93.6	-53.2	0.0
600	915	-246.5	-222.2	-189.4	-140.9	-77.0
0	2200	-1.5	-1.5	-1.5	-1.5	-1.5
100	1834	0.0	4.6	10.4	19.2	31.0
200	1517	-9.2	0.0	11.7	29.3	52.8
300	1234	-31.2	-17.5	0.0	26.5	61.6
400	1052	-77.0	-58.6	-35.3	0.0	46.9
500	969	-154.8	-131.9	-102.7	-58.6	0.0
600	897	-268.0	-240.6	-205.6	-152.6	-82.3
0	2100	-1.5	-1.5	-1.5	-1.5	-1.5
100	1742	0.0	5.2	12.6	22.1	34.4
200	1439	-10.4	0.0	14.8	33.8	58.4
300	1199	-37.7	-22.2	0.0	28.5	65.4
400	1037	-88.3	-67.6	-38.0	0.0	49.1
500	950	-171.8	-145.9	-109.0	-61.4	0.0
600	880	-290.1	-259.0	-214.7	-157.7	-83.9

Universal Table for Rifle Bullets with Ballistic Coefficients of .220-.229						
					Sight Above Bore (in.)	+1.5
			Bullet Path			
Range (yds)	Velocity (fps)	Zero (100 yds)	Zero (200 yds)	Zero (300 yds)	Zero (400 yds)	Zero (500 yds)
0	4400	-1.5	-1.5	-1.5	-1.5	-1.5
100	3889	0.0	0.4	1.5	3.0	4.9
200	3434	-0.8	0.0	2.2	5.1	9.0
300	3025	-4.6	-3.4	0.0	4.3	10.1
400	2650	-11.9	-10.2	-5.8	0.0	7.7
500	2306	-24.5	-22.5	-16.9	-9.7	0.0
600	1988	-44.0	-41.6	-34.8	-26.2	-14.6
0	4300	-1.5	-1.5	-1.5	-1.5	-1.5
100	3800	0.0	0.5	1.7	3.3	5.2
200	3354	-0.9	0.0	2.4	5.6	9.4
300	2952	-5.0	-3.6	0.0	4.8	10.6
400	2585	-13.0	-11.2	-6.4	0.0	7.7
500	2244	-25.9	-23.6	-17.6	-9.7	0.0
600	1931	-46.5	-43.8	-36.6	-27.0	-15.4
0	4200	-1.5	-1.5	-1.5	-1.5	-1.5
100	3710	0.0	0.5	1.8	3.5	5.5
200	3274	-1.0	0.0	2.5	5.9	9.9
300	2879	-5.4	-3.8	0.0	5.1	11.1
400	2517	-13.9	-11.9	-6.8	0.0	8.0
500	2182	-27.4	-24.8	-10.4	-9.9	0.0
600	1874	-49.2	-46.1	-38.5	-28.3	-16.4
0	4100	-1.5	-1.5	-1.5	-1.5	-1.5
100	3619	0.0	0.6	1.9	3.7	6.0
200	3193	-1.1	0.0	2.7	6.3	10.8
300	2806	-5.8	-4.0	0.0	5.4	12.1
400	2450	-14.9	-12.6	-7.2	0.0	9.0
500	2121	-29.8	-27.0	-20.2	-11.3	0.0
600	1816	-52.1	-48.6	-40.5	-29.8	-16.3
0	4000	-1.5	-1.5	-1.5	-1.5	-1.5
100	3529	0.0	0.6	2.1	4.0	6.4
200	3111	-1.3	0.0	2.9	6.7	11.5
300	2732	-6.2	-4.3	0.0	5.7	12.9
400	2382	-15.9	-13.3	-7.6	0.0	9.6
500	2058	-31.9	-28.6	-21.5	-12.0	0.0
600	1757	-55.2	-51.3	-42.8	-31.3	-16.9
0	3900	-1.5	-1.5	-1.5	-1.5	-1.5
100	3443	0.0	0.7	2.1	4.2	6.7
200	3033	-1.4	0.0	2.8	6.9	12.0
300	2658	-6.4	-4.3	0.0	6.1	13.7
400	2313	-16.7	-13.8	-8.1	0.0	10.2
500	1995	-33.6	-30.0	-22.9	-12.8	0.0
600	1705	-59.5	-55.2	-46.6	-34.5	-19.2
0	3800	-1.5	-1.5	-1.5	-1.5	-1.5
100	3353	0.0	0.8	2.4	4.4	7.2
200	2951	-1.6	0.0	3.2	7.3	12.7
300	2584	-7.3	-4.8	0.0	6.1	14.3
400	2243	-17.8	-14.6	-8.1	0.0	10.9
500	1931	-35.8	-31.8	-23.8	-13.6	0.0
600	1649	-63.6	-58.7	-49.1	-36.9	-20.6
0	3700	-1.5	-1.5	-1.5	-1.5	-1.5
100	3263	0.0	0.9	2.6	4.7	7.7
200	2870	-1.8	0.0	3.5	7.7	13.5
300	2508	-7.9	-5.2	0.0	6.3	15.1
400	2173	-18.9	-15.4	-8.4	0.0	11.7
500	1866	-38.3	-33.8	-25.1	-14.6	0.0
600	1592	-68.1	-62.8	-52.4	-39.7	-22.2

Universal Table for Rifle Bullets with Ballistic Coefficients of .220-.229						
					Sight Above Bore (in.)	+1.5
Bullet Path						
Range (yds)	Velocity (fps)	Zero (100 yds)	Zero (200 yds)	Zero (300 yds)	Zero (400 yds)	Zero (500 yds)
0	3600	-1.5	-1.5	-1.5	-1.5	-1.5
100	3172	0.0	1.0	2.8	5.3	8.2
200	2787	-2.0	0.0	3.7	8.5	14.3
300	2433	-8.5	-5.5	0.0	7.3	16.0
400	2105	-21.0	-17.0	-9.7	0.0	11.6
500	1801	-40.8	-35.8	-26.6	-14.5	0.0
600	1536	-73.0	-67.0	-55.9	-41.4	-24.0
0	3500	-1.5	-1.5	-1.5	-1.5	-1.5
100	3081	0.0	1.1	3.1	5.7	8.7
200	2704	-2.2	0.0	4.0	9.2	15.2
300	2356	-9.3	-6.0	0.0	7.8	16.9
400	2034	-22.8	-18.3	-10.4	0.0	12.1
500	1735	-43.6	-38.1	-28.2	-15.2	0.0
600	1479	-78.3	-71.7	-59.8	-44.2	-26.0
0	3400	-1.5	-1.5	-1.5	-1.5	-1.5
100	2993	0.0	1.2	3.2	6.0	9.5
200	2622	-2.4	0.0	4.0	9.6	16.7
300	2279	-9.7	-6.1	0.0	8.4	19.0
400	1963	-24.1	-19.3	-11.2	0.0	14.1
500	1677	-47.7	-41.7	-31.6	-17.6	0.0
600	1423	-83.2	-76.0	-63.9	-47.1	-26.0
0	3300	-1.5	-1.5	-1.5	-1.5	-1.5
100	2902	0.0	1.3	3.5	6.5	10.3
200	2538	-2.7	0.0	4.2	10.3	17.9
300	2201	-10.4	-6.4	0.0	9.0	20.6
400	1892	-25.9	-20.5	-12.0	0.0	15.4
500	1614	-51.6	-44.9	-34.3	-19.2	0.0
600	1366	-89.2	-81.1	-68.4	-50.3	-27.3
0	3200	-1.5	-1.5	-1.5	-1.5	-1.5
100	2811	0.0	1.5	3.9	7.0	11.1
200	2454	-3.0	0.0	4.9	10.9	19.3
300	2125	-11.8	-7.3	0.0	9.1	21.6
400	1820	-27.8	-21.9	-12.1	0.0	16.7
500	1552	-55.7	-48.2	-36.0	-20.9	0.0
600	1309	-95.5	-86.6	-71.9	-53.7	-28.7
0	3100	-1.5	-1.5	-1.5	-1.5	-1.5
100	2719	0.0	1.6	4.3	7.5	12.1
200	2370	-3.3	0.0	5.3	11.7	20.8
300	2047	-12.9	-8.0	0.0	9.6	23.3
400	1747	-30.0	-23.4	-12.8	0.0	18.3
500	1489	-60.3	-52.1	-38.8	-22.8	0.0
600	1273	-106.3	-96.4	-80.5	-61.3	-33.9
0	3000	-1.5	-1.5	-1.5	-1.5	-1.5
100	2627	0.0	1.8	4.7	8.4	13.0
200	2284	-3.7	0.0	5.8	13.2	22.4
300	1968	-14.2	-8.7	0.0	11.1	25.0
400	1682	-33.7	-26.4	-14.8	0.0	18.4
500	1427	-65.2	-56.1	-41.6	-23.1	0.0
600	1224	-115.6	-104.6	-87.3	-65.0	-37.3
0	2900	-1.5	-1.5	-1.5	-1.5	-1.5
100	2536	0.0	1.8	5.0	9.1	14.0
200	2199	-3.7	0.0	6.3	14.5	24.3
300	1890	-14.9	-9.5	0.0	12.3	26.9
400	1613	-36.3	-29.0	-16.3	0.0	19.6
500	1364	-69.8	-60.7	-44.9	-24.5	0.0
600	1177	-124.4	-113.4	-94.5	-70.0	-40.6

Universal Table for Rifle Bullets with Ballistic Coefficients of .220-.229

Sight Above Bore (in.) +1.5

		Bullet Path				
Range (yds)	Velocity (fps)	Zero (100 yds)	Zero (200 yds)	Zero (300 yds)	Zero (400 yds)	Zero (500 yds)
0	2800	-1.5	-1.5	-1.5	-1.5	-1.5
100	2444	0.0	2.2	5.4	9.9	15.0
200	2115	-4.4	0.0	6.4	15.4	25.6
300	1810	-16.2	-9.6	0.0	13.5	28.8
400	1544	-39.6	-30.8	-18.0	0.0	20.5
500	1302	-75.1	-64.1	-48.0	-25.6	0.0
600	1131	-134.5	-121.3	-102.0	-75.1	-44.4
0	2700	-1.5	-1.5	-1.5	-1.5	-1.5
100	2351	0.0	2.5	5.8	10.8	17.0
200	2030	-4.9	0.0	6.8	16.7	29.1
300	1731	-17.5	-10.1	0.0	14.9	33.5
400	1476	-43.2	-33.4	-19.9	0.0	24.8
500	1262	-85.0	-72.7	-55.8	-31.0	0.0
600	1085	-145.7	-130.9	-110.6	-80.9	-43.7
0	2600	-1.5	-1.5	-1.5	-1.5	-1.5
100	2258	0.0	2.8	6.8	11.8	18.6
200	1944	-5.5	0.0	8.1	18.1	31.7
300	1660	-20.4	-12.1	0.0	15.1	35.4
400	1407	-47.3	-36.3	-20.1	0.0	27.1
500	1210	-93.1	-79.2	-59.0	-33.9	0.0
600	1067	-162.4	-145.9	-121.6	-91.4	-50.8
0	2500	-1.5	-1.5	-1.5	-1.5	-1.5
100	2165	0.0	3.1	7.6	12.9	20.4
200	1858	-6.2	0.0	9.0	19.6	34.6
300	1585	-22.8	-13.5	0.0	15.9	38.4
400	1340	-51.6	-39.2	-21.1	0.0	30.0
500	1159	-102.0	-86.5	-64.0	-37.5	0.0
600	1032	-177.2	-158.5	-131.5	-99.8	-54.7
0	2400	-1.5	-1.5	-1.5	-1.5	-1.5
100	2074	0.0	3.1	8.1	13.7	22.0
200	1773	-6.2	0.0	10.1	21.3	37.8
300	1511	-24.4	-15.1	0.0	16.8	41.6
400	1272	-54.9	-42.6	-22.4	0.0	33.1
500	1109	-110.0	-94.5	-69.3	-41.3	0.0
600	1011	-194.0	-175.5	-145.2	-111.7	-62.1
0	2300	-1.5	-1.5	-1.5	-1.5	-1.5
100	1982	0.0	3.7	9.0	15.9	24.7
200	1694	-7.5	0.0	10.5	24.4	41.9
300	1437	-26.9	-15.7	0.0	20.9	47.2
400	1233	-63.7	-48.8	-27.8	0.0	35.0
500	1083	-123.4	-104.7	-78.6	-43.8	0.0
600	989	-211.5	-189.1	-157.7	-115.9	-63.4
0	2200	-1.5	-1.5	-1.5	-1.5	-1.5
100	1889	0.0	4.3	9.8	17.6	27.0
200	1612	-8.5	0.0	11.1	26.7	45.5
300	1364	-29.4	-16.6	0.0	23.4	51.6
400	1177	-70.4	-53.3	-31.2	0.0	37.7
500	1045	-135.1	-113.8	-86.1	-47.1	0.0
600	966	-232.4	-206.8	-173.5	-126.7	-70.2
0	2100	-1.5	-1.5	-1.5	-1.5	-1.5
100	1796	0.0	4.9	10.8	19.4	30.5
200	1531	-9.8	0.0	11.8	29.0	51.3
300	1290	-32.4	-17.7	0.0	25.8	59.2
400	1123	-77.6	-58.0	-34.4	0.0	44.5
500	1015	-152.6	-128.2	-98.7	-55.7	0.0
600	946	-254.5	-225.2	-189.8	-138.2	-71.4

Universal Table for Rifle Bullets with Ballistic Coefficients of .230-.239

Sight Above Bore (in.) +1.5

Range (yds)	Velocity (fps)	Bullet Path Zero (100 yds)	Zero (200 yds)	Zero (300 yds)	Zero (400 yds)	Zero (500 yds)
0	4400	-1.5	-1.5	-1.5	-1.5	-1.5
100	3910	0.0	0.4	1.5	2.9	4.8
200	3471	-0.8	0.0	2.2	5.0	8.8
300	3075	-4.5	-3.3	0.0	4.2	9.9
400	2712	-11.6	-10.0	-5.7	0.0	7.5
500	2377	-23.9	-21.9	-16.4	-9.4	0.0
600	2067	-42.4	-40.1	-33.5	-25.0	-13.8
0	4300	-1.5	-1.5	-1.5	-1.5	-1.5
100	3820	0.0	0.5	1.6	3.1	5.0
200	3390	-0.9	0.0	2.3	5.3	9.2
300	3002	-4.8	-3.5	0.0	4.4	10.3
400	2644	-12.3	-10.5	-5.9	0.0	7.9
500	2315	-25.2	-23.0	-17.2	-9.8	0.0
600	2009	-44.9	-42.2	-35.2	-26.4	-14.6
0	4200	-1.5	-1.5	-1.5	-1.5	-1.5
100	3730	0.0	0.5	1.7	3.4	5.3
200	3309	-1.0	0.0	2.5	5.8	9.7
300	2928	-5.2	-3.7	0.0	4.9	10.8
400	2577	-13.6	-11.5	-6.6	0.0	7.8
500	2251	-26.7	-24.2	-18.0	-9.8	0.0
600	1951	-47.5	-44.5	-37.1	-27.2	-15.5
0	4100	-1.5	-1.5	-1.5	-1.5	-1.5
100	3639	0.0	0.6	1.9	3.6	5.7
200	3228	-1.1	0.0	2.6	6.1	10.2
300	2854	-5.6	-3.9	0.0	5.2	11.3
400	2509	-14.5	-12.2	-7.0	0.0	8.1
500	2187	-28.3	-25.4	-18.9	-10.2	0.0
600	1892	-50.4	-47.0	-39.1	-28.7	-16.5
0	4000	-1.5	-1.5	-1.5	-1.5	-1.5
100	3548	0.0	0.6	2.0	3.9	6.2
200	3146	-1.3	0.0	2.8	6.5	11.1
300	2779	-6.1	-4.2	0.0	5.5	12.4
400	2440	-15.5	-13.0	-7.4	0.0	9.2
500	2125	-30.9	-27.7	-20.7	-11.5	0.0
600	1832	-53.5	-49.7	-41.3	-30.2	-16.4
0	3900	-1.5	-1.5	-1.5	-1.5	-1.5
100	3462	0.0	0.7	2.1	4.1	6.5
200	3067	-1.4	0.0	2.8	6.8	11.6
300	2704	-6.4	-4.2	0.0	5.9	13.2
400	2370	-16.3	-13.5	-7.9	0.0	9.7
500	2061	-32.6	-29.1	-22.0	-12.2	0.0
600	1772	-56.3	-52.0	-43.6	-31.7	-17.1
0	3800	-1.5	-1.5	-1.5	-1.5	-1.5
100	3371	0.0	0.8	2.4	4.3	7.0
200	2984	-1.6	0.0	3.1	7.1	12.3
300	2629	-7.1	-4.7	0.0	6.0	13.8
400	2300	-17.4	-14.2	-7.9	0.0	10.4
500	1996	-34.8	-30.8	-23.0	-13.0	0.0
600	1717	-61.1	-56.4	-46.9	-35.0	-19.4
0	3700	-1.5	-1.5	-1.5	-1.5	-1.5
100	3281	0.0	0.9	2.6	4.6	7.4
200	2902	-1.8	0.0	3.4	7.5	13.1
300	2553	-7.7	-5.1	0.0	6.2	14.6
400	2229	-18.6	-15.0	-8.3	0.0	11.2
500	1930	-37.2	-32.8	-24.4	-14.0	0.0
600	1659	-65.5	-60.2	-50.1	-37.6	-20.8

Universal Table for Rifle Bullets with Ballistic Coefficients of .230-.239

Sight Above Bore (in.) +1.5

Range (yds)	Velocity (fps)	Bullet Path Zero (100 yds)	Zero (200 yds)	Zero (300 yds)	Zero (400 yds)	Zero (500 yds)
0	3600	-1.5	-1.5	-1.5	-1.5	-1.5
100	3190	0.0	1.0	2.8	5.1	8.0
200	2819	-2.0	0.0	3.6	8.3	13.9
300	2476	-8.4	-5.4	0.0	7.0	15.5
400	2159	-20.5	-16.6	-9.3	0.0	11.3
500	1864	-39.8	-34.9	-25.8	-14.2	0.0
600	1601	-70.3	-64.4	-53.5	-39.5	-22.5
0	3500	-1.5	-1.5	-1.5	-1.5	-1.5
100	3098	0.0	1.1	3.0	5.5	8.5
200	2736	-2.2	0.0	3.9	8.9	14.9
300	2399	-9.1	-5.9	0.0	7.5	16.4
400	2088	-22.1	-17.8	-10.0	0.0	11.9
500	1797	-42.5	-37.1	-27.4	-14.9	0.0
600	1543	-75.5	-69.0	-57.3	-42.3	-24.4
0	3400	-1.5	-1.5	-1.5	-1.5	-1.5
100	3010	0.0	1.1	3.2	5.9	9.0
200	2651	-2.2	0.0	4.2	9.6	15.8
300	2321	-9.6	-6.3	0.0	8.1	17.4
400	2016	-23.5	-19.1	-10.8	0.0	12.4
500	1730	-44.9	-39.4	-29.0	-15.5	0.0
600	1485	-80.5	-73.9	-61.3	-45.2	-26.6
0	3300	-1.5	-1.5	-1.5	-1.5	-1.5
100	2918	0.0	1.3	3.4	6.3	9.9
200	2568	-2.6	0.0	4.2	10.0	17.3
300	2243	-10.2	-6.3	0.0	8.8	19.6
400	1943	-25.3	-20.1	-11.7	0.0	14.5
500	1671	-49.7	-43.2	-32.7	-18.1	0.0
600	1427	-86.3	-78.4	-65.8	-48.3	-26.6
0	3200	-1.5	-1.5	-1.5	-1.5	-1.5
100	2827	0.0	1.5	3.9	6.8	10.8
200	2483	-2.9	0.0	4.8	10.7	18.6
300	2165	-11.6	-7.1	0.0	8.9	20.8
400	1870	-27.3	-21.4	-11.9	0.0	15.8
500	1607	-53.9	-46.5	-34.6	-19.7	0.0
600	1369	-92.5	-83.7	-69.4	-51.6	-27.9
0	3100	-1.5	-1.5	-1.5	-1.5	-1.5
100	2734	0.0	1.6	4.2	7.4	11.7
200	2398	-3.2	0.0	5.2	11.5	20.1
300	2087	-12.6	-7.7	0.0	9.5	22.4
400	1796	-29.4	-22.9	-12.6	0.0	17.3
500	1543	-58.3	-50.2	-37.3	-21.6	0.0
600	1310	-99.4	-89.7	-74.2	-55.3	-29.4
0	3000	-1.5	-1.5	-1.5	-1.5	-1.5
100	2642	0.0	1.8	4.6	7.9	12.7
200	2313	-3.6	0.0	5.6	12.3	21.7
300	2008	-13.9	-8.5	0.0	9.9	24.2
400	1722	-31.7	-24.5	-13.3	0.0	19.0
500	1479	-63.4	-54.4	-40.3	-23.7	0.0
600	1273	-110.9	-100.1	-83.2	-63.3	-34.8
0	2900	-1.5	-1.5	-1.5	-1.5	-1.5
100	2551	0.0	1.8	4.9	8.8	13.6
200	2227	-3.6	0.0	6.2	14.0	23.6
300	1928	-14.7	-9.3	0.0	11.7	26.0
400	1658	-35.3	-28.0	-15.6	0.0	19.1
500	1415	-67.9	-58.9	-43.4	-23.9	0.0
600	1223	-119.8	-109.0	-90.4	-67.0	-38.3

Universal Table for Rifle Bullets with Ballistic Coefficients of .230-.239						
					Sight Above Bore (in.)	+1.5
			Bullet Path			
Range (yds)	Velocity (fps)	Zero (100 yds)	Zero (200 yds)	Zero (300 yds)	Zero (400 yds)	Zero (500 yds)
0	2800	-1.5	-1.5	-1.5	-1.5	-1.5
100	2458	0.0	2.2	5.3	9.6	14.7
200	2142	-4.3	0.0	6.3	14.9	25.0
300	1848	-16.0	-9.5	0.0	12.9	28.1
400	1588	-38.5	-29.9	-17.2	0.0	20.2
500	1351	-73.4	-62.6	-46.8	-25.3	0.0
600	1175	-129.9	-117.0	-98.0	-72.2	-41.9
0	2700	-1.5	-1.5	-1.5	-1.5	-1.5
100	2365	0.0	2.4	5.8	10.6	15.8
200	2056	-4.8	0.0	6.7	16.3	26.8
300	1768	-17.3	-10.0	0.0	14.4	30.2
400	1518	-42.2	-32.5	-19.2	0.0	21.1
500	1288	-79.1	-67.0	-50.3	-26.3	0.0
600	1128	-140.8	-126.3	-106.3	-77.5	-45.9
0	2600	-1.5	-1.5	-1.5	-1.5	-1.5
100	2272	0.0	2.7	6.6	11.5	18.1
200	1970	-5.4	0.0	7.8	17.7	30.7
300	1695	-19.9	-11.7	0.0	14.7	34.3
400	1449	-46.2	-35.3	-19.7	0.0	26.0
500	1249	-90.3	-76.7	-57.1	-32.5	0.0
600	1081	-153.0	-136.7	-113.2	-83.7	-44.7
0	2500	-1.5	-1.5	-1.5	-1.5	-1.5
100	2179	0.0	3.1	7.4	12.7	19.8
200	1884	-6.1	0.0	8.7	19.2	33.5
300	1619	-22.3	-13.1	0.0	15.7	37.1
400	1380	-50.6	-38.4	-20.9	0.0	28.6
500	1197	-99.0	-83.7	-61.9	-35.7	0.0
600	1063	-171.4	-153.1	-126.9	-95.5	-52.7
0	2400	-1.5	-1.5	-1.5	-1.5	-1.5
100	2088	0.0	3.1	8.0	13.5	21.4
200	1797	-6.2	0.0	9.8	20.8	36.7
300	1543	-24.0	-14.7	0.0	16.6	40.3
400	1311	-54.1	-41.7	-22.1	0.0	31.7
500	1145	-107.2	-91.7	-67.2	-39.6	0.0
600	1028	-184.9	-166.3	-136.9	-103.8	-56.2
0	2300	-1.5	-1.5	-1.5	-1.5	-1.5
100	1995	0.0	3.7	8.8	15.5	23.3
200	1717	-7.3	0.0	10.2	23.7	39.3
300	1469	-26.3	-15.3	0.0	20.2	43.7
400	1265	-62.1	-47.4	-27.0	0.0	31.2
500	1095	-116.7	-98.3	-72.8	-39.0	0.0
600	1009	-203.9	-181.9	-151.2	-110.7	-63.8
0	2200	-1.5	-1.5	-1.5	-1.5	-1.5
100	1902	0.0	4.2	9.7	17.2	26.4
200	1634	-8.4	0.0	11.0	26.1	44.3
300	1394	-29.1	-16.4	0.0	22.6	50.0
400	1207	-69.0	-52.1	-30.2	0.0	36.5
500	1071	-131.8	-110.7	-83.3	-45.6	0.0
600	984	-224.7	-199.4	-166.5	-121.3	-66.6
0	2100	-1.5	-1.5	-1.5	-1.5	-1.5
100	1809	0.0	4.8	10.7	19.0	29.1
200	1553	-9.6	0.0	11.7	28.4	48.5
300	1320	-32.0	-17.5	0.0	25.1	55.2
400	1152	-76.1	-56.8	-33.5	0.0	40.1
500	1032	-145.3	-121.2	-92.0	-50.2	0.0
600	964	-247.1	-218.3	-183.2	-133.0	-72.8

Universal Table for Rifle Bullets with Ballistic Coefficients of .240-.249

Sight Above Bore (in.) +1.5

Range (yds)	Velocity (fps)	Bullet Path Zero (100 yds)	Zero (200 yds)	Zero (300 yds)	Zero (400 yds)	Zero (500 yds)
0	4400	-1.5	-1.5	-1.5	-1.5	-1.5
100	3886	0.0	0.4	1.5	3.0	4.9
200	3429	-0.8	0.0	2.3	5.1	9.0
300	3018	-4.6	-3.4	0.0	4.3	10.2
400	2642	-11.9	-10.3	-5.8	0.0	7.8
500	2296	-24.6	-22.6	-17.0	-9.8	0.0
600	1978	-44.2	-41.7	-35.0	-26.3	-14.6
0	4300	-1.5	-1.5	-1.5	-1.5	-1.5
100	3797	0.0	0.5	1.7	3.3	5.2
200	3349	-0.9	0.0	2.4	5.6	9.5
300	2946	-5.0	-3.6	0.0	4.8	10.6
400	2576	-13.1	-11.2	-6.5	0.0	7.7
500	2235	-26.0	-23.7	-17.7	-9.6	0.0
600	1921	-46.7	-43.9	-36.8	-27.1	-15.5
0	4200	-1.5	-1.5	-1.5	-1.5	-1.5
100	3707	0.0	0.5	1.8	3.5	5.5
200	3269	-1.0	0.0	2.5	6.0	10.0
300	2873	-5.4	-3.8	0.0	5.1	11.1
400	2509	-14.0	-11.9	-6.8	0.0	8.0
500	2172	-27.5	-24.9	-18.5	-10.0	0.0
600	1864	-49.4	-46.3	-38.7	-28.4	-16.4
0	4100	-1.5	-1.5	-1.5	-1.5	-1.5
100	3617	0.0	0.6	1.9	3.7	6.0
200	3188	-1.2	0.0	2.7	6.3	10.8
300	2799	-5.8	-4.1	0.0	5.4	12.2
400	2442	-14.9	-12.6	-7.2	0.0	9.1
500	2112	-30.0	-27.1	-20.3	-11.3	0.0
600	1806	-52.3	-48.8	-40.7	-29.9	-16.3
0	4000	-1.5	-1.5	-1.5	-1.5	-1.5
100	3526	0.0	0.6	2.1	4.0	6.4
200	3107	-1.3	0.0	2.9	6.7	11.5
300	2726	-6.2	-4.3	0.0	5.7	13.0
400	2374	-16.0	-13.4	-7.7	0.0	9.7
500	2049	-32.0	-28.8	-21.6	-12.1	0.0
600	1747	-55.4	-51.5	-43.0	-31.5	-17.0
0	3900	-1.5	-1.5	-1.5	-1.5	-1.5
100	3441	0.0	0.7	2.2	4.2	6.8
200	3029	-1.5	0.0	2.9	6.9	12.1
300	2652	-6.5	-4.3	0.0	6.1	13.8
400	2305	-16.8	-13.9	-8.2	0.0	10.3
500	1986	-33.8	-30.2	-23.0	-12.8	0.0
600	1696	-59.8	-55.5	-46.9	-34.7	-19.3
0	3800	-1.5	-1.5	-1.5	-1.5	-1.5
100	3351	0.0	0.8	2.4	4.5	7.2
200	2947	-1.6	0.0	3.2	7.3	12.8
300	2578	-7.3	-4.8	0.0	6.1	14.3
400	2236	-17.8	-14.6	-8.1	0.0	11.0
500	1922	-36.0	-32.0	-23.9	-13.7	0.0
600	1639	-64.0	-59.2	-49.5	-37.3	-20.8
0	3700	-1.5	-1.5	-1.5	-1.5	-1.5
100	3260	0.0	0.9	2.6	4.7	7.7
200	2865	-1.8	0.0	3.5	7.7	13.5
300	2502	-7.9	-5.2	0.0	6.3	15.1
400	2166	-18.9	-15.4	-8.4	0.0	11.7
500	1858	-38.4	-33.9	-25.2	-14.7	0.0
600	1583	-68.5	-63.1	-52.7	-40.1	-22.5

Universal Table for Rifle Bullets with Ballistic Coefficients of .240-.249

Sight Above Bore (in.) +1.5

		Bullet Path				
Range (yds)	Velocity (fps)	Zero (100 yds)	Zero (200 yds)	Zero (300 yds)	Zero (400 yds)	Zero (500 yds)
0	3600	-1.5	-1.5	-1.5	-1.5	-1.5
100	3170	0.0	1.0	2.9	5.3	8.2
200	2783	-2.0	0.0	3.7	8.6	14.4
300	2427	-8.6	-5.6	0.0	7.3	16.0
400	2098	-21.1	-17.1	-9.7	0.0	11.6
500	1793	-40.9	-35.9	-26.7	-14.6	0.0
600	1527	-73.4	-67.4	-56.3	-41.7	-24.3
0	3500	-1.5	-1.5	-1.5	-1.5	-1.5
100	3079	0.0	1.1	3.1	5.7	8.7
200	2700	-2.2	0.0	4.0	9.2	15.3
300	2350	-9.3	-6.0	0.0	7.8	16.9
400	2027	-22.8	-18.4	-10.4	0.0	12.1
500	1727	-43.7	-38.2	-28.2	-15.2	0.0
600	1471	-78.6	-72.0	-60.0	-44.4	-26.2
0	3400	-1.5	-1.5	-1.5	-1.5	-1.5
100	2991	0.0	1.2	3.2	6.0	9.6
200	2618	-2.4	0.0	4.1	9.7	16.7
300	2273	-9.7	-6.1	0.0	8.4	19.0
400	1956	-24.2	-19.4	-11.2	0.0	14.1
500	1670	-47.9	-41.8	-31.7	-17.6	0.0
600	1414	-83.8	-76.5	-64.3	-47.5	-26.3
0	3300	-1.5	-1.5	-1.5	-1.5	-1.5
100	2900	0.0	1.3	3.5	6.5	10.3
200	2534	-2.7	0.0	4.2	10.3	18.0
300	2196	-10.4	-6.3	0.0	9.1	20.7
400	1885	-26.0	-20.6	-12.1	0.0	15.4
500	1607	-51.7	-45.0	-34.4	-19.3	0.0
600	1358	-89.5	-81.5	-68.8	-50.6	-27.5
0	3200	-1.5	-1.5	-1.5	-1.5	-1.5
100	2809	0.0	1.5	3.9	7.0	11.2
200	2450	-3.0	0.0	4.9	11.0	19.4
300	2119	-11.8	-7.4	0.0	9.1	21.8
400	1813	-27.9	-22.0	-12.2	0.0	16.9
500	1544	-56.0	-48.5	-36.3	-21.1	0.0
600	1301	-95.9	-86.9	-72.2	-54.0	-28.7
0	3100	-1.5	-1.5	-1.5	-1.5	-1.5
100	2717	0.0	1.6	4.3	7.5	12.1
200	2366	-3.3	0.0	5.3	11.7	20.9
300	2041	-13.0	-8.0	0.0	9.6	23.4
400	1740	-30.1	-23.5	-12.8	0.0	18.4
500	1482	-60.5	-52.3	-38.9	-22.9	0.0
600	1266	-106.9	-97.1	-81.0	-61.8	-34.3
0	3000	-1.5	-1.5	-1.5	-1.5	-1.5
100	2625	0.0	1.8	4.7	8.5	13.1
200	2281	-3.7	0.0	5.8	13.3	22.5
300	1963	-14.2	-8.7	0.0	11.2	25.1
400	1676	-33.9	-26.5	-14.9	0.0	18.5
500	1420	-65.5	-56.4	-41.8	-23.2	0.0
600	1218	-116.1	-105.1	-87.7	-65.3	-37.5
0	2900	-1.5	-1.5	-1.5	-1.5	-1.5
100	2534	0.0	1.8	5.0	9.1	14.0
200	2195	-3.7	0.0	6.4	14.5	24.4
300	1884	-15.0	-9.5	0.0	12.3	27.0
400	1607	-36.4	-29.1	-16.4	0.0	19.7
500	1357	-70.1	-60.9	-45.0	-24.6	0.0
600	1171	-124.9	-113.9	-94.9	-70.3	-40.9

Universal Table for Rifle Bullets with Ballistic Coefficients of .240-.249

Sight Above Bore (in.) +1.5

Range (yds)	Velocity (fps)	Bullet Path Zero (100 yds)	Zero (200 yds)	Zero (300 yds)	Zero (400 yds)	Zero (500 yds)
0	2800	-1.5	-1.5	-1.5	-1.5	-1.5
100	2442	0.0	2.2	5.4	9.9	15.1
200	2111	-4.4	0.0	6.4	15.5	25.7
300	1805	-16.3	-9.6	0.0	13.6	29.0
400	1538	-39.8	-30.9	-18.1	0.0	20.6
500	1295	-75.4	-64.3	-48.3	-25.7	0.0
600	1125	-135.2	-121.8	-102.6	-75.5	-44.7
0	2700	-1.5	-1.5	-1.5	-1.5	-1.5
100	2349	0.0	2.5	5.8	10.8	17.1
200	2026	-4.9	0.0	6.8	16.8	29.3
300	1726	-17.5	-10.1	0.0	15.0	33.8
400	1470	-43.4	-33.5	-20.0	0.0	25.0
500	1256	-85.5	-73.2	-56.3	-31.3	0.0
600	1080	-146.0	-131.2	-110.9	-80.9	-43.4
0	2600	-1.5	-1.5	-1.5	-1.5	-1.5
100	2256	0.0	2.8	6.8	11.8	18.7
200	1941	-5.5	0.0	8.1	18.2	32.0
300	1656	-20.4	-12.2	0.0	15.1	35.8
400	1402	-47.4	-36.4	-20.1	0.0	27.6
500	1204	-93.7	-79.9	-59.6	-34.4	0.0
600	1063	-163.0	-146.4	-122.1	-91.9	-50.6
0	2500	-1.5	-1.5	-1.5	-1.5	-1.5
100	2163	0.0	3.1	7.6	13.0	20.5
200	1855	-6.2	0.0	9.0	19.7	34.8
300	1581	-22.8	-13.5	0.0	16.0	38.6
400	1334	-51.8	-39.4	-21.4	0.0	30.1
500	1154	-102.4	-86.9	-64.3	-37.6	0.0
600	1028	-177.8	-159.3	-132.2	-100.1	-55.0
0	2400	-1.5	-1.5	-1.5	-1.5	-1.5
100	2073	0.0	3.1	8.2	13.8	22.1
200	1769	-6.2	0.0	10.1	21.3	38.0
300	1507	-24.5	-15.1	0.0	16.8	41.8
400	1267	-55.0	-42.5	-22.4	0.0	33.4
500	1104	-110.5	-94.9	-69.7	-41.7	0.0
600	1008	-194.6	-175.9	-145.7	-112.1	-62.0
0	2300	-1.5	-1.5	-1.5	-1.5	-1.5
100	1980	0.0	3.7	9.0	16.0	24.8
200	1691	-7.5	0.0	10.5	24.6	42.1
300	1433	-26.9	-15.7	0.0	21.2	47.5
400	1228	-64.1	-49.1	-28.2	0.0	35.1
500	1079	-124.0	-105.4	-79.2	-43.9	0.0
600	986	-212.4	-190.0	-158.6	-116.3	-63.6
0	2200	-1.5	-1.5	-1.5	-1.5	-1.5
100	1887	0.0	4.3	9.8	17.6	27.2
200	1609	-8.5	0.0	11.1	26.7	45.8
300	1360	-29.4	-16.6	0.0	23.5	52.1
400	1173	-70.6	-53.5	-31.3	0.0	38.2
500	1041	-135.9	-114.6	-86.8	-47.7	0.0
600	964	-232.3	-206.7	-173.4	-126.4	-69.2
0	2100	-1.5	-1.5	-1.5	-1.5	-1.5
100	1794	0.0	4.9	10.8	19.4	30.6
200	1528	-9.8	0.0	11.8	29.1	51.5
300	1286	-32.4	-17.7	0.0	25.9	59.5
400	1119	-77.8	-58.2	-34.6	0.0	44.8
500	1012	-153.2	-128.7	-99.2	-56.0	0.0
600	943	-256.6	-227.3	-191.9	-140.0	-72.8

Universal Table for Rifle Bullets with Ballistic Coefficients of .250-.259						
					Sight Above Bore (in.) +1.5	
			Bullet Path			
Range (yds)	Velocity (fps)	Zero (100 yds)	Zero (200 yds)	Zero (300 yds)	Zero (400 yds)	Zero (500 yds)
0	4400	-1.5	-1.5	-1.5	-1.5	-1.5
100	3946	0.0	0.4	1.4	2.8	4.5
200	3537	-0.7	0.0	2.0	4.8	8.3
300	3163	-4.1	-3.0	0.0	4.2	9.5
400	2823	-11.2	-9.7	-5.7	0.0	7.0
500	2506	-22.6	-20.8	-15.8	-8.7	0.0
600	2209	-39.0	-36.8	-30.7	-22.3	-11.8
0	4300	-1.5	-1.5	-1.5	-1.5	-1.5
100	3856	0.0	0.4	1.5	3.0	4.8
200	3455	-0.9	0.0	2.2	5.1	8.8
300	3090	-4.6	-3.4	0.0	4.3	9.8
400	2753	-11.9	-10.2	-5.7	0.0	7.4
500	2441	-24.1	-21.9	-16.3	-9.2	0.0
600	2150	-42.1	-39.5	-32.8	-24.3	-13.2
0	4200	-1.5	-1.5	-1.5	-1.5	-1.5
100	3765	0.0	0.5	1.7	3.2	5.1
200	3373	-1.0	0.0	2.4	5.3	9.2
300	3015	-5.0	-3.6	0.0	4.4	10.3
400	2683	-12.6	-10.7	-5.9	0.0	7.8
500	2376	-25.5	-23.1	-17.1	-9.7	0.0
600	2090	-44.6	-41.7	-34.6	-25.7	-14.0
0	4100	-1.5	-1.5	-1.5	-1.5	-1.5
100	3673	0.0	0.5	1.8	3.5	5.4
200	3290	-1.1	0.0	2.5	5.8	9.7
300	2939	-5.4	-3.8	0.0	4.9	10.8
400	2614	-13.8	-11.7	-6.6	0.0	7.8
500	2311	-27.0	-24.3	-17.9	-9.7	0.0
600	2029	-47.4	-44.1	-36.5	-26.6	-15.0
0	4000	-1.5	-1.5	-1.5	-1.5	-1.5
100	3582	0.0	0.6	2.0	3.7	5.8
200	3207	-1.2	0.0	2.7	6.2	10.3
300	2863	-5.9	-4.1	0.0	5.3	11.3
400	2543	-14.9	-12.4	-7.0	0.0	8.1
500	2244	-28.8	-25.7	-18.9	-10.1	0.0
600	1967	-50.5	-46.8	-38.6	-28.1	-16.0
0	3900	-1.5	-1.5	-1.5	-1.5	-1.5
100	3495	0.0	0.7	2.1	3.9	6.2
200	3126	-1.4	0.0	2.7	6.5	11.0
300	2787	-6.2	-4.1	0.0	5.6	12.4
400	2473	-15.7	-12.9	-7.4	0.0	9.1
500	2179	-30.9	-27.5	-20.6	-11.3	0.0
600	1906	-53.2	-49.0	-40.8	-29.7	-16.1
0	3800	-1.5	-1.5	-1.5	-1.5	-1.5
100	3403	0.0	0.8	2.2	4.2	6.6
200	3043	-1.5	0.0	2.9	6.8	11.7
300	2709	-6.6	-4.3	0.0	5.9	13.2
400	2401	-16.7	-13.7	-7.9	0.0	9.7
500	2112	-33.0	-29.2	-22.0	-12.1	0.0
600	1843	-56.4	-51.9	-43.2	-31.3	-16.8
0	3700	-1.5	-1.5	-1.5	-1.5	-1.5
100	3312	0.0	0.9	2.5	4.5	7.1
200	2959	-1.7	0.0	3.2	7.3	12.4
300	2633	-7.4	-4.8	0.0	6.0	13.8
400	2328	-17.9	-14.5	-8.1	0.0	10.3
500	2045	-35.3	-31.1	-23.0	-12.9	0.0
600	1779	-60.1	-54.9	-45.3	-33.2	-17.7

Universal Table for Rifle Bullets with Ballistic Coefficients of .250-.259

Sight Above Bore (in.) +1.5

Range (yds)	Velocity (fps)	Bullet Path Zero (100 yds)	Zero (200 yds)	Zero (300 yds)	Zero (400 yds)	Zero (500 yds)
0	3600	-1.5	-1.5	-1.5	-1.5	-1.5
100	3220	0.0	1.0	2.7	4.8	7.6
200	2875	-1.9	0.0	3.5	7.7	13.2
300	2555	-8.0	-5.2	0.0	6.3	14.7
400	2255	-19.2	-15.3	-8.4	0.0	11.1
500	1977	-37.9	-33.1	-24.4	-13.9	0.0
600	1721	-65.5	-59.8	-49.4	-36.8	-20.1
0	3500	-1.5	-1.5	-1.5	-1.5	-1.5
100	3128	0.0	1.1	2.9	5.3	8.1
200	2791	-2.1	0.0	3.8	8.4	14.1
300	2476	-8.8	-5.6	0.0	7.0	15.5
400	2183	-21.1	-16.9	-9.4	0.0	11.3
500	1909	-40.6	-35.3	-25.9	-14.2	0.0
600	1661	-70.3	-64.0	-52.7	-38.6	-21.6
0	3400	-1.5	-1.5	-1.5	-1.5	-1.5
100	3039	0.0	1.1	3.1	5.6	8.6
200	2706	-2.1	0.0	4.0	9.1	15.1
300	2397	-9.3	-6.1	0.0	7.6	16.5
400	2109	-22.5	-18.3	-10.2	0.0	11.9
500	1840	-43.0	-37.7	-27.5	-14.8	0.0
600	1600	-75.2	-68.8	-56.6	-41.4	-23.6
0	3300	-1.5	-1.5	-1.5	-1.5	-1.5
100	2947	0.0	1.3	3.3	6.1	9.2
200	2621	-2.5	0.0	4.1	9.6	15.8
300	2317	-10.0	-6.2	0.0	8.2	17.6
400	2035	-24.3	-19.2	-11.0	0.0	12.5
500	1770	-45.9	-39.6	-29.3	-15.6	0.0
600	1539	-80.9	-73.3	-60.9	-44.5	-25.8
0	3200	-1.5	-1.5	-1.5	-1.5	-1.5
100	2855	0.0	1.4	3.6	6.6	10.1
200	2536	-2.8	0.0	4.3	10.3	17.5
300	2237	-10.7	-6.5	0.0	8.9	19.7
400	1961	-26.2	-20.5	-11.9	0.0	14.4
500	1707	-50.7	-43.7	-32.8	-18.0	0.0
600	1478	-87.2	-78.7	-65.7	-47.9	-26.3
0	3100	-1.5	-1.5	-1.5	-1.5	-1.5
100	2762	0.0	1.6	4.1	7.1	11.0
200	2450	-3.1	0.0	5.0	11.0	18.9
300	2158	-12.2	-7.4	0.0	9.1	20.9
400	1885	-28.4	-22.1	-12.2	0.0	15.7
500	1640	-55.1	-47.3	-34.9	-19.6	0.0
600	1417	-94.0	-84.5	-69.7	-51.4	-27.8
0	3000	-1.5	-1.5	-1.5	-1.5	-1.5
100	2669	0.0	1.8	4.5	7.7	12.0
200	2363	-3.5	0.0	5.4	11.8	20.5
300	2077	-13.4	-8.1	0.0	9.6	22.7
400	1810	-30.7	-23.6	-12.8	0.0	17.4
500	1573	-60.1	-51.3	-37.8	-21.8	0.0
600	1356	-101.3	-90.7	-74.5	-55.3	-29.2
0	2900	-1.5	-1.5	-1.5	-1.5	-1.5
100	2577	0.0	1.8	4.7	8.4	12.9
200	2276	-3.6	0.0	5.9	13.2	22.3
300	1997	-14.2	-8.9	0.0	11.0	24.5
400	1739	-33.6	-26.4	-14.6	0.0	18.1
500	1507	-64.6	-55.7	-40.9	-22.6	0.0
600	1296	-107.9	-97.2	-79.5	-57.5	-30.4

Universal Table for Rifle Bullets with Ballistic Coefficients of .250-.259

Sight Above Bore (in.) +1.5

		Bullet Path				
Range (yds)	Velocity (fps)	Zero (100 yds)	Zero (200 yds)	Zero (300 yds)	Zero (400 yds)	Zero (500 yds)
0	2800	-1.5	-1.5	-1.5	-1.5	-1.5
100	2484	0.0	2.1	5.2	9.2	14.0
200	2190	-4.2	0.0	6.1	14.2	23.8
300	1916	-15.5	-9.2	0.0	12.1	26.6
400	1667	-36.7	-28.3	-16.1	0.0	19.4
500	1441	-70.1	-59.6	-44.3	-24.2	0.0
600	1258	-121.5	-108.9	-90.6	-66.4	-37.4
0	2700	-1.5	-1.5	-1.5	-1.5	-1.5
100	2391	0.0	2.4	5.6	10.1	15.2
200	2103	-4.7	0.0	6.5	15.4	25.6
300	1834	-16.9	-9.8	0.0	13.4	28.6
400	1595	-40.3	-30.9	-17.8	0.0	20.3
500	1376	-75.8	-64.0	-47.7	-25.4	0.0
600	1207	-132.5	-118.4	-98.8	-72.0	-41.5
0	2600	-1.5	-1.5	-1.5	-1.5	-1.5
100	2297	0.0	2.6	6.1	11.1	16.4
200	2016	-5.3	0.0	6.9	16.9	27.6
300	1752	-18.3	-10.4	0.0	14.9	31.0
400	1523	-44.4	-33.8	-19.9	0.0	21.4
500	1310	-82.2	-69.0	-51.6	-26.7	0.0
600	1157	-144.6	-128.7	-107.9	-78.0	-45.9
0	2500	-1.5	-1.5	-1.5	-1.5	-1.5
100	2203	0.0	3.0	7.2	12.2	18.8
200	1929	-6.0	0.0	8.3	18.4	31.6
300	1678	-21.5	-12.5	0.0	15.2	34.9
400	1452	-48.8	-36.9	-20.2	0.0	26.3
500	1266	-93.9	-79.0	-58.1	-32.9	0.0
600	1109	-157.2	-139.3	-114.3	-83.9	-44.5
0	2400	-1.5	-1.5	-1.5	-1.5	-1.5
100	2111	0.0	3.0	7.7	13.1	20.3
200	1842	-6.0	0.0	9.4	20.1	34.6
300	1601	-23.1	-14.0	0.0	16.1	37.9
400	1382	-52.2	-40.2	-21.4	0.0	29.1
500	1212	-101.7	-86.6	-63.2	-36.4	0.0
600	1084	-174.0	-155.9	-127.8	-95.7	-52.0
0	2300	-1.5	-1.5	-1.5	-1.5	-1.5
100	2018	0.0	3.3	8.5	14.2	22.4
200	1754	-6.5	0.0	10.5	21.8	38.2
300	1525	-25.5	-15.7	0.0	17.0	41.6
400	1312	-56.7	-43.7	-22.7	0.0	32.7
500	1158	-111.8	-95.5	-69.3	-40.9	0.0
600	1046	-190.7	-171.1	-139.7	-105.6	-56.5
0	2200	-1.5	-1.5	-1.5	-1.5	-1.5
100	1924	0.0	4.1	9.4	16.4	24.4
200	1674	-8.1	0.0	10.7	24.7	40.8
300	1449	-28.1	-16.0	0.0	21.1	45.1
400	1264	-65.6	-49.4	-28.1	0.0	32.1
500	1106	-122.2	-101.9	-75.2	-40.1	0.0
600	1021	-211.5	-187.2	-155.2	-113.1	-64.9
0	2100	-1.5	-1.5	-1.5	-1.5	-1.5
100	1831	0.0	4.7	10.4	18.3	27.6
200	1592	-9.3	0.0	11.5	27.3	46.0
300	1373	-31.2	-17.3	0.0	23.7	51.7
400	1205	-73.2	-54.6	-31.6	0.0	37.4
500	1079	-138.2	-115.0	-86.2	-46.7	0.0
600	996	-232.7	-204.8	-170.3	-122.9	-66.8

Universal Table for Rifle Bullets with Ballistic Coefficients of .260-.269

Sight Above Bore (in.) +1.5

Range (yds)	Velocity (fps)	Bullet Path Zero (100 yds)	Zero (200 yds)	Zero (300 yds)	Zero (400 yds)	Zero (500 yds)
0	4400	-1.5	-1.5	-1.5	-1.5	-1.5
100	3963	0.0	0.4	1.4	2.7	4.4
200	3567	-0.7	0.0	2.0	4.8	8.1
300	3204	-4.1	-3.0	0.0	4.2	9.2
400	2873	-11.0	-9.5	-5.5	0.0	6.7
500	2564	-22.1	-20.3	-15.4	-8.4	0.0
600	2275	-38.1	-35.9	-30.0	-21.6	-11.5
0	4300	-1.5	-1.5	-1.5	-1.5	-1.5
100	3872	0.0	0.4	1.5	2.9	4.7
200	3484	-0.8	0.0	2.2	5.0	8.6
300	3130	-4.5	-3.3	0.0	4.2	9.5
400	2803	-11.7	-10.0	-5.6	0.0	7.1
500	2499	-23.5	-21.4	-15.9	-8.9	0.0
600	2214	-40.9	-38.4	-31.8	-23.4	-12.7
0	4200	-1.5	-1.5	-1.5	-1.5	-1.5
100	3781	0.0	0.5	1.6	3.1	5.0
200	3401	-1.0	0.0	2.3	5.2	9.0
300	3054	-4.9	-3.5	0.0	4.4	10.0
400	2732	-12.4	-10.5	-5.8	0.0	7.5
500	2433	-25.0	-22.5	-16.7	-9.4	0.0
600	2153	-43.4	-40.5	-33.5	-24.8	-13.4
0	4100	-1.5	-1.5	-1.5	-1.5	-1.5
100	3689	0.0	0.5	1.8	3.4	5.3
200	3318	-1.1	0.0	2.5	5.7	9.5
300	2978	-5.3	-3.7	0.0	4.8	10.6
400	2662	-13.5	-11.4	-6.4	0.0	7.7
500	2366	-26.5	-23.8	-17.6	-9.6	0.0
600	2091	-46.1	-42.9	-35.4	-25.8	-14.3
0	4000	-1.5	-1.5	-1.5	-1.5	-1.5
100	3597	0.0	0.6	1.9	3.6	5.6
200	3235	-1.2	0.0	2.7	6.1	10.1
300	2901	-5.8	-4.0	0.0	5.1	11.1
400	2590	-14.6	-12.2	-6.8	0.0	8.0
500	2299	-28.2	-25.2	-18.5	-10.0	0.0
600	2029	-49.0	-45.4	-37.4	-27.2	-15.2
0	3900	-1.5	-1.5	-1.5	-1.5	-1.5
100	3509	0.0	0.6	2.0	3.8	5.9
200	3151	-1.2	0.0	2.8	6.5	10.6
300	2824	-6.1	-4.3	0.0	5.4	11.7
400	2519	-15.4	-12.9	-7.3	0.0	8.3
500	2232	-29.6	-26.5	-19.4	-10.4	0.0
600	1966	-51.8	-48.2	-39.6	-28.8	-16.3
0	3800	-1.5	-1.5	-1.5	-1.5	-1.5
100	3418	0.0	0.8	2.2	4.1	6.4
200	3069	-1.5	0.0	2.9	6.7	11.4
300	2746	-6.5	-4.3	0.0	5.8	12.8
400	2446	-16.5	-13.5	-7.8	0.0	9.3
500	2165	-32.2	-28.5	-21.3	-11.6	0.0
600	1902	-55.1	-50.6	-42.1	-30.4	-16.5
0	3700	-1.5	-1.5	-1.5	-1.5	-1.5
100	3326	0.0	0.8	2.4	4.4	6.9
200	2985	-1.7	0.0	3.2	7.1	12.1
300	2668	-7.3	-4.8	0.0	5.9	13.4
400	2373	-17.6	-14.3	-7.9	0.0	10.0
500	2097	-34.5	-30.3	-22.3	-12.5	0.0
600	1838	-58.6	-53.6	-44.0	-32.2	-17.2

Universal Table for Rifle Bullets with Ballistic Coefficients of .260-.269						
					Sight Above Bore (in.)	+1.5
			Bullet Path			
Range (yds)	Velocity (fps)	Zero (100 yds)	Zero (200 yds)	Zero (300 yds)	Zero (400 yds)	Zero (500 yds)
0	3600	-1.5	-1.5	-1.5	-1.5	-1.5
100	3234	0.0	0.9	2.6	4.7	7.4
200	2900	-1.9	0.0	3.4	7.5	12.9
300	2590	-7.9	-5.1	0.0	6.2	14.3
400	2299	-18.9	-15.1	-8.3	0.0	10.8
500	2028	-37.0	-32.3	-23.8	-13.4	0.0
600	1773	-62.4	-56.7	-46.6	-34.1	-18.0
0	3500	-1.5	-1.5	-1.5	-1.5	-1.5
100	3142	0.0	1.0	2.9	5.0	7.9
200	2816	-2.1	0.0	3.7	8.0	13.8
300	2511	-8.7	-5.5	0.0	6.5	15.2
400	2225	-20.2	-16.0	-8.6	0.0	11.6
500	1959	-39.7	-34.5	-25.3	-14.5	0.0
600	1714	-68.3	-62.0	-51.0	-38.0	-20.6
0	3400	-1.5	-1.5	-1.5	-1.5	-1.5
100	3052	0.0	1.1	3.0	5.5	8.4
200	2730	-2.1	0.0	4.0	8.9	14.8
300	2431	-9.1	-6.0	0.0	7.4	16.2
400	2151	-22.1	-17.8	-9.9	0.0	11.7
500	1889	-42.2	-36.9	-26.9	-14.6	0.0
600	1652	-73.0	-66.6	-54.7	-39.9	-22.4
0	3300	-1.5	-1.5	-1.5	-1.5	-1.5
100	2960	0.0	1.3	3.3	5.9	9.0
200	2645	-2.5	0.0	4.1	9.4	15.5
300	2351	-9.8	-6.1	0.0	8.0	17.2
400	2077	-23.8	-18.8	-10.7	0.0	12.3
500	1819	-45.1	-38.8	-28.7	-15.3	0.0
600	1590	-78.6	-71.1	-58.9	-42.9	-24.5
0	3200	-1.5	-1.5	-1.5	-1.5	-1.5
100	2867	0.0	1.4	3.5	6.4	9.6
200	2559	-2.8	0.0	4.3	10.1	16.5
300	2270	-10.6	-6.4	0.0	8.7	18.3
400	2001	-25.8	-20.2	-11.6	0.0	12.8
500	1748	-48.2	-41.2	-30.5	-16.0	0.0
600	1528	-84.7	-76.4	-63.5	-46.1	-26.9
0	3100	-1.5	-1.5	-1.5	-1.5	-1.5
100	2774	0.0	1.6	4.0	7.0	10.8
200	2472	-3.1	0.0	4.8	10.8	18.4
300	2190	-11.9	-7.2	0.0	9.0	20.3
400	1925	-27.9	-21.6	-12.0	0.0	15.1
500	1684	-53.8	-45.9	-33.9	-18.9	0.0
600	1466	-91.4	-82.0	-67.6	-49.6	-26.9
0	3000	-1.5	-1.5	-1.5	-1.5	-1.5
100	2681	0.0	1.7	4.4	7.6	11.7
200	2386	-3.5	0.0	5.3	11.6	19.9
300	2109	-13.2	-8.0	0.0	9.5	21.9
400	1849	-30.2	-23.3	-12.7	0.0	16.6
500	1617	-58.5	-49.8	-36.5	-20.7	0.0
600	1404	-98.8	-88.3	-72.4	-53.4	-28.6
0	2900	-1.5	-1.5	-1.5	-1.5	-1.5
100	2589	0.0	1.8	4.7	8.0	12.6
200	2298	-3.5	0.0	5.8	12.4	21.7
300	2027	-14.1	-8.7	0.0	9.9	23.8
400	1773	-32.0	-24.9	-13.2	0.0	18.5
500	1549	-63.1	-54.2	-39.7	-23.1	0.0
600	1342	-105.6	-95.0	-77.5	-57.7	-29.9

Universal Table for Rifle Bullets with Ballistic Coefficients of .260-.269

Sight Above Bore (in.) +1.5

Range (yds)	Velocity (fps)	Bullet Path Zero (100 yds)	Zero (200 yds)	Zero (300 yds)	Zero (400 yds)	Zero (500 yds)
0	2800	-1.5	-1.5	-1.5	-1.5	-1.5
100	2496	0.0	2.1	5.1	9.0	13.7
200	2211	-4.2	0.0	6.0	13.9	23.3
300	1946	-15.3	-9.1	0.0	11.7	25.8
400	1702	-36.0	-27.7	-15.7	0.0	18.8
500	1482	-68.6	-58.2	-43.1	-23.5	0.0
600	1297	-117.8	-105.3	-87.2	-63.7	-35.5
0	2700	-1.5	-1.5	-1.5	-1.5	-1.5
100	2402	0.0	2.3	5.5	9.9	14.9
200	2124	-4.6	0.0	6.4	15.1	25.1
300	1864	-16.6	-9.7	0.0	13.0	27.9
400	1629	-39.6	-30.3	-17.4	0.0	19.8
500	1416	-74.3	-62.6	-46.5	-24.8	0.0
600	1244	-128.6	-114.7	-95.4	-69.3	-39.5
0	2600	-1.5	-1.5	-1.5	-1.5	-1.5
100	2308	0.0	2.6	6.1	10.9	16.1
200	2037	-5.2	0.0	6.9	16.5	27.1
300	1781	-18.2	-10.3	0.0	14.4	30.3
400	1557	-43.5	-33.1	-19.3	0.0	21.1
500	1340	80.7	67.7	50.4	26.4	0.0
600	1193	-140.3	-124.7	-104.0	-75.1	-43.5
0	2500	-1.5	-1.5	-1.5	-1.5	-1.5
100	2214	0.0	3.0	7.0	12.0	18.3
200	1949	-5.9	0.0	8.1	18.1	30.6
300	1705	-21.1	-12.2	0.0	14.9	33.7
400	1485	-47.9	-36.1	-19.8	0.0	25.1
500	1299	-91.4	-76.6	-56.2	-31.4	0.0
600	1143	-153.3	-135.5	-111.1	-81.3	-43.6
0	2400	-1.5	-1.5	-1.5	-1.5	-1.5
100	2122	0.0	3.0	7.6	12.9	19.9
200	1862	-6.0	0.0	9.2	19.7	33.7
300	1627	-22.8	-13.8	0.0	15.8	36.8
400	1414	-51.5	-39.5	-21.1	0.0	27.9
500	1243	-99.3	-84.3	-61.3	-34.9	0.0
600	1110	-169.7	-151.7	-124.2	-92.5	-50.6
0	2300	-1.5	-1.5	-1.5	-1.5	-1.5
100	2028	0.0	3.2	8.4	14.0	21.8
200	1773	-6.5	0.0	10.3	21.5	37.1
300	1550	-25.1	-15.4	0.0	16.8	40.3
400	1343	-55.9	-43.0	-22.4	0.0	31.3
500	1188	-109.1	-92.8	-67.2	-39.1	0.0
600	1071	-185.6	-166.1	-135.4	-101.7	-54.8
0	2200	-1.5	-1.5	-1.5	-1.5	-1.5
100	1935	0.0	4.0	9.3	16.1	24.0
200	1692	-8.0	0.0	10.6	24.1	40.0
300	1473	-27.9	-15.9	0.0	20.3	44.1
400	1290	-64.3	-48.3	-27.1	0.0	31.7
500	1134	-120.0	-99.9	-73.4	-39.6	0.0
600	1039	-207.1	-183.0	-151.2	-110.6	-63.1
0	2100	-1.5	-1.5	-1.5	-1.5	-1.5
100	1840	0.0	4.6	10.3	18.0	27.0
200	1609	-9.1	0.0	11.4	26.8	44.8
300	1397	-30.8	-17.0	0.0	23.2	50.2
400	1229	-71.9	-53.6	-30.9	0.0	36.0
500	1101	-134.9	-112.0	-83.6	-45.0	0.0
600	1012	-227.6	-200.1	-166.0	-119.7	-65.7

Universal Table for Rifle Bullets with Ballistic Coefficients of .270-.279						
					Sight Above Bore (in.) +1.5	
		Bullet Path				
Range (yds)	Velocity (fps)	Zero (100 yds)	Zero (200 yds)	Zero (300 yds)	Zero (400 yds)	Zero (500 yds)
0	4400	-1.5	-1.5	-1.5	-1.5	-1.5
100	3978	0.0	0.4	1.3	2.7	4.3
200	3594	-0.7	0.0	2.0	4.7	7.9
300	3242	-4.0	-3.0	0.0	4.1	9.0
400	2920	-10.8	-9.3	-5.4	0.0	6.5
500	2619	-21.6	-19.8	-14.9	-8.2	0.0
600	2336	-37.3	-35.2	-29.3	-21.2	-11.4
0	4300	-1.5	-1.5	-1.5	-1.5	-1.5
100	3887	0.0	0.4	1.4	2.9	4.6
200	3511	-0.8	0.0	2.1	4.9	8.4
300	3165	-4.3	-3.1	0.0	4.3	9.5
400	2849	-11.5	-9.8	-5.7	0.0	6.9
500	2553	-23.0	-20.9	-15.8	-8.6	0.0
600	2274	-39.3	-36.9	-30.7	-22.1	-11.8
0	4200	-1.5	-1.5	-1.5	-1.5	-1.5
100	3795	0.0	0.5	1.6	3.1	4.9
200	3428	-0.9	0.0	2.3	5.2	8.8
300	3090	-4.9	-3.5	0.0	4.3	9.8
400	2777	-12.2	-10.4	-5.8	0.0	7.3
500	2486	-24.4	-22.1	-16.3	-9.1	0.0
600	2212	-42.2	-39.4	-32.5	-23.9	-12.9
0	4100	-1.5	-1.5	-1.5	-1.5	-1.5
100	3703	0.0	0.5	1.8	3.2	5.2
200	3344	-1.1	0.0	2.4	5.4	9.3
300	3014	-5.3	-3.7	0.0	4.5	10.3
400	2706	-13.0	-10.9	-6.0	0.0	7.8
500	2419	-26.0	-23.3	-17.2	-9.7	0.0
600	2150	-44.9	-41.7	-34.3	-25.4	-13.7
0	4000	-1.5	-1.5	-1.5	-1.5	-1.5
100	3611	0.0	0.6	1.9	3.6	5.5
200	3260	-1.2	0.0	2.6	5.9	9.9
300	2936	-5.7	-3.9	0.0	5.0	10.9
400	2634	-14.3	-11.9	-6.7	0.0	7.8
500	2351	-27.7	-24.6	-18.1	-9.8	0.0
600	2086	-47.9	-44.2	-36.4	-26.4	-14.7
0	3900	-1.5	-1.5	-1.5	-1.5	-1.5
100	3523	0.0	0.6	2.0	3.8	5.8
200	3176	-1.2	0.0	2.8	6.3	10.4
300	2859	-6.0	-4.2	0.0	5.3	11.4
400	2562	-15.1	-12.7	-7.1	0.0	8.1
500	2283	-29.1	-26.0	-19.1	-10.2	0.0
600	2023	-50.5	-46.9	-38.5	-27.9	-15.7
0	3800	-1.5	-1.5	-1.5	-1.5	-1.5
100	3431	0.0	0.7	2.2	4.1	6.3
200	3093	-1.5	0.0	2.8	6.6	11.1
300	2780	-6.5	-4.2	0.0	5.7	12.4
400	2488	-16.2	-13.2	-7.6	0.0	9.0
500	2215	-31.5	-27.8	-20.7	-11.2	0.0
600	1958	-53.8	-49.4	-40.9	-29.5	-16.1
0	3700	-1.5	-1.5	-1.5	-1.5	-1.5
100	3339	0.0	0.8	2.3	4.3	6.7
200	3009	-1.7	0.0	3.0	7.0	11.8
300	2701	-6.9	-4.4	0.0	6.1	13.3
400	2415	-17.3	-14.0	-8.1	0.0	9.6
500	2146	-33.7	-29.6	-22.2	-12.1	0.0
600	1893	-57.3	-52.4	-43.5	-31.3	-16.9

Universal Table for Rifle Bullets with Ballistic Coefficients of .270-.279

Sight Above Bore (in.) +1.5

Range (yds)	Velocity (fps)	Bullet Path Zero (100 yds)	Zero (200 yds)	Zero (300 yds)	Zero (400 yds)	Zero (500 yds)
0	3600	-1.5	-1.5	-1.5	-1.5	-1.5
100	3247	0.0	0.9	2.6	4.7	7.2
200	2924	-1.9	0.0	3.3	7.4	12.6
300	2623	-7.8	-5.0	0.0	6.2	13.9
400	2340	-18.6	-14.9	-8.2	0.0	10.4
500	2076	-36.2	-31.6	-23.2	-13.0	0.0
600	1828	-61.0	-55.4	-45.4	-33.1	-17.5
0	3500	-1.5	-1.5	-1.5	-1.5	-1.5
100	3154	0.0	1.0	2.8	5.0	7.8
200	2839	-2.1	0.0	3.6	7.9	13.5
300	2543	-8.5	-5.4	0.0	6.4	14.8
400	2265	-19.9	-15.8	-8.6	0.0	11.2
500	2006	-38.9	-33.8	-24.7	-14.0	0.0
600	1762	-65.0	-58.8	-47.9	-35.1	-18.3
0	3400	-1.5	-1.5	-1.5	-1.5	-1.5
100	3064	0.0	1.0	3.0	5.4	8.3
200	2753	-2.1	0.0	3.9	8.7	14.4
300	2463	-9.0	-5.9	0.0	7.2	15.8
400	2191	-21.6	-17.4	-9.6	0.0	11.4
500	1936	-41.3	-36.1	-26.3	-14.3	0.0
600	1702	-70.9	-64.6	-52.9	-38.5	-21.3
0	3300	-1.5	-1.5	-1.5	-1.5	-1.5
100	2972	0.0	1.2	3.2	5.9	8.8
200	2667	-2.5	0.0	4.0	9.2	15.2
300	2382	-9.7	-6.0	0.0	7.8	16.8
400	2115	-23.4	-18.5	-10.4	0.0	12.0
500	1865	-44.2	-38.0	-28.0	-15.0	0.0
600	1638	-76.5	-69.1	-57.0	-41.4	-23.4
0	3200	-1.5	-1.5	-1.5	-1.5	-1.5
100	2879	0.0	1.4	3.5	6.3	9.5
200	2581	-2.8	0.0	4.2	9.9	16.2
300	2301	-10.5	-6.4	0.0	8.5	18.0
400	2039	-25.3	-19.8	-11.4	0.0	12.6
500	1793	-47.4	-40.5	-30.0	-15.7	0.0
600	1575	-82.5	-74.3	-61.6	-44.5	-25.6
0	3100	-1.5	-1.5	-1.5	-1.5	-1.5
100	2786	0.0	1.5	3.9	6.9	10.5
200	2494	-3.1	0.0	4.8	10.6	17.9
300	2220	-11.8	-7.1	0.0	8.8	19.7
400	1963	-27.4	-21.3	-11.8	0.0	14.6
500	1726	-52.5	-44.8	-32.9	-18.2	0.0
600	1512	-89.2	-79.9	-65.7	-48.0	-26.2
0	3000	-1.5	-1.5	-1.5	-1.5	-1.5
100	2692	0.0	1.7	4.3	7.4	11.4
200	2406	-3.5	0.0	5.2	11.4	19.4
300	2138	-13.0	-7.8	0.0	9.3	21.4
400	1886	-29.7	-22.8	-12.4	0.0	16.0
500	1657	-57.2	-48.6	-35.6	-20.0	0.0
600	1449	-96.4	-86.0	-70.5	-51.8	-27.7
0	2900	-1.5	-1.5	-1.5	-1.5	-1.5
100	2600	0.0	1.8	4.6	7.9	12.3
200	2319	-3.5	0.0	5.7	12.3	21.1
300	2056	-13.9	-8.6	0.0	9.8	23.1
400	1809	-31.6	-24.6	-13.1	0.0	17.7
500	1589	-61.6	-52.8	-38.5	-22.1	0.0
600	1386	-103.3	-92.8	-75.6	-55.9	-29.4

Universal Table for Rifle Bullets with Ballistic Coefficients of .270-.279

Sight Above Bore (in.) +1.5

Range (yds)	Velocity (fps)	Bullet Path Zero (100 yds)	Zero (200 yds)	Zero (300 yds)	Zero (400 yds)	Zero (500 yds)
0	2800	-1.5	-1.5	-1.5	-1.5	-1.5
100	2506	0.0	1.9	5.0	8.8	13.4
200	2231	-3.7	0.0	6.3	13.9	23.0
300	1974	-15.1	-9.5	0.0	11.3	25.1
400	1736	-35.2	-27.7	-15.1	0.0	18.3
500	1521	-67.0	-57.6	-41.8	-22.9	0.0
600	1323	-111.2	-100.0	-81.1	-58.4	-30.9
0	2700	-1.5	-1.5	-1.5	-1.5	-1.5
100	2412	0.0	2.3	5.5	9.7	14.6
200	2144	-4.5	0.0	6.4	14.8	24.6
300	1891	-16.5	-9.7	0.0	12.6	27.3
400	1662	-38.7	-29.6	-16.7	0.0	19.6
500	1453	-72.9	-61.5	-45.4	-24.5	0.0
600	1280	-124.8	-111.2	-91.8	-66.7	-37.3
0	2600	-1.5	-1.5	-1.5	-1.5	-1.5
100	2318	0.0	2.6	6.0	10.7	15.8
200	2056	-5.1	0.0	6.8	16.2	26.5
300	1808	-18.0	-10.3	0.0	14.0	29.5
400	1589	-42.6	-32.3	-18.6	0.0	20.7
500	1386	-79.2	-66.3	-49.2	-25.9	0.0
600	1226	-137.0	-121.6	-101.1	-73.1	-42.0
0	2500	-1.5	-1.5	-1.5	-1.5	-1.5
100	2224	0.0	2.9	6.9	11.8	17.3
200	1968	-5.8	0.0	7.9	17.7	28.7
300	1731	-20.6	-11.9	0.0	14.7	31.2
400	1516	-47.1	-35.4	-19.6	0.0	22.0
500	1318	-86.3	-71.7	-51.9	-27.5	0.0
600	1175	-149.7	-132.2	-108.5	-79.1	-46.1
0	2400	-1.5	-1.5	-1.5	-1.5	-1.5
100	2132	0.0	3.0	7.5	12.7	19.4
200	1880	-6.0	0.0	8.9	19.4	32.9
300	1652	-22.4	-13.4	0.0	15.7	36.0
400	1444	-50.7	-38.7	-20.9	0.0	27.0
500	1272	-97.2	-82.2	-59.9	-33.8	0.0
600	1124	-162.3	-144.3	-117.6	-86.2	-45.7
0	2300	-1.5	-1.5	-1.5	-1.5	-1.5
100	2038	0.0	3.2	8.3	13.8	21.4
200	1791	-6.5	0.0	10.0	21.2	36.2
300	1574	-24.8	-15.0	0.0	16.7	39.3
400	1372	-55.3	-42.3	-22.3	0.0	30.1
500	1216	-106.8	-90.6	-65.5	-37.7	0.0
600	1095	-181.1	-161.6	-131.6	-98.2	-52.9
0	2200	-1.5	-1.5	-1.5	-1.5	-1.5
100	1944	0.0	3.9	9.2	15.8	23.5
200	1709	-7.9	0.0	10.5	23.7	39.1
300	1496	-27.6	-15.7	0.0	19.9	43.0
400	1314	-63.2	-47.5	-26.5	0.0	30.8
500	1161	-117.5	-97.9	-71.6	-38.5	0.0
600	1056	-198.6	-174.9	-143.5	-103.7	-57.5
0	2100	-1.5	-1.5	-1.5	-1.5	-1.5
100	1850	0.0	4.6	10.2	17.6	25.9
200	1625	-9.1	0.0	11.3	26.1	42.7
300	1419	-30.6	-16.9	0.0	22.3	47.1
400	1253	-70.4	-52.2	-29.7	0.0	33.1
500	1107	-129.4	-106.6	-78.4	-41.3	0.0
600	1028	-221.8	-194.5	-160.6	-116.1	-66.5

Universal Table for Rifle Bullets with Ballistic Coefficients of .280-.289

Sight Above Bore (in.) +1.5

Range (yds)	Velocity (fps)	Bullet Path Zero (100 yds)	Zero (200 yds)	Zero (300 yds)	Zero (400 yds)	Zero (500 yds)
0	4000	-1.5	-1.5	-1.5	-1.5	-1.5
100	3576	0.0	0.6	2.0	3.7	5.8
200	3197	-1.2	0.0	2.7	6.2	10.3
300	2849	-5.9	-4.1	0.0	5.3	11.4
400	2526	-15.0	-12.5	-7.1	0.0	8.1
500	2224	-28.9	-25.8	-19.0	-10.2	0.0
600	1944	-50.9	-47.2	-39.1	-28.5	-16.3
0	3900	-1.5	-1.5	-1.5	-1.5	-1.5
100	3489	0.0	0.7	2.1	3.9	6.2
200	3116	-1.4	0.0	2.8	6.5	11.1
300	2773	-6.2	-4.1	0.0	5.6	12.5
400	2455	-15.8	-13.0	-7.5	0.0	9.1
500	2159	-31.2	-27.7	-20.8	-11.4	0.0
600	1882	-53.8	-49.7	-41.4	-30.1	-16.4
0	3800	-1.5	-1.5	-1.5	-1.5	-1.5
100	3398	0.0	0.8	2.2	4.2	6.7
200	3033	-1.5	0.0	2.9	6.9	11.8
300	2695	-6.7	-4.4	0.0	6.0	13.3
400	2383	-16.9	-13.8	-8.0	0.0	9.8
500	2092	-33.3	-29.5	-22.2	-12.2	0.0
600	1820	-57.0	-52.4	-43.7	-31.7	-17.0
0	3700	-1.5	-1.5	-1.5	-1.5	-1.5
100	3307	0.0	0.9	2.5	4.5	7.1
200	2949	-1.7	0.0	3.2	7.3	12.5
300	2619	-7.5	-4.9	0.0	6.1	13.9
400	2311	-18.0	-14.6	-8.1	0.0	10.5
500	2025	-35.7	-31.4	-23.2	-13.1	0.0
600	1757	-60.6	-55.4	-45.6	-33.5	-17.7
0	3600	-1.5	-1.5	-1.5	-1.5	-1.5
100	3215	0.0	1.0	2.7	4.8	7.6
200	2866	-1.9	0.0	3.5	7.7	13.4
300	2541	-8.1	-5.3	0.0	6.3	14.8
400	2238	-19.3	-15.5	-8.4	0.0	11.2
500	1958	-38.1	-33.4	-24.6	-14.0	0.0
600	1701	-66.1	-60.4	-49.9	-37.2	-20.4
0	3500	-1.5	-1.5	-1.5	-1.5	-1.5
100	3123	0.0	1.1	2.9	5.3	8.2
200	2781	-2.1	0.0	3.8	8.5	14.2
300	2463	-8.8	-5.6	0.0	7.2	15.7
400	2166	-21.3	-17.1	-9.5	0.0	11.4
500	1890	-40.9	-35.5	-26.1	-14.2	0.0
600	1640	-71.3	-64.9	-53.6	-39.3	-22.2
0	3400	-1.5	-1.5	-1.5	-1.5	-1.5
100	3034	0.0	1.1	3.1	5.7	8.7
200	2697	-2.1	0.0	4.1	9.2	15.2
300	2384	-9.3	-6.1	0.0	7.7	16.7
400	2093	-22.7	-18.4	-10.3	0.0	12.0
500	1821	-43.3	-38.0	-27.8	-15.0	0.0
600	1580	-76.1	-69.7	-57.4	-42.1	-24.1
0	3300	-1.5	-1.5	-1.5	-1.5	-1.5
100	2942	0.0	1.3	3.3	6.1	9.3
200	2612	-2.6	0.0	4.1	9.7	16.0
300	2305	-10.0	-6.2	0.0	8.4	17.8
400	2019	-24.5	-19.4	-11.1	0.0	12.6
500	1751	-46.3	-39.9	-29.7	-15.7	0.0
600	1519	-81.9	-74.3	-61.9	-45.2	-26.4

Universal Table for Rifle Bullets with Ballistic Coefficients of .280-.289

Sight Above Bore (in.) +1.5

		Bullet Path				
Range (yds)	Velocity (fps)	Zero (100 yds)	Zero (200 yds)	Zero (300 yds)	Zero (400 yds)	Zero (500 yds)
0	3200	-1.5	-1.5	-1.5	-1.5	-1.5
100	2850	0.0	1.4	3.6	6.6	10.3
200	2527	-2.8	0.0	4.3	10.4	17.7
300	2225	-10.7	-6.5	0.0	9.1	20.1
400	1945	-26.4	-20.7	-12.1	0.0	14.7
500	1689	-51.3	-44.2	-33.4	-18.3	0.0
600	1459	-88.1	-79.6	-66.6	-48.5	-26.5
0	3100	-1.5	-1.5	-1.5	-1.5	-1.5
100	2757	0.0	1.6	4.1	7.1	11.1
200	2441	-3.2	0.0	5.0	11.1	19.1
300	2146	-12.2	-7.5	0.0	9.2	21.2
400	1870	-28.5	-22.2	-12.3	0.0	16.0
500	1623	-55.7	-47.8	-35.3	-20.0	0.0
600	1399	-94.8	-85.3	-70.3	-52.0	-27.9
0	3000	-1.5	-1.5	-1.5	-1.5	-1.5
100	2665	0.0	1.8	4.5	7.7	12.1
200	2354	-3.6	0.0	5.5	11.9	20.7
300	2065	-13.5	-8.2	0.0	9.6	22.9
400	1795	-30.9	-23.7	-12.8	0.0	17.7
500	1557	-60.7	-51.8	-38.1	-22.1	0.0
600	1338	-102.3	-91.6	-75.2	-56.0	-29.5
0	2900	-1.5	-1.5	-1.5	-1.5	-1.5
100	2573	0.0	1.8	4.8	8.5	13.0
200	2268	-3.6	0.0	6.0	13.4	22.5
300	1985	-14.3	-9.0	0.0	11.1	24.8
400	1725	-33.9	-26.7	-14.8	0.0	18.3
500	1491	-65.2	-56.3	-41.3	-22.9	0.0
600	1295	-112.9	-102.2	-84.3	-62.1	-34.7
0	2800	-1.5	-1.5	-1.5	-1.5	-1.5
100	2480	0.0	2.1	5.2	9.3	14.1
200	2182	-4.2	0.0	6.2	14.3	24.0
300	1904	-15.6	-9.3	0.0	12.2	26.7
400	1653	-37.1	-28.7	-16.3	0.0	19.4
500	1426	-70.6	-60.0	-44.6	-24.2	0.0
600	1243	-123.1	-110.5	-91.9	-67.5	-38.4
0	2700	-1.5	-1.5	-1.5	-1.5	-1.5
100	2386	0.0	2.4	5.6	10.2	15.3
200	2095	-4.7	0.0	6.5	15.6	25.8
300	1823	-16.9	-9.8	0.0	13.6	28.9
400	1581	-40.7	-31.3	-18.2	0.0	20.3
500	1361	-76.3	-64.5	-48.1	-25.4	0.0
600	1193	-134.0	-119.8	-100.2	-72.9	-42.4
0	2600	-1.5	-1.5	-1.5	-1.5	-1.5
100	2293	0.0	2.7	6.1	11.2	16.6
200	2008	-5.3	0.0	6.9	17.0	27.8
300	1741	-18.4	-10.4	0.0	15.1	31.3
400	1510	-44.8	-34.1	-20.2	0.0	21.5
500	1295	-82.8	-69.5	-52.1	-26.9	0.0
600	1144	-145.9	-129.9	-109.1	-78.8	-46.5
0	2500	-1.5	-1.5	-1.5	-1.5	-1.5
100	2199	0.0	3.0	7.2	12.3	18.9
200	1921	-6.0	0.0	8.4	18.5	31.9
300	1668	-21.6	-12.6	0.0	15.2	35.3
400	1440	-49.1	-37.1	-20.3	0.0	26.7
500	1254	-94.7	-79.7	-58.8	-33.4	0.0
600	1096	-158.8	-140.7	-115.6	-85.1	-45.1

Universal Table for Rifle Bullets with Ballistic Coefficients of .280-.289

Sight Above Bore (in.) +1.5

Range (yds)	Velocity (fps)	Bullet Path Zero (100 yds)	Zero (200 yds)	Zero (300 yds)	Zero (400 yds)	Zero (500 yds)
0	2400	-1.5	-1.5	-1.5	-1.5	-1.5
100	2107	0.0	3.0	7.8	13.1	20.6
200	1834	-6.1	0.0	9.4	20.2	35.0
300	1591	-23.3	-14.1	0.0	16.1	38.4
400	1370	-52.5	-40.4	-21.5	0.0	29.7
500	1200	-102.8	-87.6	-64.0	-37.1	0.0
600	1074	-176.0	-157.7	-129.5	-97.2	-52.7
0	2300	-1.5	-1.5	-1.5	-1.5	-1.5
100	2014	0.0	3.3	8.6	14.3	22.5
200	1747	-6.5	0.0	10.6	22.0	38.5
300	1515	-25.7	-15.9	0.0	17.1	41.9
400	1300	-57.0	-44.0	-22.8	0.0	33.1
500	1147	-112.6	-96.4	-69.8	-41.4	0.0
600	1037	-192.2	-172.7	-140.8	-106.7	-57.1
0	2200	-1.5	-1.5	-1.5	-1.5	-1.5
100	1921	0.0	4.1	9.5	16.6	24.6
200	1668	-8.1	0.0	10.8	25.0	41.0
300	1439	-28.5	-16.3	0.0	21.3	45.3
400	1254	-66.3	-50.0	-28.4	0.0	32.0
500	1096	-122.9	-102.5	76.4	-40.0	0.0
600	1015	-213.3	-188.9	-156.3	-113.8	-65.8
0	2100	-1.5	-1.5	-1.5	-1.5	-1.5
100	1827	0.0	4.7	10.4	18.4	27.8
200	1585	-9.4	0.0	11.5	27.4	46.3
300	1364	-31.3	-17.3	0.0	23.9	52.2
400	1196	-73.6	-54.9	-31.9	0.0	37.8
500	1071	-139.2	-115.8	-87.0	-47.2	0.0
600	989	-235.4	-207.3	-172.8	-125.0	-68.4
0	2000	-1.5	-1.5	-1.5	-1.5	-1.5
100	1737	0.0	5.0	11.1	20.1	31.0
200	1503	-10.0	0.0	12.3	30.1	52.1
300	1288	-33.4	-18.4	0.0	26.8	59.7
400	1139	-80.3	-60.3	-35.7	0.0	43.9
500	1037	-155.2	-130.2	-99.5	-54.8	0.0
600	967	-258.7	-228.7	-191.8	-138.3	-72.5

Universal Table for Rifle Bullets with Ballistic Coefficients of .290-.299

Sight Above Bore (in.) +1.5

Range (yds)	Velocity (fps)	Bullet Path Zero (100 yds)	Zero (200 yds)	Zero (300 yds)	Zero (400 yds)	Zero (500 yds)
0	4000	-1.5	-1.5	-1.5	-1.5	-1.5
100	3590	0.0	0.6	2.0	3.7	5.7
200	3222	-1.2	0.0	2.7	6.1	10.2
300	2883	-5.9	-4.0	0.0	5.2	11.2
400	2568	-14.7	-12.3	-6.9	0.0	8.0
500	2273	-28.5	-25.4	-18.7	-10.1	0.0
600	2000	-49.7	-46.0	-37.9	-27.6	-15.5
0	3900	-1.5	-1.5	-1.5	-1.5	-1.5
100	3502	0.0	0.6	2.0	3.9	6.1
200	3138	-1.2	0.0	2.9	6.5	11.0
300	2806	-6.1	-4.3	0.0	5.5	12.2
400	2497	-15.5	-13.1	-7.3	0.0	8.9
500	2207	-30.5	-27.5	-20.3	-11.1	0.0
600	1937	-52.5	-48.8	-40.2	-29.2	-15.9
0	3800	-1.5	-1.5	-1.5	-1.5	-1.5
100	3411	0.0	0.8	2.2	4.2	6.5
200	3057	-1.5	0.0	2.9	6.8	11.5
300	2729	-6.6	-4.3	0.0	5.9	13.0
400	2424	-16.6	-13.6	-7.9	0.0	9.4
500	2140	-32.6	-28.8	-21.7	-11.8	0.0
600	1874	-55.8	-51.2	-42.6	-30.8	-16.6
0	3700	-1.5	-1.5	-1.5	-1.5	-1.5
100	3319	0.0	0.8	2.4	4.4	7.0
200	2973	-1.7	0.0	3.2	7.2	12.3
300	2651	-7.3	-4.8	0.0	6.0	13.6
400	2352	-17.7	-14.4	-7.9	0.0	10.2
500	2072	-34.9	-30.7	-22.7	-12.7	0.0
600	1810	-59.3	-54.2	-44.6	-32.7	-17.4
0	3600	-1.5	-1.5	-1.5	-1.5	-1.5
100	3228	0.0	0.9	2.7	4.8	7.5
200	2889	-1.9	0.0	3.5	7.6	13.1
300	2573	-8.0	-5.2	0.0	6.3	14.4
400	2278	-19.0	-15.3	-8.3	0.0	10.9
500	2004	-37.4	-32.7	-24.1	-13.6	0.0
600	1746	-63.0	-57.4	-47.0	-34.5	-18.1
0	3500	-1.5	-1.5	-1.5	-1.5	-1.5
100	3135	0.0	1.0	2.9	5.2	8.0
200	2804	-2.1	0.0	3.7	8.4	14.0
300	2495	-8.7	-5.5	0.0	7.0	15.4
400	2205	-20.9	-16.7	-9.3	0.0	11.2
500	1935	-40.1	-34.9	-25.7	-14.0	0.0
600	1689	-69.2	-62.9	-51.8	-37.8	-21.0
0	3400	-1.5	-1.5	-1.5	-1.5	-1.5
100	3046	0.0	1.1	3.1	5.6	8.5
200	2719	-2.1	0.0	4.0	9.0	14.9
300	2415	-9.2	-6.0	0.0	7.5	16.3
400	2131	-22.3	-18.1	-10.0	0.0	11.7
500	1866	-42.5	-37.2	-27.2	-14.7	0.0
600	1627	-74.1	-67.7	-55.7	-40.6	-23.0
0	3300	-1.5	-1.5	-1.5	-1.5	-1.5
100	2954	0.0	1.3	3.3	6.0	9.1
200	2634	-2.5	0.0	4.1	9.5	15.7
300	2335	-9.9	-6.1	0.0	8.1	17.4
400	2057	-24.0	-19.0	-10.8	0.0	12.3
500	1796	-45.5	-39.2	-29.0	-15.4	0.0
600	1566	-79.7	-72.1	-59.8	-43.6	-25.1

Universal Table for Rifle Bullets with Ballistic Coefficients of .290-.299

						Sight Above Bore (in.) +1.5
			Bullet Path			
Range (yds)	Velocity (fps)	Zero (100 yds)	Zero (200 yds)	Zero (300 yds)	Zero (400 yds)	Zero (500 yds)
0	3200	-1.5	-1.5	-1.5	-1.5	-1.5
100	2861	0.0	1.4	3.6	6.5	10.0
200	2548	-2.8	0.0	4.3	10.2	17.2
300	2254	-10.7	-6.5	0.0	8.8	19.3
400	1982	-26.0	-20.3	-11.7	0.0	14.0
500	1731	-50.0	-43.0	-32.2	-17.5	0.0
600	1504	-85.9	-77.5	-64.6	-47.0	-25.9
0	3100	-1.5	-1.5	-1.5	-1.5	-1.5
100	2769	0.0	1.6	4.0	7.0	10.9
200	2462	-3.1	0.0	4.9	10.9	18.6
300	2175	-12.1	-7.3	0.0	9.0	20.6
400	1907	-28.1	-21.8	-12.0	0.0	15.5
500	1663	-54.5	-46.6	-34.4	-19.4	0.0
600	1443	-92.6	-83.2	-68.5	-50.5	-27.3
0	3000	-1.5	-1.5	-1.5	-1.5	-1.5
100	2676	0.0	1.8	4.4	7.6	11.9
200	2375	-3.5	0.0	5.3	11.7	20.2
300	2094	-13.3	-8.0	0.0	9.5	22.3
400	1831	-30.4	-23.4	-12.7	0.0	17.0
500	1596	-59.3	-50.5	-37.2	-21.3	0.0
600	1381	-100.1	-89.5	-73.5	-54.5	-28.9
0	2900	-1.5	-1.5	-1.5	-1.5	-1.5
100	2583	0.0	1.8	4.7	8.1	12.8
200	2288	-3.5	0.0	5.9	12.6	22.0
300	2013	-14.1	-8.8	0.0	10.1	24.2
400	1754	-32.2	-25.2	-13.4	0.0	18.8
500	1529	-63.8	-55.0	-40.3	-23.5	0.0
600	1320	-106.7	-96.1	-78.5	-58.4	-30.2
0	2800	-1.5	-1.5	-1.5	-1.5	-1.5
100	2490	0.0	2.1	5.1	9.1	13.8
200	2201	-4.2	0.0	6.1	14.0	23.5
300	1931	-15.4	-9.2	0.0	11.9	26.1
400	1685	-36.4	-28.1	-15.8	0.0	18.9
500	1463	-69.1	-58.7	-43.5	-23.7	0.0
600	1278	-119.6	-107.1	-88.8	-65.1	-36.6
0	2700	-1.5	-1.5	-1.5	-1.5	-1.5
100	2397	0.0	2.3	5.6	10.0	15.0
200	2114	-4.7	0.0	6.5	15.3	25.3
300	1850	-16.7	-9.7	0.0	13.2	28.3
400	1613	-39.9	-30.5	-17.6	0.0	20.1
500	1397	-75.0	-63.3	-47.1	-25.1	0.0
600	1226	-130.6	-116.6	-97.1	-70.8	-40.6
0	2600	-1.5	-1.5	-1.5	-1.5	-1.5
100	2303	0.0	2.6	6.1	11.0	16.3
200	2027	-5.3	0.0	6.9	16.7	27.3
300	1768	-18.2	-10.3	0.0	14.7	30.6
400	1541	-43.9	-33.4	-19.6	0.0	21.2
500	1331	-81.3	-68.2	-51.0	-26.4	0.0
600	1176	-142.3	-126.5	-105.9	-76.4	-44.7
0	2500	-1.5	-1.5	-1.5	-1.5	-1.5
100	2209	0.0	3.0	7.1	12.1	18.5
200	1940	-5.9	0.0	8.2	18.2	31.0
300	1693	-21.2	-12.3	0.0	15.0	34.2
400	1470	-48.3	-36.4	-20.0	0.0	25.6
500	1284	-92.3	-77.5	-57.0	-32.0	0.0
600	1126	-155.6	-137.8	-113.2	-83.2	-44.8

Universal Table for Rifle Bullets with Ballistic Coefficients of .290-.299

Sight Above Bore (in.) +1.5

Range (yds)	Velocity (fps)	Bullet Path Zero (100 yds)	Zero (200 yds)	Zero (300 yds)	Zero (400 yds)	Zero (500 yds)
0	2400	-1.5	-1.5	-1.5	-1.5	-1.5
100	2117	0.0	3.0	7.6	13.0	20.1
200	1852	-6.1	0.0	9.2	19.9	34.2
300	1615	-22.9	-13.8	0.0	16.0	37.4
400	1399	-51.8	-39.7	-21.3	0.0	28.6
500	1228	-100.5	-85.4	-62.4	-35.8	0.0
600	1097	-172.2	-154.0	-126.4	-94.5	-51.6
0	2300	-1.5	-1.5	-1.5	-1.5	-1.5
100	2023	0.0	3.2	8.4	14.1	22.0
200	1764	-6.5	0.0	10.4	21.7	37.6
300	1538	-25.3	-15.6	0.0	16.9	40.8
400	1328	-56.3	-43.4	-22.6	0.0	31.8
500	1174	-110.2	-93.9	-68.0	-39.8	0.0
600	1059	-188.0	-168.5	-137.4	-103.5	-55.8
0	2200	-1.5	-1.5	-1.5	-1.5	-1.5
100	1930	0.0	4.0	9.3	16.3	24.2
200	1684	-8.0	0.0	10.6	24.5	40.3
300	1462	-28.0	-15.9	0.0	20.9	44.5
400	1277	-65.2	-49.1	-27.8	0.0	31.6
500	1121	-120.9	-100.8	-74.2	-39.4	0.0
600	1029	-209.5	-185.4	-153.5	-111.8	-64.5
0	2100	-1.5	-1.5	-1.5	-1.5	-1.5
100	1836	0.0	4.6	10.4	18.1	27.4
200	1601	-9.2	0.0	11.5	27.0	45.5
300	1385	-31.1	-17.3	0.0	23.2	51.0
400	1218	-72.4	-53.9	-30.9	0.0	37.1
500	1090	-136.9	-113.8	-85.0	-46.4	0.0
600	1005	-230.0	-202.3	-167.8	-121.4	-65.8
0	2000	-1.5	-1.5	-1.5	-1.5	-1.5
100	1754	0.0	4.9	11.0	19.5	29.3
200	1533	-9.9	0.0	12.1	29.1	48.8
300	1330	-33.0	-18.1	0.0	25.5	55.0
400	1181	-78.0	-58.2	-34.1	0.0	39.3
500	1068	-146.6	-121.9	-91.7	-49.1	0.0
600	994	-246.5	-216.9	-180.6	-129.5	-70.6

Universal Table for Rifle Bullets with Ballistic Coefficients of .300-.309

Sight Above Bore (in.) +1.5

Range (yds)	Velocity (fps)	Bullet Path Zero (100 yds)	Zero (200 yds)	Zero (300 yds)	Zero (400 yds)	Zero (500 yds)
0	3500	-1.5	-1.5	-1.5	-1.5	-1.5
100	3147	0.0	1.0	2.9	5.0	7.9
200	2825	-2.1	0.0	3.6	8.0	13.7
300	2524	-8.6	-5.5	0.0	6.5	15.1
400	2241	-20.1	-15.9	-8.6	0.0	11.4
500	1978	-39.4	-34.2	-25.1	-14.3	0.0
600	1735	-67.5	-61.2	-50.3	-37.3	-20.2
0	3400	-1.5	-1.5	-1.5	-1.5	-1.5
100	3057	0.0	1.1	3.0	5.5	8.4
200	2739	-2.1	0.0	3.9	8.8	14.6
300	2444	-9.1	-5.9	0.0	7.3	16.0
400	2167	-21.9	-17.7	-9.8	0.0	11.6
500	1908	-41.8	-36.5	-26.7	-14.5	0.0
600	1672	-72.2	-65.8	-54.0	-39.3	-22.0
0	3300	-1.5	-1.5	-1.5	-1.5	-1.5
100	2965	0.0	1.2	3.3	5.9	9.0
200	2654	-2.5	0.0	4.0	9.3	15.4
300	2364	-9.8	-6.0	0.0	8.0	17.1
400	2092	-23.7	-18.7	-10.6	0.0	12.2
500	1837	-44.8	-38.6	-28.5	-15.2	0.0
600	1610	-77.7	-70.2	-58.1	-42.1	-23.9
0	3200	-1.5	-1.5	-1.5	-1.5	-1.5
100	2872	0.0	1.4	3.5	6.4	9.6
200	2568	-2.8	0.0	4.3	10.0	16.4
300	2282	-10.6	-6.4	0.0	8.6	18.2
400	2017	-25.5	-20.0	-11.4	0.0	12.8
500	1766	-47.9	-41.0	-30.3	-16.0	0.0
600	1547	-83.8	-75.5	-62.7	-45.5	-26.3
0	3100	-1.5	-1.5	-1.5	-1.5	-1.5
100	2779	0.0	1.6	4.0	6.9	10.6
200	2481	-3.1	0.0	4.8	10.7	18.2
300	2202	-11.9	-7.2	0.0	8.9	20.1
400	1941	-27.7	-21.4	-11.8	0.0	14.9
500	1701	-53.2	-45.5	-33.5	-18.7	0.0
600	1485	-90.4	-81.1	-66.7	-48.9	-26.5
0	3000	-1.5	-1.5	-1.5	-1.5	-1.5
100	2686	0.0	1.7	4.4	7.5	11.6
200	2394	-3.5	0.0	5.2	11.5	19.7
300	2121	-13.1	-7.9	0.0	9.5	21.7
400	1864	-30.1	-23.1	-12.6	0.0	16.4
500	1633	-58.0	-49.3	-36.2	-20.5	0.0
600	1422	-97.9	-87.4	-71.7	-52.8	-28.3
0	2900	-1.5	-1.5	-1.5	-1.5	-1.5
100	2593	0.0	1.8	4.6	8.0	12.5
200	2307	-3.5	0.0	5.8	12.4	21.5
300	2039	-13.9	-8.7	0.0	9.9	23.6
400	1787	-31.8	-24.8	-13.3	0.0	18.1
500	1565	-62.5	-53.7	-39.3	-22.7	0.0
600	1360	-104.6	-94.1	-76.7	-56.8	-29.6
0	2800	-1.5	-1.5	-1.5	-1.5	-1.5
100	2500	0.0	1.9	5.1	8.9	13.6
200	2219	-3.8	0.0	6.4	14.1	23.4
300	1957	-15.2	-9.6	0.0	11.5	25.5
400	1716	-35.7	-28.1	-15.4	0.0	18.6
500	1498	-67.9	-58.4	-42.5	-23.3	0.0
600	1312	-116.4	-105.1	-85.9	-62.8	-34.9

Universal Table for Rifle Bullets with Ballistic Coefficients of .300-.309

Sight Above Bore (in.) +1.5

Range (yds)	Velocity (fps)	Bullet Path Zero (100 yds)	Zero (200 yds)	Zero (300 yds)	Zero (400 yds)	Zero (500 yds)
0	2700	-1.5	-1.5	-1.5	-1.5	-1.5
100	2406	0.0	2.3	5.5	9.8	14.7
200	2132	-4.6	0.0	6.4	15.0	24.9
300	1875	-16.6	-9.7	0.0	12.8	27.7
400	1643	-39.1	-29.9	-17.0	0.0	19.8
500	1431	-73.7	-62.2	-46.1	-24.8	0.0
600	1258	-127.3	-113.5	-94.2	-68.6	-38.8
0	2600	-1.5	-1.5	-1.5	-1.5	-1.5
100	2312	0.0	2.6	6.0	10.8	16.0
200	2045	-5.2	0.0	6.9	16.4	26.9
300	1792	-18.1	-10.3	0.0	14.3	30.0
400	1570	-43.1	-32.8	-19.0	0.0	21.0
500	1364	-80.1	-67.2	-50.0	-26.2	0.0
600	1206	-139.1	-123.6	-103.0	-74.5	-43.0
0	2500	-1.5	-1.5	-1.5	-1.5	-1.5
100	2218	0.0	2.9	6.9	11.9	18.1
200	1957	-5.9	0.0	8.0	17.9	30.2
300	1716	-20.8	-12.0	0.0	14.8	33.3
400	1498	-47.5	-35.8	-19.7	0.0	24.7
500	1312	-90.3	-75.6	-55.6	-30.9	0.0
600	1156	-151.8	-134.2	-110.1	-80.5	-43.4
0	2400	-1.5	-1.5	-1.5	-1.5	-1.5
100	2126	0.0	3.0	7.5	12.8	19.7
200	1869	-6.0	0.0	9.1	19.6	33.3
300	1637	-22.6	-13.6	0.0	15.8	36.4
400	1426	-51.2	-39.2	-21.0	0.0	27.5
500	1255	-98.3	-83.3	-60.6	-34.3	0.0
600	1106	-164.2	-146.1	-118.9	-87.4	-46.2
0	2300	-1.5	-1.5	-1.5	-1.5	-1.5
100	2032	0.0	3.2	8.3	13.9	21.7
200	1781	-6.4	0.0	10.2	21.4	36.9
300	1560	-25.0	-15.3	0.0	16.8	40.0
400	1355	-55.6	-42.7	-22.4	0.0	31.0
500	1199	-108.3	-92.2	-66.7	-38.7	0.0
600	1081	-183.5	-164.2	-133.6	-100.1	-53.6
0	2200	-1.5	-1.5	-1.5	-1.5	-1.5
100	1938	0.0	4.0	9.2	15.9	23.8
200	1699	-7.9	0.0	10.5	23.9	39.6
300	1483	-27.6	-15.7	0.0	20.2	43.7
400	1300	-63.7	-47.9	-26.9	0.0	31.4
500	1145	-118.9	-99.1	-72.8	-39.2	0.0
600	1042	-201.7	-178.0	-146.5	-106.2	-59.1
0	2100	-1.5	-1.5	-1.5	-1.5	-1.5
100	1844	0.0	4.5	10.2	17.8	26.8
200	1616	-9.1	0.0	11.4	26.5	44.6
300	1406	-30.7	-17.1	0.0	22.7	49.8
400	1239	-71.2	-53.0	-30.3	0.0	36.1
500	1109	-134.1	-111.4	-83.0	-45.1	0.0
600	1019	-223.6	-196.4	-162.2	-116.8	-62.6

Universal Table for Rifle Bullets with Ballistic Coefficients of .310-.319

Sight Above Bore (in.) +1.5

		Bullet Path				
Range (yds)	Velocity (fps)	Zero (100 yds)	Zero (200 yds)	Zero (300 yds)	Zero (400 yds)	Zero (500 yds)
0	3500	-1.5	-1.5	-1.5	-1.5	-1.5
100	3197	0.0	1.0	2.7	4.8	7.2
200	2917	-2.0	0.0	3.4	7.5	12.5
300	2654	-8.1	-5.1	0.0	6.2	13.6
400	2404	-19.0	-15.0	-8.2	0.0	10.0
500	2169	-36.2	-31.2	-22.7	-12.4	0.0
600	1947	-60.4	-54.4	-44.3	-31.9	-17.0
0	3400	-1.5	-1.5	-1.5	-1.5	-1.5
100	3105	0.0	1.0	2.9	5.0	7.7
200	2830	-2.0	0.0	3.7	8.0	13.4
300	2571	-8.6	-5.5	0.0	6.5	14.5
400	2326	-20.1	-16.0	-8.6	0.0	10.8
500	2096	-38.5	-33.4	-24.2	-13.4	0.0
600	1877	-64.2	-58.1	-47.0	-34.1	-18.0
0	3300	-1.5	-1.5	-1.5	-1.5	-1.5
100	3012	0.0	1.1	3.1	5.5	8.3
200	2742	-2.2	0.0	4.0	8.8	14.4
300	2489	-9.3	-6.0	0.0	7.1	15.6
400	2249	-21.9	-17.5	-9.5	0.0	11.3
500	2022	-41.5	-36.0	-26.0	-14.1	0.0
600	1807	-68.5	-61.9	-49.9	-35.7	-18.8
0	3200	-1.5	-1.5	-1.5	-1.5	-1.5
100	2918	0.0	1.3	3.4	5.9	8.9
200	2655	-2.6	0.0	4.1	9.3	15.2
300	2405	-10.1	-6.2	0.0	7.7	16.6
400	2170	-23.8	-18.6	-10.3	0.0	11.8
500	1948	-44.5	-38.0	-27.7	-14.8	0.0
600	1741	-75.0	-67.2	-54.9	-39.4	-21.6
0	3100	-1.5	-1.5	-1.5	-1.5	-1.5
100	2824	0.0	1.5	3.6	6.5	9.6
200	2566	-3.0	0.0	4.3	10.0	16.2
300	2322	-10.9	-6.5	0.0	8.5	17.9
400	2091	-25.9	-20.0	-11.4	0.0	12.5
500	1873	-48.0	-40.6	-29.8	-15.6	0.0
600	1673	-81.4	-72.6	-59.6	-42.6	-23.8
0	3000	-1.5	-1.5	-1.5	-1.5	-1.5
100	2730	0.0	1.7	4.1	7.0	10.3
200	2478	-3.3	0.0	4.9	10.8	17.4
300	2238	-12.3	-7.3	0.0	8.8	18.7
400	2012	-28.2	-21.6	-11.8	0.0	13.1
500	1798	-51.7	-43.4	-31.2	-16.4	0.0
600	1606	-88.4	-78.5	-63.8	-46.1	-26.4
0	2900	-1.5	-1.5	-1.5	-1.5	-1.5
100	2637	0.0	1.7	4.4	7.5	11.4
200	2389	-3.4	0.0	5.4	11.6	19.4
300	2154	-13.2	-8.1	0.0	9.4	21.0
400	1933	-30.1	-23.2	-12.5	0.0	15.5
500	1727	-56.9	-48.4	-34.9	-19.3	0.0
600	1539	-95.3	-85.1	-68.9	-50.2	-27.0
0	2800	-1.5	-1.5	-1.5	-1.5	-1.5
100	2542	0.0	1.8	4.8	8.1	12.4
200	2299	-3.7	0.0	5.9	12.5	21.1
300	2070	-14.4	-8.9	0.0	9.9	22.8
400	1853	-32.4	-25.1	-13.2	0.0	17.2
500	1655	-62.0	-52.9	-38.1	-21.5	0.0
600	1472	-103.3	-92.3	-74.5	-54.7	-28.9

Universal Table for Rifle Bullets with Ballistic Coefficients of .310-.319

Sight Above Bore (in.) +1.5

Range (yds)	Velocity (fps)	Bullet Path Zero (100 yds)	Zero (200 yds)	Zero (300 yds)	Zero (400 yds)	Zero (500 yds)
0	2700	-1.5	-1.5	-1.5	-1.5	-1.5
100	2448	0.0	2.2	5.3	9.0	13.6
200	2210	-4.4	0.0	6.1	13.7	22.7
300	1986	-15.8	-9.2	0.0	11.3	24.9
400	1776	-36.1	-27.3	-15.1	0.0	18.1
500	1584	-67.8	-56.8	-41.5	-22.6	0.0
600	1406	-111.9	-98.7	-80.3	-57.7	-30.5
0	2600	-1.5	-1.5	-1.5	-1.5	-1.5
100	2353	0.0	2.5	5.8	10.0	14.8
200	2121	-4.9	0.0	6.6	15.0	24.7
300	1901	-17.3	-9.9	0.0	12.6	27.2
400	1699	-39.9	-30.0	-16.8	0.0	19.5
500	1513	-74.2	-61.8	-45.3	-24.3	0.0
600	1340	-121.1	-106.2	-86.4	-61.3	-32.1
0	2500	-1.5	-1.5	-1.5	-1.5	-1.5
100	2258	0.0	2.8	6.3	11.0	16.3
200	2032	-5.6	0.0	7.0	16.5	26.9
300	1816	-19.0	-10.6	0.0	14.1	29.8
400	1623	-44.2	-32.9	-18.9	0.0	20.9
500	1442	-81.3	-67.3	-49.7	-26.1	0.0
600	1291	-137.0	-120.2	-99.0	-70.7	-39.4
0	2400	-1.5	-1.5	-1.5	-1.5	-1.5
100	2165	0.0	2.9	7.1	12.0	17.5
200	1942	-5.9	0.0	8.2	18.2	29.1
300	1736	-21.2	-12.4	0.0	14.9	31.3
400	1547	-48.1	-36.3	-19.8	0.0	21.9
500	1372	-87.6	-72.8	-52.2	-27.4	0.0
600	1235	-149.5	-131.8	-107.1	-77.3	-44.4
0	2300	-1.5	-1.5	-1.5	-1.5	-1.5
100	2070	0.0	3.1	7.9	13.2	19.8
200	1853	-6.3	0.0	9.4	20.1	33.2
300	1655	-23.6	-14.1	0.0	16.0	35.7
400	1472	-52.8	-40.2	-21.4	0.0	26.3
500	1316	-98.8	-83.0	-59.5	-32.8	0.0
600	1181	-163.5	-144.6	-116.4	-84.3	-44.9
0	2200	-1.5	-1.5	-1.5	-1.5	-1.5
100	1976	0.0	3.8	8.8	14.5	21.9
200	1766	-7.5	0.0	10.0	21.4	36.2
300	1575	-26.3	-15.0	0.0	17.0	39.3
400	1398	-57.8	-42.7	-22.7	0.0	29.8
500	1256	-109.5	-90.6	-65.6	-37.2	0.0
600	1127	-179.7	-157.1	-127.0	-93.0	-48.4
0	2100	-1.5	-1.5	-1.5	-1.5	-1.5
100	1881	0.0	4.4	9.8	16.5	24.3
200	1680	-8.7	0.0	10.8	24.3	40.0
300	1496	-29.3	-16.2	0.0	20.2	43.7
400	1335	-66.1	-48.6	-27.0	0.0	31.3
500	1197	-121.7	-99.9	-72.8	-39.1	0.0
600	1095	-202.5	-176.3	-143.9	-103.4	-56.5

Universal Table for Rifle Bullets with Ballistic Coefficients of .320-.329

Sight Above Bore (in.) +1.5

Range (yds)	Velocity (fps)	Bullet Path Zero (100 yds)	Zero (200 yds)	Zero (300 yds)	Zero (400 yds)	Zero (500 yds)
0	3500	-1.5	-1.5	-1.5	-1.5	-1.5
100	3206	0.0	1.0	2.7	4.7	7.1
200	2934	-2.0	0.0	3.4	7.4	12.2
300	2677	-8.0	-5.0	0.0	6.1	13.3
400	2434	-18.8	-14.8	-8.1	0.0	9.7
500	2205	-35.6	-30.6	-22.2	-12.1	0.0
600	1987	-59.5	-53.5	-43.4	-31.3	-16.8
0	3400	-1.5	-1.5	-1.5	-1.5	-1.5
100	3114	0.0	1.0	2.8	5.0	7.6
200	2846	-2.0	0.0	3.6	7.9	13.1
300	2595	-8.5	-5.4	0.0	6.4	14.3
400	2356	-19.8	-15.8	-8.5	0.0	10.5
500	2131	-37.9	-32.8	-23.8	-13.1	0.0
600	1917	-63.1	-57.0	-46.2	-33.4	-17.6
0	3300	-1.5	-1.5	-1.5	-1.5	-1.5
100	3021	0.0	1.1	3.1	5.3	8.2
200	2758	-2.2	0.0	3.9	8.4	14.2
300	2512	-9.2	-5.9	0.0	6.7	15.3
400	2277	-21.2	-16.8	-9.0	0.0	11.5
500	2056	-40.9	-35.4	-25.5	-14.4	0.0
600	1846	-67.5	-60.9	-49.1	-35.7	-18.5
0	3200	-1.5	-1.5	-1.5	-1.5	-1.5
100	2927	0.0	1.3	3.3	5.9	8.8
200	2670	-2.6	0.0	4.0	9.1	15.0
300	2428	-10.0	-6.1	0.0	7.6	16.4
400	2198	-23.5	-18.3	-10.2	0.0	11.7
500	1981	-44.0	-37.5	-27.4	-14.6	0.0
600	1778	-73.5	-65.7	-53.5	-38.3	-20.7
0	3100	-1.5	-1.5	-1.5	-1.5	-1.5
100	2833	0.0	1.5	3.6	6.4	9.5
200	2582	-2.9	0.0	4.3	9.9	16.0
300	2344	-10.8	-6.4	0.0	8.3	17.6
400	2119	-25.6	-19.7	-11.1	0.0	12.4
500	1906	-47.4	-40.1	-29.3	-15.5	0.0
600	1709	-79.8	-71.0	-58.1	-41.4	-22.9
0	3000	-1.5	-1.5	-1.5	-1.5	-1.5
100	2738	0.0	1.6	4.0	7.0	10.2
200	2493	-3.3	0.0	4.8	10.6	17.2
300	2260	-12.1	-7.2	0.0	8.8	18.5
400	2039	-27.9	-21.3	-11.7	0.0	13.0
500	1830	-51.1	-42.9	-30.9	-16.3	0.0
600	1641	-86.6	-76.8	-62.4	-44.8	-25.3
0	2900	-1.5	-1.5	-1.5	-1.5	-1.5
100	2645	0.0	1.7	4.4	7.5	11.2
200	2403	-3.4	0.0	5.3	11.5	18.9
300	2175	-13.1	-7.9	0.0	9.3	20.4
400	1959	-29.8	-23.0	-12.4	0.0	14.9
500	1758	-55.9	-47.3	-34.1	-18.6	0.0
600	1573	-93.5	-83.2	-67.4	-48.7	-26.5
0	2800	-1.5	-1.5	-1.5	-1.5	-1.5
100	2550	0.0	1.8	4.7	8.0	12.2
200	2314	-3.7	0.0	5.8	12.4	20.7
300	2091	-14.2	-8.8	0.0	9.9	22.3
400	1879	-32.2	-24.9	-13.2	0.0	16.6
500	1685	-60.9	-51.8	-37.2	-20.7	0.0
600	1505	-101.5	-90.5	-73.0	-53.2	-28.4

Universal Table for Rifle Bullets with Ballistic Coefficients of .320-.329

Sight Above Bore (in.) +1.5

Range (yds)	Velocity (fps)	Bullet Path: Zero (100 yds)	Zero (200 yds)	Zero (300 yds)	Zero (400 yds)	Zero (500 yds)
0	2700	-1.5	-1.5	-1.5	-1.5	-1.5
100	2455	0.0	2.1	5.2	8.7	13.3
200	2225	-4.3	0.0	6.1	13.1	22.4
300	2006	-15.6	-9.2	0.0	10.4	24.4
400	1798	-34.8	-26.2	-13.9	0.0	18.6
500	1612	-66.7	-56.1	-40.7	-23.3	0.0
600	1438	-110.0	-97.2	-78.8	-57.9	-29.9
0	2600	-1.5	-1.5	-1.5	-1.5	-1.5
100	2361	0.0	2.5	5.7	9.8	14.7
200	2135	-4.9	0.0	6.5	14.7	24.4
300	1921	-17.2	-9.8	0.0	12.3	26.8
400	1723	-39.3	-29.5	-16.4	0.0	19.3
500	1540	-73.3	-61.0	-44.6	-24.1	0.0
600	1371	-119.5	-104.7	-85.1	-60.5	-31.5
0	2500	-1.5	-1.5	-1.5	-1.5	-1.5
100	2265	0.0	2.8	6.3	10.9	16.0
200	2045	-5.6	0.0	6.9	16.2	26.5
300	1836	-18.8	-10.4	0.0	13.9	29.3
400	1646	-43.6	-32.4	-18.5	0.0	20.6
500	1469	-80.2	-66.2	-48.9	-25.7	0.0
600	1317	-134.4	-117.6	-96.8	-69.0	-38.2
0	2400	-1.5	-1.5	-1.5	-1.5	-1.5
100	2172	0.0	2.9	6.9	11.8	17.3
200	1956	-5.8	0.0	8.1	17.9	28.8
300	1755	-20.8	-12.1	0.0	14.7	31.1
400	1570	-47.4	-35.8	-19.6	0.0	21.9
500	1398	-86.6	-72.1	-51.9	-27.4	0.0
600	1260	-146.6	-129.2	-104.9	-75.5	-42.7
0	2300	-1.5	-1.5	-1.5	-1.5	-1.5
100	2077	0.0	3.1	7.7	13.0	19.5
200	1866	-6.3	0.0	9.2	19.8	32.6
300	1673	-23.2	-13.8	0.0	15.9	35.1
400	1494	-52.1	-39.6	-21.2	0.0	25.7
500	1338	-97.3	-81.6	-58.6	-32.1	0.0
600	1204	-161.1	-142.3	-114.6	-82.9	-44.3
0	2200	-1.5	-1.5	-1.5	-1.5	-1.5
100	1982	0.0	3.7	8.7	14.3	21.5
200	1779	-7.3	0.0	10.0	21.2	35.7
300	1592	-26.0	-15.0	0.0	16.9	38.6
400	1419	-57.2	-42.5	-22.5	0.0	28.9
500	1277	-107.6	-89.3	-64.3	-36.2	0.0
600	1150	-176.5	-154.5	-124.6	-90.8	-47.4
0	2100	-1.5	-1.5	-1.5	-1.5	-1.5
100	1887	0.0	4.3	9.7	15.7	24.0
200	1692	-8.6	0.0	10.8	22.8	39.4
300	1512	-29.1	-16.2	0.0	18.0	42.8
400	1344	-62.7	-45.6	-24.0	0.0	33.1
500	1217	-119.8	-98.4	-71.4	-41.4	0.0
600	1113	-198.6	-172.9	-140.4	-104.5	-54.8

Universal Table for Rifle Bullets with Ballistic Coefficients of .330-.339

Sight Above Bore (in.) +1.5

Range (yds)	Velocity (fps)	Bullet Path: Zero (100 yds)	Zero (200 yds)	Zero (300 yds)	Zero (400 yds)	Zero (500 yds)
0	3900	-1.5	-1.5	-1.5	-1.5	-1.5
100	3548	0.0	0.6	2.0	3.7	5.6
200	3222	-1.2	0.0	2.7	6.1	10.0
300	2923	-5.9	-4.1	0.0	5.1	11.0
400	2642	-14.6	-12.2	-6.8	0.0	7.9
500	2378	-28.1	-25.1	-18.3	-9.8	0.0
600	2129	-48.2	-44.7	-36.5	-26.3	-14.5
0	3800	-1.5	-1.5	-1.5	-1.5	-1.5
100	3456	0.0	0.7	2.1	3.9	6.0
200	3139	-1.4	0.0	2.8	6.4	10.5
300	2843	-6.3	-4.2	0.0	5.4	11.6
400	2567	-15.7	-12.8	-7.3	0.0	8.2
500	2307	-29.9	-26.3	-19.3	-10.3	0.0
600	2063	-51.4	-47.1	-38.8	-27.9	-15.5
0	3700	-1.5	-1.5	-1.5	-1.5	-1.5
100	3363	0.0	0.8	2.3	4.2	6.5
200	3053	-1.6	0.0	2.9	6.8	11.3
300	2763	-6.8	-4.4	0.0	5.8	12.6
400	2492	-16.8	-13.6	-7.8	0.0	9.0
500	2237	-32.3	-28.3	-21.0	-11.3	0.0
600	1996	-54.9	-50.0	-41.3	-29.6	-16.1
0	3600	-1.5	-1.5	-1.5	-1.5	1.5
100	3270	0.0	0.9	2.5	4.5	6.9
200	2967	-1.8	0.0	3.2	7.2	12.1
300	2684	-7.5	-4.8	0.0	6.0	13.3
400	2417	-18.0	-14.4	-8.0	0.0	9.7
500	2166	-34.7	-30.2	-22.1	-12.2	0.0
600	1929	-58.6	-53.1	-43.5	-31.5	-16.9
0	3500	-1.5	-1.5	-1.5	-1.5	-1.5
100	3177	0.0	1.0	2.8	4.8	7.5
200	2881	-2.0	0.0	3.5	7.7	12.9
300	2603	-8.3	-5.2	0.0	6.3	14.2
400	2341	-19.4	-15.3	-8.3	0.0	10.6
500	2094	-37.4	-32.4	-23.6	-13.2	0.0
600	1862	-62.5	-56.4	-45.9	-33.4	-17.6
0	3400	-1.5	-1.5	-1.5	-1.5	-1.5
100	3086	0.0	1.0	2.9	5.1	8.0
200	2795	-2.0	0.0	3.8	8.2	13.9
300	2522	-8.7	-5.7	0.0	6.6	15.1
400	2264	-20.4	-16.3	-8.8	0.0	11.4
500	2022	-39.8	-34.7	-25.2	-14.3	0.0
600	1793	-66.3	-60.2	-48.8	-35.7	-18.5
0	3300	-1.5	-1.5	-1.5	-1.5	-1.5
100	2994	0.0	1.2	3.2	5.7	8.5
200	2708	-2.4	0.0	3.9	8.9	14.7
300	2440	-9.5	-5.9	0.0	7.5	16.1
400	2187	-22.6	-17.8	-10.0	0.0	11.6
500	1950	-42.7	-36.7	-26.9	-14.4	0.0
600	1730	-72.5	-65.3	-53.5	-38.6	-21.3
0	3200	-1.5	-1.5	-1.5	-1.5	-1.5
100	2900	0.0	1.3	3.4	6.1	9.2
200	2621	-2.7	0.0	4.2	9.6	15.7
300	2357	-10.3	-6.3	0.0	8.1	17.2
400	2110	-24.5	-19.1	-10.8	0.0	12.2
500	1877	-45.8	-39.2	-28.7	-15.2	0.0
600	1664	-78.4	-70.4	-57.8	-41.7	-23.4

Universal Table for Rifle Bullets with Ballistic Coefficients of .330-.339

Sight Above Bore (in.) +1.5

Range (yds)	Velocity (fps)	Bullet Path Zero (100 yds)	Zero (200 yds)	Zero (300 yds)	Zero (400 yds)	Zero (500 yds)
0	3100	-1.5	-1.5	-1.5	-1.5	-1.5
100	2807	0.0	1.5	3.7	6.7	9.9
200	2533	-3.0	0.0	4.4	10.3	16.7
300	2275	-11.1	-6.5	0.0	8.9	18.5
400	2032	-26.6	-20.6	-11.9	0.0	12.8
500	1803	-49.3	-41.8	-30.9	-16.0	0.0
600	1599	-84.8	-75.8	-62.7	-44.9	-25.6
0	3000	-1.5	-1.5	-1.5	-1.5	-1.5
100	2713	0.0	1.7	4.2	7.2	10.9
200	2445	-3.4	0.0	5.0	11.1	18.5
300	2192	-12.6	-7.5	0.0	9.1	20.2
400	1954	-29.0	-22.2	-12.1	0.0	14.8
500	1734	-54.7	-46.2	-33.6	-18.5	0.0
600	1533	-92.2	-82.0	-66.9	-48.8	-26.6
0	2900	-1.5	-1.5	-1.5	-1.5	-1.5
100	2620	0.0	1.7	4.5	7.7	11.8
200	2357	-3.5	0.0	5.5	11.9	20.2
300	2109	-13.5	-8.3	0.0	9.6	21.9
400	1876	-30.8	-23.9	-12.8	0.0	16.5
500	1663	-59.1	-50.4	-36.6	-20.6	0.0
600	1468	-99.1	-88.7	-72.1	-52.9	-28.2
0	2800	-1.5	-1.5	-1.5	-1.5	-1.5
100	2526	0.0	1.9	4.9	8.3	12.9
200	2268	-3.7	0.0	6.1	12.9	22.0
300	2026	-14.7	-9.2	0.0	10.1	23.9
400	1797	-33.2	-25.7	-13.5	0.0	18.3
500	1593	-64.4	-55.1	-39.8	-22.9	0.0
600	1403	-107.1	-95.9	-77.6	-57.3	-29.9
0	2700	-1.5	-1.5	-1.5	-1.5	-1.5
100	2432	0.0	2.2	5.4	9.3	14.1
200	2180	-4.5	0.0	6.3	14.2	23.7
300	1943	-16.1	-9.4	0.0	11.9	26.1
400	1723	-37.4	-28.4	-15.9	0.0	18.9
500	1523	-70.3	-59.1	-43.5	-23.6	0.0
600	1339	-115.5	-102.1	-83.3	-59.4	-31.1
0	2600	-1.5	-1.5	-1.5	-1.5	-1.5
100	2337	0.0	2.5	5.9	10.3	15.3
200	2091	-5.1	0.0	6.7	15.5	25.6
300	1859	-17.6	-10.0	0.0	13.3	28.4
400	1648	-41.2	-31.1	-17.7	0.0	20.1
500	1454	-76.6	-64.0	-47.3	-25.1	0.0
600	1291	-130.2	-115.0	-95.0	-68.4	-38.3
0	2500	-1.5	-1.5	-1.5	-1.5	-1.5
100	2243	0.0	2.9	6.4	11.4	16.7
200	2003	-5.7	0.0	7.1	17.0	27.7
300	1775	-19.3	-10.6	0.0	14.9	31.0
400	1574	-45.5	-34.1	-19.9	0.0	21.4
500	1385	-83.7	-69.4	-51.6	-26.8	0.0
600	1237	-142.8	-125.6	-104.3	-74.5	-42.3

Universal Table for Rifle Bullets with Ballistic Coefficients of .340-.349

Sight Above Bore (in.) +1.5

Range (yds)	Velocity (fps)	Bullet Path Zero (100 yds)	Zero (200 yds)	Zero (300 yds)	Zero (400 yds)	Zero (500 yds)
0	3500	-1.5	-1.5	-1.5	-1.5	-1.5
100	3186	0.0	1.0	2.7	4.8	7.4
200	2898	-2.0	0.0	3.4	7.6	12.7
300	2627	-8.2	-5.2	0.0	6.2	13.9
400	2370	-19.2	-15.2	-8.3	0.0	10.2
500	2129	-36.8	-31.8	-23.2	-12.8	0.0
600	1901	-61.6	-55.6	-45.3	-32.8	-17.4
0	3400	-1.5	-1.5	-1.5	-1.5	-1.5
100	3095	0.0	1.0	2.9	5.1	7.8
200	2811	-2.0	0.0	3.7	8.1	13.6
300	2545	-8.7	-5.6	0.0	6.5	14.8
400	2293	-20.2	-16.1	-8.7	0.0	11.1
500	2057	-39.2	-34.0	-24.7	-13.8	0.0
600	1832	-65.3	-59.2	-48.0	-35.0	-18.3
0	3300	-1.5	-1.5	-1.5	-1.5	-1.5
100	3002	0.0	1.1	3.1	5.6	8.4
200	2724	-2.2	0.0	4.1	8.9	14.7
300	2463	-9.4	-6.1	0.0	7.3	15.9
400	2216	-22.3	-17.9	0.8	0.0	11.5
500	1983	-42.2	-36.7	-26.5	-14.3	0.0
600	1767	-71.0	-64.4	-52.2	-37.6	-20.4
0	3200	-1.5	-1.5	-1.5	-1.5	-1.5
100	2909	0.0	1.3	3.4	6.0	9.1
200	2637	-2.7	0.0	4.1	9.4	15.5
300	2380	-10.2	-6.2	0.0	7.9	17.0
400	2138	-24.2	-18.9	-10.6	0.0	12.0
500	1910	-45.3	-38.6	-28.3	-15.1	0.0
600	1700	-76.8	-68.8	-56.4	-40.5	-22.5
0	3100	-1.5	-1.5	-1.5	-1.5	-1.5
100	2815	0.0	1.5	3.7	6.6	9.8
200	2549	-3.0	0.0	4.4	10.1	16.5
300	2296	-11.1	-6.6	0.0	8.6	18.2
400	2060	-26.2	-20.3	-11.5	0.0	12.8
500	1835	-48.8	-41.4	-30.4	-16.0	0.0
600	1633	-83.3	-74.4	-61.2	-43.9	-24.8
0	3000	-1.5	-1.5	-1.5	-1.5	-1.5
100	2721	0.0	1.7	4.1	7.2	10.7
200	2460	-3.4	0.0	4.9	10.9	18.1
300	2214	-12.4	-7.4	0.0	9.0	19.8
400	1981	-28.6	-21.9	-12.0	0.0	14.3
500	1765	-53.7	-45.2	-32.9	-17.9	0.0
600	1567	-90.4	-80.3	-65.5	-47.5	-26.0
0	2900	-1.5	-1.5	-1.5	-1.5	-1.5
100	2628	0.0	1.7	4.5	7.6	11.6
200	2372	-3.4	0.0	5.5	11.8	19.8
300	2130	-13.4	-8.2	0.0	9.5	21.4
400	1902	-30.5	-23.6	-12.7	0.0	15.9
500	1693	-58.0	-49.4	-35.7	-19.9	0.0
600	1501	-97.3	-87.0	-70.6	-51.5	-27.7
0	2800	-1.5	-1.5	-1.5	-1.5	-1.5
100	2534	0.0	1.9	4.9	8.2	12.7
200	2282	-3.7	0.0	6.0	12.7	21.6
300	2047	-14.6	-9.0	0.0	10.1	23.4
400	1823	-32.9	-25.4	-13.5	0.0	17.7
500	1622	-63.3	-53.9	-39.0	-22.2	0.0
600	1435	-105.4	-94.2	-76.3	-56.1	-29.5

Universal Table for Rifle Bullets with Ballistic Coefficients of .340-.349

Sight Above Bore (in.) +1.5

		Bullet Path				
Range (yds)	Velocity (fps)	Zero (100 yds)	Zero (200 yds)	Zero (300 yds)	Zero (400 yds)	Zero (500 yds)
0	2700	-1.5	-1.5	-1.5	-1.5	-1.5
100	2439	0.0	2.2	5.3	9.2	13.8
200	2194	-4.4	0.0	6.2	13.9	23.3
300	1963	-15.9	-9.3	0.0	11.6	25.6
400	1748	-36.7	-27.9	-15.5	0.0	18.6
500	1551	-69.1	-58.1	-42.6	-23.3	0.0
600	1370	-113.8	-100.6	-82.0	-58.8	-30.8
0	2600	-1.5	-1.5	-1.5	-1.5	-1.5
100	2345	0.0	2.5	5.8	10.1	15.1
200	2105	-5.0	0.0	6.7	15.2	25.2
300	1878	-17.5	-10.0	0.0	12.9	27.8
400	1672	-40.5	-30.5	-17.2	0.0	19.9
500	1481	-75.6	-63.0	-46.4	-24.9	0.0
600	1318	-127.4	-112.3	-92.3	-66.6	-36.6
0	2500	-1.5	-1.5	-1.5	-1.5	-1.5
100	2250	0.0	2.9	6.4	11.3	16.5
200	2016	-5.7	0.0	7.0	16.8	27.3
300	1794	-19.1	-10.6	0.0	14.6	30.3
400	1596	-45.0	-33.6	-19.5	0.0	20.9
500	1412	-82.5	-68.2	-50.5	-26.1	0.0
600	1262	-140.0	-122.9	-101.7	-72.5	-41.1
0	2400	-1.5	-1.5	-1.5	-1.5	-1.5
100	2157	0.0	3.0	7.1	12.2	17.8
200	1927	-5.9	0.0	8.4	18.4	29.6
300	1716	-21.4	-12.5	0.0	15.1	31.9
400	1522	-48.7	-36.9	-20.2	0.0	22.4
500	1342	-88.8	-74.0	-53.1	-27.9	0.0
600	1207	-152.7	-135.0	-109.9	-79.7	-46.1
0	2300	-1.5	-1.5	-1.5	-1.5	-1.5
100	2062	0.0	3.2	8.0	13.3	20.1
200	1838	-6.3	0.0	9.6	20.3	33.9
300	1635	-23.9	-14.4	0.0	16.1	36.5
400	1448	-53.2	-40.6	-21.4	0.0	27.3
500	1291	-100.6	-84.8	-60.9	-34.1	0.0
600	1154	-166.9	-147.9	-119.2	-87.0	-46.1
0	2200	-1.5	-1.5	-1.5	-1.5	-1.5
100	1968	0.0	3.8	8.9	14.6	22.3
200	1752	-7.6	0.0	10.1	21.6	37.0
300	1556	-26.6	-15.1	0.0	17.2	40.4
400	1374	-58.4	-43.1	-23.0	0.0	30.9
500	1232	-111.5	-92.5	-67.3	-38.6	0.0
600	1117	-187.0	-164.2	-133.9	-99.5	-53.2
0	2100	-1.5	-1.5	-1.5	-1.5	-1.5
100	1873	0.0	4.4	9.9	16.8	24.7
200	1667	-8.7	0.0	11.0	24.9	40.6
300	1477	-29.6	-16.5	0.0	20.8	44.4
400	1314	-67.2	-49.8	-27.8	0.0	31.4
500	1175	-123.3	-101.5	-74.0	-39.2	0.0
600	1076	-205.7	-179.5	-146.5	-104.8	-57.7

Universal Table for Rifle Bullets with Ballistic Coefficients of .350-.359

Sight Above Bore (in.) +1.5

		Bullet Path				
Range (yds)	Velocity (fps)	Zero (100 yds)	Zero (200 yds)	Zero (300 yds)	Zero (400 yds)	Zero (500 yds)
0	3500	-1.5	-1.5	-1.5	-1.5	-1.5
100	3230	0.0	1.0	2.6	4.5	6.7
200	2979	-1.9	0.0	3.2	7.2	11.5
300	2741	-7.8	-4.9	0.0	5.9	12.4
400	2515	-18.2	-14.3	-7.9	0.0	8.7
500	2300	-33.7	-28.8	-20.7	-10.9	0.0
600	2096	-56.8	-51.0	-41.3	-29.5	-16.4
0	3400	-1.5	-1.5	-1.5	-1.5	-1.5
100	3137	0.0	1.0	2.7	4.8	7.3
200	2890	-2.0	0.0	3.5	7.7	12.6
300	2657	-8.2	-5.3	0.0	6.2	13.6
400	2435	-19.3	-15.3	-8.3	0.0	9.8
500	2224	-36.4	-31.4	-22.7	-12.3	0.0
600	2024	-60.4	-54.4	-43.9	-31.5	-16.7
0	3300	-1.5	-1.5	-1.5	-1.5	-1.5
100	3043	0.0	1.1	3.0	5.2	7.8
200	2801	-2.2	0.0	3.8	8.2	13.5
300	2573	-8.9	-5.7	0.0	6.5	14.6
400	2355	-20.6	-16.3	-8.7	0.0	10.7
500	2148	-39.2	-33.8	-24.3	-13.4	0.0
600	1951	-64.8	-58.3	46.9	-33.8	-17.7
0	3200	-1.5	-1.5	-1.5	-1.5	-1.5
100	2949	0.0	1.3	3.3	5.6	8.5
200	2713	-2.5	0.0	4.0	8.8	14.4
300	2488	-9.8	-6.0	0.0	7.2	15.6
400	2275	-22.6	-17.6	-9.6	0.0	11.2
500	2072	-42.3	-36.0	-26.0	-14.0	0.0
600	1878	-69.5	-61.9	-49.9	-35.6	-18.7
0	3100	-1.5	-1.5	-1.5	-1.5	-1.5
100	2855	0.0	1.4	3.5	6.2	9.1
200	2623	-2.9	0.0	4.2	9.4	15.4
300	2403	-10.6	-6.3	0.0	7.9	16.8
400	2194	-24.6	-18.9	-10.5	0.0	11.9
500	1995	-45.7	-38.5	-28.0	-14.9	0.0
600	1807	-75.6	-67.0	-54.3	-38.6	-20.7
0	3000	-1.5	-1.5	-1.5	-1.5	-1.5
100	2760	0.0	1.6	3.9	6.7	9.9
200	2533	-3.3	0.0	4.5	10.2	16.5
300	2317	-11.6	-6.7	0.0	8.6	18.1
400	2113	-26.9	-20.4	-11.5	0.0	12.7
500	1917	-49.5	-41.4	-30.2	-15.9	0.0
600	1735	-82.4	-72.6	-59.2	-42.0	-22.9
0	2900	-1.5	-1.5	-1.5	-1.5	-1.5
100	2666	0.0	1.7	4.2	7.2	10.6
200	2443	-3.4	0.0	5.1	11.1	17.8
300	2232	-12.7	-7.6	0.0	9.0	19.1
400	2031	-28.9	-22.2	-12.0	0.0	13.4
500	1839	-52.9	-44.4	-31.8	-16.7	0.0
600	1665	-88.8	-78.7	-63.5	-45.4	-25.4
0	2800	-1.5	-1.5	-1.5	-1.5	-1.5
100	2571	0.0	1.8	4.6	7.8	11.7
200	2353	-3.6	0.0	5.6	12.1	19.7
300	2146	-13.9	-8.5	0.0	9.6	21.1
400	1949	-31.4	-24.1	-12.8	0.0	15.2
500	1765	-58.3	-49.2	-35.1	-19.0	0.0
600	1595	-96.6	-85.6	-68.7	-49.5	-26.6

Universal Table for Rifle Bullets with Ballistic Coefficients of .350-.359

Sight Above Bore (in.) +1.5

Range (yds)	Velocity (fps)	Bullet Path: Zero (100 yds)	Zero (200 yds)	Zero (300 yds)	Zero (400 yds)	Zero (500 yds)
0	2700	-1.5	-1.5	-1.5	-1.5	-1.5
100	2476	0.0	2.1	5.1	8.5	12.8
200	2263	-4.2	0.0	6.0	12.8	21.3
300	2060	-15.3	-9.0	0.0	10.2	23.0
400	1867	-34.0	-25.6	-13.6	0.0	17.1
500	1690	-63.8	-53.3	-38.3	-21.3	0.0
600	1525	-105.2	-92.6	-74.6	-54.2	-28.6
0	2600	-1.5	-1.5	-1.5	-1.5	-1.5
100	2380	0.0	2.4	5.6	9.4	14.0
200	2172	-4.8	0.0	6.4	14.1	23.2
300	1974	-16.8	-9.6	0.0	11.5	25.2
400	1788	-37.8	-28.2	-15.4	0.0	18.3
500	1616	-70.0	-58.1	-42.0	-22.8	0.0
600	1455	-114.8	-100.4	-81.2	-58.1	-30.8
0	2500	-1.5	-1.5	-1.5	-1.5	-1.5
100	2285	0.0	2.7	6.2	10.5	15.5
200	2082	-5.5	0.0	6.9	15.5	25.4
300	1888	-18.5	-10.3	0.0	13.0	27.9
400	1709	-42.0	-31.0	-17.3	0.0	19.8
500	1542	-77.3	-63.6	-46.4	-24.8	0.0
600	1386	-125.1	-108.7	-88.1	-62.1	-32.4
0	2400	-1.5	-1.5	-1.5	-1.5	-1.5
100	2190	0.0	2.8	6.7	11.4	16.7
200	1992	-5.7	0.0	7.8	17.2	27.8
300	1804	-20.2	-11.7	0.0	14.1	30.0
400	1631	-45.6	-34.3	-18.8	0.0	21.3
500	1469	-83.6	-69.5	-50.0	-26.6	0.0
600	1330	-138.7	-121.7	-98.4	-70.2	-38.3
0	2300	-1.5	-1.5	-1.5	-1.5	-1.5
100	2095	0.0	3.1	7.5	12.6	18.2
200	1901	-6.2	0.0	8.9	19.1	30.2
300	1720	-22.6	-13.3	0.0	15.3	32.0
400	1553	-50.5	-38.1	-20.4	0.0	22.3
500	1397	-91.0	-75.5	-53.4	-27.9	0.0
600	1270	-153.2	-134.6	-108.0	-77.5	-44.0
0	2200	-1.5	-1.5	-1.5	-1.5	-1.5
100	2000	0.0	3.3	8.4	13.9	20.6
200	1810	-6.7	0.0	10.1	21.2	34.6
300	1638	-25.2	-15.2	0.0	16.5	36.7
400	1476	-55.7	-42.3	-22.1	0.0	26.9
500	1335	-103.2	-86.5	-61.2	-33.6	0.0
600	1212	-169.2	-149.2	-118.8	-85.7	-45.4
0	2100	-1.5	-1.5	-1.5	-1.5	-1.5
100	1905	0.0	4.2	9.5	15.3	22.9
200	1724	-8.4	0.0	10.6	22.3	37.5
300	1556	-28.5	-15.9	0.0	17.5	40.3
400	1400	-61.3	-44.6	-23.3	0.0	30.5
500	1273	-114.7	-93.8	-67.2	-38.1	0.0
600	1156	-186.6	-161.5	-129.6	-94.7	-49.0

Universal Table for Rifle Bullets with Ballistic Coefficients of .360-.369

Sight Above Bore (in.) +1.5

Range (yds)	Velocity (fps)	Bullet Path Zero (100 yds)	Zero (200 yds)	Zero (300 yds)	Zero (400 yds)	Zero (500 yds)
0	3500	-1.5	-1.5	-1.5	-1.5	-1.5
100	3237	0.0	1.0	2.6	4.5	6.7
200	2992	-1.9	0.0	3.2	7.1	11.4
300	2760	-7.7	-4.8	0.0	5.8	12.3
400	2539	-18.0	-14.2	-7.8	0.0	8.6
500	2329	-33.3	-28.5	-20.5	-10.8	0.0
600	2129	-55.9	-50.2	-40.6	-28.9	-16.0
0	3400	-1.5	-1.5	-1.5	-1.5	-1.5
100	3144	0.0	1.0	2.7	4.8	7.2
200	2904	-2.0	0.0	3.5	7.6	12.4
300	2676	-8.2	-5.2	0.0	6.2	13.4
400	2459	-19.1	-15.2	-8.2	0.0	9.6
500	2253	-35.9	-31.0	-22.3	-12.0	0.0
600	2056	-59.6	-53.8	-43.3	-31.0	-16.6
0	3300	-1.5	-1.5	-1.5	-1.5	-1.5
100	3050	0.0	1.1	3.0	5.1	7.7
200	2814	-2.2	0.0	3.8	8.1	13.3
300	2591	-8.9	-5.7	0.0	6.5	14.3
400	2378	-20.5	-16.2	-8.7	0.0	10.4
500	2176	-38.7	-33.3	-23.9	-13.0	0.0
600	1983	-63.9	-57.5	-46.1	-33.1	-17.5
0	3200	-1.5	-1.5	-1.5	-1.5	-1.5
100	2956	0.0	1.3	3.2	5.5	8.4
200	2725	-2.5	0.0	3.9	8.5	14.2
300	2506	-9.7	-5.9	0.0	6.8	15.4
400	2297	-22.0	-17.0	-9.1	0.0	11.4
500	2099	-41.8	-35.5	-25.6	-14.3	0.0
600	1909	-68.7	-61.1	-49.3	-35.7	-18.5
0	3100	-1.5	-1.5	-1.5	-1.5	-1.5
100	2861	0.0	1.4	3.5	6.1	9.1
200	2636	-2.8	0.0	4.2	9.4	15.3
300	2421	-10.5	-6.3	0.0	7.8	16.6
400	2216	-24.4	-18.8	-10.4	0.0	11.8
500	2021	-45.3	-38.2	-27.7	-14.8	0.0
600	1835	-73.7	-65.2	-52.6	-37.1	-19.4
0	3000	-1.5	-1.5	-1.5	-1.5	-1.5
100	2766	0.0	1.6	3.8	6.6	9.8
200	2546	-3.2	0.0	4.4	10.1	16.4
300	2335	-11.4	-6.7	0.0	8.5	18.0
400	2135	-26.6	-20.2	-11.3	0.0	12.6
500	1943	-49.0	-41.0	-29.9	-15.8	0.0
600	1764	-81.0	-71.4	-58.1	-41.1	-22.2
0	2900	-1.5	-1.5	-1.5	-1.5	-1.5
100	2672	0.0	1.7	4.2	7.1	10.5
200	2455	-3.4	0.0	5.0	10.9	17.6
300	2249	-12.5	-7.5	0.0	8.9	18.9
400	2053	-28.6	-21.9	-11.9	0.0	13.3
500	1865	-52.3	-43.9	-31.4	-16.6	0.0
600	1693	-87.4	-77.3	-62.3	-44.5	-24.6
0	2800	-1.5	-1.5	-1.5	-1.5	-1.5
100	2577	0.0	1.8	4.6	7.7	11.5
200	2365	-3.6	0.0	5.6	11.9	19.4
300	2163	-13.7	-8.3	0.0	9.5	20.8
400	1971	-31.0	-23.8	-12.7	0.0	15.0
500	1789	-57.5	-48.5	-34.6	-18.8	0.0
600	1622	-95.1	-84.3	-67.7	-48.7	-26.1

Universal Table for Rifle Bullets with Ballistic Coefficients of .360-.369

Sight Above Bore (in.) +1.5

Range (yds)	Velocity (fps)	Bullet Path Zero (100 yds)	Zero (200 yds)	Zero (300 yds)	Zero (400 yds)	Zero (500 yds)
0	2700	-1.5	-1.5	-1.5	-1.5	-1.5
100	2482	0.0	2.1	5.0	8.4	12.6
200	2274	-4.2	0.0	5.9	12.7	21.0
300	2077	-15.1	-8.8	0.0	10.2	22.6
400	1888	-33.7	-25.3	-13.6	0.0	16.6
500	1714	-62.9	-52.4	-37.7	-20.8	0.0
600	1551	-103.9	-91.3	-73.7	-53.3	-28.4
0	2600	-1.5	-1.5	-1.5	-1.5	-1.5
100	2386	0.0	2.4	5.6	9.3	13.8
200	2183	-4.8	0.0	6.4	13.9	22.9
300	1990	-16.7	-9.5	0.0	11.3	24.8
400	1808	-37.3	-27.8	-15.1	0.0	18.0
500	1639	-69.1	-57.2	-41.3	-22.5	0.0
600	1481	-113.2	-98.9	-79.9	-57.3	-30.3
0	2500	-1.5	-1.5	-1.5	-1.5	-1.5
100	2291	0.0	2.7	6.1	10.4	15.3
200	2093	-5.5	0.0	6.8	15.3	25.1
300	1904	-18.4	-10.2	0.0	12.8	27.5
400	1728	-41.5	-30.6	-17.0	0.0	19.6
500	1564	-76.4	-62.8	-45.8	-24.5	0.0
600	1411	-123.7	-107.4	-87.0	-61.4	-32.0
0	2400	-1.5	-1.5	-1.5	-1.5	-1.5
100	2196	0.0	2.9	6.5	11.3	16.5
200	2002	-5.7	0.0	7.2	16.9	27.3
300	1817	-19.4	-10.8	0.0	14.6	30.2
400	1649	-45.3	-33.9	-19.4	0.0	20.8
500	1491	-82.6	-68.4	-50.3	-26.1	0.0
600	1351	-136.9	-119.7	-98.1	-69.0	-37.7
0	2300	-1.5	-1.5	-1.5	-1.5	-1.5
100	2101	0.0	3.1	7.4	12.5	18.1
200	1911	-6.2	0.0	8.6	18.8	30.0
300	1735	-22.3	-12.9	0.0	15.2	32.0
400	1571	-50.0	-37.5	-20.3	0.0	22.4
500	1417	-90.5	-74.9	-53.4	-28.0	0.0
600	1290	-151.1	-132.5	-106.6	-76.2	-42.5
0	2200	-1.5	-1.5	-1.5	-1.5	-1.5
100	2006	0.0	3.4	8.3	13.8	20.3
200	1820	-6.7	0.0	10.0	21.0	33.9
300	1652	-25.0	-14.9	0.0	16.5	35.9
400	1493	-55.4	-41.9	-22.0	0.0	25.8
500	1354	-101.5	-84.7	-59.8	-32.3	0.0
600	1232	-166.5	-146.3	-116.4	-83.4	-44.7
0	2100	-1.5	-1.5	-1.5	-1.5	-1.5
100	1910	0.0	4.1	9.4	15.2	22.6
200	1734	-8.2	0.0	10.5	22.3	37.0
300	1570	-28.1	-15.8	0.0	17.7	39.8
400	1416	-61.0	-44.6	-23.5	0.0	29.5
500	1290	-113.1	-92.6	-66.3	-36.9	0.0
600	1175	-183.7	-159.0	-127.5	-92.2	-47.9

Universal Table for Rifle Bullets with Ballistic Coefficients of .370-.379

Sight Above Bore (in.) +1.5

Range (yds)	Velocity (fps)	Bullet Path Zero (100 yds)	Zero (200 yds)	Zero (300 yds)	Zero (400 yds)	Zero (500 yds)
0	3500	-1.5	-1.5	-1.5	-1.5	-1.5
100	3244	0.0	1.0	2.5	4.5	6.6
200	3005	-1.9	0.0	3.0	7.0	11.3
300	2778	-7.4	-4.5	0.0	6.0	12.4
400	2562	-17.9	-14.1	-8.0	0.0	8.6
500	2356	-33.1	-28.3	-20.7	-10.7	0.0
600	2160	-55.3	-49.5	-40.4	-28.4	-15.5
0	3400	-1.5	-1.5	-1.5	-1.5	-1.5
100	3151	0.0	1.0	2.7	4.7	7.1
200	2916	-2.0	0.0	3.4	7.5	12.2
300	2694	-8.1	-5.1	0.0	6.1	13.2
400	2482	-18.9	-15.0	-8.2	0.0	9.4
500	2280	-35.5	-30.5	-22.0	-11.8	0.0
600	2087	-58.9	-52.9	-42.7	-30.5	-16.3
0	3300	-1.5	-1.5	-1.5	-1.5	-1.5
100	3057	0.0	1.1	2.9	5.1	7.7
200	2827	-2.1	0.0	3.7	8.0	13.2
300	2609	-8.8	-5.6	0.0	6.4	14.2
400	2401	-20.3	-16.1	-8.6	0.0	10.3
500	2202	-38.3	-33.0	-23.6	-12.9	0.0
600	2013	-63.2	-56.8	-45.6	32.7	-17.2
0	3200	-1.5	-1.5	-1.5	-1.5	-1.5
100	2962	0.0	1.2	3.2	5.5	8.3
200	2737	-2.5	0.0	3.9	8.4	14.0
300	2523	-9.6	-5.9	0.0	6.7	15.2
400	2319	-21.8	-16.8	-9.0	0.0	11.2
500	2125	-41.3	-35.1	-25.3	-14.0	0.0
600	1939	-67.8	-60.4	-48.6	-35.1	-18.3
0	3100	-1.5	-1.5	-1.5	-1.5	-1.5
100	2867	0.0	1.4	3.5	6.0	8.9
200	2647	-2.8	0.0	4.2	9.3	15.1
300	2437	-10.5	-6.3	0.0	7.6	16.3
400	2237	-24.2	-18.5	-10.2	0.0	11.6
500	2047	-44.7	-37.6	-27.2	-14.5	0.0
600	1864	-72.9	-64.5	-51.9	-36.7	-19.3
0	3000	-1.5	-1.5	-1.5	-1.5	-1.5
100	2772	0.0	1.6	3.8	6.6	9.7
200	2557	-3.2	0.0	4.4	10.0	16.2
300	2351	-11.4	-6.6	0.0	8.4	17.7
400	2155	-26.4	-20.1	-11.2	0.0	12.4
500	1968	-48.5	-40.5	-29.5	-15.5	0.0
600	1792	-79.7	-70.2	-56.9	-40.1	-21.5
0	2900	-1.5	-1.5	-1.5	-1.5	-1.5
100	2678	0.0	1.7	4.1	7.1	10.4
200	2467	-3.3	0.0	5.0	10.9	17.4
300	2265	-12.4	-7.5	0.0	8.8	18.6
400	2073	-28.4	-21.8	-11.8	0.0	13.0
500	1890	-51.8	-43.5	-31.0	-16.3	0.0
600	1720	-86.1	-76.1	-61.2	-43.5	-23.9
0	2800	-1.5	-1.5	-1.5	-1.5	-1.5
100	2583	0.0	1.8	4.5	7.7	11.3
200	2376	-3.6	0.0	5.5	11.8	19.1
300	2179	-13.6	-8.2	0.0	9.4	20.4
400	1991	-30.8	-23.6	-12.6	0.0	14.6
500	1813	-56.7	-47.7	-34.0	-18.2	0.0
600	1648	-93.8	-83.0	-66.6	-47.7	-25.8

Universal Table for Rifle Bullets with Ballistic Coefficients of .370-.379

Sight Above Bore (in.) +1.5

Range (yds)	Velocity (fps)	Bullet Path Zero (100 yds)	Zero (200 yds)	Zero (300 yds)	Zero (400 yds)	Zero (500 yds)
0	2700	-1.5	-1.5	-1.5	-1.5	-1.5
100	2487	0.0	2.1	5.0	8.3	12.4
200	2285	-4.1	0.0	5.9	12.6	20.7
300	2092	-15.0	-8.8	0.0	10.0	22.3
400	1908	-33.4	-25.1	-13.4	0.0	16.3
500	1736	-62.1	-51.8	-37.1	-20.4	0.0
600	1576	-102.5	-90.1	-72.5	-52.4	-27.9
0	2600	-1.5	-1.5	-1.5	-1.5	-1.5
100	2392	0.0	2.4	5.5	9.1	13.7
200	2194	-4.7	0.0	6.3	13.4	22.7
300	2005	-16.6	-9.5	0.0	10.6	24.5
400	1825	-36.2	-26.8	-14.1	0.0	18.6
500	1660	-68.5	-56.7	-40.8	-23.2	0.0
600	1506	-111.7	-97.5	-78.5	-57.3	-29.5
0	2500	-1.5	-1.5	-1.5	-1.5	-1.5
100	2296	0.0	2.7	6.1	10.3	15.1
200	2103	-5.4	0.0	6.8	15.1	24.8
300	1918	-18.3	-10.2	0.0	12.5	27.0
400	1746	-41.1	-30.2	-16.6	0.0	19.4
500	1585	-75.5	-62.0	-45.0	-24.2	0.0
600	1435	-122.2	-106.0	-85.6	-60.6	-31.6
0	2400	-1.5	-1.5	-1.5	-1.5	-1.5
100	2201	0.0	2.8	6.4	11.2	16.4
200	2012	-5.7	0.0	7.2	16.8	27.1
300	1831	-19.3	-10.8	0.0	14.3	29.8
400	1666	-44.9	-33.5	-19.1	0.0	20.6
500	1511	-81.9	-67.7	-49.6	-25.8	0.0
600	1365	-131.7	-114.7	-93.0	-64.4	-33.5
0	2300	-1.5	-1.5	-1.5	-1.5	-1.5
100	2106	0.0	3.1	7.4	12.4	17.9
200	1921	-6.2	0.0	8.6	18.5	29.7
300	1748	-22.2	-12.9	0.0	14.9	31.6
400	1588	-49.5	-37.1	-19.9	0.0	22.2
500	1437	-89.6	-74.2	-52.7	-27.8	0.0
600	1310	-148.7	-130.1	-104.3	-74.5	-41.1
0	2200	-1.5	-1.5	-1.5	-1.5	-1.5
100	2011	0.0	3.3	8.3	13.7	19.5
200	1830	-6.7	0.0	9.9	20.7	32.4
300	1665	-24.8	-14.8	0.0	16.2	33.8
400	1510	-54.7	-41.4	-21.6	0.0	23.4
500	1364	-97.7	-81.0	-56.3	-29.3	0.0
600	1250	-164.6	-144.5	-114.9	-82.5	-47.4
0	2100	-1.5	-1.5	-1.5	-1.5	-1.5
100	1915	0.0	4.1	9.3	15.1	22.4
200	1743	-8.2	0.0	10.5	22.1	36.6
300	1582	-28.0	-15.8	0.0	17.4	39.1
400	1432	-60.6	-44.2	-23.2	0.0	29.0
500	1306	-111.9	-91.5	-65.2	-36.2	0.0
600	1192	-181.9	-157.5	-125.9	-91.1	-47.6

Universal Table for Rifle Bullets with Ballistic Coefficients of .380-.389

						Sight Above Bore (in.) +1.5
		Bullet Path				
Range (yds)	Velocity (fps)	Zero (100 yds)	Zero (200 yds)	Zero (300 yds)	Zero (400 yds)	Zero (500 yds)
0	3500	-1.5	-1.5	-1.5	-1.5	-1.5
100	3250	0.0	0.9	2.5	4.4	6.6
200	3017	-1.9	0.0	3.0	7.0	11.2
300	2795	-7.4	-4.5	0.0	5.9	12.3
400	2584	-17.7	-13.9	-7.9	0.0	8.5
500	2382	-32.8	-28.1	-20.5	-10.7	0.0
600	2190	-54.5	-48.8	-39.8	-28.0	-15.1
0	3400	-1.5	-1.5	-1.5	-1.5	-1.5
100	3157	0.0	1.0	2.7	4.7	6.9
200	2928	-2.0	0.0	3.4	7.5	11.9
300	2711	-8.0	-5.1	0.0	6.1	12.8
400	2503	-18.8	-14.9	-8.1	0.0	8.9
500	2305	-34.6	-29.7	-21.3	-11.1	0.0
600	2116	-58.2	-52.3	-42.2	-30.0	-16.6
0	3300	-1.5	-1.5	-1.5	-1.5	-1.5
100	3063	0.0	1.1	2.9	5.1	7.6
200	2838	-2.1	0.0	3.7	8.0	13.0
300	2625	-8.8	-5.6	0.0	6.4	13.9
400	2422	-20.2	-15.9	-8.5	0.0	10.0
500	2228	-37.8	-32.5	-23.2	-12.6	0.0
600	2042	-62.5	-56.0	-44.9	-32.2	-17.1
0	3200	-1.5	-1.5	-1.5	-1.5	-1.5
100	2968	0.0	1.2	3.2	5.4	8.2
200	2749	-2.4	0.0	3.9	8.4	13.9
300	2540	-9.5	-5.8	0.0	6.7	15.0
400	2340	-21.7	-16.8	-9.0	0.0	11.0
500	2150	-40.8	-34.7	-25.0	-13.7	0.0
600	1968	-67.0	-59.6	-47.9	-34.5	-18.0
0	3100	-1.5	-1.5	-1.5	-1.5	-1.5
100	2873	0.0	1.4	3.5	6.0	8.9
200	2659	-2.8	0.0	4.2	9.2	14.9
300	2453	-10.4	-6.3	0.0	7.5	16.1
400	2258	-23.9	-18.3	-10.0	0.0	11.5
500	2071	-44.3	-37.3	-26.9	-14.4	0.0
600	1892	-72.1	-63.9	-51.3	-36.3	-19.0
0	3000	-1.5	-1.5	-1.5	-1.5	-1.5
100	2778	0.0	1.6	3.8	6.5	9.6
200	2568	-3.2	0.0	4.4	9.9	16.0
300	2367	-11.3	-6.6	0.0	8.3	17.5
400	2175	-26.2	-19.8	-11.1	0.0	12.2
500	1992	-48.0	-40.1	-29.1	-15.3	0.0
600	1819	-78.6	-69.0	-55.9	-39.3	-20.9
0	2900	-1.5	-1.5	-1.5	-1.5	-1.5
100	2683	0.0	1.7	4.1	7.0	10.3
200	2477	-3.3	0.0	4.9	10.7	17.2
300	2280	-12.3	-7.4	0.0	8.7	18.5
400	2093	-28.0	-21.4	-11.6	0.0	13.0
500	1913	-51.3	-43.1	-30.8	-16.3	0.0
600	1745	-85.0	-75.1	-60.3	-42.9	-23.4
0	2800	-1.5	-1.5	-1.5	-1.5	-1.5
100	2588	0.0	1.8	4.5	7.6	11.1
200	2386	-3.6	0.0	5.4	11.7	18.6
300	2194	-13.5	-8.1	0.0	9.4	19.7
400	2010	-30.5	-23.3	-12.5	0.0	13.8
500	1833	-55.4	-46.4	-32.9	-17.3	0.0
600	1673	-92.4	-81.7	-65.4	-46.7	-26.0

Universal Table for Rifle Bullets with Ballistic Coefficients of .380-.389

Sight Above Bore (in.) +1.5

Range (yds)	Velocity (fps)	Bullet Path Zero (100 yds)	Zero (200 yds)	Zero (300 yds)	Zero (400 yds)	Zero (500 yds)
0	2700	-1.5	-1.5	-1.5	-1.5	-1.5
100	2493	0.0	2.1	5.0	8.3	12.3
200	2295	-4.1	0.0	5.8	12.4	20.4
300	2107	-14.9	-8.7	0.0	10.0	21.9
400	1927	-33.1	-24.9	-13.3	0.0	15.9
500	1758	-61.3	-51.0	-36.5	-19.9	0.0
600	1601	-100.9	-88.5	-71.1	-51.2	-27.3
0	2600	-1.5	-1.5	-1.5	-1.5	-1.5
100	2397	0.0	2.3	5.5	9.0	13.5
200	2204	-4.7	0.0	6.3	13.3	22.4
300	2020	-16.4	-9.4	0.0	10.6	24.2
400	1843	-36.0	-26.7	-14.2	0.0	18.0
500	1681	-67.6	-55.9	-40.3	-22.5	0.0
600	1529	-110.4	-96.4	-77.6	-56.4	-29.3
0	2500	-1.5	-1.5	-1.5	-1.5	-1.5
100	2301	0.0	2.7	6.1	10.2	14.9
200	2113	-5.4	0.0	6.8	15.0	24.4
300	1932	-18.2	-10.2	0.0	12.3	26.4
400	1763	-40.7	-30.0	-16.4	0.0	18.9
500	1606	-74.4	-61.0	-44.1	-23.6	0.0
600	1458	-120.8	-104.7	-84.3	-59.8	-31.5
0	2400	-1.5	-1.5	-1.5	-1.5	-1.5
100	2206	0.0	2.8	6.4	11.1	16.2
200	2022	-5.6	0.0	7.2	16.6	26.8
300	1845	-19.2	-10.8	0.0	14.0	29.3
400	1683	-44.4	-33.1	-18.7	0.0	20.4
500	1531	-80.9	-66.9	-48.9	-25.5	0.0
600	1387	-130.6	-113.7	-92.1	-64.0	-33.4
0	2300	-1.5	-1.5	-1.5	-1.5	-1.5
100	2111	0.0	3.1	7.3	12.3	17.8
200	1931	-6.1	0.0	8.5	18.4	29.4
300	1761	-22.0	-12.8	0.0	14.8	31.3
400	1604	-49.1	-36.8	-19.8	0.0	22.0
500	1456	-88.9	-73.6	-52.2	-27.5	0.0
600	1329	-146.6	-128.2	-102.6	-73.0	-40.0
0	2200	-1.5	-1.5	-1.5	-1.5	-1.5
100	2015	0.0	3.3	8.1	13.6	19.4
200	1839	-6.6	0.0	9.7	20.6	32.2
300	1678	-24.4	-14.5	0.0	16.3	33.7
400	1525	-54.3	-41.1	-21.8	0.0	23.2
500	1382	-97.0	-80.4	-56.2	-29.0	0.0
600	1268	-162.1	-142.3	-113.2	-80.6	-45.8
0	2100	-1.5	-1.5	-1.5	-1.5	-1.5
100	1920	0.0	4.1	9.2	15.0	22.1
200	1751	-8.2	0.0	10.3	21.8	36.0
300	1595	-27.7	-15.4	0.0	17.3	38.6
400	1448	-59.9	-43.6	-23.0	0.0	28.5
500	1322	-110.5	-90.0	-64.4	-35.6	0.0
600	1209	-179.8	-155.2	-124.5	-89.9	-47.2

Universal Table for Rifle Bullets with Ballistic Coefficients of .390-.399

Sight Above Bore (in.) +1.5

Range (yds)	Velocity (fps)	Bullet Path: Zero (100 yds)	Zero (200 yds)	Zero (300 yds)	Zero (400 yds)	Zero (500 yds)
0	3500	-1.5	-1.5	-1.5	-1.5	-1.5
100	3256	0.0	0.9	2.4	4.4	6.5
200	3029	-1.9	0.0	3.0	6.9	11.1
300	2812	-7.3	-4.5	0.0	5.8	12.1
400	2605	-17.5	-13.8	-7.8	0.0	8.4
500	2408	-32.4	-27.8	-20.2	-10.5	0.0
600	2219	-53.8	-48.2	-39.2	-27.5	-14.9
0	3400	-1.5	-1.5	-1.5	-1.5	-1.5
100	3163	0.0	1.0	2.7	4.7	6.9
200	2939	-2.0	0.0	3.4	7.4	11.8
300	2727	-8.0	-5.0	0.0	6.0	12.6
400	2524	-18.6	-14.7	-8.0	0.0	8.8
500	2330	-34.3	-29.4	-21.1	-11.0	0.0
600	2145	-57.4	-51.5	-41.4	-29.4	-16.2
0	3300	-1.5	-1.5	-1.5	-1.5	-1.5
100	3069	0.0	1.1	2.9	5.0	7.5
200	2849	-2.1	0.0	3.7	7.9	12.8
300	2641	-8.7	-5.5	0.0	6.3	13.7
400	2442	-20.1	-15.8	-8.5	0.0	9.9
500	2252	-37.4	-32.1	-22.9	-12.3	0.0
600	2070	-61.7	-55.3	-44.3	-31.6	-16.8
0	3200	-1.5	-1.5	-1.5	-1.5	-1.5
100	2974	0.0	1.2	3.2	5.4	8.1
200	2760	-2.4	0.0	3.9	8.3	13.8
300	2555	-9.5	-5.8	0.0	6.6	14.8
400	2360	-21.5	-16.7	-8.9	0.0	10.9
500	2173	-40.5	-34.4	-24.7	-13.6	0.0
600	1995	-66.3	-59.0	-47.3	-34.0	-17.7
0	3100	-1.5	-1.5	-1.5	-1.5	-1.5
100	2879	0.0	1.4	3.4	5.9	8.8
200	2669	-2.8	0.0	4.1	9.1	14.8
300	2469	-10.3	-6.2	0.0	7.5	16.0
400	2277	-23.7	-18.2	-9.9	0.0	11.4
500	2094	-43.9	-36.9	-26.7	-14.2	0.0
600	1919	-71.4	-63.0	-50.7	-35.8	-18.7
0	3000	-1.5	-1.5	-1.5	-1.5	-1.5
100	2784	0.0	1.6	3.8	6.5	9.5
200	2578	-3.2	0.0	4.3	9.8	15.8
300	2382	-11.3	-6.5	0.0	8.2	17.3
400	2194	-26.0	-19.6	-10.9	0.0	12.1
500	2015	-47.6	-39.6	-28.8	-15.1	0.0
600	1843	-76.9	-67.3	-54.3	-37.9	-19.8
0	2900	-1.5	-1.5	-1.5	-1.5	-1.5
100	2689	0.0	1.7	4.1	7.0	10.2
200	2487	-3.3	0.0	4.8	10.6	17.1
300	2295	-12.3	-7.3	0.0	8.7	18.4
400	2111	-27.9	-21.3	-11.6	0.0	12.9
500	1935	-51.0	-42.7	-30.6	-16.1	0.0
600	1770	-83.9	-73.9	-59.4	-42.0	-22.6
0	2800	-1.5	-1.5	-1.5	-1.5	-1.5
100	2593	0.0	1.8	4.5	7.6	11.0
200	2396	-3.6	0.0	5.4	11.6	18.4
300	2208	-13.4	-8.0	0.0	9.3	19.6
400	2028	-30.3	-23.1	-12.4	0.0	13.7
500	1855	-54.9	-46.0	-32.6	-17.1	0.0
600	1697	-91.1	-80.4	-64.4	-45.7	-25.2

Universal Table for Rifle Bullets with Ballistic Coefficients of .390-.399

Sight Above Bore (in.) +1.5

Range (yds)	Velocity (fps)	Bullet Path Zero (100 yds)	Zero (200 yds)	Zero (300 yds)	Zero (400 yds)	Zero (500 yds)
0	2700	-1.5	-1.5	-1.5	-1.5	-1.5
100	2479	0.0	2.1	5.0	8.5	12.7
200	2269	-4.2	0.0	5.9	12.7	21.2
300	2070	-15.1	-8.8	0.0	10.2	22.9
400	1879	-33.8	-25.4	-13.6	0.0	16.9
500	1703	-63.4	-52.9	-38.2	-21.2	0.0
600	1540	-104.4	-91.8	-74.1	-53.7	-28.3
0	2600	-1.5	-1.5	-1.5	-1.5	-1.5
100	2384	0.0	2.4	5.6	9.4	13.9
200	2179	-4.8	0.0	6.4	14.1	23.1
300	1983	-16.8	-9.6	0.0	11.4	25.0
400	1799	-37.6	-28.1	-15.2	0.0	18.1
500	1629	-69.6	-57.7	-41.6	-22.6	0.0
600	1470	-113.9	-99.6	-80.4	-57.5	-30.4
0	2500	-1.5	-1.5	-1.5	-1.5	-1.5
100	2288	0.0	2.7	6.1	10.4	15.3
200	2088	-5.5	0.0	6.8	15.4	25.2
300	1897	-18.4	-10.2	0.0	12.8	27.6
400	1720	-41.7	-30.7	-17.1	0.0	19.6
500	1555	-76.6	-63.0	-45.9	-24.5	0.0
600	1401	-124.0	-107.6	-87.2	-61.5	-32.0
0	2400	-1.5	-1.5	-1.5	-1.5	-1.5
100	2194	0.0	2.9	6.7	11.4	16.6
200	1998	-5.7	0.0	7.8	17.1	27.5
300	1812	-20.2	-11.6	0.0	14.0	29.6
400	1641	-45.6	-34.2	-18.6	0.0	20.9
500	1482	-83.0	-68.8	-49.4	-26.1	0.0
600	1342	-137.7	-120.6	-97.3	-69.4	-38.1
0	2300	-1.5	-1.5	-1.5	-1.5	-1.5
100	2099	0.0	3.1	7.5	12.6	18.1
200	1907	-6.2	0.0	8.7	18.9	30.0
300	1729	-22.4	-13.1	0.0	15.4	32.0
400	1563	-50.3	-37.9	-20.5	0.0	22.1
500	1409	-90.6	-75.0	-53.3	-27.7	0.0
600	1282	-151.8	-133.1	-107.0	-76.3	-43.0
0	2200	-1.5	-1.5	-1.5	-1.5	-1.5
100	2003	0.0	3.3	8.4	13.8	20.4
200	1816	-6.6	0.0	10.1	21.0	34.2
300	1646	-25.1	-15.1	0.0	16.5	36.2
400	1486	-55.4	-42.1	-21.9	0.0	26.3
500	1346	-102.1	-85.4	-60.3	-32.9	0.0
600	1223	-167.8	-147.9	-117.7	-84.8	-45.4
0	2100	-1.5	-1.5	-1.5	-1.5	-1.5
100	1908	0.0	4.1	9.4	15.3	22.8
200	1730	-8.3	0.0	10.6	22.3	37.4
300	1564	-28.3	-15.9	0.0	17.6	40.2
400	1409	-61.2	-44.7	-23.5	0.0	30.2
500	1282	-114.2	-93.6	-67.1	-37.7	0.0
600	1167	-184.8	-160.1	-128.3	-93.1	-47.8

Universal Table for Rifle Bullets with Ballistic Coefficients of .400-.409

Sight Above Bore (in.) +1.5

Range (yds)	Velocity (fps)	Bullet Path Zero (100 yds)	Zero (200 yds)	Zero (300 yds)	Zero (400 yds)	Zero (500 yds)
0	3500	-1.5	-1.5	-1.5	-1.5	-1.5
100	3240	0.0	0.9	2.5	4.5	6.6
200	2999	-1.9	0.0	3.2	7.1	11.4
300	2769	-7.6	-4.8	0.0	5.8	12.3
400	2550	-18.0	-14.2	-7.8	0.0	8.6
500	2342	-33.2	-28.5	-20.5	-10.8	0.0
600	2144	-55.6	-49.9	-40.3	-28.7	-15.7
0	3400	-1.5	-1.5	-1.5	-1.5	-1.5
100	3148	0.0	1.0	2.7	4.8	7.1
200	2910	-2.0	0.0	3.4	7.6	12.3
300	2685	-8.1	-5.2	0.0	6.2	13.3
400	2470	-19.1	-15.1	-8.2	0.0	9.5
500	2266	-35.7	-30.8	-22.2	-11.9	0.0
600	2071	-59.3	-53.4	-43.1	-30.7	-16.5
0	3300	-1.5	-1.5	-1.5	-1.5	-1.5
100	3053	0.0	1.1	2.9	5.1	7.7
200	2820	-2.1	0.0	3.7	8.1	13.2
300	2600	-8.8	-5.6	0.0	6.5	14.2
400	2389	-20.4	-16.1	-8.7	0.0	10.3
500	2189	-38.5	-33.1	-23.7	-12.9	0.0
600	1998	-63.5	-57.1	-45.8	-32.9	-17.4
0	3200	-1.5	-1.5	-1.5	-1.5	-1.5
100	2959	0.0	1.3	3.2	5.5	8.3
200	2731	-2.5	0.0	3.9	8.4	14.2
300	2515	-9.6	-5.9	0.0	6.8	15.4
400	2308	-21.9	-16.9	-9.1	0.0	11.4
500	2111	-41.6	-35.4	-25.6	-14.3	0.0
600	1924	-68.2	-60.7	-49.0	-35.4	-18.2
0	3100	-1.5	-1.5	-1.5	-1.5	-1.5
100	2864	0.0	1.4	3.5	6.1	9.0
200	2641	-2.8	0.0	4.2	9.3	15.1
300	2429	-10.5	-6.2	0.0	7.7	16.5
400	2226	-24.3	-18.6	-10.3	0.0	11.6
500	2034	-44.9	-37.9	-27.4	-14.5	0.0
600	1849	-73.3	-64.8	-52.3	-36.9	-19.4
0	3000	-1.5	-1.5	-1.5	-1.5	-1.5
100	2769	0.0	1.6	3.8	6.6	9.7
200	2551	-3.2	0.0	4.4	10.0	16.3
300	2343	-11.4	-6.6	0.0	8.5	17.8
400	2145	-26.5	-20.1	-11.3	0.0	12.4
500	1956	-48.6	-40.6	-29.6	-15.5	0.0
600	1778	-80.3	-70.7	-57.5	-40.6	-21.9
0	2900	-1.5	-1.5	-1.5	-1.5	-1.5
100	2675	0.0	1.7	4.2	7.1	10.4
200	2461	-3.3	0.0	5.0	10.9	17.5
300	2257	-12.5	-7.5	0.0	8.9	18.8
400	2063	-28.5	-21.8	-11.8	0.0	13.2
500	1877	-52.1	-43.8	-31.3	-16.5	0.0
600	1706	-86.8	-76.8	-61.8	-44.1	-24.3
0	2800	-1.5	-1.5	-1.5	-1.5	-1.5
100	2580	0.0	1.8	4.6	7.7	11.4
200	2370	-3.6	0.0	5.5	11.9	19.2
300	2171	-13.7	-8.2	0.0	9.6	20.6
400	1980	-31.0	-23.7	-12.8	0.0	14.7
500	1801	-57.1	-48.0	-34.3	-18.3	0.0
600	1634	-94.7	-83.8	-67.3	-48.2	-26.2

Universal Table for Rifle Bullets with Ballistic Coefficients of .400-.409

Sight Above Bore (in.) +1.5

		Bullet Path				
Range (yds)	Velocity (fps)	Zero (100 yds)	Zero (200 yds)	Zero (300 yds)	Zero (400 yds)	Zero (500 yds)
0	2700	-1.5	-1.5	-1.5	-1.5	-1.5
100	2484	0.0	2.1	5.0	8.4	12.5
200	2279	-4.2	0.0	5.9	12.6	20.8
300	2084	-15.0	-8.8	0.0	10.0	22.4
400	1898	-33.5	-25.2	-13.4	0.0	16.5
500	1725	-62.4	-52.0	-37.3	-20.6	0.0
600	1563	-103.2	-90.7	-73.1	-53.0	-28.3
0	2600	-1.5	-1.5	-1.5	-1.5	-1.5
100	2389	0.0	2.4	5.5	9.3	13.8
200	2189	-4.7	0.0	6.3	13.9	22.8
300	1998	-16.6	-9.5	0.0	11.3	24.7
400	1817	-37.1	-27.7	-15.1	0.0	17.9
500	1649	-68.8	-57.1	-41.2	-22.4	0.0
600	1493	-112.5	-98.4	-79.4	-56.8	-29.9
0	2500	-1.5	-1.5	-1.5	-1.5	-1.5
100	2293	0.0	2.7	6.1	10.3	15.1
200	2098	-5.4	0.0	6.8	15.2	24.9
300	1911	-18.3	-10.2	0.0	12.6	27.1
400	1737	-41.2	-30.4	-16.8	0.0	19.4
500	1575	-75.7	-62.2	-45.2	-24.2	0.0
600	1423	-122.8	-106.6	-86.2	-61.0	-31.9
0	2400	-1.5	-1.5	-1.5	-1.5	-1.5
100	2199	0.0	2.9	6.5	11.3	16.5
200	2007	-5.7	0.0	7.2	16.9	27.2
300	1824	-19.4	-10.8	0.0	14.5	30.0
400	1657	-45.2	-33.8	-19.3	0.0	20.6
500	1501	-82.3	-68.0	-49.9	-25.8	0.0
600	1353	-132.6	-115.4	-93.8	-64.8	-33.8
0	2300	-1.5	-1.5	-1.5	-1.5	-1.5
100	2103	0.0	3.1	7.4	12.4	18.0
200	1916	-6.2	0.0	8.7	18.7	29.8
300	1741	-22.2	-13.0	0.0	15.1	31.7
400	1579	-49.7	-37.4	-20.1	0.0	22.2
500	1427	-89.9	-74.5	-52.8	-27.7	0.0
600	1300	-149.7	-131.2	-105.2	-75.1	-41.8
0	2200	-1.5	-1.5	-1.5	-1.5	-1.5
100	2008	0.0	3.3	8.3	13.8	19.6
200	1825	-6.7	0.0	10.0	20.9	32.5
300	1658	-24.9	-15.0	0.0	16.4	33.9
400	1501	-55.1	-41.7	-21.8	0.0	23.3
500	1354	-98.0	-81.4	-56.4	-29.2	0.0
600	1241	-165.3	-145.3	-115.4	-82.7	-47.6
0	2100	-1.5	-1.5	-1.5	-1.5	-1.5
100	1912	0.0	4.1	9.3	15.2	22.6
200	1738	-8.2	0.0	10.5	22.1	37.0
300	1576	-28.0	-15.7	0.0	17.5	39.7
400	1424	-60.6	-44.3	-23.4	0.0	29.6
500	1297	-112.8	-92.4	-66.2	-37.1	0.0
600	1183	-182.9	-158.3	-127.0	-91.9	-47.5

Universal Table for Rifle Bullets with Ballistic Coefficients of .410-.419

Sight Above Bore (in.) +1.5

Range (yds)	Velocity (fps)	Bullet Path Zero (100 yds)	Zero (200 yds)	Zero (300 yds)	Zero (400 yds)	Zero (500 yds)
0	3500	-1.5	-1.5	-1.5	-1.5	-1.5
100	3246	0.0	0.9	2.5	4.4	6.6
200	3010	-1.9	0.0	3.0	7.0	11.3
300	2785	-7.4	-4.5	0.0	6.0	12.4
400	2571	-17.8	-14.0	-7.9	0.0	8.5
500	2367	-32.9	-28.2	-20.6	-10.7	0.0
600	2172	-54.9	-49.3	-40.2	-28.3	-15.5
0	3400	-1.5	-1.5	-1.5	-1.5	-1.5
100	3153	0.0	1.0	2.7	4.7	7.1
200	2921	-2.0	0.0	3.4	7.5	12.2
300	2700	-8.1	-5.2	0.0	6.1	13.1
400	2490	-18.9	-15.0	-8.1	0.0	9.3
500	2290	-35.3	-30.4	-21.8	-11.7	0.0
600	2099	-58.5	-52.6	-42.3	-30.1	-16.1
0	3300	-1.5	-1.5	-1.5	-1.5	-1.5
100	3059	0.0	1.1	2.9	5.1	7.6
200	2831	-2.1	0.0	3.7	8.0	13.1
300	2615	-8.8	-5.6	0.0	6.4	14.1
400	2409	-20.3	-16.0	-8.5	0.0	10.2
500	2212	-38.1	-32.8	-23.5	-12.8	0.0
600	2025	-62.8	-56.4	-45.2	-32.4	-17.1
0	3200	-1.5	-1.5	-1.5	-1.5	-1.5
100	2965	0.0	1.2	3.2	5.5	8.2
200	2742	-2.5	0.0	3.9	8.4	14.0
300	2530	-9.6	-5.9	0.0	6.8	15.1
400	2327	-21.8	-16.9	-9.0	0.0	11.1
500	2135	-41.1	-34.9	-25.2	-13.9	0.0
600	1950	-67.6	-60.2	-48.5	-34.9	-18.3
0	3100	-1.5	-1.5	-1.5	-1.5	-1.5
100	2870	0.0	1.4	3.5	6.0	8.9
200	2652	-2.8	0.0	4.2	9.2	15.1
300	2444	-10.4	-6.2	0.0	7.6	16.3
400	2246	-24.0	-18.4	-10.1	0.0	11.7
500	2056	-44.6	-37.6	-27.2	-14.6	0.0
600	1876	-72.5	-64.1	-51.6	-36.5	-18.9
0	3000	-1.5	-1.5	-1.5	-1.5	-1.5
100	2775	0.0	1.6	3.8	6.6	9.7
200	2561	-3.2	0.0	4.4	10.0	16.1
300	2357	-11.4	-6.6	0.0	8.3	17.5
400	2163	-26.4	-19.9	-11.1	0.0	12.3
500	1978	-48.3	-40.3	-29.2	-15.3	0.0
600	1803	-79.3	-69.6	-56.4	-39.7	-21.3
0	2900	-1.5	-1.5	-1.5	-1.5	-1.5
100	2680	0.0	1.7	4.1	7.1	10.3
200	2471	-3.3	0.0	5.0	10.8	17.3
300	2271	-12.4	-7.4	0.0	8.8	18.6
400	2081	-28.2	-21.6	-11.7	0.0	13.1
500	1899	-51.6	-43.4	-31.0	-16.3	0.0
600	1730	-85.6	-75.7	-60.8	-43.3	-23.7
0	2800	-1.5	-1.5	-1.5	-1.5	-1.5
100	2585	0.0	1.8	4.5	7.7	11.3
200	2380	-3.6	0.0	5.5	11.8	19.0
300	2185	-13.6	-8.2	0.0	9.5	20.3
400	1998	-30.7	-23.5	-12.6	0.0	14.4
500	1822	-56.4	-47.4	-33.8	-18.0	0.0
600	1658	-93.2	-82.5	-66.1	-47.1	-25.6

Universal Table for Rifle Bullets with Ballistic Coefficients of .410-.419

Sight Above Bore (in.) +1.5

Range (yds)	Velocity (fps)	Bullet Path Zero (100 yds)	Zero (200 yds)	Zero (300 yds)	Zero (400 yds)	Zero (500 yds)
0	2700	-1.5	-1.5	-1.5	-1.5	-1.5
100	2489	0.0	2.0	5.0	8.3	12.3
200	2289	-4.1	0.0	5.8	12.6	20.6
300	2098	-14.9	-8.8	0.0	10.1	22.1
400	1915	-33.3	-25.1	-13.4	0.0	16.1
500	1745	-61.7	-51.5	-36.8	-20.1	0.0
600	1586	-101.8	-89.5	-71.9	-51.8	-27.7
0	2600	-1.5	-1.5	-1.5	-1.5	-1.5
100	2394	0.0	2.3	5.5	9.0	13.6
200	2198	-4.7	0.0	6.3	13.4	22.5
300	2011	-16.5	-9.5	0.0	10.6	24.3
400	1832	-36.2	-26.8	-14.2	0.0	18.2
500	1669	-68.0	-56.2	-40.4	-22.7	0.0
600	1515	-111.2	-97.1	-78.2	-57.0	-29.7
0	2500	-1.5	-1.5	-1.5	-1.5	-1.5
100	2298	0.0	2.7	6.1	10.2	15.0
200	2107	-5.4	0.0	6.8	15.0	24.6
300	1924	-18.2	-10.1	0.0	12.4	26.7
400	1753	-40.9	-30.1	-16.6	0.0	19.1
500	1594	-74.9	-61.4	-44.5	-23.8	0.0
600	1444	-121.7	-105.5	-85.2	-60.4	-31.8
0	2400	-1.5	-1.5	-1.5	-1.5	-1.5
100	2203	0.0	2.8	6.4	11.2	16.3
200	2016	-5.6	0.0	7.2	16.7	27.0
300	1837	-19.3	-10.8	0.0	14.2	29.6
400	1673	-44.6	-33.3	-19.0	0.0	20.6
500	1519	-81.5	-67.4	-49.4	-25.7	0.0
600	1374	-131.2	-114.2	-92.7	-64.2	-33.4
0	2300	-1.5	-1.5	-1.5	-1.5	-1.5
100	2108	0.0	3.1	7.3	12.4	17.8
200	1925	-6.2	0.0	8.5	18.5	29.5
300	1754	-22.0	-12.7	0.0	15.1	31.5
400	1594	-49.4	-37.1	-20.1	0.0	21.9
500	1445	-89.2	-73.8	-52.6	-27.4	0.0
600	1318	-147.6	-129.1	-103.6	-73.5	-40.6
0	2200	-1.5	-1.5	-1.5	-1.5	-1.5
100	2012	0.0	3.3	8.2	13.6	19.5
200	1833	-6.7	0.0	9.8	20.6	32.3
300	1670	-24.6	-14.6	0.0	16.2	33.8
400	1516	-54.5	-41.2	-21.6	0.0	23.4
500	1371	-97.3	-80.7	-56.3	-29.2	0.0
600	1257	-163.6	-143.6	-114.3	-81.8	-46.7
0	2100	-1.5	-1.5	-1.5	-1.5	-1.5
100	1917	0.0	4.1	9.3	15.1	22.3
200	1746	-8.2	0.0	10.4	22.0	36.4
300	1587	-27.9	-15.6	0.0	17.4	39.0
400	1438	-60.4	-44.0	-23.2	0.0	28.8
500	1312	-111.5	-91.1	-65.0	-36.0	0.0
600	1199	-180.9	-156.3	-125.1	-90.3	-47.0

Universal Table for Rifle Bullets with Ballistic Coefficients of .420-.429

Sight Above Bore (in.) +1.5

Range (yds)	Velocity (fps)	Bullet Path Zero (100 yds)	Zero (200 yds)	Zero (300 yds)	Zero (400 yds)	Zero (500 yds)
0	3500	-1.5	-1.5	-1.5	-1.5	-1.5
100	3252	0.0	0.9	2.5	4.4	6.5
200	3021	-1.9	0.0	3.0	6.9	11.2
300	2800	-7.4	-4.6	0.0	5.8	12.2
400	2591	-17.6	-13.9	-7.8	0.0	8.5
500	2390	-32.7	-28.0	-20.4	-10.7	0.0
600	2199	-54.3	-48.7	-39.5	-27.9	-15.1
0	3400	-1.5	-1.5	-1.5	-1.5	-1.5
100	3159	0.0	1.0	2.7	4.7	6.9
200	2932	-1.9	0.0	3.4	7.4	11.9
300	2716	-8.0	-5.1	0.0	6.0	12.7
400	2510	-18.7	-14.9	-8.1	0.0	8.9
500	2313	-34.5	-29.7	-21.2	-11.1	0.0
600	2125	-58.0	-52.1	-41.9	-29.8	-16.5
0	3300	-1.5	-1.5	-1.5	-1.5	-1.5
100	3065	0.0	1.1	2.9	5.0	7.6
200	2842	-2.1	0.0	3.7	8.0	13.0
300	2630	-8.8	-5.6	0.0	6.4	13.9
400	2428	-20.2	-15.9	-8.5	0.0	10.0
500	2235	-37.8	-32.4	-23.1	-12.5	0.0
600	2051	-62.2	-55.8	41.7	-31.9	-16.9
0	3200	-1.5	-1.5	-1.5	-1.5	-1.5
100	2970	0.0	1.2	3.2	5.4	8.1
200	2752	-2.5	0.0	3.9	8.4	13.8
300	2544	-9.6	-5.9	0.0	6.7	14.9
400	2346	-21.6	-16.7	-8.9	0.0	10.9
500	2157	-40.7	-34.6	-24.8	-13.7	0.0
600	1976	-66.8	-59.4	-47.7	-34.4	-17.9
0	3100	-1.5	-1.5	-1.5	-1.5	-1.5
100	2875	0.0	1.4	3.5	6.0	8.8
200	2662	-2.8	0.0	4.2	9.1	14.9
300	2458	-10.4	-6.2	0.0	7.5	16.1
400	2264	-23.8	-18.3	-9.9	0.0	11.5
500	2078	-44.2	-37.2	-26.8	-14.4	0.0
600	1901	-71.8	-63.5	-51.0	-36.1	-18.8
0	3000	-1.5	-1.5	-1.5	-1.5	-1.5
100	2780	0.0	1.6	3.8	6.5	9.6
200	2571	-3.2	0.0	4.4	9.9	16.0
300	2372	-11.3	-6.5	0.0	8.3	17.4
400	2181	-26.1	-19.8	-11.0	0.0	12.2
500	1999	-47.9	-40.0	-29.1	-15.3	0.0
600	1827	-78.2	-68.7	-55.6	-39.1	-20.7
0	2900	-1.5	-1.5	-1.5	-1.5	-1.5
100	2685	0.0	1.7	4.1	7.0	10.2
200	2480	-3.3	0.0	4.9	10.7	17.2
300	2285	-12.3	-7.3	0.0	8.7	18.4
400	2099	-28.0	-21.3	-11.6	0.0	13.0
500	1920	-51.2	-42.9	-30.7	-16.3	0.0
600	1753	-84.6	-74.6	-60.0	-42.6	-23.1
0	2800	-1.5	-1.5	-1.5	-1.5	-1.5
100	2590	0.0	1.8	4.5	7.6	11.1
200	2389	-3.6	0.0	5.4	11.7	18.5
300	2198	-13.5	-8.1	0.0	9.4	19.6
400	2015	-30.5	-23.3	-12.5	0.0	13.7
500	1840	-55.3	-46.2	-32.7	-17.1	0.0
600	1680	-92.2	-81.4	-65.2	-46.4	-25.9

Universal Table for Rifle Bullets with Ballistic Coefficients of .420-.429						
					Sight Above Bore (in.)	+1.5
			Bullet Path			
Range (yds)	Velocity (fps)	Zero (100 yds)	Zero (200 yds)	Zero (300 yds)	Zero (400 yds)	Zero (500 yds)
0	2700	-1.5	-1.5	-1.5	-1.5	-1.5
100	2494	0.0	2.0	4.9	8.3	12.2
200	2298	-4.1	0.0	5.8	12.5	20.4
300	2111	-14.8	-8.7	0.0	10.0	21.8
400	1932	-33.1	-24.9	-13.3	0.0	15.8
500	1764	-61.1	-50.9	-36.4	-19.8	0.0
600	1608	-100.5	-88.2	-70.8	-50.9	-27.1
0	2600	-1.5	-1.5	-1.5	-1.5	-1.5
100	2399	0.0	2.3	5.5	9.0	13.5
200	2207	-4.7	0.0	6.3	13.4	22.3
300	2024	-16.4	-9.4	0.0	10.6	24.0
400	1848	-36.1	-26.7	-14.2	0.0	17.9
500	1687	-67.5	-55.8	-40.1	-22.4	0.0
600	1536	-110.1	-96.0	-77.2	-56.0	-29.1
0	2500	-1.5	-1.5	-1.5	-1.5	-1.5
100	2303	0.0	2.7	6.0	10.2	14.8
200	2116	-5.4	0.0	6.7	14.9	24.3
300	1937	-18.1	-10.1	0.0	12.3	26.4
400	1768	-40.6	-29.9	-16.5	0.0	18.8
500	1612	-74.2	-60.8	-44.0	-23.5	0.0
600	1465	-120.4	-104.3	-84.1	-59.4	-31.3
0	2400	-1.5	-1.5	-1.5	-1.5	-1.5
100	2208	0.0	2.8	6.4	11.1	16.1
200	2025	-5.6	0.0	7.2	16.5	26.7
300	1849	-19.3	-10.8	0.0	13.9	29.2
400	1688	-44.3	-33.0	-18.6	0.0	20.3
500	1537	-80.7	-66.6	-48.6	-25.3	0.0
600	1394	-130.1	-113.2	-91.5	-63.7	-33.2
0	2300	-1.5	-1.5	-1.5	-1.5	-1.5
100	2112	0.0	3.1	7.3	12.2	17.7
200	1933	-6.1	0.0	8.4	18.3	29.2
300	1765	-21.9	-12.6	0.0	14.8	31.2
400	1609	-48.8	-36.5	-19.7	0.0	21.9
500	1462	-88.4	-73.0	-52.0	-27.4	0.0
600	1335	-145.7	-127.2	-102.0	-72.5	-39.6
0	2200	-1.5	-1.5	-1.5	-1.5	-1.5
100	2017	0.0	3.3	8.2	13.6	19.3
200	1842	-6.6	0.0	9.8	20.5	32.0
300	1681	-24.6	-14.6	0.0	16.1	33.4
400	1530	-54.2	-40.9	-21.4	0.0	23.1
500	1388	-96.7	-80.1	-55.7	-28.9	0.0
600	1273	-161.9	-142.0	-112.7	-80.6	-45.9
0	2100	-1.5	-1.5	-1.5	-1.5	-1.5
100	1921	0.0	4.0	9.2	15.0	22.1
200	1754	-8.1	0.0	10.4	21.9	36.1
300	1598	-27.7	-15.6	0.0	17.2	38.6
400	1452	-59.9	-43.7	-23.0	0.0	28.4
500	1326	-110.4	-90.2	-64.3	-35.5	0.0
600	1214	-179.1	-154.9	-123.8	-89.3	-46.6

Universal Table for Rifle Bullets with Ballistic Coefficients of .430-.439

Sight Above Bore (in.) +1.5

		Bullet Path				
Range (yds)	Velocity (fps)	Zero (100 yds)	Zero (200 yds)	Zero (300 yds)	Zero (400 yds)	Zero (500 yds)
0	3500	-1.5	-1.5	-1.5	-1.5	-1.5
100	3258	0.0	0.9	2.4	4.4	6.5
200	3031	-1.9	0.0	3.0	6.9	11.1
300	2815	-7.3	-4.5	0.0	5.8	12.1
400	2610	-17.5	-13.7	-7.7	0.0	8.5
500	2413	-32.4	-27.7	-20.2	-10.6	0.0
600	2225	-53.7	-48.0	-39.0	-27.5	-14.8
0	3400	-1.5	-1.5	-1.5	-1.5	-1.5
100	3164	0.0	1.0	2.7	4.7	6.9
200	2942	-1.9	0.0	3.4	7.4	11.8
300	2730	-8.0	-5.1	0.0	6.0	12.6
400	2528	-18.6	-14.8	-8.0	0.0	8.8
500	2335	-34.3	-29.5	-21.0	-11.0	0.0
600	2151	-57.2	-51.4	-41.3	-29.3	-16.0
0	3300	-1.5	-1.5	-1.5	-1.5	-1.5
100	3070	0.0	1.1	2.9	5.0	7.5
200	2852	-2.1	0.0	3.7	7.9	12.8
300	2645	-8.7	-5.5	0.0	6.4	13.7
400	2446	-20.1	-15.8	-8.5	0.0	9.8
500	2257	-37.3	-32.1	-22.9	-12.3	0.0
600	2076	-61.5	-55.2	-44.2	31.4	-16.7
0	3200	-1.5	-1.5	-1.5	-1.5	-1.5
100	2975	0.0	1.2	3.2	5.4	8.1
200	2762	-2.4	0.0	3.9	8.3	13.7
300	2558	-9.5	-5.8	0.0	6.6	14.7
400	2364	-21.5	-16.6	-8.8	0.0	10.8
500	2178	-40.4	-34.3	-24.6	-13.5	0.0
600	2000	-66.3	-59.0	-47.3	-34.0	-17.8
0	3100	-1.5	-1.5	-1.5	-1.5	-1.5
100	2880	0.0	1.4	3.4	5.9	8.8
200	2671	-2.8	0.0	4.1	9.1	14.7
300	2472	-10.3	-6.2	0.0	7.4	15.9
400	2281	-23.7	-18.1	-9.9	0.0	11.3
500	2099	-43.8	-36.8	-26.5	-14.2	0.0
600	1925	-71.2	-62.8	-50.5	-35.6	-18.7
0	3000	-1.5	-1.5	-1.5	-1.5	-1.5
100	2785	0.0	1.6	3.8	6.5	9.5
200	2581	-3.1	0.0	4.4	9.8	15.8
300	2385	-11.3	-6.6	0.0	8.2	17.2
400	2198	-25.9	-19.6	-10.9	0.0	12.0
500	2020	-47.5	-39.6	-28.6	-15.0	0.0
600	1849	-76.6	-67.2	-54.0	-37.7	-19.7
0	2900	-1.5	-1.5	-1.5	-1.5	-1.5
100	2690	0.0	1.6	4.1	7.0	10.2
200	2490	-3.3	0.0	4.9	10.7	17.1
300	2298	-12.2	-7.3	0.0	8.7	18.3
400	2115	-27.9	-21.3	-11.6	0.0	12.8
500	1940	-50.9	-42.7	-30.5	-16.0	0.0
600	1775	-83.7	-73.8	-59.2	-41.9	-22.6
0	2800	-1.5	-1.5	-1.5	-1.5	-1.5
100	2594	0.0	1.8	4.5	7.5	10.9
200	2398	-3.6	0.0	5.3	11.5	18.3
300	2211	-13.4	-8.0	0.0	9.3	19.5
400	2032	-30.2	-23.1	-12.4	0.0	13.6
500	1860	-54.7	-45.8	-32.5	-17.0	0.0
600	1702	-90.9	-80.2	-64.2	-45.6	-25.2

Universal Table for Rifle Bullets with Ballistic Coefficients of .430-.439

Sight Above Bore (in.) +1.5

Range (yds)	Velocity (fps)	Bullet Path Zero (100 yds)	Zero (200 yds)	Zero (300 yds)	Zero (400 yds)	Zero (500 yds)
0	2700	-1.5	-1.5	-1.5	-1.5	-1.5
100	2499	0.0	2.0	4.9	8.2	12.1
200	2307	-4.1	0.0	5.8	12.4	20.1
300	2124	-14.7	-8.6	0.0	10.0	21.5
400	1948	-32.9	-24.8	-13.3	0.0	15.4
500	1783	-60.5	-50.3	-35.9	-19.3	0.0
600	1629	-99.4	-87.2	-69.9	-50.0	-26.8
0	2600	-1.5	-1.5	-1.5	-1.5	-1.5
100	2403	0.0	2.3	5.4	9.0	13.3
200	2216	-4.6	0.0	6.3	13.3	22.0
300	2036	-16.3	-9.4	0.0	10.5	23.5
400	1864	-35.8	-26.6	-14.0	0.0	17.4
500	1706	-66.5	-55.0	-39.2	-21.7	0.0
600	1556	-109.0	-95.2	-76.3	-55.2	-29.2
0	2500	-1.5	-1.5	-1.5	-1.5	-1.5
100	2307	0.0	2.7	6.0	10.1	14.7
200	2124	-5.3	0.0	6.7	14.8	24.1
300	1949	-18.0	-10.0	0.0	12.1	26.1
400	1783	-40.2	-29.5	-16.2	0.0	18.7
500	1629	-73.6	-60.2	-43.6	-23.3	0.0
600	1485	-119.0	-103.0	-83.0	-58.7	-30.7
0	2400	-1.5	-1.5	-1.5	-1.5	-1.5
100	2212	0.0	2.8	6.4	10.9	16.0
200	2033	-5.6	0.0	7.2	16.3	26.4
300	1861	-19.2	-10.8	0.0	13.6	28.8
400	1703	-43.7	-32.5	-18.2	0.0	20.2
500	1554	-79.9	-65.9	-48.0	-25.2	0.0
600	1413	-129.0	-112.2	-90.7	-63.4	-33.1
0	2300	-1.5	-1.5	-1.5	-1.5	-1.5
100	2117	0.0	3.1	7.3	12.1	17.5
200	1941	-6.2	0.0	8.3	18.1	28.9
300	1776	-21.8	-12.5	0.0	14.6	30.8
400	1623	-48.5	-36.1	-19.4	0.0	21.6
500	1479	-87.7	-72.2	-51.3	-27.0	0.0
600	1351	-144.5	-125.9	-100.8	-71.7	-39.3
0	2200	-1.5	-1.5	-1.5	-1.5	-1.5
100	2021	0.0	3.3	8.1	13.4	19.2
200	1849	-6.7	0.0	9.5	20.1	31.7
300	1692	-24.4	-14.3	0.0	15.9	33.2
400	1544	-53.7	-40.3	-21.2	0.0	23.1
500	1404	-96.0	-79.2	-55.4	-28.9	0.0
600	1289	-159.7	-139.6	-111.0	-79.2	-44.6
0	2100	-1.5	-1.5	-1.5	-1.5	-1.5
100	1925	0.0	4.0	9.1	14.9	21.8
200	1761	-8.1	0.0	10.2	21.7	35.6
300	1609	-27.4	-15.3	0.0	17.3	38.1
400	1465	-59.6	-43.4	-23.1	0.0	27.7
500	1340	-109.1	-88.9	-63.5	-34.7	0.0
600	1229	-177.0	-152.8	-122.3	-87.7	-46.1

Universal Table for Rifle Bullets with Ballistic Coefficients of .440-.449

Sight Above Bore (in.) +1.5

		Bullet Path				
Range (yds)	Velocity (fps)	Zero (100 yds)	Zero (200 yds)	Zero (300 yds)	Zero (400 yds)	Zero (500 yds)
0	3500	-1.5	-1.5	-1.5	-1.5	-1.5
100	3263	0.0	0.9	2.4	4.3	6.4
200	3041	-1.9	0.0	3.0	6.8	11.0
300	2829	-7.3	-4.5	0.0	5.7	12.0
400	2628	-17.4	-13.6	-7.6	0.0	8.4
500	2435	-32.1	-27.5	-20.0	-10.5	0.0
600	2250	-53.1	-47.5	-38.5	-27.1	-14.5
0	3400	-1.5	-1.5	-1.5	-1.5	-1.5
100	3170	0.0	1.0	2.6	4.6	6.8
200	2952	-1.9	0.0	3.4	7.3	11.7
300	2744	-7.9	-5.0	0.0	6.0	12.6
400	2546	-18.5	-14.7	-7.9	0.0	8.8
500	2356	-34.2	-29.3	-20.9	-11.0	0.0
600	2175	-56.7	-50.9	-40.9	-28.9	-15.7
0	3300	-1.5	-1.5	-1.5	-1.5	-1.5
100	3075	0.0	1.1	2.9	5.0	7.4
200	2861	-2.1	0.0	3.6	7.8	12.7
300	2658	-8.6	-5.5	0.0	6.3	13.5
400	2464	-19.9	-15.6	-8.4	0.0	9.7
500	2278	-37.0	-31.7	-22.6	-12.1	0.0
600	2100	-60.9	-54.5	-43.6	-31.1	-16.5
0	3200	-1.5	-1.5	-1.5	-1.5	-1.5
100	2980	0.0	1.2	3.1	5.4	8.0
200	2771	-2.4	0.0	3.8	8.3	13.5
300	2572	-9.4	-5.7	0.0	6.7	14.6
400	2381	-21.4	-16.5	-8.9	0.0	10.6
500	2199	-40.0	-33.9	-24.3	-13.2	0.0
600	2024	-65.5	-58.2	-46.7	-33.4	-17.6
0	3100	-1.5	-1.5	-1.5	-1.5	-1.5
100	2885	0.0	1.4	3.4	5.9	8.7
200	2681	-2.7	0.0	4.1	9.0	14.6
300	2485	-10.3	-6.2	0.0	7.3	15.7
400	2298	-23.5	-18.0	-9.8	0.0	11.2
500	2119	-43.4	-36.6	-26.2	-14.1	0.0
600	1948	-70.6	-62.4	-50.0	-35.3	-18.5
0	3000	-1.5	-1.5	-1.5	-1.5	-1.5
100	2789	0.0	1.6	3.7	6.4	9.4
200	2590	-3.1	0.0	4.4	9.7	15.7
300	2398	-11.2	-6.6	0.0	8.0	17.0
400	2215	-25.6	-19.4	-10.7	0.0	11.9
500	2040	-47.0	-39.2	-28.3	-14.9	0.0
600	1871	-76.1	-66.8	-53.7	-37.6	-19.7
0	2900	-1.5	-1.5	-1.5	-1.5	-1.5
100	2694	0.0	1.6	4.0	6.9	10.1
200	2498	-3.3	0.0	4.8	10.5	16.8
300	2311	-12.1	-7.2	0.0	8.6	18.1
400	2131	-27.6	-21.1	-11.5	0.0	12.6
500	1960	-50.3	-42.1	-30.2	-15.8	0.0
600	1797	-82.5	-72.7	-58.4	-41.1	-22.2
0	2800	-1.5	-1.5	-1.5	-1.5	-1.5
100	2599	0.0	1.8	4.5	7.5	10.9
200	2407	-3.6	0.0	5.3	11.4	18.2
300	2223	-13.4	-8.0	0.0	9.1	19.3
400	2048	-30.0	-22.9	-12.2	0.0	13.5
500	1879	-54.4	-45.5	-32.2	-16.9	0.0
600	1723	-89.9	-79.2	-63.2	-44.9	-24.6

Universal Table for Rifle Bullets with Ballistic Coefficients of .440-.449						
					Sight Above Bore (in.)	+1.5
			Bullet Path			
Range (yds)	Velocity (fps)	Zero (100 yds)	Zero (200 yds)	Zero (300 yds)	Zero (400 yds)	Zero (500 yds)
0	2700	-1.5	-1.5	-1.5	-1.5	-1.5
100	2503	0.0	1.9	4.9	8.2	12.0
200	2315	-3.8	0.0	5.9	12.5	20.1
300	2136	-14.6	-8.9	0.0	9.8	21.3
400	1964	-32.6	-25.0	-13.1	0.0	15.2
500	1801	-59.8	-50.3	-35.4	-19.1	0.0
600	1649	-98.3	-86.8	-69.0	-49.4	-26.5
0	2600	-1.5	-1.5	-1.5	-1.5	-1.5
100	2407	0.0	2.3	5.4	8.9	13.2
200	2224	-4.6	0.0	6.2	13.2	21.8
300	2048	-16.2	-9.3	0.0	10.5	23.3
400	1879	-35.6	-26.5	-14.0	0.0	17.1
500	1723	-65.8	-54.4	-38.8	-21.3	0.0
600	1576	-107.6	-93.9	-75.2	-54.2	-28.6
0	2500	-1.5	-1.5	-1.5	-1.5	-1.5
100	2311	0.0	2.7	6.0	9.9	14.6
200	2132	-5.3	0.0	6.7	14.6	23.8
300	1960	-18.0	-10.0	0.0	11.8	25.7
400	1798	-39.7	-29.1	-15.8	0.0	18.5
500	1646	-72.8	-59.5	-42.8	-23.2	0.0
600	1504	-117.8	-101.9	-81.9	-58.3	-30.5
0	2400	-1.5	-1.5	-1.5	-1.5	-1.5
100	2216	0.0	2.8	6.4	10.8	15.9
200	2041	-5.6	0.0	7.2	16.1	26.2
300	1872	-19.1	-10.8	0.0	13.4	28.5
400	1717	-43.3	-32.2	-17.8	0.0	20.1
500	1570	-79.3	-65.4	-47.4	-25.1	0.0
600	1432	-127.7	-111.0	-89.4	-62.7	-32.5
0	2300	-1.5	-1.5	-1.5	-1.5	-1.5
100	2121	0.0	3.1	7.2	12.1	17.5
200	1949	-6.2	0.0	8.3	18.0	28.8
300	1787	-21.6	-12.4	0.0	14.6	30.7
400	1636	-48.3	-35.9	-19.4	0.0	21.6
500	1494	-87.3	-71.9	-51.2	-26.9	0.0
600	1368	-142.3	-123.8	-99.1	-69.9	-37.6
0	2200	-1.5	-1.5	-1.5	-1.5	-1.5
100	2025	0.0	3.3	8.1	13.3	19.1
200	1857	-6.6	0.0	9.5	20.0	31.5
300	1702	-24.3	-14.3	0.0	15.7	33.0
400	1557	-53.3	-40.0	-20.9	0.0	23.1
500	1419	-95.5	-78.9	-55.0	-28.9	0.0
600	1304	-158.1	-138.1	-109.5	-78.1	-43.5
0	2100	-1.5	-1.5	-1.5	-1.5	-1.5
100	1929	0.0	4.0	9.1	14.8	21.6
200	1768	-8.0	0.0	10.1	21.5	35.2
300	1619	-27.2	-15.1	0.0	17.1	37.7
400	1478	-59.1	-43.0	-22.8	0.0	27.5
500	1353	-108.2	-88.1	-62.9	-34.4	0.0
600	1243	-175.5	-151.4	-121.1	-86.9	-45.6

Universal Table for Rifle Bullets with Ballistic Coefficients of .450-.459

Sight Above Bore (in.) +1.5

Range (yds)	Velocity (fps)	Bullet Path Zero (100 yds)	Zero (200 yds)	Zero (300 yds)	Zero (400 yds)	Zero (500 yds)
0	3500	-1.5	-1.5	-1.5	-1.5	-1.5
100	3268	0.0	0.9	2.4	4.3	6.4
200	3051	-1.8	0.0	3.0	6.8	11.0
300	2843	-7.3	-4.5	0.0	5.7	11.9
400	2645	-17.3	-13.6	-7.6	0.0	8.3
500	2455	-32.0	-27.4	-19.9	-10.4	0.0
600	2274	-52.5	-47.0	-38.0	-26.7	-14.1
0	3400	-1.5	-1.5	-1.5	-1.5	-1.5
100	3175	0.0	1.0	2.6	4.6	6.8
200	2961	-1.9	0.0	3.3	7.3	11.6
300	2758	-7.9	-4.9	0.0	6.0	12.5
400	2563	-18.4	-14.5	-7.9	0.0	8.7
500	2377	-33.9	-29.0	-20.8	-10.9	0.0
600	2198	-56.3	-50.4	-40.5	-28.6	-15.6
0	3300	-1.5	-1.5	-1.5	-1.5	-1.5
100	3080	0.0	1.0	2.9	4.9	7.3
200	2871	-2.1	0.0	3.6	7.8	12.6
300	2671	-8.6	-5.5	0.0	6.2	13.4
400	2481	-19.8	-15.6	-8.3	0.0	9.6
500	2290	-36.7	-31.4	-22.3	-12.0	0.0
600	2123	-60.3	-54.0	-43.1	-30.7	-16.3
0	3200	-1.5	-1.5	-1.5	-1.5	-1.5
100	2985	0.0	1.2	3.1	5.3	7.9
200	2780	-2.4	0.0	3.8	8.2	13.5
300	2585	-9.4	-5.7	0.0	6.7	14.5
400	2397	-21.4	-16.5	-8.9	0.0	10.4
500	2218	-39.7	-33.6	-24.1	-13.0	0.0
600	2047	-64.9	-57.6	-46.2	-32.9	-17.2
0	3100	-1.5	-1.5	-1.5	-1.5	-1.5
100	2889	0.0	1.4	3.4	5.8	8.6
200	2689	-2.7	0.0	4.1	8.9	14.4
300	2498	-10.2	-6.1	0.0	7.3	15.5
400	2314	-23.3	-17.8	-9.7	0.0	11.0
500	2139	-42.9	-36.0	-25.9	-13.8	0.0
600	1970	-69.9	-61.7	-49.6	-35.0	-18.5
0	3000	-1.5	-1.5	-1.5	-1.5	-1.5
100	2794	0.0	1.6	3.7	6.4	9.4
200	2598	-3.1	0.0	4.3	9.6	15.6
300	2410	-11.2	-6.5	0.0	7.9	16.9
400	2231	-25.4	-19.2	-10.5	0.0	12.0
500	2058	-46.8	-38.9	-28.1	-15.0	0.0
600	1893	-75.5	-66.1	-53.1	-37.3	-19.3
0	2900	-1.5	-1.5	-1.5	-1.5	-1.5
100	2699	0.0	1.6	3.9	6.9	10.0
200	2507	-3.3	0.0	4.6	10.4	16.8
300	2323	-11.8	-6.9	0.0	8.8	18.3
400	2147	-27.4	-20.9	-11.7	0.0	12.7
500	1978	-50.1	-41.9	-30.5	-15.8	0.0
600	1818	-81.7	-71.9	-58.1	-40.5	-21.5
0	2800	-1.5	-1.5	-1.5	-1.5	-1.5
100	2603	0.0	1.8	4.4	7.4	10.8
200	2415	-3.5	0.0	5.3	11.3	18.1
300	2235	-13.2	-7.9	0.0	9.1	19.2
400	2063	-29.8	-22.7	-12.2	0.0	13.5
500	1897	-54.1	-45.2	-32.0	-16.8	0.0
600	1743	-88.9	-78.2	-62.4	-44.2	-24.0

Universal Table for Rifle Bullets with Ballistic Coefficients of .450-.459

Sight Above Bore (in.) +1.5

Range (yds)	Velocity (fps)	Bullet Path Zero (100 yds)	Zero (200 yds)	Zero (300 yds)	Zero (400 yds)	Zero (500 yds)
0	2700	-1.5	-1.5	-1.5	-1.5	-1.5
100	2508	0.0	1.9	4.9	8.1	11.9
200	2323	-3.9	0.0	5.9	12.4	19.9
300	2147	-14.7	-8.9	0.0	9.7	21.0
400	1979	-32.5	-24.7	-12.9	0.0	15.1
500	1818	-59.4	-49.8	-35.0	-18.8	0.0
600	1668	-97.5	-85.9	-68.2	-48.8	-26.2
0	2600	-1.5	-1.5	-1.5	-1.5	-1.5
100	2412	0.0	2.3	5.4	8.9	13.1
200	2232	-4.6	0.0	6.2	13.1	21.5
300	2060	-16.1	-9.2	0.0	10.5	23.0
400	1894	-35.4	-26.3	-14.0	0.0	16.8
500	1740	-65.3	-53.8	-38.4	-21.0	0.0
600	1595	-106.6	-92.9	-74.4	-53.4	-28.3
0	2500	-1.5	-1.5	-1.5	-1.5	-1.5
100	2315	0.0	2.6	5.9	9.8	14.4
200	2140	-5.3	0.0	6.6	14.4	23.6
300	1972	-17.8	-9.9	0.0	11.7	25.5
400	1812	-39.4	-28.9	-15.7	0.0	18.4
500	1662	-72.2	-59.0	-42.6	-23.0	0.0
600	1522	-116.8	-101.1	-81.3	-57.8	-30.2
0	2400	-1.5	-1.5	-1.5	-1.5	-1.5
100	2220	0.0	2.8	6.4	10.8	15.7
200	2048	-5.6	0.0	7.1	15.9	25.8
300	1883	-19.1	-10.7	0.0	13.3	28.1
400	1730	-43.1	-31.9	-17.7	0.0	19.8
500	1586	-78.5	-64.5	-46.8	-24.7	0.0
600	1449	-127.0	-110.2	-88.9	-62.4	-32.7
0	2300	-1.5	-1.5	-1.5	-1.5	-1.5
100	2124	0.0	3.0	7.1	11.9	17.2
200	1957	-6.0	0.0	8.2	17.8	28.4
300	1798	-21.3	-12.2	0.0	14.5	30.4
400	1649	-47.7	-35.7	-19.4	0.0	21.2
500	1510	-86.1	-71.1	-50.7	-26.5	0.0
600	1377	-138.0	-120.0	-95.5	-66.5	-34.7
0	2200	-1.5	-1.5	-1.5	-1.5	-1.5
100	2029	0.0	3.3	8.0	13.3	19.0
200	1864	-6.7	0.0	9.4	19.9	31.2
300	1712	-24.1	-14.1	0.0	15.7	32.8
400	1569	-53.1	-39.7	-20.9	0.0	22.7
500	1434	-94.8	-78.1	-54.6	-28.4	0.0
600	1319	-156.2	-136.2	-108.0	-76.6	-42.4
0	2100	-1.5	-1.5	-1.5	-1.5	-1.5
100	1933	0.0	4.0	9.1	14.7	21.4
200	1775	-8.0	0.0	10.1	21.4	34.9
300	1628	-27.2	-15.2	0.0	16.9	37.1
400	1490	-58.8	-42.8	-22.5	0.0	27.0
500	1366	-107.2	-87.2	-61.9	-33.7	0.0
600	1257	-173.7	-149.7	-119.3	-85.5	-45.0

Universal Table for Rifle Bullets with Ballistic Coefficients of .460-.469

					Sight Above Bore (in.)	+1.5
		Bullet Path				
Range (yds)	Velocity (fps)	Zero (100 yds)	Zero (200 yds)	Zero (300 yds)	Zero (400 yds)	Zero (500 yds)
0	3500	-1.5	-1.5	-1.5	-1.5	-1.5
100	3273	0.0	0.9	2.4	4.3	6.4
200	3060	-1.8	0.0	3.0	6.7	10.9
300	2856	-7.2	-4.5	0.0	5.6	11.8
400	2662	-17.1	-13.4	-7.5	0.0	8.3
500	2475	-31.8	-27.2	-19.7	-10.4	0.0
600	2297	-52.1	-46.5	-37.6	-26.4	-13.9
0	3400	-1.5	-1.5	-1.5	-1.5	-1.5
100	3179	0.0	1.0	2.6	4.6	6.7
200	2970	-1.9	0.0	3.3	7.2	11.6
300	2770	-7.8	-5.0	0.0	5.8	12.4
400	2580	-18.2	-14.4	-7.8	0.0	8.7
500	2396	-33.7	-28.9	-20.6	-10.9	0.0
600	2221	-55.6	-49.9	-39.9	-28.3	-15.2
0	3300	-1.5	-1.5	-1.5	-1.5	-1.5
100	3084	0.0	1.0	2.8	4.9	7.3
200	2879	-2.1	0.0	3.6	7.7	12.5
300	2684	-8.5	-5.4	0.0	6.2	13.3
400	2497	-19.6	-15.4	-8.3	0.0	9.5
500	2317	-36.4	-31.1	-22.2	-11.8	0.0
600	2145	-59.7	-53.4	-42.7	-30.3	-16.1
0	3200	-1.5	-1.5	-1.5	-1.5	-1.5
100	2989	0.0	1.2	3.1	5.3	7.9
200	2789	-2.4	0.0	3.8	8.2	13.4
300	2597	-9.3	-5.7	0.0	6.6	14.3
400	2413	-21.2	-16.4	-8.8	0.0	10.3
500	2237	-39.3	-33.4	-23.8	-12.9	0.0
600	2068	-64.4	-57.3	-45.8	-32.6	-17.2
0	3100	-1.5	-1.5	-1.5	-1.5	-1.5
100	2894	0.0	1.4	3.4	5.7	8.5
200	2698	-2.7	0.0	4.1	8.7	14.3
300	2510	-10.2	-6.1	0.0	7.0	15.4
400	2329	-22.8	-17.4	-9.3	0.0	11.3
500	2157	-42.7	-35.9	-25.7	-14.1	0.0
600	1991	-69.5	-61.3	-49.1	-35.2	-18.3
0	3000	-1.5	-1.5	-1.5	-1.5	-1.5
100	2798	0.0	1.5	3.7	6.3	9.2
200	2607	-3.1	0.0	4.4	9.6	15.4
300	2422	-11.1	-6.5	0.0	7.8	16.6
400	2246	-25.2	-19.1	-10.4	0.0	11.7
500	2077	-46.2	-38.6	-27.7	-14.7	0.0
600	1914	-74.8	-65.6	-52.6	-37.0	-19.4
0	2900	-1.5	-1.5	-1.5	-1.5	-1.5
100	2703	0.0	1.6	3.9	6.8	9.9
200	2515	-3.3	0.0	4.6	10.3	16.6
300	2334	-11.8	-6.9	0.0	8.6	18.1
400	2162	-27.2	-20.7	-11.5	0.0	12.6
500	1996	-49.7	-41.6	-30.1	-15.7	0.0
600	1838	-80.8	-71.1	-57.3	-40.0	-21.2
0	2800	-1.5	-1.5	-1.5	-1.5	-1.5
100	2607	0.0	1.8	4.4	7.4	10.7
200	2423	-3.5	0.0	5.2	11.3	17.9
300	2247	-13.1	-7.8	0.0	9.2	19.1
400	2077	-29.7	-22.6	-12.2	0.0	13.2
500	1915	-53.6	-44.8	-31.8	-16.5	0.0
600	1762	-88.0	-77.5	-61.9	-43.5	-23.7

Universal Table for Rifle Bullets with Ballistic Coefficients of .460-.469

Sight Above Bore (in.) +1.5

		Bullet Path				
Range (yds)	Velocity (fps)	Zero (100 yds)	Zero (200 yds)	Zero (300 yds)	Zero (400 yds)	Zero (500 yds)
0	2700	-1.5	-1.5	-1.5	-1.5	-1.5
100	2512	0.0	1.9	4.8	8.1	11.8
200	2331	-3.8	0.0	5.8	12.3	19.7
300	2159	-14.5	-8.7	0.0	9.8	20.8
400	1993	-32.3	-24.6	-13.0	0.0	14.8
500	1835	-58.9	-49.3	-34.7	-18.4	0.0
600	1687	-96.4	-84.9	-67.4	-47.9	-25.8
0	2600	-1.5	-1.5	-1.5	-1.5	-1.5
100	2416	0.0	2.3	5.4	8.8	12.9
200	2239	-4.6	0.0	6.2	13.0	21.3
300	2070	-16.1	-9.2	0.0	10.3	22.7
400	1908	-35.3	-26.1	-13.8	0.0	16.5
500	1756	-64.7	-53.2	-37.8	-20.6	0.0
600	1613	-105.6	-91.8	-73.4	-52.7	-28.0
0	2500	-1.5	-1.5	-1.5	-1.5	-1.5
100	2319	0.0	2.6	5.9	9.8	14.3
200	2147	-5.3	0.0	6.6	14.3	23.3
300	1982	-17.8	-9.9	0.0	11.6	25.1
400	1825	-39.1	-28.6	-15.4	0.0	18.0
500	1678	-71.4	-58.3	-41.8	-22.5	0.0
600	1540	-115.7	-99.9	-80.1	-57.0	-29.9
0	2400	-1.5	-1.5	-1.5	-1.5	-1.5
100	2224	0.0	2.8	6.3	10.7	15.6
200	2056	-5.5	0.0	7.1	15.8	25.7
300	1894	-18.9	-10.6	0.0	13.1	27.8
400	1743	-42.7	-31.7	-17.5	0.0	19.6
500	1601	-78.0	-64.1	-46.4	-24.5	0.0
600	1467	-125.6	-109.0	-87.8	-61.5	-32.1
0	2300	-1.5	-1.5	-1.5	-1.5	-1.5
100	2128	0.0	3.0	7.1	11.9	17.2
200	1964	-6.0	0.0	8.2	17.7	28.3
300	1807	-21.3	-12.3	0.0	14.3	30.2
400	1661	-47.5	-35.5	-19.1	0.0	21.1
500	1524	-85.8	-70.7	-50.3	-26.4	0.0
600	1394	-137.0	-118.9	-94.4	-65.7	-34.0
0	2200	-1.5	-1.5	-1.5	-1.5	-1.5
100	2032	0.0	3.3	7.9	13.2	18.8
200	1871	-6.6	0.0	9.3	19.7	31.1
300	1722	-23.8	-13.9	0.0	15.7	32.7
400	1581	-52.6	-39.4	-20.9	0.0	22.7
500	1448	-94.1	-77.6	-54.5	-28.4	0.0
600	1333	-154.6	-134.8	-107.0	-75.7	-41.6
0	2100	-1.5	-1.5	-1.5	-1.5	-1.5
100	1936	0.0	3.9	9.0	14.6	20.7
200	1782	-7.8	0.0	10.2	21.4	33.5
300	1637	-27.0	-15.3	0.0	16.9	35.0
400	1501	-58.6	-42.9	-22.6	0.0	24.1
500	1372	-103.4	-83.8	-58.3	-30.1	0.0
600	1270	-172.2	-148.7	-118.2	-84.3	-48.2

Universal Table for Rifle Bullets with Ballistic Coefficients of .470-.479

Sight Above Bore (in.) +1.5

Range (yds)	Velocity (fps)	Bullet Path Zero (100 yds)	Zero (200 yds)	Zero (300 yds)	Zero (400 yds)	Zero (500 yds)
0	3500	-1.5	-1.5	-1.5	-1.5	-1.5
100	3278	0.0	0.9	2.4	4.3	6.3
200	3069	-1.8	0.0	3.0	6.7	10.8
300	2869	-7.2	-4.5	0.0	5.6	11.7
400	2678	-17.0	-13.4	-7.4	0.0	8.2
500	2495	-31.5	-26.9	-19.5	-10.2	0.0
600	2319	-51.6	-46.1	-37.2	-26.1	-13.8
0	3400	-1.5	-1.5	-1.5	-1.5	-1.5
100	3184	0.0	1.0	2.6	4.5	6.7
200	2979	-1.9	0.0	3.3	7.2	11.5
300	2783	-7.8	-4.9	0.0	5.9	12.3
400	2595	-18.2	-14.4	-7.8	0.0	8.6
500	2415	-33.5	-28.7	-20.6	-10.8	0.0
600	2243	-55.1	-49.4	-39.6	-27.8	-14.9
0	3300	-1.5	-1.5	-1.5	-1.5	-1.5
100	3089	0.0	1.0	2.8	4.9	7.2
200	2888	-2.1	0.0	3.6	7.7	12.2
300	2696	-8.5	-5.4	0.0	6.2	13.0
400	2512	-19.6	-15.4	-8.3	0.0	9.1
500	2335	-35.8	-30.6	-21.7	-11.3	0.0
600	2166	-59.3	-53.0	-42.3	-29.9	-10.4
0	3200	-1.5	-1.5	-1.5	-1.5	-1.5
100	2994	0.0	1.2	3.1	5.3	7.8
200	2797	-2.4	0.0	3.8	8.2	13.2
300	2609	-9.3	-5.7	0.0	6.6	14.2
400	2428	-21.1	-16.3	-8.8	0.0	10.1
500	2255	-39.1	-33.1	-23.6	-12.7	0.0
600	2089	-63.9	-56.7	-45.4	-32.2	-17.0
0	3100	-1.5	-1.5	-1.5	-1.5	-1.5
100	2898	0.0	1.3	3.4	5.7	8.5
200	2706	-2.7	0.0	4.0	8.7	14.2
300	2522	-10.1	-6.0	0.0	7.0	15.3
400	2344	-22.7	-17.3	-9.3	0.0	11.1
500	2175	-42.3	-35.5	-25.5	-13.9	0.0
600	2012	-68.8	-60.7	-48.6	-34.7	-18.1
0	3000	-1.5	-1.5	-1.5	-1.5	-1.5
100	2802	0.0	1.5	3.7	6.3	9.2
200	2614	-3.1	0.0	4.3	9.5	15.3
300	2434	-11.0	-6.4	0.0	7.8	16.5
400	2260	-25.1	-19.0	-10.4	0.0	11.6
500	2094	-46.0	-38.2	-27.6	-14.6	0.0
600	1934	-74.3	-65.0	-52.2	-36.6	-19.1
0	2900	-1.5	-1.5	-1.5	-1.5	-1.5
100	2707	0.0	1.6	3.9	6.8	9.9
200	2523	-3.2	0.0	4.6	10.3	16.5
300	2345	-11.7	-6.9	0.0	8.5	17.9
400	2176	-27.0	-20.6	-11.4	0.0	12.5
500	2013	-49.4	-41.3	-29.8	-15.6	0.0
600	1856	-79.4	-69.7	-56.0	-38.9	-20.2
0	2800	-1.5	-1.5	-1.5	-1.5	-1.5
100	2611	0.0	1.7	4.3	7.4	10.6
200	2431	-3.5	0.0	5.2	11.2	17.8
300	2258	-13.0	-7.8	0.0	9.1	18.9
400	2091	-29.5	-22.5	-12.2	0.0	13.1
500	1932	-53.2	-44.5	-31.6	-16.4	0.0
600	1781	-87.1	-76.6	-61.1	-42.9	-23.2

Universal Table for Rifle Bullets with Ballistic Coefficients of .470-.479

					Sight Above Bore (in.)	+1.5
Bullet Path						
Range (yds)	**Velocity (fps)**	**Zero (100 yds)**	**Zero (200 yds)**	**Zero (300 yds)**	**Zero (400 yds)**	**Zero (500 yds)**
0	2700	-1.5	-1.5	-1.5	-1.5	-1.5
100	2515	0.0	1.9	4.8	8.0	11.5
200	2338	-3.8	0.0	5.8	12.2	19.2
300	2169	-14.4	-8.7	0.0	9.6	20.1
400	2007	-32.0	-24.4	-12.8	0.0	14.0
500	1850	-57.5	-48.0	-33.5	-17.5	0.0
600	1705	-95.3	-83.9	-66.5	-47.3	-26.3
0	2600	-1.5	-1.5	-1.5	-1.5	-1.5
100	2419	0.0	2.2	5.3	8.8	12.8
200	2247	-4.5	0.0	6.1	13.1	21.2
300	2081	-15.9	-9.2	0.0	10.4	22.6
400	1921	-35.1	-26.1	-13.9	0.0	16.3
500	1771	-64.2	-53.0	-37.6	-20.3	0.0
600	1631	-104.3	-90.9	-72.5	-51.7	-27.3
0	2500	-1.5	-1.5	-1.5	-1.5	-1.5
100	2323	0.0	2.6	5.9	9.7	14.2
200	2155	-5.2	0.0	6.7	14.2	23.2
300	1992	-17.8	-10.0	0.0	11.4	24.8
400	1838	-38.9	-28.5	-15.2	0.0	17.8
500	1693	-70.9	-57.9	-41.3	-22.3	0.0
600	1557	-114.7	-99.1	-79.2	-56.4	-29.6
0	2400	-1.5	-1.5	-1.5	-1.5	-1.5
100	2228	0.0	2.8	6.3	10.6	15.5
200	2063	-5.5	0.0	7.1	15.7	25.5
300	1904	-18.9	-10.6	0.0	13.0	27.6
400	1755	-42.6	-31.5	-17.3	0.0	19.5
500	1615	-77.6	-63.8	-46.1	-24.4	0.0
600	1483	-125.0	-108.4	-87.1	-61.1	-31.8
0	2300	-1.5	-1.5	-1.5	-1.5	-1.5
100	2132	0.0	3.1	7.1	11.8	17.1
200	1970	-6.1	0.0	8.0	17.5	28.0
300	1817	-21.2	-12.0	0.0	14.3	30.0
400	1673	-47.3	-35.0	-19.0	0.0	21.0
500	1538	-85.3	-70.0	-50.0	-26.3	0.0
600	1410	-136.2	-117.8	-93.9	-65.3	-33.8
0	2200	-1.5	-1.5	-1.5	-1.5	-1.5
100	2036	0.0	3.3	7.9	13.1	18.7
200	1878	-6.6	0.0	9.2	19.5	30.8
300	1731	-23.8	-13.9	0.0	15.4	32.4
400	1593	-52.3	-39.1	-20.6	0.0	22.6
500	1462	-93.5	-77.0	-54.0	-28.2	0.0
600	1347	-153.1	-133.3	-105.6	-74.7	-40.9
0	2100	-1.5	-1.5	-1.5	-1.5	-1.5
100	1939	0.0	3.9	8.9	14.5	20.5
200	1788	-7.8	0.0	10.1	21.2	33.2
300	1646	-26.8	-15.1	0.0	16.6	34.7
400	1513	-57.9	-42.3	-22.2	0.0	24.0
500	1386	-102.4	-82.9	-57.8	-30.0	0.0
600	1284	-169.7	-146.3	-116.1	-82.8	-46.8

Universal Table for Rifle Bullets with Ballistic Coefficients of .480-.489

Sight Above Bore (in.) +1.5

Range (yds)	Velocity (fps)	Bullet Path Zero (100 yds)	Zero (200 yds)	Zero (300 yds)	Zero (400 yds)	Zero (500 yds)
0	3500	-1.5	-1.5	-1.5	-1.5	-1.5
100	3282	0.0	0.9	2.4	4.2	6.3
200	3077	-1.8	0.0	2.9	6.6	10.7
300	2881	-7.2	-4.4	0.0	5.5	11.6
400	2693	-16.9	-13.3	-7.4	0.0	8.1
500	2513	-31.3	-26.8	-19.4	-10.2	0.0
600	2340	-50.8	-45.3	-36.4	-25.4	-13.2
0	3400	-1.5	-1.5	-1.5	-1.5	-1.5
100	3188	0.0	1.0	2.6	4.5	6.6
200	2987	-1.9	0.0	3.2	7.1	11.4
300	2795	-7.7	-4.9	0.0	5.9	12.2
400	2610	-18.1	-14.3	-7.8	0.0	8.5
500	2434	-33.2	-28.5	-20.4	-10.6	0.0
600	2264	-54.6	-48.9	-39.2	-27.5	-14.7
0	3300	-1.5	-1.5	-1.5	-1.5	-1.5
100	3093	0.0	1.0	2.8	4.9	7.1
200	2896	-2.1	0.0	3.5	7.6	12.2
300	2708	-8.4	-5.3	0.0	6.2	12.9
400	2527	-19.4	-15.3	-8.2	0.0	9.0
500	2353	-35.6	-30.4	-21.6	-11.3	0.0
600	2187	-58.7	-52.5	-41.9	-29.5	-16.0
0	3200	-1.5	-1.5	-1.5	-1.5	-1.5
100	2998	0.0	1.2	3.1	5.2	7.7
200	2805	-2.4	0.0	3.8	8.1	13.1
300	2620	-9.2	-5.7	0.0	6.5	14.0
400	2443	-21.0	-16.2	-8.7	0.0	10.0
500	2273	-38.7	-32.8	-23.3	-12.5	0.0
600	2109	-63.4	-56.3	-44.9	-31.9	-16.9
0	3100	-1.5	-1.5	-1.5	-1.5	-1.5
100	2902	0.0	1.3	3.3	5.6	8.4
200	2714	-2.7	0.0	4.0	8.6	14.1
300	2533	-10.0	-6.0	0.0	6.9	15.1
400	2359	-22.5	-17.2	-9.1	0.0	11.0
500	2192	-42.0	-35.3	-25.2	-13.8	0.0
600	2032	-68.2	-60.2	-48.1	-34.4	-17.8
0	3000	-1.5	-1.5	-1.5	-1.5	-1.5
100	2806	0.0	1.5	3.7	6.2	9.1
200	2622	-3.1	0.0	4.3	9.4	15.2
300	2445	-11.0	-6.4	0.0	7.7	16.4
400	2274	-25.0	-18.9	-10.3	0.0	11.5
500	2111	-45.6	-38.0	-27.3	-14.4	0.0
600	1954	-73.6	-64.5	-51.7	-36.2	-18.9
0	2900	-1.5	-1.5	-1.5	-1.5	-1.5
100	2711	0.0	1.6	3.9	6.7	9.8
200	2530	-3.2	0.0	4.5	10.2	16.3
300	2356	-11.7	-6.8	0.0	8.4	17.7
400	2190	-26.8	-20.4	-11.3	0.0	12.3
500	2030	-49.0	-40.9	-29.5	-15.4	0.0
600	1875	-78.9	-69.2	-55.6	-38.7	-20.2
0	2800	-1.5	-1.5	-1.5	-1.5	-1.5
100	2615	0.0	1.8	4.3	7.3	10.6
200	2438	-3.5	0.0	5.1	11.1	17.7
300	2268	-13.0	-7.7	0.0	8.9	18.8
400	2105	-29.2	-22.2	-11.9	0.0	13.1
500	1948	-53.0	-44.2	-31.3	-16.4	0.0
600	1799	-86.3	-75.8	-60.3	-42.4	-22.7

Universal Table for Rifle Bullets with Ballistic Coefficients of .480-.489

Sight Above Bore (in.) +1.5

Range (yds)	Velocity (fps)	Bullet Path Zero (100 yds)	Zero (200 yds)	Zero (300 yds)	Zero (400 yds)	Zero (500 yds)
0	2700	-1.5	-1.5	-1.5	-1.5	-1.5
100	2519	0.0	1.9	4.8	8.0	11.4
200	2346	-3.8	0.0	5.7	12.2	19.1
300	2180	-14.3	-8.6	0.0	9.6	20.1
400	2020	-31.9	-24.3	-12.8	0.0	13.9
500	1866	-57.2	-47.8	-33.5	-17.4	0.0
600	1723	-94.3	-83.0	-65.8	-46.5	-25.6
0	2600	-1.5	-1.5	-1.5	-1.5	-1.5
100	2423	0.0	2.2	5.3	8.7	12.7
200	2254	-4.5	0.0	6.1	13.0	21.0
300	2091	-15.9	-9.2	0.0	10.3	22.3
400	1934	-34.9	-26.0	-13.8	0.0	16.0
500	1786	-63.7	-52.5	-37.2	-20.0	0.0
600	1648	-103.4	-90.0	-71.6	-51.0	-27.0
0	2500	-1.5	-1.5	-1.5	-1.5	-1.5
100	2327	0.0	2.6	5.9	9.5	14.1
200	2161	-5.3	0.0	6.6	13.8	22.9
300	2002	-17.7	-9.8	0.0	10.8	24.6
400	1849	-38.1	-27.6	-14.5	0.0	18.3
500	1707	-70.5	-57.4	-41.0	-22.9	0.0
600	1573	-114.0	-98.2	-78.5	-56.9	-29.4
0	2400	-1.5	-1.5	-1.5	-1.5	-1.5
100	2231	0.0	2.8	6.3	10.5	15.4
200	2069	-5.5	0.0	7.0	15.5	25.2
300	1914	-18.8	-10.4	0.0	12.9	27.4
400	1767	-42.2	-31.1	-17.2	0.0	19.4
500	1629	-76.9	-63.1	-45.7	-24.2	0.0
600	1499	-123.9	-107.2	-86.4	-60.6	-31.6
0	2300	-1.5	-1.5	-1.5	-1.5	-1.5
100	2135	0.0	3.0	7.0	11.7	16.9
200	1977	-6.0	0.0	8.0	17.3	27.8
300	1826	-21.0	-12.0	0.0	14.0	29.7
400	1685	-46.7	-34.7	-18.7	0.0	20.9
500	1552	-84.5	-69.4	-49.4	-26.1	0.0
600	1426	-134.9	-116.8	-92.9	-64.8	-33.5
0	2200	-1.5	-1.5	-1.5	-1.5	-1.5
100	2039	0.0	3.3	7.8	13.0	18.6
200	1884	-6.6	0.0	9.1	19.3	30.6
300	1740	-23.5	-13.6	0.0	15.4	32.2
400	1604	-51.9	-38.7	-20.5	0.0	22.5
500	1475	-93.0	-76.5	-53.7	-28.1	0.0
600	1361	-151.2	-131.4	-104.1	-73.4	-39.6
0	2100	-1.5	-1.5	-1.5	-1.5	-1.5
100	1943	0.0	3.9	8.9	14.5	20.5
200	1794	-7.8	0.0	9.9	21.2	33.1
300	1655	-26.6	-14.9	0.0	16.9	34.8
400	1523	-58.0	-42.3	-22.5	0.0	24.0
500	1398	-102.4	-82.8	-58.0	-29.9	0.0
600	1296	-168.9	-145.4	-115.6	-81.9	-45.9

Universal Table for Rifle Bullets with Ballistic Coefficients of .490-.499

Sight Above Bore (in.) +1.5

Range (yds)	Velocity (fps)	Bullet Path Zero (100 yds)	Zero (200 yds)	Zero (300 yds)	Zero (400 yds)	Zero (500 yds)
0	3500	-1.5	-1.5	-1.5	-1.5	-1.5
100	3286	0.0	0.9	2.4	4.2	6.2
200	3085	-1.8	0.0	2.9	6.6	10.6
300	2893	-7.1	-4.4	0.0	5.5	11.6
400	2708	-16.8	-13.2	-7.3	0.0	8.1
500	2531	-31.1	-26.6	-19.3	-10.1	0.0
600	2361	-50.4	-44.9	-36.1	-25.2	-13.0
0	3400	-1.5	-1.5	-1.5	-1.5	-1.5
100	3192	0.0	0.9	2.6	4.5	6.6
200	2995	-1.9	0.0	3.2	7.1	11.3
300	2806	-7.7	-4.8	0.0	5.8	12.2
400	2625	-17.9	-14.2	-7.7	0.0	8.5
500	2451	-33.1	-28.3	-20.3	-10.7	0.0
600	2284	-54.1	-48.5	-38.8	-27.2	-14.5
0	3300	-1.5	-1.5	-1.5	-1.5	-1.5
100	3097	0.0	1.0	2.8	4.8	7.1
200	2904	-2.1	0.0	3.5	7.6	12.1
300	2719	-8.4	-5.3	0.0	6.2	12.9
400	2541	-19.4	-15.2	-8.2	0.0	9.0
500	2370	-35.4	-30.3	-21.5	-11.2	0.0
600	2207	-58.2	-52.0	-41.4	-29.1	-15.7
0	3200	-1.5	-1.5	-1.5	-1.5	-1.5
100	3002	0.0	1.1	3.1	5.2	7.7
200	2812	-2.3	0.0	3.9	8.2	13.1
300	2631	-9.2	-5.8	0.0	6.5	13.9
400	2457	-20.9	-16.4	-8.6	0.0	9.8
500	2290	-38.4	-32.8	-23.1	-12.3	0.0
600	2129	-62.8	-56.0	-44.4	-31.4	-16.7
0	3100	-1.5	-1.5	-1.5	-1.5	-1.5
100	2906	0.0	1.3	3.3	5.6	8.3
200	2721	-2.7	0.0	4.0	8.5	14.0
300	2544	-10.0	-6.0	0.0	6.8	15.0
400	2373	-22.4	-17.1	-9.1	0.0	10.9
500	2209	-41.6	-34.9	-25.0	-13.6	0.0
600	2051	-67.7	-59.7	-47.7	-34.1	-17.8
0	3000	-1.5	-1.5	-1.5	-1.5	-1.5
100	2810	0.0	1.5	3.7	6.2	9.1
200	2629	-3.1	0.0	4.3	9.3	15.1
300	2455	-11.0	-6.4	0.0	7.6	16.2
400	2288	-24.8	-18.6	-10.1	0.0	11.5
500	2127	-45.3	-37.7	-27.0	-14.4	0.0
600	1972	-73.3	-64.1	-51.3	-36.2	-18.9
0	2900	-1.5	-1.5	-1.5	-1.5	-1.5
100	2715	0.0	1.6	3.9	6.7	9.8
200	2537	-3.3	0.0	4.5	10.1	16.3
300	2367	-11.6	-6.7	0.0	8.4	17.7
400	2203	-26.7	-20.2	-11.2	0.0	12.4
500	2045	-48.9	-40.7	-29.5	-15.4	0.0
600	1894	-78.3	-68.6	-55.1	-38.2	-19.7
0	2800	-1.5	-1.5	-1.5	-1.5	-1.5
100	2619	0.0	1.8	4.3	7.3	10.6
200	2445	-3.5	0.0	5.1	11.0	17.6
300	2278	-13.0	-7.7	0.0	8.9	18.7
400	2118	-29.1	-22.1	-11.8	0.0	13.2
500	1963	-52.8	-44.0	-31.2	-16.5	0.0
600	1816	-85.7	-75.2	-59.8	-42.1	-22.3

Universal Table for Rifle Bullets with Ballistic Coefficients of .490-.499

Sight Above Bore (in.) +1.5

Range (yds)	Velocity (fps)	Bullet Path Zero (100 yds)	Zero (200 yds)	Zero (300 yds)	Zero (400 yds)	Zero (500 yds)
0	2700	-1.5	-1.5	-1.5	-1.5	-1.5
100	2523	0.0	1.9	4.8	8.0	11.4
200	2352	-3.8	0.0	5.7	12.1	19.0
300	2189	-14.3	-8.6	0.0	9.6	19.9
400	2032	-31.9	-24.2	-12.8	0.0	13.8
500	1881	-57.1	-47.5	-33.2	-17.2	0.0
600	1740	-93.5	-82.0	-64.8	-45.7	-25.0
0	2600	-1.5	-1.5	-1.5	-1.5	-1.5
100	2426	0.0	2.2	5.2	8.7	12.6
200	2260	-4.5	0.0	6.0	12.8	20.7
300	2101	-15.7	-9.0	0.0	10.2	22.1
400	1947	-34.6	-25.7	-13.7	0.0	15.8
500	1801	-63.0	-51.8	-36.8	-19.7	0.0
600	1664	-102.5	-89.1	-71.1	-50.6	-26.9
0	2500	-1.5	-1.5	-1.5	-1.5	-1.5
100	2330	0.0	2.6	5.8	9.5	14.0
200	2168	-5.2	0.0	6.5	13.8	22.8
300	2012	-17.5	-9.8	0.0	10.9	24.4
400	1861	-37.9	-27.6	-14.5	0.0	18.0
500	1721	-69.9	-57.0	-40.7	-22.5	0.0
600	1589	-112.9	-97.4	-77.8	-56.0	-29.0
0	2400	-1.5	-1.5	-1.5	-1.5	-1.5
100	2234	0.0	2.7	6.2	10.5	15.2
200	2076	-5.4	0.0	7.0	15.5	25.0
300	1923	-18.7	-10.5	0.0	12.8	27.0
400	1778	-42.0	-31.1	-17.0	0.0	18.9
500	1643	-76.1	-62.5	-45.0	-23.7	0.0
600	1515	-122.6	-106.2	-85.2	-59.6	-31.2
0	2300	-1.5	-1.5	-1.5	-1.5	-1.5
100	2138	0.0	3.0	6.9	11.6	16.8
200	1983	-6.0	0.0	7.9	17.2	27.5
300	1835	-20.8	-11.8	0.0	14.0	29.5
400	1696	-46.4	-34.4	-18.6	0.0	20.7
500	1565	-83.9	-68.8	-49.1	-25.9	0.0
600	1441	-133.9	-115.8	-92.2	-64.3	-33.3
0	2200	-1.5	-1.5	-1.5	-1.5	-1.5
100	2042	0.0	3.3	7.8	12.9	18.4
200	1890	-6.6	0.0	9.1	19.3	30.3
300	1748	-23.5	-13.6	0.0	15.3	31.9
400	1614	-51.8	-38.6	-20.5	0.0	22.0
500	1488	-92.2	-75.8	-53.1	-27.5	0.0
600	1374	-149.8	-130.1	-102.9	-72.2	-39.2
0	2100	-1.5	-1.5	-1.5	-1.5	-1.5
100	1946	0.0	3.9	8.8	14.4	20.3
200	1800	-7.8	0.0	9.9	21.0	32.9
300	1663	-26.5	-14.9	0.0	16.6	34.4
400	1534	-57.4	-41.9	-22.1	0.0	23.8
500	1411	-101.6	-82.2	-57.4	-29.8	0.0
600	1308	-167.5	-144.3	-114.5	-81.4	-45.6

Universal Table for Rifle Bullets with Ballistic Coefficients of .500-.509

Sight Above Bore (in.) +1.5

Range (yds)	Velocity (fps)	Bullet Path Zero (100 yds)	Zero (200 yds)	Zero (300 yds)	Zero (400 yds)	Zero (500 yds)
0	3400	-1.5	-1.5	-1.5	-1.5	-1.5
100	3196	0.0	1.0	2.5	4.5	6.6
200	3002	-1.9	0.0	3.1	7.0	11.2
300	2817	-7.5	-4.6	0.0	5.9	12.3
400	2639	-17.8	-14.0	-7.9	0.0	8.5
500	2468	-32.9	-28.1	-20.4	-10.6	0.0
600	2303	-53.8	-48.1	-38.9	-27.0	-14.3
0	3300	-1.5	-1.5	-1.5	-1.5	-1.5
100	3101	0.0	1.0	2.8	4.8	7.0
200	2911	-2.1	0.0	3.5	7.6	12.0
300	2730	-8.3	-5.2	0.0	6.1	12.8
400	2555	-19.3	-15.1	-8.2	0.0	8.9
500	2387	-35.2	-30.0	-21.3	-11.1	0.0
600	2226	-57.7	-51.5	-41.1	-28.8	-15.5
0	3200	-1.5	-1.5	-1.5	-1.5	-1.5
100	3005	0.0	1.1	3.0	5.2	7.6
200	2820	-2.2	0.0	3.8	8.1	13.0
300	2642	-9.1	-5.8	0.0	6.4	13.8
400	2471	-20.7	-16.3	-8.6	0.0	9.8
500	2300	-38.1	-32.6	-23.0	-12.3	0.0
600	2148	-62.1	-55.5	-44.0	-31.1	-16.4
0	3100	-1.5	-1.5	-1.5	-1.5	-1.5
100	2910	0.0	1.3	3.3	5.6	8.3
200	2728	-2.7	0.0	4.0	8.5	13.8
300	2554	-10.0	-6.0	0.0	6.8	14.8
400	2386	-22.4	-17.0	-9.1	0.0	10.7
500	2225	-41.3	-34.6	-24.7	-13.3	0.0
600	2069	-67.3	-59.3	-47.4	-33.7	-17.7
0	3000	-1.5	-1.5	-1.5	-1.5	-1.5
100	2814	0.0	1.5	3.7	6.2	9.0
200	2636	-3.1	0.0	4.3	9.3	14.9
300	2465	-11.0	-6.4	0.0	7.5	16.0
400	2301	-24.6	-18.5	-10.0	0.0	11.4
500	2143	-45.0	-37.3	-26.7	-14.2	0.0
600	1991	-72.6	-63.4	-50.6	-35.7	-18.6
0	2900	-1.5	-1.5	-1.5	-1.5	-1.5
100	2718	0.0	1.6	3.8	6.6	9.7
200	2544	-3.2	0.0	4.5	10.0	16.1
300	2377	-11.5	-6.7	0.0	8.3	17.5
400	2216	-26.5	-20.1	-11.1	0.0	12.2
500	2061	-48.3	-40.3	-29.1	-15.2	0.0
600	1911	-77.9	-68.3	-54.9	-38.2	-19.9
0	2800	-1.5	-1.5	-1.5	-1.5	-1.5
100	2622	0.0	1.7	4.3	7.2	10.5
200	2452	-3.5	0.0	5.1	11.0	17.5
300	2288	-12.8	-7.6	0.0	8.9	18.7
400	2130	-29.0	-22.0	-11.9	0.0	13.0
500	1978	-52.5	-43.8	-31.1	-16.3	0.0
600	1833	-84.9	-74.5	-59.3	-41.5	-22.0
0	2700	-1.5	-1.5	-1.5	-1.5	-1.5
100	2526	0.0	1.9	4.7	7.9	11.3
200	2359	-3.8	0.0	5.7	12.0	18.9
300	2199	-14.2	-8.5	0.0	9.5	19.8
400	2045	-31.5	-24.0	-12.6	0.0	13.8
500	1896	-56.7	-47.2	-33.0	-17.2	0.0
600	1756	-92.7	-81.4	-64.4	-45.4	-24.7

Universal Table for Rifle Bullets with Ballistic Coefficients of .500-.509						
					Sight Above Bore (in.)	+1.5
			Bullet Path			
Range (yds)	Velocity (fps)	Zero (100 yds)	Zero (200 yds)	Zero (300 yds)	Zero (400 yds)	Zero (500 yds)
0	2600	-1.5	-1.5	-1.5	-1.5	-1.5
100	2430	0.0	2.2	5.2	8.6	12.5
200	2267	-4.4	0.0	6.0	12.8	20.6
300	2110	-15.7	-9.1	0.0	10.2	21.8
400	1959	-34.5	-25.6	-13.5	0.0	15.6
500	1815	-62.6	-51.5	-36.4	-19.5	0.0
600	1680	-101.7	-88.3	-70.2	-49.9	-26.5
0	2500	-1.5	-1.5	-1.5	-1.5	-1.5
100	2333	0.0	2.6	5.8	9.4	13.8
200	2174	-5.2	0.0	6.5	13.7	22.5
300	2021	-17.4	-9.7	0.0	10.8	24.0
400	1873	-37.7	-27.4	-14.4	0.0	17.7
500	1735	-69.2	-56.3	-40.1	-22.1	0.0
600	1604	-112.0	-96.6	-77.1	-55.5	-29.1
0	2400	-1.5	-1.5	-1.5	-1.5	-1.5
100	2238	0.0	2.8	6.2	10.4	15.2
200	2082	-5.5	0.0	7.0	15.3	24.8
300	1932	-18.7	-10.5	0.0	12.5	26.7
400	1790	-41.6	-30.6	-16.6	0.0	19.0
500	1656	-75.8	-62.0	-44.5	-23.8	0.0
600	1529	-122.3	-105.7	-84.8	-59.9	-31.4
0	2300	-1.5	-1.5	-1.5	-1.5	-1.5
100	2141	0.0	3.0	6.9	11.5	16.6
200	1989	-6.0	0.0	7.8	17.0	27.3
300	1844	-20.6	-11.6	0.0	13.9	29.3
400	1707	-46.0	-34.0	-18.5	0.0	20.5
500	1578	-83.2	-68.2	-48.8	-25.7	0.0
600	1455	-133.2	-115.2	-92.0	-64.2	-33.4
0	2200	-1.5	-1.5	-1.5	-1.5	-1.5
100	2045	0.0	3.3	7.7	12.8	18.4
200	1896	-6.6	0.0	8.9	19.1	30.2
300	1757	-23.2	-13.3	0.0	15.3	31.9
400	1625	-51.2	-38.1	-20.3	0.0	22.2
500	1500	-91.8	-75.4	-53.2	-27.7	0.0
600	1381	-145.4	-125.8	-99.1	-68.6	-35.3
0	2100	-1.5	-1.5	-1.5	-1.5	-1.5
100	1949	0.0	3.9	8.8	14.3	20.2
200	1805	-7.8	0.0	9.8	20.8	32.6
300	1671	-26.4	-14.7	0.0	16.5	34.3
400	1544	-57.1	-41.5	-22.0	0.0	23.7
500	1423	-101.0	-81.5	-57.1	-29.6	0.0
600	1320	-166.1	-142.7	-113.3	-80.3	-44.8
0	2000	-1.5	-1.5	-1.5	-1.5	-1.5
100	1853	0.0	4.4	9.8	15.8	23.0
200	1715	-8.9	0.0	10.8	22.7	37.1
300	1586	-29.5	-16.1	0.0	17.8	39.5
400	1463	-63.1	-45.3	-23.8	0.0	28.8
500	1354	-114.9	-92.7	-65.8	-36.0	0.0
600	1257	-184.0	-157.3	-125.0	-89.3	-46.1

Universal Table for Rifle Bullets with Ballistic Coefficients of .510-.519

Sight Above Bore (in.) +1.5

		Bullet Path				
Range (yds)	Velocity (fps)	Zero (100 yds)	Zero (200 yds)	Zero (300 yds)	Zero (400 yds)	Zero (500 yds)
0	3400	-1.5	-1.5	-1.5	-1.5	-1.5
100	3200	0.0	0.9	2.5	4.4	6.5
200	3010	-1.9	0.0	3.1	7.0	11.2
300	2827	-7.5	-4.7	0.0	5.8	12.2
400	2653	-17.7	-13.9	-7.7	0.0	8.5
500	2484	-32.7	-28.0	-20.3	-10.6	0.0
600	2322	-53.4	-47.7	-38.4	-26.8	-14.1
0	3300	-1.5	-1.5	-1.5	-1.5	-1.5
100	3105	0.0	1.0	2.8	4.8	7.0
200	2918	-2.1	0.0	3.5	7.5	11.9
300	2740	-8.3	-5.2	0.0	6.1	12.7
400	2568	-19.2	-15.0	-8.1	0.0	8.8
500	2403	-35.0	-29.8	-21.2	-11.0	0.0
600	2244	-57.4	-51.1	-40.8	-28.6	-15.4
0	3200	-1.5	-1.5	-1.5	-1.5	-1.5
100	3009	0.0	1.1	3.0	5.2	7.6
200	2827	-2.2	0.0	3.8	8.1	12.9
300	2652	-9.1	-5.7	0.0	6.4	13.6
400	2484	-20.6	-16.2	-8.5	0.0	9.7
500	2322	-37.8	-32.3	-22.7	-12.1	0.0
600	2166	-61.7	-55.1	-43.6	-30.8	16.3
0	3100	-1.5	-1.5	-1.5	-1.5	-1.5
100	2913	0.0	1.3	3.3	5.6	8.2
200	2735	-2.6	0.0	4.0	8.5	13.8
300	2564	-9.9	-5.9	0.0	6.8	14.7
400	2399	-22.2	-16.9	-9.0	0.0	10.6
500	2240	-41.0	-34.4	-24.5	-13.3	0.0
600	2087	-66.7	-58.8	-46.9	-33.4	-17.5
0	3000	-1.5	-1.5	-1.5	-1.5	-1.5
100	2817	0.0	1.5	3.6	6.1	8.9
200	2643	-3.0	0.0	4.2	9.2	14.8
300	2475	-10.9	-6.4	0.0	7.5	15.9
400	2313	-24.5	-18.5	-10.0	0.0	11.2
500	2158	-44.7	-37.1	-26.5	-14.0	0.0
600	2008	-72.2	-63.1	-50.4	-35.4	-18.6
0	2900	-1.5	-1.5	-1.5	-1.5	-1.5
100	2722	0.0	1.6	3.9	6.6	9.6
200	2551	-3.2	0.0	4.5	10.0	16.0
300	2386	-11.6	-6.8	0.0	8.2	17.3
400	2228	-26.4	-20.0	-11.0	0.0	12.0
500	2076	-48.1	-40.1	-28.8	-15.0	0.0
600	1929	-77.3	-67.7	-54.1	-37.7	-19.6
0	2800	-1.5	-1.5	-1.5	-1.5	-1.5
100	2626	0.0	1.8	4.3	7.2	10.4
200	2458	-3.5	0.0	5.0	10.9	17.3
300	2297	-12.9	-7.6	0.0	8.8	18.4
400	2142	-28.9	-21.8	-11.7	0.0	12.9
500	1993	-52.2	-43.3	-30.7	-16.1	0.0
600	1850	-84.2	-73.6	-58.5	-40.9	-21.6
0	2700	-1.5	-1.5	-1.5	-1.5	-1.5
100	2529	0.0	1.9	4.7	7.9	11.3
200	2365	-3.8	0.0	5.6	11.9	18.8
300	2208	-14.1	-8.4	0.0	9.5	19.7
400	2056	-31.5	-23.9	-12.7	0.0	13.6
500	1910	-56.4	-46.9	-32.9	-17.0	0.0
600	1772	-91.8	-80.5	-63.6	-44.6	-24.2

Universal Table for Rifle Bullets with Ballistic Coefficients of .510-.519

Sight Above Bore (in.) +1.5

Range (yds)	Velocity (fps)	Bullet Path: Zero (100 yds)	Zero (200 yds)	Zero (300 yds)	Zero (400 yds)	Zero (500 yds)
0	2600	-1.5	-1.5	-1.5	-1.5	-1.5
100	2433	0.0	2.2	5.2	8.6	12.4
200	2273	-4.4	0.0	6.0	12.8	20.4
300	2119	-15.6	-9.0	0.0	10.2	21.6
400	1970	-34.4	-25.6	-13.6	0.0	15.2
500	1829	-62.0	-50.9	-35.9	-19.0	0.0
600	1695	-100.9	-87.6	-69.6	-49.3	-26.5
0	2500	-1.5	-1.5	-1.5	-1.5	-1.5
100	2337	0.0	2.6	5.8	9.4	13.8
200	2180	-5.2	0.0	6.4	13.7	22.3
300	2030	-17.4	-9.6	0.0	10.8	23.9
400	1884	-37.7	-27.3	-14.5	0.0	17.4
500	1748	-68.8	-55.8	-39.8	-21.7	0.0
600	1619	-111.3	-95.7	-76.4	-54.7	-28.7
0	2400	-1.5	-1.5	-1.5	-1.5	-1.5
100	2241	0.0	2.7	6.3	10.4	15.0
200	2088	-5.5	0.0	7.0	15.3	24.6
300	1940	-18.8	-10.5	0.0	12.4	26.3
400	1800	-41.5	-30.5	-16.5	0.0	18.6
500	1669	-75.1	-61.4	-43.8	-23.2	0.0
600	1544	-121.1	-104.6	-83.6	-58.8	-31.0
0	2300	-1.5	-1.5	-1.5	-1.5	-1.5
100	2145	0.0	3.0	6.9	11.5	16.6
200	1995	-6.1	0.0	7.7	16.9	27.1
300	1852	-20.7	-11.6	0.0	13.8	29.1
400	1717	-46.0	-33.9	-18.4	0.0	20.4
500	1590	-82.9	-67.8	-48.4	-25.5	0.0
600	1469	-132.6	-114.5	-91.2	-63.7	-33.1
0	2200	-1.5	-1.5	-1.5	-1.5	-1.5
100	2048	0.0	3.3	7.7	12.7	18.2
200	1902	-6.5	0.0	8.8	19.0	30.0
300	1765	-23.0	-13.3	0.0	15.2	31.7
400	1635	-50.9	-37.9	-20.2	0.0	22.0
500	1512	-91.2	-74.9	-52.8	-27.5	0.0
600	1395	-144.4	-124.8	-98.3	-68.0	-34.9
0	2100	-1.5	-1.5	-1.5	-1.5	-1.5
100	1952	0.0	3.8	8.8	14.3	20.2
200	1811	-7.7	0.0	9.9	20.9	32.6
300	1678	-26.4	-14.8	0.0	16.4	34.1
400	1553	-57.1	-41.7	-21.9	0.0	23.5
500	1434	-100.8	-81.6	-56.8	-29.4	0.0
600	1332	-164.4	-141.3	-111.6	-78.7	-43.4
0	2000	-1.5	-1.5	-1.5	-1.5	-1.5
100	1855	0.0	4.4	9.8	15.7	22.6
200	1720	-8.8	0.0	10.8	22.6	36.5
300	1593	-29.3	-16.1	0.0	17.8	38.6
400	1472	-62.8	-45.2	-23.7	0.0	27.8
500	1365	-113.2	-91.3	-64.4	-34.7	0.0
600	1268	-182.2	-155.8	-123.6	-88.0	-46.3

Universal Table for Rifle Bullets with Ballistic Coefficients of .520-.529

Sight Above Bore (in.) +1.5

Range (yds)	Velocity (fps)	Bullet Path Zero (100 yds)	Zero (200 yds)	Zero (300 yds)	Zero (400 yds)	Zero (500 yds)
0	3400	-1.5	-1.5	-1.5	-1.5	-1.5
100	3197	0.0	0.9	2.5	4.5	6.6
200	3004	-1.9	0.0	3.1	7.0	11.3
300	2819	-7.5	-4.6	0.0	5.9	12.3
400	2642	-17.8	-14.0	-7.8	0.0	8.5
500	2471	-32.9	-28.2	-20.4	-10.6	0.0
600	2307	-53.7	-48.0	-38.8	-27.0	-14.2
0	3300	-1.5	-1.5	-1.5	-1.5	-1.5
100	3102	0.0	1.0	2.8	4.8	7.0
200	2913	-2.1	0.0	3.5	7.6	12.0
300	2732	-8.3	-5.2	0.0	6.1	12.7
400	2558	-19.2	-15.1	-8.1	0.0	8.8
500	2391	-35.1	-29.9	-21.2	-11.0	0.0
600	2230	-57.6	-51.4	-41.0	-28.8	-15.5
0	3200	-1.5	-1.5	-1.5	-1.5	-1.5
100	3006	0.0	1.1	3.0	5.2	7.6
200	2821	-2.2	0.0	3.8	8.1	13.0
300	2644	-9.1	-5.7	0.0	6.4	13.8
400	2474	-20.7	-16.2	-8.5	0.0	9.8
500	2309	-38.1	-32.6	-23.0	-12.3	0.0
600	2152	-62.0	-55.3	-43.9	-31.1	16.3
0	3100	-1.5	-1.5	-1.5	-1.5	-1.5
100	2911	0.0	1.3	3.3	5.6	8.3
200	2730	-2.7	0.0	4.0	8.5	13.9
300	2556	-10.0	-6.0	0.0	6.8	14.8
400	2389	-22.3	-17.0	-9.0	0.0	10.7
500	2228	-41.3	-34.7	-24.7	-13.4	0.0
600	2073	-67.2	-59.2	-47.2	-33.7	-17.6
0	3000	-1.5	-1.5	-1.5	-1.5	-1.5
100	2815	0.0	1.5	3.6	6.1	9.0
200	2638	-3.0	0.0	4.2	9.2	15.0
300	2468	-10.9	-6.3	0.0	7.5	16.1
400	2304	-24.6	-18.5	-10.0	0.0	11.4
500	2146	-45.0	-37.4	-26.8	-14.3	0.0
600	1994	-72.7	-63.6	-50.9	-35.8	-18.7
0	2900	-1.5	-1.5	-1.5	-1.5	-1.5
100	2719	0.0	1.6	3.8	6.6	9.7
200	2546	-3.2	0.0	4.5	10.1	16.1
300	2379	-11.5	-6.8	0.0	8.4	17.5
400	2218	-26.6	-20.2	-11.2	0.0	12.1
500	2064	-48.3	-40.4	-29.1	-15.1	0.0
600	1915	-77.8	-68.2	-54.7	-37.9	-19.8
0	2800	-1.5	-1.5	-1.5	-1.5	-1.5
100	2623	0.0	1.8	4.3	7.2	10.5
200	2453	-3.5	0.0	5.0	10.9	17.4
300	2290	-12.8	-7.6	0.0	8.8	18.5
400	2133	-28.9	-21.9	-11.8	0.0	12.9
500	1982	-52.3	-43.5	-30.9	-16.1	0.0
600	1837	-84.7	-74.1	-59.0	-41.3	-22.0
0	2700	-1.5	-1.5	-1.5	-1.5	-1.5
100	2527	0.0	1.9	4.7	7.9	11.3
200	2360	-3.8	0.0	5.6	12.0	18.8
300	2201	-14.2	-8.4	0.0	9.5	19.8
400	2047	-31.6	-24.0	-12.7	0.0	13.7
500	1899	-56.6	-47.1	-33.0	-17.1	0.0
600	1759	-92.7	-81.3	-64.4	-45.3	-24.7

Universal Table for Rifle Bullets with Ballistic Coefficients of .520-.529						
					Sight Above Bore (in.)	+1.5
		Bullet Path				
Range (yds)	Velocity (fps)	Zero (100 yds)	Zero (200 yds)	Zero (300 yds)	Zero (400 yds)	Zero (500 yds)
0	2600	-1.5	-1.5	-1.5	-1.5	-1.5
100	2431	0.0	2.2	5.2	8.7	12.5
200	2268	-4.5	0.0	6.0	12.8	20.5
300	2112	-15.7	-9.0	0.0	10.2	21.8
400	1961	-34.6	-25.6	-13.6	0.0	15.4
500	1818	-62.5	-51.3	-36.3	-19.3	0.0
600	1683	-101.6	-88.2	-70.2	-49.7	-26.6
0	2500	-1.5	-1.5	-1.5	-1.5	-1.5
100	2334	0.0	2.6	5.8	9.4	13.8
200	2176	-5.1	0.0	6.5	13.8	22.5
300	2023	-17.4	-9.8	0.0	10.9	24.0
400	1875	-37.8	-27.5	-14.5	0.0	17.5
500	1738	-69.0	-56.3	-40.0	-21.8	0.0
600	1607	-112.0	-96.6	-77.1	-55.3	-29.1
0	2400	-1.5	-1.5	-1.5	-1.5	-1.5
100	2238	0.0	2.7	6.2	10.4	15.1
200	2083	-5.5	0.0	6.9	15.3	24.7
300	1934	-18.6	-10.4	0.0	12.5	26.6
400	1792	-41.5	-30.6	-16.7	0.0	18.8
500	1659	-75.4	-61.7	-44.4	-23.5	0.0
600	1532	-122.0	-105.5	-84.7	-59.7	-31.5
0	2300	-1.5	-1.5	-1.5	-1.5	-1.5
100	2142	0.0	3.0	6.9	11.5	16.7
200	1991	-6.0	0.0	7.9	17.1	27.4
300	1845	-20.8	-11.9	0.0	13.7	29.2
400	1709	-46.0	-34.1	-18.3	0.0	20.6
500	1580	-83.4	-68.5	-48.7	-25.8	0.0
600	1458	-133.1	-115.3	-91.5	-64.1	-33.1
0	2200	-1.5	-1.5	-1.5	-1.5	-1.5
100	2046	0.0	3.3	7.8	12.8	18.4
200	1898	-6.5	0.0	9.1	19.1	30.3
300	1758	-23.3	-13.6	0.0	15.1	31.9
400	1627	-51.3	-38.3	-20.2	0.0	22.3
500	1502	-92.0	-75.7	-53.1	-27.9	0.0
600	1384	-145.3	-125.8	-98.6	-68.4	-34.9
0	2100	-1.5	-1.5	-1.5	-1.5	-1.5
100	1949	0.0	3.8	8.8	14.2	20.2
200	1807	-7.6	0.0	10.0	20.9	32.8
300	1672	-26.4	-15.0	0.0	16.3	34.2
400	1546	-57.0	-41.7	-21.8	0.0	23.8
500	1425	-101.0	-81.9	-57.0	-29.8	0.0
600	1323	-165.1	-142.3	-112.4	-79.7	-44.0
0	2000	-1.5	-1.5	-1.5	-1.5	-1.5
100	1853	0.0	4.4	9.8	15.7	22.8
200	1716	-8.8	0.0	10.8	22.6	36.8
300	1587	-29.5	-16.3	0.0	17.7	38.9
400	1465	-62.9	-45.2	-23.6	0.0	28.3
500	1357	-114.0	-91.9	-64.8	-35.4	0.0
600	1259	-183.7	-157.2	-124.7	-89.4	-46.9

Universal Table for Rifle Bullets with Ballistic Coefficients of .530-.539

Sight Above Bore (in.) +1.5

Range (yds)	Velocity (fps)	Bullet Path Zero (100 yds)	Zero (200 yds)	Zero (300 yds)	Zero (400 yds)	Zero (500 yds)
0	3400	-1.5	-1.5	-1.5	-1.5	-1.5
100	3201	0.0	0.9	2.5	4.4	6.5
200	3011	-1.9	0.0	3.1	7.0	11.2
300	2829	-7.5	-4.6	0.0	5.8	12.1
400	2655	-17.7	-13.9	-7.8	0.0	8.4
500	2487	-32.7	-28.0	-20.2	-10.5	0.0
600	2325	-53.4	-47.7	-38.4	-26.8	-14.1
0	3300	-1.5	-1.5	-1.5	-1.5	-1.5
100	3105	0.0	1.0	2.8	4.8	7.0
200	2920	-2.0	0.0	3.5	7.5	11.9
300	2742	-8.3	-5.2	0.0	6.1	12.7
400	2571	-19.1	-15.0	-8.1	0.0	8.9
500	2406	-34.9	-29.8	-21.2	-11.1	0.0
600	2247	-57.3	-51.2	-40.8	-28.7	-15.4
0	3200	-1.5	-1.5	-1.5	-1.5	-1.5
100	3010	0.0	1.1	3.0	5.2	7.6
200	2828	-2.2	0.0	3.8	8.1	12.9
300	2654	-9.1	-5.7	0.0	6.4	13.7
400	2486	-20.7	-16.2	-8.6	0.0	9.7
500	2324	-37.9	-32.4	-22.8	-12.1	0.0
600	2169	-61.7	-55.0	-43.6	-30.7	-16.2
0	3100	-1.5	-1.5	-1.5	-1.5	-1.5
100	2914	0.0	1.3	3.3	5.6	8.2
200	2736	-2.7	0.0	3.9	8.5	13.7
300	2566	-9.9	-5.9	0.0	6.8	14.7
400	2401	-22.3	-16.9	-9.1	0.0	10.5
500	2243	-41.0	-34.3	-24.5	-13.2	0.0
600	2090	-66.7	-58.7	-46.9	-33.3	-17.5
0	3000	-1.5	-1.5	-1.5	-1.5	-1.5
100	2818	0.0	1.5	3.6	6.1	9.0
200	2644	-3.0	0.0	4.2	9.2	14.9
300	2477	-10.9	-6.3	0.0	7.4	16.0
400	2316	-24.4	-18.3	-9.9	0.0	11.4
500	2160	-44.8	-37.2	-26.6	-14.2	0.0
600	2011	-72.1	-63.0	-50.3	-35.5	-18.4
0	2900	-1.5	-1.5	-1.5	-1.5	-1.5
100	2722	0.0	1.6	3.8	6.6	9.6
200	2552	-3.2	0.0	4.5	10.0	16.0
300	2388	-11.5	-6.7	0.0	8.3	17.3
400	2230	-26.4	-20.0	-11.0	0.0	12.1
500	2078	-48.1	-40.1	-28.9	-15.1	0.0
600	1931	-77.4	-67.8	-54.3	-37.8	-19.7
0	2800	-1.5	-1.5	-1.5	-1.5	-1.5
100	2626	0.0	1.8	4.3	7.2	10.4
200	2459	-3.5	0.0	5.0	10.9	17.3
300	2299	-12.8	-7.5	0.0	8.8	18.5
400	2144	-28.8	-21.8	-11.8	0.0	12.9
500	1995	-52.1	-43.4	-30.9	-16.1	0.0
600	1853	-83.9	-73.4	-58.4	-40.7	-21.3
0	2700	-1.5	-1.5	-1.5	-1.5	-1.5
100	2530	0.0	1.9	4.7	7.9	11.3
200	2367	-3.7	0.0	5.6	12.0	18.8
300	2210	-14.1	-8.4	0.0	9.5	19.8
400	2058	-31.5	-24.0	-12.7	0.0	13.7
500	1912	-56.5	-47.1	-33.0	-17.1	0.0

Universal Table for Rifle Bullets with Ballistic Coefficients of .530-.539

Sight Above Bore (in.) +1.5

Range (yds)	Velocity (fps)	Bullet Path Zero (100 yds)	Zero (200 yds)	Zero (300 yds)	Zero (400 yds)	Zero (500 yds)
0	2600	-1.5	-1.5	-1.5	-1.5	-1.5
100	2434	0.0	2.2	5.2	8.6	12.4
200	2274	-4.5	0.0	6.0	12.8	20.4
300	2120	-15.7	-9.1	0.0	10.1	21.5
400	1972	-34.5	-25.5	-13.5	0.0	15.2
500	1831	-62.1	-50.9	-35.9	-19.0	0.0
600	1698	-100.7	-87.3	-69.2	-49.0	-26.2
0	2500	-1.5	-1.5	-1.5	-1.5	-1.5
100	2337	0.0	2.6	5.8	9.4	13.7
200	2181	-5.2	0.0	6.5	13.6	22.3
300	2031	-17.4	-9.7	0.0	10.8	23.8
400	1886	-37.6	-27.3	-14.3	0.0	17.4
500	1750	-68.7	-55.8	-39.6	-21.7	0.0
600	1622	-110.8	-95.3	-75.9	-54.4	-28.3
0	2400	-1.5	-1.5	-1.5	-1.5	-1.5
100	2241	0.0	2.7	6.2	10.3	15.0
200	2089	-5.4	0.0	7.0	15.2	24.5
300	1942	-18.6	-10.4	0.0	12.4	26.4
400	1802	-41.3	-30.5	-16.5	0.0	18.6
500	1671	-74.9	-61.4	-43.9	-23.3	0.0
600	1546	-121.0	-104.7	-83.8	-59.0	-31.1
0	2300	-1.5	-1.5	-1.5	-1.5	-1.5
100	2145	0.0	3.0	6.9	11.4	16.6
200	1996	-6.0	0.0	7.8	16.9	27.1
300	1853	-20.7	-11.7	0.0	13.6	29.0
400	1719	-45.8	-33.8	-18.2	0.0	20.4
500	1592	-82.8	-67.7	-48.3	-25.5	0.0
600	1471	-132.5	-114.5	-91.2	-63.9	-33.2
0	2200	-1.5	-1.5	-1.5	-1.5	-1.5
100	2049	0.0	3.3	7.7	12.8	18.2
200	1903	-6.6	0.0	8.9	19.0	29.9
300	1766	-23.2	-13.3	0.0	15.2	31.6
400	1636	-51.2	-38.1	-20.3	0.0	21.8
500	1514	-91.2	-74.9	-52.7	-27.2	0.0
600	1397	-144.6	-124.9	-98.3	-67.8	-35.1
0	2100	-1.5	-1.5	-1.5	-1.5	-1.5
100	1952	0.0	3.8	8.7	14.2	20.1
200	1812	-7.6	0.0	9.8	20.8	32.6
300	1680	-26.1	-14.7	0.0	16.5	34.2
400	1555	-56.8	-41.6	-22.0	0.0	23.6
500	1436	-100.6	-81.5	-57.0	-29.5	0.0
600	1334	-164.0	-141.1	-111.8	-78.8	-43.3
0	2000	-1.5	-1.5	-1.5	-1.5	-1.5
100	1856	0.0	4.4	9.8	15.7	22.7
200	1721	-8.8	0.0	10.8	22.5	36.7
300	1594	-29.5	-16.2	0.0	17.6	38.8
400	1474	-62.7	-45.1	-23.4	0.0	28.3
500	1366	-113.7	-91.6	-64.6	-35.3	0.0
600	1269	-182.9	-156.4	-124.0	-88.9	-46.5

Universal Table for Rifle Bullets with Ballistic Coefficients of .540-.549

Sight Above Bore (in.) +1.5

		Bullet Path				
Range (yds)	Velocity (fps)	Zero (100 yds)	Zero (200 yds)	Zero (300 yds)	Zero (400 yds)	Zero (500 yds)
0	3400	-1.5	-1.5	-1.5	-1.5	-1.5
100	3204	0.0	0.9	2.5	4.4	6.5
200	3018	-1.9	0.0	3.1	6.9	11.1
300	2839	-7.4	-4.6	0.0	5.8	12.1
400	2668	-17.6	-13.8	-7.7	0.0	8.4
500	2502	-32.5	-27.8	-20.1	-10.5	0.0
600	2343	-52.4	-46.8	-37.5	-26.0	-13.4
0	3300	-1.5	-1.5	-1.5	-1.5	-1.5
100	3109	0.0	1.0	2.8	4.8	7.0
200	2926	-2.1	0.0	3.4	7.5	11.8
300	2751	-8.3	-5.2	0.0	6.0	12.6
400	2583	-19.1	-14.9	-8.0	0.0	8.8
500	2421	-34.8	-29.6	-21.0	-11.0	0.0
600	2265	-56.8	-50.5	-40.2	-28.2	-15.0
0	3200	-1.5	-1.5	-1.5	-1.5	-1.5
100	3013	0.0	1.1	3.0	5.1	7.5
200	2834	-2.2	0.0	3.8	8.0	12.8
300	2663	-9.0	-5.7	0.0	6.4	13.5
400	2498	-20.5	-16.1	-8.5	0.0	9.5
500	2339	-37.6	-32.0	-22.6	-11.9	0.0
600	2186	-61.2	-54.5	-43.1	-30.4	-16.1
0	3100	-1.5	-1.5	-1.5	-1.5	-1.5
100	2917	0.0	1.3	3.3	5.5	8.1
200	2743	-2.6	0.0	4.0	8.5	13.7
300	2575	-9.8	-5.9	0.0	6.7	14.6
400	2413	-22.1	-16.9	-9.0	0.0	10.5
500	2257	-40.7	-34.2	-24.3	-13.1	0.0
600	2107	-66.1	-58.3	-46.4	-32.9	-17.2
0	3000	-1.5	-1.5	-1.5	-1.5	-1.5
100	2821	0.0	1.5	3.6	6.1	8.9
200	2651	-3.0	0.0	4.2	9.2	14.8
300	2486	-10.8	-6.4	0.0	7.4	15.8
400	2327	-24.4	-18.4	-9.9	0.0	11.2
500	2174	-44.5	-37.0	-26.4	-14.0	0.0
600	2027	-71.7	-62.7	-50.0	-35.1	-18.3
0	2900	-1.5	-1.5	-1.5	-1.5	-1.5
100	2726	0.0	1.6	3.8	6.6	9.6
200	2558	-3.2	0.0	4.5	9.9	15.9
300	2397	-11.5	-6.7	0.0	8.2	17.2
400	2241	-26.4	-19.9	-11.0	0.0	11.9
500	2092	-47.8	-39.8	-28.6	-14.9	0.0
600	1947	-77.0	-67.3	-54.0	-37.5	-19.6
0	2800	-1.5	-1.5	-1.5	-1.5	-1.5
100	2629	0.0	1.7	4.3	7.2	10.3
200	2465	-3.5	0.0	5.0	10.8	17.2
300	2307	-12.8	-7.5	0.0	8.7	18.3
400	2155	-28.6	-21.7	-11.6	0.0	12.7
500	2009	-51.7	-43.0	-30.5	-15.9	0.0
600	1867	-82.7	-72.2	-57.2	-39.7	-20.7
0	2700	-1.5	-1.5	-1.5	-1.5	-1.5
100	2533	0.0	1.9	4.7	7.8	11.2
200	2372	-3.8	0.0	5.6	11.9	18.6
300	2218	-14.1	-8.4	0.0	9.4	19.6
400	2069	-31.3	-23.7	-12.6	0.0	13.5
500	1926	-56.0	-46.5	-32.6	-16.9	0.0
600	1789	-91.2	-79.8	-63.1	-44.2	-24.0

Universal Table for Rifle Bullets with Ballistic Coefficients of .540-.549

Sight Above Bore (in.) +1.5

Range (yds)	Velocity (fps)	Bullet Path Zero (100 yds)	Zero (200 yds)	Zero (300 yds)	Zero (400 yds)	Zero (500 yds)
0	2600	-1.5	-1.5	-1.5	-1.5	-1.5
100	2437	0.0	2.2	5.2	8.6	12.4
200	2280	-4.4	0.0	6.0	12.7	20.3
300	2129	-15.6	-9.0	0.0	10.1	21.5
400	1983	-34.3	-25.4	-13.5	0.0	15.2
500	1843	-61.9	-50.8	-35.9	-19.0	0.0
600	1712	-100.0	-86.8	-68.8	-48.6	-25.8
0	2500	-1.5	-1.5	-1.5	-1.5	-1.5
100	2340	0.0	2.6	5.8	9.3	13.7
200	2187	-5.1	0.0	6.5	13.6	22.2
300	2039	-17.4	-9.7	0.0	10.6	23.6
400	1897	-37.4	-27.1	-14.2	0.0	17.3
500	1762	-68.3	-55.5	-39.3	-21.6	0.0
600	1635	-110.3	-95.0	-75.5	-54.3	-28.4
0	2400	-1.5	-1.5	-1.5	-1.5	-1.5
100	2244	0.0	2.7	6.2	10.3	14.9
200	2094	-5.5	0.0	6.9	15.1	24.4
300	1950	-18.6	-10.3	0.0	12.3	26.3
400	1812	-41.1	-30.2	-16.4	0.0	18.7
500	1682	-74.7	-61.0	-43.8	-23.3	0.0
600	1560	-120.0	-103.5	-82.8	-58.3	-30.3
0	2300	-1.5	-1.5	-1.5	-1.5	-1.5
100	2148	0.0	3.0	6.7	11.4	16.5
200	2001	-6.1	0.0	7.4	16.8	26.9
300	1860	-20.1	-11.0	0.0	14.2	29.3
400	1728	-45.7	-33.6	-18.9	0.0	20.2
500	1603	-82.4	-67.3	-48.9	-25.3	0.0
600	1484	-131.8	-113.6	-91.5	-63.2	-32.9
0	2200	-1.5	-1.5	-1.5	-1.5	-1.5
100	2051	0.0	3.3	7.7	12.6	18.1
200	1908	-6.5	0.0	8.8	18.8	29.7
300	1773	-23.0	-13.2	0.0	14.9	31.3
400	1646	-50.5	-37.5	-19.9	0.0	21.9
500	1525	-90.6	-74.3	-52.2	-27.4	0.0
600	1409	-144.0	-124.4	-98.0	-68.1	-35.3
0	2100	-1.5	-1.5	-1.5	-1.5	-1.5
100	1955	0.0	3.8	8.7	14.2	20.0
200	1817	-7.6	0.0	9.8	20.7	32.4
300	1687	-26.1	-14.6	0.0	16.4	34.0
400	1564	-56.6	-41.4	-21.8	0.0	23.4
500	1447	-100.0	-81.0	-56.6	-29.3	0.0
600	1345	-162.7	-139.9	-110.6	-77.8	-42.7
0	2000	-1.5	-1.5	-1.5	-1.5	-1.5
100	1858	0.0	4.3	9.7	15.6	22.5
200	1726	-8.7	0.0	10.8	22.6	36.3
300	1601	-29.2	-16.1	0.0	17.8	38.2
400	1482	-62.6	-45.2	-23.7	0.0	27.3
500	1376	-112.4	-90.7	-63.7	-34.1	0.0
600	1280	-180.7	-154.6	-122.3	-86.8	-45.8

Universal Table for Rifle Bullets with Ballistic Coefficients of .550-.559

Sight Above Bore (in.) +1.5

Range (yds)	Velocity (fps)	Bullet Path: Zero (100 yds)	Zero (200 yds)	Zero (300 yds)	Zero (400 yds)	Zero (500 yds)
0	3400	-1.5	-1.5	-1.5	-1.5	-1.5
100	3208	0.0	1.0	2.5	4.4	6.5
200	3024	-1.9	0.0	3.1	6.9	11.0
300	2848	-7.5	-4.6	0.0	5.7	11.9
400	2680	-17.5	-13.7	-7.6	0.0	8.3
500	2517	-32.3	-27.6	-19.9	-10.4	0.0
600	2360	-52.2	-46.5	-37.2	-25.9	-13.4
0	3300	-1.5	-1.5	-1.5	-1.5	-1.5
100	3112	0.0	1.0	2.7	4.7	6.9
200	2933	-2.0	0.0	3.4	7.4	11.8
300	2761	-8.2	-5.1	0.0	6.0	12.6
400	2595	-18.9	-14.9	-8.0	0.0	8.8
500	2435	-34.6	-29.5	-21.0	-11.0	0.0
600	2281	-56.5	-50.4	-40.1	-28.1	-14.9
0	3200	-1.5	-1.5	-1.5	-1.5	-1.5
100	3016	0.0	1.1	3.0	5.1	7.4
200	2841	-2.2	0.0	3.8	8.0	12.6
300	2672	-9.0	-5.7	0.0	6.3	13.2
400	2510	-20.4	-16.0	-8.4	0.0	9.2
500	2353	-37.0	-31.5	-22.0	-11.5	0.0
600	2202	-60.8	-54.2	-42.8	-30.2	-16.4
0	3100	-1.5	-1.5	-1.5	-1.5	-1.5
100	2921	0.0	1.3	3.3	5.5	8.1
200	2749	-2.6	0.0	3.9	8.4	13.6
300	2584	-9.9	-5.9	0.0	6.8	14.5
400	2424	-22.2	-16.9	-9.0	0.0	10.3
500	2271	-40.5	-33.9	-24.1	-12.8	0.0
600	2123	-65.7	-57.8	-46.0	-32.5	-17.1
0	3000	-1.5	-1.5	-1.5	-1.5	-1.5
100	2824	0.0	1.5	3.6	6.1	8.8
200	2657	-3.0	0.0	4.2	9.2	14.7
300	2495	-10.8	-6.3	0.0	7.4	15.7
400	2338	-24.2	-18.3	-9.9	0.0	11.0
500	2188	-44.1	-36.7	-26.1	-13.8	0.0
600	2043	-71.1	-62.2	-49.6	-34.8	-18.2
0	2900	-1.5	-1.5	-1.5	-1.5	-1.5
100	2729	0.0	1.6	3.9	6.6	9.5
200	2564	-3.2	0.0	4.5	9.9	15.8
300	2405	-11.6	-6.7	0.0	8.1	17.0
400	2252	-26.2	-19.8	-10.8	0.0	11.9
500	2105	-47.6	-39.6	-28.4	-14.8	0.0
600	1963	-76.4	-66.8	-53.3	-37.1	-19.3
0	2800	-1.5	-1.5	-1.5	-1.5	-1.5
100	2632	0.0	1.7	4.2	7.1	10.3
200	2471	-3.5	0.0	5.0	10.8	17.1
300	2316	-12.6	-7.4	0.0	8.7	18.2
400	2166	-28.5	-21.5	-11.6	0.0	12.7
500	2022	-51.4	-42.8	-30.4	-15.9	0.0
600	1882	-82.3	-71.9	-57.1	-39.6	-20.6
0	2700	-1.5	-1.5	-1.5	-1.5	-1.5
100	2536	0.0	1.9	4.7	7.8	11.2
200	2378	-3.8	0.0	5.6	11.8	18.6
300	2226	-14.0	-8.4	0.0	9.3	19.5
400	2080	-31.1	-23.6	-12.4	0.0	13.7
500	1938	-55.9	-46.5	-32.6	-17.1	0.0
600	1804	-90.3	-79.0	-62.3	-43.7	-23.2

Universal Table for Rifle Bullets with Ballistic Coefficients of .550-.559

Sight Above Bore (in.) +1.5

Range (yds)	Velocity (fps)	Bullet Path Zero (100 yds)	Zero (200 yds)	Zero (300 yds)	Zero (400 yds)	Zero (500 yds)
0	2600	-1.5	-1.5	-1.5	-1.5	-1.5
100	2440	0.0	2.2	5.2	8.5	12.3
200	2285	-4.5	0.0	5.9	12.6	20.1
300	2137	-15.5	-8.9	0.0	10.1	21.2
400	1993	-34.2	-25.3	-13.5	0.0	14.8
500	1856	-61.3	-50.1	-35.4	-18.5	0.0
600	1726	-99.2	-85.9	-68.2	-47.9	-25.7
0	2500	-1.5	-1.5	-1.5	-1.5	-1.5
100	2343	0.0	2.6	5.8	9.3	13.6
200	2192	-5.1	0.0	6.4	13.5	22.0
300	2047	-17.3	-9.6	0.0	10.6	23.4
400	1907	-37.3	-27.0	-14.2	0.0	17.0
500	1774	-67.8	-55.0	-38.9	-21.2	0.0
600	1649	-109.3	-93.9	-74.6	-53.4	-27.9
0	2400	-1.5	-1.5	-1.5	-1.5	-1.5
100	2247	0.0	2.7	6.2	10.2	14.8
200	2100	-5.4	0.0	6.9	15.0	24.2
300	1958	-18.5	-10.4	0.0	12.2	26.0
400	1822	-40.9	-30.0	-16.2	0.0	18.4
500	1694	-74.1	-60.5	-43.3	-23.0	0.0
600	1573	-119.2	-102.9	-82.2	-57.9	-30.3
0	2300	-1.5	-1.5	-1.5	-1.5	-1.5
100	2150	0.0	2.9	6.7	11.3	16.4
200	2007	-5.9	0.0	7.5	16.7	26.8
300	1867	-20.1	-11.2	0.0	13.8	29.0
400	1738	-45.1	-33.3	-18.4	0.0	20.3
500	1614	-81.8	-67.1	-48.3	-25.4	0.0
600	1497	-130.6	-112.9	-90.4	-62.9	-32.4
0	2200	-1.5	-1.5	-1.5	-1.5	-1.5
100	2054	0.0	3.3	7.7	12.6	18.1
200	1913	-6.5	0.0	8.8	18.6	29.6
300	1780	-23.0	-13.2	0.0	14.8	31.3
400	1655	-50.3	-37.3	-19.7	0.0	22.0
500	1535	-90.5	-74.1	-52.2	-27.5	0.0
600	1422	-142.9	-123.3	-96.9	-67.4	-34.3
0	2100	-1.5	-1.5	-1.5	-1.5	-1.5
100	1957	0.0	3.7	8.6	14.0	19.9
200	1822	-7.5	0.0	9.7	20.5	32.3
300	1694	-25.8	-14.6	0.0	16.3	33.9
400	1573	-56.1	-41.1	-21.7	0.0	23.6
500	1457	-99.6	-80.8	-56.6	-29.5	0.0
600	1356	-161.0	-138.5	-109.4	-76.9	-41.5
0	2000	-1.5	-1.5	-1.5	-1.5	-1.5
100	1861	0.0	4.3	9.7	15.6	22.4
200	1731	-8.7	0.0	10.7	22.5	36.1
300	1608	-29.1	-16.1	0.0	17.6	38.1
400	1491	-62.3	-44.9	-23.5	0.0	27.3
500	1385	-111.9	-90.2	-63.5	-34.1	0.0
600	1290	-179.5	-153.5	-121.4	-86.1	-45.2

Universal Table for Rifle Bullets with Ballistic Coefficients of .560-.569

Sight Above Bore (in.) +1.5

		Bullet Path				
Range (yds)	Velocity (fps)	Zero (100 yds)	Zero (200 yds)	Zero (300 yds)	Zero (400 yds)	Zero (500 yds)
0	3400	-1.5	-1.5	-1.5	-1.5	-1.5
100	3211	0.0	0.9	2.5	4.4	6.4
200	3031	-1.9	0.0	3.1	6.9	11.0
300	2858	-7.4	-4.6	0.0	5.8	11.9
400	2691	-17.5	-13.8	-7.7	0.0	8.2
500	2531	-32.2	-27.5	-19.9	-10.3	0.0
600	2376	-51.9	-46.4	-37.2	-25.7	-13.3
0	3300	-1.5	-1.5	-1.5	-1.5	-1.5
100	3115	0.0	1.0	2.7	4.7	6.9
200	2939	-2.0	0.0	3.4	7.4	11.7
300	2770	-8.1	-5.1	0.0	6.0	12.5
400	2607	-18.8	-14.7	-7.9	0.0	8.8
500	2449	-34.4	-29.4	-20.9	-11.0	0.0
600	2297	-56.1	-50.0	-39.8	-27.9	-14.7
0	3200	-1.5	-1.5	-1.5	-1.5	-1.5
100	3020	0.0	1.1	3.0	5.1	7.4
200	2847	-2.2	0.0	3.8	8.0	12.5
300	2681	-9.0	-5.6	0.0	6.3	13.2
400	2521	-20.4	-16.0	-8.5	0.0	9.1
500	2367	-36.9	-31.3	-21.9	-11.4	0.0
600	2218	-60.4	-53.8	-42.5	-29.8	-16.2
0	3100	-1.5	-1.5	-1.5	-1.5	-1.5
100	2924	0.0	1.3	3.3	5.5	8.1
200	2755	-2.6	0.0	4.0	8.4	13.5
300	2592	-9.9	-5.9	0.0	6.7	14.4
400	2435	-22.1	-16.8	-8.9	0.0	10.2
500	2284	-40.3	-33.8	-23.9	-12.7	0.0
600	2138	-65.4	-57.5	-45.7	-32.3	-17.0
0	3000	-1.5	-1.5	-1.5	-1.5	-1.5
100	2827	0.0	1.5	3.6	5.9	8.8
200	2663	-2.9	0.0	4.2	8.9	14.6
300	2503	-10.8	-6.4	0.0	7.1	15.5
400	2349	-23.8	-17.9	-9.4	0.0	11.3
500	2201	-43.9	-36.5	-25.9	-14.1	0.0
600	2058	-70.7	-61.9	-49.2	-35.1	-18.1
0	2900	-1.5	-1.5	-1.5	-1.5	-1.5
100	2732	0.0	1.6	3.8	6.5	9.5
200	2570	-3.2	0.0	4.5	9.8	15.7
300	2414	-11.5	-6.7	0.0	8.1	16.9
400	2263	-26.0	-19.7	-10.8	0.0	11.8
500	2118	-47.3	-39.4	-28.2	-14.8	0.0
600	1978	-76.0	-66.4	-53.0	-36.9	-19.2
0	2800	-1.5	-1.5	-1.5	-1.5	-1.5
100	2635	0.0	1.7	4.2	7.1	10.3
200	2477	-3.4	0.0	5.0	10.7	17.1
300	2324	-12.6	-7.4	0.0	8.6	18.2
400	2177	-28.3	-21.4	-11.5	0.0	12.8
500	2034	-51.3	-42.7	-30.3	-16.0	0.0
600	1897	-81.8	-71.5	-56.6	-39.4	-20.2
0	2700	-1.5	-1.5	-1.5	-1.5	-1.5
100	2539	0.0	1.9	4.7	7.8	11.1
200	2384	-3.7	0.0	5.6	11.8	18.5
300	2234	-14.0	-8.4	0.0	9.3	19.4
400	2090	-31.0	-23.5	-12.4	0.0	13.4
500	1951	-55.6	-46.2	-32.3	-16.8	0.0
600	1818	-89.7	-78.4	-61.7	-43.1	-23.0

Universal Table for Rifle Bullets with Ballistic Coefficients of .560-.569						
					Sight Above Bore (in.)	+1.5
			Bullet Path			
Range (yds)	Velocity (fps)	Zero (100 yds)	Zero (200 yds)	Zero (300 yds)	Zero (400 yds)	Zero (500 yds)
0	2600	-1.5	-1.5	-1.5	-1.5	-1.5
100	2442	0.0	2.2	5.1	8.5	12.0
200	2291	-4.3	0.0	5.9	12.6	19.7
300	2145	-15.4	-8.9	0.0	10.1	20.7
400	2003	-34.0	-25.3	-13.5	0.0	14.2
500	1867	-60.2	-49.3	-34.6	-17.7	0.0
600	1739	-98.5	-85.5	-67.8	-47.6	-26.3
0	2500	-1.5	-1.5	-1.5	-1.5	-1.5
100	2346	0.0	2.5	5.7	9.3	13.5
200	2198	-5.1	0.0	6.4	13.5	21.8
300	2055	-17.2	-9.6	0.0	10.6	23.1
400	1917	-37.1	-27.0	-14.2	0.0	16.7
500	1786	-67.3	-54.6	-38.5	-20.8	0.0
600	1662	-108.6	-93.4	-74.1	-52.9	-27.9
0	2400	-1.5	-1.5	-1.5	-1.5	-1.5
100	2249	0.0	2.7	6.1	10.2	14.7
200	2105	-5.4	0.0	6.9	14.9	24.0
300	1965	-18.4	-10.4	0.0	12.0	25.7
400	1831	-40.6	-29.9	-16.1	0.0	18.2
500	1705	-73.5	-60.1	-42.8	-22.7	0.0
600	1585	-118.5	-102.4	-81.7	-57.6	-30.3
0	2300	-1.5	-1.5	-1.5	-1.5	-1.5
100	2153	0.0	3.0	6.7	11.2	16.3
200	2012	-5.9	0.0	7.4	16.6	26.6
300	1875	-20.0	-11.1	0.0	13.7	28.8
400	1747	-45.0	-33.1	-18.3	0.0	20.1
500	1625	-81.3	-66.5	-48.0	-25.1	0.0
600	1509	-130.0	-112.3	-90.1	-62.6	-32.5
0	2200	-1.5	-1.5	-1.5	-1.5	-1.5
100	2057	0.0	3.3	7.6	12.6	18.0
200	1918	-6.6	0.0	8.7	18.6	29.4
300	1787	-22.9	-13.1	0.0	14.9	31.0
400	1663	-50.4	-37.3	-19.9	0.0	21.5
500	1546	-89.8	-73.4	-51.6	-26.8	0.0
600	1434	-142.2	-122.5	-96.4	-66.6	-34.4
0	2100	-1.5	-1.5	-1.5	-1.5	-1.5
100	1960	0.0	3.8	8.6	14.0	19.9
200	1826	-7.6	0.0	9.7	20.4	32.1
300	1700	-25.9	-14.5	0.0	16.1	33.6
400	1581	-56.1	-40.8	-21.5	0.0	23.3
500	1467	-99.3	-80.2	-56.1	-29.2	0.0
600	1366	-160.3	-137.4	-108.4	-76.1	-41.1
0	2000	-1.5	-1.5	-1.5	-1.5	-1.5
100	1863	0.0	4.3	9.7	15.5	22.2
200	1735	-8.7	0.0	10.7	22.3	35.7
300	1614	-29.0	-16.0	0.0	17.5	37.6
400	1499	-62.0	-44.6	-23.3	0.0	26.9
500	1394	-111.1	-89.4	-62.7	-33.6	0.0
600	1299	-178.8	-152.8	-120.8	-85.8	-45.5

Universal Table for Rifle Bullets with Ballistic Coefficients of .570-.579

Sight Above Bore (in.) +1.5

Range (yds)	Velocity (fps)	Bullet Path Zero (100 yds)	Zero (200 yds)	Zero (300 yds)	Zero (400 yds)	Zero (500 yds)
0	3400	-1.5	-1.5	-1.5	-1.5	-1.5
100	3214	0.0	0.9	2.5	4.3	6.4
200	3037	-1.9	0.0	3.1	6.8	10.9
300	2866	-7.4	-4.6	0.0	5.6	11.8
400	2703	-17.3	-13.6	-7.5	0.0	8.2
500	2545	-32.0	-27.3	-19.6	-10.3	0.0
600	2392	-51.7	-46.1	-36.9	-25.6	-13.3
0	3300	-1.5	-1.5	-1.5	-1.5	-1.5
100	3119	0.0	1.0	2.7	4.7	6.9
200	2945	-2.1	0.0	3.4	7.3	11.7
300	2778	-8.2	-5.1	0.0	5.9	12.4
400	2618	-18.8	-14.7	-7.9	0.0	8.7
500	2463	-34.3	-29.2	-20.6	-10.8	0.0
600	2313	-55.7	-49.6	-39.4	-27.6	-14.6
0	3200	-1.5	-1.5	-1.5	-1.5	-1.5
100	3023	0.0	1.1	3.0	5.1	7.4
200	2853	-2.2	0.0	3.7	7.9	12.5
300	2690	-8.9	-5.6	0.0	6.3	13.2
400	2532	-20.3	-15.9	-8.5	0.0	9.1
500	2380	-36.8	-31.2	-21.9	-11.4	0.0
600	2233	-60.1	-53.5	-42.3	-29.6	-16.0
0	3100	-1.5	-1.5	-1.5	-1.5	-1.5
100	2927	0.0	1.3	3.3	5.5	8.0
200	2761	-2.6	0.0	3.9	8.4	13.4
300	2601	-9.8	-5.9	0.0	6.7	14.3
400	2446	-22.0	-16.8	-9.0	0.0	10.1
500	2297	-40.1	-33.6	-23.8	-12.6	0.0
600	2153	-65.0	-57.2	-45.5	-32.0	-16.8
0	3000	-1.5	-1.5	-1.5	-1.5	-1.5
100	2830	0.0	1.5	3.6	5.9	8.7
200	2668	-3.0	0.0	4.2	8.9	14.5
300	2511	-10.8	-6.3	0.0	7.0	15.4
400	2360	-23.6	-17.7	-9.3	0.0	11.2
500	2214	-43.6	-36.1	-25.6	-14.0	0.0
600	2073	-70.2	-61.3	-48.7	-34.8	-18.0
0	2900	-1.5	-1.5	-1.5	-1.5	-1.5
100	2734	0.0	1.6	3.8	6.5	9.4
200	2575	-3.2	0.0	4.4	9.8	15.7
300	2422	-11.4	-6.6	0.0	8.0	16.9
400	2273	-25.9	-19.5	-10.7	0.0	11.8
500	2130	-47.0	-39.1	-28.1	-14.7	0.0
600	1992	-75.5	-66.0	-52.8	-36.7	-19.1
0	2800	-1.5	-1.5	-1.5	-1.5	-1.5
100	2638	0.0	1.7	4.2	7.0	10.2
200	2482	-3.5	0.0	4.9	10.6	16.9
300	2332	-12.5	-7.3	0.0	8.6	18.0
400	2187	-28.2	-21.2	-11.4	0.0	12.6
500	2047	-50.9	-42.2	-30.0	-15.7	0.0
600	1911	-81.5	-71.1	-56.4	-39.2	-20.4
0	2700	-1.5	-1.5	-1.5	-1.5	-1.5
100	2542	0.0	1.9	4.6	7.7	11.1
200	2389	-3.8	0.0	5.5	11.7	18.4
300	2242	-13.9	-8.2	0.0	9.3	19.3
400	2100	-30.9	-23.3	-12.4	0.0	13.4
500	1963	-55.3	-45.9	-32.2	-16.7	0.0
600	1831	-89.3	-77.9	-61.5	-42.9	-22.9

Universal Table for Rifle Bullets with Ballistic Coefficients of .570-.579

Sight Above Bore (in.) +1.5

Range (yds)	Velocity (fps)	Bullet Path: Zero (100 yds)	Zero (200 yds)	Zero (300 yds)	Zero (400 yds)	Zero (500 yds)
0	2600	-1.5	-1.5	-1.5	-1.5	-1.5
100	2445	0.0	2.2	5.1	8.4	12.0
200	2296	-4.3	0.0	5.9	12.5	19.7
300	2152	-15.4	-8.9	0.0	10.0	20.7
400	2013	-33.8	-25.1	-13.3	0.0	14.3
500	1878	-60.2	-49.3	-34.5	-17.9	0.0
600	1752	-97.9	-84.8	-67.1	-47.2	-25.7
0	2500	-1.5	-1.5	-1.5	-1.5	-1.5
100	2348	0.0	2.5	5.7	9.3	13.4
200	2203	-5.0	0.0	6.4	13.5	21.7
300	2062	-17.2	-9.6	0.0	10.6	22.9
400	1926	-37.0	-27.0	-14.1	0.0	16.4
500	1797	-66.8	-54.2	-38.2	-20.5	0.0
600	1674	-108.0	-92.9	-73.6	-52.4	-27.8
0	2400	-1.5	-1.5	-1.5	-1.5	-1.5
100	2252	0.0	2.7	6.2	10.1	14.6
200	2110	-5.4	0.0	6.9	14.9	23.8
300	1972	-18.5	-10.4	0.0	11.9	25.4
400	1840	-40.5	-29.7	-15.9	0.0	17.9
500	1716	-73.1	-59.6	-42.3	-22.4	0.0
600	1597	-118.0	-101.8	-81.1	-57.2	-30.3
0	2300	-1.5	-1.5	-1.5	-1.5	-1.5
100	2156	0.0	3.0	6.7	11.3	16.2
200	2016	-6.0	0.0	7.3	16.5	26.4
300	1882	-20.0	-10.9	0.0	13.8	28.6
400	1755	-45.0	-32.9	-18.3	0.0	19.8
500	1635	-81.1	-66.0	-47.7	-24.8	0.0
600	1521	-129.3	-111.2	-89.3	-61.8	-32.1
0	2200	-1.5	-1.5	-1.5	-1.5	-1.5
100	2059	0.0	3.2	7.6	12.5	17.8
200	1923	-6.5	0.0	8.6	18.4	29.2
300	1794	-22.7	-12.9	0.0	14.7	30.9
400	1672	-49.8	-36.9	-19.6	0.0	21.5
500	1556	-89.2	-73.0	-51.5	-26.9	0.0
600	1445	-141.6	-122.2	-96.3	-66.8	-34.5
0	2100	-1.5	-1.5	-1.5	-1.5	-1.5
100	1962	0.0	3.7	8.5	13.9	19.7
200	1831	-7.5	0.0	9.6	20.4	32.0
300	1707	-25.6	-14.4	0.0	16.2	33.6
400	1589	-55.8	-40.8	-21.7	0.0	23.1
500	1477	-98.6	-79.9	-55.9	-28.9	0.0
600	1376	-159.1	-136.7	-107.9	-75.4	-40.8
0	2000	-1.5	-1.5	-1.5	-1.5	-1.5
100	1866	0.0	4.3	9.7	15.5	21.7
200	1740	-8.6	0.0	10.8	22.3	34.7
300	1620	-29.1	-16.1	0.0	17.3	36.0
400	1507	-61.9	-44.6	-23.1	0.0	24.8
500	1398	-108.4	-86.9	-59.9	-31.1	0.0
600	1309	-177.3	-151.4	-119.2	-84.5	-47.2

Universal Table for Rifle Bullets with Ballistic Coefficients of .580-.589

Sight Above Bore (in.) +1.5

Range (yds)	Velocity (fps)	Bullet Path Zero (100 yds)	Zero (200 yds)	Zero (300 yds)	Zero (400 yds)	Zero (500 yds)
0	3400	-1.5	-1.5	-1.5	-1.5	-1.5
100	3217	0.0	0.9	2.4	4.3	6.4
200	3043	-1.8	0.0	3.1	6.8	10.9
300	2875	-7.3	-4.6	0.0	5.6	11.8
400	2714	-17.3	-13.6	-7.5	0.0	8.2
500	2558	-31.8	-27.2	-19.6	-10.3	0.0
600	2408	-51.3	-45.8	-36.6	-25.4	-13.1
0	3300	-1.5	-1.5	-1.5	-1.5	-1.5
100	3122	0.0	1.0	2.7	4.7	6.8
200	2951	-2.0	0.0	3.4	7.3	11.6
300	2787	-8.1	-5.0	0.0	5.9	12.4
400	2629	-18.7	-14.6	-7.9	0.0	8.7
500	2476	-34.2	-29.0	-20.6	-10.8	0.0
600	2328	-55.4	-49.3	-39.2	-27.4	-14.4
0	3200	-1.5	-1.5	-1.5	-1.5	-1.5
100	3026	0.0	1.1	3.0	5.1	7.3
200	2859	-2.2	0.0	3.7	7.9	12.4
300	2698	-8.9	-5.6	0.0	6.3	13.1
400	2543	-20.2	-15.8	-8.4	0.0	9.1
500	2393	-36.6	-31.1	-21.8	-11.3	0.0
600	2248	-59.7	-53.1	-41.9	-29.4	-15.0
0	3100	-1.5	-1.5	-1.5	-1.5	-1.5
100	2930	0.0	1.3	3.2	5.5	8.0
200	2766	-2.6	0.0	3.9	8.3	13.4
300	2609	-9.7	-5.8	0.0	6.6	14.3
400	2457	-21.8	-16.6	-8.8	0.0	10.2
500	2309	-40.0	-33.5	-23.8	-12.7	0.0
600	2168	-64.5	-56.6	-45.0	-31.7	-16.5
0	3000	-1.5	-1.5	-1.5	-1.5	-1.5
100	2833	0.0	1.5	3.6	5.9	8.7
200	2674	-2.9	0.0	4.2	8.9	14.4
300	2519	-10.7	-6.3	0.0	7.0	15.3
400	2370	-23.6	-17.7	-9.3	0.0	11.1
500	2226	-43.4	-36.1	-25.5	-13.9	0.0
600	2087	-69.9	-61.1	-48.5	-34.5	-17.8
0	2900	-1.5	-1.5	-1.5	-1.5	-1.5
100	2737	0.0	1.6	3.8	6.4	9.4
200	2581	-3.1	0.0	4.5	9.8	15.6
300	2429	-11.4	-6.7	0.0	7.9	16.7
400	2283	-25.8	-19.5	-10.5	0.0	11.7
500	2142	-46.8	-39.0	-27.8	-14.6	0.0
600	2006	-75.1	-65.8	-52.3	-36.5	-18.9
0	2800	-1.5	-1.5	-1.5	-1.5	-1.5
100	2641	0.0	1.7	4.2	7.0	10.2
200	2488	-3.4	0.0	5.0	10.6	16.9
300	2339	-12.6	-7.4	0.0	8.4	17.9
400	2197	-28.0	-21.2	-11.2	0.0	12.7
500	2058	-50.9	-42.3	-29.9	-15.9	0.0
600	1925	-81.0	-70.7	-55.9	-39.0	-20.0
0	2700	-1.5	-1.5	-1.5	-1.5	-1.5
100	2544	0.0	1.9	4.6	7.6	11.0
200	2394	-3.7	0.0	5.5	11.6	18.3
300	2249	-13.8	-8.2	0.0	9.1	19.2
400	2110	-30.6	-23.1	-12.2	0.0	13.5
500	1974	-55.1	-45.8	-32.1	-16.9	0.0
600	1844	-88.6	-77.4	-60.9	-42.7	-22.4

Universal Table for Rifle Bullets with Ballistic Coefficients of .580-.589

Sight Above Bore (in.) +1.5

Range (yds)	Velocity (fps)	Bullet Path Zero (100 yds)	Zero (200 yds)	Zero (300 yds)	Zero (400 yds)	Zero (500 yds)
0	2600	-1.5	-1.5	-1.5	-1.5	-1.5
100	2448	0.0	2.2	5.1	8.4	12.0
200	2301	-4.4	0.0	5.9	12.5	19.6
300	2159	-15.4	-8.9	0.0	9.9	20.5
400	2022	-33.8	-25.1	-13.2	0.0	14.1
500	1890	-59.8	-48.9	-34.1	-17.6	0.0
600	1765	-97.2	-84.1	-66.3	-46.5	-25.4
0	2500	-1.5	-1.5	-1.5	-1.5	-1.5
100	2351	0.0	2.5	5.7	9.3	13.3
200	2208	-5.0	0.0	6.4	13.5	21.7
300	2069	-17.2	-9.6	0.0	10.6	22.8
400	1935	-37.0	-27.0	-14.1	0.0	16.4
500	1807	-66.7	-54.1	-38.1	-20.4	0.0
600	1687	-107.0	-92.0	-72.7	-51.5	-27.0
0	2400	-1.5	-1.5	-1.5	-1.5	-1.5
100	2254	0.0	2.7	6.1	10.0	14.5
200	2115	-5.3	0.0	6.9	14.8	23.8
300	1979	-18.3	-10.4	0.0	11.8	25.3
400	1849	-40.2	-29.6	-15.7	0.0	18.0
500	1726	-72.7	-59.4	-42.1	-22.4	0.0
600	1609	-117.0	-101.1	-80.4	-56.8	-29.9
0	2300	-1.5	-1.5	-1.5	-1.5	-1.5
100	2158	0.0	3.0	6.6	11.1	16.1
200	2021	-6.0	0.0	7.3	16.3	26.3
300	1889	-19.9	-10.9	0.0	13.5	28.5
400	1764	-44.5	-32.6	-18.1	0.0	19.9
500	1645	-80.5	-65.6	-47.4	-24.8	0.0
600	1532	-128.8	-110.9	-89.0	-61.9	-32.1
0	2200	-1.5	-1.5	-1.5	-1.5	-1.5
100	2061	0.0	3.2	7.5	12.4	17.8
200	1928	-6.4	0.0	8.7	18.4	29.2
300	1800	-22.6	-13.0	0.0	14.6	30.8
400	1680	-49.6	-36.8	-19.4	0.0	21.6
500	1565	-89.0	-73.1	-51.3	-27.0	0.0
600	1456	-140.9	-121.8	-95.7	-66.5	-34.1
0	2100	-1.5	-1.5	-1.5	-1.5	-1.5
100	1965	0.0	3.8	8.6	13.9	19.7
200	1835	-7.6	0.0	9.5	20.2	31.8
300	1713	-25.7	-14.3	0.0	16.1	33.5
400	1597	-55.7	-40.5	-21.4	0.0	23.2
500	1486	-98.6	-79.6	-55.8	-29.0	0.0
600	1386	-158.1	-135.4	-106.8	-74.6	-39.8
0	2000	-1.5	-1.5	-1.5	-1.5	-1.5
100	1868	0.0	4.3	9.6	15.5	21.6
200	1744	-8.6	0.0	10.7	22.3	34.6
300	1626	-28.9	-16.0	0.0	17.5	35.9
400	1514	-61.9	-44.7	-23.3	0.0	24.6
500	1407	-108.1	-86.5	-59.8	-30.7	0.0
600	1318	-176.3	-150.5	-118.4	-83.5	-46.6

Universal Table for Rifle Bullets with Ballistic Coefficients of .590-.599

Sight Above Bore (in.) +1.5

Range (yds)	Velocity (fps)	Bullet Path Zero (100 yds)	Zero (200 yds)	Zero (300 yds)	Zero (400 yds)	Zero (500 yds)
0	3400	-1.5	-1.5	-1.5	-1.5	-1.5
100	3220	0.0	0.9	2.5	4.3	6.3
200	3049	-1.8	0.0	3.1	6.8	10.8
300	2883	-7.4	-4.6	0.0	5.6	11.7
400	2724	-17.3	-13.6	-7.5	0.0	8.1
500	2571	-31.7	-27.1	-19.4	-10.1	0.0
600	2422	-51.2	-45.7	-36.5	-25.4	-13.2
0	3300	-1.5	-1.5	-1.5	-1.5	-1.5
100	3125	0.0	1.0	2.7	4.7	6.8
200	2956	-2.1	0.0	3.3	7.3	11.6
300	2795	-8.1	-5.0	0.0	5.9	12.4
400	2639	-18.6	-14.5	-7.9	0.0	8.6
500	2488	-34.1	-28.9	-20.6	-10.8	0.0
600	2343	-55.0	-48.8	-38.8	-27.1	-14.1
0	3200	-1.5	-1.5	-1.5	-1.5	-1.5
100	3029	0.0	1.1	3.0	5.0	7.3
200	2864	-2.2	0.0	3.7	7.9	12.4
300	2706	-8.9	-5.5	0.0	6.3	13.1
400	2553	-20.2	-15.7	-8.4	0.0	9.1
500	2405	-36.5	-31.0	-21.8	-11.3	0.0
600	2262	-59.4	-52.8	-41.7	-29.2	-15.6
0	3100	-1.5	-1.5	-1.5	-1.5	-1.5
100	2932	0.0	1.3	3.2	5.4	7.9
200	2772	-2.5	0.0	3.9	8.3	13.3
300	2617	-9.6	-5.8	0.0	6.6	14.1
400	2467	-21.7	-16.6	-8.8	0.0	10.0
500	2322	-39.6	-33.2	-23.5	-12.4	0.0
600	2182	-64.0	-56.4	-44.7	-31.5	-16.6
0	3000	-1.5	-1.5	-1.5	-1.5	-1.5
100	2836	0.0	1.5	3.6	5.9	8.6
200	2679	-3.0	0.0	4.2	8.8	14.3
300	2527	-10.7	-6.2	0.0	7.0	15.2
400	2380	-23.6	-17.6	-9.3	0.0	11.0
500	2238	-43.2	-35.8	-25.4	-13.7	0.0
600	2101	-69.5	-60.7	-48.2	-34.2	-17.7
0	2900	-1.5	-1.5	-1.5	-1.5	-1.5
100	2740	0.0	1.6	3.8	6.4	9.3
200	2586	-3.1	0.0	4.4	9.7	15.5
300	2437	-11.4	-6.7	0.0	7.9	16.6
400	2293	-25.6	-19.3	-10.5	0.0	11.6
500	2154	-46.6	-38.7	-27.6	-14.5	0.0
600	2020	-74.7	-65.2	-51.9	-36.2	-18.8
0	2800	-1.5	-1.5	-1.5	-1.5	-1.5
100	2643	0.0	1.7	4.1	7.0	10.1
200	2493	-3.4	0.0	4.9	10.6	16.8
300	2347	-12.4	-7.3	0.0	8.5	17.8
400	2206	-27.9	-21.1	-11.3	0.0	12.5
500	2070	-50.4	-42.0	-29.7	-15.6	0.0
600	1938	-80.6	-70.5	-55.8	-38.8	-20.1
0	2700	-1.5	-1.5	-1.5	-1.5	-1.5
100	2547	0.0	1.9	4.6	7.6	11.0
200	2399	-3.8	0.0	5.4	11.5	18.2
300	2257	-13.7	-8.1	0.0	9.2	19.3
400	2119	-30.6	-23.1	-12.3	0.0	13.4
500	1985	-55.0	-45.6	-32.2	-16.8	0.0
600	1857	-88.0	-76.8	-60.6	-42.2	-22.0

Universal Table for Rifle Bullets with Ballistic Coefficients of .590-.599						
					Sight Above Bore (in.)	+1.5
			Bullet Path			
Range (yds)	Velocity (fps)	Zero (100 yds)	Zero (200 yds)	Zero (300 yds)	Zero (400 yds)	Zero (500 yds)
0	2600	-1.5	-1.5	-1.5	-1.5	-1.5
100	2450	0.0	2.1	5.1	8.4	11.9
200	2306	-4.3	0.0	5.9	12.5	19.5
300	2166	-15.3	-8.9	0.0	9.9	20.4
400	2031	-33.6	-25.0	-13.2	0.0	14.0
500	1901	-59.5	-48.8	-34.0	-17.5	0.0
600	1777	-96.5	-83.7	-65.9	-46.1	-25.1
0	2500	-1.5	-1.5	-1.5	-1.5	-1.5
100	2353	0.0	2.5	5.7	9.2	13.2
200	2212	-5.0	0.0	6.3	13.4	21.4
300	2076	-17.0	-9.5	0.0	10.5	22.6
400	1944	-36.8	-26.7	-14.1	0.0	16.1
500	1818	-66.1	-53.5	-37.7	-20.1	0.0
600	1699	-106.2	-91.1	-72.1	-51.0	-26.9
0	2400	-1.5	-1.5	-1.5	-1.5	-1.5
100	2257	0.0	2.7	6.1	10.1	14.5
200	2119	-5.4	0.0	6.8	14.7	23.6
300	1986	-18.3	-10.2	0.0	11.8	25.1
400	1857	-40.2	-29.4	-15.8	0.0	17.7
500	1736	-72.4	-58.9	-41.9	-22.1	0.0
600	1621	-116.3	-100.0	-79.6	-55.9	-29.4
0	2300	-1.5	-1.5	-1.5	-1.5	-1.5
100	2160	0.0	2.9	6.6	11.1	16.0
200	2026	-5.9	0.0	7.4	16.3	26.1
300	1895	-19.9	-11.1	0.0	13.4	28.1
400	1772	-44.3	-32.6	-17.8	0.0	19.6
500	1655	-79.9	-65.3	-46.8	-24.5	0.0
600	1544	-127.5	-109.9	-87.7	-61.0	-31.6
0	2200	-1.5	-1.5	-1.5	-1.5	-1.5
100	2064	0.0	3.2	7.5	12.4	17.7
200	1932	-6.5	0.0	8.5	18.2	28.9
300	1807	-22.5	-12.7	0.0	14.6	30.6
400	1688	-49.5	-36.5	-19.5	0.0	21.3
500	1575	-88.5	-72.2	-51.1	-26.7	0.0
600	1467	-140.3	-120.8	-95.4	-66.1	-34.1
0	2100	-1.5	-1.5	-1.5	-1.5	-1.5
100	1967	0.0	3.8	8.5	13.8	19.6
200	1839	-7.6	0.0	9.5	20.1	31.7
300	1719	-25.5	-14.2	0.0	15.9	33.4
400	1605	-55.2	-40.1	-21.2	0.0	23.3
500	1495	-98.1	-79.2	-55.6	-29.1	0.0
600	1396	-156.7	-134.1	-105.7	-73.9	-39.0
0	2000	-1.5	-1.5	-1.5	-1.5	-1.5
100	1870	0.0	4.3	9.6	15.3	21.5
200	1748	-8.6	0.0	10.6	22.1	34.5
300	1632	-28.8	-15.9	0.0	17.3	35.8
400	1522	-61.4	-44.2	-23.0	0.0	24.7
500	1416	-107.6	-86.1	-59.6	-30.9	0.0
600	1327	-175.1	-149.4	-117.5	-83.0	-46.0

Universal Table for Rifle Bullets with Ballistic Coefficients of .600-.609

Sight Above Bore (in.) +1.5

Range (yds)	Velocity (fps)	Bullet Path Zero (100 yds)	Zero (200 yds)	Zero (300 yds)	Zero (400 yds)	Zero (500 yds)
0	3400	-1.5	-1.5	-1.5	-1.5	-1.5
100	3246	0.0	0.9	2.4	4.2	6.1
200	3097	-1.8	0.0	3.0	6.5	10.4
300	2954	-7.2	-4.4	0.0	5.3	11.1
400	2815	-16.7	-13.0	-7.1	0.0	7.7
500	2681	-30.5	-25.9	-18.5	-9.6	0.0
600	2550	-49.2	-43.7	-34.8	-24.2	-12.6
0	3300	-1.5	-1.5	-1.5	-1.5	-1.5
100	3149	0.0	1.0	2.6	4.5	6.6
200	3004	-2.0	0.0	3.1	7.0	11.1
300	2864	-7.7	-4.7	0.0	5.8	12.0
400	2728	-17.9	-14.0	-7.7	0.0	8.3
500	2596	-32.8	-27.8	-20.0	-10.3	0.0
600	2468	-52.5	-46.5	-37.2	-25.6	-13.2
0	3200	-1.5	-1.5	-1.5	-1.5	-1.5
100	3053	0.0	1.1	2.8	4.9	7.1
200	2911	-2.2	0.0	3.5	7.6	11.9
300	2774	-8.5	-5.3	0.0	6.1	12.6
400	2640	-19.5	-15.2	-8.2	0.0	8.7
500	2511	-35.3	-29.9	-21.1	-10.9	0.0
600	2385	-56.2	-49.7	-39.2	-27.0	-13.9
0	3100	-1.5	-1.5	-1.5	-1.5	-1.5
100	2956	0.0	1.3	3.1	5.3	7.6
200	2817	-2.5	0.0	3.7	8.1	12.7
300	2683	-9.4	-5.6	0.0	6.5	13.4
400	2552	-21.2	-16.1	-8.6	0.0	9.2
500	2425	-38.0	-31.7	-22.3	-11.5	0.0
600	2302	-61.1	-53.5	-42.3	-29.3	-15.5
0	3000	-1.5	-1.5	-1.5	-1.5	-1.5
100	2859	0.0	1.4	3.5	5.7	8.2
200	2724	-2.8	0.0	4.1	8.6	13.7
300	2592	-10.4	-6.1	0.0	6.8	14.4
400	2464	-22.9	-17.2	-9.1	0.0	10.1
500	2340	-41.2	-34.1	-23.9	-12.6	0.0
600	2219	-66.2	-57.7	-45.4	-31.8	-16.7
0	2900	-1.5	-1.5	-1.5	-1.5	-1.5
100	2763	0.0	1.5	3.7	6.1	8.9
200	2630	-3.1	0.0	4.3	9.2	14.7
300	2501	-11.1	-6.5	0.0	7.3	15.6
400	2375	-24.5	-18.3	-9.7	0.0	11.1
500	2254	-44.6	-36.8	-26.0	-13.9	0.0
600	2135	-71.6	-62.3	-49.3	-34.8	-18.1
0	2800	-1.5	-1.5	-1.5	-1.5	-1.5
100	2666	0.0	1.7	4.0	6.8	9.7
200	2536	-3.4	0.0	4.7	10.1	16.1
300	2409	-12.0	-7.0	0.0	8.2	17.1
400	2286	-27.1	-20.3	-11.0	0.0	11.9
500	2167	-48.7	-40.2	-28.6	-14.8	0.0
600	2052	-77.3	-67.2	-53.2	-36.7	-18.9
0	2700	-1.5	-1.5	-1.5	-1.5	-1.5
100	2568	0.0	1.8	4.4	7.4	10.5
200	2441	-3.7	0.0	5.2	11.1	17.4
300	2317	-13.3	-7.8	0.0	8.8	18.3
400	2197	-29.5	-22.2	-11.8	0.0	12.7
500	2081	-52.7	-43.6	-30.5	-15.8	0.0
600	1968	-83.5	-72.5	-56.8	-39.2	-20.2

Universal Table for Rifle Bullets with Ballistic Coefficients of .600-.609

Sight Above Bore (in.) +1.5

Range (yds)	Velocity (fps)	Bullet Path: Zero (100 yds)	Zero (200 yds)	Zero (300 yds)	Zero (400 yds)	Zero (500 yds)
0	2600	-1.5	-1.5	-1.5	-1.5	-1.5
100	2471	0.0	2.1	4.9	8.1	11.5
200	2347	-4.1	0.0	5.7	12.0	18.9
300	2226	-14.7	-8.5	0.0	9.5	19.9
400	2109	-32.3	-24.0	-12.6	0.0	13.9
500	1994	-57.7	-47.3	-33.1	-17.3	0.0
600	1884	-91.2	-78.8	-61.8	-42.8	-22.1
0	2500	-1.5	-1.5	-1.5	-1.5	-1.5
100	2374	0.0	2.5	5.5	8.9	12.6
200	2252	-4.9	0.0	6.2	12.9	20.2
300	2134	-16.6	-9.3	0.0	10.1	21.0
400	2020	-35.7	-25.8	-13.5	0.0	14.6
500	1908	-62.8	-50.5	-35.0	-18.2	0.0
600	1801	-100.8	-86.0	-67.5	-47.3	-25.4
0	2400	-1.5	-1.5	-1.5	-1.5	-1.5
100	2277	0.0	2.6	6.0	9.7	13.8
200	2158	-5.3	0.0	6.6	14.0	22.2
300	2043	-17.9	-10.0	0.0	11.1	23.4
400	1930	-38.6	-28.0	-14.8	0.0	16.5
500	1823	-68.9	-55.6	-39.0	-20.6	0.0
600	1720	-110.2	-94.3	-74.4	-52.3	-27.6
0	2300	-1.5	-1.5	-1.5	-1.5	-1.5
100	2180	0.0	2.9	6.5	10.7	15.2
200	2064	-5.8	0.0	7.2	15.5	24.6
300	1951	-19.5	-10.8	0.0	12.5	26.1
400	1842	-42.7	-31.0	-16.6	0.0	18.2
500	1739	-76.1	-61.5	-43.5	-22.7	0.0
600	1640	-121.3	-103.8	-82.2	-57.3	-30.0
0	2200	-1.5	-1.5	-1.5	-1.5	-1.5
100	2083	0.0	3.2	7.2	11.9	16.9
200	1969	-6.4	0.0	8.0	17.3	27.3
300	1860	-21.7	-12.0	0.0	14.0	29.0
400	1755	-47.6	-34.7	-18.7	0.0	20.0
500	1656	-84.4	-68.3	-48.3	-25.0	0.0
600	1560	-134.1	-114.7	-90.7	-62.7	-32.7
0	2100	-1.5	-1.5	-1.5	-1.5	-1.5
100	1986	0.0	3.7	8.2	13.3	18.9
200	1875	-7.3	0.0	9.1	19.3	30.5
300	1770	-24.7	-13.7	0.0	15.2	32.0
400	1670	-53.2	-38.6	-20.3	0.0	22.4
500	1573	-94.5	-76.2	-53.4	-28.0	0.0
600	1481	-148.0	-126.0	-98.7	-68.2	-34.6
0	2000	-1.5	-1.5	-1.5	-1.5	-1.5
100	1888	0.0	4.2	9.3	14.9	20.9
200	1782	-8.4	0.0	10.3	21.3	33.5
300	1681	-28.0	-15.4	0.0	16.6	34.8
400	1585	-59.4	-42.6	-22.1	0.0	24.3
500	1491	-104.6	-83.7	-58.0	-30.4	0.0
600	1406	-165.7	-140.6	-109.8	-76.6	-40.1

Universal Table for Rifle Bullets with Ballistic Coefficients of .720-.729

Sight Above Bore (in.) +1.5

Range (yds)	Velocity (fps)	Bullet Path Zero (100 yds)	Zero (200 yds)	Zero (300 yds)	Zero (400 yds)	Zero (500 yds)
0	2900	-1.5	-1.5	-1.5	-1.5	-1.5
100	2783	0.0	1.5	3.6	6.0	8.6
200	2670	-3.0	0.0	4.2	8.9	14.1
300	2560	-10.9	-6.3	0.0	7.1	14.9
400	2452	-23.9	-17.9	-9.4	0.0	10.4
500	2347	-42.9	-35.3	-24.8	-13.0	0.0
600	2245	-68.1	-59.1	-46.4	-32.3	-16.7
0	2800	-1.5	-1.5	-1.5	-1.5	-1.5
100	2686	0.0	1.7	3.9	6.5	9.3
200	2575	-3.4	0.0	4.5	9.6	15.3
300	2467	-11.8	-6.8	0.0	7.7	16.1
400	2362	-26.0	-19.3	-10.2	0.0	11.3
500	2259	-46.6	-38.2	-26.9	-14.1	0.0
600	2159	-74.0	-63.9	-50.4	-35.0	-18.1
0	2700	-1.5	-1.5	-1.5	-1.5	-1.5
100	2588	0.0	1.8	4.3	7.1	10.1
200	2480	-3.6	0.0	5.0	10.6	16.7
300	2374	-12.9	-7.5	0.0	8.4	17.5
400	2271	-28.4	-21.2	-11.2	0.0	12.1
500	2171	-50.6	-41.7	-29.1	-15.1	0.0
600	2073	-80.2	-69.4	-54.4	-37.6	-19.4
0	2600	-1.5	-1.5	-1.5	-1.5	-1.5
100	2491	0.0	2.0	4.8	7.8	11.2
200	2385	-4.1	0.0	5.5	11.6	18.2
300	2281	-14.4	-8.3	0.0	9.0	19.0
400	2181	-31.3	-23.2	-12.1	0.0	13.3
500	2082	-55.8	-45.6	-31.7	-16.6	0.0
600	1986	-87.7	-75.5	-58.8	-40.7	-20.8
0	2500	-1.5	-1.5	-1.5	-1.5	-1.5
100	2393	0.0	2.3	5.3	8.7	12.2
200	2290	-4.7	0.0	6.0	12.7	19.7
300	2189	-16.0	-9.0	0.0	10.0	20.5
400	2090	-34.7	-25.4	-13.4	0.0	14.0
500	1994	-60.9	-49.2	-34.2	-17.5	0.0
600	1900	-95.9	-81.9	-63.9	-43.8	-22.8
0	2400	-1.5	-1.5	-1.5	-1.5	-1.5
100	2296	0.0	2.6	5.9	9.4	13.2
200	2195	-5.1	0.0	6.6	13.6	21.2
300	2096	-17.6	-9.9	0.0	10.6	22.0
400	2000	-37.6	-27.3	-14.1	0.0	15.2
500	1905	-66.0	-53.1	-36.6	-19.0	0.0
600	1815	-104.9	-89.5	-69.7	-48.6	-25.8
0	2300	-1.5	-1.5	-1.5	-1.5	-1.5
100	2198	0.0	2.8	6.4	10.1	14.6
200	2100	-5.6	0.0	7.1	14.7	23.6
300	2003	-19.1	-10.7	0.0	11.3	24.6
400	1909	-40.6	-29.4	-15.1	0.0	17.8
500	1818	-73.0	-58.9	-41.1	-22.2	0.0
600	1731	-115.7	-98.8	-77.4	-54.8	-28.2

ENERGY CHART FOR RIFLE BULLETS

Example: A 110 grain bullet at 2300 FPS will develop 1,292 ft-lbs of energy.

VELOCITY IN FEET PER SECOND

BULLET WEIGHT IN GRAINS	3,100	3,000	2,900	2,800	2,700	2,600	2,500	2,400	2,300	2,200	2,100	2,000	1,900	1,800	1,700	1,600	1,500	1,400	1,300	1,200	1,100	1,000
500	10,668	9,991	9,336	8,703	8,093	7,504	6,938	6,394	5,873	5,373	4,896	4,440	4,008	3,597	3,208	2,842	2,498	2,176	1,876	1,599	1,343	1,110
450	9,601	8,992	8,403	7,833	7,284	6,754	6,244	5,755	5,285	4,836	4,406	3,996	3,607	3,237	2,887	2,558	2,248	1,958	1,688	1,439	1,209	999
400	8,535	7,993	7,469	6,963	6,474	6,004	5,551	5,115	4,698	4,298	3,917	3,552	3,206	2,877	2,567	2,274	1,998	1,741	1,501	1,279	1,075	888
300	6,401	5,995	5,602	5,222	4,856	4,503	4,163	3,837	3,524	3,224	2,937	2,664	2,405	2,158	1,925	1,705	1,499	1,306	1,126	959	806	666
286	6,102	5,715	5,340	4,978	4,629	4,293	3,969	3,658	3,359	3,073	2,800	2,540	2,292	2,057	1,835	1,626	1,429	1,245	1,073	914	768	635
260	5,548	5,195	4,855	4,526	4,208	3,902	3,608	3,325	3,054	2,794	2,546	2,309	2,084	1,870	1,668	1,478	1,299	1,131	976	831	698	577
250	5,334	4,996	4,668	4,352	4,046	3,752	3,469	3,197	2,936	2,687	2,448	2,220	2,004	1,798	1,604	1,421	1,249	1,088	938	799	672	555
240	5,121	4,796	4,481	4,178	3,885	3,602	3,330	3,069	2,819	2,579	2,350	2,131	1,924	1,726	1,540	1,364	1,199	1,044	901	767	645	533
225	4,801	4,496	4,201	3,917	3,642	3,377	3,122	2,877	2,643	2,418	2,203	1,998	1,803	1,619	1,444	1,279	1,124	979	844	719	604	500
220	4,694	4,396	4,108	3,829	3,561	3,302	3,053	2,813	2,584	2,364	2,154	1,954	1,763	1,583	1,412	1,250	1,099	957	825	703	591	488
210	4,481	4,196	3,921	3,655	3,399	3,152	2,914	2,686	2,466	2,257	2,056	1,865	1,683	1,511	1,347	1,194	1,049	914	788	671	564	466
200	4,267	3,996	3,734	3,481	3,237	3,002	2,775	2,558	2,349	2,149	1,958	1,776	1,603	1,439	1,283	1,137	999	870	750	639	537	444
180	3,841	3,597	3,361	3,133	2,913	2,702	2,498	2,302	2,114	1,934	1,762	1,599	1,443	1,295	1,155	1,023	899	783	675	575	484	400
175	3,734	3,497	3,268	3,046	2,832	2,627	2,428	2,238	2,055	1,881	1,713	1,554	1,403	1,259	1,123	995	874	762	657	560	470	389
170	3,627	3,397	3,174	2,959	2,752	2,552	2,359	2,174	1,997	1,827	1,665	1,510	1,363	1,223	1,091	966	849	740	638	544	457	377
168	3,585	3,357	3,137	2,924	2,719	2,521	2,331	2,148	1,973	1,805	1,645	1,492	1,347	1,209	1,078	955	839	731	630	537	451	373
165	3,521	3,297	3,081	2,872	2,671	2,476	2,290	2,110	1,938	1,773	1,616	1,465	1,322	1,187	1,059	938	824	718	619	528	443	366
160	3,414	3,197	2,988	2,785	2,590	2,401	2,220	2,046	1,879	1,719	1,567	1,421	1,282	1,151	1,027	909	799	696	600	512	430	355
155	3,307	3,097	2,894	2,698	2,509	2,326	2,151	1,982	1,820	1,666	1,518	1,377	1,242	1,115	995	881	774	675	582	496	416	344
150	3,200	2,997	2,801	2,611	2,428	2,251	2,081	1,918	1,762	1,612	1,469	1,332	1,202	1,079	962	853	749	653	563	480	403	333
140	2,987	2,798	2,614	2,437	2,266	2,101	1,943	1,790	1,644	1,504	1,371	1,243	1,122	1,007	898	796	699	609	525	448	376	311
130	2,774	2,598	2,427	2,263	2,104	1,951	1,804	1,663	1,527	1,397	1,273	1,155	1,042	935	834	739	649	566	488	416	349	289
125	2,667	2,498	2,334	2,176	2,023	1,876	1,735	1,599	1,468	1,343	1,224	1,110	1,002	899	802	710	624	544	469	400	336	278
120	2,560	2,398	2,241	2,089	1,942	1,801	1,665	1,535	1,409	1,290	1,175	1,066	962	863	770	682	599	522	450	384	322	266
115	2,454	2,298	2,147	2,002	1,861	1,726	1,596	1,471	1,351	1,236	1,126	1,021	922	827	738	654	574	500	432	368	309	255
110	2,347	2,198	2,054	1,915	1,780	1,651	1,526	1,407	1,292	1,182	1,077	977	882	791	706	625	550	479	413	352	296	244
100	2,134	1,998	1,867	1,741	1,619	1,501	1,388	1,279	1,175	1,075	979	888	802	719	642	568	500	435	375	320	269	222
95	2,027	1,898	1,774	1,654	1,538	1,426	1,318	1,215	1,116	1,021	930	844	761	683	610	540	475	413	356	304	255	211
90	1,920	1,798	1,681	1,567	1,457	1,351	1,249	1,151	1,057	967	881	799	721	647	577	512	450	392	338	288	242	200
85	1,814	1,698	1,587	1,480	1,376	1,276	1,180	1,087	998	913	832	755	681	611	545	483	425	370	319	272	228	189
80	1,707	1,599	1,494	1,393	1,295	1,201	1,110	1,023	940	860	783	710	641	575	513	455	400	348	300	256	215	178
77	1,643	1,539	1,438	1,340	1,246	1,156	1,068	985	904	827	754	684	617	554	494	438	385	335	289	246	207	171
70	1,494	1,399	1,307	1,218	1,133	1,051	971	895	822	752	685	622	561	504	449	398	350	305	263	224	188	155
69	1,472	1,379	1,288	1,201	1,117	1,036	957	882	810	741	676	613	553	496	443	392	345	300	259	221	185	153
60	1,280	1,199	1,120	1,044	971	901	833	767	705	645	587	533	481	432	385	341	300	261	225	192	161	133
55	1,174	1,099	1,027	957	890	825	763	703	646	591	539	488	441	396	353	313	275	239	206	176	148	122
52	1,110	1,039	971	905	842	780	722	665	611	559	509	462	417	374	334	296	260	226	195	166	140	115
50	1,067	999	934	870	809	750	694	639	587	537	490	444	401	360	321	284	250	218	188	160	134	111
45	960	899	840	783	728	675	624	575	529	484	441	400	361	324	289	256	225	196	169	144	121	100
40	853	799	747	696	647	600	555	512	470	430	392	355	321	288	257	227	200	174	150	128	107	89
35	747	699	654	609	566	525	486	448	411	376	343	311	281	252	225	199	175	152	131	112	94	78
32	683	639	598	557	518	480	444	409	376	344	313	284	256	230	205	182	160	139	120	102	86	71
20	427	400	373	348	324	300	278	256	235	215	196	178	160	144	128	114	100	87	75	64	54	44
17	363	340	317	296	275	255	236	217	200	183	166	151	136	122	109	97	85	74	64	54	46	38

ENERGY CHART FOR RIFLE BULLETS

Example: A 90 grain bullet at 3400 FPS will develop 2,310 ft-lbs of energy.

VELOCITY IN FEET PER SECOND

BULLET WEIGHT IN GRAINS	4,900	4,800	4,700	4,600	4,500	4,400	4,300	4,200	4,100	4,000	3,900	3,800	3,700	3,600	3,500	3,400	3,300	3,200	3,100
500	26,654	25,577	24,523	23,490	22,480	21,492	20,526	19,583	18,661	17,762	16,385	16,030	15,198	14,387	13,599	12,833	12,089	11,368	10,668
450	23,989	23,020	22,070	21,141	20,232	19,343	18,474	17,624	16,795	15,986	15,196	14,427	13,678	12,948	12,239	11,550	10,880	10,231	9,601
400	21,323	20,462	19,618	18,792	17,984	17,194	16,421	15,666	14,929	14,210	13,508	12,824	12,158	11,510	10,879	10,266	9,671	9,094	8,535
300	15,992	15,346	14,714	14,094	13,488	12,895	12,316	11,750	11,197	10,657	10,131	9,618	9,119	8,632	8,159	7,700	7,254	6,821	6,401
286	15,246	14,630	14,027	13,436	12,859	12,293	11,741	11,201	10,674	10,160	9,558	9,169	8,693	8,229	7,779	7,340	6,915	6,502	6,102
260	13,860	13,300	12,752	12,215	11,690	11,176	10,674	10,183	9,704	9,236	8,780	8,336	7,903	7,481	7,071	6,673	6,286	5,911	5,548
250	13,327	12,789	12,261	11,745	11,240	10,746	10,263	9,791	9,331	8,881	8,442	8,015	7,599	7,194	6,800	6,417	6,045	5,684	5,334
240	12,794	12,277	11,771	11,275	10,790	10,316	9,853	9,400	8,957	8,526	8,105	7,694	7,295	6,906	6,528	6,160	5,803	5,456	5,121
225	11,994	11,510	11,035	10,571	10,116	9,671	9,237	8,812	8,398	7,993	7,598	7,214	6,839	6,474	6,120	5,775	5,440	5,115	4,801
220	11,728	11,254	10,790	10,336	9,891	9,456	9,032	8,616	8,211	7,815	7,429	7,053	6,687	6,330	5,984	5,647	5,319	5,002	4,694
210	11,195	10,742	10,300	9,866	9,442	9,027	8,621	8,225	7,838	7,460	7,092	6,733	6,383	6,043	5,712	5,390	5,077	4,774	4,481
200	10,662	10,231	9,809	9,396	8,992	8,597	8,210	7,833	7,464	7,105	6,754	6,412	6,079	5,755	5,440	5,133	4,836	4,547	4,267
180	9,595	9,208	8,828	8,456	8,093	7,737	7,389	7,050	6,718	6,394	6,079	5,771	5,471	5,179	4,896	4,620	4,352	4,092	3,841
175	9,329	8,952	8,583	8,222	7,868	7,522	7,184	6,854	6,531	6,217	5,910	5,611	5,319	5,036	4,760	4,492	4,231	3,979	3,734
170	9,062	8,696	8,338	7,987	7,643	7,307	6,979	6,658	6,345	6,039	5,741	5,450	5,167	4,892	4,624	4,363	4,110	3,865	3,627
168	8,956	8,594	8,240	7,893	7,553	7,221	6,897	6,580	6,270	5,968	5,673	5,386	5,106	4,834	4,569	4,312	4,062	3,820	3,585
165	8,796	8,440	8,092	7,752	7,418	7,092	6,774	6,462	6,158	5,861	5,572	5,290	5,015	4,748	4,488	4,235	3,989	3,751	3,521
160	8,529	8,185	7,847	7,517	7,194	6,877	6,568	6,266	5,972	5,684	5,403	5,130	4,863	4,604	4,352	4,107	3,869	3,638	3,414
155	8,263	7,929	7,602	7,282	6,969	6,663	6,363	6,071	5,785	5,506	5,234	4,969	4,711	4,460	4,216	3,978	3,748	3,524	3,307
150	7,996	7,673	7,357	7,047	6,744	6,448	6,158	5,875	5,598	5,329	5,065	4,809	4,559	4,316	4,080	3,850	3,627	3,410	3,200
140	7,463	7,162	6,866	6,577	6,294	6,018	5,747	5,483	5,225	4,973	4,728	4,488	4,255	4,028	3,808	3,593	3,385	3,183	2,987
130	6,930	6,650	6,376	6,107	5,845	5,588	5,337	5,091	4,852	4,618	4,390	4,168	3,951	3,741	3,536	3,337	3,143	2,956	2,774
125	6,664	6,394	6,131	5,873	5,620	5,373	5,132	4,896	4,665	4,440	4,221	4,008	3,799	3,597	3,400	3,208	3,022	2,842	2,667
120	6,397	6,139	5,885	5,638	5,395	5,158	4,926	4,700	4,479	4,263	4,052	3,847	3,647	3,453	3,264	3,080	2,901	2,728	2,560
115	6,130	5,883	5,640	5,403	5,170	4,943	4,721	4,504	4,292	4,085	3,884	3,687	3,495	3,309	3,128	2,952	2,781	2,615	2,454
110	5,864	5,627	5,395	5,168	4,946	4,728	4,516	4,308	4,105	3,908	3,715	3,527	3,343	3,165	2,992	2,823	2,660	2,501	2,347
100	5,331	5,115	4,905	4,698	4,496	4,298	4,105	3,917	3,732	3,552	3,377	3,206	3,040	2,877	2,720	2,567	2,418	2,274	2,134
95	5,064	4,860	4,659	4,463	4,271	4,083	3,900	3,721	3,546	3,375	3,208	3,046	2,888	2,734	2,584	2,438	2,297	2,160	2,027
90	4,798	4,604	4,414	4,228	4,046	3,869	3,695	3,525	3,359	3,197	3,039	2,885	2,736	2,590	2,448	2,310	2,176	2,046	1,920
85	4,531	4,348	4,169	3,993	3,822	3,654	3,489	3,329	3,172	3,020	2,870	2,725	2,584	2,446	2,312	2,182	2,055	1,933	1,814
80	4,265	4,092	3,924	3,758	3,597	3,439	3,284	3,133	2,986	2,842	2,702	2,565	2,432	2,302	2,176	2,053	1,934	1,819	1,707
77	4,105	3,939	3,776	3,617	3,462	3,310	3,161	3,016	2,874	2,735	2,600	2,469	2,340	2,216	2,094	1,976	1,862	1,751	1,643
70	3,732	3,581	3,433	3,289	3,147	3,009	2,874	2,742	2,613	2,487	2,364	2,244	2,128	2,014	1,904	1,797	1,692	1,591	1,494
69	3,678	3,530	3,384	3,242	3,102	2,966	2,833	2,702	2,575	2,451	2,330	2,212	2,097	1,985	1,877	1,771	1,668	1,569	1,472
60	3,198	3,069	2,943	2,819	2,698	2,579	2,463	2,350	2,239	2,131	2,026	1,924	1,824	1,726	1,632	1,540	1,451	1,364	1,280
55	2,932	2,813	2,697	2,584	2,473	2,364	2,258	2,154	2,053	1,954	1,857	1,763	1,672	1,583	1,496	1,412	1,330	1,250	1,174
52	2,772	2,660	2,550	2,443	2,338	2,235	2,135	2,037	1,941	1,847	1,756	1,667	1,581	1,496	1,414	1,335	1,257	1,182	1,110
50	2,665	2,558	2,452	2,349	2,248	2,149	2,053	1,958	1,866	1,776	1,688	1,603	1,520	1,439	1,360	1,283	1,209	1,137	1,067
45	2,399	2,302	2,207	2,114	2,023	1,934	1,847	1,762	1,680	1,599	1,520	1,443	1,368	1,295	1,224	1,155	1,088	1,023	960
40	2,132	2,046	1,962	1,879	1,798	1,719	1,642	1,567	1,493	1,421	1,351	1,282	1,216	1,151	1,088	1,027	967	909	853
35	1,866	1,790	1,717	1,644	1,574	1,504	1,437	1,371	1,306	1,243	1,182	1,122	1,064	1,007	952	898	846	796	747
32	1,706	1,637	1,569	1,503	1,439	1,375	1,314	1,253	1,194	1,137	1,081	1,026	973	921	870	821	774	728	683
20	1,066	1,023	981	940	899	860	821	783	746	710	675	641	608	575	544	513	484	455	427
17	906	870	834	799	764	731	698	666	634	604	574	545	517	489	462	436	411	387	363

ENERGY CHART FOR HANDGUN BULLETS

Example: A 240 grain bullet at 1100 FPS will develop 645 ft-lbs of energy.

VELOCITY IN FEET PER SECOND

WEIGHT IN GRAINS	2,500	2,400	2,300	2,200	2,100	2,000	1,900	1,800	1,700	1,600	1,500	1,400	1,300	1,200	1,100	1,000	900	800	700	600	500
300	4,163	3,837	3,524	3,224	2,937	2,664	2,405	2,158	1,925	1,705	1,499	1,306	1,126	959	806	666	540	426	326	240	167
260	3,608	3,325	3,054	2,794	2,546	2,309	2,084	1,870	1,668	1,478	1,299	1,131	976	831	698	577	468	369	283	208	144
250	3,469	3,197	2,936	2,687	2,448	2,220	2,004	1,798	1,604	1,421	1,249	1,088	938	799	672	555	450	355	272	200	139
240	3,330	3,069	2,819	2,579	2,350	2,131	1,924	1,726		1,364	1,199	1,044	901	767	645	533	432	341	261	192	133
230	3,192	2,941	2,701	2,472	2,252	2,043	1,843	1,655	1,476	1,307	1,149	1,001	863	735	618	511	414	327	250	184	128
210	2,914	2,686	2,466	2,257	2,056	1,865	1,683	1,511	1,347	1,194	1,049	914	788	671	564	466	378	298	228	168	117
200	2,775	2,558	2,349	2,149	1,958	1,776	1,603	1,439	1,283	1,137	999	870	750	639	537	444	360	284	218	160	111
185	2,567	2,366	2,173	1,988	1,811	1,643	1,483	1,331	1,187	1,052	924	805	694	591	497	411	333	263	201	148	103
180	2,498	2,302	2,114	1,934	1,762	1,599	1,443	1,295	1,155	1,023	899	783	675	575	484	400	324	256	196	144	100
170	2,359	2,174	1,997	1,827	1,665	1,510	1,363	1,223	1,091	966	849	740	638	544	457	377	306	242	185	136	94
158	2,192	2,021	1,856	1,698	1,547	1,403	1,266	1,137	1,014	898	789	688	593	505	424	351	284	225	172	126	88
150	2,081	1,918	1,762	1,612	1,469	1,332	1,202	1,079	962	853	749	653	563	480	403	333	270	213	163	120	83
135	1,873	1,726	1,586	1,451	1,322	1,199	1,082	971	866	767	674	587	507	432	363	300	243	192	147	108	75
125	1,735	1,599	1,468	1,343	1,224	1,110	1,002	899	802	710	624	544	469	400	336	278	225	178	136	100	69
115	1,596	1,471	1,351	1,236	1,126	1,021	922	827	738	654	574	500	432	368	309	255	207	163	125	92	64

Nosler® Ballistics App

Nosler Ballistics is a trajectory calculator and range log for Apple and Android devices. This application calculates trajectory, windage, velocity, energy, and bullet flight time for any valid range, and can compensate for atmospheric conditions such as temperature, barometric pressure, humidity, and altitude (it can also accept density of air or density altitude input).

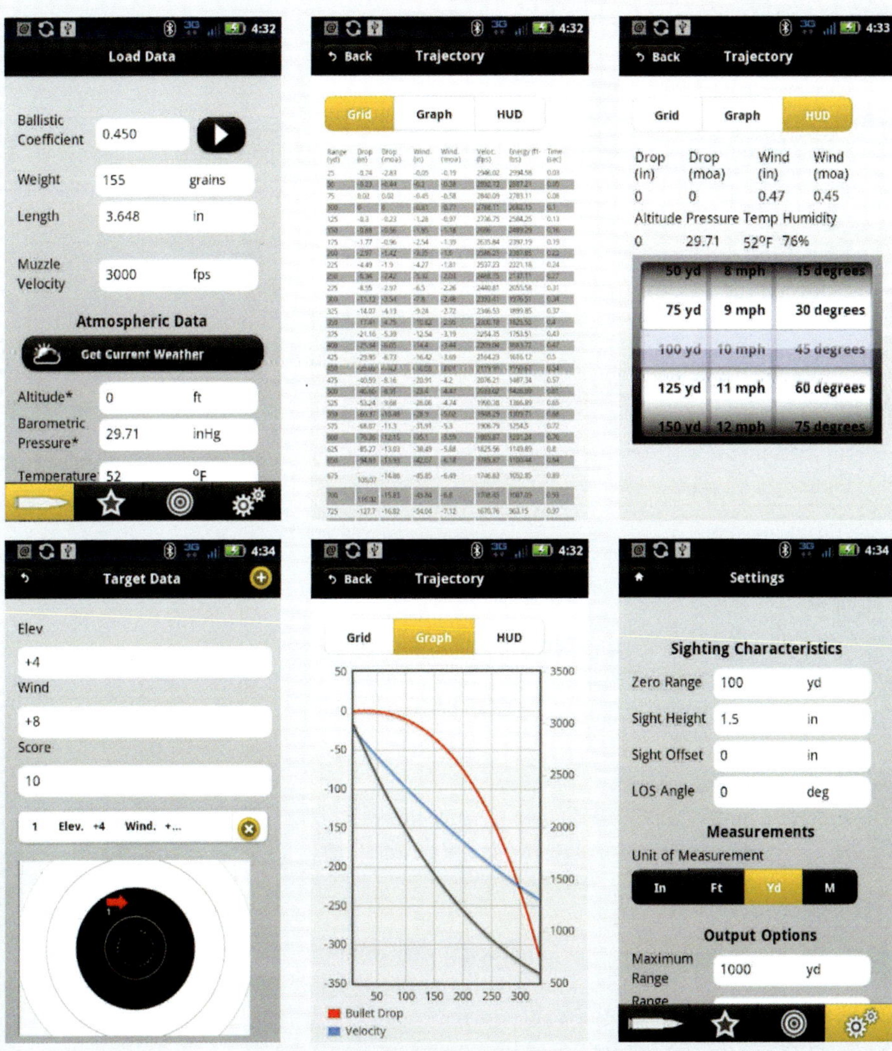

Reloading Guide Glossary

Action
The part of a firearm by which it is loaded, fired and unloaded.

Annealing
The process by which metal is softened by exposure to heat. Case necks which have become brittle as a result of repeated resizing can be annealed to prevent cracking and yield longer life. Case neck annealing must be done with care to prevent softening of the case head.

Antimony
A metallic element which increases the hardness of lead when combined as an alloy.

Anvil
The part of the primer which offers resistance to the primer compound as the firing pin strikes the primer. In firing, the primer compound is compressed sharply against the anvil and the resulting shearing force sets off the ignition constituents contained in the primer compound.

Ball
A term used by the military to denote bullet. Also used by Olin Industries to describe their spherical powder.

Ballistics
The science and study of projectiles in motion. Divided into three parts: INTERIOR ballistics is used to describe the events which take place from ignition until the projectile leaves the muzzle. EXTERIOR ballistics describes the projectile's flight from muzzle to target, and TERMINAL ballistics is the study of events after the bullet reaches the target.

Ballistic Coefficient
The ratio of a bullet's sectional density to its coefficient of form, used to describe the bullet's effectiveness in overcoming air resistance during flight.

Ballistic Tip®
A unique rifle bullet designed by John Nosler that combines the qualities of the Nosler® Solid Base® (see SOLID BASE) bullet with a polycarbonate tip that resists deformation in the magazine or feed ramp of a rifle during recoil. These bullets have a higher ballistic coefficient than most other bullets of the equivalent size and weight due to the sharp, aerodynamic tip.

Barrel Cylinder Gap
The gap or space between the cylinder and forcing cone of the barrel in a revolver.

Bearing surface
The portion of a projectile which comes in contact with the bore.

Bedding
The part of the stock which forms the contact with the action and/or barrel. The act of fitting the action to the stock.

Belted Case
A cartridge case with a reinforcing belt forward of the extractor groove, characteristics of most modern "magnum" cartridges.

Berdan Primer
Normally found in only European cartridges, the Berdan Primer requires a protrusion in the bottom of the primer pocket to serve as an anvil (see ANVIL).

Boat Tail
The pronounced taper at the heel of a bullet. Used frequently on military and match bullets and found on all Nosler AccuBond®, Ballistic Tip® and E-Tip® rifle bullets.

Body
The portion of a case between the head and the point where the shoulder begins to taper.

Bolt
The part of the action which contains the firing pin assembly, extractor mechanism and the locking system.

Bore
The interior of a barrel. When used to describe a rifled barrel, the diameter before the rifling is cut. Also called "bore diameter."

Bore Sight
The technique of aligning the sights of a rifle with the target by sighting through the barrel.

Boxer Primer
The primer type used in American centerfire ammunition, both sporting and military.

Brass
A common term used by handloaders in referring to empty cartridge cases. An alloy of copper and zinc.

Bullet
The cartridge component which, when in flight, becomes a projectile.

Bullet Path
The vertical distance of the bullet's trajectory above or below the line of sight.

Bullet Puller
Tool used to remove bullet from case.

Burning Rate
A relative term used to define the rate at which a powder burns in comparison with other powders in individual cartridges.

Caliber
The nominal diameter of the groove or bore, often modified to show the date of adoption (.30-06) or the designer (.257 Roberts), etc. Caliber is expressed either in decimals of an inch (.308") or in metric (7mm) measurements. Rifle or cartridge designations are often merely approximated; for example: .30, .308 and .300 cartridges all use a .308" diameter bullet.

Cannelure
A groove around a bullet that is used for crimping, lubricating or identification.

Cap
An early form of today's primer, used to ignite black powder charges. The term is still sometimes used to denote primer.

Cartridge
A complete round of ammunition with all components intact.

Case
The salvageable brass component of a cartridge which serves as a container for the expendable components.

Case Trimmer
Tool used for trimming case length back to original.

Centerfire
A cartridge with the primer located in the center of the case head.

Chamber
The section of the barrel which encloses the cartridge to restrict case expansion during fire.

Chamber Cast
A molding taken of the chamber by pouring a low-melting-point metal or compound into the chamber. The casting permits accurate measurement of critical chamber dimensions.

Chamfer
The process of reaming or beveling a taper on the inside or outside of a case mouth to remove burrs left by the case trimming operation.

Charge
The quantity of powder loaded into a case. Usually expressed in grains and/or tenths of a grain.

Chronograph
An instrument which measures a projectile's velocity electronically, based on the elapsed time between two measured reference points.

Components
The items necessary to load a cartridge (brass, primer, powder, bullet).

Compressed Charge
A condition where powder is [packed more densely in a case than would normally occur. Caused by filling the case to a point where seating a bullet compresses the powder.

Core
The center, lead section(s) of a bullet.

Corrosive Primer
An older type primer in which the compound, if not removed after firing, attracted moisture and resulted in rust. Use of this compound was generally discontinued about 1950.

Crimp
To turn the case mouth slightly inward, usually into a cannelure, to grip a bullet firmly.

Cylindrical Powder
Tubular shaped smokeless powder which is manufactured by an extrusion process.

Deburr
To remove the rough and jagged edges left when a case is trimmed to length.

Deburring Tool
A tool for removing the rough metal burrs from a trimmed case mouth.

Decap (Deprime)
To eject the spent primer from the pocket of a fired case.

Die
In reloading, a tool used to re-form a fired brass case or to seat a fresh bullet into the case mouth. In bullet making, a tool used to form the bullet.

Double Base Powder
Smokeless powder manufactured of nitrocellulose and nitroglycerine.

Drift
A term used in exterior ballistics to describe the deviation of a projectile laterally from the line of departure. Typically caused by wind.

Drop
The effect of gravity on the projectile. The distance between the line of departure and the trajectory at a given distance.

Elevation
The vertical adjustment of a sight to bring the point of aim into coincidence with the point of impact.

Energy
In ballistics, the amount of work (see FOOT POUND) a projectile is capable of accomplishing. Usually measured at the muzzle and as remaining energy at various distances from the muzzle.

Engraving
The indentations and other marks on the bullet surface caused by the rifling.

Erosion
The wearing away of the rifling due to hot powder gases and/or bullet friction. The erosion is usually heaviest in the riflings closest to the throat area.

Expander Ball (Button)
A thicker section of the decapping stem, used to expand the case mouth to the precise diameter required to hold the bullet firmly.

Extruder Primer
An indication of excessive pressure shown by a ring or "cratering" around the primer indent where the primer has been extruded into the firing pin recess on the bolt face.

FPS
Feet per second. The measurement used to describe the velocity of the projectile.

Fire-forming
Firing the cartridge to achieve full expansion of the case to the dimensions of the chamber.

Firing Pin (Striker)
A spring-loaded, round-tipped pin which is held under compression after cocking. Pulling the trigger activates a sear which allows the firing pin to snap sharply forward to strike the primer.

Flake Powder
Thin, round, flake-type of smokeless powder which is usually of a fast-burning nature.

Flash Hole
An opening that provides access through the web for the primer flash to ignite the powder granules in the case interior.

Foot Pound
A unit of energy. The force required to lift one pound of weight to a height of one foot.

Freebore
(See THROAT)

Gas
In handloading, the vapor form of burning powder. This heavy gas is capable of expanding rapidly, creating sufficient pressure to propel the bullet.

Gilding Metal
An alloy of copper and zinc, commonly used for bullet jacket material. Nosler bullets are made from an alloy of copper and zinc.

Grain
The weight unit used in measuring powder charges or bullet weights. One pound avoirdupois is equal to 7,000 grains. 437.5 grains is equivalent to one ounce.

Grooves
The spiral cut that is removed from the bore to leave the "lands." The grooves of a .30 caliber rifle are approximately .308" in diameter.

Hangfire
A condition where ignition is delayed after the strike of the firing pin.

Headspace
In handloading, the slight gap that is permitted between the bolt face and the case head to facilitate closure of the bolt. Headspace is actually the distance between the bolt face and the part of the chamber that acts as the cartridge stop.

Holdover
The distance above the target a shooter must aim when the rifle is zeroed at a lesser range. The Nosler Ballistic Tables are designed to be helpful to the shooter in estimation holdover at various ranges.

Hollow Point
A bullet design in which the lead core does not come all the way to the bullet tip.

Hydrostatic Shock
A highly destructive shock wave created by a bullet passing through animal tissue, which is high in water content.

IMR
Improved Military Rifle. Single base rifle powder brand name used in reloading.

Ignition
The initial combustion of the powder caused by the flame of the primer.

Impact Extrusion
The shaping process used in the manufacture of Nosler bullet jackets. The jacket material is formed in precise dies to maintain consistency of jacket wall thickness.

Improved Cartridge
A standard cartridge that has been modified to fit a specially chambered rifle, with less body taper and/or a more sharply angled shoulder. Modification is done by fire-forming in the chamber in which it is intended to be used.

Ingalls Tables
Ballistic tables computed by the late Col. James M. Ingalls. The projectiles used by Col. Ingalls server as the basis for his computations which are regarded as one of the standards for comparison of all small arms projectiles.

Jacket
The outer covering of a bullet that is used to contain the lead core and improve the "mushrooming" characteristics. Gilding metal is considered best for this purpose.

Keyhole
The keyhole-shaped print of a bullet on the target which indicates it was not stabilized.

Lands
The raised portion of the spiral rifling that remains after the groove cuts have been made.

Leade
(See THROAT)

Line of Departure
A projection of the axis of the bore, which is straight to infinity. Line of Departure is coincident with the path the projectile up to the time the projectile leaves the muzzle.

Line of Sight
An imaginary line, straight to infinity, which passes through the sights. Coincident with the point of aim.

Load Density

The weight of the powder charge, expressed in grains, divided by the volume of the cartridge case, also expressed in grains. The cartridge case volume is obtained by determining the amount of water the case will hold when the bullet is seated to the SAAMI suggested maximum cartridge length.

Locking Lugs

Protrusions on the bolt that engage a mating recess inside the receiver ring when the bolt is closed. This feature prevents the bolt from moving rearward when the rifle is fired.

Lock Time

The elapsed time between release of the firing pin and its strike on the primer.

Magnum

A firearm or cartridge case capable of greater power than is normal for the bullet diameter.

Max Ord

(Maximum Ordinate) or Mid Range Trajectory. The point at which the bullet reaches its greatest vertical distance above the line of sight. Usually occurring between 50% to 60% downrange.

Meplat

The diameter of the blunt section of a bullet tip.

Mercuric Primer

A primer in which a mercuric compound is used. The corrosive effect of this type of compound has discouraged its use in recent years (see CORROSIVE PRIMER).

Metal Fouling

Metal deposited from the projectile as it passes through the bore. This condition is not common with bullet jackets constructed of gilding metal.

Micrometer

A caliper-type instrument used to measure thickness or diameter.

Mid-range Trajectory

(See MAX ORD)

Minute of Angle (MOA)

A unit of angular measurement equal to 1/60th of a degree. Usually approximated as 1" (actually 1.047") at 100 yards, 2" at 200 yards, etc.

Misfire
The failure of a cartridge to fire after the strike of the firing pin.

Mushroom
The ideal shape of a bullet following impact with animal tissue.

Muzzle
The end of the barrel where the projectile leaves the bore.

Muzzle Blast
The release of gas from the muzzle immediately following the projectile's exit.

Muzzle Energy
The energy of a projectile, usually measure in foot pounds, at the muzzle.

Muzzle Velocity
The velocity of a projectile, usually measured in fps, at the muzzle.

Neck
The narrow portion of the case, forward of the shoulder, which grips the bullet.

Neck Reaming
A technique used to achieve uniform case wall thickness at the neck by removing metal from the inside of the case wall.

Neck Sizing
A resizing technique in which only the neck portion of the case is resized.

Neck Turning
A technique used to achieve uniform case wall thickness at the neck by removing metal from the outside of the case wall.

Non-corrosive Primer
A primer containing a priming compound which does not induce rust or corrosion in the bore or case. Most military ammunition prior to 1950 contained a corrosive compound. This practice was discontinued in 1950.

O.A.L.
Overall Cartridge Length

Ogive
The curved, forward portion of a bullet.

Partition®

The unique bullet concept developed by John Nosler, in which a part of the jacket material is used to form a barrier between the front and rear lead cores of a bullet. This design yields the most consistent performance possible from a weight-retaining game bullet. A registered trademark of Nosler, Inc.

Pierced Primer

A primer that has been punctured, usually by a defective firing pin.

Point of aim

The point where the bullet path intersects the line of sight.

Powder

The highly combustible substance that generates a heavy gas upon ignition. The pressure of this gas acts as a propellant to drive the projectile down the bore.

Powder Measure

A measuring device designed to throw uniform charges of powder.

Powder Scale

A precision measuring device designed for weighting powder; calibrated in grains and tenths of grains.

Pressure

The force exerted by burning powder. Expressed as "peak pressure" and measured in pounds per square inch.

Primer

(see BERDAN, BOXER) – A small metal cup containing a priming compound which, when crushed against the anvil by the firing pin, creates an intense spark that ignites the powder granules inside the case. Also commonly referred to as a "cap".

Primer Indent

A depression in the primer resulting from the strike of the firing pin.

Primer Pocket

A recess in the head of a centerfire cartridge case which holds the primer.

Projectile

A bullet when it is in motion.

Protected Point

A Nosler bullet design that is intended to prevent nose damage caused by battering in the magazines of heavy magnums. This design is particularly adaptable to cartridges that require deep seating to cut down overall length.

Protruding (Popped) Primer
A primer that has partially backed out of the primer pocket.

Ram
The section of a reloading press, supporting the shellholder, which is used to raise or lower the case into the die during resizing and bullet seating.

Ream
To remove material from a cavity with a rotary cutting tool.

Rebated Case
A case with a rim that is of smaller diameter than the case body.

Remaining Velocity
The velocity of a projectile, usually measured in fps, at a given distance from the muzzle.

Remaining Energy
The residual energy of a projectile, measured in foot pounds, at a given distance from the muzzle.

Resizing Die
A precisely chambered die used to re-form a fired case to the proper dimensions. When properly used, it will allow the case to be chambered in any rifle of the same caliber.

Rifling
Parallel spiral grooves, cut into the bore of the firearm, which impart a spin to the projectile.

Rim
The flange behind the extractor groove on a cartridge case. Used as a means of extracting a case from the chamber after firing.

Rimfire
A relatively low-velocity cartridge in which the priming compound is located under the rim, inside the case. The firing pin strikes the edge of the case rim, crushing the priming compound and igniting the powder. These cases are not reloadable.

Rimmed Case
A case with a rim of greater diameter than the case body.

Rimless Case
A case with a rim of the same diameter as the case body.

Round
A completely assembled cartridge with all components unused.

Round Nose
A bullet design that features a blunt, spherical shape at the tip.

Rupture
A split or separation in the case wall or neck.

Seating Depth
The depth to which a bullet is inserted into the case mouth.

Seating Die
A die that can be adjusted to control seating depth of a bullet.

Sectional density
The ratio of a bullet's weight, in pounds, to the square of its diameter, in inches.

Semi-rimmed Case
A case with a rim of slightly larger diameter than the case body.

Shank
The cylindrical, straight section of a bullet behind the ogive.

Shellholder
The device that holds the case upright and guides it into the die cavity. A given cartridge requires a shellholder of a specific size.

Shoulder
That portion of a case that slopes from the body to the neck.

Single Base Powder
Smokeless powder manufactured of nitrocellulose.

Sight Radius
The distance between the front and rear sights.

Solid Base®
A bullet design, developed by John Nosler, which features a thick base section that improves expansion characteristics and resists deformation during firing. Used in both Nosler Solid Base® and Ballistic Tip® bullets. A registered trademark of Nosler, Inc.

Spherical Powder (Ball Type)
Smokeless powder shaped into flattened ball granules or round balls.

Spent Primer
A primer that has been fired and in which the priming compound has been expended.

Spin
The rotation of the projectile, caused by the spiral rifling of the bore.

Stabilize
To cause a projectile to rotate rapidly enough around its long axis to keep it from tumbling or yawing in flight.

Swaging
A process used to form a material under pressure.

Throat
The unrifled section of the bore immediately ahead of the chamber.

Time of Flight
The amount of time a projectile is in free flight.

Twist
The rate of spiral of the rifling. Given as the barrel length required to make one full revolution: i.e., "One Turn in Ten Inches," etc.

Velocity
The speed of a projectile in flight. Usually measured in fps at a specific distance.

Vernier Caliper
A finely graduated, precision measuring instrument of much use to the handloader.

Web
The solid portion of a cartridge case between the bottom of the primer pocket and the case interior. A flash hole through the web provides a means for igniting the powder by the primer.

Wildcat
A cartridge that is not commercially available. Produced from standard cases through the use of special forming dies, the altered case must be fire-formed in the chamber of the rifle to complete the reforming process.

Windage
The horizontal adjustment of a sight to bring the point of aim into coincidence with the point of impact.

Wind Deflection
A lateral change in bullet trajectory caused by crosswind.

Zero

In shooting, the point at which the line of the sight is coincident with the bullet trajectory at a given distance from the muzzle. In "zeroing" a rifle, sight adjustments are made until this condition exists at the range.

NOTES

NOTES

NOTES

NOTES

NOTES

NOTES

NOTES